Ecological Developmental Biology
The Environmental Regulation of Development, Health, and Evolution

Ecological Developmental Biology

The Environmental Regulation of Development, Health, and Evolution

Second Edition

Scott F. Gilbert

University of Helsinki
Swarthmore College

David Epel

Hopkins Marine Station,
Stanford University

Sinauer Associates, Inc. Publishers
Sunderland, Massachusetts U.S.A.

The Cover

Bicyclus anynana, a butterfly native to Malawi, has two forms, each determined by temperature. Warm weather (front cover) induces brightly colored eyespots that provide the flying butterfly with the ability to deflect predators. During cooler weather (back cover), the butterfly remains on the ground. The butterfly does not form such obvious eyespots, and it can be protected by blending into the brown-colored leaf litter. The hormone ecdysone is made during the warm weather, and it stabilizes the transcription of the *Distal-less* gene (left-most panel), which organizes each eyespot. The concentric circles of pigmented scales then develop onto the nascent wing according to the factors organized by *Distal-less*. Photographs courtesy of Patricia Beldade, Sean Carroll, Steve Paddock, Leila T. Shirai, Steven Woodhall, and Oskar Brattström.

Ecological Developmental Biology: The Environmental Regulation of Development, Health, and Evolution
Second Edition
Copyright © 2015 by Sinauer Associates, Inc. All rights reserved. This book may not be reproduced in whole or in part without permission from the publisher.

Sinauer Associates, Inc.
P.O. Box 407
Sunderland, MA 01375 USA
PHONE: 413-549-4300
FAX: 413-549-1118
Email: publish@sinauer.com, orders@sinauer.com
Website: www.sinauer.com

Library of Congress Cataloging-in-Publication Data

Gilbert, Scott F., 1949-
  Ecological developmental biology : the environmental regulation of development, health, and evolution / Scott F. Gilbert, University of Helsinki, Swarthmore College, David Epel, Hopkins Marine Station, Stanford University. -- Second edition.
      pages cm
  Includes bibliographical references and index.
   ISBN 978-1-60535-344-9 (alk. paper)
  1. Phenotypic plasticity. 2. Developmental biology. 3. Epigenesis. 4. Evolution (Biology) 5. Symbiosis. I. Epel, David. II. Title.
   QH438.5.G55 2015
   576.5'3--dc23
                                              2015009872

Printed in China
5 4 3 2 1

To Anne and Talia
and
To Lois

Brief Contents

Contents

3 Developmental Symbiosis
Co-Development as a Strategy for Life 79

Part 2: Ecological Developmental Biology and Disease States 135

4 Developmental Physiology for Survival in Changing Environments 137

Teratogenesis

Environmental Assaults on Development 185

Endocrine Disruptors 221

Part 3: Toward a Developmental Evolutionary Synthesis   345

Environment, Development, and Evolution
Toward a New Evolutionary Synthesis   435

Coda
Philosophical Concerns Raised by Ecological Developmental Biology   477

Appendix A
Lysenko, Kammerer, and the Truncated Tradition of Ecological Developmental Biology   509

Appendix B
The Molecular Mechanisms of Epigenetic Change   521

Preface

A quiet biological revolution, driven by new technologies in molecular, cell, and developmental biology and ecology has made the biology of the twenty-first century a different science than that of the twentieth.

This revolution was not the one expected. Rather than confirm and deepen what we already knew, these new technologies uncovered new layers of inheritance, development, and evolution. They have given us a new humility. There is plenty we don't know, and many of our assumptions about the mechanisms of development, inheritance, physiology, disease, and evolution have to be questioned.

These unexpected challenges have given rise to Ecological Developmental Biology, the science seeking to understand how environments interact with developing organisms to produce new phenotypes, and how these interactions affect disease and evolution. It is a science that may be transforming our thinking about life as profoundly as evolution, the cell theory, or the gene theory.

Some unexpected ideas must be integrated into our new thinking about inheritance, development, evolution, and health. These include:

- **Symbiosis**. Once thought of as the exception to the rules of life, symbiosis is now recognized as a signature of life, including its development and evolution. We function, develop, and evolve as consortia.

- **Developmental plasticity.** Also thought of as an exception to the rules of life, developmental plasticity is also ubiquitous. A single genome can generate numerous phenotypes, depending on environmental conditions.

- **Epialleles, environmentally induced modifications of the genome.** Formerly considered impossible, such environmentally modified chromatin not only exists but can be inherited for many generations.

These new discoveries alter the way we think about the world:

- **Evolutionary biology** must more fully integrate genetics with epialleles, symbiosis, and plasticity to generate a new evolutionary framework for the origins and maintenance of biodiversity.

- **Disease susceptibilities**, especially to diseases such as cancer, diabetes, asthma, autism, and obesity, may be caused by environmental toxins, mismatches in developmental plasticity, particular combinations of symbionts, or epialleles, Our understanding of disease has to change.

- **Global climate change** and **endocrine disruptors** are affecting how organisms develop and how they behave. These often concern changes in symbionts, the limits of plasticity, and the generation of epialleles.

Ecological Developmental Biology was published almost six years ago, and the above challenges and new ideas form an important part of our revision. The resulting book is organized into four parts. The first part concerns *the ways by which the developing organism interacts with its environment during normal development*. It focuses on the three newly appreciated mechanisms of development that were mentioned above: developmental plasticity, inherited epigenetic modification, and developmental symbioses. Evidence will be given that these three phenomena are crucial to understanding the generation of phenotypes. In each case, the environment is not a merely permissive agent, but one that helps instruct development.

The second part of the book examines *how the environment can cause development to go awry*. First, it looks at the physiological functions and strategies that have evolved to protect the embryo before the developing organism has its adult defense systems. It also details how climate change can circumvent some of those strategies. The next chapters looks at those chemicals—teratogens and endocrine disruptors—that can disrupt normal development, and the final two chapters of this section look at the ways that developmental information coming from the environment can predispose us to develop diseases later in life..

The third part of the book presents *the evidence for a new evolutionary synthesis*. Sometimes called "the expanded synthesis," "eco-evo-devo," or "the developmental synthesis," this synthesis seeks to bring into evolutionary biology the rules by which an organism's genes, environment, and development interact to create the variation and selective pressures needed for evolution. It starts with traditional evolutionary biology, proceeds through evolutionary developmental biology, and then builds on this foundation with ecological evolutionary developmental biology.

A philosophical coda and a series of appendices that go more deeply into the historical, philosophical, and scientific matters discussed in the body of the text follow the final chapter. This book has not hesitated to discuss public policy issues. Indeed, it would be a caricature of science to discuss modern science as if it were done without an eye on funding questions, government regulations, economic considerations, and even ideological issues.

Given that we live in a world characterized by the accelerating reduction of species, the sudden increase in non-infectious diseases, and the breakdown of ecological communities, ecological developmental biology is a critically important science. When we sent the chapters out for review, we found that we were not alone in this feeling. One reviewer, calling eco-devo "the most important field of science at the moment," noted that:

> Ecology isn't prepared to analyze at the molecular level the ills of
> our present world, genetics doesn't contain the background in tissue

interactions, and developmental biology has the tools but is only just now turning its attention to an environment outside of cells and the individual organism. Eco-devo is the synthesis that combines all of the above, and we and our students desperately need to have a basic understanding of this new field to become proper stewards of our planet.

Ecological developmental biology integrates molecular biology, ecology, developmental biology, evolutionary biology, physiology, cell biology, and genetics into a syncytial science that is at the core of twenty-first century concerns.

This book is intended both for students and for our scientific colleagues. While it would help students to have had courses in developmental biology, cell biology, ecology, or evolution, a good first-year biology course should be adequate. This book is also for specialists who would like to learn something about how their particular subdiscipline might interact with other biological sciences.

We hope that the examples presented here will reinforce the sense of wonder that biologists find in the world, and at the same time be a jumping-off point for discussions about both the integration of different areas of biology as well as the increasingly critical question of biology's relation to public policy. While we have tried to be integrative, we realize that we are still bound by our past history and training. So we hope that college students, still relatively undifferentiated, will come up with their own connections and syntheses and that they will see patterns that we haven't yet imagined. We are extremely glad that this book has been used in senior and graduate seminars to unite students of different backgrounds. Indeed, we have to thank the students in Jeannette Wyneken's seminar at Florida Atlantic University for giving us a running commentary as they went through the book, augmenting and critiquing the text as they discussed its contents.

Finally, we hope the ideas in this book evoke a way of approaching nature, an approach exemplified in the banner that hangs over the library of the Woods Hole Marine Biology Laboratory, reminding us that we should "Study Nature, Not Books." One is constantly surprised by the wonderful improvisation of development. The photograph, sent by Dr. Bill Bates, a friend of both authors, shows a clutch of toad eggs developing in a small pond in north India. Only, the "pond" is rainwater collected in the footprint of an elephant. Who would have thought that elephants might be necessary for the completion of a toad's life cycle? As Dr. Ian Malcolm says in Jurassic Park, "I'm simply saying that life finds a way."

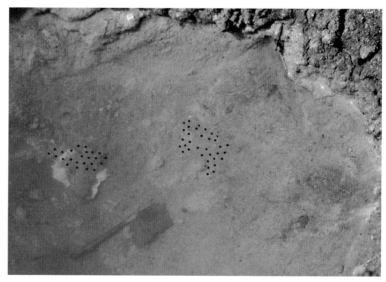

Toad eggs developing in the footprint of an Indian elephant. (Courtesy of W. Bates.)

Acknowledgments

We would like to express our thanks to our team at Sinauer Associates: Andy Sinauer, Azelie Fortier, Danna Lockwood, Lou Doucette, Chris Small, Jen Basil-Whitaker, Johannah Walkowicz, and David McIntyre. We would also like to thank Elizabeth Morales for transforming our sketches into the figures you see in this book.

Scott Gilbert gratefully acknowledges that he's been incredibly fortunate to have the members of the Swarthmore College biology department and the University of Helsinki Developmental Biology Center of Excellence to give him new ideas and to be sounding boards for him. Attending meetings has been crucial for getting new information, and Scott wishes to thank Swarthmore College, the Academy of Finland, NESCent, the NSF, AURA, and others for their support. Also important for this volume were discussions with Anna Cha, Ruston Hogness, O of Serenity House, Anna Tsing, and many others, including members of the FAU seminar, whose respectful comments and criticisms have been invaluable in framing these ideas. There are now entirely new sources of information coming from dedicated and knowledgeable biology bloggers such as Faye Flamm, P. Z. Meyers, Ed Yong, and Carl Zimmer, and Scott thanks them for having brought to his attention many papers he would not have seen.

David Epel thanks his mentors: Daniel Mazia, a modeler of questioning dogma; C. B. Van Niel, an exemplar of precise communication; and Donald P. Abbott, David's guide to the rabbit hole of invertebrate development. He thanks colleagues Sam Dupont, Amro Hamdoun, Donal Manahan, and George Somero for insights into adaptive cell biology and Mark Denny, Michael Hadfield, Chris Lowe, and John Pearse for melding development with ecology.

Special thanks go to the chapter reviewers: Ehab Abouheif, Wallace Arthur, Vince Formica, Mark Hanson, Donna Haraway, Randall Jirtle, Cris Ledon-Rettig, Armin Moczek, Michael Skinner, Carlos Sonnenschein, Ana Soto, Kathy Sulik, Alan Vajda, and Jeannette Wyneken. They provided remarkably insightful feedback (and stopped us from writing things we'd regret), and they were extremely supportive. Any errors, or course, are ours. Finally, we want to thank the scientists who reviewed the first edition of this textbook, as well. Their comments have been very important both personally and professionally, and they have allowed us to revise this book in the way it's been done.

Part 1

Environmental Signals and Normal Development

Ecological developmental biology studies the interactions between developing organisms and their environments. Part 1 identifies three major ways through which these interactions occur. **Chapter 1** details the phenomenon of *developmental plasticity*, documenting that the genome is a repertoire for the production of numerous different phenotypes. Various environmental agents—such as temperature and diet—elicit a particular phenotype from the possible range. **Chapter 2** details the mechanisms by which the environment might elicit those particular phenotypes. This chapter thus introduces us to the concept of the *environmentally induced epiallele*, inherited differences in chromatin structure rather than in DNA sequence, which can produce phenotypic differences. **Chapter 3** demonstrates that *symbiotic organisms*—usually commensal bacteria—are important sources of chemical signals that enable normal development. Thus, symbionts, epialleles, and developmental plasticity allow the environment to help construct the phenotypes of organisms.

Developmental Plasticity

The Environment as a Normal Agent in Producing Phenotypes

My soul is wrought to sing of forms transformed to bodies new and strange.

Ovid, 1 CE

A single genotype can produce many phenotypes, depending on many contingencies encountered during development. That is, phenotype is an outcome of a complex series of developmental processes that are influenced by environmental factors as well as genes.

H. F. Nijhout, 1999

Imagine a young aquatic organism developing in a particular pond. This organism has the ability to sense soluble biochemicals in the water—chemicals given off in the saliva or urine of its major predator. In the presence of these chemical signals, the organism's pattern of development changes, resulting in a phenotype that is less likely to be eaten by its predator. For instance, in the presence of the dragonfly larvae that feed on them, tadpoles of the gray tree frogs *Hyla chrysoscelis* and *H. versicolor* develop bright-red tails that deflect the predators' attention, and a set of trunk muscles that enables them to make "ice hockey turns" to escape being eaten (McCollum and Van Buskirk 1996; Relyea 2003a; Figure 1.1A).

Imagine an organism that develops different phenotypes depending on the season. *Nemoria arizonaria* larvae hatching on oak trees in the spring have a form that blends remarkably with young oak flowers ("catkins"). But caterpillars that hatch in the summer would be very conspicuous if they looked like the long-fallen oak flowers; thus the summer caterpillars resemble newly formed twigs (Figure 1.1B). Here, it is the larva's diet that determines its phenotype. Larvae who feed on young oak leaves will look like the catkins, while larvae eating older leaves (which have a different chemical composition) will develop to resemble twigs (Greene 1989).

Figure 1.1 Environmental cues can result in the development of completely different ▶ phenotypes in individuals of the same species. (A) Tadpoles of the tree frog *Hyla chrysoscelis* developing in the presence of cues from a predator's larvae (left) develop strong trunk muscles, and a red "warning" coloration. When predator cues are absent (right), the tadpoles grow longer and sleeker. (B) *Nemoria arizonaria* caterpillars that hatch in the spring (left) eat young oak leaves and develop a cuticle that resembles the oak's flowers (catkins). Caterpillars that hatch in the summer (right), after the catkins are gone, eat mature oak leaves and develop a cuticle that resembles young twigs. (C) A single male blue-headed wrasse (*Thalassoma bifasciatum*) swims with a cohort of the less colorful females. Should the male die, one of the females will grow testes, changing phenotype completely to become a male. (A courtesy of T. Johnson/USGS; B courtesy of E. Greene.)

Next, imagine an organism whose sex is determined not by its chromosomes, but by the environment the embryo experiences during a particular time during its development. In many species of fish, turtles, and alligators, sex is determined by the temperature of incubation. The same egg developing at one temperature will be male, but at another temperature it will be female. The blue-headed wrasse (*Thalassoma bifasciatum*), a Caribbean reef fish, is one of several fish species whose sex depends on the other fish it encounters (Figure 1.1C). When an immature wrasse reaches a reef where a single male lives and defends a territory with many females, the newcomer develops into a female. If the same immature wrasse had reached a reef that was undefended by a male, it would have developed into a male (Warner 1984). If the territorial male dies, one of the females (usually the largest) becomes a male; within a day, its ovaries shrink and testes grow (Godwin et al. 2000, 2003).

Consider now an organism with a set of cells that can recognize and attack invading viruses and bacterial cells. It has billions of such immune cells, each of which will divide and produce antibodies only when it binds to a particular virus or bacterium. While its genetic repertoire allows this organism to form billions of different types of immune cells, the actual number of different antibody-producing cell types in a given individual is only a fraction of this potential and will depend on which bacteria and viruses infect that individual. This same organism has the ability to regulate its muscular phenotype such that continued physical stress on a particular muscle will cause that muscle to grow. Furthermore, the brain development of this organism can be altered by experience, making learning possible. Moreover, parts of its digestive system develop in response to the many different bacteria residing symbiotically in its gut. This species, in which so much of the phenotype is due to environmental circumstances, is *Homo sapiens*.

Plasticity Is a Normal Part of Development

In each of the above instances, the environment has profound effects on the animal's phenotype. In other words, everything one needs for phenotype production is *not* packaged in the fertilized egg. This ability of a single individual to develop into more than one phenotype has been called **phenotypic plasticity** (Nilsson-Ehle 1914). Phenotypic plasticity was well known to late nineteenth-century embryologists, who showed that different environmental conditions produced different phenotypes during normal

(A) *Hyla chrysoscelis*

Predator present

Predator absent

(B) *Nemoria arizonaria*

Spring morph among catkins

Summer morph on twig

(C) *Thalassoma bifasciatum*

development (Nyhart 1995). Today we define phenotypic plasticity as the ability of an organism to react to an environmental input with a change in form, state, movement, behavior, or rate of activity (West-Eberhard 2003; Duckworth 2009). This plasticity is the property of the trait, not the individual; indeed, most individuals have several plastic traits. When seen in embryonic or larval stages of animals or plants, phenotypic plasticity is often referred to as **developmental plasticity**.

A century of studies

In his 1894 volume *The Biological Problem of Today: Preformation or Epigenesis?*, Oscar Hertwig summarized the studies demonstrating that development involved not only the interactions between embryonic cells, but also important interactions between developing organisms and their environments. He cited numerous cases of developmental plasticity, especially instances in which the sex of an organism was determined by the environment. These included the well-known case of *Bonellia viridis* (see the box), as well as temperature-dependent sex determination in rotifers, nutrition-dependent production of workers and queens in ant colonies (see Figure 1.8B), and temperature-dependent pigmentation patterns of butterfly wings (see Figure 1.5). Hertwig wrote (1894, p. 122), "These seem to me to show how very different final results may grow from identical rudiments, if these, in their early stages of development, be subjected to different external influences."

In 1909, two publications brought the concept of phenotypic plasticity to the awareness of many biologists. The Danish biologist Wilhelm Johannsen's *Elemente der Exakten Erblichkeitslehre* made clear the distinction between the genotype and phenotype (Figure 1.2). Rejecting August Weismann's 1893 proposal that all the causes of an embryo's development were compressed into the nucleus of the egg, Johannsen stated specifically that phenotype (what the organism looks like and how it behaves) is not merely the expression or actualization of the genotype (the set of inherited genes), but rather depends on the interactions of inherited genes with components of the environment. Like Hertwig, Johannsen felt that early development was genetically controlled but that the environment could effect changes in the later developmental stages (Moss 2003; Roll-Hansen 2007). Johannsen

Figure 1.2 One hundred years ago, Wilhelm Johannsen noted that the phenotype is the product of both the genome and environmental circumstances. Here he writes on the board that *Anlaegspraeg* ("genotype") + *Kaar* (Danish for "conditions" or "circumstances") gives *Fremtonnigspraeg* ("phenotype"). (Photograph from a movie of Professor Johannsen at www.wjc.ku.dk/library/video/original.avi.)

also believed that Weismann's refutation of the inheritance of acquired characteristics had not been complete.

Another important paper published in 1909 was from the German biologist Richard Woltereck, who reported that genetically identical lines of *Daphnia* (a water flea that reproduces asexually) could produce different phenotypes during different times of the year (see Figure 1.16). Woltereck argued that what actually was inherited was the *potential* to generate an almost infinite number of small variations in phenotype in response to environmental cues. He called this potential the **Reaktionsnorm (reaction norm** or **norm of reaction).*** Moreover, different pure lines (genotypes) responded differently to seasonal cues. This suggested to Woltereck that the reaction norm was heritable and that, as with any other trait, natural populations would harbor genetic variation in this potential.

However, after the 1920s and the rise of genetics, plasticity dropped out of the study of animal development. In the laboratory, we usually study development of only a few organisms and over a very narrow set of conditions. *Drosophila* are bred at 18°C, chick embryos at 37°C. Nutrition is similarly controlled. Indeed, as Bolker (1995, 2014) has pointed out, our "model organisms" for studying development have been selected for traits (early separation of germline and somatic lines; rapid generation) that would suppress the environmental contributions to the study of development. Since most of the contemporary developmental biology has focused on the genetic causation of cell differentiation and morphogenesis, the environmental effects have been seen as "noise." As evolution and developmental biology became sciences of gene frequency and gene expression, respectively, nongenetic sources of variation became marginalized and stopped being taught or discussed. So for most of the past half century, developmental biologists have not studied plasticity.

A contextually integrated view of life

The view expressed in the preceding examples is that the environment is not merely a filter that selects existing variations. Rather, it is a *source of variation*. The environment contains signals that can enable a developing organism to produce a phenotype that will increase its fitness in that particular environment. This isn't the view of life usually presented in today's textbooks or popular presentations of biology.

Since World War II, the dominant paradigm for explaining biodiversity has been genetics. Indeed, there has been marked antipathy against the notion of phenotypic plasticity and the inheritance of nonallelic phenotypic variation (Sarkar 2006; see Appendices A and C). The architects of twentieth-century biology (including such remarkable people as Ernst Mayr, Eric Davidson, and Jacques Monod) have emphasized the gene as

---

*Sarkar (1999) has pointed out that Woltereck actually concluded the *Reaktionsnorm* is what is inherited and that hereditary change consists of an alteration of the *Reaktionsnorm*. Woltereck identified the *Reaktionsnorm* with the genotype: "*Der 'Genotypus' ... eines Quantitativmerkmals ist die vereberte Reaktionsnorm*" (Woltereck 1909, p. 136). Johannsen agreed, saying that Woltereck's *Reaktionsnorm* was "nearly synonymous" with his own conception of genotype (Johannsen 1911, p. 133). As we will see later in this chapter, the ability of enzymes to have different properties at different temperatures ensures that all organisms have plasticity during their development.

Bonellia viridis: When the Environment Determines Sex

Some early studies of the effect of the environment on development included the fascinating case of sex determination in echiuroid worms, specifically *Bonellia viridis* (the green spoonworm). Females of this marine species have a deep-green, round body that burrows into rock crevasses and gravel on the seafloor, and from which is extended a projection (proboscis) that can grow up to a meter long. Males of *B. viridis* have an amazingly different phenotype; indeed, the males are rarely seen, being colorless, only about 3 mm long, and living parasitically inside the female's genital sac, where their sole function is to produce sperm to fertilize the female's eggs.

It has been known since the nineteenth century that the sex of a *B. viridis* individual depends solely on the environment in which the larva develops (Baltzer 1914; Leutert 1974; Jaccarini et al. 1983). Fertilized *B. viridis* eggs are expelled into the seawater. Larvae that settle and develop on the seafloor become female, maturing over several years as the proboscis extends. The female's cells, especially those of its proboscis, generate a powerful attractant to *B. viridis* larvae. Larvae passing within range of these signals will land on the proboscis of the sessile female and then crawl up into her mantle and/or be sucked into her gut, where they develop into the miniscule males.

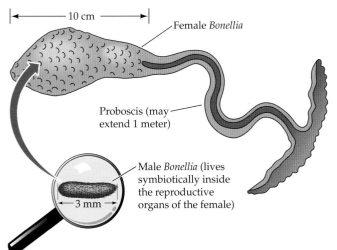

Sexual dimorphism in *Bonellia viridis*. While the body of the adult female remains buried in rocks or ocean sediments, her proboscis extends widely across the seafloor, where it is used for feeding. The proboscis also produces chemical signals that attract other *B. viridis* larvae, which, upon landing on the female, develop into males.

This mode of environmental sex determination enables the optimum use of space by females and the prevention of further competition for the limited burrowing regions (Beree et al. 2005).

being the core of animal identities and the "master molecule" of life. James Watson (1989) claimed, "We used to think our fate was in the stars. Now we know in large measure, our fate is in our genes." In his popular book *The Selfish Gene*, Richard Dawkins (1976) wrote of the genome as "the book of life" and proposed that our bodies are merely transient vehicles for the survival and propagation of our immortal DNA. In 1995, Nelkin and Lindee reviewed the popular accounts of DNA and concluded that DNA is being perceived as the secular equivalent of the soul. It is thought of as the essence of our being and that which determines our behaviors.* Richard Lewontin (1993) has also documented the dominance acquired by genetic determinism as an explanation of behavioral phenotypes.

But as we have seen (and will see much more of), genes are not the only explanation for animal diversity. During the past decade, interest in

*This is still a popular theme in the anti-Choice websites, where one is told that at fertilization, we receive our new DNA that pre-determines our physical and psychological traits for the rest of our lives (Gilbert 2008).

the environmental mechanisms of variation proposed early in the twentieth century has been renewed, fueled by new findings in conservation biology, developmental biology, public health, and evolutionary biology. The breakthroughs in molecular biology that have led to our exponentially expanded genomic knowledge also led to our current understanding of molecular signaling, casting a brilliant new light on the work of a century earlier. Just as in the 1600s the microscope revolutionized our view of life by revealing a previously invisible world, so in the twenty-first century technologies such as PCR (the polymerase chain reaction) and high-throughput RNA analysis have allowed us glimpses of a hitherto unsuspected world of interactions and interrelationships between genes and the environment. The result is a new perspective on life, its origins, and its interconnections.

"Eco-Devo": Embryology Meets Developmental Plasticity

Ecological developmental biology, casually known as **eco-devo**, is an approach to embryonic development that studies the interactions between a developing organism and its environment (Gilbert 2001; Sultan 2007). It focuses on how animals have evolved to integrate signals from the environment into their normal developmental trajectories. As we detail in this section, ecological developmental biology has three major sources: developmental plasticity (this chapter), environmentally induced gene configurations (Chapter 2), and developmental symbiosis (Chapter 3). Each of these phenomena is a means of producing phenotypic variation.

In many ways, ecological developmental biology is the extension of embryology to levels above that of the individual. In standard embryology, the focus has always been on the internal dynamics through which the genes of an individual's cell nuclei produce the phenotype of the organism. Within the past century, we have discovered that cell-cell communication is key to this phenomenon. By itself, the genetic information in a cell's nucleus cannot directly produce the many differentiated cell types in a multicellular organism; cells must interact, reciprocally instructing each other as they differentiate. Molecular signals called **paracrine factors** are released by one set of cells and induce gene expression changes in the cells adjacent to them. These neighboring cells, with their newly acquired characteristics, then produce their own paracrine factors that can change the gene expression of *their* neighbors—sometimes including the cell that originally induced them! By such cooperative signaling between cells, organs are formed.

But as Paul Weiss (1970) and others hypothesized, such molecular signals are not limited to the internally generated paracrine factors, but can also come from sources outside the organism. Oscar Hertwig (1894), Curt Herbst (1901), and others catalogued these environmental agents and discussed them as normal components in determining the phenotype of the embryo. Thus the same genotype can generate different phenotypes depending on what cues are present in the environment, allowing the embryo to change its developmental trajectory in response to environmental input.

Sonia Sultan (2007, p. 575) summarized the modern status of ecological developmental biology:

> Ecological developmental biology ("eco-devo") examines how organisms develop in "real-life" environments... [and] aims to provide an integrated framework for investigating development in its ecological context.... Eco-devo is not simply a repackaging of plasticity studies under a new name.... Whereas plasticity studies draw on quantitative genetic and phenotypic selection analyses to examine developmental outcomes and their evolution as adaptive traits, eco-devo adds an explicit focus on the molecular and cellular mechanisms of environmental perception and gene regulation underlying these responses, and how these signaling pathways operate in genetically and/or ecologically distinct individuals, populations, communities, and taxa.

Developmental plasticity is usually adaptive—that is, it makes the organism more fit for its environment. This idea that the developing organism has evolved mechanisms to receive and to respond to environmental cues to produce particular phenotypes has important evolutionary implications, which will be discussed more fully in Chapter 11. We will see that for evolutionary change to occur, the phenotypically plastic change may come first, followed by the genetic change. Moreover, as we will see in later chapters, there are times when plasticity is maladaptive—either when environmental cues alter development in a pathological manner, or when there is a mismatch between the phenotype induced by the embryonic environment and the environment experienced by the organism later in life. In both instances, developmental plasticity can give rise to disease, as will be discussed in the second section of the book, Chapters 4–8.

In most developmental interactions, the genome provides specific instructions, while the environment is permissive. That is to say, the genes determine what structures get made, and the only requirement of the environment is that it support and not disturb the developmental processes. Dogs will generate dogs and cats will beget cats, even if the animals live in the same house. However, in most species, there are instances in development when the *environment* plays the instructive role and the genome is merely permissive. In these instances, the environment determines what type of structure is made—but the genetic repertoire has to be capable of building that structure. The genetic ability to respond to environmental factors has to be inherited, of course, but in these cases it is the environment that directs the formation of the specific phenotype (Sarkar 1998; Gilbert 2001; Jablonka and Lamb 2005).

Reaction norms and polyphenisms

Two main types of phenotypic plasticity are currently recognized: reaction norms and polyphenisms (Woltereck 1909; Schmalhausen 1949; Stearns et al. 1991; West-Eberhard 2003). As mentioned earlier, in a reaction norm, the genome encodes a *continuous range* of potential phenotypes, and the environment the individual encounters determines the phenotype. One obvious example is muscle hypertrophy in humans. The size of our muscle depends on environmental conditions—how much load it experiences.

(A)   (B)

Figure 1.3   Density-induced polyphenism in the desert (or "plague") locust *Schistocerca gregaria*. (A) The low-density morph has green pigmentation and miniature wings. (B) Triggered by crowding, the high-density morph develops with deep pigmentation and wing and leg development suitable for migration. (From Tawfik et al. 1999, courtesy of S. Tanaka.)

Those people who exercise have larger muscles—but only within hereditarily defined limits. Reaction norms allow developing organisms to "titrate" their responses to the strength of a signal. Tadpoles of the wood frog *Rana sylvatica*, for instance, respond to the presence of predators by developing deeper tails and shorter bodies. Moreover, the larger the predation risk (as measured by the chemical secreted by the predators), the deeper the tail and the shorter the body (Relyea 2004).

The second type of phenotypic plasticity, **polyphenism**, refers to *discontinuous* (either/or) phenotypes elicited by the environment. One obvious example is sex determination in turtles, where one range of temperatures induces female development in the embryo, while a different range of temperatures elicits male development. Between these two ranges is a small band of temperatures that will produce different proportions of males and females—but these intermediate temperatures do not induce intersexual animals.

An important example of polyphenism is seen in the migratory locust *Schistocerca gregaria*. These plant-eating grasshoppers exist in two mutually exclusive forms (Figure 1.3): they are either short winged, uniformly green, and solitary or they are long winged, brightly colored, and gregarious (Pener 1991; Rogers et al. 2003, 2004). The phenotypic differences between these two morphs are so striking that only in 1921 did the Russian biologist Boris Usarov finally realize they were the same species. Cues in the environment determine which morphology a young locust will develop. The major stimulus appears to be population density, as measured by the rubbing of legs. When locust nymphs get crowded enough that a certain neuron in the hind femur is stimulated by other nymphs, their developmental pattern changes, and the next time they molt, they emerge with long, brightly colored wings, as well as with gregarious (flocking) and migratory behaviors.

The different phenotypes induced by the environment are sometimes called **morphs** or **ecomorphs**. Genetically identical animals can have different morphs depending on the season, their larval diet, or other signals present in the environment. Confusingly, the phenotypes produced by different genetic alleles are called either "mutants" (if rare) or "polymorphisms" (if common—arbitrarily defined as found in more than 5% of the population). Polymorphism is therefore a condition where variation is the product of genetic differences, while polyphenism is a condition where variation (different morphs) is the product of environmental signals (Mayr 1963).

Epigenetics

In 1968, Waddington coined the term "epigenetics" to describe a way of integrating the series of ordered interactions in development (epigenesis) with genetics (see Van Speybroeck 2002). Since then, epigenetics has been redefined as the set of mechanisms involved in regulating gene activity during development. For instance, the methylation of certain regions of genes suppresses their expression. This is very important in preventing genes from being expressed in the wrong types of cells or at the wrong times. By these epigenetic mechanisms, hemoglobin is made in red blood cell precursors and in no other type of cell, and insulin is made only in the beta cells of the pancreas. This is discussed more fully in Appendix B. **Epigenetics** is defined in this book as those genetic mechanisms that create phenotypic variation without altering the base-pair nucleotide sequence of the genes. Specifically, we use this term to refer to those mechanisms that cause variation by altering the *expression* of genes rather than their sequence.

As we will see in the next chapter, epigenetic mechanisms can integrate genomic and environmental inputs to generate the instructions for producing a particular phenotype. Epigenetic investigations have become focused on the mechanisms of phenotypic plasticity and on how changes in gene expression patterns mediated by the environment can cause diseases (such as cancers and hypertension).

The term **epigenetic inheritance** has been used to denote heritable phenotypes that are not encoded in the genome (Jablonka and Lamb 2005; Jablonka and Raz 2008). Epigenetic inheritance includes:

- Variations inherited over *cell* generations, such as changes in chromatin that stabilize a particular cell type during normal development. For instance, during the course of their differentiation, mammalian liver precursor cells obtain a chromatin configuration that instructs their liver-specific genes to function, and henceforth all liver cells "remember" this chromatin configuration and maintain it over progressive cell divisions.

- Variations inherited from one *organismal* generation to the next. For instance, certain drugs can induce changes in the chromatin structure in the nuclei of mouse cells, including the mouse's germ cells. The progeny of such a mouse can inherit the drug-induced chromatin change from its parents, even if the drug is no longer present (see Chapter 6). There are also some variations that can be inherited from one generation to the next by modifying maternal nursing behavior (see Chapter 2).

- The inheritance of symbionts from one generation to another. Chapters 3 and 11 will document that these microorganism help construct the animal body and can be a source of heritable variation.

Among humans (and possibly among other animals; see Avital and Jablonka 2000), variations in cultural inheritance represent another inheritance pattern that is not mediated by changes in DNA.

Agents of developmental plasticity

Possibly, every organism has environmentally determined components in its phenotype. When asked, "Where does polyphenism occur among the insects?" Simpson and colleagues (2011) reply, "'Everywhere' is the brief answer." Therefore, a complete list of organisms with phenotypic plasticity would resemble a survey of all the planet's eukaryotes. Examples of how the environment acts in normal development are given in Table 1.1,

---

TABLE 1.1 Some environmental contributors to phenotype development

Context-dependent normal development (Chs. 1, 2)

A. Morphological polyphenisms
1. Nutrition-dependent (*Nemoria*, hymenoptera castes)
2. Temperature-dependent (*Arachnia*, *Bicyclus*)
3. Density-dependent (locusts)
4. Stress-dependent (*Scaphiopus*)

B. Sex determination polyphenisms
1. Location-dependent (*Bonellia*)
2. Temperature-dependent (*Menidia*, turtles)
3. Social-dependent (wrasses, gobys)

C. Predator-induced polyphenisms
1. Adaptive predator-avoidance morphologies (*Daphnia*, *Hyla*)
2. Adaptive immunological responses (mammals)
3. Adaptive reproductive allocations (ant colonies)

D. Stress-induced bone formation
1. Prenatal (fibular crest in birds)
2. Postnatal (patella in mammals; lower jaw in humans?)

Context-dependent life cycle progression (Chs. 2, 3)

A. Larval settlement
1. Substrate-induced metamorphosis (bivalves, gastropods)
2. Prey-induced metamorphosis (gastropods, chitons)
3. Temperature/photoperiod-dependent metamorphosis

B. Diapause
1. Overwintering in insects
2. Delayed implantation in mammals

C. Sexual/asexual progression
1. Temperature/photoperiod-induced (aphids, *Megoura*)
2. Temperature/colony-induced (*Volvox*)

D. Symbioses/parasitism
1. Blood meals (*Rhodnius*, *Aedes*)
2. Commensalism (*Euprymna/Vibrio*, eggs/algae, mammalian gut microbiota)
3. Parasites (*Wolbachia*)

E. Developmental plant-insect interactions

Adaptations of embryos and larvae to environments (Ch. 4)

A. Egg protection
1. Sunscreens against radiation (*Rana*, sea urchins)
2. Plant-derived protection (*Utetheisa*)

B. Larval protection
1. Plant-derived protection (*Danaus*, tortoise beetles)

Source: Gilbert 2001.

Note: This list should not be thought to be inclusive. For example, the list is limited to animals; plant developmental plasticity and many plant-animal interactions have not been included here.

which shows some of the numerous environmental agents that contribute to producing normal phenotypes, including:

- Temperature
- Nutrition
- Pressure and gravity
- Light
- The presence of dangerous conditions (predators or stress)
- The presence or absence of conspecifics (other members of the same species)

The remainder of this chapter describes how environmental cues affect the course of normal development in a variety of species. In subsequent chapters, more specific details about the mechanisms of developmental plasticity will be discussed.

Temperature-Dependent Phenotypes

Temperature is the causal factor in a number of phenotypes: when temperature differences cause amino acid chains to fold differently, traits (determined by enzymes) are turned on or off. Often, temperature will cause a suite of morphological and behavioral changes. Indeed, in some species, different temperatures cause embryos to develop as either male or female, each with its own set of behavioral characteristics. Thus, the organism is born with the possibility for both sets of organs and behaviors, and the temperature will select both the organ and the behavior that goes with it.

Enzyme activity as a function of temperature

Nearly all enzyme activity is temperature-dependent. This concept is often expressed as the enzyme's Q_{10}, or the ratio of its activities at two temperatures, one 10°C higher than the other. Temperature can cause changes in the way a protein folds and thereby determine the shape of an enzyme's active site and the sites of interaction with other proteins. One example of such a protein is the tyrosinase enzyme variant found in Siamese cats and Himalayan rabbits (Figure 1.4). Tyrosinase is critical for making melanin, the dark pigment of vertebrate skin. (Indeed, mutations that block melanin production result in albinism, the lack of dark pigment throughout the body.) The mutation that creates the phenotype of Siamese cats and Himalayan rabbits transforms tyrosinase from an enzyme that is not temperature-dependent (in the physiological ranges expected in an organism) into a temperature-dependent enzyme. In these animals, tyrosinase folds properly at relatively cold temperatures but does not fold properly—and thus does not work—at warmer temperatures. Cooler temperatures are normally found at the extremities (the tips of the ears, the paws and tail, and part of the snout), with warmer temperatures throughout the major parts of the body (Schmalhausen 1949). Thus, tyrosinase functions (and melanin pigment is made) only in the extremities of Siamese cats and Himalayan

Figure 1.4   The dark pigment melanin is synthesized only in the colder areas of the vertebrate skin in Siamese cats and Himalayan rabbits. This is due to a mutation in the gene for tyrosinase—the rate-limiting enzyme of melanin synthesis—that renders the protein heat-sensitive. In the colder extremities of the body, tyrosinase folds properly and melanin (dark pigment) is produced. In warmer regions of the body, however, the enzyme folds improperly and cannot function, thus limiting the production of melanin.

rabbits, demonstrating that enzymes are affected by temperature and that their subsequent responses can have large impacts on phenotype.*

There are analogous conditions in humans in which only the hair at the extremities is pigmented (Berson et al. 2000). These conditions result from a single G → A mutation that replaces the positively charged amino acid arginine at position 402 of tyrosinase with the uncharged glutamine.

The induction of melanin pigment by the environment is part of our plastic response to the environment and is the basis for suntanning[†] (D'Orazio et al. 2006; April and Barsh 2007). By increasing epidermal melanin content, tanning is the skin's major response against acute and ultraviolet light-induced damage (Chen et al. 2014). Plants also have an inducible system for melanin production. When certain fruits are cut, melanin is induced, creating a dark, protective meshwork that prevents bacterial and fungal penetration. This is why apples, potatoes, and bananas turn brown

---

*Both Siamese and Burmese cats possess mutations of the tyrosinase gene, but they occur at slightly different sites within the gene. This apparently produces different thresholds for gene activity, allowing the Burmese breeds to have darker body color (Schmidt-Küntzel et al. 2005).

[†]There is an unexpected side-effect to this induction. UV light induces the epidermal cells to produce the pro-hormone proopiomelanocortin. This protein is processed into several other smaller proteins. One of these is melanocyte-stimulating hormone, which induces the production of melanin that is characteristic of tanning. Another product, however, is β-endorphin, which gives pleasurable sensations. Fell and colleagues (2014) have evidence that this β-endorphin can cause addiction to tanning and a subsequent increase in melanoma tumors.

when sliced. (See Szent-Györgyi 1966 and Bachem et al. 2004 for discussion of the importance of this reaction.)

Seasonal polyphenism

Since enzymes (and presumably other proteins, such as transcription factors) can be influenced by temperature, it is not surprising that animals have evolved such that thermal cues can cause different phenotypes at different seasons. Ecologists have long known that in North America, the pigmentation of many butterfly species follows a seasonal pattern. Throughout much of the Northern Hemisphere, one can see such a polyphenism in butterflies of the family Pieridae (the cabbage whites), with phenotypes that differ between individuals that eclose from their pupa during the long days of summer and those that eclose at the beginning of the season, in the shorter, cooler days of spring. The hindwing pigments of the spring forms are darker than those of the summer butterflies (Figure 1.5). Pigmentation has a functional advantage during the cooler months: darker pigments absorb sunlight more efficiently than lighter ones, raising the body temperature more rapidly (Shapiro 1968; Watt 1968; see also Nijhout 1991). As we will see later, temperature may effect these changes in color by affecting the production of hormones needed for growth and differentiation.

Seasonal changes in fur color are typical for many mammals that thrive in winter snow. Their summer pelage is brown or gray, blending into the trees and grasses, but when the average amount of daylight (photoperiod) gets progressively less, hormones in their bodies activate those genes producing the white fur that camouflages them in the snow. The photoperiod,

Figure 1.5 Polyphenic variation in *Pontia* (Pieridae) butterflies. The top row shows summer morphs: *P. protodice* female (left) and male (center), *P. occidentalis* male (right). The bottom row shows spring morphs, which have a more highly pigmented ventral hindwing: *P. protodice* female (left) and male (center), *P. occidentalis* male (right). (Photograph courtesy of T. Valente.)

not temperature, is the cue, and it has been shown to be a very accurate one (Grange 1932; Flux 1970). Global climate change, however, can cause mismatch between the coat color and background. As snow is coming much later to areas of the northern United States, snowshoe hares are turning white long before the first snowfall (Mills et al. 2013; Zimova et al. 2014). Conservation biologists are concerned that this might lead to the elimination of this species in many areas of its range.

Temperature and sex

Aristotle—a noteworthy naturalist and history's first embryologist—made few major errors in his embryological descriptions. One of these, however, was to attribute human sex determination to temperature (Aristotle 355 BCE). He felt that maleness was generated through the heat of the semen, and he encouraged elderly men to mate in the summertime if they desired male heirs.

Although Aristotle was wrong about temperature having a role in human sex determination, in many species, temperature *does* control whether an embryo develops testes or ovaries. Indeed, among certain reptile groups (turtles and crocodilians), there are many species in which the temperature at which an embryo develops determines whether an individual is male or female (Figure 1.6). This type of environmental sex determination, which also is found in certain fishes, has advantages and disadvantages.

One probable advantage is that it gives the species the benefits of sexual reproduction without tying the species to a 1:1 sex ratio. In crocodiles, in which temperature extremes produce females while moderate temperatures produce males, the sex ratio may be as great as 10 females to each male (Woodward and Murray 1993). In instances where the population size is limited by the number of females, such a ratio is more advantageous than the 1:1 ratio usually resulting from genotypic sex determination.

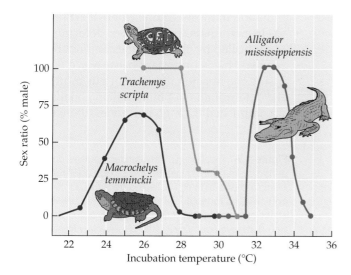

Figure 1.6 Temperature-dependent sex determination in three different reptilian species: the American alligator (*Alligator mississippiensis*), the red-eared slider turtle (*Trachemys scripta*), and the alligator snapping turtle (*Macrochelys temminckii*). (After Crain and Guillette 1998.)

The major disadvantage of temperature-dependent sex determination may involve its narrowing of the temperature range within which a species can persist. Thus, thermal pollution (either local or due to global warming) could conceivably eliminate a species from a given area (Janzen and Paukstis 1991). Researchers have speculated that dinosaurs may have had temperature-dependent sex determination, and that their sudden demise may have been the result of a slight change in temperature that created conditions wherein only males or only females hatched (Ferguson and Joanen 1982; Miller et al. 2004). Unlike turtles, which have long reproductive lives, can hibernate for years, and whose females can store sperm, dinosaurs may have had a relatively narrow window of time in which to reproduce and lacked the ability to hibernate through long stretches of bad times.

Charnov and Bull (1977) argued that environmental sex determination would be adaptive in those habitats characterized by patchiness—that is, habitats having some regions where it is more advantageous to be male and other regions where it is more advantageous to be female. Conover and Heins (1987) provided evidence for this hypothesis. In certain fish species, females benefit from being larger, since larger size translates into higher fecundity. If you are a female Atlantic silverside (*Menidia menidia*), it is advantageous to be born early in the breeding season, because you have a longer feeding season and thus can grow larger. (The size of males in this species doesn't influence mating success or outcomes.) In the southern range of *M. menidia*, females are indeed born early in the breeding season, and temperature appears to play a major role in this pattern. However, in the northern reaches of its range, the species shows no environmental sex determination. Rather, a 1:1 sex ratio is generated at all temperatures (Figure 1.7). Conover and Heins speculated that the more northern populations have such a short feeding season, there is no reproductive advantage for females in being born earlier. Thus, this fish has environmental sex determination in those regions where it is adaptive, and genotypic sex determination in those regions where it is not.

In mammals, primary sex determination is controlled by chromosomes and not by hormones. This is important because we develop inside the hormonal milieu of our mothers. If the determination of mammalian gonads were accomplished through hormones, there would

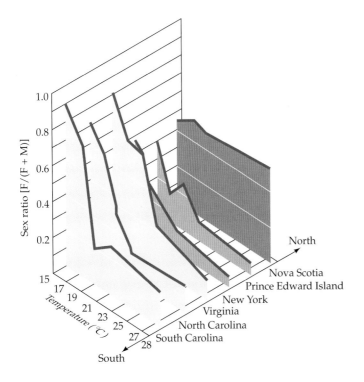

Figure 1.7 Relationship between temperature during the period of sex determination and sex ratio [F/(F + M)] in *Menidia menidia*. In fish collected from the northernmost portion of its range (Nova Scotia), temperature had little effect on sex determination. However, among fish collected at more southerly locations (especially from Virginia through South Carolina), temperature had a large effect. (After Conover and Heins 1987.)

be no males. In mammals, two stages of sex determination have evolved. Primary sex determination is controlled by the X and Y chromosomes, which determine whether the gonads differentiate as ovaries or as testes. Secondary sex determination is accomplished by the hormones (testosterone, estrogen, and others) made by the gonads. This second stage is responsible for the male- and female-specific external genitalia, as well as for the differentiation of the uterus and oviducts in females and the development of the spermatic ducts in males.

In other vertebrates (including fishes, amphibians, and birds), the hormones estrogen and testosterone appear to be responsible for making ovaries or testes, respectively. The enzyme responsible for controlling the ratio of these hormones is **aromatase**, which converts testosterone into estrogen. Aromatase has been found to be temperature-regulated in several vertebrate species (Kroon et al. 2005). As we will see in later chapters, the enzyme is a target for environmental mutagens that can seriously alter the sexual development of a number of vertebrate species.*

Nutritional Polyphenism: What You Eat Becomes You

The food an organism eats may contain powerful chemical signals that induce phenotypic changes. We saw at the start of the chapter that the larval phenotype of the moth *Nemoria arizonaria* depends on its diet (see Figure 1.1B). Such effects are not uncommon among insects.

Royal jelly and egg-laying queens

In hymenopteran insects (bees, wasps, and ants), the determination of queen and worker castes can be effected by several factors, including genes, nutrition, temperature, and even volatile chemicals secreted by other members of the hive. In the honeybee, new queens are generated within 2 weeks after the death of the preceding queen (or in anticipation of the colony's splitting and a second queen being needed); they are almost never produced otherwise. Queen formation is dependent almost entirely on diet. A larva fed "royal jelly" (a protein-rich food that contains secretions from the workers' salivary glands) for most of its larval life will be a queen (with functional ovaries), while a larva fed a poorer diet (and given royal jelly for only a brief time late in larval development) will become a sterile worker (Figure 1.8A). In addition to nutritive proteins, the royal jelly contains a relatively small protein dubbed "royalactin" that increases the juvenile hormone titer, increases body size, and speeds up development (Kamakura 2011).

---

*This sex-altering property of aromatase was useful in an experiment that demonstrated the adaptive value of temperature-dependent sex determination. In the jacky dragon lizard (*Amphibolurus muricatus*), males are produced at intermediate temperatures (around 27°C), whereas both higher and lower temperatures produce females. By using aromatase inhibitors to block the conversion of testosterone to estrogen, Warner and Shine (2008) were able to produce males throughout the temperature range 23°C–33°C. There were no morphological differences among the males produced at any of these temperatures, but males produced at the intermediate temperatures had significantly better fitness (i.e., they sired more progeny) than the males produced at the extreme (normally female-producing) temperatures. The reason for this increased fitness as yet remains unknown.

(A)

(B)

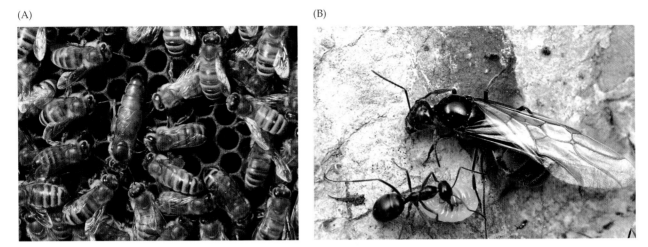

Figure 1.8   Reproductive queens in hymenopteran colonies. (A) The queen of the honeybee (*Apis mellifera*) with her sister workers. (B) Gyne (reproductive queen) and worker of the ant *Pheidologeton*. This picture shows the remarkable dimorphism between the large queen and the small worker (seen near the queen's antennae). The difference between these two sisters involves larval feeding and juvenile hormone synthesis.

The increase in juvenile hormone is very important for the development of fertile queens. Larvae can become queens only if they reach a certain size before metamorphosis. A larva continually fed royal jelly from a relatively early stage retains the activity of a structure called the corpora allatum throughout its larval stages. The corpora allatum secretes juvenile hormone (JH), which delays metamorphosis, allowing the larva to grow larger and to have functional ovaries (Brian 1974, 1980; Plowright and Pendrel 1977). The rate of JH synthesis in the "queen larvae" is 25 times greater than the rate of synthesis in larvae not fed royal jelly. Applying large amounts of JH to worker larvae late in life can transform them into queens (Wirtz 1973; Rachinsky and Hartfelder 1990). Thus, the queen does not achieve her large and fertile status due to a genetic predisposition, but from nutritional supplementation.

Similarly, ant colonies are predominantly female, and the females can be very different in size and function. The much larger reproductive females ("gynes" or "queens") have fully functional ovaries and wings; the workers do not (Figure 1.8B). These striking differences in anatomy and physiology are also regulated through juvenile hormone (Wheeler 1991). The influence of the environment on hormone levels and gene expression in ants was analyzed by Abouheif and Wray (2002), who found that nutrition-induced JH levels regulated wing formation. In the queen, both the forewing and the hindwing disc undergo normal development, expressing the same genes as *Drosophila* wing discs. However, in the wing imaginal discs of workers, some of these genes remain unexpressed, and the wings fail to form.

Horn length in the male dung beetle

The structural and behavioral male phenotypes of some male dung beetles depend on the quality and quantity of the nutrition—in the form of maternally provided dung—that they have access to during development (Emlen 1997; Moczek and Emlen 2000). In dung beetle species such as *Onthophagus taurus* and *O. acuminatus*, males have the ability to grow horns while females do not. The hornless female beetle gathers manure, digs tunnels, and places balls of dung in brood chambers that she constructs at the ends of the tunnels. She then lays a single egg on each cluster of dung, and when the larvae hatch, they eat the dung. Metamorphosis occurs when the dung cluster is consumed.

The amount of food affects the titer of juvenile hormone present during the developing beetle's last larval molt. In the males, the last organs to form are the horns. The size of the larva at metamorphosis determines the titer of JH, and the titer of JH affects the growth of the ectodermal regions that make the horns (Emlen and Nijhout 1999; Moczek 2005; Figure 1.9A). If juvenile hormone is added to an *Onthophagus* male larva during the sensitive period of his last molt, the cuticle in its head expands to produce a horn. The male horn does not grow unless the beetle larva reaches a certain size. Above this threshold body size, horn size is proportional to body size. Thus, although body size has a normal distribution, there is a bimodal distribution of horn sizes. About half the males (the small-bodied ones) have no horns, while the other half have horns of considerable length (Figure 1.9B).

(A)   Horned male   Hornless male

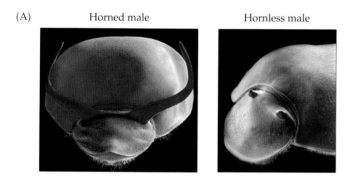

(B)

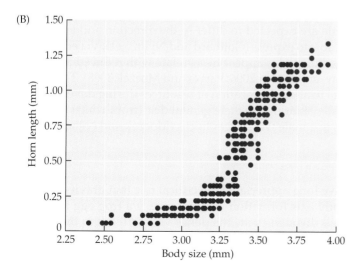

Figure 1.9 Diet and *Onthophagus* horn size. (A) Horned and hornless males of the dung beetle *Onthophagus acuminatus* (horns have been artificially colored). (B) Whether a male of this species is horned or hornless is determined by the titer of juvenile hormone at the last molt. This hormone titer depends in turn on the size of the larva. There is a sharp threshold of body size, before which horns fail to form and after which horn growth is linearly correlated with the size of the beetle. This threshold effect produces populations with no horns and with large horns, but very few with horns of intermediate size. (After Emlen 2000, photographs courtesy of D. Emlen.)

Figure 1.10 The presence or absence of horns determines the male reproductive strategy in some dung beetle species. Females dig tunnels in the soil beneath a pile of dung and bring dung fragments into the tunnels. These will be the food supply of the larvae. Horned males guard the entrances to the tunnels and mate repeatedly with the females. They fight to prevent other males from entering the tunnels, and those males with long horns usually win such contests. Smaller, hornless males do not guard tunnels. Rather, they dig their own tunnels to connect with those of females, mate, and exit. (After Emlen 2000.)

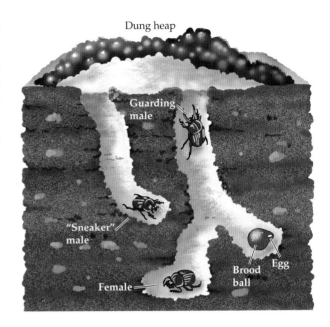

The behaviors of horned and hornless males also differ. Horned males guard the females' tunnels and use their horns to prevent other males from mating with the females; the male with the biggest horns wins such contests. But what about the males with no horns? Hornless males do not fight with the horned males for mates. Since they, like the females, lack horns, they are able to dig their own tunnels without the horns getting in the way. These "sneaker" males dig tunnels that intersect with those of the females and mate, while the horned males stand guard at the tunnel entrances (Figure 1.10; Emlen 2000; Moczek and Emlen 2000). Both strategies appear to be highly successful.

The heritability of horn length is zero; it is a phenotype that is environmentally determined by the response of the endocrine system to food intake. However, the size threshold a male larva must reach in order to produce a horn is a property of the genome that can be selected. Different species of beetle are expected to differ in the direction and amount of plasticity they are able to express (Gotthard and Nylin 1995; Via et al. 1995). Interestingly, large horns also appear to correlate with reduced penis and testis size (Simmons and Emlen 2006; Parzer and Moczek 2008). This **trade-off** is probably due to altered allocation of resources during development, since experimentally ablating the male genital disc (from which the penis originates) results in a male with larger horns (Moczek and Nijhout 2004).

Gravity and Pressure

Embryologists have long appreciated the critical role that gravity plays in frog and chick body axis formation. For instance, if a frog egg is rotated during the first cell division cycle, the dense yolk will fall to the bottom

(A)   (B)

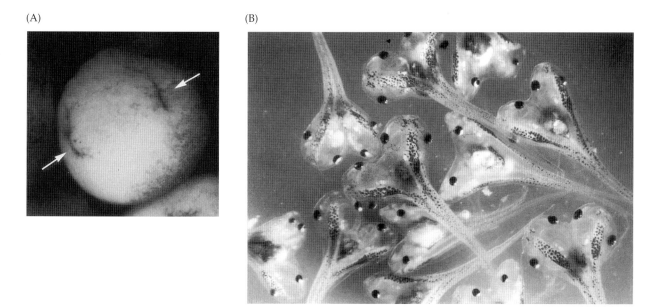

Figure 1.11   Expectation of a 1-G gravitational field in frog development. The dense yolk of a frog egg instructs a single axis to form, defined by the gravitational force on the yolk. If a frog egg is rotated during first cleavage, the yolk travels into a new area of the egg, displacing its contents. As a result, gastrulation (the beginning of organ formation) is initiated at two sites rather than one site (A) and two axes form, each with a fully developed head (B). (Photographs courtesy of J. Gerhart.)

of the rotated egg and displace the proteins and mRNA molecules there. This can result in the formation of two heads, one defined by the old axis, and one defined by the new axis (Kirschner et al. 1980; Figure 1.11). The axes will not form correctly, or several axes will form—in which case the embryo will have more than one head. Bird eggs use the force of gravity to form their anterior-posterior (head-to-tail) axis (see Gilbert 2013).

More recent experiments have shown that the human body requires a 1-G gravitational field for the proper development and maintenance of bones and muscles. Astronauts experiencing weightless conditions undergo severe muscle atrophy. As Figure 1.12 shows, weightlessness results in dramatic structural changes in the muscles, leading to tears and loss of strength and coordination. Spending 11 days in microgravity (without exercising) can cause a 30% shrinkage in the mass of certain muscles (NASA 2003). Several genes necessary for muscle differentiation and maintenance—including those genes encoding the transcription factors MyoD and myogenin—are not expressed in microgravity conditions (Inobe et al. 2002). Moreover, in mice and rats, genes encoding proteins that support mitochondria and muscle growth also fail to function without normal gravity (Nikawa et al. 2004; Allen et al. 2009).

In addition to muscle, the formation of several vertebrate bones is dependent on gravity (or on pressure from the environment). Such stresses are known to be responsible for the formation of the human patella (kneecap)

(A)

(B)

Lipid droplets

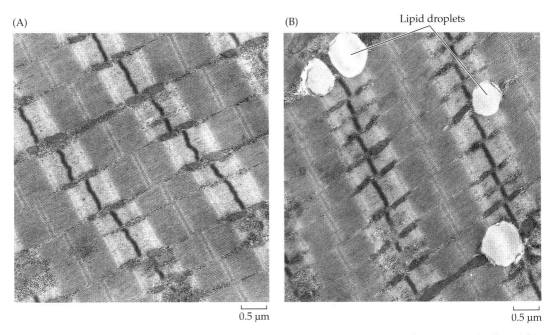

0.5 μm

0.5 μm

Figure 1.12 Human soleus muscle tissues, showing the effects of exercise in weightless conditions. (A) Exercised tissue, where gravitational load stimulates the production of proteins that keep muscle fibers strong. (B) After 17 days in microgravity, muscle protein synthesis has slowed down. The muscle cells have grown more irregular and show signs of atrophy. The prevalence of lipid droplets indicates that in microgravity, the muscle cells store fat instead of using it for energy. (From Widrick et al. 1999.)

after birth and have also been found to be critical for jaw growth in humans and fish. The jaws of cichlid fish differ enormously, depending on the food they eat (Figure 1.13; Meyer 1987). Similarly, normal human jaw development may be predicated on expected tension due to grinding food: mechanical tension appears to stimulate the expression of the *indian hedgehog* gene in mammalian mandibular cartilage, and this paracrine factor stimulates cartilage growth (Tang et al. 2004). If an infant monkey is given

(A)

(B)

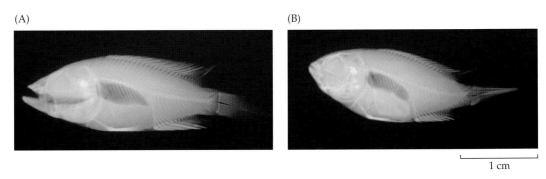

1 cm

Figure 1.13 X-rays of cichlid fish fed different diets for 8 months. (A) Fish fed shrimp larvae developed a narrow-angled jaw. (B) Fish fed commercial flaked food and nematodes had a wider-angled jaw. (From Meyer 1987, photographs courtesy of Axel Meyer.)

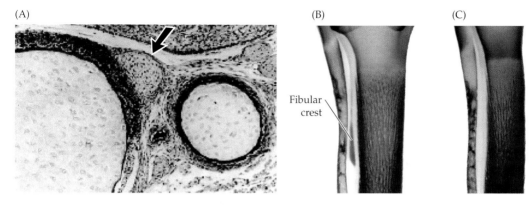

Figure 1.14   Activity-induced formation of the fibular crest. The fibular crest (syndesmosis tibiofibularis) is formed when the movement of the embryo in the egg puts stress on the tibia. (A) Transverse section through the 10-day embryonic chick limb, showing the condensation (arrow) that will become the fibular crest. (B) A 13-day chick embryo, showing fibular crest forming between the tibia and fibula. (C) Absence of fibular crest in the connective tissue of a 13-day embryo whose movement was inhibited. The blue dye stains cartilage; the red dye stains the bone elements. (From Müller 2003, photographs courtesy of G. Müller.)

soft food, its lower jaw is smaller than normal. Corruccini and Beecher (1982, 1984) and Varela (1992) have shown that people in cultures where infants are fed hard food have jaws that "fit" better, and they speculate that soft infant food explains why so many children in Western societies need braces on their teeth. Indeed, the notion that mechanical tension can change jaw size and shape is the basis of the functional hypothesis of modern orthodontics (Moss 1962, 1997).

In the chick, several bones do not form if the embryo's movement inside the egg is suppressed. One of these bones is the fibular crest, which connects the tibia directly to the fibula. This direct connection is believed to be important in the evolution of birds, and the fibular crest is a universal feature of the bird hindlimb (Müller and Steicher 1989). When the chick is prevented from moving within its egg, the fibular crest fails to develop (Figure 1.14; Wu et al. 2001; Müller 2003).

Predator-Induced Polyphenisms

At the beginning of this chapter, we asked you to imagine an animal that is frequently confronted by a particular predator, that could recognize soluble molecules secreted by that predator, and that could use those molecules to activate the development of structures that would make this individual less likely to be eaten. This ability to modulate development in the presence of predators is called predator-induced defense, or **predator-induced polyphenism**.

To demonstrate predator-induced polyphenism, one has to show that the phenotypic modification is caused by the presence of the predator. In addition, many investigators say, the modification should increase the

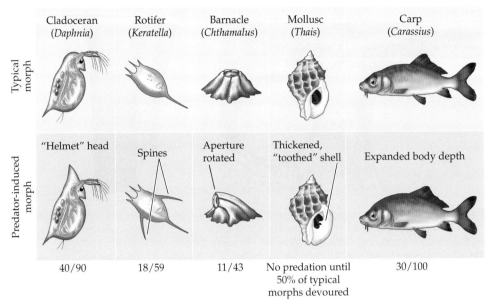

Figure 1.15   Predator-induced defenses. Typical (upper row) and predator-induced (lower row) morphs of various organisms are shown. The numbers beneath each column represent the percentages of organisms surviving predation when both induced and uninduced individuals were presented with predators (in various assays). (Data from Adler and Harvell 1990 and references cited therein.)

fitness of its bearers when the predator is present (see Adler and Harvell 1990; Tollrian and Harvell 1999). Figure 1.15 shows both the typical and predator-induced morphs for several species. In each case, the induced morph is more successful at surviving the predator, and soluble filtrate from water surrounding the predator is able to induce the changes. The chemicals that are released by a predator and that induce defenses in its prey are called **kairomones**.

One important concept to remember is that, as with the larger horns of male dung beetles mentioned earlier, there is usually a trade-off. That is, the energy and material used to produce the adaptation to the predator often come at the expense of making other organs or cells. Thus, what is adaptive in one environment is less adaptive in another. *Daphnia*, for instance, make a spiked "helmet" in the presence of kairomones (Figure 1.16). However, helmeted *Daphnia* individuals produce fewer eggs than their smaller counterparts. Similarly, tadpoles that develop quickly in order to escape predators are usually less robust than those that take the full time developing.

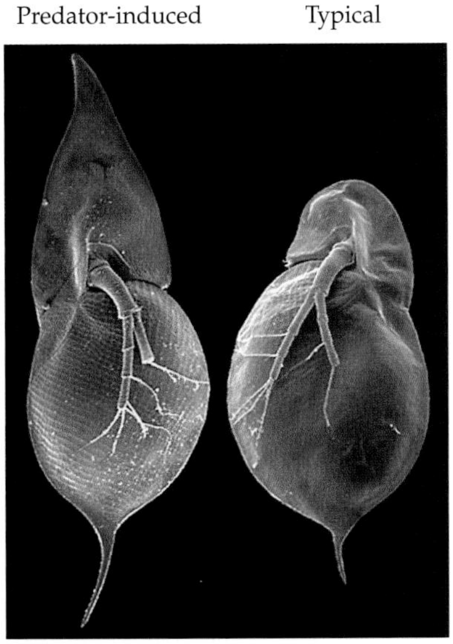

Predator-induced   Typical

Figure 1.16   Scanning electron micrographs showing predator-induced and typical morphs of genetically identical individuals of the water flea genus *Daphnia*. (Photographs by C. Laforsch and R. Tollrian, courtesy of A. A. Agrawal.)

Predator-induced polyphenism in invertebrates

Several rotifer species will alter their morphology when they develop in pond water in which their predators were cultured (Dodson 1989; Adler and Harvell 1990). The predatory rotifer *Asplanchna* releases a soluble compound that induces the eggs of a prey rotifer species, *Keratella slacki*, to develop into individuals with slightly larger bodies and anterior spines 130% longer than they otherwise would be (see Figure 1.15), making the prey more difficult to eat. Also shown in Figure 1.15, the snail *Thais lamellosa* develops a thickened shell and a "tooth" in its aperture when exposed to water that once contained the crab species that preys on it. In a mixed snail population, crabs will not attack the thicker snails until more than half of the typical-morph snails are devoured (Palmer 1985).

When parthenogenetic water fleas (*Daphnia cucullata*) encounter the predatory larvae of the fly *Chaoborus*, the heads of the *Daphnia* grow to twice the normal size, becoming long and helmet shaped (see Figure 1.16). This increase in size lessens the chances that *Daphnia* will be eaten by the fly larvae. This same helmet induction occurs if the *Daphnia* are exposed to extracts of water in which the fly larvae have been swimming. Further, the predator-induced polyphenism of the *Daphnia* is beneficial not only to itself, but also to its offspring. Agrawal and colleagues (1999) have shown that the offspring of such induced *Daphnia* are born with this same altered head morphology. It is possible that the *Chaoborus* kairomone regulates gene expression both in the adult and in the developing embryo.

Predator-induced polyphenism in vertebrates

Predator-induced polyphenism is not limited to invertebrates. Indeed, predator-induced polyphenisms are abundant among amphibians. Tadpoles found in ponds or reared in the presence of other species may differ significantly from tadpoles reared by themselves in aquaria. For instance, newly hatched wood frog tadpoles (*Rana sylvatica*) reared in tanks containing the predatory larval dragonfly *Anax* (confined in mesh cages so that they cannot eat the tadpoles) grow smaller than those reared in similar tanks without predators. Moreover, as with the *Hyla* species shown in Figure 1.1A, the wood frog tadpoles' tail musculature deepens, allowing faster turning and swimming speeds (Van Buskirk and Relyea 1998). The addition of more predators to the tank causes a continuously deeper tail fin and tail musculature, and in fact what initially appeared to be a polyphenism may be a reaction norm that responds to the number (and type) of predators. In some species, phenotypic plasticity is even reversible, and removing the predators can restore the original phenotype (Relyea 2003a).

Predator-induced defensive reactions in some other frogs involve responding to specific vibrational cues produced by predators. The embryos of the Costa Rican red-eyed tree frog (*Agalychnis callidryas*) use vibrations transmitted through the egg mass to escape egg-eating snakes. These egg masses are laid on leaves that overhang ponds. The embryos usually develop into tadpoles within 7 days, at which time the tadpoles wiggle out of the eggs and fall into the pond. However, when snakes attempt to feed on the frog eggs (Figure 1.17A), the vibrations from the snakes' movements cue any embryos remaining inside the eggs to begin the twitching

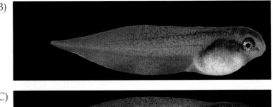

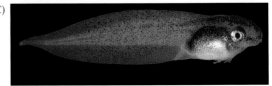

Figure 1.17   Predator-induced polyphenism in the red-eyed tree frog *Agalychnis callidryas*. (A) A snake eats a clutch of *A. callidryas* eggs. As the snake eats the egg mass, some of the embryos inside respond to the vibrations by hatching prematurely (arrow) and falling into the water. (A movie of this phenomenon is on the Internet at sites.bu.edu/warkentinlab/video-library/.) (B) Immature tadpole, induced to hatch at day 5. (C) Normal tadpoles such as this one hatch at day 7. (Photographs courtesy of Karen Warkentin.)

movements that initiate their hatching (within seconds!), and they drop into the pond, escaping the snakes. Embryos are competent to begin these hatching movements as early as day 5 (Figure 1.17B). Interestingly, the frog embryos respond this way only to vibrations given at a certain frequency and interval, and research has shown that these vibrations alone (and not sight or smell) cue the hatching movements (Warkentin 2005; Warkentin et al. 2006). Up to 80% of the remaining embryos can escape snake predation in this way. However, though these embryos escape their snake predators, they are at greater risk from water-borne predators than are "full-term" embryos, since the musculature of the early hatchers has not fully developed (Figure 1.17C).

The Presence of Conspecifics: It's Who You Know

Cues to change phenotype can come not only from predators but also from conspecifics (organisms of the same species); an individual in a large population can have a markedly different phenotype from that of an individual of the same species that is solitary. As mentioned above, the presence of predators induces wood frog (*Rana sylvatica*) tadpoles to develop thicker trunk muscles and shorter bodies. In contrast, when the tadpoles are raised together in high population density (in the absence of predators), development is slowed down, resulting in shallower tails and longer bodies relative to those raised in isolation. Thus, predation results in the development of short, muscular tadpoles, while competition from conspecifics results in long, sleek tadpoles (Relyea 2004). A developing tadpole apparently integrates competing signals from predators and conspecifics to produce a body shape that will optimize its performance.

A swarm of locusts: Polyphenism through touch

Crowding among conspecifics can produce remarkably different phenotypes; this phenomenon is especially obvious in migratory desert locusts, *Schistocerca gregaria*. In this species, mechanoreceptors are responsible for the induction of the crowding phenotype. Locusts in the low-density "solitary" phase are usually green, have short wings and large abdomens, and avoid each other. However, when forced to crowd together (as in small areas of patchy food), they actively aggregate to form a high-density "gregarious" or "migratory" phase. Individuals change their behavior (from avoidance to attraction), change their color (from green to brown, black, and orange), and molt into adults with longer wings and more slender abdomens (see Figure 1.3). These profound changes in color and behavior can be accomplished by subjecting solitary locust nymphs to buffeting with small paper mache balls (Roessingh et al. 1998).

The behavioral phase of the phenotypic change is mediated by direct physical contact among locusts, and the major sites of this mechanosensory input are the femurs of the hind legs. Repeatedly touching a minute region of the outer surface area of a hind femur with a fine paintbrush produces full behavioral gregarization within 4 hours (Simpson et al. 2001; Rogers et al. 2003). The colorization of the gregarious phenotype, however, may come from different cues. The smell of other locusts appears able to induce dark coloration in solitary nymphs by inducing the secretion of the neuropeptide hormone corazonin (Lester et al. 2005).

The green coloration of the solitary stage blends in with the background, making the grasshopper harder for predators to see. In contrast, the black-and-orange pattern on the gregarious locusts functions as a warning, telling potential predators that these nymphs have been feeding on toxicity-conferring plants. Moreover, gregariousness is thought to enhance the efficiency of this protective coloration. Thus the cryptic (hiding) coloration of the solitary morph and the aposematic (warning) coloration of the gregarious morph both serve as predator-avoidance strategies (Sword et al. 2000).

As in so many polyphenisms, a suite of behavioral practices are intimately connected with the morphological changes. In this locust, the changes associated with predator avoidance can be observed prior to the morphological changes. Within a few hours of crowding (and before molting), the solitary nymphs that are to change their phenotype start eating plants that had been distasteful previously. The plant-derived chemicals are responsible for make the migrating locusts unpalatable to their predators (Roessingh et al. 1993; Simões et al. 2013).

Sexual polyphenism induced by the community environment

Fish of many species change their sex based on social interactions; the blue-headed wrasse described at the start of the chapter is one good example (see Figure 1.1C). Marine goby fish are among the few fishes that can change their sex more than once—and in either direction. A female goby can become male if the male of the group dies. However, if a larger male enters the group, such males revert to being female (Black et al. 2005). Grober and Sunobe (1996) induced females to become males, males to become females, and females to become males and then females again, merely by changing their companions. A goby can change its sex in about 4 days.

In both the goby and the blue-headed wrasse, the shift of sex is mediated by hormonal changes caused by the environmental conditions. Interestingly, the sex changes in behaviors take place within hours, whereas the female-to-male color changes and gamete production take about a week to complete. The male behavioral changes may arise so quickly from the inactivation of the aromatase gene in certain hypothalamic neurons in the brain. This would enable testosterone to accumulate in these neurons and would prevent the synthesis of new estrogens. Indeed, the addition of estrogen implants into the brains of these fish prevents the female-to-male transition (and raises the levels of aromatase). These neurons are thought to mediate the sex-specific competitive and mating behaviors (Godwin et al. 2003; Perry and Grober 2003; Kroon et al. 2005; Marsh-Hunkin et al. 2013). The gonadal and color changes involved in such sex reversals also seem to be mediated by estrogens, wherein the gonadal aromatases are inhibited and elevated serum estrogen levels are seen. At this point, the ovaries start forming testes and sperm (and estrogen implants into the ovaries can also block these changes) (Horiguchi et al. 2013).

Cannibalism: An extreme phenotype for extreme times

One of the most impressive of these stress-induced phenotypes is cannibalism in some spadefoot toads of the genus *Spea*. These amphibians have a remarkable strategy for coping with a very harsh environment. The toads are called out of hibernation by the thunder that accompanies the first spring storms in Arizona's Sonoran Desert. (Unfortunately, motorcycles produce much the same sound, causing the toads to come out of hibernation only to die in the scorching sun.) The toads breed in temporary ponds formed by the rain, and the embryos develop quickly into larvae. After the larvae metamorphose into toads, the young toads burrow into the sand until the next year's storms bring them out.

Desert ponds are ephemeral pools that can either dry up quickly or persist, depending on the initial depth and the frequency of the rainfall. One might envision two alternative scenarios confronting a tadpole in such a pond: (1) the pond persists until tadpoles have time to metamorphose, and they live; or (2) the pond dries up before the tadpoles' metamorphosis is complete, and they die. *Spea*, however, has evolved a third alternative. The timing of its metamorphosis is controlled by the pond. If the pond persists at a viable level, development continues at its normal rate, and the algae-eating tadpoles develop into juvenile toads. However, if the pond is

Polyphenisms and Conservation Biology

Phenotypic plasticity means that animals in the wild may develop differently than those in the laboratory. Embryos and larvae in the wild develop in the presence of particular plants, predators, and conspecifics, and they experience variations of temperature and day length. In contrast, animals developing in a laboratory are usually grown in a monoculture of conspecific organisms, under a single particular temperature regime. This has important consequences when we apply knowledge gained in the laboratory to a field science such as conservation biology.

For instance, the metabolism of predator-induced morphs may differ significantly from that of the uninduced morphs, and this phenomenon has important consequences. Relyea (2003b, 2004) has found that in the presence of the chemical cues emitted by predators, pesticides such as carbaryl (Sevin®) can become up to 46 times more lethal than they are without the predator cues. Bullfrog and green frog tadpoles were especially sensitive to carbaryl when they were exposed simultaneously to predator chemicals. Relyea (2003b) has related these findings to the global decline of amphibian populations, saying that governments should test the toxicity of the chemicals under natural conditions, including that of predator stress. He concluded that "ignoring the relevant ecology can cause incorrect estimates of a pesticide's lethality in nature, yet it is the lethality of pesticides under natural conditions that is of utmost interest."

Temperature-dependent polyphenisms are also important for conservation biology and are likely to become more so with global warming. The significance of thermal polyphenisms was highlighted by Morreale and colleagues (1982) in a paper documenting temperature-dependent sex determination in a range of sea turtle species. Prior to that time, conservation biologists interested in restoring sea turtle populations had been growing the eggs in laboratory incubators set at a certain temperature, or culturing them in a single area of a beach. But these practices result in turtles of only one sex. Thus, Morreale and colleagues concluded that "current practices threaten conservation of sea turtles" rather than enhance them. They suggested protecting existing nests from predators, thereby maintaining the normal sex ratio.

Developmental plasticity also allows invasive species to alter their development to enable them to consume new prey (Bernays 1986; Kishida et al. 2006). The ability of such predators to display such plasticity may be a critically important factor in their success or failure to expand their ranges (Baldridge and Smith 2008).

One of the most interesting types of polyphenisms involves larval cues for metamorphosis. Environmental cues are critical to metamorphosis in many species, and some of the best-studied examples are the settlement cues used by marine larvae. A free-swimming marine larva often needs to settle near a source of food or on a firm substrate on which it can metamorphose. If prey, conspecifics, or substrates give off soluble molecules, these molecules can be used by the larvae as cues to settle and begin metamorphosis. Among the mollusks, there are often very specific cues for settlement (Hadfield 1977). In some cases, the prey supply the cues, while in other cases the substrate itself gives off molecules used by the larvae to initiate settlement. These cues may not be constant, but they need to be part of the environment if normal development is to occur (Pechenik et al. 1998).

The importance of substrates for larval settlement and metamorphosis was demonstrated in 1880 when William Keith Brooks, an embryologist at Johns Hopkins University, was asked to help the ailing oyster industry of Chesapeake Bay. For decades, oysters had been dredged from the bay, and there had always been a new crop to take their place. But suddenly, each year brought fewer oysters. What was responsible for the decline? Experimenting with larval oysters, Brooks discovered that the American oyster (*Crassostrea virginica*) requires a hard substrate on which to metamorphose. For years, oystermen had simply thrown the mollusks' shells back into the water, but with the advent of suburban sidewalks, they began selling the shells to cement factories. Brooks's solution: go back to returning the oyster shells to the bay. The oyster population responded, and Baltimore wharves still sell their descendants.

Knowledge of phenotypic plasticity is critical in conservation biology and is necessary for making informed decisions that will benefit the environment. We will revisit this theme several times in this book. For more information on plasticity and conservation biology, see the video "Race for Survival" (Baressi et al. 2013).

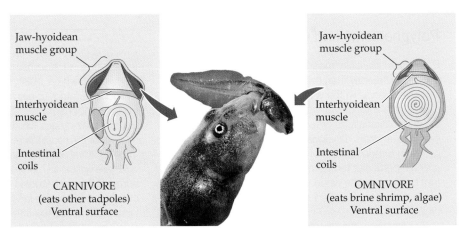

Figure 1.18 Polyphenism in tadpoles of a spadefoot toad (*Spea*). The typical morph (right) is an omnivore, feeding on insects and algae. When ponds are drying out quickly, however, a carnivorous (cannibalistic) morph forms. It develops a wider mouth, larger jaw muscles, and an intestine modified for a carnivorous diet. The center photograph shows a cannibalistic tadpole eating a smaller pondmate. (Photograph © Thomas Wiewandt; drawings courtesy of R. Ruibel.)

drying out and getting smaller, some of the tadpoles embark on an alternative developmental pathway. They develop a wider mouth and powerful jaw muscles, which enable them to eat (among other things) other *Spea* tadpoles (Figure 1.18). These carnivorous tadpoles metamorphose quickly, albeit into a smaller version of the juvenile spadefoot toad. But they survive while other *Spea* tadpoles perish from desiccation (Newman 1989, 1992; Denver 1997).

Convergence on Favorable Phenotypes

One principle of environmentally induced polyphenisms worth stressing is the fact that *different* environmental cues can produce the same favorable phenotype. For instance, the helmet-and-spiked-tailed *Daphnia* morph can be induced by different predators, and the chemical signals eliciting this phenotype are probably different. The water conditioned by dragonfly larvae can induce this phenotype, but so can chemicals released by dead *Daphnia* individuals being digested inside a fish's gut (Stabell et al. 2003). Similarly, the hatch-early-into-the-pond behavior of the red-eyed tree frog can be induced not only by snake vibrations, but also by wasp predation and by fungal infection (Warkentin 2000; Warkentin et al. 2001). The gregarious phase change of desert locusts can be induced by mechanostimulation of the hindlimb neurons (as described on p. 29), but it can also take place in the presence of a combination of visual and olfactory stimuli. In these developing locusts, either cause will induce a rise in the levels of serotonin in the thoracic ganglia of solitary individuals and the subsequent components of the phase change. Drugs that block the action or synthesis of serotonin will prevent the phase change in both cases (Simpson and Sword 2009; Anstey et al. 2009).

Looking Ahead

Plasticity had been considered as an exception to the rule, something seen in odd critters such as *Daphnia*. Now, plasticity is seen as the norm, something common to all animals. Environment is therefore considered to play a role in the generation of phenotypes, in addition to its well-established role in the natural selection of which phenotypes will survive and reproduce. This opens up numerous questions concerning both health and evolution, and we will be dealing with these throughout the book.

But one of the major questions concerns the mechanisms by which several plastic traits in an organism are regulated. This is especially important when looking at morphs differing in both *anatomical* and *behavioral* traits. For instance, the polyphenisms in the male dung beetle involve not only the presence or absence of horns, but also the type of behavior males can express—fighter or sneaker. It seems that the different components of the dung beetle response to diet is controlled by different sets of genes (Snell-Rood et al. 2011). Similarly, wood frog tadpoles not only develop tails of have different shapes and colors, depending on the sensing of predators; their behavior also changes. The first component of the sex change in fish and the gregarious phenotype in locusts involves their new behaviors. Indeed, behavioral plasticity may be more common than we thought. According to John Dupre (personal comm.), plants exhibit their plasticity through morphological change, while animals may manifest their plasticity primarily through behaviors and the morphological structures that enable them. The ability of organisms to be plastic in both their behaviors and their anatomies opens up important areas of study that will be critically important for studying evolution and psychological diseases. Much of the remainder of the book will concern the ramifications of plasticity for disease and evolution.

References

Abouheif, E. and G. A. Wray. 2002. Evolution of the gene network underlying wing polyphenism in ants. *Science* 297: 249–252.

Adler, F. R. and C. D. Harvell. 1990. Inducible defenses, phenotypic variability, and biotic environments. *Trends Ecol. Evol.* 5: 407–410.

Agrawal, A. A., C. Laforsch and R. Tollrian. 1999. Transgenerational induction of defenses in animals and plants. *Nature* 401: 60–63.

Allen, D. L. and 12 others. 2009. Effects of spaceflight on murine skeletal muscle gene expression. *J. Appl. Physiol.* 106: 582–595.

Anstey, M. L., S. M. Rogers, S. R. Ott, M. Burrows and S. J. Simpson. 2009. Serotonin mediates behavioral gregarization underlying swarm formation in desert locusts. *Science* 323: 627–630.

April, C. S. and G. S. Barsh. 2007. Distinct pigmentary and melanocortin-1 receptor-induced components of cutaneous defense against ultraviolet radiation. *PLoS Genet.* 3(1): e9.

Aristotle. 355 BCE. *Generation of Animals*. 737a 27-29; 766B 29-767A 2.

Avital, E. and E. Jablonka. 2000. *Animal Traditions: Behavioural Inheritance in Evolution*. Cambridge University Press, Cambridge.

Bachem, C. W. B, G.-J. Speckmann, P. C. G. van der Linde, F. T. M. Verheggen, M. D. Hunt, J. C. Steffens and M. Zabeau. 2004. Antisense expression of polyphenol oxidase genes inhibits enzymatic browning in potato tubers. *Biotechnology* 12: 1101–1105.

Baldridge, A. K. and L. D. Smith. 2008. Temperature constraints on phenotypic plasticity explain biogeographic patterns in predator trophic morphology. *Mar. Ecol. Prog. Ser.* 365: 25–34.

Baltzer, F. 1914. Die Bestimmung und der Dimorphismus des Geschlechtes bei Bonellia. *Sber. Phys.-Med. Ges. Würzb.* 43: 1–4.

Baressi, M., A. Chong, K. Dymek, S. Gaffney, J. Hopson and G. Lyons. 2013. *Race for Survival: Phenotypic Plasticity in a Changing World*. www.sophia.smith.edu/~mbaressi/lab/documentaries.html

Beree, L., P. J. Schembri and D. S. Boukal. 2005. Sex determination in *Bonellia viridis* (Echiura: Bonelliidae). *OIKOS* 108: 473–484.

Bernays, E. A. 1986. Diet-induced head allometry among foliage-chewing insects and its importance for graminivores. *Science* 231: 495–497.

Berson, J. F., D. W. Frank, P. A. Calvo, B. M. Bieler and M. S. Marks. 2000. A common temperature-sensitive allelic form of human tyrosinase is retained in the endoplasmic reticulum at the nonpermissive temperature. *J. Biol. Chem*. 275: 12281–12289.

Black, M. P., T. B. Moore, A. V. M. Canario, D. Ford, R. H. Reavis and M. S. Grober. 2005. Reproduction in context: Field-testing a lab model of socially controlled sex change in *Lythrypnus dalli*. *J. Exp. Marine Biol. Ecol*. 318: 127–143.

Bolker, J. A. 1995. Model systems in developmental biology. *Bioessays*. 17: 451–455.

Bolker, J. A. 2014. Model species in evo-devo: A philosophical perspective. *Evol. Dev*. 16: 49–56.

Brian, M. V. 1974. Caste differentiation in *Myrmica rubra*: The role of hormones. *J. Insect Physiol*. 20: 1351–1365.

Brian, M. V. 1980. Social control over sex and caste in bees, wasps, and ants. *Biol. Rev*. 55: 379–415.

Brooks, W. K. 1880. *Development of the American Oyster: Report of the Commission of Fisheries of Maryland*. Inglehart, Annapolis, MD.

Charnov, E. L. and J. J. Bull. 1977. When is sex environmentally determined? *Nature* 266: 828–830.

Chen, H., Q. Y. Weng and D. E. Fisher. 2014. UV signaling pathways within the skin. *J. Invest. Dermatol*. 34: 2080–2085.

Conover, D. O. and S. W. Heins. 1987. Adaptive variation in environmental and genetic sex determination in a fish. *Nature* 326: 496–498.

Corruccini, R. S. and C. L. Beecher. 1982. Occlusal variation related to soft diet in a nonhuman primate. *Science* 218: 74–76.

Corruccini, R. S. and C. L. Beecher. 1984. Occlusofacial morphological integration lowered in baboons raised on soft diet. *J. Craniofacial Genet. Dev. Biol*. 4: 135–142.

Crain, D. A. and L. L. Guillette, Jr. 1998. Reptiles as models of contaminant-induced endocrine disruption. *Anim. Reprod. Sci*. 53: 77–86.

Dawkins, R. 1976. *The Selfish Gene*. Oxford University Press, New York.

Denver, R. J. 1997. Proximate mechanism of phenotypic plasticity in amphibian metamorphosis. *Am. Zool*. 37: 172–184.

Dodson, S. 1989. Predator-induced reaction norms. *BioScience* 39: 447–452.

D'Orazio, J. A. and 11 others. 2006. Topical drug rescue strategy and skin protection based on the role of Mc1r in UV-induced tanning. *Nature* 443: 340–344.

Duckworth, R. A. 2009. The role of behavior in evolution: A search for mechanism. *Evol. Ecol*. 23: 513–531.

Emlen, D. J. 1997. Alternative reproductive tactics and male dimorphism in the horned beetle *Onthophagus acuminatus* (Coleoptera: Scarabaeidae). *Behav. Ecol. Sociobiol*. l41: 335–341.

Emlen, D. J. 2000. Integrating development with evolution: A case study with beetle horns. *BioScience* 50: 403–418.

Emlen, D. J. and H. F. Nijhout. 1999. Hormonal control of male horn length dimorphism in the horned beetle *Onthophagus taurus*. *J. Insect Physiol*. 45: 45–53.

Fell, G. L., K. C. Robinson, J. Mao, C. J. Woolf and D. E. Fisher. 2014. Skin β-endorphin mediates addiction to UV light. *Cell* 157: 1527–1534.

Ferguson, M. W. J. and T. Joanen. 1982. Temperature of egg incubation determines sex in *Alligator mississippiensis*. *Nature* 296: 850–853.

Flux, J. E. C. 1970. Colour change of mountain hares (*Lepus timidus scoticus*) in north-east Scotland. *J. Zool*. 162: 345–358.

Gilbert, S. F. 2001. Ecological developmental biology: Developmental biology meets the real world. *Dev. Biol*. 233: 1–12.

Gilbert, S. F. 2008. When "Personhood" begins in the embryo: Avoiding a syllabus of errors. *Birth Def. Res. C Embryo Today* 84: 164–173.

Gilbert, S. F. 2013. *Developmental Biology*, 10th Ed. Sinauer Associates, Sunderland, MA.

Godwin, J., J. A. Luckenbach and R. J. Borski. 2003. Ecology meets endocrinology: Environmental sex determination in fishes. *Evol. Dev*. 5: 40–49.

Godwin, J., R. Sawby, R. R. Warner, D. Crews and M. S. Grober. 2000. Hypothalamic arginine vasotocin mRNA abundance variation across sexes and with sex change in a coral reef fish. *Brain Behav. Evol*. 55: 77–84.

Gotthard, K. and S. Nylin. 1995. Adaptive plasticity and plasticity as an adaptation: A selective review of plasticity in animal morphology and life history. *Oikos* 74: 3–17.

Grange, W. B. 1932. The pelages and color changes of the snowshoe hare, *Lepus americanus phaeonotus*. *J. Mammal*. 13: 99–116.

Greene, E. 1989. A diet-induced developmental polymorphism in a caterpillar. *Science* 243: 643–646.

Grober, M. S. and T. Sunobe. 1996. Serial adult sex change involves rapid and reversible changes in forebrain neurochemistry. *Neuroreport* 7: 2945–2949.

Hadfield, M. G. 1977. Metamorphosis in marine molluscan larvae: An analysis of stimulus and response. In R.-S. Chia and M. E. Rice (eds.), *Settlement and Metamorphosis of Marine Invertebrate Larvae*. Elsevier, New York, pp. 165–175.

Herbst, C. 1901. Formative Reize in der tierischen Ontogenese, Ein Beitrag zum Verständnis der tierischen Embryonalentwicklung. Arthur Georgi, Leipzig. Quoted in J. M. Oppenheimer. 1991. Curt Herbst's contribution to the concept of embryonic induction. In S. F. Gilbert (ed.), *A Conceptual History of Modern Embryology*. Plenum, New York, pp. 63–89.

Hertwig, O. 1894. Zeit- und Streitfragen der Biologie I. Präformation oder Epigenese? Grundzüge einer Entwicklungstheorie der Organismen. Gustav Fischer, Jena. Translated as *The Biological Problem of Today: Preformation or Epigenesis?* P. C. Mitchell (trans.). Macmillan, New York.

Horiguchi, R., R. Nozu, T. Hirai, Y. Kobayashi, Y. Nagahama and M. Nakamura. 2013. Characterization of gonadal soma-derived factor expression during sex change in the protogynous wrasse, *Halichoeres trimaculatus*. *Dev. Dyn*. 242: 388–399.

Inobe, M., I. Inobe, G. R. Adams, K. M. Baldwin and S. Takeda. 2002. Effects of microgravity on myogenic factor expressions during postnatal development of rat skeletal muscle. *J. Appl. Physiol*. 92: 1936–1942.

Jablonka, E. and M. J. Lamb. 2005. *Evolution in Four Dimensions: Genetic, Epigenetic, Behavioral, and Symbolic Variation in the History of Life*. MIT Press, Cambridge, MA.

Jablonka, E. and G. Raz. 2008. Transgenerational epigenetic inheritance: Prevalence, mechanisms, and implications for the study of heredity. *Q. Rev. Biol*. 84: 131–176.

Jaccarini, V., L. Agius, P. J. Schembri and M. Rizzo. 1983. Sex determination and larval sexual interaction in *Bonellia viridis* Rolando (Echiura: Bonelliidae). *J. Exp. Mar. Biol. Ecol*. 66: 25–40.

Janzen, F. J. and G. L. Paukstis. 1991. Environmental sex determination in reptiles: Ecology, evolution, and experimental design. *Q. Rev. Biol*. 66: 149–179.

Johannsen, W. 1909. *Elemente der Exakten Erblichkeitslehre*. Gustav Fischer, Jena.

Johannsen, W. 1911. The genotype conception of heredity. *Amer. Nat*. 45: 129–159.

Kamakura, M. 2011. Royalactin induces queen differentiation in honeybees. *Nature* 473: 478–483.

Keiner, C. 1998. W. K. Brooks and the oyster question: Science, politics, and resource management in Maryland, 1880–1930. *J. Hist. Biol.* 31: 383–424.

Kirschner W. M., J. C. Gerhart, K. Hara and G. Ubbels. 1980. Initiation of the cell cycle and establishment of bilateral symmetry in *Xenopus* eggs. *Symp. Soc. Dev. Biol.* 38: 187–215.

Kishida, O., Y. Mizuta and K. Nishimura. 2006. Reciprocal phenotypic plasticity in a predator-prey interaction between larval amphibians. *Ecology* 87: 1599–1604.

Kroon, F. J., P. L. Munday, D. A. Westcott, J. P. Hobbs and N. R. Liley. 2005. Aromatase pathway mediates sex change in each direction. *Proc. Biol. Sci.* 272: 1399–1405.

Lester, R. L., C. Grach, M. P. Pener and S. J. Simpson. 2005. Stimuli inducing gregarious colouration and behaviour in nymphs of *Schistocerca gregaria*. *J. Insect Physiol.* 51: 737–747.

Leutert, T. R. 1974. Zur Geschlechtsbestimmung und Gametogenese von *Bonellia viridis* Rolando. *J. Embryol. Exp. Morphol.* 32: 169–193.

Lewontin, R. C. 1993. *Biology as Ideology: The Doctrine of DNA*. Harper Collins, New York.

Marsh-Hunkin, K. E., H. M. Heinz, M. B. Hawkins and J. Godwin. 2013. Estrogenic control of behavioral sex change in the bluehead wrasse, *Thalassoma bifasciatum*. *Integr. Comp. Biol.* 53: 951- 959.

Mayr, E. 1963. *Animal Species and Evolution*. Harvard University Press, Cambridge, MA.

McCollum, S. A. and J. Van Buskirk. 1996. Costs and benefits of a predator-induced polyphenism in the gray treefrog *Hyla chrysoscelis*. *Evolution* 50: 583–593.

Meyer, A. 1987. Phenotypic plasticity and heterochrony in *Cichlasoma managues* (Pisces, Cichlidae) and their implications for speciation in cichlid fishes. *Evolution* 41: 1357–1369.

Miller, D., J. Summers and S. Silber. 2004. Environmental versus genetic sex determination: A possible factor in dinosaur extinction? *Fertil. Steril.* 81: 954–964.

Mills, L. S., M. Zimova, J. Oyler, S. Running, J. T. Abatzoglou and P. M. Lukacs. 2013. Camouflage mismatch in seasonal coat color due to decreased snow duration. *Proc. Natl. Acad. Sci. USA* 110: 7360–7365.

Moczek, A. P. 2005. The evolution of development of novel traits, or how beetles got their horns. *BioScience* 55: 937–951.

Moczek, A. P. and D. J. Emlen. 2000. Male horn dimorphism in the scarab beetle *Onthophagus taurus*: Do alternative tactics favor alternative phenotypes? *Anim. Behav.* 59: 459–466.

Moczek, A. P. and H. F. Nijhout. 2004. Trade-offs during development of primary and secondary sexual traits in a horned beetle. *Amer. Nat.* 163: 184–191.

Morreale, S. J., G. J. Ruiz, J. R. Spotila and E. A. Standora. 1982. Temperature-dependent sex determination: Current practices threaten conservation of sea turtles. *Science* 216: 1245–1247.

Moss, L. 2003. *What Genes Can't Do*. MIT Press, Cambridge, MA.

Moss, M. L. 1962. The functional matrix. In B. Kraus and R. Reidel (eds.), *Vistas in Orthodontics*. Lea & Febiger, Philadelphia, pp. 85–98.

Moss, M. L. 1997. The functional matrix hypothesis revisited. IV. The epigenetic antithesis and the resolving synthesis. *Amer. J. Orthod. Dentofac. Orthop.* 112: 410–417.

Müller, G. B. 2003. Embryonic motility: Environmental influences on evolutionary innovation. *Evo. Dev.* 5: 56–60.

Müller, G. B. and J. Steicher. 1989. Ontogeny of the syndesmosis tibiofibularis and the evolution of the bird hindlimb: A caenogenetic feature triggers phenotypic novelty. *Anat. Embryol.* 179: 327–339.

National Aeronautics and Space Administration (NASA). 2003. Pumping iron in microgravity. Space Research Office Biol. Physical Res. Newsltr. spaceresearch.nasa.gov/general_info/pumpingiron.html

Nelkin, D. and M. S. Lindee. 1996. *The DNA Mystique: The Gene as a Cultural Icon*. W. H. Freeman, New York.

Newman, R. A. 1989. Developmental plasticity of *Scaphiopus couchii* tadpoles in an unpredictable environment. *Ecology* 70: 1775–1787.

Newman, R. A. 1992. Adaptive plasticity in amphibian metamorphosis. *BioScience* 42: 671–678.

Nijhout, H. F. 1991. *The Development and Evolution of Butterfly Wing Patterns*. Smithsonian Institution Press, Washington, DC.

Nijhout, H. F. 1999. Control mechanisms of polyphenic development in insects. *BioScience* 49: 181–192

Nikawa, T. and 13 others. 2004. Skeletal muscle gene expression in space-flown rats. *FASEB J.* 18: 522–524.

Nilsson-Ehle, H. 1914. Vilka erfarenheter hava hittills vunnits roerande mojligheten av vaexters acklimatisering? *Kungl. Landtbruks-Akademiens Handlinger och Tidskrift* 53: 537–572.

Nyhart, L. 1995. *Biology Takes Form*. University of Chicago Press, Chicago.

Ovid. 1 CE. *Metamorphoses*. B. Moore (trans.), Cornhill Publishers, Boston, MA, line I1.

Palmer, A. R. 1985. Adaptive value of shell variation in *Thais lamellosa*: Effect of thick shells on vulnerability to and preference by crabs. *Veliger* 27: 349–356.

Parzer, H. F. and A. P. Moczek. 2008. Rapid antagonistic coevolution between primary and secondary sexual characters in horned beetles. *Evolution* 62: 2423–2428.

Pechenik, J. A., D. E. Wendt and J. N. Jarrett. 1998. Metamorphosis is not a new beginning. *BioScience* 48: 901–910.

Pener, M. P. 1991. Locust phase polymorphism and its endocrine relations. *Adv. Insect Physiol.* 3: 1–79.

Perry, A. N. and M. S. Grober. 2003. A model for social control of sex change: Interactions of behavior, neuropeptides, glucocorticoids, and sex steroids. *Horm. Behav.* 43: 31–38.

Plowright, R. C. and B. A. Pendrel. 1977. Larval growth in bumble-bees. *Can. Entomol.* 109: 967–973.

Rachinsky, A. and K. Hartfelder. 1990. Corpora allata activity, a prime regulating element for caste-specific juvenile hormone titre in honey bee larvae (*Apis mellifera carnica*). *J. Insect Physiol.* 36: 329–349.

Relyea, R. A. 2003a. Predators come and predators go: The reversibility of predator-induced traits. *Ecology* 84: 1840–1848.

Relyea, R. A. 2003b. Predator cues and pesticides: A double dose of danger for amphibians. *Ecol. Applic.* 13: 1515–1521.

Relyea, R. 2004. Fine-tuned phenotypes: Tadpole plasticity under 16 combinations of predators and competitors. *Ecology* 85: 172–179.

Roessingh, P., A. Bouaïchi and S. J. Simpson. 1998. Effects of sensory stimuli on the behavioural phase state of the desert locust, *Schistocerca gregaria*. *J. Insect Physiol.* 44: 883–893.

Roessingh, P., S. J. Simpson and S. James. 1993. Analysis of phase-related changes in behavior of desert locust nymphs. *Proc. Biol. Sci.* 252: 43–49.

Rogers, S. M., T. Matheson, E. Despland, T. Dodgson, M. Burrows and S. J. Simpson. 2003. Mechanosensory-induced behavioural gregarization in the desert locust *Schistocerca gregaria*. *J. Exp. Biol.* 206: 3991–4002.

Rogers, S. M., T. Matheson, K. Sasaki, K. Kendsrick, S. J. Simpson and M. Burrows. 2004. Substantial changes in central nervous system neurotransmitters and neuromodulators accompany phase change in the locust. *J. Exp. Biol.* 207: 3603–3617.

Roll-Hansen, N. 2007. Sources of Johannsen's genotype theory. Meeting of the International Society for the History, Philosophy, and Social Studies of Biology, Exeter, England.

Sarkar, S. 1998. *Genetics and Reductionism*. Cambridge University Press, Cambridge.

Sarkar, S. 1999. From the Reaktionsnorm to the adaptive norm: The norm of reaction, 1906–1960. *Biol. Philos.* 14: 235–252.

Sarkar, S. 2006. From genes as determinants to DNA as resource: Historical notes on development and genetics. In E. Neumann-Held and C. Rehmann-Sutter (eds.), *Genes in Development: Re-Reading the Molecular Paradigm*. Duke University Press, Durham, NC, pp. 77–95.

Schmalhausen, I. I. 1949. *Factors of Evolution: The Theory of Stabilizing Selection*. University of Chicago Press, Chicago.

Schmidt-Küntzel, A., E. Eizirik, S. J. O'Brien and M. Menotti-Raymond. 2005. Tyrosinase and tyrosinase-related protein 1 genetic variants specify domestic cat coat color alleles of the albino and brown loci. *J. Hered.* 96: 289–301.

Shapiro, A. M. 1968. Photoperiodic induction of vernal phenotype in *Pieris protodice* Boisduval and Le Conta (Lepidoptera: Pieridae). *Wasmann J. Biol.* 26: 137–149.

Simmons, L. W. and D. J. Emlen. 2006. Evolutionary trade-off between weapons and testes. *Proc. Natl. Acad. Sci. USA* 103: 16346–16351.

Simões, P. M., J. E. Niven and S. R. Ott. 2013. Phenotypic transformation affects associative learning in the desert locust. *Curr. Biol.* 23: 2407–2412.

Simpson, S. J., E. Despland, B. F. Hägele and T. Dodgson. 2001. Gregarious behavior in desert locusts is evoked by touching their back legs. *Proc. Natl. Acad. Sci. USA* 98: 3895–3897.

Simpson, S. J. and G. A. Sword. 2009. Phase polyphenism in locusts: Mechanisms, population consequences, adaptive significance and evolution. In T. Ananthakrishnan and D. Whitman (eds.), *Phenotypic Plasticity of Insects: Mechanisms and Consequences*. Science Publishers, Plymouth, UK, pp. 147–189.

Simpson, S. J., G. A. Sword and N. Lo. 2011. Polyphenism in insects. *Curr. Biol.* 21: R738–R749.

Snell-Rood, E. C., A. Cash, M. V. Han, T. Kijimoto, J. Andrews and A. P. Moczek. 2011. Developmental decoupling of alternative phenotypes: Insights from the transcriptomes of horn-polyphenic beetles. *Evolution* 65: 231–245.

Stabell, O. B., F. Ogbeto and R. Primicerio. 2003. Inducible defenses in *Daphnia* depend on latent alarm signals from conspecific prey activated in predators. *Chem. Senses* 28: 141–153.

Stearns, S. C., G. de Jong and R. A. Newman. 1991. The effects of phenotypic plasticity on genetic correlations. *Trends Ecol. Evol.* 6: 122–126.

Sultan, S. E. 2007. Development in context: The timely emergence of eco-devo. *Trends Ecol. Evol.* 22: 575–582.

Sword, G. A., S. J. Simpson, O. T. M. El Hadi and H. Wilps. 2000. Density-dependent aposematism in the desert locust. *Proc. R. Soc. Lond. B Biol. Sci.* 267: 63–68.

Szent-Györgyi, A. 1966. In search of simplicity and generalizations: Fifty years of poaching in science. In N. O. Kaplan and E. P. Kennedy (eds), *Current Aspects of Biochemical Energetics*. Academic Press, New York, pp. 63–76.

Tang, G. H., A. B. Rabie and U. Hagg. 2004. Indian hedgehog: A mechanotransduction mediator in condylar cartilage. *J. Dent. Res.* 83: 434–438.

Tawfik, A. I. and 9 others. 1999. Identification of the gregarization-associated dark pigmentotropin in locusts through an albino mutant. *Proc. Natl. Acad. Sci. USA* 96: 7083–7087.

Tollrian, R. and C. D. Harvell. 1999. *The Ecology and Evolution of Inducible Defenses*. Princeton University Press, Princeton, NJ.

Usarov, B. P. 1921. A revision of the genus *Locusta* L. (= *Patchytylus* Fieb.), with a new theory as to the periodicity and migrations of locusts. *Bull. Entomolog. Res.* 12: 135–163.

Van Buskirk, J. and R. A. Relyea. 1998. Natural selection for phenotypic plasticity: Predator-induced morphological responses in tadpoles. *Biol. J. Linn. Soc.* 65: 301–328.

Van Speybroeck, L. 2002. From epigenesis to epigenetics: The case of C. H. Waddington. *Ann. N.Y. Acad. Sci.* 981: 61–81.

Varrela, J. 1992. Dimensional variation of craniofacial structures in relation to changing masticatory-functional demands. *Eur. J. Orthod.* 14: 31–36.

Via, S., R. Gomulkiewicz, G. De Jong, S. M. Scheiner, C. D. Schlichting and P. H. Van Tienderen. 1995. Adaptive phenotypic plasticity: Consensus and controversy. *Trends Ecol. Evol.* 10: 212–217

Waddington, C. H. 1968. The basic ideas of biology. In C. H. Waddington (ed.), *Towards a Theoretical Biology*. Edinburgh University Press, Edinburgh, pp. 1–32.

Warkentin, K. M. 2000. Wasp predation and wasp-induced hatching of red-eyed treefrog eggs. *Anim. Behav.* 60: 503–510.

Warkentin, K. M. 2005. How do embryos assess risk? Vibrational cues in predator-induced hatching of red-eyed treefrogs. *Anim. Behav.* 70: 59–71.

Warkentin, K. M., M. S. Caldwell and J. G. McDaniel. 2006. Temporal pattern cues in vibrational risk assessment by embryos of the red-eyed treefrog, *Agalychnis callidryas*. *J. Exp. Biol.* 209: 1376–1384.

Warkentin, K. M., C. C. Currie and S. A. Rehner. 2001. Egg-killing fungus induces early hatching of red-eyed treefrog eggs. *Ecology* 82: 2860–2869.

Warner, D. A. and R. Shine. 2008. The adaptive significance of temperature-dependent sex determination in a reptile. *Nature* 451: 566–568.

Warner, R. R. 1984. Deferred reproduction as a response to sexual selection in a coral reef fish: A test of the life historical consequences. *Evolution* 38: 148–162.

Watson, J. 1989. Quoted in L. Jaroff, "The gene hunt." *Time Magazine*, March 20, 1989, p. 67.

Watt, W. B. 1968. Adaptive significance of pigment polymorphism in *Colias* butterflies. I. Variation of melanin in relation to thermoregulation. *Evolution* 22: 437–458.

Weismann, A. 1893. *The Germ Plasm*. W. N. Parker and H. Rönnfeldt (trans.). Charles Scribners Sons, New York.

Weiss, P. A. 1970. *Life, Order, and Understanding: A Theme in Three Variations*. University of Texas Press, Austin.

West-Eberhard, M. J. 2003. *Developmental Plasticity and Evolution*. Oxford University Press, Oxford.

Wheeler, D. 1991. The developmental basis of worker caste polymorphism in ants. *Amer. Nat.* 138: 1218–1238.

Widrick, J. J. and 10 others. 1999. Effect of a 17-day spaceflight on contractile properties of human soleus muscle fibres. *J. Physiol.* 516: 915–930.

Wirtz, P. 1973. Differentiation in the honeybee larva. Meded. Landb. Hogesch. *Wagningen* 73–75, 1–66.

Woltereck, R. 1909. Weitere experimentelle Untersuchungen über Artveränderung, speziell über das Wesen quantitativer

Artunderscheide bei Daphniden. *Versuch. Deutsch. Zool. Ges.* 1909: 110–172.

Woodward, D. E. and J. D. Murray. 1993. On the effect of temperature-dependent sex determination on sex ratio and survivorship in crocodilians. *Proc. R. Soc. Lond. B Biol. Sci.* 252: 149–155.

Wu, K. C., J. Streicher, M. L. Lee, B. I. Hall and G. B. Müller. 2001. Role of motility in embryonic development. I. Embryo movements and amnion contractions in the chick and the influence of illumination. *J. Exp. Zool.* 291: 186–194.

Zimova, M., L. S. Mills, P. M. Lukacs and M. S. Mitchell. 2014. Snowshoe hares display limited phenotypic plasticity to mismatch in seasonal camouflage. *Proc. Biol. Sci.* 281(1782): 20140029.

2

Environmental Epigenetics

How Agents in the Environment Effect Molecular Changes in Development

The guiding motto in the life of every natural philosopher should be, "Seek simplicity and distrust it."

Alfred North Whitehead, 1926

Genotypes are only as good as the environments that they find themselves in.

Greg Gibson, 2008

One cannot reduce phenotype completely to the inherited genes. Experience must be added to endowment. In the preceding chapter, we saw that the natural environment has an incredibly rich repertoire of ways to generate phenotypes. Diet affects whether a bee or ant larva develops into a huge, fertile queen or a small, sterile worker. Temperature determines the sex of many reptiles and affects wing color and pattern in many butterflies. The presence of predators or competitors alters the developmental trajectories of numerous organisms, helping them survive in a potentially hostile environment. "Virtually all organisms," write Denver and Middlemis Maher (2010), "display some form of developmental plasticity."

But exactly *how* do environmental agents instruct a particular phenotype to emerge from the inherited repertoire of possible phenotypes? What are these factors doing to alter gene expression or cell division in the embryo? This chapter surveys the mechanisms by which environmental cues can instruct changes in normal development. Moreover, it will demonstrate some of the ways by which such phenotypes can become stabilized such that these altered patterns of development are inherited from one generation to the next.

The environmental agents involved in altering phenotypes usually do so by altering gene expression during development. This can be accomplished by at least three major routes (Gilbert 2005):

- *Direct transcriptional regulation*. Environmental factors can alter gene expression patterns by chemically modifying particular regions of DNA. These modifications regulate whether particular genes are activated (transcribed) or turned off (repressed).
- *The neuroendocrine system*. In most of these cases, the nervous system receives signals from the environment, and chemical signals from the nervous system cause changes in the hormone (i.e., endocrine) milieu within the organism. Hormones produced in response to such neural signals can alter gene expression patterns and regulate phenotype production.
- *Direct induction*. The environmental factor may interact directly with the cell to activate or repress the signal transduction cascades that activate gene expression and alter cell behaviors.

Regulation of Gene Transcription

Because every non-microbial cell in an individual's body is derived by cell division from the same fertilized egg, all these cells (with a few interesting exceptions such as mammalian lymphocytes) have the same set of genes; this set of genes comprises one's **genome**. Thus, the genes in pancreas cells are the same as the genes in the lens precursor cells of the eye. What makes the pancreatic cells different from the lens cells are the genes that are activated (i.e., *expressed*). In the lens precursor cell, a cascade of tissue-specific events activates genes producing crystalline proteins that allow the lens tissue to be transparent, whereas in certain cells in the pancreas, the gene for insulin is activated (and the lens-specific genes are turned off). To understand how such **differential gene expression** is accomplished, we must briefly review some molecular biology.

Differential gene expression

In addition to the structural portion of the gene—the DNA that encodes the amino acid sequence of the protein—there are regions of the gene that regulate where, when, and how efficiently the gene will be active. These regulatory regions (sometimes called *cis*-regulatory elements) include the promoter and the enhancers (Figure 2.1):

- Every gene has a **promoter**—a region of DNA containing a specific nucleotide sequence (usually TATA) that binds the enzyme **RNA polymerase**. This enzyme unwinds the double helix of DNA and initiates the synthesis of nuclear RNA. In exiting the nucleus and entering the cytoplasm, the nuclear RNA is processed into messenger RNA (mRNA), which in turn specifies a sequence of amino acids that fold to form a functional protein.
- **Enhancers** are DNA sequences that bind a set of proteins called **transcription factors**. The resulting complex of transcription

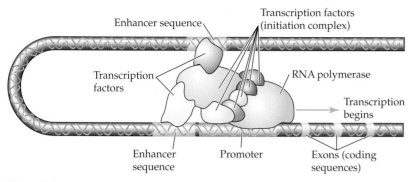

Figure 2.1   RNA polymerase is stabilized on the promoter site of the DNA by transcription factors recruited by the enhancers. The TATA sequence at the promoter binds a protein that serves as a "saddle" for RNA polymerase. However, RNA polymerase would not remain bound long enough to initiate transcription were it not for the stabilization provided by the transcription factors. The transcription factors collected by the enhancers and promoter interact with RNA polymerase to form a stable initiation complex that begins transcription.

factors interacts with the promoter to tell the gene when to be active, where to be active, and how much mRNA it should synthesize.

RNA polymerase makes mRNA that is complementary to the DNA strand, or **template**, it is reading from. The process whereby RNA polymerase synthesizes this complementary mRNA is called **transcription**. The promoter sequence orients RNA polymerase so that it reads in the correct direction and binds at the beginning of the gene. This makes sense, since if RNA polymerase could bind anywhere on the DNA, mRNAs would have no way of defining the starting point of a protein. The promoter can thus be thought of as a punctuation mark that tells RNA polymerase (the "reader") where to start its transcription of the structural gene. Once RNA polymerase binds to the promoter and transcription begins, the gene is said to be activated.

If every cell contains the same genes, how does each cell type get its own unique constellation of messenger RNAs? This differential gene transcription is accomplished by the enhancers. The enhancer sequences of the DNA (see Figure 2.1) are the same in every cell type; what differs is the combination of transcription factor proteins present. Enhancers can bind several transcription factors, and it is the specific combination of transcription factors that allows a gene to be active in a particular cell type. That is, the same transcription factor, in conjunction with different other transcription factors, will activate different promoters in different cells. Moreover, the same gene can have several enhancers, with each enhancer binding transcription factors that enable that same gene to be expressed in different cell types. Figure 2.2 illustrates this phenomenon for expression of the mammalian *Pax6* gene. First, there are several enhancer elements that can activate the *Pax6* gene differentially in several different cell types (Figure 2.2A,B). Second, within each of these enhancer regions there are sites that bind specific transcription factors (Figure 2.2C).

Transcription factor complexes generally activate gene expression in one of two ways. First, the complex can clear a space in the chromosome,

(A)

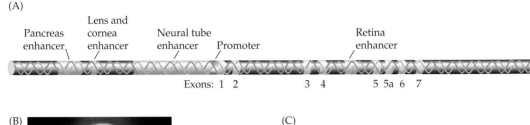

Pancreas enhancer

Lens and cornea enhancer

Neural tube enhancer   Promoter

Retina enhancer

Exons: 1 2   3 4   5 5a 6 7

(B)

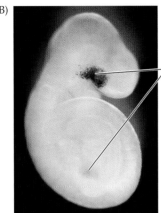

β-galactosidase

(C)

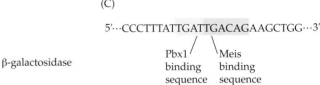

5′···CCCTTTATTGATTGACAGAAGCTGG···3′

Pbx1 binding sequence

Meis binding sequence

Figure 2.2 Enhancer region modularity. (A) The gene for Pax6, a protein critical in the development of a number of widely different tissues, has several enhancer elements (green). These enhancers direct *Pax6* expression (yellow exons 1–7) differentially in the pancreas, the lens and cornea of the eye, the retina, and the neural tube. (B) When the gene for bacterial β-galactosidase is fused to *Pax6* enhancers for expression in the pancreas and the lens/cornea, this enzyme (which is easily stained) can be seen in those tissues. (C) The pancreas-specific enhancer element has binding sites for the Pbx1 and Meis transcription factors; both of these factors must be present in order to activate the *Pax6* gene in the pancreas. (Photograph from Williams et al. 1998.)

enabling RNA polymerase to find the promoter. Second, and most simply, the transcription factors bound on an enhancer can stabilize the RNA polymerase so that it will stay on the promoter and begin transcription.

Finding a promoter is not easy, because the DNA is usually so wound up that the promoter sites are not accessible. Indeed, over 6 *feet* of DNA is packaged into chromosomes of each human cell nucleus (Schones and Zhao 2008). The protein particles that wrap the DNA are called **nucleosomes**. In most cases, the nucleosomes wrap the DNA so tightly they prevent any other protein (including RNA polymerase) from binding to the DNA. However, some enhancers bind transcription factors that modify the nucleosomes such that they unwind and the promoter site becomes accessible (North et al. 2012; Barozzi et al. 2014).

Each nucleosome is made of a core of eight **histone proteins** (two molecules each of histones H2A, H2B, H3, and H4; Figure 2.3A). Adjacent nucleosomes are packed together into tight arrays that prevent transcription factors and RNA polymerases from gaining access to the genes (Thoma et al. 1979; Schlissel and Brown 1984). It is generally thought that the "default" condition of chromatin is a repressed (compressed) state, and that cell- and tissue-specific genes become activated by local interruption of this repression (Weintraub 1985). Repression can be locally strengthened (so that it is very difficult to transcribe those genes within the nucleosomes) or relieved (so that it becomes relatively easy to transcribe them) by modifying the histones.

Even in compressed nucleosomes, "tails" projecting from the histones are accessible to important enzymes that can place specific chemical groups on the DNA. In general, the addition of acetyl groups (**acetylation**) to histone tails will loosen the nucleosome and allow transcription, while the addition of methyl groups (**methylation**) to histone tails tends to recruit proteins that will aggregate the nucleosomes more tightly and repress transcription (Figure 2.3B). More specifically, enzymes (**histone acetyltransferases**) that place acetyl groups on histones (especially on lysines in histones H3 and H4) destabilize the nucleosome so that it unwraps and the DNA is exposed. Enzymes that remove acetyl groups (**histone deacetylases**)

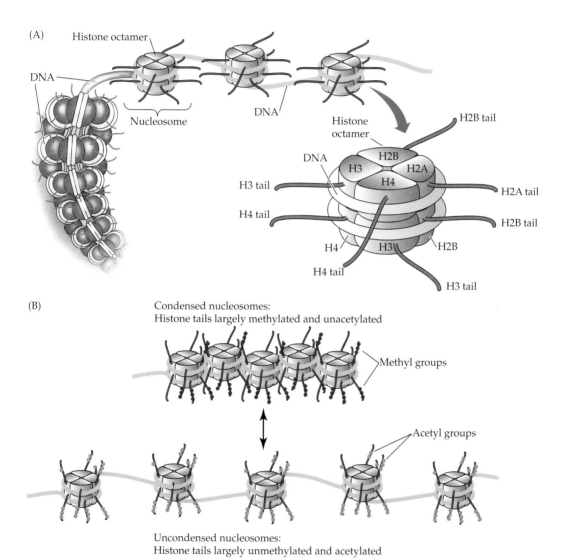

Figure 2.3 Nucleosomes condense the DNA into tight configurations on the chromosomes. (A) Nucleosomes are composed of histone subunits wound tightly by DNA. Histone "tails" protruding from the subunits allow for the attachment of chemical groups. (B) Methyl groups condense nucleosomes more tightly, preventing access to promoter sites and thus preventing gene transcription. Acetylation loosens the nucleosome packing, exposing the DNA to RNA polymerase and other transcription factors that will activate the genes.

stabilize nucleosomes and prevent transcription. The addition of methyl groups to histones by **histone methyltransferases** generally represses transcription even further, whereas removing methyl groups brings the nucleosomes back to their baseline stability (see Strahl and Allis 2000; Cosgrove et al. 2004).

These changes in the chromatin can be transmitted from cell to cell, leading to certain genes being turned on in particular groups of cells. Since these stable changes regulate which genes are expressed, they are critically important in producing differentiated cell types and in the physiological states of cells. **Epigenetics** is the study of such chromatin alterations that result in changing the stable, long-term changes in the transcriptional potential of cells. **Epigenetic changes** have been broadly defined as being "molecular processes around DNA that regulate genome activity independent of DNA sequence and are mitotically stable" (Skinner et al. 2010). Note that in epigenetic changes—DNA methylation, histone methylation or acetylation, and other long-term modifications of chromatin—the nucleotide sequence of the DNA does not change.

An overwhelming majority of environmental factors and toxicants are not able to alter DNA sequence or cause genetic mutations (McCarrey 2012). However, numerous environmental factors can dramatically influence epigenetic processes to change gene expression and thus alter development and physiology. Therefore, epigenetics provides molecular mechanisms whereby the environment can affect the biology of an organism. As DNA sequence changes are called genetic mutations, environmentally altered epigenetic sites that influence genome activity are often referred to as **epimutations** (Skinner et al. 2010). Two different forms of a gene, having the same DNA sequence but different epigenetic marks, are often called **epialleles**.

DNA methylation

Methylation is one of the most important and best studied mechanisms of keeping genes inactive. Certain cytosine (C) bases on DNA are especially susceptible to methylation. These are CpG (i.e., cytosine-phosphate-guanine) sequences in animals and CpNpGp triplets in plants (where N can be any of the four bases). Many transcription factor proteins recognize a specific DNA sequence but cannot find that sequence if the DNA is methylated. Therefore, one way of keeping a gene inactive is to methylate the promoter and enhancer regions of that gene. This inactivation can be transmitted at each cell division by enzymes called DNA methyltransferases that recognize a methylated C on a CpG sequence of the template strand and then methylate the C on the newly synthesized complementary strand.

Methylated cytosines in DNA can further inhibit transcription by binding other proteins such as MeCP2. Once connected to a methylated cytosine, MeCP2 binds to histone deacetylases and histone methyltransferases, which remove acetyl groups from (Figure 2.4A) or add methyl groups to (Figure 2.4B) the histones. As a result, the nucleosomes form tight complexes with the DNA and don't allow other transcription factors and RNA

(A)

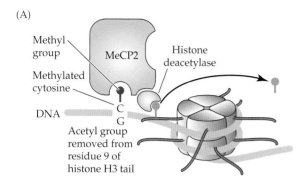

(B)

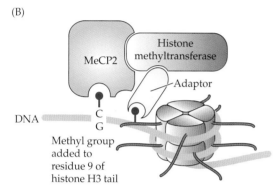

Figure 2.4 Modifying nucleosomes through methylated DNA. The MeCP2 protein recognizes methylated cytosines on DNA. It binds to the DNA and is thereby able to recruit (A) histone deacetylases (which remove acetyl groups) or (B) histone methyltransferases (which add more methyl groups). Both of these modifications promote the stability of the nucleosome and the tight packing of DNA, thereby repressing gene expression. (After Fuks 2005.)

polymerases to find the genes. Other proteins, such as HP1, also bind and aggregate methylated histones (Fuks 2005).

Another enzyme recruited to the DNA by MeCP2 is DNA methyltransferase 3 (Dnmt3). This enzyme methylates previously unmethylated cytosines on the DNA. The newly established methylation pattern is then promulgated by DNA methyltransferase 1 (Dnmt1), which recognizes methyl groups on one strand of DNA and places methyl groups on the newly synthesized strand, as mentioned above (see Bird 2002; Burdge et al. 2007a). In this way, once the DNA methylation pattern is established in a cell, it can be stably inherited by all the progeny of that cell. Such heritable changes in methylation state are epigenetic.

Thus, during normal development, differences in cytosine methylation are critical in telling a nucleus which genes can be expressed, and which genes are expressed determines what type of cell it will become. Moreover, thanks to Dnmt1, this pattern of gene expression is stably transmitted to that cell's progeny. Appendix B provides further details on how such chromatin alterations are passed on epigenetically from one cell generation to another.

Vernalization: Temperature-Dependent Chromatin Changes

One reason why Western scientists have ignored the roles of the environment in normal development is that the idea of environmentally induced phenotypes formed the center of an ideologically biased version of heredity. From the 1930s through the mid-1960s, the Soviet biologist Trofim Lysenko fought for an environmentally based, anti-Mendelian science of heredity, and in 1948 he succeeded in making the study of genetics illegal in the Soviet Union (see Appendix A).

The paradigm for Lysenko's approach was vernalization, a phenomenon whereby the seeds of certain plants need to be exposed to cold temperatures in order for the plants to flower. The vernalization requirement of winter strains of cereals such as wheat means that their seeds are planted in the fall and experience the winter cold, making the plant competent to flower the following spring. Lysenko's fervent claims that the vernalized state could be inherited by seeds that had never been exposed to cold temperatures failed miserably as a model of heredity (and probably led to crop failures in the Ukraine and China). But vernalization is a very real natural phenomenon in which the plant's development is altered by the environment. Moreover, the phenomenon has a genetic basis.

The genetic basis of vernalization appears to be straightforward: Cold temperatures repress a repressor of flowering. In wheat, for instance, expression of the *VNR1* (*vernalization-1*) gene promotes flowering. *VNR1* encodes a transcription factor that *activates* genes that promote flowering while at the same time repressing genes that maintain vegetative (nonreproductive) growth. The activity of *VNR1*, however, is repressed by the transcription factor encoded by the *VNR2* gene. Cold conditions repress *VNR2*, thus releasing the inhibition of *VNR1* and allowing flowering to occur (see figure). This repression continues through the spring, allowing the wheat to become competent to flower as early as possible.

In the wild mustard *Arabidopsis* (the model system widely used in plant genetics because of its rapid life cycle and the ability to obtain mutants of flowering in the laboratory), the *FLC* ("flowering control") gene is equivalent to wheat's *VRN2* gene. Although the details of

General pathway for vernalization in wheat. Cold temperatures inhibit the expression of the *VRN2* gene. *VRN2* encodes a transcription factor that actively represses the *VRN1* gene, the products of which promote flowering and suppress vegetative growth. Thus, cold temperatures inhibit the inhibitor of flowering.

vernalization differ between cereals and *Arabidopsis*, the mechanisms are similar (see Sung and Amasino 2005; Baulcombe and Dean 2014). Using genetic screens of *Arabidopsis* to find mutations that interfere with vernalization has led to the discovery of genes whose protein products are critical in retaining the repression of the *FLC* gene. This repression is accomplished by methylating and deacetylating the gene's regulatory region. Moreover, we now know that further epigenetic events, specifically the reactivation of the *FLC* gene in the early seed by a DNA demethylase, prevents the inheritance of the vernalized condition (Crevillén et al. 2014). Thus, while it is true that Lysenko was an ideologue and a poor scientist, his concept that the environment could give plants new properties—and that cold treatment might induce a longer and safer growing season in wheat—is something that is being actively revisited today.

DNA methylation and plasticity: Environmental control through direct DNA transcription

There are numerous cases where the environmental agent regulates the production of different phenotypes by altering DNA methylation. This causes different genes to be expressed, depending on whether the environmental factor is present or absent. Not all animals methylate their transcription by methylating their DNA. But for those that use this process, differential methylation of promoters and enhancers is a well used mechanism for responding to environmental cues.

YOU ARE WHAT YOU EAT: DIET-INDUCED PHENOTYPIC CHANGE   Environmental agents can significantly alter gene expression and the resulting phenotype by changing DNA methylation patterns. One study by Waterland and Jirtle (2003) used a dominant mutation of the *agouti* gene in mice to show that changes in maternal diet can produce changes in DNA methylation, and that these methylation changes can affect offspring phenotype.

The normal (wild-type) allele of the mouse *agouti* gene results in brown-pigmented fur and normal body size. The variant *Agouti* allele, however, is dominant over *agouti* and produces proteins that lead to obese mice with yellowish fur (Figure 2.5A). Waterland and Jirtle's study made use

(A)

(B)

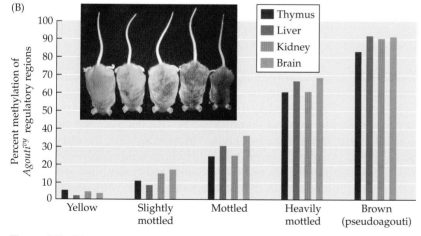

Figure 2.5   Maternal diet can affect phenotype. (A) These two adult mice are genetically identical; both are heterozygous for the *viable yellow* mutation of the *Agouti* allele, whose protein product converts brown pigment to yellow and accelerates fat storage. The obese yellow mouse is the offspring of a mother whose diet was not supplemented with methyl donors (e.g., folic acid) during her pregnancy. The embryo's *Agouti* gene was not methylated, and Agouti protein was made. The sleek brown mouse was born of a mother whose prenatal diet was supplemented with methyl donors. The *Agouti* gene was turned off, and no Agouti protein was made. (B) Norm of reaction for the *viable yellow Agouti* gene. As methyl donor supplementation increased, genetically identical mice became progressively more brown. The histogram shows coat color correlated with the amount of methylation of this gene in all tissues studied. (A photograph courtesy of R. L. Jirtle; B after Waterland and Jirtle 2003, photograph from Dolinoy et al. 2006, courtesy of R. L. Jirtle.)

of a mutant allele of *Agouti* called *viable yellow* that contains an inserted sequence that can be methylated, thus blocking transcription of *Agouti*.

Because the mutant *Agouti* allele is dominant, we would predict that all the offspring having one copy of this dominant gene (which they inherit from their fathers) would be yellow-furred and tend to obesity. However, when Waterland and Jirtle fed pregnant mice different amounts of methyl-donor dietary supplements (folate, choline, and betaine), their *viable yellow Agouti* offspring did not show the predicted phenotype (Figure 2.5B). Even though the offspring were genetically identical, they had strikingly different phenotypes due to their mothers' diets. The researchers found that the more methyl donors in the mother's diet, the greater the methylation of the *viable yellow* insertion in the offspring, and the darker (i.e., more normal) their pigmentation. Additionally, the obesity characteristic of the *Agouti* phenotype was reduced, and other methylation differences were seen throughout the body (Jirtle 2013). In other words, the dietary methyl groups fed to the mother turned off the *Agouti* gene in her offspring.

There are several studies showing that DNA methylation is affected by diet and other environmental agents experienced during early development. As we will discuss in Chapter 7, feeding a pregnant rat a low-protein diet causes her offspring to develop phenotypes that predispose them to diabetes, hypertension, and obesity later in life. It has been proposed that this susceptibility to disease is caused by the inappropriate methylation of the genes encoding enzymes used in liver metabolism and kidney development. For example, rats starved for protein in utero produced much more *PPARα* mRNA in their livers than did rats whose mothers were fed a normal diet (Lillycrop et al. 2007). This increase is due to lower amounts of promoter methylation (Burdge et al. 2007b; Lillycrop et al. 2007). The PPARα protein leads to greater fat storage.

RUNNING HOT AND COLD: SEX SPECIFICATION IN TURTLES   Unlike the situation in mammals, where primary sex determination (i.e., whether gonads are ovaries or testes) is specified by the chromosomes, gonadal specification in many fishes, amphibians, and reptiles can be affected by hormones. While the detailed mechanisms are not known in most species, the enzyme **aromatase** appears to play an important role in both temperature-dependent sex determination in turtles and crocodilians and community-dependent sex determination in certain fishes. Aromatase, as mentioned in Chapter 1, can convert testosterone into estrogens (Figure 2.6A).

Environmentally regulated hormones may directly control the temperature-sensitive sex determination cascades of some species of turtles. When turtle embryos develop below a certain temperature, they are one sex; above that temperature, they develop into the other sex. The sex and set point differ between species (see Bull 1980). In such cases of environmental sex determination, hormones can determine the direction of gonad development, and the environment controls the production of different hormones. Injection of turtle eggs with estrogen organizes the gonadal rudiment into ovaries, no matter what their temperature of incubation. Similarly, injecting turtle eggs with inhibitors of estrogen synthesis produces male offspring, even if the eggs are incubated at temperatures that usually produce females (Dorizzi et al. 1994; Rhen and Lang 1994). The

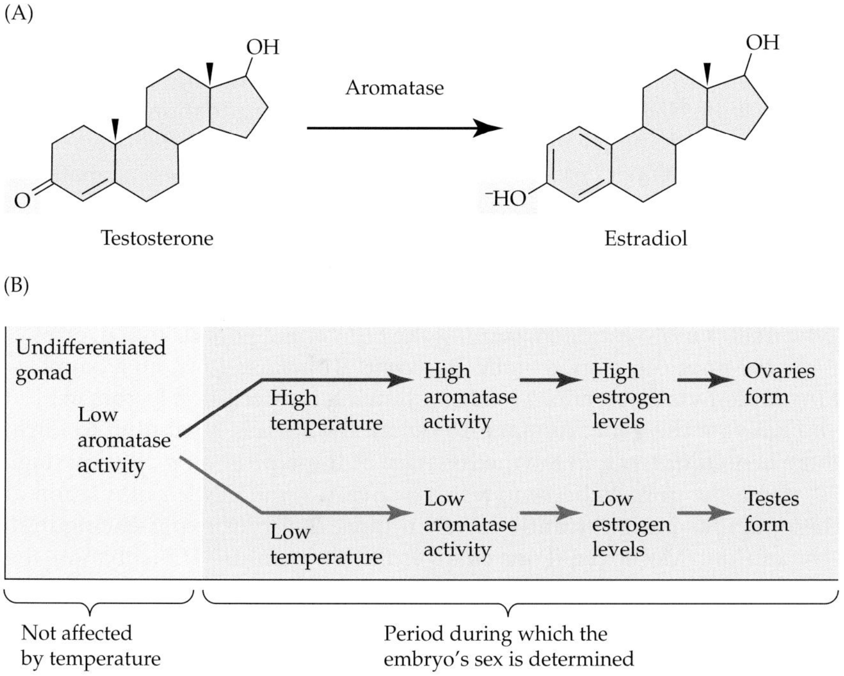

(A)

Testosterone

Aromatase

Estradiol

(B)

Undifferentiated gonad

Low aromatase activity

High temperature

Low temperature

High aromatase activity → High estrogen levels → Ovaries form

Low aromatase activity → Low estrogen levels → Testes form

Not affected by temperature

Period during which the embryo's sex is determined

Figure 2.6 The enzyme aromatase converts androgens (such as testosterone) into estrogens (such as estradiol). (A) The name "aromatase" comes from its ability to aromatize the six-membered carbon ring by reducing the ring-stabilizing keto group (=O) to a hydroxyl group (–OH⁻). This allows the hormones to bind to different receptors (dormant transcription factors) and activate different genes. (B) Scheme for sex determination in the turtle *Emys orbicularis* (the European pond turtle). High temperatures during the sex-determining period cause high aromatase levels, leading to the production of estrogen and ovary formation. Aromatase is not promoted at low temperatures, thereby preventing ovaries from forming.

sensitive time for the effects of estrogens and their inhibitors coincides with the time when sex determination usually occurs (Bull et al. 1988; Gutzke and Chymiy 1988).

High levels of aromatase are associated with the generation of ovaries in many species having temperature-dependent sex determination. The aromatase activity of the European pond turtle (*Emys*), for instance, is very low at the male-promoting temperature of 25°C (Figure 2.6B). At the female-promoting temperature of 30°C, aromatase activity increases dramatically during the critical period for sex determination (Desvages et al. 1993; Pieau et al. 1994; Ramsey et al. 2007). Temperature-dependent aromatase activity is also seen in diamondback terrapins, where its inhibition masculinizes their gonads (Jeyasuria et al. 1994), and it also appears to be involved in the temperature-dependent differentiation of lizards and salamanders (Sakata et al. 2005).*

It is possible that the expression of aromatase is activated differently in different species. In some species, the aromatase *protein* itself may be

*What about mammals? Although we don't change gonadal sex when injected with hormones or hormone inhibitors, aromatase may still be important. It is known that the mammalian testes make aromatase and that this aromatase is used to synthesize estrogen from testosterone. (Estrogen is needed in male gonads for the proper functioning of the duct system.) It is also well known that if human or mouse testes don't descend, sperm production is abnormal. One explanation for the low sperm count in undescended testes is that the higher temperatures experienced within the abdomen increase aromatase function, thereby converting testosterone—required for efficient sperm production—into estrogen (Bilinska et al. 2003). Thus, the scrotum may have evolved because it holds the testes outside the main body wall, where these male gonads experience lower temperatures and the correct ratios of testosterone and estrogen can be maintained.

temperature-sensitive. In other species, the expression of the aromatase *gene* may be differentially activated at high temperatures.

In *Trachemys scripta*, the red-eared slider turtle (commonly seen in pet stores), temperature plays a major role in sex determination. Clutches of *Trachemys* eggs incubated at 26°C produce 100% male offspring, while temperatures of 31°C produce all female offspring. Clutches incubated at temperatures between these two extremes have different percentages of male and female hatchlings. The critical gene for *Trachemys* sex determination encodes the aromatase enzyme. So, in these turtles, high temperature is thought to activate the aromatase gene, and the protein encoded by that gene makes more estrogen, turning the bipotential gonads into ovaries.

In the promoter region of the *Trachemys* aromatase gene, high temperature demethylates certain cytosines, allowing transcription factors to bind and activate the gene. At stage 16 (when the shell is beginning to form) there is no difference in the methylation of these promoters. But by stage 20, when the gonads become testes or ovaries, the DNA methylation at these regions is significantly lower in those embryos experiencing high temperature. Moreover, if one changes the temperature of incubation, the DNA methylation of this region changes, as does the sex of the gonad. Thus the aromatase gene is active in those gonads becoming ovaries and is not active in those gonads that will be testes (Matsumoto et al. 2013). In addition one of the DNA sites that becomes accessible is a binding site for the estrogen receptor. So, once estrogen is made, it may feed back positively to keep the aromatase gene active. This allows the gonads to become ovaries even after the initial temperature conditions are gone. The mechanism by which temperature regulates DNA methylation has still to be delineated, and other transcription factors may also be involved (Murdock and Wibbels 2006; Shoemaker et al. 2007).

JELLY IN THE RECIPE: CASTE SPECIFICATION IN HONEYBEES   In many species of ants and bees, the "queen" is the only female having functional ovaries (see Figure 1.8). In the honeybee (*Apis mellifera*), the queen appears to be determined by the amount of royalactin, a 57kDa protein, in the larva's diet (Figure 2.7). Kamakura (2011) showed that this protein binds to the receptor for epidermal growth factor, which in turn activates the TOR protein (which increases body size), mitogen-activated protein kinase (which accelerates development), and juvenile hormone (JH) titers (which promote ovary development). The JH titer is increased during the fourth instar (larval molt) stage, which is when ovary growth would occur (Wirtz and Beetsma 1972; Patel et al. 2007).

Like vertebrates, honeybees have methylated DNA. Indeed, in the larval heads of worker bees,

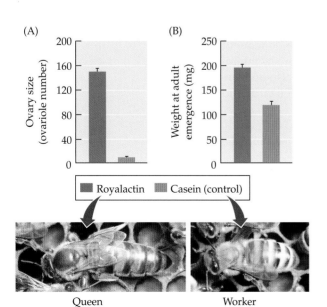

(A)

Ovary size (ovariole number)

(B)

Weight at adult emergence (mg)

■ Royalactin   ▦ Casein (control)

Queen   Worker

Figure 2.7 Diet-induced developmental changes can produce either reproductively competent queens or sterile workers. Royalactin induces functional ovaries (A) and increased body weight (B) in the honeybee *Apis mellifera*. (After Kamakura 2011.)

more than 80% of the differentially methylated cytosines are methylated. These methylated genes include those involved in the synthesis and reception of JH and insulin, as well as several genes involved in regulating metabolism (Foret et al. 2012). Moreover, like vertebrates, honeybees use the methyltransferase enzyme Dnmt3 to methylate cytosines in the DNA without a template. Reducing the levels of Dnmt3 (by inhibitor RNAs that bind to the Dnmt3 mRNA) reduces the amount of DNA methylation and promotes the formation of queens at the expense of workers. In control experiments, very few bee larvae became queens (Kucharski et al. 2008).

POWER AND CONTROL: PARASITES AT THE SWITCH   Parasites and mutualistic symbionts in the environment live in intimate association with their hosts, and they often manage their life cycles by altering the expression of host genes. Indeed, they can cause significant phenotypic changes through changing DNA methylation patterns. As will be discussed in Chapter 3, in the leafhopper insect *Zyginidia pullula*, genomic sex can be overridden by *Wolbachia* bacterial symbionts that turn males (which cannot transmit *Wolbachia*) into females (which can). Methylation-sensitive DNA sequencing studies have show that *Z. pullula* males that have been completely feminized by *Wolbachia* bacteria no longer have a male pattern of genome methylation. The *Wolbachia* have changed it to a female pattern (Negri et al. 2009).

Similarly, when male rats are infected with *Toxoplasma gondii*, the parasitic protist takes control. *T. gondii* causes toxoplasmosis, a flu-like disease that can be fatal to human fetuses. Although the parasite infects many species of mammals, the reproductive portion of its life cycle can only be accomplished inside cats. *T. gondii* alters the fear behavior of infected rodents in manners thought to increase the rodents' chances of being eaten by cats and thereby completing the parasite's life cycle. It appears that *T. gondii* infection causes the demethylation of the promoter region of the arginine-vasopressin gene in the rat brain after exposure to cat urine. This greatly reduces the threat-signaling capacity of the urine, allowing the rat to enter more dangerous areas and be eaten. By altering the methylation of the host's genome, the parasite becomes better able to complete its life cycle (Hari Dass and Vyas 2014).

We've thus seen that certain environmental agents—diet, temperature, and microbes—can directly change DNA transcription through altering the epigenetic contours of the chromatin. These transcriptional changes alter the phenotype of the organism. This direct alteration of transcription is one pathway through which environmental factors can alter phenotype.

Signal Transduction from the Environment to the Genome Via the Neuroendocrine System

Another phenotype-altering pathway involves the nerves and hormones. In studying the metamorphosis of insects, H. Fred Nijhout (1999, 2003) came to formulate one of the most important principles of developmental plasticity: namely, that actions of the neuroendocrine system can change gene expression by transducing sensory information from the environment into the body. This seems reasonable, since the nervous system monitors

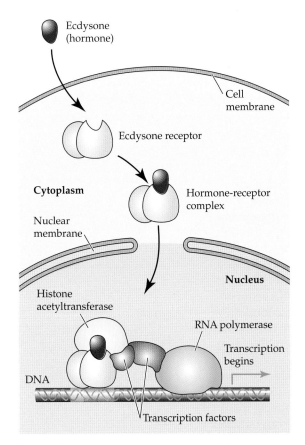

Figure 2.8 Simplified representation of the proposed mechanism of ecdysone action. Like other steroid hormones, ecdysone enters the cell readily (it is a lipid). Once inside, it binds to its receptor, a protein that is always present in the cytoplasm but is nonfunctional until it binds ecdysone. Once the receptor binds the steroid, this complex can enter the nucleus and activate gene expression by recruiting histone acetyltransferase enzymes and other transcription factors.

the outside environment through our sensory neurons. Signals from these neurons can then inform the endocrine glands to secrete (or to not secrete) hormones—and one of the functions of hormones is the regulation gene expression. In this manner, an environmental stimulus can cause changes in gene expression. It is important to recognize that such responses to the environment must already be part of the genetic repertoire of the organism. Expression of such genes, however, requires an environmental signal; otherwise the genes will remain inactive.

Insect hormones work largely by activating transcription factors. Hormones such as the metamorphic molting agent ecdysone (from *ecdysis*, "to shed") induce many changes in the developing insect. Ecdysone functions by binding to an inactive transcription factor (the ecdysone receptor). Once the hormone binds, it converts the receptor to an active form that can enter the nucleus, interact with other transcription factors, and initiate gene expression on particular promoters (Figure 2.8). Juvenile hormone (which prevents metamorphosis into an adult, hence its name) probably works in a similar manner.*

Neuroendocrine regulation of temperature-dependent polyphenism in insects

One of the earliest documented cases of environmental determination of phenotype came from studies on the European map butterfly (*Araschnia levana*). The butterflies eclosing from their pupal cases in the spring are so different from the butterflies eclosing in the summer that Linnaeus's *Systema Naturae* (1758) classified them as two different species. The spring morph is bright orange with black spots, while the summer form is mostly black with a white band. In 1875, August Weismann demonstrated that these butterflies were the same species but that temperature conditions during their larval development resulted in different phenotypes. By incubating caterpillars and pupae in different conditions, Weismann could produce the spring form in the summer.

It is now known that the normal change from spring to summer morph is controlled by changes in both day length and temperature during the larval period (Koch and Buckmann 1987; Nijhout 1991). Moreover, we now know that the temperature and daylight signals produce the phenotype by regulating the amount of the hormone 20- hydroxyecdysone (20E; the active form of ecdysone) during the larval stage of development. Larvae

---

*Indeed, several lipid hormones, including testosterone, estrogens, ecdysone, juvenile hormone, retinoic acid, and glucocorticoids, work in this remarkable way: they convert their receptors into transcription factors. Once the receptor protein has bound the hormone, it can enter the nucleus and bind to promoters or enhancers to regulate gene expression. Hormones are known to regulate phenotypic plasticity in many insects (Table 2.1).

TABLE 2.1 Examples of endocrine-mediated polyphenism in insects, their mechanisms, and environmental cues

Organism	Polymorphism	Mechanism	Cue
Ants, bees	Castes	JH titer[a]	Nutrition; pheromones
Aphids	Winged forms (alates)	JH titer	Seasonal: photoperiod; temperature; population density
Butterflies	Hindwing melanism and eyespot size	Ecdysteroid timing and duration	Seasonal: photoperiod; temperature
	Eyespot presence	Ecdysteroid receptor expression	Seasonal: photoperiod; temperature
Crickets	Wing length	JH titer	Temperature
		Ecdysteroid titer	Photoperiod; diet; population density
Daphnia spp.	Sex determination	JH titer	Seasonal: photoperiod; temperature; nutrition; population density
	Defense morphology	JH titer	Kairomones
Manduca sexta larvae	Larval color green/black	JH titer	Temperature
Onthophagus beetles	Horn growth	Ecdysteroid pulse	Body size
	Horn size	JH titer	Nutrition
Tadpoles	Size, shape, coloration	Stress hormones	Predators; stress
Termites	Caste differentiation	JH titer	Pheromone

Sources: Beldade et al. 2011, Dennis et al. 2014.
[a]JH = juvenile hormone

that develop in the spring do not experience this pulse of hormone; larvae developing in the longer, warmer days of summer do. When 20E is injected into larvae, the spring form can be converted to a summer form, and intermediate amounts of 20E produce intermediate phenotypes not found in any natural conditions (Figure 2.9; Nijhout 2003).

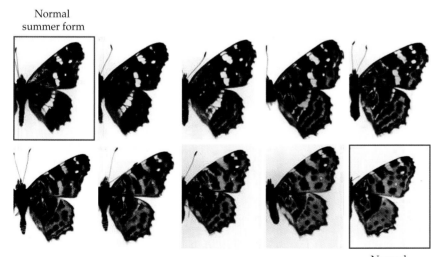

Normal summer form

Normal spring form

Figure 2.9 Hormonal regulation mediates the environmentally controlled pigmentation of *Araschnia levana*. In the wild, different generations experience significantly different photoperiods. In the short photoperiod (below the critical day length), there is no pulse of 20-hydroxyecdysone (20E) during early pupation, and the spring form of the butterfly is generated. When these spring butterflies mate, the larvae experience a long photoperiod and generate the summer pigmentation. In the laboratory, injections of 20E at different times during pupation can induce both phenotypes, as well as intermediate phenotypes not seen in the wild. (From Nijhout 2003, photographs courtesy of H. F. Nijhout.)

There is evidence that in some butterflies, the presence of eyespots is controlled by the temperature-dependent hormonal regulation of the *Distal-less* gene. In the African country of Malawi, where there is a hot wet season and a cooler dry season, the butterfly *Bicyclus anynana* occurs in two environmentally induced phenotypes ("morphs"; see the cover of this book). The dry-season morph is somewhat sluggish, with cryptic coloration resembling the dead brown leaves of its habitat. The wet-season morph is an active flier and has ventral hindwing eyespots that deflect attacks from predatory birds, mantids, and lizards (Brakefield and Frankino 2009; Olofsson et al. 2010; Prudic et al. 2015). The determining factor in this seasonal polyphenism appears to be the temperature during pupation: low temperatures produce the dry-season morph, whereas high temperatures produce the wet-season morph (Figure 2.10; Brakefield and Reitsma 1991). The development of butterfly eyespots begins in the late larval stages, when the transcription of the *Distal-less* gene is restricted to a small focus that will become the center of each eyespot. Paracrine factors from the central eyespot cells and local sensitivities to ecydysone may regulate the size and color of the eyespot in a complex manner not yet completely understood (Mateus et al. 2014). Monteiro and colleagues (2013) have shown that overexpressing *Distal-less* in the larval wing disc induces the formation of new eyespots, while down-regulating this gene decreases the size of those pigmented areas.

The seasonal *B. anynana* morphs appear to diverge at the later stages of signal activation and color differentiation, and this also appears to be regulated by the hormone ecdysone. When 20E is present early in pupal development, *Distal-less* function is retained and the hot-and-wet season

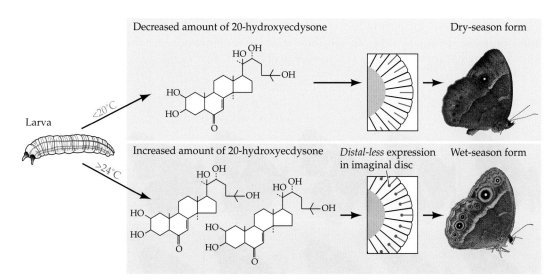

Figure 2.10 Phenotypic plasticity in *Bicyclus anynana* is regulated by temperature. High temperature allows the rapid synthesis of 20-hydroxyecdysone (20E), a hormone that is able to sustain *Distal-less* expression in the pupal imaginal disc. The region of *Distal-less* expression becomes the focus of each eyespot. In cooler weather, 20E is not formed as rapidly, other transcription factors are upregulated, which may suppress *Distal-less* activity, and eyespots fail to form. (Photographs courtesy of S. Carroll and P. Brakefield.)

form of the wing, with its large eyespot, is formed (Oliver et al. 2013; Beldade et al. 2002; Oostra et al. 2014). Conversely, in the dry season, 20E levels will not be sufficient to retain the expression of *Distal-less* in early wing primordia, and thus no eyespots will form. Thus, environmental temperature can be sensed and converted into an endocrine signal that regulates gene expression. The developmental effects of these hormones are at the level of DNA and epigenetic regulation.

Daphnia, the crab-like arthropod whose predator-induced polyphenism was discussed in Chapter 1, also has a hormone-mediated defensive phenotype. Like the nutritionally induced horns in dung beetles, the elongation of the head and tail in *Daphnia* is induced through juvenile hormones. The kairo-mones released by the predators of *Daphnia* appear to activate the JH pathway in the juveniles, thereby activating those genes responsible for the defensive phenotype (Dennis et al. 2014).

Sex and social interaction: Joy to the fishes in the deep blue sea

As mentioned in Chapter 1, many fishes can change sex based on social interactions, and these changes are mediated by the neuroendocrine system (Godwin 2010). Interestingly, although the trigger of the sex change may be stress hormones (such as cortisol) that induce sex-specific neuropeptides, the effector of the change may once again be aromatase.

There are two aromatase genes in many animals; one is expressed in the brain, and one is expressed in the gonads. Black and colleagues (2005) have shown a striking correlation between changes in brain aromatase levels and changes in behavior during the sexual transitions of goby fish. The removal of the male from a stable group caused a rapid increase (more than 200%) in the aggressive behavior of the largest female, which is destined to become a male in about a week's time. This aggression is thought to have resulted from an increase in brain testosterone levels, since within hours upon removal of the male, dominant females developed lower brain aromatase levels than the other females (Figure 2.11). Gonadal aromatase levels, however, stayed the same, and gonadal sex change came later.

Aromatase-induced estrogen can work quickly because the gonad, itself, is ready to change. For instance, the goby *Trimma okinawae* can change sex in both directions repeatedly. It has both testicular and ovarian

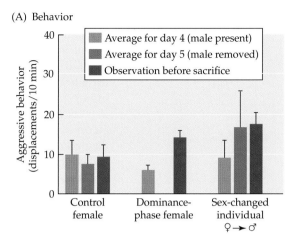

(A) Behavior

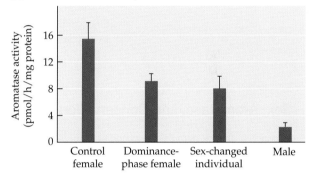

(B) Brain aromatase activity

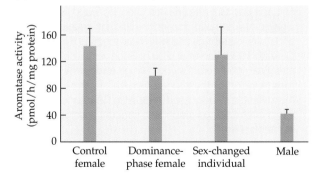

(C) Gonadal aromatase activity

Figure 2.11 Aggressive behavior and aromatase activity (AA) in the brain and gonads of *Lythrypnus dalli*, a goby fish species in which individuals can change sex (with dominant females becoming males if the group's lone male is removed; see Chapter 1). (A) On day 4 (prior to male removal), there was no statistical difference in average daily aggressive behaviors (displacements) between the largest females. On day 5, the male was removed and dominant females increased their aggressive behavior. (There are no day 5 data for dominance-phase fish because they were sacrificed during day 5 or just after.) (B,C) Brain but not gonadal AA was significantly lower in dominance-phase and sex-changed individuals compared with control females. Established males had lower brain AA than all other groups, and lower gonadal AA than all groups except dominance-phase females. (After Black et al. 2005.)

tissue in the same gonad, but only one of these tissues is active at any time. Social conditions will stimulate the transcription factors that activate aromatase and also activate the genes making the gonadotropin receptors (for luteinizing hormone and follicle-stimulating hormone) in either of these tissues. Induced social changes will alter the mRNA levels within a day (Kobayashi et al. 2005, 2009). Changes in the social group, perceived by the nervous system, become expressed by the hormonal system within hours, thereby changing the behavioral phenotype of a female fish. Interestingly, when it comes to behaviors in these fish, sex is in the brain before it is in the gonads.

In porgy fish, where the young normally start off as males and become females as they grow older (a phenomenon called protandrous hermaphroditism), aromatase inhibitors (which elevate testosterone levels) block the natural sex change from male to female and also induce male development in those older, female, fish (Lee et al. 2002). Lower aromatase activity in the gonad and brain paralleled the suppression of sex change, strongly suggesting that aromatase is directly involved in the mechanism of natural porgy sex change.

Sex, Aromatase, and Conservation Biology

Environmental sex determination has become a large issue in conservation biology because numerous man-made chemicals can alter hormone production. Atrazine, for instance, is the most widely used herbicide in the world; the United States alone uses 80 million pounds of it annually. But atrazine has effects beyond killing weeds. Atrazine can also induce aromatase expression and, thus, disrupt normal hormonal function (Crain and Guillette 1998; Fan et al. 2007). Such hormone-altering chemical compounds are known as **endocrine disruptors**. A variety of these environmental substances and their effects are discussed in detail in Chapter 6, but the effect of atrazine on amphibian gonadal development is of interest here.

One such case involves the development of hermaphroditic and demasculinized frogs after exposure to extremely low doses of atrazine. Hayes and colleagues (2002a) found that exposing tadpoles to atrazine concentrations as low as 0.1 part per billion (ppb) induced higher levels of aromatase and produced gonadal and other sexual anomalies in male frogs. Many male frogs developing at atrazine doses of 0.1 ppb and higher had ovaries in addition to testes. At concentrations of 1 ppb, the male vocal sacs (which the male frog must have in order to signal potential mates) failed to develop properly. The testosterone levels of adult male *Xenopus* were reduced nearly 90% (to the same level as control females) by exposure to 25 ppb atrazine for 46 days (Figure A).

These levels of atrazine are ecologically very relevant.

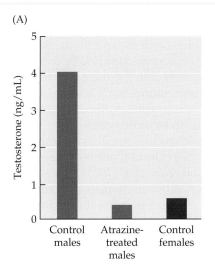

(A) Effect of a 46-day exposure to 25 ppb atrazine on plasma testosterone levels in sexually mature male *Xenopus* frogs. Testosterone levels in control males were 10 times higher than in control females; levels in atrazine-treated males were at or below those of the control females. (After Hayes et al. 2002a.)

The allowable amount of atrazine in our drinking water is 3 ppb, but levels are as high as 224 ppb in some streams in the agricultural heartland of the American Midwest (Battaglin et al. 2000; Barbash et al. 2001). Given the amount of atrazine in the water and the sensitivity of

Sex, Aromatase, and Conservation Biology (*continued*)

(B)

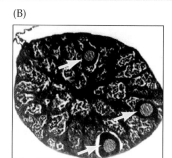

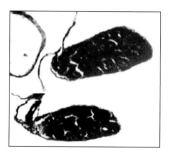

(B) Demasculinization of frogs by low amounts of atrazine. (Top) Testis of a frog from a natural site having 0.5 ppb atrazine. The testis contains three lobules that are developing oocytes (arrows) as well as sperm. (Bottom) Two testes of a frog from a natural site containing 0.8 ppb atrazine. These organs show the severe testicular atrophy that characterized 28% of the frogs at that site. (From Hayes et al. 2003, photographs courtesy of T. Hayes.)

frogs to this compound, the herbicide could be devastating to wild populations. Hayes and his colleagues collected leopard frogs (*Lithobates pipiens*) and water samples at eight sites across the central United States (Hayes et al. 2002b, 2003). They sent the water samples to two separate laboratories for the determination of atrazine, and they coded the frog specimens so that the technicians dissecting the gonads did not know from which site an animal came. The results showed that all but one site contained atrazine—and that was the only site from which the frog specimens had no gonadal abnormalities. At concentrations as low as 0.1 ppb, leopard frogs displayed testicular dysgenesis (abnormally developed testes) or conversion of the testes into ovaries. In many examples, oocytes were found in the testes (Figure B).

These studies have obvious ramifications for humans. Indeed, in subsequent chapters we will see that mice and rats exposed to endocrine disruptors develop certain diseases (notably infertility syndromes and cancers) similar to those of humans. Many agricultural and social concerns mediate the amount of atrazine use, and the corporation that manufactures atrazine has lobbied against the work of some independent researchers whose results suggest a link between the chemical and reproductive malfunctions and cancers in wildlife and humans* (Blumenstyk 2003; Aviv 2014). However, concern over atrazine's apparent ability to disrupt sex hormones has resulted in a ban on this compound by the European Union in 2004.

*Indeed, atrazine has been shown to lower the sperm count and testicular function in male rats, and some recent studies have shown that, in areas with higher amounts of pesticides, human sperm counts are low (Kniewald et al. 2000; Swan et al. 2003). In women, aromatase *inhibitors* are used clinically to lower the amount of estrogen in postmenopausal women who have estrogen-responsive breast cancer. The Environmental Protection Agency (2013) says that atrazine does not pose a reproductive health risk for humans; but the methods leading to this conclusion have been criticized by several researchers (Vandenberg et al. 2012; see Chapter 6).

Stress hormones and the induced phenotype

Stress is a great inducer of polyphenisms, and it works through the production of adrenal corticosteroid hormones such as cortisol or corticosterone. Like insect juvenile hormone and ecdysone (and vertebrate sex hormones), corticosterone binds to an inactive transcription factor, converting the transcription factor to an active form that can enter the nucleus and regulate gene expression (Figure 2.12). Corticosteroids are produced by the adrenal glands (or, in amphibians, by the interrenal glands) in response to the pituitary gland's secretion of adrenocorticotropic hormone (ACTH). ACTH is synthesized by corticotropin-releasing hormone (CRH), a small

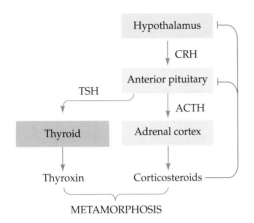

Figure 2.12 Hypothalamus-pituitary-adrenal-thyroid axis. The hypothalamus responds to internal or external stress signals by making corticotropin-releasing hormone (CRH), which tells the anterior portion of the pituitary gland to secrete adrenocorticotropic hormone (ACTH). ACTH circulates through the blood and tells the adrenal cortex to make corticosteroids. Corticosteroids circulate in the blood to activate various stress-response and polyphenism pathways; they also go to the pituitary and to the hypothalamus, where they bind to receptors that, in a negative feedback loop, down-regulate CRH and ACTH production. The CRH stimulus from the hypothalamus to the pituitary is also responsible for the production of thyroid-stimulating hormone (TSH), resulting in the elevated levels of thyroid hormones that trigger metamorphosis.

(41 amino acids) protein made in the hypothalamus. In amphibians and fish, CRH appears to induce the synthesis and secretion of both ACTH (for corticosteroids) and thyroid-stimulating hormone (for thyroxine; see Figure 2.11; see Denver 1999).

Levels of corticosterone have been directly linked to several polyphenisms. Wood frog (*Rana sylvatica*) tadpoles elongate their tails in response to predators. In the wild, those tadpoles undergoing this predator-induced response have higher levels of corticosterone. Moreover, if exposed to caged predators in controlled conditions, the tadpoles develop their new morphologies in parallel with higher concentrations of corticosterone, and the morphological changes can be halted when metyrapone, an inhibitor of corticosterone synthesis, is added to the water (Figure 2.13; Middlemis Maher et al. 2013). Corticosteroids also mediate polyphenisms in response to intraspecies competition. When leopard frog (*Lithobates pipiens*, formerly known as *Rana pipiens*) tadpoles grow at high densities, they tend to be smaller and develop more slowly. Their size and development are negatively correlated with the corticosteroid levels in their blood, suggesting that corticosteroid synthesis is responsible for slowing down development when population

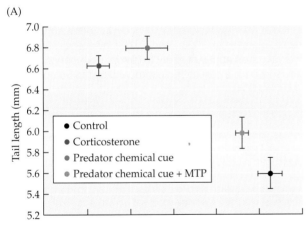

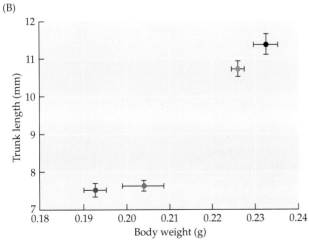

Figure 2.13 Exposure of wood frog tadpoles to predator chemical cues or to corticosterone generates similar antipredator morphologies, and the effect of the predator cues can be blocked by co-treatment with metyrapone (MTP), an inhibitor of corticosteroid synthesis. Tadpoles were treated with a predator chemical cue, corticosterone (125 n*M*), or predator chemical cue plus MTP (110 m*M*). At the end of the treatment period, tail length (A), trunk length (B), and body weight were measured. (After Middlemis Maher et al. 2013.)

density is high (Glennemeier and Denver 2002). Corticosterone is also involved in the plasticity of metamorphosis. Metamorphosis occurs earlier in times of stress, and this hormone makes cells more sensitive to thyroid hormone and more able to convert thyroid hormone into more active forms (Bonett et al. 2010).

In Chapter 1, we discussed the need for spadefoot toad tadpoles to metamorphose quickly to get out of their drying ponds. The signal for accelerated metamorphosis appears to be the change in water volume. In the laboratory, these tadpoles are able to sense the removal of water from aquaria, and their acceleration of metamorphosis depends on the rate at which the water is removed. The stress-induced CRH signaling system appears to mediate this effect (Denver et al. 1998; Middlemis Maher et al. 2013). This increase in brain CRH is thought to be responsible for the subsequent elevation of the thyroid hormones that initiate metamorphosis (see Figure 2.11) and the corticosteroids that accelerate it (Denver 1997; Boorse and Denver 2003). Thus, as in many other cases of polyphenism, the developmental changes in the spadefoot tadpoles are mediated through the neuroendocrine system—sensory organs monitor the external environment and send a neural signal to regulate hormone release. The hormones then can alter gene expression in a coordinated and relatively rapid fashion. Moreover, hormonally mediated change means the environmental cue is able to affect several different target tissues, systematically altering the phenotype of the entire organism.

The effects of maternal behavior on gene methylation

A synthesis of the work on stress hormones and on DNA methylation was realized in the discovery that adult behavioral phenotypes such as anxiety could be generated by environmental stress soon after birth. Such stress can affect the expression of genes involved in the rat corticosteroid response pathway.

There are anxious laboratory rats and there are nonanxious laboratory rats, and even if they have nearly the same inbred genome, and have been given the same food and environment, they experience stress differently. It turned out that the anxious rats had received little maternal care during the first week after birth, while the nonanxious rats had experienced more maternal licking and grooming. But how did this relate to anxiety? Tactile stimulation during this time regulates the synthesis of the glucocorticoid receptor in the hippocampus of the brain. This is the receptor that binds the glucocorticoid hormones (such as corticosterone) and down-regulates the stress response.

How is the adult phenotype regulated by these perinatal (i.e., near the time of birth) experiences? Weaver and colleagues (2004, 2007) have shown that the difference involves the methylation of the binding site for a particular transcription factor (NGF1-A) in an enhancer region of the glucocorticoid receptor (*GR*) gene. Before birth, there is no methylation at this site; one day after birth, the site is methylated in all rat pups. However, in those pups that experience intensive grooming and licking during the first week after birth, this site *loses* its methylation, while methylation is

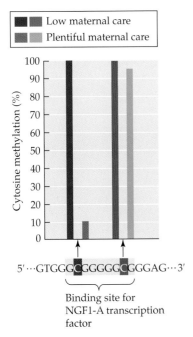

Low maternal care
Plentiful maternal care

5′ ···GTGGG**C**GGGGG**C**GGGAG···3′

Binding site for
NGF1-A transcription
factor

Figure 2.14 Differential DNA methylation due to behavioral differences in parental care. A portion of an enhancer sequence of the rat glucocorticoid receptor gene is shown, indicating the binding site for the NGF1-A transcription factor. Two cytosine residues within this site have the potential to be methylated. The cytosine at the 5′ end is completely methylated in the brains of pups that did not receive extensive licking and grooming from their mothers (red bar). The transcription factor did not bind these methylated sites, and thus the *GR* gene remained inactive. In pups that received sufficient maternal care, this same site was largely unmethylated (orange bar), and the gene was transcribed in the brain. The cytosine at the 3′ end of the enhancer (blue bars) was observed as a control; it was always methylated and had no effect on binding. (After Weaver et al. 2004.)

retained in those rats that do not receive extensive care. Moreover, this methylation difference is not seen at other sites in or near the gene (Figure 2.14). When unmethylated, the hippocampal enhancer DNA binds the NGF1-A transcription factor and is associated with active (i.e., acetylated) nucleosomes. NGF1-A does not bind when this DNA has been methylated, and in such cases the surrounding chromatin (containing the *GR* gene) is not activated. Moreover, without the binding of NGF1-A, the chromatin is not made into an active conformation and remains repressive. These chromatin differences, established during the first week after birth, appear to be retained throughout the life of the rat. Thus, adult rats that received extensive perinatal grooming have more glucocorticoid receptors in their brains and are able to deal with stress better than rats that received less care.*

The developmental pathway from prenatal care to these adult behaviors has many steps. It starts with tactile stimulation. The tactile stimulation of licking and grooming, mediated by neurons, increases the levels of thyroid hormone in the blood. (This tactile stimulation can also be accomplished by a scientist stroking the rat pup. It is not specifically a tongue or a mother thing.) The thyroid hormones stimulate the hippocampal cells to make the neurotransmitter serotonin (5-hydroxytryptamine). Serotonin binds to its receptors on neighboring hippocampal cells, stimulating the activity of the membrane-bound adenylate cyclase protein, which converts ATP into cyclic AMP (cAMP). This cAMP stimulates protein kinase A (PKA), an enzyme that activates numerous proteins, including transcription factor NGF1-A, histone acetyltransferase CBP, and transcription factor Sp1, resulting in DNA demethylation (Figure 2.15; Hellstrom et al. 2012). Thus, the tactile stimulation both induces a DNA region that allows the binding of NGF1-A and also promotes the activation of NGF1-A. This enables those rats that

---

*Of course, one big question is "Do these experiments on caged rats relate to humans?" The results are, of course, statistical, and humans live in an environment full of different stimuli, so extrapolations are risky. Still, using an "extreme" set of cases, McGowan and colleagues (2009) and Labonte and colleagues (2012) have shown that the promoter of the human hippocampal-specific glucocorticoid receptor is more highly methylated in the brains of suicide victims with a history of childhood abuse than the promoter of the same gene from either suicide victims with no childhood abuse or controls. There is also evidence that depression and anxiety in humans can be influenced by maternal care and tactile experiences (Hellstrom et al. 2012; Zhang et al. 2013).

received sufficient maternal care to lower their glucocorticoid hormones after a stressful situation, while the rats who did not experience this maternal care do not lower their glucocorticoid titers as quickly.

By switching pups and parents, Michael Meaney's laboratory (see Laplante et al. 2002) also demonstrated that the hippocampal DNA methylation difference depended on the mother rat's care and was not the result of differences in the pups themselves. One of the fascinating parts of this pathway is that the female rats that did not receive maternal licking and grooming are themselves not prone to give their offspring much licking and grooming. Consequently, their female offspring perpetuate the anxious phenotype. The lack of maternal licking and grooming results in the maintenance of DNA methylation, and this DNA methylation prevents the synthesis of hormone receptors, which predispose the offspring toward certain behaviors. Thus, this pathway shows how behavioral differences can be inherited without different genes.

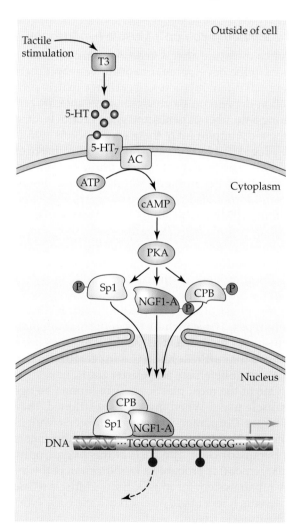

Figure 2.15 Schematic of the proposed mechanism for tactile stimulation effects on *GR* gene expression. Licking and grooming provide tactile stimulation, which increases plasma levels of thyroid hormone (T3), which in turn stimulates serotonin (5-HT) signaling in the hippocampus. Increased 5-HT levels stimulate adenylate cyclase (AC) to make cAMP, which in turn stimulates protein kinase A (PKA). PKA phosphorylates and activates the histone acetyltransferase CBP and transcription factors NGF1-A and Sp1. These proteins form a complex that binds to specific genomic regions such as the promoter of the glucocorticoid gene. The binding of this complex initiates the demethylation of the cytosines in this region, thus providing a permissive environment for gene expression. (After Hellstrom et al. 2012.)

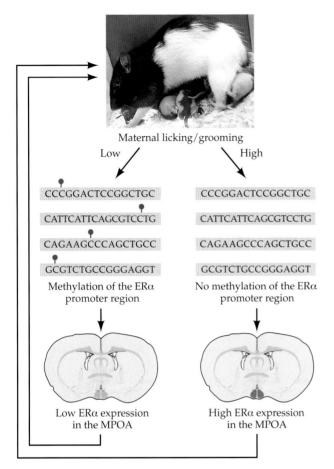

Maternal licking/grooming

Low — High

CCCGGACTCCGGCTGC

CATTCATTCAGCGTCCTG

CAGAAGCCCAGCTGCC

GCGTCTGCCGGGAGGT

Methylation of the ERα
promoter region

CCCGGACTCCGGCTGC

CATTCATTCAGCGTCCTG

CAGAAGCCCAGCTGCC

GCGTCTGCCGGGAGGT

No methylation of the ERα
promoter region

Low ERα expression
in the MPOA

High ERα expression
in the MPOA

Figure 2.16 Environmental regulation of DNA methylation and the transmission of behavior. Female rats vary in the amount of time and effort spent licking and grooming their newborn pups. Pups that experience high levels of licking and grooming have a relatively unmethylated promoter region on the gene encoding the α estrogen receptor (ERα). This gene becomes active in the MPOA and promotes the oxytocin-based licking and grooming in females. Among pups that do not experience high levels of licking and grooming, the ERα promoter is relatively highly methylated, causing the gene to function at lower levels. The result is a decrease in oxytocin-induced licking and grooming behavior. Thus, the behavior modifies the DNA, and this in turn affects the behavior. (After Champagne 2006, photograph from Meaney and Szyf 2005.)

One explanation for why these female rats were not as physical as other mothers was that these rats were anxious; but this behavior could also be due to the effects that licking and grooming have on the estrogen receptor (Figure 2.16). In rats, high levels of maternal care also cause demethylation of the regulatory regions of the estrogen receptor genes, enabling their expression in the MPOA region of the brain that is associated with sex-specific behaviors. Those female rats with fewer estrogen receptors in the MPOA region of the brain have a more receptive sexual phenotype than the rats who had been licked and groomed thoroughly when young, but they do not provide as much licking and grooming to their offspring (Champagne et al. 2006; Cameron et al. 2008).

"We will pump you up": Muscle hypertrophy

Physical pressure affects gene expression in many places in the developing organism. Vertebrate cartilage cell differentiation and cartilage matrix production depend on mechanosensitive interactions among a number of genes and gene products. One of the most important of these genes is *Sox9*, which is up-regulated by compressive force (Takahashi et al. 1998). The Sox9 protein is a transcription factor that activates numerous bone-forming genes. Tension forces also activate bone morphogenetic proteins (BMPs) and align chondrocytes (Bard 1990; Sato et al. 1999; Ikegame et al. 2001; Young and Badyaev 2007). Several studies implicate *indian hedgehog* as a key signaling molecule that is stimulated by stress and which activates the BMPs (Wu et al. 2001). Such examples of normal development's needing physical stress may explain certain human phenotypes. These include the very common malocclusion of the jaw as well as the numerous bone changes that afflict children with cerebral palsy after birth (Shefelbine and Carter 2004).

One of the most plastic traits of human beings is muscle mass. When muscles are strained, they get larger. Carpenters get enlarged biceps, weight lifters develop large pectoral muscles, and runners get big quadriceps. The basis of this hypertrophy (enlargement) is probably a modification of the injury repair system for muscle tissue. Muscle tissue represents fibers made of fused muscle cells. Next to the muscle tissue, usually within the muscular tissue matrix, are satellite cells—unfused muscle cells that have the ability to proliferate, grow, and fuse with muscle tissue upon stress. In other words, these satellites constitute a population of cells—usually held in reserve—that can be used to repair muscle tissue.

When first stressed, muscles respond by increased protein synthesis. Each nucleus in the muscle cell (a muscle cell has hundreds of nuclei because it was formed by a series of cell fusion events) has a region in which new protein synthesis occurs. However, if the physical stress continues, it appears that the muscle fibers cannot keep up with the amount of protein synthesis needed (Kadi et al. 2004). This may cause injury, or it may activate the same pathway that is used to protect against injury. In either case, the muscle starts making a hormone-like protein called insulin-like growth factor-1 (IGF-1).

IGF-1 is made by both liver and muscle tissues. In physically stressed muscle, however, it appears that IGF-1 is produced in high amounts and remains near the muscle, where it can be used. Yang and colleagues (1996) found that in stressed (but not in unstressed) rabbit muscles, IGF-1 mRNA appeared. The protein products of this mRNA message could take many forms (because of alternative splicing of the nuclear RNA into mRNA*), and one variant, seen chiefly in stressed muscles, gave rise to an IGF-1 protein called mechanogrowth factor (MGF). When extra DNA for this protein (cDNA), derived from the mRNA for MGF, was injected into mice, the animals showed a 25% increase in lean muscle within 3 weeks.

IGF-1/MGF appears to cause the satellite cells to proliferate and to fuse with the existing muscle fibers (Goldspink 2004; Kandalla et al. 2011). IGF-1 also acts on the muscle fibers through the pathway to further increase protein synthesis (Figure 2.17A). These would be the bases of the muscle hypertrophy. Lai and colleagues (2004) showed that when IGF-1 bound to its receptor, it initiated a cascade via a kinase called Akt. If Akt is mutated so that it doesn't function, the mice have reduced musculature. Conversely, if Akt (without IGF-1) is artificially activated in mice, the mice develop three times the usual muscle mass, at the expense of the fat tissue (Figure 2.17B,C).

Interestingly for bodybuilders, the molecular responses to different forms of exercise—endurance or resistance training—seem to be overlapping. Both exercise and passive stretching activate Akt (Sakamoto et al. 2003; Garma et al. 2007). Moreover, both types of exercise elevate the levels of a transcriptional coactivator called PGC-1α1. This transcription factor regulates many of the effects of training, such as angiogenesis, mitochondria production, and muscle fiber-type switching. However, gain-of-function mutations of this gene in mice have not shown any increases in muscle mass or strength (Lin et al. 2002). An increase in strength and

---

*Alternative splicing of nuclear RNA (nRNA) can make different mRNAs from the same nRNA. The mRNAs will be translated into different proteins. This is another important mechanism of differential gene expression. Interestingly, the dietary differences in honeybees also cause changes in nRNA splicing as well as DNA methylation. Indeed, Dnmt3 has been implicated in both processes (Li-Byarlay et al. 2013).

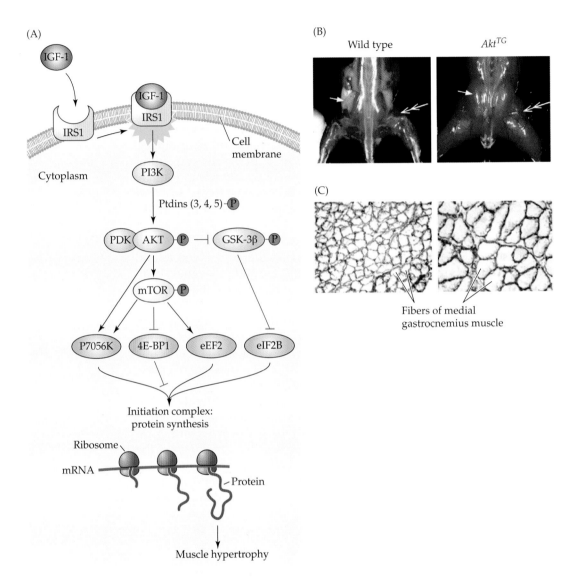

Figure 2.17 The actions of IGF-1 and Akt stimulate muscle development. (A) IGF-1 binds to its receptor (IRS1) on the cell membrane. After binding IGF-1, IRS1 becomes activated and it in turn activates compounds that phosphorylate Akt. Now Akt becomes active and can phosphorylate several target proteins, including GSK-3β. Another protein phosphorylated by Akt is mTOR, which promotes the activity of the protein synthesis initiators P7056K and eEF2; mTOR also blocks another protein synthesis inhibitor, 4E-BP1. As a result, a protein synthesis initiation complex can be made on the ribosome and many more proteins can be produced, resulting in muscle growth. (B,C) Constitutive activation of the *Akt* gene in transgenic mice results in skeletal muscle hypertrophy. (B) The double-headed arrows point to a muscle that is significantly larger in the *Akt*^{TG} mouse than in its wild-type littermate. The single arrow points to a fat pad that is present in the wild type but absent in *Akt*^{TG} animals. (C) The micrographs show transverse sections of medial gastrocnemius muscle immunostained with an anti-laminin antibody to outline the perimeters of muscle fibers. In comparison with its wild-type littermate, the *Akt*^{TG} mouse clearly has enlarged fibers. (B,C from Lai et al. 2004.)

muscle mass comes from resistance training (Figure 2.18). These resistance regimens induce the production of another, closely related transcription factor, PGC-1α4. This protein is formed from the same gene as PGC-1α1, but the RNA is spliced differently (Ruas et al. 2012). The amino-terminal region of the protein differs, allowing it to regulate a different set of genes. Importantly, it activates the gene for IGF-1, and it suppresses the gene for myostatin. IGF-1, as we've seen above, is critical for activating AKT. Myostatin is an inhibitor of muscle cell growth.

(A)

Figure 2.18 Phenotypic plasticity and resistance training. (A) Charles Atlas (born Angelo Siciliano in Calabria, Italy) won the title of "The World's Most Perfectly Developed Man" in 1922 and became an icon of male body-building. Atlas pioneered resistance training for muscle hypertrophy. His "dynamic tension technique" was based on exercises that could be done within the confines of a small apartment and did not rely on equipment or drugs. While biologists in the 1920s were saying that your genes were your destiny, Atlas's system, which pitted muscles against one another, delivered on his promise that "I can make you a new man!" (B) Alternative RNA splicing makes two proteins. Endurance training activates the proximal promoter (P.P.), closest to the structural gene, of the *PGC-1α* gene and generates a transcription factor protein, PGC-1α1, that activates certain types of myosin heavy chains (MyHC; for strength), mitochondrial oxidative phosphorylation proteins (OXPHOS; for energy), fatty acid oxidases (FAO; for energy), and VEGF (to make more blood vessels supplying oxygen). Resistance training, on the other hand, uses another promoter (A.P.) and forms another protein, PGC-1α4, that has a different N-terminal region. This protein can down-regulate myostatin (thus enlarging muscles) and up-regulate insulin-like growth factor-1 (causing more protein synthesis and proliferation). (B after Ruas et al. 2012.)

(B)

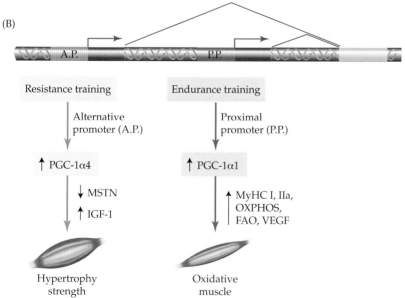

Therefore, by suppressing myostatin, PGC-1α4 enhances muscle cell growth. It inhibits the inhibitor. Mice that have an added gene that causes an overproduction of PGC-1α4 have an increase in muscle force that is proportional to their increase in muscle mass. Moreover, resistance training seems to be more effective for young men than for young women or the elderly (Petrella et al. 2006). This is reflected both in the muscle mass produced by training and in the amount of IGF-1 made by the cells. It is thought that older subjects lack satellite cells (Kandalla et al. 2011).

Thus we see that exercise activates the developmental mechanisms that promote the formation of muscles. Moreover, new studies show that without exercise, mitochondria fail to proliferate and they produce fewer mRNA transcripts for the genes that promote energy metabolism. This leads to a vicious cycle in which lack of exercise diminishes mitochondrial function, which leads to lethargy and lack of exercise. Researchers at the University of Helsinki (Mustelin et al. 2008; Pietiläinen et al. 2008) have studied monozygotic twin pairs (who have identical genotypes) in which one of the twins was obese and the other was not. The number of mitochondria in the muscles of the obese twins was only about half that of their thinner twins. Moreover, the obese twins' mitochondria had significantly less expression of those genes responsible for oxidative phosphorylation. The obese twins' mitochondria also synthesized different enzymes, leading to their having lower energy metabolism, less ability to respond to insulin, and more fat accumulation. The researchers concluded that physical activity is important for establishing normal mitochondrial transcript and enzyme levels (Pietiläinen et al. 2008). Indeed, PGC-1α1 appears to promote the production of new mitochondria and its oxidative phosphorylation enzymes (Chan and Arany 2014).

Exercise changes many aspects of adult anatomy and physiology (Rottensteiner et al. 2015). Moreover, such exercise may act by altering DNA methylation patterns. The enhancers of many genes encoding proteins involved in muscle development, insulin responsiveness, and energy metabolism are differentially methylated by exercise. When volunteers had an exercise regimen involving pedaling a bicycle with only one of their legs, the DNA methylation changes were seen only in the exercised leg (Lindholm et al. 2014).

Thus, one of the most obvious examples of the environmental regulation of human phenotype—muscle hypertrophy—is beginning to be understood at a molecular level.

Signal Transduction from the Environment to the Genome by Direct Induction: The Immune System

When one tissue induces the differentiation of another during normal embryonic development (as when the presumptive neural retina tells the outer ectoderm to become the lens of the eye, or when the ureteric bud tells the mass of intermediate mesoderm next to it to become the kidney tubules), soluble factors are sent from the inducing tissue and are received by receptors in the tissue adjacent to it. In a similar manner, microbes are able to direct the induction of antibodies in our own predator-induced polyphenism.

Microbial induction of the vertebrate immune response

If predator-induced polyphenism is the alteration of development to counter potential predators, the mammalian immune system may be the acme of such polyphenic responses (see Frost 1999). The major predators of humans have never been lions, tigers, and bears, but rather bacteria, viruses, and protists. The mammalian immune system is an incredibly elaborate mechanism for sensing and destroying these microbes (unless they have remained compartmentalized in the gut or on the skin, where they may be essential partners).

Our immune system recognizes and destroys most material that is foreign to the body. When we are exposed to a foreign molecule (called an **antigen**), we manufacture **antibodies** and secrete them into our blood serum. These antibodies recognize and then inactivate or eliminate the antigen. When viruses or bacteria enter our bodies, they are seen as a collection of antigens, and an immune response is mounted against them. The basis for the antibody immune response is summarized in the five major postulates of the clonal selection hypothesis shown in Figure 2.19 (Burnett 1959):

1. Each B lymphocyte (B cell) can make one, and only one, type of antibody. It is specific for one shape of antigen only.

2. Each B cell places the antibodies it makes into its cell membrane with the specificity-bearing side pointing outward.

3. Bacteria and viruses are digested and internalized by macrophage cells, and the digested fragments (now considered antigens) are placed on their cell membranes. These antigens are presented by the macrophages to the antibodies on the B-cell membranes.

4. Only those B cells that bind to the antigen can complete their development into antibody-secreting plasma cells. These B cells divide repeatedly, produce an extensive rough endoplasmic reticulum, and synthesize enormous amounts of antibody molecules. These antibodies are secreted into the blood.

5. The specificity of the antibody made by the plasma cell is exactly the same as that which was on the cell surface of the B cells.

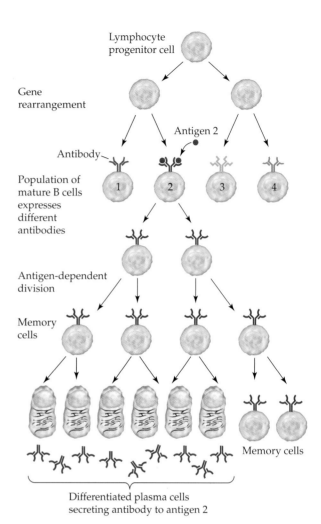

Figure 2.19 The clonal selection model. Each B cell produces only one type of the millions of potential antibody proteins. If and when the antigen to its specific antibody is presented, the B cell is stimulated to divide. (The steps of macrophage digestion and its presentation to the antibody are omitted for simplicity.) Most cells that result from this antigen-stimulated B-cell division differentiate, becoming plasma cells that secrete the antibody unique to that B-cell clone; some of the dividing cells become memory cells that will respond rapidly if the body is again exposed to the original antigen.

Out of the approximately 10 million types of antibody proteins a B cell can potentially synthesize, each B cell will make only one type. That is, one B cell may be making antibodies that bind to poliovirus, while a neighboring B cell might be making antibodies to diphtheria toxin. The type of antibody molecule on the B-cell surface is determined by chance.

B cells are continually being created and destroyed. However, when a specific antigen binds to a set of B cells, these cells are stimulated to divide and differentiate (see Figure 2.18). Most differentiate to become **plasma cells** that secrete the antibody; some divide to become **memory cells** that populate lymph nodes and can respond rapidly when exposed to the same antigen at a later time. The antigen activates the signal transduction cascade, which in turn activates transcription factors that activate the expression of those genes involved with lymphocyte differentiation and division. Thus, each person's constellation of plasma cells and memory cells differs depending on the antigens he or she has encountered. Even identical twins will have different populations of B-cell descendants in their spleens and lymph nodes.

This is, of course, a very simplified version of what actually happens. For our purposes here, the crucial point to understand is that the actual substances of the bacterial cell or virus—or anything else that is not seen as being a normal component of our bodies—can initiate the differentiation of a set of B cells into plasma cells. This not only shows that an environmental agent can induce differentiation directly, it also shows how sensitive our predator-induced polyphenism is.

Transgenerational Epigenetic Transmission

That acquired characteristics cannot be inherited is one of the key concepts of genetics. And it is almost always the case. Children of weight lifters don't inherit their parents' physiques, and accident victims who have lost limbs can rest assured that their children will be born with normal arms and legs. August Weismann cut off the tails of mice for several generations, "de-tailing" over 900 mice in all, but the following generations of mice were always born with normal tails. Darwin's colleague Thomas Huxley (see Richards 2004) remarked that these experiments demonstrated the truth of Hamlet's remark, "There's a divinity that shapes our ends / rough-hew them how we will." However, there have always been exceptions to this rule (see Appendix D), and some noteworthy exceptions involve the transgenerational continuity of polyphenisms.

In order to document the inheritance of environmentally induced epigenetic changes in mammals, one must look at least at the F_3 (grandpup) generation. If an original (F_0) female animal is exposed to an environmental agent, so are the fetuses (F_1) in the uterus. But if the agent persists in the pregnant animal, the germ cells (that produce F_2) in the fetuses are also exposed to the agent. So before it can be claimed that the environmental agent caused hereditary effects, the F_3 generations must be the first examined. Epigenetic transgenerational inheritance is defined (Skinner et al. 2010) as "germline (sperm or egg) transmission of epigenetic information between generations in the absence of any direct exposures or genetic manipulations."

Transgenerational polyphenism in locusts

The brown, gregarious, and long-winged migratory morph of the locust *Schistocerca gregaria* is retained for several generations after the crowding stimulus initiates the transformation from the solitary morph (see Chapter 1). This transgenerational effect is now known to be mediated during oviposition (egg laying) by a chemical agent introduced into the foam surrounding the eggs. Part of the gregarious female's phenotype involves the production of a particular chemical that she introduces into the foam surrounding her eggs. This chemical agent, which appears to be a modified form of the neurotransmitter L-dopa, is thought to be synthesized in the accessory glands of the female reproductive tract and to act during the time of egg laying (Hägele et al. 2000; Miller et al. 2008). If foam is transferred from egg masses laid by gregarious females to egg masses produced by solitary females, the solitary eggs turn into gregarious locusts. If the foam is washed off, the gregarious state reverts to the solitary phenotype after a few molts (Figure 2.20; McCaffery and Simpson 1998; Simpson and Sword 2009). Thus, transgenerational inheritance does not have to rely solely on genes.

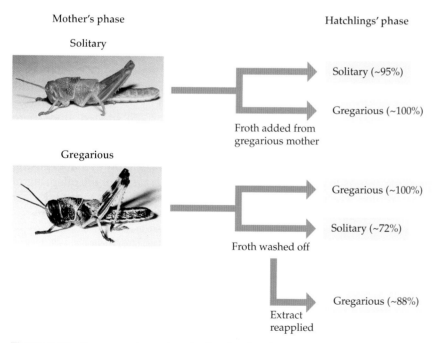

Figure 2.20   Summary of experiments showing that the source of the gregarizing agent in *Schistocerca gregaria* is an aqueous substance found in the froth surrounding the egg masses of gregarious females. (After McCaffery and Simpson 1998 and Simpson and Sword 2009.)

Transgenerational predator-induced polyphenisms

Ralph Tollrian's laboratory demonstrated that predator-induced polyphenism also can be stably transmitted from generation to generation in both plants and animals (Agrawal et al. 1999). The investigators showed that, among plants, the wild radish has a very powerful predator-induced polyphenism. Compared with control plants, radish plants damaged by the caterpillar of the cabbage white butterfly (*Pieris rapae*) produce 10 times the concentration of mustard oil glycosides (which make the plant very distasteful), as well as 30% higher densities of trichomes (hooked, hairlike structures on the leaf, which stab the caterpillar's belly). Once these inducible defenses are generated, they are also found in the plant's offspring. Seedlings from induced plants were shown to germinate with more trichomes and higher levels of glycosides, even when the new generation was not exposed to herbivory by caterpillars.

Agrawal and colleagues also demonstrated that the predator-induced polyphenisms established in *Daphnia* were also transmitted to their progeny. *Daphnia cucullata* is a parthenogenetic (all-female) species whose eggs develop within the parent's body and need not be fertilized. When the eggs of a *Daphnia* that underwent predator-induced polyphenism (exaggerated helmet and tail; see Figure 1.15) were grown in the absence of predators, they still developed the induced phenotype. The mechanisms for the transgenerational inheritance of the induced phenotype is presently unknown.

Among the aphids, another group of parthenogenetic arthropods, winged morphs are induced either when the wingless form is at high density or when the larvae are exposed to the scent of a predator. When mature induced aphids are placed on clean plants to reproduce, they tend to give rise to winged offspring for several subsequent generations, even though the inducer is not present. In the pea aphid (*Acyrthosiphon pisum*), this transgenerational response is probably induced by the alarm pheromone β-farnesene (Mondor et al. 2005; Podjesak et al. 2005). Indeed, since there is no genetic variability between clone mates, and since alarm pheromones dissipate over time, aphids are an excellent model for looking at the inheritance of induced traits across generations. It has been found that exposing juvenile aphids to β-farnesene also changes the *behavior* of the offspring of these aphids. The newly born aphids will seek safer places to feed (Keiser and Mondor 2013).

Methylation and transgenerational continuity: Toadflax

Epigenetic inheritance is rather common in plants (since the plant germline arises from somatic cells that are exposed to environmental cues; see Heard and Martienssen 2014). And in 1999, studies of DNA methylation solved a long-standing mystery that had been vexing plant taxonomists since 1742. This enigma was the origin of the symmetrical species of the toadflax plant, *Linaria vulgaris*. In the mid-eighteenth century, Linnaeus encountered some specimens of toadflax that were so different, they had to be classified as a new species (see Coen 1999). Normal *L. vulgaris* has flowers with definite upper and lower petals (Figure 2.21A). In the new form, the petals were symmetrically arranged, all on the same level—but every other part of the

(A)

(B)

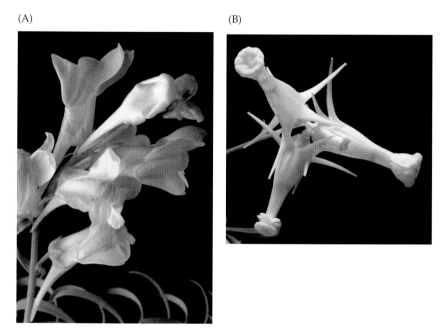

Figure 2.21 Epigenetic forms of toadflax. (A) *Linaria vulgaris* with its relatively unmethylated *cycloidea* gene. (B) *Peloria* has a relatively heavily methylated *cycloidea* gene. Despite their different phenotypes, these two plants are the same species; *Peloria* is now considered to be a variant of *Linaria*. (Photographs courtesy of R. Grant-Downton.)

plant was obviously toadflax (Figure 2.21B). Moreover, the symmetrical toadflax continued to generate such symmetrical forms, generation after generation. Since Linnaeus's binomial classification system for plants was based on petal and stamen parts, this caused a crisis: it meant that by Linnaean definitions, this was a newly formed species. For Linnaeus and his colleagues, species were invariant entities formed by God at the time of Creation. The concept of a new species was a dangerous thing to consider, and Linnaeus called the new form *Peloria* (from the Greek *pelor*, "monstrosity"), writing that "this is certainly no less remarkable than if a cow had given birth to a calf with a wolf's head."

In modern times, Coen's laboratory discovered that in fact the peloric form of toadflax is not a separate species (Coen 1999). Nor is there any mutation in the DNA between the linarian and peloric forms; rather, the DNA methylation patterns on their respective *cycloidea* genes are different. Compared with the same gene in *Linaria*, the *cycloidea* gene of *Peloria* is heavily methylated, and thereby inactivated. This hypermethylation, retaining the symmetrical phenotype, has been inherited for hundreds of generations (Cubas et al. 1999).

Methylation and transgenerational continuity: Rodents and insects

Earlier in this chapter, we described experiments that showed that a female mouse's diet during pregnancy controlled the coat color and obesity phenotypes of offspring carrying the *viable yellow* allele of the *Agouti* gene (see Figure 2.5B). The expression of this gene was regulated by the differential

DNA methylation in an inserted piece of DNA near the *Agouti* promoter site. What is just as amazing is that the epigenetic state of the *Agouti* gene (whether methylated or unmethylated) can be inherited transgenerationally: the coat colors of the grandoffspring resembled that of their mother, indicating that the *grandmother's* diet affected the coat color of her grandpups (Morgan et al. 1999). This same inheritance pattern has been noted in another mouse allele (*Axin*) that produces a polyphenism (kinky tail) due to DNA methylation differences (Rakyan et al. 2003). In this latter case, the transgenerational effect can arise through either the male or the female lineages, suggesting that the DNA methylation is on the germline cells, and that it is not erased during gamete production or during early development.* These transgenerational effects may have important considerations in medicine and in evolutionary biology.

As we will see in subsequent chapters, mammalian embryos react to the maternal diet in utero, and levels of gene expression are set accordingly. If the maternal diet is poor in proteins, some of the genes involved in metabolism become more highly methylated, so the body will store food rather than utilize it. For instance, as mentioned earlier, the offspring of pregnant rats fed a protein-restricted diet will have an unmethylated CG at a specific site in the promoter of the *PPARα* gene in the liver. This gene encodes a protein that regulations fatty acid metabolism, and the methylation pattern established in the embryo persists into adulthood (Lillycrop et al. 2008). Moreover, the methylation state of these genes appears to be transmitted transgenerationally.

Rats fed a low-protein diet for 12 generations demonstrated progressively slower fetal growth (Stewart et al. 1975); even when a normal diet was restored, the rats' growth and development did not normalize until 3 generations had passed. Moreover, as we will discuss in Chapter 7, rats exposed to low-protein diets in utero have the propensity to develop elevated blood pressure, blood vessel dysfunction, and insulin resistance. Even when a normal diet was fed to subsequent generations, Burdge and colleagues (2007b) found that these adverse effects persisted in the F_1 and F_2 generations (Figure 2.22). Burdge's study demonstrated that the methylation of gene promoters was probably the mechanism for the transgenerational stability of this polyphenism. This will become an important concept, and we will see that diet, stress, and certain environmental chemicals can alter DNA methylation patterns and thereby alter the gene expression pattern and phenotype of subsequent generations.

Simple experiments on the brine shrimp *Artemia* have shown that even in arthropods (many of which don't have DNA methylation), chromatin and DNA modifications may enable the transgenerational inheritance of environmentally induced phenotypes. Like *Daphnia*, certain *Artemia*

---

*In mammals (at least in mice, where it has best been studied), there appear to be two waves of methylation erasure; the first takes place in the germ cell precursors, and the second, in the early embryo prior to implantation. Thus, most DNA methylation has vanished by the time of implantation, except in a relatively small number of genes that are said to be *imprinted*. For more on this subject, see Appendix D.

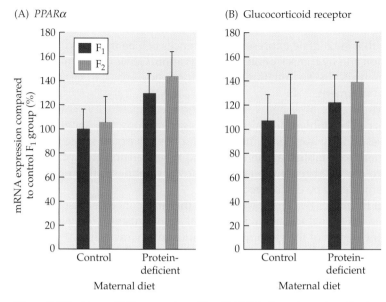

Figure 2.22   Liver mRNA expression of the (A) *PPARα* (peroxisomal proliferators-activated receptor alpha) and (B) glucocorticoid receptor (*GR*) genes. Though diet was normalized, altered promoter methylation effects persisted in the F_1 and F_2 generations. (After Burdge et al. 2007b.)

species in certain environments produce clonal, all-female populations. A mild heat shock (38°C) given to larval females causes an increase in their levels of Hsp70, a chaperonin (heat shock) protein that protects other proteins against heat-induced denaturation. This Hsp70 provides tolerance of lethal heat stress, and resistance against pathogenic *Vibrio campbellii* bacteria. Interestingly, these acquired phenotypic traits were transmitted to three successive generations of *Artemia*, even though these generations had not been exposed to the parental stressor. This transgenerational inheritance of these environmentally induced traits was associated with altered levels of global DNA methylation and acetylated histones H3 and H4 in the heat-shocked group compared with the control group, where both the parental and successive generations were reared at a standard temperature (Norouzitallab et al. 2014). These results indicated that epigenetic mechanisms, such as DNA methylation and histone modification, might be critical in the heritability of environmentally induced adaptive phenotypic traits across generations.

Looking Ahead

We now know of several mechanisms whereby the environment can alter development in perfectly understandable ways. Diet, temperature, maternal conditions, and even social interactions can change development. In many instances, this occurs through the altering of gene expression. Gene expression can be altered by direct induction (as when bacteria induce changes in the expression of intestinal genes) or by the neuroendocrine system (as in butterfly wing colors or muscle hypertrophy). There is also evidence that such gene expression differences can be mediated by DNA methylation and by transcription factor activation. Genes can be inactivated by mutation and by methylation. Methylation, however, can be regulated by the environment.

In addition, we now know some of the mechanisms by which polyphenisms can be stably transmitted across generations. DNA methylation has been especially important in explaining how an organism can use the environment to establish a phenotype and then transmit that phenotype stably across several generations, and there are doubtless other epigenetic mechanisms yet to be discovered.

One of the major research objectives in this area concerns the biochemical mechanisms by which an environmental agent causes *specific* alterations in a specific region of the genome or chromosome. How is it that a certain cytosine gets methylated, and not neighboring cytosines? What instructs the demethylases and methyltransferases? How do environmental agents activate the nucleosome remodeling enzymes? One of the mechanisms being investigated concerns **RNA-dependent DNA methylation**. In plants and yeasts, small non-coding RNAs can bind histone methyltransferases and DNA methyltransferase enzymes and bring them to particular genes, thereby silencing them (see Heard and Martienssen 2014). As discussed in Appendix B, such small non-coding RNAs (apparently synthesized by an RNA-dependent RNA polymerase) can induce transcriptional silencing for more than 20 generations (Burton et al. 2011; Ashe et al. 2012; Gu et al. 2012).

Environmental agents certainly affect the phenotype of animals, and they do so by altering the animal's development. This can be accomplished directly or through the neuroendocrine system. Thus, we know of mechanisms that empower the environment to play instructive roles in development. We will see throughout this book how epigenetic developmental changes create the conditions that enable health, disease, behaviors, and evolution.

References

Agrawal, A. A., C. Laforsch and R. Tollrian. 1999. Transgenerational induction of defences in animals and plants. *Nature* 401: 60–63.

Ashe, A. and 14 others. 2012. piRNAs can trigger a multigenerational epigenetic memory in the germline of *C. elegans*. Cell 150: 88–99.

Aviv, R. 2014. A valuable reputation. *The New Yorker* (Feb.10): www.newyorker.com/reporting/2014/02/10/140210fa_fact_aviv?currentPage=all

Barbash, J. E., G. P. Thelin, D. W. Kolpin and R. J. Gilliom. 2001. Major herbicides in ground water: Results from the National Water-Quality Assessment. *J. Environ. Qual.* 30: 831–845.

Bard, J. B. L. 1990. Traction and the formation of mesenchymal condensations *in vivo*. *BioEssays* 12: 389–395.

Barozzi, I., M. Simonatto, S. Bonifacio, L. Yang, R. Rohs, S. Ghisletti and G. Natoli. 2014. Coregulation of transcription factor binding and nucleosome occupancy through DNA features of mammalian enhancers. *Mol. Cell.* 54: 844–857.

Battaglin, W. A., E. T. Furlong, M. R. Burkhardt and C. J. Peter. 2000. Occurrence of sulfonylurea, sulfonamide, imidazolinone, and other herbicides in rivers, reservoirs, and ground water in the midwestern United States 1998. *Sci. Total Environ.* 248: 123–133.

Baulcombe, D. C. and C. Dean. 2014. Epigenetic regulation in plant responses to the environment. *Cold Spring Harb. Perspect. Biol.* 6(9): a019471.

Beck, S., A. Olek and J. Walter. 1999. From genomics to epigenomics: A loftier view of life. *Nature Biotechnol.* 17: 1144.

Beldade, P., P. M. Brakefield and A. D. Long. 2002. Contribution of Distal-less to quantitative variation in butterfly eyespots. *Nature* 415: 315–318.

Beldade, P., A. R. Mateus and R. A. Keller. 2011. Evolution and molecular mechanisms of adaptive developmental plasticity. *Mol. Ecol.* 20: 1347–1363.

Bilinska, B., M. Kotula-Balak, M. Gancarczyk, J. Sadowska, Z. Tabarowskii and A. Wojtusiak. 2003. Androgen aromatization in cryptorchid mouse testis. *Acta. Histochem.* 105: 57–65.

Bird, A. 2002. DNA methylation patterns and epigenetic memory. *Genes Dev.* 16: 6–21.

Black, M. P., M. Baillien, J. Balthazart and M. S. Grober. 2005. Socially induced and rapid increases in aggression are inversely related to brain aromatase activity in a sex-changing fish, *Lythrypnus dalli*. *Proc. Biol. Sci.* 272: 2435–2440.

Blumenstyk, G. 2003. The story of Syngenta and Tyrone Hayes at UC Berkeley: The price of research. *Chron. Higher Educ.* 50: 110. Available at www.mindfully.org/Pesticide/2003/Syngenta-Tyrone-Hayes31oct03.htm.

Bonett, R. M., E. D. Hoopfer and R. J. Denver. 2010. Molecular mechanisms of corticosteroid synergy with thyroid hormone during tadpole metamorphosis. *Gen. Comp. Endocrinol.* 168: 209–219

Boorse, G. C. and R. J. Denver. 2003. Endocrine mechanisms underlying plasticity in metamorphic timing in spadefoot toads. *Integr. Comp. Biol.* 43: 646–657.

Brakefield, P. M. and W. A. Frankino. 2009. Polyphenisms in Lepidoptera: Multidisciplinary approaches to studies of evolution. In *Phenotypic Plasticity in Insects: Mechanisms and Consequences*, pp. 281–312. Science Publishers, Plymouth, MA.

Brakefield, P. M. and N. Reitsma. 1991. Phenotypic plasticity, seasonal climate, and the population biology of *Bicyclus* butterflies (Satyridae) in Malawi. *Ecol. Entomol.* 16: 291–303.

Bull, J. J. 1980. Sex determination in reptiles. *Q. Rev. Biol.* 55: 3–21.

Bull, J. J., W. H. N. Gutzke and D. Crews. 1988. Sex reversal by estradiol in three reptilian orders. *Gen. Comp. Endocrinol.* 70: 425–428.

Burdge, G. C., M. A. Hanson, J. L. Stater-Jefferies and K. A. Lillycrop. 2007a. Epigenetic regulation of transcription: A mechanism for inducing variations in phenotype (fetal programming) by differences in nutrition during early life? *Brit. J. Nutrit.* 97: 1036–1046.

Burdge, G. C., J. L. Slater-Jefferies, C. Torrens, E. S. Phillips, M. A. Hanson and K. A. Lillycrop. 2007b. Dietary protein restriction of pregnant rats in the F_0 generation induces altered methylation of hepatic gene promoters in the adult male offspring in the F_1 and F_2 generations. *Brit. J. Nutrit.* 97: 435–439.

Burnett, F. M. 1959. *The Clonal Selection Theory of Immunity*. Vanderbilt University Press, Nashville.

Burton, N.O., K. B. Burkhart and S. Kennedy. 2011. Nuclear RNAi maintains heritable gene silencing in *Caenorhabditis elegans*. *Proc. Natl. Acad. Sci. USA* 108: 19683–19688.

Cameron, N. M., D. Shahrokh, A. Del Corpo, S. K. Dhir , M. Szyf, F. A. Champagne and M. J. Meaney. 2008. Epigenetic programming of phenotypic variations in reproductive strategies in the rat through maternal care. *J. Neuroendocrinol.* 20: 795–801.

Champagne, F. A., I. C. Weaver, J. Diorio, S. Dymov, M. Szyf and M. J. Meaney. 2006. Maternal care associated with methylation of the estrogen receptor-alpha1b promoter and estrogen receptor-alpha expression in the medial preoptic area of female offspring. *Endocrinology* 147: 2909–2915.

Chan, M. C. and Z. Arany. 2014. The many roles of PGC-1α in muscle—Recent developments. *Metabolism* 63: 441–451.

Coen, E. 1999. *The Art of Genes*. Oxford University Press, Oxford.

Cosgrove, M. S., J. D. Boeke and C. Wolberger. 2004. Regulated nucleosome mobility and the histone code. *Nature Struct. Mol. Biol.* 11: 1037–1043.

Crain, D. A. and L. J. Guillette, Jr. 1998. Reptiles as models of contaminant-induced endocrine disruption. *Anim. Reprod. Sci.* 53: 77–86.

Crevillén, P. and 7 others. 2014. Epigenetic reprogramming that prevents transgenerational inheritance of the vernalized state. *Nature* 515: 587–590.

Cubas, P., C. Vincent and E. Coen. 1999. An epigenetic mutation responsible for natural variation in floral symmetry. *Nature* 401: 157–161.

Dennis, S. R., G. A. LeBlanc and A. P. Beckerman. 2014. Endocrine regulation of predator-induced phenotypic plasticity. *Oecologia*. 176: 625–35.

Denver, R. J. 1997. Proximate mechanism of phenotypic plasticity in amphibian metamorphosis. *Am. Zool.* 37: 172–184.

Denver, R. J. 1999. Evolution of the corticotropin-releasing hormone signaling system and its role in stress-induced developmental plasticity. *Ann. NY Acad. Sci.* 897: 46–53.

Denver, R. J. and J. Middlemis-Maher. 2010. Lessons from evolution: Developmental plasticity in vertebrates with complex life cycles. *J. Dev. Orig. Health Dis.* 1: 282–291.

Denver, R. J., N. Mirhadi and M. Phillips. 1998. Adaptive plasticity in amphibian metamorphosis: Response of *Scaphiopus hammondii* tadpoles to habitat desiccation. *Ecology* 79: 1859–1872.

Desvages, G., M. Girondot and C. Pieau. 1993. Sensitive stages for the effects of temperature on gonadal aromatase activity in embryos of the marine turtle *Dermochelys coriacea*. *Gen. Comp. Endocrinol.* 92: 54–61.

Dolinoy, D. C., J. R. Weidman, R. A. Waterland and R. L. Jirtle. 2006. Maternal genistein alters coat color and protects A^{vy} mouse offspring from obesity by modifying the fetal epigenome. *Environ. Health Perspect.* 114: 567–572.

Dorizzi, M., G. Richard-Mercier, G. Desvages and C. Pieau. 1994. Masculinization of gonads by aromatase inhibitors in a turtle with temperature-dependent sex determination. *Differentiation* 58: 1–8.

EPA (Environmental Protection Agency). 2013. Atrazine updates. www.epa.gov/pesticides/reregistration/atrazine/atrazine_update.htm#atrazine

Fan, W. and 10 others. 2007. Atrazine-induced aromatase expression is SF1-dependent: Implications for endocrine disruption in wildlife and reproductive cancers in humans. *Environ. Health Persp.* 115: 720–727.

Ferguson-Smith, A. C., H. Sasaki, B. M. Cattanach and M. A. Surani. 1993. Paternal-origin-specific epigenetic modification of the mouse *H19* gene. *Nature* 362: 751–755.

Foret, S., R. Kucharski, M. Pellegrini, S. Feng, S. E. Jacobsen, G. E. Robinson and R. Maleszka. 2012. DNA methylation dynamics, metabolic fluxes, gene splicing, and alternative phenotypes in honey bees. *Proc. Natl. Acad. Sci. USA* 109: 4968–4973.

Frost, S. D. W. 1999. The immune system as an inducible defense. In R. Tollrian and and C. D. Harvell (eds.), *The Ecology and Evolution of Inducible Defenses*. Princeton University Press, Princeton, NJ, pp. 104–126.

Fuks, F. 2005. DNA methylation and histone modifications: Teaming up to silence genes. *Curr. Opin. Genet. Devel.* 15: 490–495.

Garma, T., C. Kobayashi, F. Haddad, G. R. Adams, P. W. Bodell and K. M. Baldwin. 2007. Similar acute molecular responses to equivalent volumes of isometric lengthening, or shortening mode resistance exercise. *J. Appl. Physiol.* 102: 135–143.

Gilbert, S. F. 2005. Mechanisms for the environmental regulation of gene expression: Ecological aspects of animal development. *J. Biosci.* 30: 101–110.

Glennemeier, K. A. and R. J. Denver. 2002. Small changes in whole-body corticosterone content affect larval *Rana pipiens* fitness components. *Gen. Comp. Endocrinol.* 127: 16–25.

Godwin, J. 2010. Neuroendocrinology of sexual plasticity in teleost fishes. *Front. Neuroendocrinol.* 31: 203–216.

Goldspink, G. 2004. Mechanical signals, *IGF-1* gene splicing, and muscle adaptation. *J. Physiol.* 20: 232–238.

Gu, S.G., J. Pak, S. Guang, J. M. Maniar, S. Kennedy and A. Fire. 2012. Amplification of siRNA in *Caenorhabditis elegans* generates a transgenerational sequence-targeted histone H3 lysine 9 methylation footprint. *Nat. Genet.* 44: 157–164.

Gutzke, W. H. N. and D. B. Chymiy. 1988. Sensitive periods during embryology for hormonally induced sex determination in turtles. *Gen. Comp. Endocrinol.* 71: 265–267.

Hägele, B. F., V. Oag, A. Bouaïchi, A. R. McCaffery and S. J. Simpson. 2000. The role of female accessory glands in maternal inheritance of phase in the desert locust *Schistocerca gregaria*. *J. Insect Physiol.* 46: 275–280.

Hari Dass, S. A. and A. Vyas. 2014. *Toxoplasma gondii* infection reduces predator aversion in rats through epigenetic modulation in the host medial amygdala. *Mole. Ecol.* 23: 6114–6122.

Hayes, T. B., A. Collins, M. Lee, M. Mendoza, N. Noriega, A. Stuart and A. Vonk. 2002a. Hermaphroditic, demasculinized frogs after exposure to the herbicide atrazine at low ecologically relevant doses. *Proc. Natl. Acad. Sci. USA* 99: 5476–5480.

Hayes, T., K. Haston, M. Tsui, A. Hoang, C. Haeffele and A. Vonk. 2002b. Herbicides: Feminization of male frogs in the wild. *Nature* 419: 895–896.

Hayes, T., K. Haston, M. Tsui, A. Hoang, C. Haeffele and A. Vonk. 2003. Atrazine-induced hermaphroditism at 0.1 ppb in American leopard frogs (*Rana pipiens*): Laboratory and field evidence. *Envir. Health Perspec.* 111: 568–575.

Heard, E. and R. A. Martiensson 2014. Transgenerational epigenetic inheritance: Myths and mechanisms. *Cell* 157: 95–109.

Hellstrom, I. C., S. K. Dhir, J. C. Diorio and M. J. Meaney. 2012. Maternal licking regulates hippocampal glucocorticoid receptor transcription through a thyroid hormone-serotonin-NGFI-A signalling cascade. *Philos. Trans. R Soc. Lond. B Biol. Sci.* 367: 2495–24510.

Ikegame, M., O. Ishibashi, T. Yoshizawa, J. Shimomura, T. Komori, H. Ozawa and H. Kawashima. 2001. Tensile stress induces bone morphogenetic protein 4 in preosteoblastic and fibroblastic cells, which later differentiate into osteoblasts leading to osteogenesis in the mouse calvariae in organ culture. *J. Bone Miner. Res.* 16: 24–32.

Jeyasuria, P., W. M. Roosenburg and A. R. Place. 1994. Role of P-450 aromatase in sex determination of the diamondback terrapin, *Malaclemys terrapin*. *J. Exp. Zool.* 270: 95–111.

Jirtle, R. L. 2013. Epigenetics: How genes and environment interact. In R. L. Jirtle and F. L. Tyson (eds), *Environmental Epigenomics in Health and Disease: Epigenetics and Disease Origins*, Springer, Heidelberg, Germany, pp. 3–30.

Kadi, F., P. Schjerling, L. Andersen, N. Charifi, J. Madsen, L. Christensen and J. Andersen. 2004. The effects of heavy resistance training and detraining on satellite cells in human skeletal muscles. *J. Physiol.* 558: 1005–1012.

Kamakura, M. 2011. Royalactin induces queen differentiation in honeybees. *Nature* 473: 478–483.

Kandalla, P. K, G. Goldspink, G. Butler-Browne and V. Mouly. 2011. Mechano Growth Factor E peptide (MGF-E), derived from an isoform of IGF-1, activates human muscle progenitor cells and induces an increase in their fusion potential at different ages. *Mech. Ageing Dev.* 132: 154–162.

Keiser, C. N. and E. B. Mondor. 2013. Transgenerational behavioral plasticity in a parthenogenetic insect in response to increased predation risk. *J. Insect Behav.* 26: 603–613.

Kniewald, J., M. Jakominic, A. Tomljenovic, B. Simic, P. Romac, D. Vranesic and Z. Kniewald. 2000. Disorders of male reproductive tract under the influence of atrazine. *J. Appl. Toxicol.* 20: 61–68.

Kobayashi, Y., T. Sunobe, T. Kobayashi, M. Nakamura, N. Suzuki and Y. Nagahama. 2005. Molecular cloning and expression of Ad4BP/SF-1 in the serial sex changing gobiid fish, Trimma okinawae. *Biochem. Biophys. Res. Commun.* 332: 1073–1080.

Kobayashi, Y. and 8 others. 2009. Sex change in the Gobiid fish is mediated through rapid switching of gonadotropin receptors from ovarian to testicular portion or vice versa. *Endocrinology* 150: 1503–1511.

Koch, P. B. and D. Buchmann. 1987. Hormonal control of seasonal morphs by the timing of ecdysteroid release in *Araschnia levana* (Nymphalidae: Lepidoptera). *J. Insect Physiol.* 36: 159–164.

Kucharski, R., J. Maleszka, S. Foret and R. Maleszka. 2008. Nutritional control of reproductive status in honeybees via DNA methylation. *Science* 319: 1827–1830.

Labonte, B., V. Yerko, J. Gross, N. Mechawar, M. J. Meaney, M. Szyf and G. Turecki. 2012. Differential glucocorticoid receptor exon 1(B), 1(C), and 1(H) expression and methylation in suicide completers with a history of childhood abuse. *Biol. Psychiatry* 72(1): 41–48.

Lai, K. V. and 9 others. 2004. Conditional activation of *Akt* in adult skeletal muscle induces rapid hypertrophy. *Mol. Cell. Biol.* 24: 9295–9304.

Laplante, P., J. Diorio and M. J. Meaney. 2002. Serotonin regulates hippocampal glucocorticoid receptor expression via a 5-HT7 receptor. *Dev. Brain Res.* 139: 199–203.

Lee, Y. H, W. S. Yueh, J. L. Du, L. T. Sun and C. F. Chang. 2002. Aromatase inhibitors block natural sex change and induce male function in the protandrous black porgy, *Acanthopagrus schlegeli* Bleeker: Possible mechanism of natural sex change. *Biol. Reprod.* 66: 1749–1754.

Li-Byarlay, H. and 12 others. 2013. RNA interference knockdown of DNA methyl-transferase 3 affects gene alternative splicing in the honey bee. *Proc. Natl. Acad. Sci. USA* 110: 12750–12755.

Lillycrop, K. A., J. L. Slater-Jefferies, M. A. Hanson, K. M. Godfrey, A. A. Jackson and G. C. Burdge. 2007. Induction of altered epigenetic regulation of the hepatic glucocorticoid receptor in the offspring of rats fed a protein-restricted diet during pregnancy suggests that reduced DNA methyltransferase-1 expression is involved in impaired DNA methylation and changes in histone modifications. *Br. J. Nutr.* 97: 1064–1073.

Lillycrop, K. A., E. S. Phillips, C. Torrens, M. A. Hanson, A. A. Jackson and C. C. Burdge. 2008. Feeding pregnant rats a protein-restricted diet persistently alters the methylation of specific cytosines in the hepatic PPARα. *Brit. J. Nutrit.* 100: 278–282.

Lin, J. and 12 others. 2002. Transcriptional co-activator PGC-1α drives the formation of slow-twitch muscle fibres. *Nature* 418: 797–801.

Lindholm, M. E., F. Marabita, D. Gomez-Cabrero, H. Rundqvist, T. J. Ekström, J. Tegnér and C. J. Sundberg. 2014. An integrative analysis reveals coordinated reprogramming of the epigenome and the transcriptome in human skeletal muscle after training. *Epigenetics.* 9: 1557–1569.

Mateus, A. R., M. Marques-Pita, V. Oostra, E. Lafuente, P. M. Brakefield, B. J. Zwaan and P. Beldade. 2014. Adaptive developmental plasticity: Compartmentalized responses to environmental cues and to corresponding internal signals provide phenotypic flexibility. *BMC Biol.* 12: 97.

Matsumoto, Y., A. Buemio, R. Chu, M. Vafaee and D. Crews. 2013. Epigenetic control of gonadal aromatase (cyp19a1) in temperature-dependent sex determination of red-eared slider turtles. *PLoS ONE* 8(6): e63599.

McCaffery, A. and S. J. Simpson. 1998. A gregarizing factor present in the egg pod foam of the desert locust *Schistocerca gregaria*. *J. Exp. Biol.* 201: 347–363.

McCarrey, J. R. 2012. The epigenome as a target for heritable environmental disruptions of cellular function. *Mol. Cell. Endocrinol.* 354: 9–15.

Meaney, M. J. and M. Szyf. 2005. Maternal care as a model for experience-dependent chromatin plasticity. *Trends Neurosci.* 28: 456–463.

Middlemis Maher, J., E. E. Werner and R. J. Denver. 2013. Stress hormones mediate predator-induced phenotypic plasticity in amphibian tadpoles. *Proc. Biol. Sci.* 280: 20123075.

Miller, G. A., M. S. Islam, T. D. W. Claridge, T. Dodgson and S. J. Simpson. 2008. Swarm formation in the desert locust *Schistocerca gregaria*: Isolation and NMR analaysis of the primary maternal gregarizing agent. *J. Exp. Biol.* 211: 370–376.

Mondor, E. B., J. A. Rosenheim and J. F. Addicott. 2005. Predator-induced transgenerational phenotypic plasticity in the cotton aphid. *Oecologia* 142: 104–108.

Monteiro, A. and 8 others. 2013. Distal-less regulates eyespot patterns and melanization in *Bicyclus* butterflies. *J. Exp. Zool. B Mol. Dev. Evol.* 320: 321–331.

Morgan, H. D., H. G. E. Sutherland, D. I. K. Martin and E. Whitelaw. 1999. Epigenetic inheritance at the *agouti* locus in the mouse. *Nature Genet.* 23: 314–318.

Murdoch, C. and T. Wibbels. 2006. *Dmrt1* expression in response to estrogen treatment in a reptile with temperature-dependent sex determination. *J. Exp. Zool. B Mol. Dev. Evol.* 306: 134–139.

Mustelin, L. and 8 others. 2008. Acquired obesity and poor physical fitness impair expression of genes of mitochondrial oxidative phosphorylation in monozygotic twins discordant for obesity. *Am. J. Physiol. Endocrinol. Metab.* 295: E148–E154.

Negri, I., A. Franchini, E. Gonella, D. Daffonchio, P. J. Mazzoglio, M. Mandrioli and A. Alma. 2009. Unravelling the *Wolbachia* evolutionary role: the reprogramming of the host genomic imprinting. *Proc. Biol. Sci.* 276: 2485–2491.

Nijhout, H. F. 1991. *The Development and Evolution of Butterfly Wing Patterns.* Smithsonian Institution Press, Washington, DC.

Nijhout, H. F. 1999. Control mechanisms of polyphonic development in insects. *BioScience* 49: 181–192.

Nijhout, H. F. 2003. Development and evolution of adaptive polyphenisms. *Evo. Dev.* 5: 9–18.

Norouzitallab, P. and 9 others. 2014. Environmental heat stress induces epigenetic transgenerational inheritance of robustness in parthenogenetic Artemia model. *FASEB J.* 28: 3552–3563.

North, J. A. and 9 others. 2012. Regulation of the nucleosome unwrapping rate controls DNA accessibility. *Nucleic Acids Res.* 40: 10215–10227.

Oliver, J. C., D. Ramos, K. L. Prudic and A. Monteiro. 2013. Temporal gene expression variation associated with eyespot size plasticity in *Bicyclus anynana*. *PLoS ONE* 8: e65830.

Olofsson, M., A. Vallin, S. Jakobsson and C. Wiklund. 2010. Marginal eyespots on butterfly wings deflect bird attacks under low light intensities with UV wavelengths. *PLoS ONE* 5(5): e10798.

Oostra, V. and 7 others. 2014. Ecdysteroid hormones link the juvenile environment to alternative adult life histories in a seasonal insect. *Amer. Nat.* 184(3): E79–E92.

Patel, A., M. K. Fondrk, O. Kaftanoglu, C. Emore, G. Hunt, K. Frederick and G. V. Amdam. 2007. The making of a queen: TOR pathway is a key player in diphenic caste development. *PLoS ONE* 2(6): e509.

Petrella, J., J. Kim, J. Cross, D. Kosek and M. Bamman. 2006. Efficacy of myonuclear addition may explain differential myofiber growth among resistance trained young and older men and women. *J. Physiol. Endocrinol. Metab.* 291: 937–946.

Pieau, C., N. Girondot, G. Richard-Mercier, M. Desvages, P. Dorizzi and P. Zaborski. 1994. Temperature sensitivity of sexual differentiation of gonads in the European pond turtle. *J. Exp. Zool.* 270: 86–93.

Pietiläinen, K. H. and 11 others. 2008. Global transcript profiles of fat in monozygotic twins discordant for BMI: Pathways behind acquired obesity. *PLoS Med.* 5(3): e51.

Podjasek, J. O., L. M. Bosnjak, D. J. Brooker and E. B. Mondor. 2005. Alarm pheromone induces a transgenerational wing polyphenism in the pea aphid, *Acyrthosiphon pisum*. *Canad. J. Zool.* 83: 1138–1141.

Prudic, K. L., A. M. Stoehr, B. R. Wasik and A. Monteiro. 2015. Eyespots deflect predator attack increasing fitness and promoting the evolution of phenotypic plasticity. *Proc. R. Soc. Lond. Biol. Sci.* 282: 20141531.

Rakyan, V. K., S. Chong, M. E. Champ, P. C. Cuthbert, H. D. Morgan, K. V. K. Lu and E. Whitelaw. 2003. Transgenerational inheritance of epigenetic states of the murine *Axin^{Fu}* allele occurs after maternal and paternal transmission. *Proc. Natl. Acad. Sci. USA* 100: 2538–2543.

Ramsey, M., C. Shoemaker and D. Crews. 2007. Gonadal expression of Sf1 and aromatase during sex determination in the red-eared slider turtle (*Trachemys scripta*), a reptile with temperature-dependent sex determination. *Differentiation* 75: 978–991.

Rhen, T. and J. W. Lang. 1994. Temperature-dependent sex determination in the snapping turtle: Manipulation of the embryonic sex steroid environment. *Gen. Comp. Endocrinol.* 96: 243–254.

Richards, R. J. 2004. Justification through biological faith: A rejoinder. *Biol. Philos.* 1: 337–354.

Rottensteiner, M. and 11 others. 2015. Physical activity, glucose homeostasis, and brain morphology in twins. *Med. Sci. Sports Exerc.* 47: 509–518.

Ruas, J. L. and 16 others. 2012. A PGC-1α isoform induced by resistance training regulates skeletal muscle hypertrophy. *Cell.* 151: 1319–1331.

Sakamoto, K., W. G. Aschenbach, M. F. Hirschman and L. J. Goodyear. 2003. Akt signaling in skeletal muscle: Regulation by exercise and passive stretch. *Am. J. Endocrinol. Metab.* 285: E1081–E1088.

Sakata, N., Y. Tamori and M. Wakahara. 2005. P450 aromatase expression in the temperature-sensitive sexual differentiation of salamander (*Hynobius retardatus*) gonads. *Int. J. Dev. Biol.* 49: 417–425.

Sato, M., T. Ochi, T. Nakase, S. Hirota, Y. Kitamura, S. Nomura and N. Yasui. 1999. Mechanical tension-stress induces expression of bone morphogenetic proteins BMP-2 and BMP-4, but not BMP-6, BMP-7, and GDF-5 mRNA, during distraction osteogenesis. *J. Bone Miner. Res.* 14: 1084–1095.

Schlissel, M. S. and D. D. Brown. 1984. The transcriptional regulation of *Xenopus* 5S RNA genes in chromatin: The roles of active stable transcription complex and histone H1. *Cell* 37: 903–913.

Schones, D. E. and K. Zhou. 2008. Genome-wide approaches to studying chromatin modifications. *Nature Rev. Genet.* 9: 179–191.

Shefelbine, S. J. and D. R. Carter. 2004. Mechanobiological predictions of femoral anteversion in cerebral palsy. *Ann. Biomed. Eng.* 32: 297–305.

Shoemaker, C., M. Ramsey, J. Queen and D. Crew. 2007. Expression of *Sox9*, *Mis*, and *Dmrt1* in the gonad of a species with temperature-dependent sex determination. *Dev. Dyn.* 236: 1055–1063.

Simpson, S. J. and G. A. Sword. 2009. Phase polyphenism in locusts: Mechanisms, population consequences, adaptive significance and evolution. In T. Ananthakrishnan and D. Whitman (eds.), *Phenotypic Plasticity of Insects: Mechanisms and Consequences*. Science Publishers, Plymouth, UK, pp. 147–190.

Skinner, M. K., M. Manikkam and C. Guerrero-Bosagna. 2010. Epigenetic transgenerational actions of environmental factors in disease etiology. *Trends Endocrinol. Metab.* 21: 214–222.

Stewart, R. J. C., R. F. Preece and H. G. Sheppard. 1975. Twelve generations of marginal protein deficiency. *Brit. J. Nutrit.* 33: 233–253.

Strahl, B. D. and C. D. Allis. 2000. The language of covalent histone modifications. *Nature* 403: 41–45.

Sung, S. and R. M. Amasino. 2005. Remembering winter: Toward a molecular understanding of vernalization. *Annu. Rev. Plant Biol.* 56: 491–508.

Swan, S. H. and 9 others, and the Study for the Future of Families Research Group. 2003. Semen quality in relation to biomarkers of pesticide exposure. *Envir. Health Persp.* 111: 1478–1484.

Takahashi, I. and 7 others. 1998. Compressive force promotes *Sox9*, type II collagen and aggrecan and inhibits *IL-1b* expression resulting in chondrogenesis in mouse embryonic limb bud mesenchymal cells. *J. Cell Sci.* 111: 2067–2076.

Thoma, F., T. Koller and A. Klug. 1979. Involvement of histone H1 in the organization of the nucleosome and of the salt-dependent superstructures of chromatin. *J. Cell Biol.* 83: 403–427.

Vandenberg, L. N. and 11 others. 2012. Hormones and endocrine-disrupting chemicals: Low-dose effects and nonmonotonic dose responses. *Endocr. Rev.* 33: 378–455.

Waterland, R. A. and R. L. Jirtle. 2003. Transposable elements: Targets for early nutritional effects of epigenetic gene regulation. *Mol. Cell. Biol.* 23: 5293–5300.

Weaver, I. C. and 8 others. 2004. Epigenetic programming by maternal behavior. *Nature Neurosci.* 7: 847–854.

Weaver, I. C. and 7 others. 2007. The transcription factor nerve growth factor-inducible protein A mediates epigenetic programming: Altering epigenetic marks by immediate-early genes. *J. Neurosci.* 27: 1756–1768.

Weintraub, H. 1985. Assembly and propagation of repressed and derepressed chromosomal states. *Cell* 42: 705–711.

Weismann, A. 1875. Über den Saison-Dimorphismus der Schmetterlinge. In *Studien zur Descendenz-Theorie*. Engelmann, Leipzig.

Williams, S. C., C. R. Altman, R. L. Chow, A. Hemmati-Brivanlou and R. A. Lang. 1998. A highly conserved lens transcriptional element from the *Pax-6* gene. *Mech. Dev.* 73: 225–229.

Wirtz, P. and J. Beetsma. 1972. Induction of caste determination in the honey bee (*Apis mellifera*) by juvenile hormone. *Entomol. Exp. Appl.* 15: 517–520.

Wu, K. C., J. Streicher, M. L. Lee, B. I. Hall and G. B. Müller. 2001. Role of motility in embryonic development: I. Embryo movements and amnion contractions in the chick and the influence of illumination. *J. Exp. Zool.* 291: 186–194.

Yang, S. Y., M. Alnaqeeb, H. Simpson and G. Goldspink. 1996. Cloning and characterization of an IGF-1 isoform expressed in skeletal muscle subjected to stretch. *J. Musc. Res. Cell Motil.* 17: 487–495.

Young, R. L. and A. V. Badyaev. 2007. Evolution of ontogeny: Linking epigenetic remodeling and genetic adaptation in skeletal structures. *Int. Comp. Biol.* 47: 234–244.

Zhang, T. Y., B. Labonté, X. L. Wen, G. Turecki and M. J. Meaney. 2013. Epigenetic mechanisms for the early environmental regulation of hippocampal glucocorticoid receptor gene expression in rodents and humans. *Neuropsychopharmacology* 38: 111–123.

Developmental Symbiosis
Co-Development as a Strategy for Life

There is a tendency for living things to join up, establish linkages, live inside each other, return to earlier arrangements, get along, whenever possible. This is the way of the world.

Lewis Thomas, 1974

Honor thy symbionts.

J. Xu and J. I. Gordon, 2003

As we saw in the previous chapters, an organism may respond to the presence of other organisms by altering its development in specific ways to produce a phenotype best suited to survive under the particular circumstances. But what if the "other organisms" are present not just sometimes (as in the case of predators), but always? Such is the situation when a host responds to cues from its **symbionts**, organisms of different species that live in close association with the host. Indeed, we often find that the host has "outsourced" developmental cues to these expected symbiotic partners and that normal development comprises the interactions of at least two organisms. This situation—where the complete development of an organism is dependent on symbionts with foreign genomes—developmental symbiosis—is probably not the exception; in fact, it appears to be the rule (McFall-Ngai 2002; McFall-Ngai et al. 2013; Douglas 2014).

Some of the most important examples of symbiosis in general and developmental symbiosis in particular come from the relationships between animals and the microbes that live inside them. One constant environmental factor is the presence of microbes. We may think in terms of "The Age of Reptiles" or "The Age of Mammals," but it has always been the microbes' world. As Stephen J. Gould (1996) wrote, "We live now in the

'Age of Bacteria.' Our planet has always been in the 'Age of Bacteria.'" We large animals supply food and niches for the enormous microbial population of the planet.

Symbiosis: An Overview

The word **symbiosis** (Greek, *sym*, "together" + *bios*, "life") refers to any close association between organisms of different species (see Margulis 1981; Sapp 1994). In many symbiotic relationships, one of the organisms is much larger than the other, and the smaller organism may live on the surface or inside the body of the larger. In such relationships, the larger organism is referred to as the **host** and the smaller as the **symbiont**. There are two important categories of symbiosis:

1. **Parasitism** occurs when one partner benefits at the expense of the other, as in a tapeworm that lives within and steal nutrients from a human digestive tract. Many important developmental symbioses involve **parasitoid** species that attach to or live within the host, eventually killing and consuming it. While **parasites** try to keep the host animal alive and plenishing them with benefits, parasitoids kill their prey. This strategy is typified by parasitoid wasp species that lay their eggs inside the larvae of other insects. The wasp larvae eat the insect larvae from within and undergo metamorphosis inside the hollow cuticle of the dead host.

2. **Mutualism** is a relationship that benefits both partners. A striking example of this type of symbiosis is that of the Egyptian plover (*Pluvianus aegyptius*) and the Nile crocodile (*Crocodylus niloticus*). Although it regards most birds as lunch, the crocodile allows the plover to roam its body, feeding on the harmful parasites there. Thus the bird obtains food while the crocodile gets rid of its parasites.

A third type of symbiosis, known as **commensalism**, is defined as a relationship that is beneficial to one partner and neither beneficial nor harmful to the other partner. Although many symbiotic relationships may appear on the surface to be commensal, few if any symbiotic relationships in nature are truly neutral with respect to either party. Finally, the term **endosymbiosis** ("living inside") describes the situation where one cell lives inside another cell, a circumstance thought to account for the evolution of the mitochondria and chloroplasts, which are organelles of eukaryotic cells (see Margulis 1971), and that describes the *Wolbachia* developmental symbioses discussed at length later in this chapter.

The term used to encompass the eukaryotic organism and its persistent populations of symbionts is **holobiont*** (Rosenberg et al. 2007). The cow

---

Holobiont is one of those terms that was coined independently by researchers discovering the same phenomenon in different organisms. The Rosenbergs were concerned with coral bleaching and recognized the holobiont (coral plus algae) as being the primary individual; in a series of seminars concerning vertebrate hormones, Richard Jefferson (1994) proposed the same term (see Arnold 2013).

holobiont can digest grass; the cow, alone without its symbionts, cannot. Similarly, the termite genome provides no enzymes that will allow the termite to digest the cellulose of wood. Those enzymes come from symbionts. When we talk about tropical reef-building corals, we are talking about the coral holobiont. Without its symbiotic algae, the coral dies for lack of oxygen and nutrients. This death is called coral bleaching, and it is a product of the rising ocean temperatures, which can cause the expulsion of the algae. The holobiont is the consortium, the team of organisms that makes a new individual (Figure 3.1). As we will see, all animals are, in fact, holobionts, and symbionts play major roles in directing the development, physiology, and behavior of the holobiont.

(A)

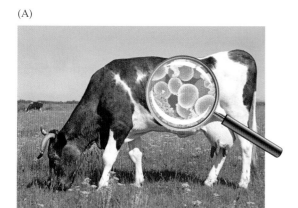

Figure 3.1 Holobionts. (A) The cow holobiont. The cow is the eukaryotic organism, but thousands of microbial species (including bacteria, fungi, and archaea) in the specialized stomach of the cow enable it to digest the cellulose of grass. (B) Coral holobiont. The coloration of living coral is due to the zooxanthellae algae that provide it with nutrition and oxygen. The white coral lacks the symbionts and is dead. (C) The termite *Mastotermes darwinensis*, whose ability to digest wood comes from the symbiotic organism *Myxotricha paradoxica*. (D) The *M. paradoxica* symbiont has a eukaryotic protist component plus at least four bacterial components. The "cilia" are actually spirochetes.

(B)   (C)

(D)

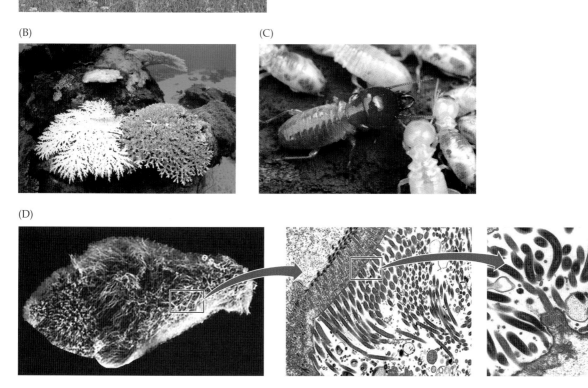

Although this chapter is concerned primarily with the developmental symbioses of animals, it is important to note that animal life as we know it rests on a variety of symbiotic relationships among plants, bacteria, and fungi.

The "Grand" Symbioses

Symbiosis, and especially mutualism, is the basis for life on Earth. Certain symbioses are essential for maintaining life as we know it on Earth. These "grand" symbioses include those relationships between bacteria and plants that make proteins possible, between bacteria and clams and between algae and corals that stabilize large marine ecosystems, and between fungi and plants that produce both the food and the oxygen that allow animal species to exist.

Nitrogen-fixing nodules

The symbiosis between *Rhizobium* bacteria and the roots of legume plants is responsible for transforming elemental chemicals from the soil or air into biologically useful molecules. Foremost among these transforming reactions is **nitrogen fixation**, in which nitrogen in the atmosphere, as the gas N_2, is transformed into the biologically useful form of nitrogen, ammonia (NH_4^+).

Most of the nitrogen fixation on Earth (about 170 million metric tons of nitrogen per year) is accomplished by living organisms, and much of this amount results from the symbioses between bacteria of the genus *Rhizobium* (and some related species) and the roots of certain green plants known as the legumes (including peas, soybeans, clover, alfalfa, and numerous tropical shrubs). Smaller amounts of nitrogen (about 20 million metric tons per year) are converted into ammonia by lightning, volcanic eruptions, and forest fires, and about 80 million metric tons are fixed industrially each year by the Haber process* (wherein nitrogen gas and hydrogen gas are heated to 500°C at a pressure of 250 atmospheres). *Rhizobium*-legume symbioses accomplish nitrogen fixation every day and at normal atmospheric temperatures and pressure. In so doing, they make all other life on Earth possible.

Neither free-living rhizobia nor uninfected legume plants can fix nitrogen; they must be in symbiotic association in order to do so. Moreover, various species of legumes are able to fix nitrogen only when partnered with a particular species of *Rhizobium*. For example, garden pea plants and soybeans are both legumes, but their nitrogen-fixing symbionts are different *Rhizobium* species.

*The Haber process is responsible for the production of fertilizers used in gardens and farms throughout the industrialized world. It is one of the most important industrial processes known and has been given major credit for the increase in world population (Smil 2001). The process uses large amounts of energy, and this energy is currently supplied by oil and petroleum. This has important economic, political, and social consequences, in that it now takes oil to grow crops. Thus, Michael Pollan (2002) writes, "Growing the vast quantities of corn used to feed livestock in this country takes vast quantities of chemical fertilizer, which in turn takes vast quantities of oil—1.2 gallons for every bushel. So the modern feedlot is really a city floating on a sea of oil."

The association between the legume root and the rhizobia begins when the root secretes flavonoid compounds that are recognized by the rhizobia (Figure 3.2A). The rhizobia respond by secreting "nod (nodulation) factors" that bind to the root hairs and induce the hairs to grow to the bacteria and

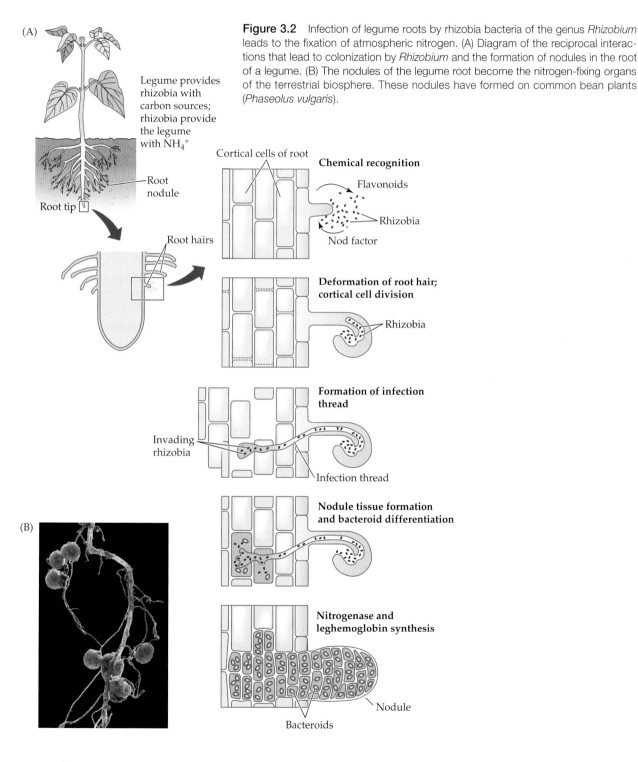

Figure 3.2 Infection of legume roots by rhizobia bacteria of the genus *Rhizobium* leads to the fixation of atmospheric nitrogen. (A) Diagram of the reciprocal interactions that lead to colonization by *Rhizobium* and the formation of nodules in the root of a legume. (B) The nodules of the legume root become the nitrogen-fixing organs of the terrestrial biosphere. These nodules have formed on common bean plants (*Phaseolus vulgaris*).

(A)

Legume provides rhizobia with carbon sources; rhizobia provide the legume with NH_4^+

Root nodule

Root tip

Root hairs

Cortical cells of root

Chemical recognition

Flavonoids

Rhizobia

Nod factor

Deformation of root hair; cortical cell division

Rhizobia

Formation of infection thread

Invading rhizobia

Infection thread

Nodule tissue formation and bacteroid differentiation

Nitrogenase and leghemoglobin synthesis

Nodule

Bacteroids

(B)

curl around them. The rhizobia then induce individual epithelial root cells to create tunnels that allow the bacteria to enter the cell. The rhizobia then migrate into the cortex (central cells) of the root. Once inside the root, the bacteria release chemicals that activate root cell division, producing nodules (Figure 3.2B). Each rhizobium then changes shape into a form called a **bacteroid** and expresses different genes. One of these bacteroidal genes encodes nitrogenase, an enzyme that binds atoms of gaseous nitrogen (N_2) and adds hydrogens to the individual atoms to make NH_4^+.

Another set of rhizobial genes encodes a heme group (similar to that found in mammalian hemoglobin) that attaches to a globin protein made by the legume. The resulting heme-containing protein, **leghemoglobin**, carries oxygen to the bacteroids to sustain their metabolism without interfering with the work of nitrogenase (which is inactivated by oxygen). In this way, atmospheric nitrogen is converted into a form that can be used as the essential material for synthesizing the bases of nucleic acids and the amino acids of proteins.

Tidal and coral ecosystems

Another major ecosystem, the worldwide coastal tidal zone, is sustained by a quadruple symbiosis involving seagrass, nitrogen-fixing bacteria, clams, and the sulfide-oxidizing bacteria living inside the clam's gills. The seagrasses, evolved from flowering land plants, provide nutrients, shelter, and protection for fish, shellfish, and even marine mammals. Interestingly, those species of seagrass preferred by the dugong (a marine mammal) have seeds that pass through the dugong's guts and remain viable. So grazing by these mammals can actually help propagate these species (Duarte 2012). Seagrasses are also critical for water clarity and sediment stabilization. As in legumes, the roots of sea grasses can support symbiotic bacteria that fix nitrogen.

The sulfide-containing sediments of decaying organic matter would be poisonous to these seagrasses living alone. However, the sulfides are detoxified by another symbiotic cycle: they are oxidized by the symbiotic bacteria of clams, while the bacteria and the clams both benefit from the oxygen produced by the seagrasses (van der Heide et al. 2012). Clams are generally underappreciated in ecosystem stabilization, because it is their symbionts that are often the major players. These clams and their symbionts seem to be critical in marine ecosystems that are under extremely stressful conditions, particularly those containing high concentrations of sulfides (such as shore ecosystems and hydrothermal vents; Roeselers and Newton 2012).

In addition to the tidal ecosystems, coral reefs are the other great incubators of life in the sea. Corals are cnidarians, and they can produce their massive and biologically diverse reefs despite growing in nutrient-poor waters because they have acquired algal symbionts. Mutualistic unicellular algae (dinoflagellates of the genus *Symbiodinium*) enter into and reside in the gastric cells of the corals, where they transport up to 95% of their photosynthetically produced carbon compounds to their hosts (Muscatine et al. 1984). The photosynthesis by the algal symbionts fuels coral growth and the calcification that creates the reef. The entry of the algae into the host changes host cell gene expression in ways that facilitate entry into the host, prevent the host immune system from destroying the algae, and enable nutrients to be exchanged between the symbiont and the host cell (Lehnert et al. 2014.) In return, the coral give their endosymbionts critical nutrients and a safe, sun-lit habitat in an otherwise nutrient-poor environment (Roth 2014).

These tidal and coral ecosystems are in peril. Globally, seagrass is disappearing at a rate of about 1% each year, and the loss of seagrass since 1980 is equivalent to losing a soccer field every 30 minutes (Waycott et al. 2009; Duarte 2012). Reef-building corals are also declining throughout the planet, largely because of human practices, including pollution, non-sustainable fishing practices, and increasing water temperatures. These conditions lead to the symbionts' being unable to function or to reside within their host corals, probably by the stress-induced activation of cell-death pathways within the symbionts (Hawkins et al. 2013; Sammarco and Strychar 2013).

Mycorrhizae

Plants also need mineral nutrients, such as phosphorus, that they must take up from the soil. About 95% of all seed plants have a symbiotic relationship with fungi, which can grow either within the roots (Figure 3.3A) or around the roots, extending the root surface area (Figure 3.3B,C). Such associations are called **mycorrhizae**, which means "fungus roots." The fungal partner receives organic compounds from the plant while it functionally extends

(A)   (B)   (C)

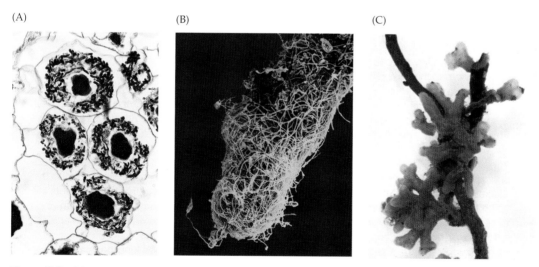

Figure 3.3 Mycorrhizae ("fungus roots") are symbiotic associations of fungi and plant roots that greatly increase the plant's ability to absorb water and nutrients. (A) Micrographic cross section of a plant root showing a mycorrhizal fungus (purple stain) living within the root cells. The fungus invades as feathery "arbuscules" that coalesce into dense vesicles. (B) Seen at twice life size in this scanning electron micrograph, "threads" of the fungus *Scleroderma geaster* surround a root of a eucalyptus tree. This mycorrhizal association greatly increases the root's absorptive surface area. (C) Mycorrhizae on these conifer roots increase absorptive area not by surrounding the root but by "blossoming" into bifurcate branches.

the plant's root system, allowing the plant to acquire more water and nutrients. Rain forest soils in particular usually have very low mineral content, and the lush growth of plants in such soil is dependent on mycorrhizae (see Martin et al. 2007; Cameron et al. 2008).

Both mycorrhizal fungi and the nitrogen-fixing bacteria in root nodules induce changes in gene expression in the plant, just as the plant changes gene expression in the fungus or bacterium. Interestingly, in *Medicago trunculata*, a legume species that can associate with both *Rhizobium* and with mycorrhizal fungi, the same gene is used for the establishment of both types of symbiosis. Mycorrhizae can be important in plant development. Orchid seeds, for example, are so small (a few milligrams) that they contain no energy reserves and cannot germinate on their own. The "dust seed" must acquire carbon from mycorrhizal fungi. All orchids need a mycorrhizal partner for their early development; an orchid may produce thousands of seeds (which will travel widely), but only those finding a fungal partner have a chance of germinating (Waterman and Bidartando 2008). This is why orchids grow best in moist tropical environments, where fungi are plentiful.

Mycorrhizae may even constitute a "wood-wide web." Work by Simard and colleagues (1997; Selosse et al. 2006) has shown that the threadlike

Endophytes

We have long known that lichens are the products of plant-fungal symbioses and have relegated them to a separate place outside the traditional eukaryotic kingdoms of plant, animal, and fungus. They are considered exceptions. However, we now grasp the possibility that most, if not all, plants are products of symbiosis. (Indeed, perhaps all animals are, too.) The idea that each "plant" is in fact a lichen-like symbiosis was driven home by the discovery of **endophytes**: fungi living inside the tissues of host plants.

While botanists had been aware of the presence of endophytes in a few species, their ubiquity was discovered accidentally. Evolutionary biologists using the polymerase chain reaction to amplify pine DNA inadvertently sequenced the DNA of fungi living within the pine needles. While at first this may have caused some consternation (since the original data made it seem that pines were really tree-shaped fungi), it soon became apparent that fungi had infiltrated the plant's tissues and become a normal part of the plant's anatomy. In some plant species, fungi have been found to infiltrate seeds, which means they have an immediate host in the next generation. In other cases, fungi generate spores that waft in the wind until they land on a host plant. When

they enter a plant, fungi usually establish themselves near the plant's vascular system and live off fluids obtained by the host.

Endophytic fungi have been found in every plant species tested, and some species have numerous fungal symbionts (Vandenkoorhuyse et al. 2002; Milius 2006). Moreover, these fungi may play an important role in plant protection (Saikkonen et al. 2006). A southwestern U.S. rye grass, *Achnatherum robustum*, is commonly called "sleepy grass" because it causes wooziness in horses and other animals that eat it. It turns out that the wooziness is caused by an LSD-like compound produced by a symbiotic fungus—a compound that also prevents insects and nematodes from eating the grass (Faeth et al. 2006). In other species, fungi provide protection against the invasion of neighboring plants.

The plants and their associated fungi generally appear to cooperate for mutual advantage. However, a fungus is a potentially dangerous thing to have around. Tanaka and colleagues (2006) found that a single gene mutation can convert a friendly, mutualistic fungus into a predatory parasite: when the fungus *Epichloë festucae* is mutated such that it cannot induce oxygen radicals in its host rye grass, its proliferation goes unchecked, and it kills the host.

mycelia of symbiotic fungi can colonize the roots of different plant species and link them together beneath the soil. Such associations may allow plants to "share the wealth," with nutrient exchange taking place among the individual plants connected by the fungi (Philip et al. 2011). Moreover, such mycorrhizal networks may help establish seedlings in understory conditions that would otherwise be hostile, and they may communicate stress signals from one plant to another (Booth 2004; Song et al. 2010, 2013). Thus, below the surface, plants may "fuse" into a cooperative network that modulates competition and provides a structure for an integrated plant community.

Life Cycle Symbiosis

The life cycles of most organisms are exquisitely coordinated with the seasons and are intimately connected to their environments. Developmental considerations are critical, since in most organisms, the adult stage of the life cycle is usually very brief and is often completely given over to reproduction. We tend to forget that the beautiful large moths that eclose from their pupae in the summer live for only 1–2 days. The males often don't even have mouthparts with which to feed; their bodies have been so formed as to be nothing more (or less) than mobile gonads. The cannibalistic practice of the female praying mantis is readily explainable from the perspective that adult male insects do not survive long, and if providing food for your developing progeny will help them survive, everyone benefits. Indeed, some male mantises die during copulation without female assistance (Johns and Maxwell, 1997; Foellmer and Fairbairn 2003).

Life cycle symbioses are also critical to marine invertebrates, whose life cycles can be regulated by mats of bacteria, called bacterial **biofilms**. Different species can take their cues from biofilms produced by different species of bacteria (Hadfield and Paul 2001; Zardus et al. 2008). These biofilms determine where and when invertebrate larvae can settle and undergo metamorphosis. These biofilms occur naturally, and they help determine the distribution of the species. However, we humans are changing this distribution by our desire to place large objects into the oceans. Such objects readily acquire biofilms and the resultant marine fauna that attach to them. As early as 1854, Charles Darwin speculated that barnacles were transported to new locales when their larvae settled on the hulls of ships. The ability of biofilms to aid invertebrate larval settlement and colony formation explains the ability of barnacles and tube worms ("biofouling invertebrates") to accumulate on ships' keels, clog sewer pipes, and deteriorate underwater structures. Such invertebrate colony formation costs billions of dollars a year in prevention, fuel consumption, and structure maintenance (Zardus et al. 2008).

Symbioses in which one organism is essential to another at a particular stage of its life cycle are thus not unusual. However, such life cycle symbioses can be exacting, because they require all the necessary players to be present and participating at precisely the right stages of development.

Life Cycle Symbioses and Human Involvement

Life cycle symbioses can be disrupted inadvertently by human involvement, even when the involvement is an attempt to preserve the species.

The large blue butterfly

The unsuccessful attempt to save the large blue butterfly *Maculinea arion* from extinction in England is an example of the need to understand a species' life cycle in all its phases (Figure A). There were hundreds of thousands of these beautiful butterflies in England during the 1950s, but censuses carried out in the early 1970s found only a few hundred individuals. In an attempt to save the species, conservation biologists tried to protect thyme plants, which were the butterflies' preferred egg-laying sites, by eliminating the grazing of cows in these areas. But the butterfly population continued to decline.

It turned out that, although the female *Macrulinea arion* does indeed lay her eggs on thyme plants, the larvae do not eat thyme. Instead the caterpillars drop to the ground, where they produce a airborn chemicals that mimics the smell of larvae of the ant species *Myrmica sabuleti*. The patrolling worker ants mistake the butterfly larvae for their own and carry the caterpillars into the ant nests (Figure B). Once there, the caterpillars eat young ants until they pupate. They undergo metamorphosis in the ant colony, surfacing as butterflies.

The conservation efforts failed because, while the grazing restrictions allowed pastures to grow thicker and taller, they changed the soil conditions to favor different species of ants. As a result, the *M. arion* larvae dropped into a predatory field rather than a nutritive one. The species went extinct in Britain in 1979, although a population introduced from Sweden, with the proper conservation management, appears to be growing (Thomas 1995; Nash et al. 2008).

But in the complex web of life histories, the parasitic *Maculinea* larvae can become the parasitized. It seems these caterpillars are the sole food

(A)

(B)

(C)

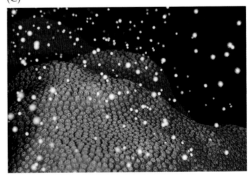

(A) A female Large Blue butterfly *Maculinea arion*, from the re-introducton program underway in Somerset, England. (B) Newly hatched caterpillar of a related species (*M. alcon*) being carried by a worker *Myrmica* ant to its nest. These fascinating interactions are shown in a series of videos on the BBC documentary *Life in the Undergrowth* (Programme 4). (C) Adult coral *Montastraea faveolata* colonies release sperm and eggs in bundles. The bundles rise to the ocean surface and the sheaths dissolve to allow the sperm and eggs to mix. The fertilized eggs become larvae that travel on ocean currents. Eventually, the larva return to the reefs by following the sound of snapping shrimp, which can be heard at www.bris.ac.uk/news/2010/7004.html. (A,B from Nash et al. 2008, photographs courtesy of D. R. Nash; C © Nature Picture Library/Alamy.)

Life Cycle Symbioses and Human Involvement (*continued*)

source for the larvae of several species of wasps of the genus *Ichneumon* (Thomas and Elmes 1993; Thomas et al. 2002). A female wasp can detect not only the ant colonies but also the presence of butterfly larva within them. She enters only colonies where caterpillars are present; once there, the wasp emits airborn chemicals that cause the ants to fight among themselves while she goes about laying a single egg in each butterfly larva. Each wasp egg hatches into a larva that eats the caterpillar as it begins pupation. Eleven months later, the pupal case is shed and there emerges not a butterfly but an adult wasp.

A righteous Caribbean dub

Most all the cues known for larval settlement and metamorphosis involve chemicals eminating from the substrate; these chemicals can signal the presence of a food source or potentially induce larval metamorphosis. However, in at least one case, vibrational cues appear to direct marine larvae to coral reefs. Coral reefs are the largest biological structures on Earth, and they grow by recruiting planktonic coral (cnidarian) larvae. While chemical cues from algae and bacterial films work within a small distance of the reef, it is the "noise of the reef"—the

snapping of the shrimp and the fish noises—that attracts coral larvae from long distances (Figure C). Vermeij and colleagues (2010) made recordings of Caribbean reefs and found that the larvae swam to the source of the sound, even in the laboratory. These findings mean that coral reefs face danger from noise pollution as well as from thermal and chemical pollution. Steve Simpson (2010), who headed the study, has warned, "Anthropogenic noise has increased dramatically in recent years, with small boats, shipping, drilling, pile driving and seismic testing now sometimes drowning out the natural sounds of fish and snapping shrimps."

Another environmental agent critical for larval settlement is the turbulent shear produced by the waves. Larvae of the sea urchin *Strongylocentrotus purpuratus* initiated settlement when exposed to turbulent shear that is typical of the wave-swept shores where adults are found. The shear force appears to make the larvae competent to settle and enable them to respond to chemical inducers that will direct them to their final place of settlement and metamorphosis (Gaylord et al. 2013). Scientists are beginning to have a new respect for the physical forces involved in relating development to the ecosystem.

The Holobiont Persepective

In addition to the "grand" symbioses discussed earlier, twenty-first-century techniques of molecular biology discovered that no organism was truly an "individual" by any biological criterion (Gilbert et al. 2012; McFall-Ngai 2013). *Anatomically*, humans are 90% composed of prokaryotic cells, and *physiologically*, about 35% of our blood metabolites come from bacteria (McFall-Ngai et al. 2013). In many organisms, essential amino acids are constructed by pathways involving enzymes encoded by both host and symbiont genomes. *Genetically*, symbionts provide important and selectable variation for the entire organism. The color, thermotolerance, and parasitoid resistance of aphids are determined by alleles of their symbionts' genomes, not of their nuclear genome (Dunbar et al. 2007; Oliver et al. 2009; see Chapter 10). Even the immune system, which had been thought of as our defensive weaponry protecting us against the outside microbial onslaught, is more like a bouncer or passport control agent, knowing which bacteria to allow in and which ones to keep out. Amazingly, the possibility even exists that natural selection sees animals as consortia, as teams, and not merely as single genetic types (Gilbert et al. 2012). This will be discussed more fully in Chapter 11, concerning evolution and ecological developmental biology.

Developmental symbiosis

But probably the most exciting idea is that we *develop* as holobionts. Remarkably, we don't develop solely from the genome we inherit from our parents; there are other genomes involved. As we will see, even normal mammalian development is predicated on developmental signals being provided by symbiotic bacteria. Such co-development, where symbionts are critical for forming the organs of their hosts is sometimes called **symbiopoiesis**. In some cases, development can barely get started without symbiotic input. For instance, in the nematode *Brugia malayi*, *Wolbachia* bacteria ride to the posterior pole on the microtubules that form the cellular mitotic spindle. Once in the posterior pole, they become essential for regulating the cell divisions that create the anterior-posterior boundary critical to the early development of the nematode. If they are removed from the egg before first cell division, that division is often abnormal, and a proper anterior-posterior polarity fails to form (Figure 3.4; Landmann et al. 2014). Here we see very starkly that the developing organism is a holobiont.

We develop as holobiont communities. In cases of obligate developmental symbiosis, the development of the holobiont can only occur if the symbionts are present. There is a co-development such that the host organism "expects" the presence of the symbiont and needs it in order to develop. For instance, studies on the parasitoid wasp *Asobara tabida* found that normal

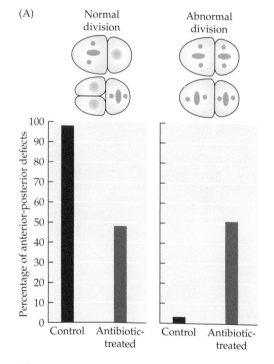

Figure 3.4 Loss of *Wolbachia* bacteria leads to defects in anterior-posterior polarity. (A) Proportion of anterior-posterior defects during the first cell division of *Brugia malayi* in the presence (control) or absence (antibiotic-treated) of *Wolbachia* symbionts. The symbionts are in the fertilized egg but can be eliminated by antibiotics. All the symbiont-containing embryos divide normally, while around half the embryos without symbionts give faulty divisions in the posterior-forming cell. (B) Control (with bacteria) and (C) antibiotic-treated (*Wolbachia*-free) embryos. The DNA is stained red and the microtubules of the mitotic spindle are stained green. (After Landmann et al. 2014, photographs courtesy of F. Landmann.)

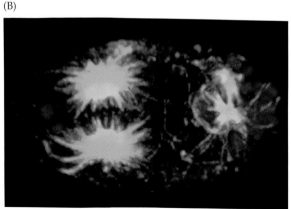

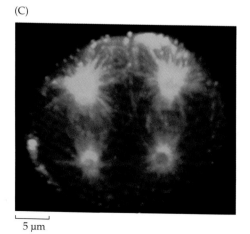

5 µm

females of this species are dependent on *Wolbachia* bacteria for successful egg production. In one experiment, *A. tabida* females carrying *Wolbachia* were treated with antibiotics that killed the symbiotic bacteria. The experimenters found this treatment had what they described as a "totally unexpected effect": these female wasps were unable to produce mature oocytes and thus could not reproduce (Figure 3.5; Dedeine et al. 2001; Pannebakker et al. 2007). Their ovaries had undergone apoptosis! The *Wolbachia* bacteria were critical in preventing the cells from undergoing programmed cell death. The results of this experiment showed that female *A. tabida* have become dependent on the presence of *Wolbachia* for successful reproduction. *Wolbachia* are part of the host's normal development.

In such developmental mutualisms, the death of the host can result from killing the symbiont. This is what's happening to many of the corals that make the reefs that support much of the world's marine species. "Coral bleaching" is seen as less thermotolerant algae die and the coral is left without its way of getting oxygen and carbon (see Figure 3.1B). This loss of algal symbionts also appears to be partly responsible for the decline to the spotted

(A)   Control   Antibiotic-treated

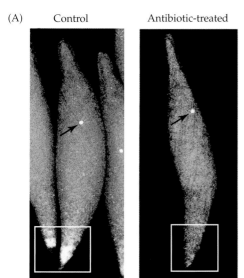

Figure 3.5   Comparison of the ovaries and oocytes (immature eggs) of the wasp *Asobara tabida* from control females and females treated with rifampicin antibiotic to remove *Wolbachia*. (A) When DNA in normal *A. tabida* oocytes was stained, the controls had a nucleus (arrow) as well as a mass of *Wolbachia* at one end (boxed area). Treated oocytes had a nucleus but no *Wolbachia*. (B) The ovaries of the wasps with Wolbachia had an average of 228 oocytes, while rifampicin-treated females had an average of 36 oocytes as their ovaries had undergone apoptosis in the absence of *Wolbachia*. (C) Ovarian cells undergo dramatic apoptosis if the symbionts are removed (antibiotic-treated), while normally (in symbiont-containing larvae; control) very few cells undergo apoptosis. ***, $p < 0.001$. (A,B from Dedeine et al. 2001; C after Pannebakker et al. 2007.)

(B)   Control   Antibiotic-treated

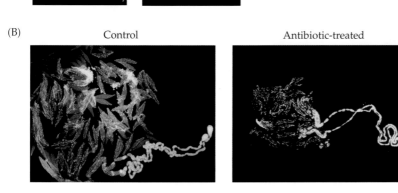

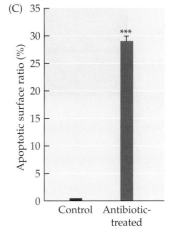

salamander (*Ambystoma maculatum*) in the American midwest. The spotted salamander lays hundreds of eggs in a dense egg mass in a pond (Figure 3.6A). The supply of oxygen limits the rate of their development, and there is a steep gradient of oxygen from the outside of the cluster to deep within it; thus embryos on the inside of the cluster develop more slowly than those near the surface (Strathmann and Strathmann 1995). The embryos seem to get around this problem by coating themselves with a thin film of photosynthetic algae, which they obtain from the pond water. In the egg clutches, photosynthesis from this algal "fouling" enables net oxygen production in the light (Bachmann et al. 1986; Pinder and Friet 1994; Cohen and Strathmann 1996). This algal symbiont is so specific that its name is *Oophilia amblystomatis* ("lover of ambysoma eggs"). The alga is actually stored in the mother's body and appears to be deposited along with the eggs (Kerney et al. 2011; Graham et al. 2013). These symbionts are being destroyed by atrazine, an herbicide widely used in the United States as a weed killer. Once applied, atrazine can remain active in the soil for more than 6 months, and it can be carried by wind and rainwater to new ponds. Concentrations of atrazine as low as 50 micrograms per liter completely eliminate this alga from the eggs, and the amphibian's hatching success is greatly lowered (Gilbert 1944; Mills and Barnhart 1999; Olivier and Moon 2010).

Attacking a symbiont can be used as a means to eradicate an unwanted host. *Mansonella*, a genus of filariasis worms that cause debilitating conditions in humans, contains *Wolbachia* bacteria as endosymbionts. The *Wolbachia* produce chemicals that enable the worm to molt, and without them, the worms die. Many species of filariasis worms have become resistant to the

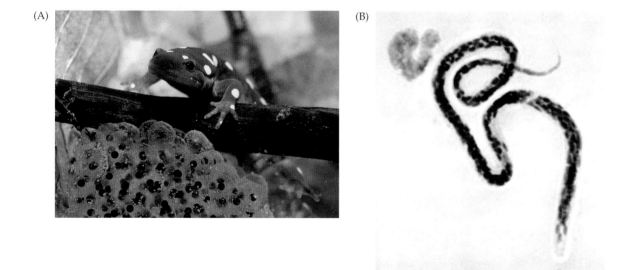

(A)

(B)

Figure 3.6 Obligate developmental symbionts. (A) Spotted salamander (*Ambystoma maculatum*) eggs at the center of the cluster cannot survive the lack of oxygen when their algal symbiont is eliminated by herbicides. (B) Filariasis worms such as *Mansonella ozzardi* cannot complete molting when their *Wolbachia* symbiont is eliminated by antibiotics.

Immunity through Developmental Symbiosis

Symbiosis may be one evolutionary path of protection for eukaryotic organisms. As we will see in Chapter 4, embryos often encounter pathogenic microorganisms before they have developed a functioning immune system. One way for embryos to gain immediate protection from these pathogens is by forming symbiotic relationships with other organisms. For example, symbioses between eggs and bacteria can protect the eggs of aquatic species from fungal pathogens. (As anyone who owns an aquarium knows, uneaten fish food is soon surrounded by a halo of filamentous fungi.)

The outer envelopes (chorions) of several crustacean eggs actually attract bacteria that produce fungicidal compounds. Embryos of the shrimp *Palaemon macrodactylus*, for instance, are extremely resistant to infection by the marine fungus *Lagenidium callinectes*, a known pathogen of many crustaceans. The chorions of these shrimp embryos are consistently found to be infected with the bacterium *Alteromonas*, which produces the antifungal compound 2,3-indolinediol. In laboratory studies,

Palaemon embryos were bred both with and without the bacterial symbionts and then exposed to the fungal pathogen (Gil-Turnes et al. 1989). When exposed to the fungus, the bacteria-free embryos died rapidly, whereas embryos reinoculated with *Alteromonas* (or treated with 2,3-indoleinedione) survived well. Therefore, it appears that the symbiotic bacteria protect the shrimp embryos from fungal infection.

In some cases, there are entire layers of symbiosis protecting the host. Aphids, for instance, have mutualistic symbioses with numerous bacteria. The pea aphid (*Acyrthosiphon pisum*) is host to the bacterium *Hamiltonella*, a symbiont that gives the aphid protection against the larvae of parasitoid wasps. But the bacterium itself is not what provides this immunity. Rather, *Hamiltonella* cells also harbor a symbiont—a type of bacteriophage (virus) that is critical for their own life cycle. It is a biochemical product of the phage that actually kills the wasp larvae (Moran et al. 2005).

drugs traditionally used to kill these parasites in humans. A new treatment strategy has been to employ antibiotics (such as doxycycline) against the *symbionts* rather than the hosts (Figure 3.6B; Hoerauf et al. 2003; Coulibaly et al. 2009). Once the antibiotic destroys the symbiont, the worms' cells undergo apoptosis and the worms die* (Landmann et al. 2011).

Constructing the Holobiont: Uniting Host and Symbionts

All symbiotic associations must meet the challenge of maintaining their partnerships over successive generations. In the partnerships that are the main subject here, in which microbes are crucial to the development of their animal hosts, the task of transmission is usually accomplished in one of two ways, either vertical or horizontal transmission. In terrestrial organisms, the symbionts (whether horizontally or vertically transmitted) are usually derived from the mother (Funkhouser and Bordenstein 2013). In aquatic environments, horizontal transmission usually occurs by the uptake of environmental bacteria, which can come from either mother or father.

---

*Symbiont therapy is also being investigated for ridding the world of malaria. The most vulnerable stage of the *Plasmodium* parasite (that causes malaria) is when it is in the gut of its mosquito hosts. Researchers have found that adding certain bacteria (such as *Wolbachia*) to mosquitoes will prevent the *Plasmodium* from maturing in the gut. Moreover, normal symbionts can be replaced with genetically engineered symbionts that secrete poisons against larval plasmodia and kill the malarial parasites (Wang et al. 2012; Bian et al. 2013).

Vertical transmission refers to the transfer of symbionts from one generation to the next through the germ cells, usually the eggs (Krueger et al. 1996). And, as we will soon learn, there are many cases in which bacteria of the genus *Wolbachia* reside in the egg cytoplasm of invertebrates and provide important signals for the development of the individuals produced by those eggs. *Wolbachia*, one of the world's most successful bacteria, can prevent development in some species and become an important agent of normal development in others. The *Wolbachia* become provisioned in the egg cytoplasm just like the mitochondria and ribosomes, traveling from nurse cells into the developing egg (Figure 3.7). There are numerous other ways that bacteria from the females body become concentrated in the developing eggs. Bacterial symbionts enter the larval gut of the *Dentalium* mollusks by having ovarian cells infect the egg. These bacteria specifically adhere to those cells destined to become the larval gut. During gastrulation, these cells (originally on the outside of the embryo) become internalized to form the gut lining, and the bacteria ride in with them (Geilenkirchen et al. 1971). In the common composting earthworm (*Eisenia fetida*), symbiotic bacteria from the nephridia ("kidneys") are deposited directly onto the capsules inside which the embryos are developing. The developing embryo makes a nephridial duct that recruits only that particular species of bacterium into its nephridia, excluding others. This bacterial species is thought to protect the worm's nephridia against colonization by other, more harmful, bacteria (Davidson and Stahl 2008).

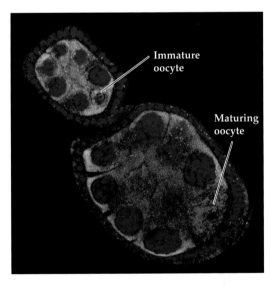

Figure 3.7   Vertical and horizontal transmission of *Wolbachia* bacteria. In *Drosophila*, the *Wolbachia* are transmitted vertically through the female germ cells. In the ovary, 15 nurse cells transport proteins, RNA, and organelles to the most distal oocyte cell. The symbiotic bacterium (stained red) is also transported by microtubules into the oocyte. Cytoplasm of the ovary is green, and blue indicates DNA. (After Ferree et al. 2005, courtesy of H. M. Frydman and E. Wieschaus.)

It has been assumed that vertical symbiotic transmission takes place solely through the female germline. The sperm of most organisms have very little cytoplasm in which symbionts can reside. However, sperm-borne symbionts have also been discovered in some insects (Watanabe et al. 2014), and these are transmitted to the offspring through the sperm head, where the bacteria appear to reside in the nucleus.

In **horizontal transmission**, the metazoan host is free of symbionts but subsequently becomes infected, either by its environment or by other members of its species. One example of horizontal transmission involves aquatic eggs that attract photosynthetic algae (see Figure 3.6A). As with the spotted salamander (*Ambystoma maculatum*) mentioned above, clutches of amphibian and snail eggs are packed together in tight masses.

As we will see, horizontal transmission is also the rule in mammals, where the newborn is colonized as it is being born. Indeed, mammalian birth can be seen as the process of going from one symbiotic environment (the mother) to another (the bacteria). Horizontal transmission also appears to be the case in the well-studied association between the bioluminescent bacterium *Vibrio fischeri* and its host, the squid *Euprymna scolopes*. In this relationship, *Vibrio* bacteria provide developmental signals for the formation of the squid's light organ. Nyholm and colleagues (2000) have shown that the newborn squid acquires *Vibrio* by horizontal transmission from seawater; the cells on the squid's nascent symbiotic organ secrete molecules that selectively attract and catch the potentially light-producing microbes.

The Squid and the Microbe: A Paradigm of Developmental Symbiosis

The Hawaiian bobtail squid, *Euprymna scolopes*, is a tiny (2 inches long) native of the shallow waters off Hawaii (Figure 3.8A). A nocturnal animal, the squid preys on shrimp. Predatory fish (who like squid hors d'oeuvres) may be alerted to the presence of the squid if the moon casts its shadow on the sea floor. *E. scolopes*, however, has developed a most creative mechanism of dealing with this potential threat: it emits light from its underside, mimicking the moonlight and hiding its shadow from potential predators. The squid cannot accomplish this feat of disguise alone, however; the presence of the symbiotic bacterium *Vibrio fischeri* in the squid's light organ is required to generate this characteristic glow (Figure 3.8B). Both the squid and the bacterium benefit from this mutualistic relationship—the squid because it gains protection from predators, and the bacterium because it is able to live safely within the host's light organ, an environment free of predators and full of nutrients.

The squid is not born with either symbionts or a developed light organ. It has to collect the bacteria, and the bacteria have to build the light organ: The *V. fischeri* symbionts are transmitted horizontally between host generations. The *V. fischeri* are present in low concentrations (between 100 and 1500 per milliliter of seawater—which is less than 0.1% of all the bacteria present in such a sample). A newly hatched squid has no *V. fischeri* on or in it, but as it swims through its environment, seawater passes through the ciliated epithelial cells on the ventral surface of the its nascent light organ (Figure 3.8C). These cells form two fields that ring the pores leading to the crypts that will eventually house the luminescent bacteria. Several species of bacteria in the sea water attempt to stick to the squid cells, but very few species can survive there because of the mucus secreted by the epithelium. *V. fischeri* can survive the chemicals in the mucus and reach the epithelium, and only about 3–5 *V. fischeri* bacteria are needed to induce the squid to respond to them in a way that facilitates their aggregation (Altura et al. 2013; Figure 3.8D).

Next, the aggregated *V. fischeri* release compounds that alter the gene expression profiles of the squid's ciliated cells, causing them to make new proteins. Some of these proteins (such as lysozyme) prevent other types of bacteria from colonizing the epithelium. Others are important in the further colonization of the squid by the *V. fischeri*. The epithelial cells (even those without *V. fischeri* on them) are instructed by the bacteria to make and secrete chitotriosidase (Kremer et al. 2013; Figure 3.8E). This is an enzyme that degrades chitin, the structural compound in the exoskeletons of most invertebates, into chitobiose (a disaccharide of *N*-acecytlglucosamine, the building block of chitin). Chitobiose acts on the bacteria, helping them to multiply, as well as causing them to become sensitive to gradients of chitobiose that extend from the ciliated appendages to the pores and ducts. The gradient develops because the majority of the chitin-degrading enzymes are generated by the cells lining the pores. The bacteria detach from the epithelial cells and, following the gradient of chiobiose from lowest to highest concentration, enter the pores leading to the ducts and eventually to the crypts of the nascent light organ (Mandel et al. 2012; Kremer et al. 2013; Figure 3.8F). Once the *V. fischeri* are inside the crypts, bacterial products (portions of the cell wall and membrane) induce apoptosis in the squid epithelial cells that had once attracted them (Koropatnick et al. 2004; Figure 3.8G).

Inside the crypts, the bacteria's bioluminescence is stimulated through a quorum-sensing feedback loop. As *V. fischeri* accumulate in concentrations far exceeding those in the external environment, the bacteria produce two

Figure 3.8 Symbiosis in the squid *Euprymna scolopes*. (A) An adult Hawaiian bobtail squid. (B) The anatomy of the host ventral surface of the squid, showing the light organ (black square) through the ventral surface of a hatchling animal. A diagram of the external (left) and internal (right) features of this organ. (C) The ventral appendages are poised to receive *V. fischeri*, creating currents and an environment (diffuse yellow stain) that attracts seaborne Gram-negative bacteria, including *V. fischeri*. (D) Aggregated *V. fischeri* (stained green) entering the pores on the light organ surface (stained red). The upper pore has a large aggregate at the pore as well as a set of cells chemotaxing down toward the pore. (E) Confocal micrograph of one side of the lateral area of a juvenile light organ. The red stain highlights the actin filaments of the squid cells, the blue stain labels nuclei, and the green stain binds to the chitobiose-producing enzymes involved in chemotaxis and nutrition. (F) Crypts (red) of both sides of the light organ become colonized by *V. fischeri* (green). (G) The appendages, which had originally trapped the microbes, degenerate by apoptosis of the epithelial cells (yellow) caused by the *V. fischeri*. (Photographs courtesy of M. McFall-Ngai.)

(A)

(B)

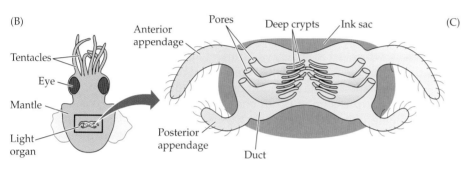

Tentacles

Eye

Mantle

Light organ

Anterior appendage

Pores

Deep crypts

Ink sac

Posterior appendage

Duct

(C)

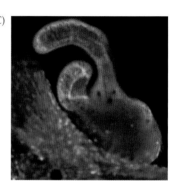

(D)

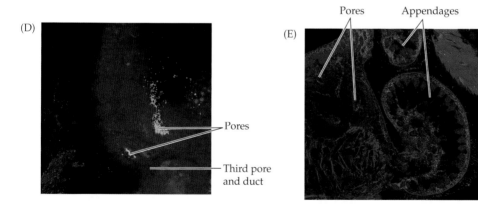

Pores

Third pore and duct

(E)

Pores

Appendages

(F)

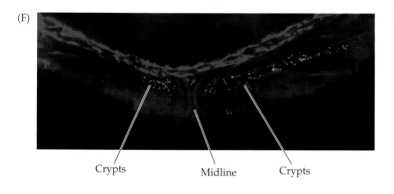

Crypts

Midline

Crypts

(G)

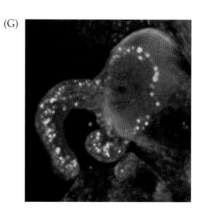

molecules that induce transcription of their own *lux* genes, which are the genes responsible for luminescence (Visick et al. 2000; Millikan and Ruby 2001). Thus, *V. fischeri* is able to luminesce *only* after it is established within the light organ of *E. scolopes*; this fundamental attribute of the symbiont is induced only in the presence of its host. It is therefore clear that the host has an important effect on the development of its symbiont and also that the symbiont is important to the normal development of its host.

How do the symbiotic bacteria complete their host's light organ development? The mechanisms are beginning to be discovered. First, the symbionts reduce the squid's production of nitric oxide (NO) by 80%; without this reduction, *E. scolope* would produce NO in concentrations toxic to the bacteria* (Nyholm and McFall-Ngai 2004). *V. fischeri* appears to be able to attenuate its host's NO production by the presentation of two sets of cell surface molecules (lipopolysaccharide and peptidoglycan derivatives), which act in synergy (Altura et al. 2011). The bacteria also induce the replacement of the ciliated epithelium by a nonciliated epithelium, the differentiation of the surrounding cells into storage sacs for the bacteria, and the expression of genes encoding opsins and other visual proteins in the light organ (Chun et al. 2008; McFall-Ngai 2008; Tong et al. 2009). Remarkably, some of the genes induced by the bacteria are the same genes that are used to make the squid's eye. Here, in the mantle (body) cavity of the squid, they make a photoreceptive area that can measure the amount of light in the environment and adjust the bacterial luminescence to match it, thereby canceling out the shadow by counterillumination (Peyer et al. 2014).

The symbiotic relationship between *E. scolopes* and *V. fischeri* is a complex one (Koch et al. 2013; Norsworthy and Visick 2013;), and many other details about the relationship—such as how the squid is able to expel those bacteria that do not luminesce—have yet to be explained. Even though our knowledge about this excellent model system is as yet incomplete, the relationship between the Hawaiian bobtail squid and its luminescent symbiont can teach us important lessons about environmental influences on the regulation of development. The squid and the bacterium have coevolved such that each plays a fundamental role in the other's development: the squid actively accumulates a high enough population density of *V. fischeri* to allow the bacterium to express its latent bioluminescence, while the bacteria trigger important morphological changes in the light organ of the host. The concrete and well-characterized impact that host and symbiont have on each other's development provides us with an excellent example of how environmental factors can play an integral role in normal development.

---

*Mary Montgomery has pointed out that this makes for interesting interaction of cooperation and competition. In order to cooperate, the bacteria must first be selected. And the bacteria are being selected by an extremely harsh mechanism, where hardly any of them survive. *V. fisheri* survives because it has evolved to respond to low levels of NO in the squid mucus by activating the genes that allow it to resist the NO "fire" of the ducts. So again, the genes of one species are being induced by the products (in this case, NO) of another species.

Mutualistic Consortia

While the squid *Euprymna* appears to harbor only one symbiont in its light organ, some animal cells and organs are home to entire ecosystems of symbionts. As we will see later in the chapter, the mammalian gut is such a complex ecosystem. Since a basic requirement of all organisms is food, consortia of photosynthesizers, gatherers, and digesters are to be found in within a "single" organism.

Providers and processors

Off the coast of Naples, burrowed in the mud near the island of Elba, resides the oligochaete worm *Olavius algarvensis*. More like a plant than an animal, this species is able to synthesize its own food. Indeed, the worm has neither mouth nor anus nor gut. Even its nephridial excretory pores have mostly degenerated.

What this animal has is an internal ecosystem of sulfur-metabolizing bacteria that can fix carbon dioxide into sugars and manufacture all the amino acids and several vitamins needed by the worm. These bacteria also appear to recycle the animal's waste products. This microbial consortium of four different bacteria species lives just beneath the worm's cuticle. As a group, they are able to fix carbon dioxide through the Calvin-Benson cycle, using the same types of enzymes (including Rubisco) as chloroplasts but coupled to the oxidation of reduced sulfur compounds. Moreover, by cycling the intermediate sulfur compounds between them, they can supply organic carbon to the worm through three different metabolic pathways. Depending on where the worm sits in the sediment (the deeper layers being more anoxic than the upper layers of mud), different pathways will be utilized (Woyke et al. 2006).

Lynn Margulis (1971, 1994) has written elegantly of symbiosis, and one of her most incredible examples involves the protozoan *Myxotricha paradoxa*, a symbiont of the Australian termite *Mastotermes darwinensis*, without which the termite would not be able to digest wood, as the termite gut by itself has no enzymes capable of digesting cellulose (see Figure 3.1C,D). But *Myxotricha paradoxa* itself is a symbiotic colony, including a protist and at least three species of bacteria. One of these bacterial species is a spirochete that attaches to the outer membrane of the protist, forming its locomotor apparatus. A second symbiotic bacterium appears to stabilize this arrangement, and a third lives inside the cytoplasm of the protist. It is the genome of this third bacterium that provides the cellulose-digesting enzyme essential to the termite. All of these bacteria appear to be transmitted through the egg yolk to the next generation (Sacchi et al. 1998).

Such **microbial consortia**—structured arrangements of microbes that function together as a unit—are found in numerous species. Some of the most important microbial corsortia are those established in coral reefs. The photosynthesis by symbiotic algae (dinoflagellates) of the genus *Symbiodinium* provides oxygen and sugars to the coral, and these algae also provide oxygen radicals that, working between the cells, block infection. (The effect is similar to placing hydrogen peroxide on a cold sore, releasing oxygen that destroys bacteria and viruses.) In addition, corals contain a resident microbiological population that includes nitrogen-fixing bacteria as well as bacteria that produce antibiotics against other, pathogenic, bacteria. Coral bleaching is caused when this symbiosis is disrupted, and the major cause of this disruption is the pathogenic bacterium *Vibrio shiloi*. Elevated temperatures cause *Vibrio* to express adhesion molecules, which allow the bacteria to bind to the coral's surface, and the substance "toxin P," which blocks photosynthesis in *Symbiodinium*.

Rosenberg and colleagues (2007) have proposed that environmental conditions select for the most beneficial symbiotic relationship between the different coral species and the different bacterial and *Symbiodinium* species. Thus, in this case, the processes of natural selection are operating not on an organism but on a network of relationships among many organisms. This is the so-called *hologenome theory of evolution*, wherein the host and its symbiont population constitute a "hologenome" that can change more rapidly than the host genome alone, thereby providing greater adaptability.

The evolution of social insects

Ants and termites might look the same (but look closely—ants have a constricted waist, termites do not; termites have equally long forewings and hindwings, ants do not). However, the two groups of insects have very different origins. Termites are basically modified cockroaches, while ants are derived from stinging wasps. Both groups have evolved eusociability—hives, where there is division of labor among the inhabitants—and symbiosis has played critical roles in allowing these eusocial groups to thrive.

(Continued on next page)

Mutualistic Consortia (*continued*)

Macrotermes natalensis, an African termite whose mounds can be 9 meters high, has evolved a symbiosis involving three kingdoms—bacteria, fungi, and animals—to extract nurients from wood. *Macrotermes* purposely grows a fungus, aptly named *Termitomyces*, in these mounds (Figure A). (Indeed, the fungus has a very narrow temperature range in which it can grow, and the termites are adept at closing and opening apertures in their mounds to retain the correct growing temperature for the fungus.) Worker termites gather, chew, and regurgitate wood onto the fungus. The fungus secretes enzymes from its hyphal processes that are able to chop the lignin and cellulose of the wood into their small oligosaccharide subunits. The termites then eat the resulting slurry, and the bacteria in their guts have the enzymes necessary to digest the oligosaccharides into glucose and other energy-producing compounds (Poulsen et al. 2014). Thus, the termites find the food, harbor the microbes, and tend the fungus. The fungus splits the large carboydrates of the wood into simpler components, and the microbes digest these into energy-forming carbohydrates for the termites and themselves.

This has been a remarkably successful symbiosis, enabling the spread of *Macrotermes* through Africa and Asia. Every species of *Macrotermes* cultivates *Termitomyces*, and the symbiosis between them is obligatory—neither lives without the other. "The fungus-growing termites thus represent a major metazoan radiation based on a tripartite life-history transition: insects became farmers, fungi became crops, and gut microbiotas adapting largely unknown complementary roles" (Poulsen et al. 2014).

While *Macrotermes* species have become incredibly important decomposers of Old World forests, the attine ants play a similar role in the New World. Similarly, their status as "prime New World pests" comes from what has been described as an "unholy alliance" with a fungus, in this case *Leucocoprinus*. These are the "leafcutters," and they have the largest colonies of any ant species. Like *Macrotermes*, they are obligate farmers, and the symbiosis between the insect and the fungus (plus the gut and cuticle bacteria) has given them remarkable success (Figure B). One of the fascinating aspects of this symbiosis is that the ants will eat fungi from the middle of the garden and concentrate in their excrement the fungal enzymes that digest wood and that destroy the plant defense compounds against the fungus. This enables the fungal hyphae to get around the plants' defense systems and to better digest plants (Schiøtt et al. 2010; De Fine Licht et al. 2013). There are numerous morphological

(A)

(B)

Fungal symbioses provide nutrition in different ways to attine ants and *Macrotermes* termites. (A) *Macrotermes* termites farm fungi to break down the cellulose and lignins in plants and wood, turning the plants into digestable products. Most termite species have their cellulose-digesting symbionts inside their guts. These termites have such digestion "outside" their bodies. (B) Leafcutter ants (*Atta texana*) distribute leaves inside their fungus colony. The white fungus will break down the leaves for its own nutrition, and the ants will eat the fungi. (A Courtesy of J. Scott Turner.)

adaptations (such as the ants' serrated mandibles) that have to develop for these symbioses to work.

Symbiosis is the way of the world. As Margulis and Sagan (1986) said, "Life did not take over the globe by combat, but by networking."

Wolbachia: Symbiotic Regulation of Sexual Development

The relationships between bacteria of the genus *Wolbachia* and their many different arthropod hosts provide clear examples of how symbionts can alter the development of their hosts and how organisms can evolve to depend on their symbionts for normal development. In the many different organisms that *Wolbachia* infects, this symbiont's effects on the sexual development of its hosts range widely, from transformation of genetically male embryos into females to the necessity of its presence for completion of normal oogenesis.

Male killing

First discovered inhabiting the reproductive organs of the mosquito *Culex pipiens*, the bacterial genus *Wolbachia* includes at least 38 different strains, though the number of species into which these strains should be classified is still uncertain. The bacteria live in the cytoplasm of cells in the gonads of many different arthropods, including, in addition to mosquitoes, parasitoid wasps, pillbugs, mites, lepidopterans (butterflies and moths), and *Drosophila* (Stouthamer et al. 1999). *Wolbachia* can be transmitted both horizontally (from one adult to another) and vertically (from one generation to the next), but vertical transmission through the female germ cells is by far the more important mechanism by which *Wolbachia* infects new hosts (Werren 1997; Charlat et al. 2003).

Because *Wolbachia* can be transmitted only by females, the bacteria benefit if they increase the ratio of females to males among the offspring of the organism that they infect. Indeed, several different methods have evolved by which *Wolbachia* can and does affect this ratio. The most straightforward way in which *Wolbachia* does this is by killing the male progeny of infected females (Charlat et al. 2003). *Wolbachia* has been found to engage in male killing in both the Asian corn borer moth *Ostrinia furnacalis* and the African butterfly *Acraea encedana* (Jiggins et al. 2000; Kageyama and Traut 2003). In their study of a Ugandan population of *A. encedana*, Jiggins and colleagues (2000) determined that over 95% of the females were infected with *Wolbachia* and that many of these infected females produced no male progeny. They also noted that the broods of infected females (that produced only female progeny) had hatch rates roughly half those of broods from uninfected females, suggesting that *Wolbachia* was actively killing the male progeny during embryogenesis rather than nonlethally altering the sex of the progeny (Figure 3.9). The impact of this male killing in the population was huge: only 6% of the adult butterflies in the population were male.

Although it is clear that male killing results in a skewed sex ratio in the host population, exactly how this skew benefits the bacteria is a bit more obscure. After all, *Wolbachia* benefits not by killing off organisms that it can't infect, but by increasing the number of host organisms available to infect, and as seen in Figure 3.9, male killing doesn't directly influence the absolute number of female progeny. Instead, the female progeny benefit because the elimination of their male siblings means reduced competition. In some species, females benefit further by eating the male embryos that

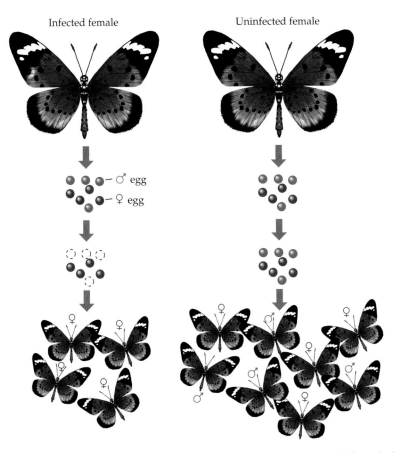

Figure 3.9   Sex ratio distortion by Wolbachia. In *Acraea encedana* infected with *Wolbachia* bacteria, male progeny die during embryogenesis (left). Because all the eggs of uninfected females hatch (right), infected females produce only half as many progeny as uninfected females. However, the females produced have better access to resources.

do not survive. Because the death of their male siblings gives the female progeny of infected females a higher survival rate than the female progeny of uninfected females, male killing does give the infecting *Wolbachia* an advantage: their potential hosts are more likely to survive and reproduce than they would be if their male siblings survived (Charlat et al. 2003). However, the absence of male progeny is so deleterious for the host population as a whole that, in insect populations with high incidences of male-killing *Wolbachia* infections, we would expect genes for resistance to either *Wolbachia* infection or its male-killing effects to be strongly selected for, and indeed there is some evidence for resistance genes in *A. encedana* (Charlat et al. 2003; Jiggins et al. 2000).

Turning males into females

A second mechanism by which *Wolbachia* may attempt to skew the sex ratio of the progeny of its host is by feminizing genetically male embryos, resulting in all-female or mostly female progenies without killing embryos.

Termed "feminization in diploids," this process has been found to occur in *Wolbachia* infections in many species of isopods, including the common pill bug, *Armadillidium vulgare* (Cordaux et al. 2004; Figure 3.10). In this species, females are usually the heterogametic sex (WZ) while males are homogametic (ZZ); however, when there is *Wolbachia* infection, genetically male ZZ embryos develop as females. This is a striking example of the power of environmental factors to influence development, for the ability of *Wolbachia* to feminize a genetically male host means that, in this case, the presence or absence of the symbiont is a more important factor in sex determination than chromosomal sex.

Interestingly, *Wolbachia* appears to cause this sex reversal by interfering with the male pattern of DNA methylation (Negri et al. 2009). *Wolbachia* feminizes genetic males in the leafhopper *Zyginidia pullula*. These feminized males were found to have the *female*-specific pattern of DNA methylation. Moreover, the extent of feminization appears to be dependent on the amount of *Wolbachia*. Since the *Wolbachia* is transmitted through the female germline, it continues to produce transgenerational epigenetic changes to any genetic male that may be generated.

Feminization of males has a clear advantage for the infecting *Wolbachia* because it leads to an increased number of females, the only sex capable of passing *Wolbachia* on to progeny. The elimination of male progeny through feminization, however, once again results in a dearth of males for the *Wolbachia*-carrying host females to mate with, and thus it is hypothesized that among arthropods carrying feminizing *Wolbachia*, selection for resistance genes will again occur (Charlat et al. 2003; Cordaux et al. 2004).

For some arthropod species (such as the hymenopteran ants, bees, and wasps), sex is determined not by chromosome composition (whether XY

Figure 3.10   Male and female *Armadillidium vulgare*. Genetically male pill bugs (right) can be transformed into phenotypic egg-producing females (left) by infection with *Wolbachia* bacteria.

or WZ), but by the number of complete chromosome sets that an organism has. In this scheme, known as *haplodiploidy*, females develop from fertilized eggs and thus are diploid, whereas males develop from unfertilized eggs and are haploid. In several species, including mites and parasitoid wasps, *Wolbachia* can interfere with this sex-determination scheme, turning haploid embryos that normally would become males into diploid females. This phenomenon is known as *parthenogenesis induction*. *Wolbachia* usually induces parthenogenesis (i.e., the ability of a female to produce female progeny without fertilization by a male) by doubling the chromosome number in unfertilized eggs to create diploid embryos with two identical sets of chromosomes (Zchori-Fein et al. 2001; Charlat et al. 2003). Such an embryo will develop into a female capable not only of passing *Wolbachia* on to its progeny, but of doing so without interaction with a male, making the reduction in male progeny caused by parthenogenesis induction a less severe problem for the host population than the shortage of males caused by male killing or feminization.

Studies (reviewed in Charlat et al. 2003) have found that if normally haplodiploid species infected with parthenogenesis-inducing *Wolbachia* are treated with antibiotics, haploid male progeny can be produced. These males, however, are often unable to successfully mate with females. A hypothesized reason for this curious fact is that *Wolbachia* infection may also be affecting the evolution of host sex determination. The theory is that because *Wolbachia*-infected populations are female-biased, any phenotype that promotes the production of more males will be favored evolutionarily. In the case of haplodiploid species, however, a phenotype that favors the production of males is one that somehow prevents or discourages the female from mating with a male, since only her unfertilized eggs will develop into males. Thus, the same genes that might favor the production of males also prevent the females carrying them from mating with males! Such "virginity genes," which in fact have been found in the parasitoid wasp *Telenomus nawai*, represent an important example of how a symbiont can profoundly influence the evolution of development in its host.

Cytoplasmic incompatibility

Perhaps the most common effect of *Wolbachia* infection, however, is **cytoplasmic incompatibility** (**CI**), in which embryos resulting from crosses between uninfected males and infected females are viable, but those resulting from crosses between infected males and uninfected females are not (Charlat et al. 2003). Cytoplasmic incompatibility benefits *Wolbachia* because it inhibits the reproduction of uninfected females, thereby increasing the relative reproductive success of infected females and promoting the spread of *Wolbachia* throughout the host population. Because in this instance females become dependent on *Wolbachia* infection for reproductive success, CI can result in selection for any phenotype that decreases resistance to *Wolbachia* infection, thus making *Wolbachia* a yet more integral component of the host's sexual development.

Recently, such cytoplasmic incompatibility became viewed as a possible engine of speciation (Brucker and Bordenstein 2013). An ancestral population of uninfected jewel wasps (*Nasonia vitripennis*) can split into two populations that are subsequently infected by different *Wolbachia* strains. These originally single infections spread throughout each group within each population by unidirectional CI, which gives a relative fitness advantage to infected females by causing embryonic death in crosses between infected males and uninfected females. Upon secondary contact of these two populations, bidirectional CI causes reciprocal incompatibility in *both* directions. In other words, the nuclear-symbiont relationship established with one of the *Wolbachia* groups cannot work with the other *Wolbachia* group, and vice versa. Thus, species can arise within a population, by its having different symbionts (Figure 3.11).

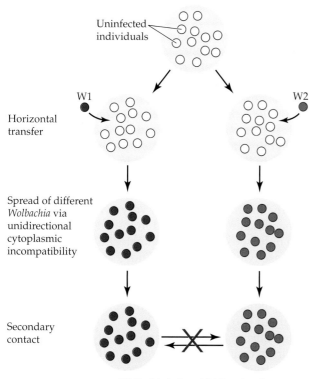

Figure 3.11 An ancestral population (large circle) of uninfected individuals (small open circles) splits into two populations that subsequently become infected with different *Wolbachia* strains (W1 and W2). The original infections spread to each individual within each population by unidirectional cytoplasmic incompatibility (CI). Because infected females (the transmitting sex of cytoplasmic bacteria) are compatible with either infected or uninfected males, they do not suffer this fitness reduction; therefore, unidirectional CI can rapidly spread the bacteria through host populations. Upon secondary contact of these populations, bidirectional CI causes reciprocal incompatibility in both cross directions. Thus, species can arise without morphological or genetic divergence. (After Brucker and Bordenstein 2013.)

The Mutualistic Bacteria of the Vertebrate Gut

Having discussed several examples of how symbionts and symbiotic consortia play important roles in the development of invertebrate host organisms, we will turn to an example much closer to home: the interaction between mammals and the bacteria that reside in the mammalian gut. This symbiotic relationship is an amazing example of mutualism: the bacteria benefit from the safe and nutrient-filled home the host's gut provides, while the host benefits from a range of functional and developmental effects induced by the bacteria. As we shall see, these bacteria not only aid the host's digestion, they also play an important role in the development of the host's intestine, capillary networks, and even the immune system and the brain.

Introduction to the gut microbiota

It is estimated that 90% of the cells in our bodies are prokaryotic. While it is difficult to determine the exact number, it is believed that the human intestinal tract may contain up to 100 trillion (10^{14}) microorganisms—about 10 times as many cells as the eukaryotic "human" cells that make up our bodies (Bäckhed et al. 2005; Ley et al. 2006a). The human gut is home to a consortium of roughly 500–1000 different species of bacteria. It is difficult to obtain a more precise estimate, in part because many human gut bacteria have thus far resisted cultivation in vitro; indeed, only about 7% have been successfully cultured in the laboratory* (Gordon et al. 2005; Rawls et al. 2006).

Compared with the microbial communities found in many other ecosystems, the bacteria of the mammalian gut are not very diverse. Only eight of the known phyla of bacteria are represented in the human gut and of these eight, species from only two divisions dominate the gut community (Bäckhed et al. 2005). These two dominant divisions are the Firmicutes (such as the *Bacillus* and *Listeria* genera) and the Cytophaga-Flavobacterium-Bacteroides complex; the genus *Bacteroides* alone (including *Bacteroides fragilis* and *B. thetaiotaomicron*) accounts for roughly 25% of the bacterial population of the human gut (Xu and Gordon 2003; Bäckhed et al. 2005; Ley et al. 2006a). Although the human gut bacteria are not terribly diverse on the phylum level, their diversity increases dramatically if analyzed on the species or subspecies (strain) level. The high species- and strain-level diversity found in the mammalian gut is indicative of a degree of selective pressure for bacteria both to occupy unique niches and to maintain overlapping functions with other members of the gut community; this point will be discussed in more detail later in the chapter.

Maintaining the gut microbial community: The biofilm model

As we will see, bacteria induce gene expression in the gut tissue, and the gut tissue reciprocates by inducing new gene expression in the microbes.

---

*The weight of the human gut microbiome is about 2 kg (4.4 lb), which is heavier than the brain. It is as metabolically active as the liver and contains 100-times more different genes than the human genome.

Both the gut and the microbial cells display phenotypic plasticity. To maintain their presence in the intestinal microbiome, the gut bacteria form **biofilms**, interconnected microbial communities that can resist the shearing forces that threaten to dislodge them and sweep them away (Sonnenburg et al. 2004). Individual bacteria utilize different genes and acquire new properties when they become part of a biofilm (Walter et al. 2003; Macfarlane et al. 2005). These genes are induced by the gut epithelia and mucus, allowing the bacteria to adhere together and to change their nutrient uptake.*

The gut microbes' use of intestinal mucus as an attachment matrix helps to explain the variable composition of the gut bacteria in different regions of the intestine. The intestinal mucus varies in thickness and composition in different regions of the gut, and thus different bacteria are able to form biofilms in different areas. This bacterial segregation allows for specialization. In one region of the gut, for example, various *Bacteroides* species produce products that are further metabolized by *Escherichia coli*, another gut inhabitant, in a more distal region of the gut (Sonnenburg et al. 2004). Interestingly, the *Bacteroides* themselves are able to metabolize the product they produce, but because the product is washed through the gut lumen along with the rest of the partially digested food while the *Bacteroides* remain attached to the slower-moving mucosal matrix, they are unable to access the product. That the *E. coli* bacteria farther along in the gut make use of the metabolite demonstrates how the differential characteristics of mucus in different regions of the gut help to maintain gut bacterial diversity.

Inheritance of the gut bacteria

The mutualistic gut bacteria of mammals are passed from host to host through some form of horizontal transmission. However, this creates a paradox, since observations and experiments in both mice and humans have indicated that the gut community is inherited. Related mice were found to have more similar gut communities than unrelated mice. Similarly, in humans, marriage partners were found to have less similar gut communities than were monozygotic twins, dizygotic twins, and normal siblings (reviewed in Bäckhed et al. 2005 and Ley et al. 2006a).

While inheritance has thus been shown to have a greater impact on the gut community than does the host's environment as an adult, the irrelevance of the type of sibship (monozygotic vs. dizygotic twins vs. non-twin siblings) to the similarity of the siblings' gut communities suggests that it is the early environment of the host, rather than the host's genotype, that has the greatest impact on the makeup of the gut bacterial community. It is currently believed that human babies acquire the beginnings of such communities largely from the reproductive and digestive tracts of their

---

*Interestingly, the presence of biofilms in the human appendix, and the location and narrow aperture of this structure, have led Bollinger and colleagues (2007) to suggest that the appendix is not a vestigial organ, as has long been supposed, but a very highly adapted structure that serves as a reservoir of symbiotic bacteria. In cases of diarrhea or other severe evacuation of the gut, the appendix would be able to maintain its bacterial population to serve as a culture that can repopulate the bowel during its recovery.

Of Colons and Colonization

When John Donne famously wrote, "No man is an island," he was fully correct from the sociological perspective. But from the bacterial perspective, a man is a remarkable island, and the rules of island colonization hold for bacteria (Costello et al. 2012). Those bacteria that arrrive first get a wide choice of options, and they restrict the conditions for the next wave of settlers.

It has long been assumed (Tissier 1900) that the fetus develops within a sterile environment, and that when the amnion (water sac) bursts during labor, the colonization could begin. The first microbes would reach the fetus as it was being born. These would be the resident microbes of the birth canal. Then, microbes from the mother's breast and skin would be in line for colonizing the newborn baby. However, we now know that a large percentage of the metabolites in a pregnant woman's blood come from symbiotic bacteria, so the fetus is not separated from the influences of microbes as it is developing. Moreover, there is new evidence that intestinal microbes from the mother actually gain entry into the fetus, bypassing the placental barriers. Strangely, bacteria have been found in normal amniotic fluid, in the umbilical cord, and in infants' first (intrauterine) bowel movements (see Funkhouser and Bordenstein 2013; Gilbert 2014). This would indicate that the bacteria were already in the fetus before birth!

To experimentally observe whether maternal gut bacteria could be transferred into the fetus, Jiménez and colleagues (2008) fed pregnant mice water that had been inoculated with genetically labeled *Enterococcus faecium* bacteria (Figure A). A day before the litters were to be naturally born, the researchers performed Caesarean sections on the mothers, delivering the pups aseptically. They found not only that the pups' first bowel movements contained bacteria, but also that some of the bacteria had the transgenic label that could only have been received through the gut of their mothers (Figure B).

How can bacteria get through the mother's gut wall, the placenta, and the amnion and into the fetal gut? One hypothesis focuses on the *dendritic cells* of the gut (Donnet-Hughes et al. 2007; Funkhouser and Bordenstein 2013). These are cells that ingest bacteria and bring the bacteria to the lymphocytes of the gut to initiate the immune response. (This is how the lymphocytes are presented with the bacteria—in the context of dendritic cell membranes.) Depending on how the bacteria are presented, the immune system may react against them or recognize them as symbionts. Dendritic cells are able to transport live bacteria from the gut to other places

of the body, and it is possible that they could bring live bacteria into the fetus. The proteins that regulate immediate ("innate") immune responses against bacteria in the adult are the Toll-like receptors. Interestingly, the activity of these receptors appears to be downregulated in the fetus. The amniotic fluid, in addition to providing suspension and anti-dessication protection to the embryo, also contains large concentrations of epidermal growth factor. This protein inhibits the function of the Toll-like receptor. So while the early digestive tract is being bathed by amniotic fluid (the mouth and anus are open and exposed to amniotic fluid), the bacteria might be accepted as colonizers (Good et al. 2012).

Once the amnion has broken, however, the fetus is exposed to a wide variety of microbes, mostly from the gut (although bacteria have now been found on normal placentae; Aagaard et al. 2014). These bacteria appear to be very important, as human babies born through Caesarean section (i.e., not passing through the birth canal) have a different bacterial colonization pattern early in life than vaginally delivered babies do (Ley et al. 2006a; Makino et al. 2013). It takes over a year for the babies born by C-section to develop a similar bacterial profile, and during this time, they have a lower microbial diversity, delayed colonization of important microbes (such as *Bacteroides*), and reduced lymphocyte responses (Guarner and Malagelada 2003; Jakobsson et al. 2014). The ability of the fetus to decide which bacteria stay and which ones must be excluded is still very much a mystery. The process of colonization appears to involve many of the same molecules that are usually used to attack bacteria. But it seems that at the core of either acceptance or rejection is recognition. The symbiotic bacteria appear to be recognized by the same sets of molecules usually used to attack bacteria, but the symbionts appear to have certain compounds on them that turn this recognition into acceptance rather than attack, and in this particular context, these bacteria are actually encouraged to settle into our guts* (Chu and Mazmanian 2013; Lee et al. 2013).

In mice, the bacterial profile of the gut on day 1 looks very much like the vaginal and gut microbiota of the mother. But as the mice are fed milk, the Lactobacilli rise

*In vitro studies have also shown that commensal bacteria can change expression of genes involved in innate immunity by causing the methylation of these genes' promotor regions. This would prevent them from making compounds that would ultimately destroy the bacteria (Takahashi et al. 2011).

Of Colons and Colonization *(continued)*

(A)

(B)

Experimental protocol and results showing that bacteria from the gut of mother mice can be found in the fetuses. (A) Experimental protocol wherein pregnant mice were fed bacteria (*Enterococcus faecium*) having a particular genetic sequence from a soy gene as a marker. The pups were isolated just before birth, and their guts were tested for the presence of the labeled bacteria. (B) The genetic sequence was found in the guts of the fetuses whose mothers had the labeled bacteria in their guts, but not in the guts of the mice whose mothers were not fed that bacteria. The primers amplified the soy transgene in the bacterium. Lane 1 shows the 100 base pair (bp) size markers for reference. Lane 2 shows the positive control (genomic DNA obtained from the soy); lanes 3–7, colonies obtained from mice whose mothers had been fed the bacteria containing the transgene; lanes 8–12, colonies obtained from mice whose mothers had not been fed that labeled bacteria; lane 13, negative control. (B after Jiménez et al. 2008.)

substantially. Upon weaning and solid food ingestion, the bacterial population changes again, to the more adult form (Pantoja-Feliciano et al. 2013). Interestingly, the types of bacteria allowed to colonize depend on (1) the prevalence of a particular species of microbe in the environment, (2) which microbes have already entered the gut, (3) the genetics of the digestive tract, and (4) the diet (Nicholson et al. 2012; Pacheco et al. 2012; Kashyap et al. 2013).

And one of the most interesting components of a newborn's diet is mother's milk. Surprisingly, some of the complex sugars found in human mother's milk are not digestible by the infant. Rather, they serve as food for certain *bacterial symbionts* that help the infant's body develop (Sela et al. 2011; Zivkovic et al. 2011; Underwood et al. 2013). Indeed, Bifidobacteria have an entire set of genes for the digestion of such milk sugars, and Bifidobacteria help structure the early gut community in a healthy manner (Makino et al. 2013). Not only do they feed the "good guys" and not the "bad guys," the

(Continued on next page)

Of Colons and Colonization *(continued)*

sugars in mother's milk "prevent the pathogens from latching onto healthy cells, routing trouble-makers into a dirty diaper instead" (Bode 2012; Manthey et al. 2014; quotation from Shugart 2014). Although each baby starts with a unique bacterial profile, within a year the types and proportions of bacteria have converged to the adult human profile that characterizes the human digestive tract (Palmer et al. 2007).

The importance of mother's milk in primates was recently shown in an experiment where laboratory-reared macaque monkeys were either fed from a bottle or given to the mother to nurse. This study found that the breast-fed infant monkeys had a remarkably different gut microbiota than did the formula-fed monkeys (Figure C; Ardeshir et al. 2014). The gut bacteria in the breast-fed macaques produced compounds such as arachidonic acid, which induced the presence of populations of helper T lymphocytes that were absent in the formula-fed macaques. These helper T lymphocytes are those that protect the newborn against *Salmonella* and other pathogens.

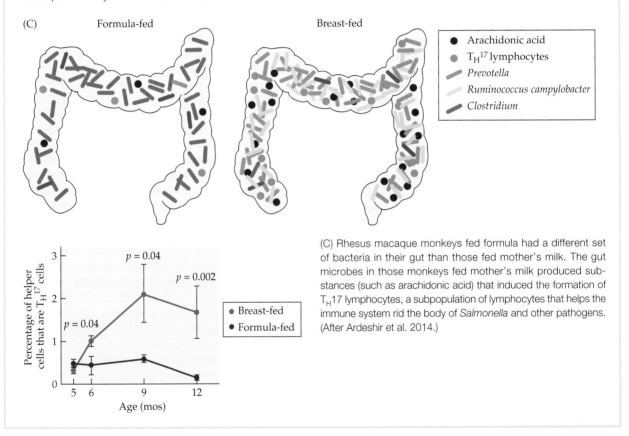

(C) Rhesus macaque monkeys fed formula had a different set of bacteria in their gut than those fed mother's milk. The gut microbes in those monkeys fed mother's milk produced substances (such as arachidonic acid) that induced the formation of T_H17 lymphocytes, a subpopulation of lymphocytes that helps the immune system rid the body of *Salmonella* and other pathogens. (After Ardeshir et al. 2014.)

mothers. In support of this idea, babies born through Caesarean section have been found to have an altered bacterial colonization pattern early in life compared with vaginally delivered babies (Ley et al. 2006a). It appears, then, that the gut microbiota are an epigenetic inheritance from the mother.

Although early environment is most important in determining the makeup of the gut community, vast differences in genotype do influence this composition. Transplantation experiments (Rawls et al. 2006) using mice and zebrafish in which a germ-free mouse (i.e., a mouse raised so that it had no microbes of any kind) was inoculated with gut bacteria from a conventionally raised zebrafish, and vice versa, showed that the lineages that take up residence in the new host are limited to those provided through inoculation (i.e., from the environment). However, the new host will remodel the proportions of the different lineages present to approach the population normally found in its species (i.e., with its genotype). Additionally, bacterial sublineages present in the donor host population but not normally present in the new host will often be eliminated. It is thus clear that large differences in genotype do indeed affect the makeup of the gut microorganism community but that the type of bacteria that the host is exposed to also has a vital, and in many cases overwhelming, influence on the community's final composition.

Gut Bacteria and Normal Mammalian Development

The enteric bacteria have been shown to induce or regulate the expression of many genes in the gut. In other words, products of the bacterial cells can induce gene expression in the mammalian intestinal cells. Moreover, this gene expression is essential for the normal development of the gut. Bacteria-induced expression of mammalian genes was first demonstrated by Umesaki (1984), who noticed that a particular fucosyl transferase enzyme characteristic of mouse intestinal villi was induced by bacteria.* More recent studies have shown that the intestines of germ-free mice (i.e., mice bred in sterile facilities and having no contact with bacteria or fungi) can initiate, but not complete, their differentiation. For complete development of the mouse gut, the microbial symbionts are needed (Hooper et al. 1998). Microarray analyses of mouse intestinal cells have shown that normally occurring gut bacteria can upregulate the transcription of numerous mouse genes, including those encoding colipase, which is important in nutrient absorption; angiogenin-4, which helps form blood vessels; and Sprr2a, a small, proline-rich protein that is thought to fortify matrices that line the intestine (Figure 3.12; Hooper et al. 2001). Thus, the "normal" amount of gene expression in the gut is that which is regulated by the microbes. In others mammals as well as mice, bacteria are expected to be in the gut, and not having bacteria is like having a loss-of-function mutation. In human cultured colon cells, bacteria (through their flagellin proteins) are responsible for activating the synthesis of transcription factors that are necessary for several intestinal functions (Becker et al. 2013). Thus, we develop as holobionts.

---

*This complex pathway of induction changes the sugar residues of the intestinal epithelial cells and enables symbiotic bacteria to utilize the cell-surface polysaccharides as food. In contrast to the mutualistic symbionts, most pathogenic bacteria lack the enzymatic ability to utilize these modified carbohydrates This is thought to help recruit and retain the symbionts (Marcobal et al. 2013; Goto et al. 2014).

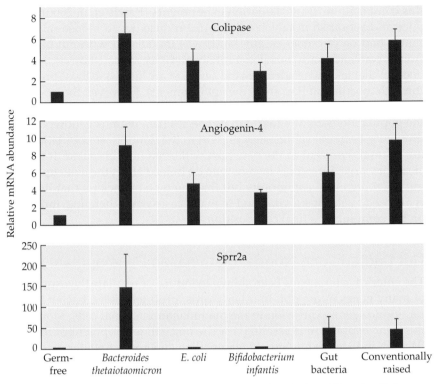

Figure 3.12   Induction of mammalian genes by symbiotic microbes. Mice raised in "germ-free" environments were either left alone or inoculated with one or more types of bacteria. After 10 days, their intestinal mRNA was isolated and tested on microarrays. Mice grown in germ-free conditions had very little expression of the genes encoding colipase, angiogenin-4, or Sprr2a. Several different bacteria—*Bacteroides thetaiotaomicron*, *Escherichia coli*, *Bifidobacterium infantis*, and an assortment of gut bacteria harvested from conventionally raised mice—induced the genes for colipase and angiogenin-4. *B. thetaiotaomicron* appeared to be totally responsible for the normal levels of angiogenin-4 and colipase mRNAs. This ecological relationship between the gut microbes and the host cells could not have been discovered without the molecular biological techniques of polymerase chain reaction and microarray analysis. (After Hooper et al. 2001.)

Mammals aren't the only vertebrates whose gut and immune system development depend on microbial symbionts. Zebrafish guts have a vast assortment of microbes (see Jemielita et al. 2014), and these microbial symbionts use the β-catenin signaling pathway to initiate cell division in the intestinal stem cells (Rawls et al. 2004; Figure 3.13). Without this stem cell division, germ-free mice have smaller intestines, with a paucity of enteroendocrine and goblet cells (Bates et al. 2006).

Symbionts and induction of gut blood vessels

Stappenbeck and colleagues (2002) have demonstrated that in the absence of particular intestinal microbes, the capillaries of the small intestinal villi fail to develop their complete vascular networks. *Bacteroides thetaiotaomicron*

(A) Germ-free

(B) Germ free + microbes

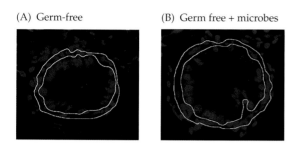

(C)

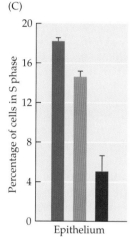

Figure 3.13 Microbial symbionts necessary for stem cell division and epithelial cell formation in zebrafish. Intestines of germ-free fish (A) and germ-free fish given normal microbes (B) were stained for nondividing cells (blue) and dividing cells (magenta). The mesenchyme and muscles outlining the gut are in white. (C) S-phase (dividing) intestinal epithelial cells in germ-free, conventionally raised, and germ-free plus added microbes specimens show significantly greater proliferation with the bacteria. (After Rawls et al. 2004.)

is vital for the induction of host angiogenesis (blood vessel formation). They demonstrated that mice raised so that they have no bacteria in their gut (germ-free mice) have reduced gut capillary network formation compared with normal mice. Colonization of the germ-free mouse intestine either with a sample of bacteria from the gut of a conventionally raised mouse or with *B. thetaiotaomicron* alone, however, completed the capillary network formation of the mice within 10 days (Figure 3.14). This result was not dependent on the age of the mouse; even adult mice were able to quickly complete capillary growth in the presence of the required symbiotic bacteria.

The experiment by Stappenbeck and colleagues also investigated the mechanism through which *B. thetaiotaomicron* was able to induce angiogenesis in its host. They determined that the Paneth cells—critical gut cells found in the intestinal crypts (at the base of the intestinal glands)—were required for the induction of the capillary network. In mice without Paneth cells, the capillary network failed to form properly even after inoculation with *B. thetaiotaomicron* or conventional gut bacteria. Other experiments showed that the Paneth cells were responding to *B. thetaiotaomicron* by transcribing the gene encoding angiogenin-4, a protein known to induce blood vessel formation (Hooper et al. 2001, 2003; Crabtree et al. 2007). These experiments made it clear that the presence of a specific environmental factor (in this case, *B. thetaiotaomicron*) is necessary to induce the expression

(A)   (B)   (C)

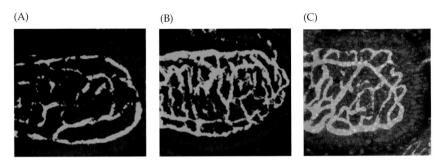

Figure 3.14   Gut microbes are necessary for mammalian capillary development. (A) The capillary network (green) of germ-free mice is severely reduced compared with (B) the capillary network in those same mice 10 days after inoculation with normal gut bacteria. (C) The addition of *Bacteroides thetaiotaomicron* alone is sufficient to complete capillary formation. (From Stappenbeck et al. 2002.)

of important factors necessary for the completion of capillary network formation in the host's gut.*

Symbionts and maturation of the GALT

Intestinal microbes also appear to be critical for the maturation of the mammalian gut-associated lymphoid tissue (GALT). GALT mediates mucosal immunity and oral immune tolerance, allowing us to eat food without making an immune response against it (see Rook and Stanford 1998; Cebra 1999; Steidler 2001). When introduced into germ-free rabbit appendices, neither *Bacteroides fragilis* nor *Bacillus subtilis* alone was capable of consistently inducing the proper formation of GALT. However, the combination of these two common mammalian gut bacteria consistently induced GALT (Figure 3.15; Rhee et al. 2004). The major inducer appears to be the bacterial protein polysaccharide-A (PSA), especially that encoded by the genome of *Bacteroides fragilis*. The PSA-deficient mutant of *B. fragilis* is not able to restore normal immune function to germ-free mice (Mazmanian et al. 2005). Thus, it appears that a bacterial compound plays a major role in inducing the host's immune system. Exposure to microbes early in life prevents the development of the T lymphocytes associated with allergies and inflammatory bowel disease, while it enhances the helper T-cell repertoire. Germ-free mice have an immunodeficiency syndrome, and the full complement of T lymphocytes is made possible only with the host's species-specific microbes (Niess et al. 2008; Duan et al. 2010; Chung et al. 2012; Olszak et al. 2012). Similarly, microbial colonization is necessary for the development of B-cells in the intestinal mucosa (Wesemann et al. 2013).

*Angiogenin-4 has another role as well. It is bacteriocidal for *Listeria monocytogenes* and *Enterococcus faecalis*, killing 99% of these bacteria within 2 hours. These bacteria are pathogens for mammals, and they are also competitors for *Bacteroides*. So while *Bacteroides* helps us make gut capillaries, we help *Bacteroides* by getting rid of its major competitors (Hooper et al. 2003). Molecules such as angiogenin-4 that are able to kill Gram-positive bacteria may serve to structure the bacterial community in the natural setting. This would be analogous to the roles that flavonoids from plant roots have in structuring the microbial communities of the rhizosphere.

(A)   (B)   (C)

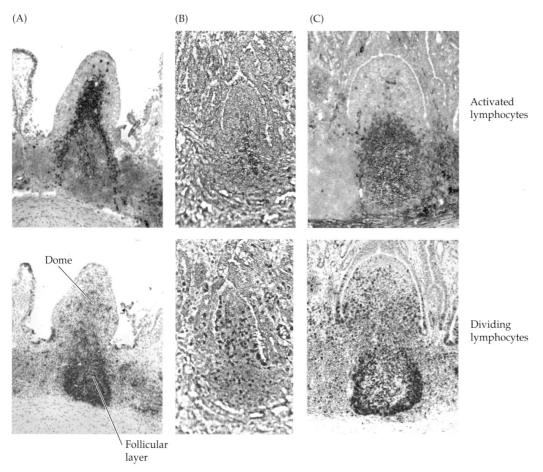

Activated
lymphocytes

Dome

Dividing
lymphocytes

Follicular
layer

Figure 3.15 Gut-associated lymphoid tissue (GALT) in serial sections of rabbit intestinal tissue. (A) Normal intestinal GALT, showing a dome of antibody-secreting lymphocytes and an activated follicular layer of dividing lymphocytes. (B) The GALT is absent in a germ-free rabbit intestine treated with bacterial protein A. (C) Partially reconstituted GALT results when *Bacteroides fragilis* plus *Bacillus subtilis* are placed into the intestines of germ-free rabbits. (From Rhee et al. 2004; photographs courtesy of Ki-Jong Rhee.)

So symbiotic bacteria are critically important in the differentiation of the lymphocytes of the mammalian immune system.

Bacteria, brain, and behavior

Although it may sound like science fiction, there is now evidence that symbiotic bacteria stimulate the postnatal development of the mammalian brain (Lorber 2005; Wang and Kasper 2014). Germ-free mice have lower levels of NGF1-A and BDNF in relevant portions of their brains than do conventionally raised mice (Figure 3.16). This correlates with behavioral differences between groups of mice, leading Diaz Heijtz and colleagues (2011) to conclude, "During evolution, the colonization of gut microbiota has become integrated into the programming of brain development, affecting motor control and anxiety-like behavior." In another investigation, a

Figure 3.16 *NGF1-A* expression in mice depends on symbiotic microbes. (A) In situ hybridization of *NGF1-A* mRNA in a section through the frontal cortex of the brain, showing high levels of *NGF1-A* in mice that have conventional microbes compared with those remaining germ-free. (B) Quantitation using radioactive probes shows the symbiont-containing mice had significantly higher levels of *NGF1-A* mRNA expression in the frontal cortex and anterior olfactory region. (After Diaz Heijtz et al. 2011.)

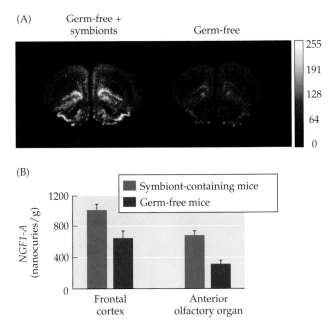

particular *Lactobacillus* strain has been reported to help regulate emotional behavior through a vagus nerve–dependent regulation of GABA receptors (Bravo et al. 2011; Forsythe et al. 2014). Thus, there may be a microbiota-gut-brain axis wherein products made by bacteria can help regulate the development of the brain (Cryan and Dinan 2012; McLean et al. 2012; Mayer et al. 2014).

Remarkably, the gut microbiota were shown to be critical for normal *social behavior* in mice. Germ-free mice have aberrant behaviors, including excessive time spent in repetitive self-grooming, social avoidance, and very little time spent in social investigation. Desbonnet and colleagues (2014) remarked that these traits appeared to be similar to those of autistic children. Moreover, many of these behavioral traits were made normal by providing the mice with gut bacteria early in life. Within the same year, Hsaio and colleagues (2013) showed that the symbiotic bacterium *Bacteroides fragilis* ameliorated the aberrant communicative, anxiety-like, and stereotypic behaviors seen in another mouse model of autism (Figure 3.17). Moreover, it did so as it altered the composition of gut microbiota, improved the integrity of the gut epithelium, and reduced the leakage of particular gut metabolites into the blood. Several investigators had seen that a subset of autistic children had altered gut bacteria, and the germ-free mice had a similarly skewed pattern of microbial species. Once the integrity of the gut epithelium had been restored and the population of bacteria changed, different metabolites were seen in the blood. Levels of indolepyruvate and ethylphenylsulfate (a chemical that induces anxiety-like behaviors) plummeted. Microbial (probiotic) treatment has sometimes been beneficial in treating psychological distress and chronic fatigue symptoms in humans

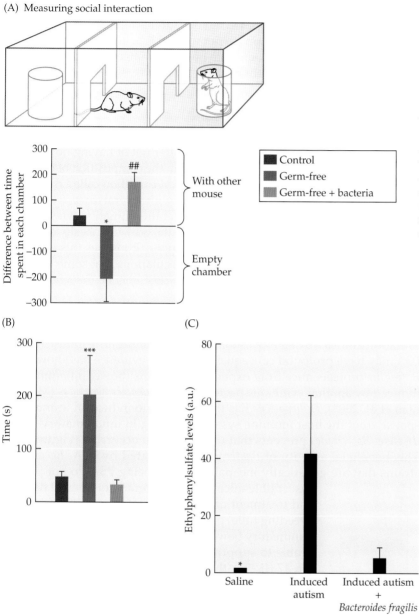

(A) Measuring social interaction

Figure 3.17 Effects of germ-free rearing and germ-free bacterial consortium on social behaviors in male mice. (A) Measuring social interaction by the difference between times spent in chambers with and without other mice. Germ-free mice had social avoidance. *, $p < 0.05$ versus controls; ##, $p < 0.001$ versus germ-free mice. (B) Time spent in repetitive self-grooming. Both autistic-like behaviors could be relieved by adding microbes later in life. ***, $p < 0.001$ versus controls. (C) Ethylphenylsulfate levels are extremely high in germ-free mice and can also be removed when epithelial integrity is restored by the addition of *Bacteroides fragilis*. *, $p < 0.05$. (A,B after Desbonnet et al. 2013; C after Hsaio et al. 2013.)

(Messaoudi et al. 2011; Rao et al. 2009). So it is possible that several of the diseases that we have come to classify as psychological have a root in bacterial metabolism. Moreover, gut microbes are responsible for controlling the permeability of the blood-brain barrier, the tight lining of the capillaries feeding the brain. This barrier shields the brain from blood-borne infections and toxins, since most large molecules cannot flow through it. However, the blood-brain barrier of germ-free mice is not impermeable to proteins, both as pups and adults. This defect can be cured by adding normal gut bacteria back to the gut of the pups (Braniste et al. 2014). It appears, then, that material from the mother's symbiotic gut bacteria are regulating the permeability of the blood brain barrier while it is being formed in the fetal mice. It is likely that neither our brain nor our behaviors develop properly without the appropriate symbionts.

Gut Bacteria and Human Health

As we have seen above, the composition of gut bacteria can have dramatic health consequences. Health becomes a matter not of having no bacteria, but of having the right bacteria. Imbalance in the composition of our symbiotic microbes is a newly recognized medical condition called **dysbiosis**, and such perturbations are being associated with several diseases. The role of gut bacteria in human obesity and bowel inflammation has been the subject of much recent study, in response to the growing obesity epidemic in industrialized countries and to the many health issues this epidemic magnifies. In addition, there have been speculations that some diseases, including asthma and rectal and gastric cancers, may be the result of improper populations of gut microbes.

Bacterial regulation of the immune response

The ability of *Bacteroides fragilis* to regulate the immune response though PSA may have profound consequences. Recent evidence has shown that *B. fragilis* protects mice from experimental colitis (bowel inflammation) induced by another symbiotic bacterium, *Helicobacter hepaticus* (Mazmanian et al. 2008). *H. hepaticus* has the potential to induce an immune response from the host immune system, resulting in inflammatory bowel disease. *Bacteroides* prevents that response from occurring (Figure 3.18). This beneficial activity of *Bacteroides* is mediated by PSA. *Bacteroides* strains that are genetically incapable of producing PSA do not protect susceptible mice against inflammatory bowel disease, and treatment of susceptible mice with PSA (through a feeding tube) prevents *Helicobacter* from inducing inflammatory bowel disease. PSA secreted by *Bacteroides* is able to suppress the inflammatory factor, interleukin-17 (IL-17), that is induced by the *Helicobacter*. The mechanism for this suppression is induction of gut CD4+ T lymphocytes that express another interleukin, IL-10, that blocks the inflammatory immune response

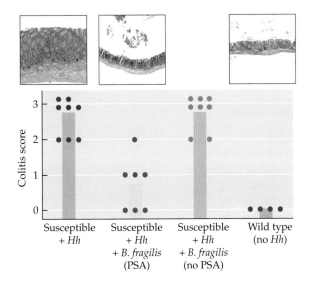

Figure 3.18 *Bacteroides fragilis* PSA protects mice from experimentally induced colitis. Susceptible mice infected with *Helicobacter hepaticus* (*Hh*) have colitis (inflammatory bowel disease), whereas wild-type mice not given *H. hepaticus* do not show disease. The incidence and severity of colitis decreases when the susceptible mice are given *B. fragilis* along with *H. hepaticus*. However, if the *B. fragilis* bacteria do not make PSA, they have no effect, and colitis ensues. (After Mazmanian et al. 2008.)

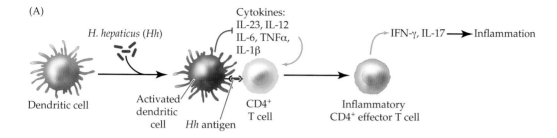

(A)

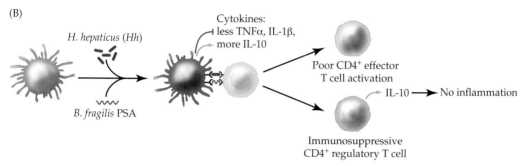

(B)

Figure 3.19 Activation of inflammatory proteins by *Helicobacter hepaticus* and their suppression by *Bacteroides fragilis*. (A) In mice that develop colitis, dendritic cells activated by *H. hepaticus* (*Hh*) produce cytokines that induce an inflammatory reaction. They also present *H. hepaticus* antigen to CD4$^+$ T cells, causing them to differentiate into inflammatory CD4$^+$ T cells that secrete the inflammatory cytokine IL-17. (B) Dendritic cells jointly stimulated with *H. hepaticus* and *B. fragilis* PSA secrete less inflammatory cytokines (such as TNFα and IL-1β) and synthesize more of the anti-inflammatory IL-10. They present both *H. hepaticus* antigen and PSA to CD4$^+$ T cells and may either produce poor T cell activation or induce the differentiation of CD4$^+$ T cells that produce IL-10. (After Kulberg 2008.)

(Figure 3.19). Mazmanian and colleagues (2008) conclude that human health depends on the proper relationship between the human hosts and their hundreds of microbial species. Moreover, they note that the finding that a single molecule from a single species of symbiotic bacteria promotes mammalian health "may provide a platform for the development of therapies based on the fundamental relationship between humans and their beneficial microbial partners."

Bacteria and the development of obesity

In the United States, 64% of the adult population is overweight or obese (Makdad et al. 2004). Clearly, there are many factors, both innate and external, that contribute to this obesity epidemic; some of the obvious ones include diet, reduced physical activity, and genes. Recently, however, yet another important environmental contributor to obesity has been discovered: the microbes living in our gut. A study by Bäckhed and colleagues (2004) showed that, in one mouse strain, adult male germ-free mice had a stunning 42% less body fat than their conventionally raised counterparts, even though the conventionally raised mice ate 29% less food per day.

Furthermore, when adult germ-free mice were colonized with conventional mouse gut bacteria, the animals underwent a whopping 57% increase in body fat in just 14 days, even though they were eating 27% less food than they had been eating prior to colonization.

Intriguingly, this huge increase in body fat content was not associated with differential weight gain in the mice, but instead with a decrease in the "lean" (nonfat) body mass; a greater percentage of the new body weight was fat. In addition, conventionalized mice (germ-free mice inoculated with normal gut bacteria) had increased metabolic rates—27% higher than those of their germ-free counterparts (Bäckhed et al. 2004; Wolf 2006).

The mechanisms by which the gut bacteria induce fat synthesis are just beginning to be elucidated. They include the introduction of specific bacteria that help the body gain or lose fat (Bäckhed et al. 2004), the interaction between members of the domains Archaea and Bacteria (Samuel and Gordon 2006), and the interaction between diet and bacteria (Turnbaugh et al. 2006). An important set of interactions involves the interplay of the microbes with diet and the host genome. In fat mice, the relative abundance of the two most common bacterial groups in the mammalian gut, *Bacteroides* and Firmicutes, was found to be altered such that the population of *Bacteroides* was decreased by 50% while that of Firmicutes increased by the same amount in Obese (leptin-deficient) mice as compared with normal mice (Ley et al. 2005; Turnbaugh et al. 2006; Figure 3.20). A similar correspondence was noted in dieting humans: as individuals lost weight, the relative abundance of members of the *Bacteroides* lineage in their guts increased (Ley et al. 2006b). When mice were made obese by feeding them a "prototypical Western diet," the high-calorie food caused a "bloom" in a single, formerly unknown, species of Firmicutes; low-calorie diets decreased the numbers of this species. Moreover, when bacteria from the intestines of mice fed high-calorie diets were transferred into lean germ-free recipients, these

Figure 3.20 Mice with a mutation in their leptin genes are genetically obese. Their gut microbes are also different from those of wild-type mice, having 50% more firmicutes and 50% less *Bacteroides*. The mix of microbes in the intestines of the obese mice is more effective at releasing calories from food.

conventionalized mice stored more fat than did mice receiving microbe transplants from mice fed low-calorie diets (Turnbaugh et al. 2008).

Experiments in germ-free mice demonstrated that certain bacteria may be major contributors to weight loss or gain. When germ-free mice are given a microbiome through a fecal transplant from a human donor, they get fatter within two days, because the bacteria help them digest their food properly. Goodrich and colleagues (2014) found that if *Christensenella minuta* were added to the feces of an obese human donor, the recipient mice became thinner than when this bacteria was not added. Moreover, thin people have a greater abundance of *C. minuta* in their gut microbe population than obese people.

This means that a high-fat diet during the last stages of pregnancy might alter the bacteria that colonize the newborn. Ma and colleagues (2014) have shown that in the Japanese macaque monkey, a high-fat maternal diet structures the offspring's gut microbiome. This microbiome, in turn, affects the metabolic health of the offspring, causing it to gain fat more rapidly.*

One of the most exciting stories of bacterial-diet interaction has come from studies of kwashiokor. This is a disease of severe acute malnutrition, characterized by generalized edema (the body swells as the capillaries become more permeable), skin ulcerations, fatty liver, and anorexia. It has been known for decades as a tropical malnutrition disorder. But new research (Smith et al. 2013; Trehan et al. 2013) has shown that it is a disease of dysbiosis. Smith and colleagues (2013) studied 317 pairs of identical twins in which one of the pair had developed kwashiorkor while the other twin had not. They found that the twin who developed kwashiorkor had different gut bacteria than the twin without kwashiorkor, and that this disease was caused by a combination of bad nutrition and bad bacteria. Kwashiorkor could actually be transmitted to germ-free mice by giving them the affected twins's bacteria as well as a poor diet. This disease did not occur when the bacteria were transferred from the healthy twin. Moreover, replacing the bacteria with "good bacteria" ameliorated the disease, and this was thought to be because of the "microbial-host co-metabolism" of the diet.

Bacterial restoration of bone density

One never stops developing. One's skeleton is being remodeled constantly by the osteoclasts (cells that degrade bone) and the osteoblasts (cells that add new bone). During menopause or other times of estrogen reduction in women, the balance of development favors the osteoclasts, whose activity

---

*Which leads to a fascinating study on diet aids. Noncaloric artificial sweeteners (such as saccharine and aspartame) may have zero calories, but they may change the microbial population in our gut in ways that *prevent* weight loss. Suez and colleagues (2014) found that mice given the artificial sweeteners did not lose weight. They gained the same weight as those mice fed glucose. Moreover, the mice fed artificial sweeteners had statistically *higher* amounts of glucose in their blood than the mice that were actually fed glucose. This was demonstrated to be due to the sweetener's ability to change the gut microbiota. The human data were not as consistent, as the mice were genetically identical while the humans had varied genotypes and varied responses to the artificial sweeteners.

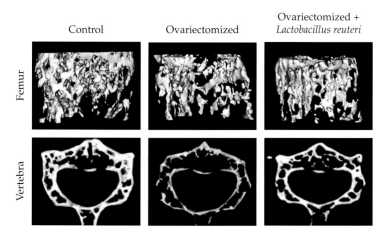

Figure 3.21   Tomographic density scans of the femur and third lumbar vertebra in control female mice, ovariectomized mice, and ovariectomized mice given *Lactobacillus reuteri* through a gastric tube for 4 weeks. The *L. reuteri* prevented much of the bone loss associated with estrogen deficiency. (From Britton et al. 2014.)

causes osteoporosis (bone density loss). Certain gut bacteria, however, can offset the loss of estrogen. In ovariectomized mice that no longer make estrogen, the addition of several different *Lactobacillus* species (which are normal gut symbionts) to their diet caused a dramatic restoration of bone density. The compounds made by *L. reuteri* appear to prevent the development of osteoclasts and the subsequent bone resorption. It has been suggested that *L. reuteri* be used as a relatively cheap probiotic for menopausal women (Figure 3.21; Britton et al. 2014; Ohlsson et al. 2014)

Bacteria and accommodation of pregnancy

One of the most exciting stories of development and symbiosis comes from investigations of human pregnancy. The gut bacteria change dramatically during human pregnancy. Indeed, they appear to respond to the hormonal status and help a pregnant woman adjust to the physiological stresses of carrying a fetus. When transferred into a germ-free mouse, bacteria from women in the early stages of pregnancy cause a normal phenotype to develop in the hosts. When bacteria from women late in their pregnancy are transferred into germ-free mice, the mice get fatter and display some of the metabolic changes (such as insulin desensitization) associated with pregnant women (Figure 3.22; Koren et al. 2012).

Bacterial protection against asthma and allergies

While most infectious diseases have been brought under control in industrialized countries, cases of allergies and asthma have actually increased with the advent of modern medicine (Braun-Fahrländer et al. 2002; MacDonald and Monteleone 2005). One widely disseminated hypothesis is that the increased prevalence of these conditions is in fact connected to the developments in medicine. This idea, known as the "hygiene hypothesis," states that we put ourselves at risk for allergies and asthma by removing bacteria

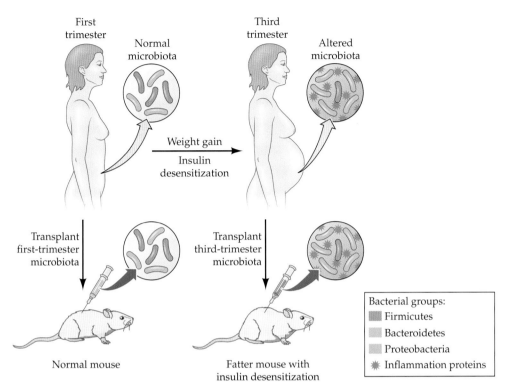

Figure 3.22 The composition of the gut microbe population changes dramatically during pregnancy. This is associated with weight gain and with the progressive insensitivity to insulin characteristic of human pregnancy. When transplanted into the guts of germ-free mice, the bacteria from women early in their pregnancy (first trimester, roughly weeks 1–12) gave a normal phenotype to the mice. However, when bacteria from women late in their pregnancy (third trimester, roughly weeks 27–40) were transplanted into the mouse gut, the bacteria induced pregnancy-like metabolism in germ-free mice, including weight gain and insulin resistance. (After Koren et al. 2012.)

from our environment. In addition, the hygiene hypothesis proposes that the level of antibiotic use prevalent in many industrialized nations may permanently disrupt the proportions of bacteria in the gut and thereby alter the direction of the host's immune responses (Noverr and Huffnagle 2004). The hygiene hypothesis is supported by studies showing that infants who go on to develop allergies later in life have statistically different enteric bacterial profiles than infants who do not develop allergies later, and that early exposure to certain microbes provides protection against having allergic responses to related microbes later in life (see Noverr and Huffnagle 2004; Debarry et al. 2007; Blaser 2014).

Antibiotics make no distinction between the good and the bad microbes. The widespread use of antibiotics has raised the alarm that "some of the useful guys are disappearing. As a consequence, human physiology is changing and therefore human health" (Blaser 2012). Cox and colleagues (2014) found that when mice received a low dose of antibiotics shortly after birth, they changed their bacterial population and gained weight. Moreover, this propensity to gain weight could be transferred through the bacteria to other mice. The new bacteria activated genes that transformed sugar into fat. It's long been known that when antibiotics are mixed into the food of pigs, cattle, and chickens, they promote faster growth and fatter bodies. (This is actually a standard agricultural practice in the United States—although banned in Europe). It is possible that the same procedure may happen in children (Blaser and Flakow 2009; Blaser 2014). Children treated with antibiotics during the first 6 months of their lives have a 22% higher probability of being obese at 3 years (Trasande et al. 2013; Blech 2012), and a comparison of the geographic data concerning obesity and antibiotic use shows stunning parallels (Figure 3.23; Petschow et al. 2013).

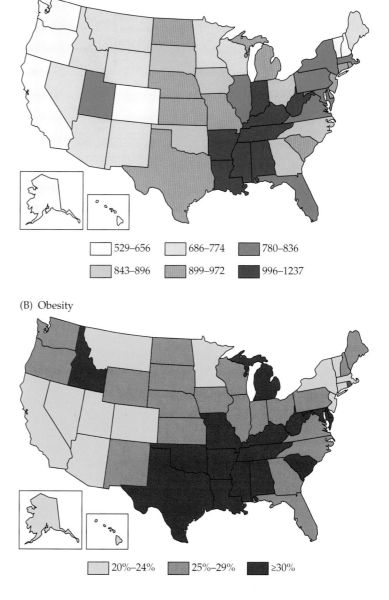

Figure 3.23   Comparisons between antibiotic use per 1000 indiviuduals (A) and the geographical distribution of obesity as a percentage of each state's population (B) in the United States, 2010. (After Petschow et al. 2013.)

Interestingly, Chen and Blaser (2007) have shown that the childhood acquisition of the *Helicobacter pylori* bacterium (which can be a dangerous pathogen, initiating gastric cancer) appears to *lower* the risk of that person developing asthma or allergies. Martin Blaser and colleagues (2008) have shown that *Helicobacter* species are relatively host-specific, and they hypothesize that for most of human history, *H. pylori* has been a normal, coevolved human gut symbiont (and that its loss in people in industrial countries is associated with acid reflux disease). Normal *H. pylori* (or a similar bacterium) might be a bacterial component regulating the immune system to block the conditions necessary for allergy and asthma induction.*

Thus, antibiotics may be destroying those microbes that have coevolved with us, which we expect for normal health and development and which afford us protection. The disappearance of ancient microbiota may be the initial cause of the "diseases of modernity."

Fecal bacteria transplantation

This opens up a new line of inexpensive therapy known as "fecal transplantation," "fecal microbiota transplation," or very simply "stool transplants." Here, dysbioses are cured by the expedient of transplanting gut bacteria from one person (often a spouse or family member) to another. The donated material is suspended in saline, checked for parasites, and administered by enema or tube into the intestine. This was first used to cure the diarrhea induced by *Clostridium difficile* infections. *Clostridium* dysbiosis probably resulted from antibiotic use in hospitals, and several particularly virulent strains have evolved. This condition kills about 300 people per day.

The first clinical trial of fecal transplantation to control *Clostridium* diarrhea had control patients being given antibiotics, whereas the experimental patients were given antibiotics and then fecal transplants. The small trial was halted ahead of schedule since the experimental patients were obviously getting better faster than the control patients (van Nood et al. 2013). Success rates have been around 90% (Kassam et al. 2013; Smith et al. 2014). Fecal transplantation has now been used to cure other dysbioses, including colitis and irritable bowel syndrome, and it is being looked at as a possible way of curing certain obese conditions (Smits et al. 2013). As in blood transfusion, though, there are dangers of transmitting infectious disease.

---

*The hygiene hypothesis claims that raising children in environments without natural symbionts (such as bacteria or parasitic worms) retards the normal development of the immune system, thereby predisposing these children to immunological diseases such as asthma, allergy, and inflammatory bowel disease. Whether or not there is a developmental window where these organisms are expected for normal development, in some cases the addition of these organisms to ill patients may help ameliorate or cure the symptoms (Elliott et al. 2000; Croese et al. 2006; Finlay et al. 2014). The ability of *H. pylori* to cause cancers in some people but not others may involve the compatibility between the strain of *H. pylori* and the host's genotype (de Sablet et al. 2011; Sheh et al. 2013). These differences may cause chronic inflammation and oxidative stress in the gut (Hardbower et al. 2013), which, as we will discuss later in the book, may predispose tissues to cancer.

Looking Ahead

It should be clear by now that most organisms are not "individuals," nor do they develop as individuals. It might be better to think of each of us as a team, or as a collection of ecosystems. In *At Home in the Universe*, Stuart Kauffman (1995) stated, "All evolution is co-evolution." We may now have to conclude that "all development is co-development."

It is safe to say that the maintenance of life is predicated on symbiotic relationships. Mycorrhizal, endophytic, and nitrogen-fixation symbioses are probably responsible for the existence of life on land. Moreover, symbiosis is an essential element of the normal development of many animals. Molecular biology has revealed that what used to be thought of as exceptions are actually the rule. Whereas microbially induced development was once seen as the province of strange arthropods and a few squids, it is now seen to be a common scheme of development. Even we mammals have "outsourced" the signals that induce normal intestinal gene expression to our gut microbes. Our life cycle is not one of a monogenetic individual. Rather, it is a holobiont life cycle—the integration of the host life cycle with that of its persistent symbionts (Figure 3.24). Thus, the notion of the "individual" becomes semipermeable. Mark Twain is quoted as saying that the only people who can use the "royal we" are kings, editors, and people with tapeworms. However, it now looks as if each of us develops as a community and can embrace that royal privilege. In *The Lives of a Cell*, Lewis

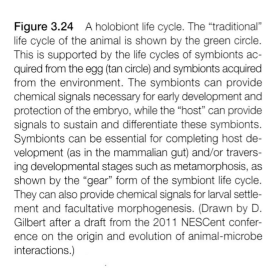

Figure 3.24 A holobiont life cycle. The "traditional" life cycle of the animal is shown by the green circle. This is supported by the life cycles of symbionts acquired from the egg (tan circle) and symbionts acquired from the environment. The symbionts can provide chemical signals necessary for early development and protection of the embryo, while the "host" can provide signals to sustain and differentiate these symbionts. Symbionts can be essential for completing host development (as in the mammalian gut) and/or traversing developmental stages such as metamorphosis, as shown by the "gear" form of the symbiont life cycle. They can also provide chemical signals for larval settlement and facultative morphogenesis. (Drawn by D. Gilbert after a draft from the 2011 NESCent conference on the origin and evolution of animal-microbe interactions.)

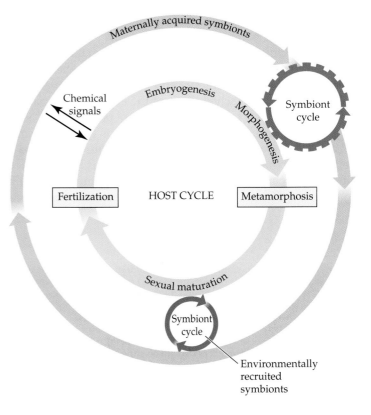

Thomas (1974, p. 142) eloquently points out the lessons of such discoveries: "This is, when you think about it, really amazing. The whole dear notion of one's own Self—marvelous, old free-willed, free-enterprising, autonomous, independent island of a Self—is a myth."

This presents us with an entirely new view of nature—a view involving holobionts—individuals made of multiple lineages. Just as the microscope introduced us to a new world of microbes, new molecular tools (including PCR, sequencing, and high-throughput RNA analysis) have introduced us to a new world of relationships. This new view of nature presents us with enormous and fundamental questions. These concern how the symbiotic microbiome is inherited, how it interacts with our nuclear inheritance, how it performs developmental tasks, and what is its medical significance. For instance, we know that (1) the majority of our resident microbes are acquired before we are three years old; (2) once acquired, most of the resident strains remain in our body for decades; (3) one can acquire new microbes from close family members; and (4) the effects of a particular strain may wait to be seen decades after its acquisition (Faith et al. 2015). So one set of new questions involves how the first colonization takes place and what the rules are for subsequent colonizers. Another set of questions involves how new probiotic bacteria can colonize in ways that might improve the health of adults. Would an entirely new microbial community be needed for them to exert healthy effects? The standardization of probiotic bacteria and probiotic bacterial communities might allow the emergence of a large area of medical practice. This use of bacteria to regulate health may be especially important if bacteria were found to induce or to be essential for certain mental states (see Gilbert et al. 2013).

The idea that early communities of probiotic bacteria might improve the health of the host has recently entered the realm of conservation biology. Certain amphibian species are being wiped out by fungal infections. However, there are symbiotic skin bacteria in some amphibians that allow the host to resist the fungus. In some endangered species, survival appears to depend largely on the initial community of symbiotic bacteria they possess (Becker et al. 2015; Yong 2015). Symbiotic protection against infectious disease may become an important way of sustaining species.

A third set of questions involve how we co-develop with the microbes. These include (1) how the resident microbes promote the formation of capillaries, gut cells, and gut lymphoid tissue; (2) if some strains are better than others at these developmental tasks; (3) if certain strains of bacteria (or sets of such bacteria) predispose us to certain diseases; (4) if there are interactions between microbial strains and certain nuclear genotypes that are pathogenic or that may provide a more fit phenotype; and (5) if there are some strains of bacteria that influence brain development and might bias behaviors.

Another set of questions involves how animals co-evolved with their respective symbiotic communities. The evolution of the symbiont community parallels the phylogenetic relationship of the host genome and closely related species can have remarkably different microbiomes (Brucker and Bordenstein 2012, Franzenburg et al. 2013). Studies are beginning to find out how this "phylosymbiosis" occurs and what its importance might be to the individual and to the species (Fraune and Bosch 2007; Chung et al. 2012). Moreover, the co-evolution can be a double-edged sword. Obligative symbiosis can open new niches. This is seen, for instance, in the ability of *Buchnera* symbionts to provide its pea aphid hosts with amino acids that are deficient in plant sap. Thus, the pea aphids containing the symbionts can exploit plant sap as a food source. However, once the aphids become adapted to using this resource, they become "addicted" to the symbionts. The pea aphids are unable to live without the symbionts, and the development of the aphid ceases if *Buchnera* is removed (Koga et al. 2007; Bennet and Moran 2015).

And a fifth set of questions concerns the importance of microbes in evolutionary transitions. In Chapter 10 we will discuss the possibility that symbionts were critical for the origin of animal multicellularity. There is also the possibility that microbes are responsible for the origin and/or maintenance of sociality (Archie and Thies 2011; Thies et al. 2012; Stilling et al. 2014). Mating phenremones in flies and hyenas can be produced by bacteria (Sharon et al. 2010, Thies et al. 2013). If *Toxoplasma* can manipulate hormones involved in social interactions (Chapter 2) perhaps the symbionts can, too. It's been proposed (Lombardo 2008) that social interactions facilitate the spread of microbes. If there is a fitness advantage for the microbes to persist in a social organism, perhaps they have evolved to make conditions possible for the sociality of their hosts.

Indeed, if we develop and function as holobionts, and if the holobiont has selectable genetic variation provided by symbiotic genomes, then our view of evolution might have to include serious discussion of how nature might select for consortia, teams of organisms. This will be the topic of Chapter 10. Developmental symbiosis—the generation of the organism as a multispecies assemblage—has profound significance for how we view life and study life.

References

Aagaard, K., J. Ma, K. M. Antony, R. Ganu, J. Petrosino and J. Versalovic. 2014. The placenta harbors a unique microbiome. *Sci. Transl. Med.* 6: 237ra65.

Altura, M. A. and 8 others 2013. The first engagement of partners in the *Euprymna scolopes-Vibrio fischeri* symbiosis is a two-step process initiated by a few environmental symbiont cells. *Environ. Microbiol.* doi: 10.1111/1462-2920.12179.

Altura, M. A., E. Stabb, W. Goldman, M. Apicella and M. J. McFall-Ngai. 2011. Attenuation of host NO production by MAMPs potentiates development of the host in the squid-vibrio symbiosis. *Cell Microbiol.* 13: 527–537.

Archie, E. A. and K. R. Theis. 2011. Animal behaviour meets microbial ecology. *Anim Behav* 82: 425–436.

Ardeshir, A. and 8 others. 2014. Breast-fed and bottle-fed infant rhesus macaques develop distinct gut microbiotas and immune systems. *Sci. Transl. Med.* 6: 252r120.

Arnold, C. 2013. The hologenome: A new view of evolution. *New Scientist* 217(2899): 30–34.

Bachmann, M. D., R. G. Carlton, J. M. Burkholder and R. G. Wetzel. 1986. Symbiosis between salamander eggs and green algae: Microelectrode measurements inside eggs demonstrate effects of photosynthesis on oxygen concentrations. *Can. Zool.* 64: 1586–1588.

Bäckhed, F. and 7 others. 2004. The gut microbiota as an environmental factor that regulates fat storage. *Proc. Natl. Acad. Sci. USA* 101: 15718–15723.

Bäckhed, F., R. E. Ley, J. L. Sonnenburg, D. A. Peterson and J. I. Gordon. 2005. Host-bacterial mutualism in the human intestine. *Science* 307: 1915–1920.

Bates, J. M., E. Mittge, J. Kuhlman, K. N. Baden, S. E. Cheesman and K. Guilemin. 2006. Distinct signals from the microbiota promote different aspects of zebrafish gut differentiation. *Dev. Biol.* 297: 374–386.

Becker, S. and 7 others. 2013. Bacteria regulate intestinal epithelial cell differentiation factors both in vitro and in vivo. *PLoS ONE* 8: e55620.

Becker, M. H. and 9 others. 2015. Composition of symbiotic bacteria predicts survival in Panamanian golden frogs infected with a lethal fungus. *Proc. Biol. Soc.* 282: doi:10.1098/rspb.2014.2881

Bennett, G. M. and N. A. Moran. 2015. Heritable symbiosis: The advantages and perils of an evolutionary rabbit hole. *Proc. Natl. Acad. Sci. USA* pii: 201421388.

Bian, G. and 8 others. 2013. *Wolbachia* invades *Anopheles stephensi* populations and induces refractoriness to *Plasmodium* infection. *Science* 340: 748–751.

Blaser, M. 2012. Quoted in J. Blech. Crucial colonizers: Western lifestyle disturbing key bacterial balance. *Spiegel Online*. www.spiegel.de/international/zeitgeist/western-lifestyle-leading-to-dangerous-bacterial-imbalances-a-856825.html

Blaser, M. J. 2014. *Missing Microbes: How the Overuse of Microbiotics is Fueling Our Modern Plagues*. Picador Press, New York.

Blaser, M. J., Y. Chen and J. Reibman. 2008. Does *Helicobacter pylori* protect against asthma and allergy? *Gut* 57: 1178–1179.

Blaser, M. J. and S. Falkow. 2009. What are the consequences of the disappearing human microbiota? *Nature Rev. Microbiol.* 7: 887–894.

Blech, J. 2012. Western lifestyle disturbing key bacterial balance. *Spiegel Online*. www.spiegel.de/international/zeitgeist/western-lifestyle-leading-to-dangerous-bacterial-imbalances-a-856825.html

Bode, L. 2012. Human milk oligosaccharides: Every baby needs a sugar mama. *Glycobiology* 22: 1147–1162.

Bollinger, R., A. S. Barbas, E. L. Bush, S. S. Lin and W. Parker. 2007. Biofilms in the large bowel suggest an apparent function of the human vermiform appendix. *J. Theoret. Biol.* 249: 826–831.

Booth, M. G. 2004. Mycorrhizal networks mediate overstorey-understorey competition in a temperate forest. *Ecol. Lett.* 7: 538–546.

Braniste, V. and 17 others. 2014. The gut microbiota influences blood-brain barrier permeability in mice. *Science Transl. Med.* 6: 263ra158.

Braun-Fahrländer, C. and 14 others. 2002. Environmental exposure to endotoxin and its relation to asthma in school-age children. *N. Engl. J. Med.* 347: 869–877.

Bravo, J. A. and 7 others. 2011. Ingestion of *Lactobacillus* strain regulates emotional behavior and central GABA receptor expression in a mouse via the vagus nerve. *Proc. Natl. Acad. Sci. USA* 108: 16050–16055.

Britton, R. A. and 7 others. 2014. Probiotic *L. reuteri* treatment prevents bone loss in a menopausal ovariectomized mouse model. *J. Cell Physiol.* 229: 1822–1830.

Brucker, R.M. and S. R. Bordenstein. 2012. The roles of host evolutionary relationships (genus: *Nasonia*) and development in structuring microbial communities. *Evol. Int. J. Org. Evol.* 66: 349–362

Brucker, R. M. and S. R. Bordenstein. 2013. The hologenomic basis of speciation: Gut bacteria cause hybrid lethality in the genus *Nasonia*. *Science* 341: 667–669.

Cameron, D. D., I. Johnson, D. J. Read and J. R. Lenke. 2008. Giving and receiving: Measuring the carbon cost in the green orchid, *Goodyera repens*. *New Phytol.* 180: 176–184.

Cebra, J. J. 1999. Influences of microbiota on intestinal immune system development. *Am. J. Clin. Nutr.* 69[Supplement]: 1046S–1051S.

Charlat, S., G. D. D. Hurst and H. Merçot. 2003. Evolutionary consequences of *Wolbachia* infections. *Trends Genet.* 19: 217–223.

Chen, Y. and M. J. Blaser. 2007. Inverse associations of *Helicobacter pylori* with asthma and allergy. *Arch. Int. Med.* 167: 821–827.

Cho, I. and 12 others. 2012. Antibiotics in early life alter the murine colonic microbiome and adiposity. *Nature* 488: 621–626.

Chu, H. and S. K. Mazmanian. 2013. Innate immune recognition of the microbiota promotes host-microbial symbiosis. *Nature Immunol.* 14: 668–675.

Chun, C. K. and 12 others. 2008. Effects of colonization, luminescence, and autoinducer on host transcription during development of the squid-*Vibrio* association. *Proc. Natl. Acad. Sci. USA* 105: 11323–11328.

Chung, H. and 14 others. 2012. Gut immune maturation depends on colonization with a host-specific microbiota. *Cell* 149: 1578–1593.

Cohen, C. S. and R. R. Strathmann. 1996. Embryos at the edge of tolerance: Effects of environment and structure of egg masses on supply of oxygen to embryos. *Biol. Bull.* 190: 8–15.

Cordaux, R., A. Michel-Salzat, M. Frelon-Raimond, T. Rigaud and D. Bouchon. 2004. Evidence for a new feminizing *Wolbachia* strain in the isopod *Armadillidium vulgare*: Evolutionary implications. *Heredity* 93: 78–84.

Costello, E. K., K. Stagaman, L. Dethlefsen, B. J. Bohannan and D. A. Relman. 2012. The application of ecological theory toward an understanding of the human microbiome. *Science* 336: 1255–1262.

Crabtree, B., D. E. Holloway, M. D. Baker, K. R. Archava and B. Subramanian. 2007. Biological and structural features of murine angiogenin-4, an angiogenic protein. *Biochemistry* 46: 2431–2443.

Croese, J., J. O'neil, J. Masson, S. Cooke, W. Melrose, D. Pritchard and R. Speare. 2006. A proof of concept study establishing *Necator americanus* in Crohn's patients and reservoir donors. *Gut* 55: 136–137.

Cryan, J. F. and T. G. Dinan. 2012. Mind-altering microorganisms: the impact of the gut microbiota on brain and behaviour. *Nature Rev. Neurosci.* 13: 701–712.

Darwin, C. R. 1854. Living Cirripedia, The Balanidæ, (or sessile cirripedes); the Verrucidæ. etc. etc. etc. Vol. 2. The Ray Society, London.

Davidson, S. K. and D. A. Stahl. 2008. Selective recruitment of bacteria during embryogenesis of an earthworm. *ISME Journal* 2: 510–518.

De Fine Licht, H. H., M. Schiøtt, A. Rogowska-Wrzesinska, S. Nygaard, P. Roepstorff and J. J. Boomsma. 2013. Laccase detoxification mediates the nutritional alliance between leaf-cutting ants and fungus-garden symbionts. *Proc. Natl. Acad. Sci. USA* 110: 583–587.

de Sablet, T. and 15 others. 2011. Phylogeographic origin of *Helicobacter pylori* is a determinant of gastric cancer risk. *Gut* 60: 1189–1195.

Debarry, J. and 10 others. 2007. *Acinetobacter lwoffii* and *Lactococcus lactis* strains isolated from farm cowsheds possess strong allergy-protective properties. *J. Allergy Clin. Immunol.* 119: 1514–1521.

Dedeine, F., F. Vavre, F. Fleury, B. Loppin, M. E. Hochberg and M. Boulétreau. 2001. Removing symbiotic *Wolbachia* specifically inhibits oogenesis in a parasitic wasp. *Proc. Natl. Acad. Sci. USA* 98: 6247–6252.

Desbonnet, L., G. Clarke, F. Shanahan, T. G. Dinan and J. F. Cryan. 2014. Microbiota is essential for social development in the mouse. *Mol. Psychiatry* 19: 146–148.

Diaz Heijtz, R. D. and 8 others. 2011. Normal gut microbiota modulates brain development and behavior. *Proc. Natl. Acad. Sci. USA* 108: 3047–3052.

Donnet-Hughes, A. and 8 others. 2010. Potential role of the intestinal microbiota of the mother in neonatal immune education. *Proc. Nutr. Soc.* 69: 407–415.

Douglas, A. E. 2014. Symbiosis as a general principle in eukaryotic evolution. *Cold Spring Harb. Perspect. Biol.* 6(2).

Duan, J., H. Chung, E. Troy and D. L. Kasper. 2010. Microbial colonization drives expansion of IL-1 receptor 1-expressing and IL-17-producing γ/δ T cells. *Cell Host Microbe* 7: 140–150.

Duarte, C. M. 2012. *Global Loss of Coastal Habitats*. Fundación BBVA, Madrid.

Elliott, D. E., J. F. Urban Jr., C. K. Argo and J. V. Weinstock. 2000. Does the failure to acquire helminthic parasites predispose to Crohn's disease? *FASEB J.* 14: 1848–1855

Faith J. J., J. F. Colombel and J. I. Gordon. 2015. Identifying strains that contribute to complex diseases through the study of microbial inheritance. *Proc. Natl. Acad. Sci. USA* 112: 633–640.

Ferree, P. M., H. M. Frydman, J. M. Li, J. Cao, E. Wieschaus and W. Sullivan. 2005 *Wolbachia* utilizes host microtubules and dynein for anterior localization in the *Drosophila* oocyte. *PLoS Pathog.* 1: e14.

Finlay, C. M., K. P. Walsh and K. H. Mills. 2014. Induction of regulatory cells by helminth parasites: Exploitation for the treatment of inflammatory diseases. *Immunol. Rev.* 259: 206– 230.

Foellmer, M. W. and D. J. Fairbairn. 2003. Spontaneous male death during copulation in an orb-weaving spider. *Proc. Roy. Soc. Lond. B Biol. Sci* 270: S183–S185.

Forsythe, P., J. Bienenstock and W. A. Kunze. 2014. Vagal pathways for microbiome-brain-gut axis communication. *Adv. Exp. Med. Biol.* 817: 115–133.

Franzenburg, S., J. Walter, S. Künzel, J. Wang, J. F. Baines, T. C. G. Bosch and S. Fraune. 2013. Distinct antimicrobial peptide expression determines host species-specific bacterial associations. *Proc. Natl. Acad. Sci. USA* 110: 3730–3738.

Fraune, S. and T. C. G. Bosch. 2007. Long-term maintenance of species-specific bacterial microbiota in the basal metazoan *Hydra. Proc. Natl. Acad. Sci. USA* 104: 13146–13151.

Funkhouser, L. J. and S. R. Bordenstein. 2013. Mom knows best: The universality of maternal microbial transmission. *PLoS Biol.* 11(8): e1001631.

Gaylord, B., J. Hodin and M. C. Ferner. 2013. Turbulent shear spurs settlement in larval sea urchins. *Proc. Natl. Acad. Sci. USA* 110: 6901–6906.

Geilenkirchen, W. L., L. P. Timmermans, C. A. van Dongen and W. J. Arnolds. 1971. Symbiosis of bacteria with eggs of Dentalium at the vegetal pole. *Exp. Cell Res.* 67: 477–479.

Gil-Turnes, M. S., M. E. Hay and W. Fenical. 1989. Symbiotic marine bacteria chemically defend crustacean embryos. *Science* 246: 116–118.

Gilbert, J. A., R. Krajmalnik-Brown, L. Dorota, D. L. Porazinska, S. J. Weiss and R. Knight. 2013. Toward effective probiotics for autism and other neurodevelopmental disorders. *Cell* 155: 1446–1448.

Gilbert, P. W. 1944. The alga-egg relationship in *Ambystoma maculatum*: A case of symbiosis. *Ecology* 25: 366–369.

Gilbert, S. F. 2014. A holobiont birth narrative: The epigenetic transmission of the human microbiome. *Front. Genet.* 5: 282.

Gilbert, S. F., J. Sapp and A. I. Tauber. 2012. A symbiotic view of life: We have never been individuals. *Q. Rev. Biol.* 87: 325–341.

Good, M. and 19 others. 2012. Amniotic fluid inhibits Toll-like receptor 4 signaling in the fetal and neonatal intestinal epithelium. *Proc. Natl. Acad. Sci. USA* 109: 11330–11335.

Goodrich, J. K and 12 others. 2014. Human genetics shape the gut microbiome. *Cell* 159: 789–799.

Gordon, J. I., R. E. Ley, R. Wilson, E. Mardis, J. Xu, C. M. Fraser, and D. A. Relman. 2005. Extending our view of self: The

human gut microbiome initiative (HGMI). Available at www.genome.gov/10002154.

Goto, Y. and 21 others. 2014. Innate lymphoid cells regulate intestinal epithelial cell glycosylation. *Science* 345: 1254009.

Gould, S. J. 1996. *Planet of the Bacteria*. Washington Post Horizon 119:(344). An essay adapted from Full House New York: Harmony Books, 1996, pp. 175–192.

Graham, E. R., S. A. Fay, A. Davey and R. W. Sanders. 2013. Intracapsular algae provide fixed carbon to developing embryos of the salamander *Ambystoma maculatum. J. Exp. Biol.* 216: 452–459.

Guarner, F. and J. R. Malagelada. 2003. Gut flora in health and disease. *Lancet* 361: 512–519.

Hadfield, M. G. and V. J. Paul. 2001. Natural chemical cues for settlement and metamorphosis of marine invertebrate larvae. In J. B. B. McClintock and B. J. Baker (eds.), *Marine Chemical Ecology*. CRC Press, Boca Raton, FL, pp. 431–461.

Hardbower, D. M., T. de Sablet, R. Chaturvedi and K. T. Wilson. 2013. Chronic inflammation and oxidative stress: the smoking gun for *Helicobacter pylori*-induced gastric cancer? *Gut Microbes* 4: 475–481.

Hatakka, K. and M. Saxelin. 2008. Probiotics in intestinal and non-intestinal infectious diseases: Clinical evidence. *Curr. Pharm. Des*. 14: 1351–1367.

Hawkins T. D., B. J. Bradley and S. K. Davy. 2013. Nitric oxide mediates coral bleaching through an apoptotic-like cell death pathway: Evidence from a model sea anemone-dinoflagellate symbiosis. *FASEB J.* 27: 4790–4798.

Hooper, L. V., L. Bry, P. G. Falk and J. I. Gordon. 1998. Host-microbial symbiosis in the mammalian intestine: Exploring an internal ecosystem. *BioEssays* 20: 336–343.

Hooper, L. V., T. S. Stappenbeck, C. V. Hong and J. I. Gordon. 2003. Angiogenins: A new class of microbicidal proteins involved in innate immunity. *Nature Immunol.* 4: 269–273.

Hooper, L. V., M. H. Wong, A. Thelin, L. Hansson, P. G. Falk and J. I. Gordon. 2001. Molecular analysis of commensal host-microbial relationships in the intestine. *Science* 291: 881–884.

Hsiao, E. Y. and 11 others. 2013. Microbiota modulate behavioral and physiological abnormalities associated with neurodevelopmental disorders. *Cell* 155: 1451–1463.

Jakobsson, H. E. and 8 others. 2014. Decreased gut microbiota diversity, delayed Bacteroidetes colonisation and reduced Th1 responses in infants delivered by Caesarean section. *Gut* 63: 559–566.

Jefferson, R. 1994. Cited in R. Jefferson, blogs.cambria,org/raj/2010/11/16/the-hologenome-theory-of-evolution.

Jemielita, M., M. J. Taormina, A. R. Burns, J. S. Hampton, A. S. Rolig, K. Guillemin and R. Parthasarathy. 2014. Spatial and temporal features of the growth of a bacterial species colonizing the zebrafish gut. *MBio* 5(6): e01751.

Jiggins, F. M., G. D. D. Hurst, C. E. Dolman and M. E. N. Majerus. 2000. High-prevalence male-killing *Wolbachia* in the butterfly *Acraea encedana. J. Evol. Biol.* 13: 495–501.

Jiménez, E. and 7 others. 2008. Is meconium from healthy newborns actually sterile? *Res. Microbiol.* 159: 187–193.

Johns, P. M. and M. R. Maxwell. 1997. Sexual cannibalism: Who benefits? *Trends Ecol. Evol.* 12: 127–128.

Kageyama, D. and W. Traut. 2003. Opposite sex-specific effects of *Wolbachia* and interference with the sex determination of its host *Ostrinia scapulalis. Proc. Roy. Soc. Lond. B Biol. Sci.* 271: 251–258.

Kashyap, P. C. and 12 others. 2013. Genetically dictated change in host mucus carbohydrate landscape exerts a diet-dependent effect on the gut microbiota. *Proc. Natl. Acad. Sci. USA* 110: 17059–17064.

Kassam, Z., C. H. Lee, Y. Yuan and R. H. Hunt. 2013. Fecal microbiota transplantation for *Clostridium difficile* infection: Systematic review and meta-analysis. *Am. J. Gastroenterol.* 108: 500–508.

Kauffman, S. A. 1995. *At Home in the Universe: The Search for the Laws of Self-Organization and Complexity.* Oxford University Press, New York.

Kerney, R., E. Kim, R. P. Hangarter, A. A. Heiss, C. D. Bishop and B. K. Hall. 2011. Intracellular invasion of green algae in a salamander host. *Proc. Natl. Acad. Sci. USA* 108: 6497–6502.

Koch, E. J., T. Miyashiro, M. J. McFall-Ngai and E. G. Ruby. 2014. Features governing symbiont persistence in the squid-vibrio association. *Mol. Ecol.* 23: 1624–1634.

Koga, R., T. Tsuchida, M. Sakurai and T. Fukatsu. 2007. Selective elimination of aphid endosymbionts: Effects of antibiotic dose and host genotype, and fitness consequences. *FEMS Microbiol Ecol.* 60: 229–239.

Koren, O. and 13 others. 2012. Host remodeling of the gut microbiome and metabolic changes during pregnancy. *Cell* 150: 470–480.

Koropatnick, T. A., J. T. Engle, M. A. Apicella, E. V. Stabb, W. E. Goldman and M. J. McFall-Ngai. 2004. Microbial factor-mediated development in a host-bacterial mutualism. *Science* 306: 1186–1188.

Kremer, N. and 14 others. 2013. Initial symbiont contact orchestrates host-organ-wide transcriptional changes that prime tissue colonization. *Cell Host Microbe* 14: 183–194.

Krueger, D. M., R. G. Gustafson and C. M. Cavanaugh. 1996. Vertical transmission of chemoautotrophic symbionts in the bivalve *Solemya velum* (Bivalvia: Protobranchia). *Biol. Bull.* 190: 195–202.

Kulberg, M. C. 2008. Soothing intestinal sugars. *Nature* 453: 602–604.

Landmann, F., D. Voronin, W. Sullivan and M. J. Taylor. 2011. Anti-filarial activity of antibiotic therapy is due to extensive apoptosis after *Wolbachia* depletion from filarial nematodes. *PLoS Pathog.* 7(11): e1002351.

Landmann, F., J. M. Foster, M. L. Michalski, B. E. Slatko and W. Sullivan. 2014. Co-evolution between a nematode and its nematode host: *Wolbachia* asymmetric localization and A-P polarity establishment. *PLoS Negl. Diseases.* 8(8): e3096.

Lee, S. M., G. P. Donaldson, Z. Mikulski, S. Boyajian, K. Ley and S. K. Mazmanian. 2013. Bacterial colonization factors control specificity and stability of the gut microbiota. *Nature* 501: 426–529.

Lehnert, E. M., M. E. Mouchka, M. S. Burriesci, N. D. Gallo, J. A. Schwarz and J. R. Pringle. 2014. Extensive differences in gene expression between symbiotic and aposymbiotic cnidarians. *G3* 4: 277–295.

Leonardo, M. P. 2008. Access to mutualistic endosymbiotic microbes: An underappreciated benefit of group living. *Behav. Ecol. Sociobiol.* 62: 479–497.

Ley, R. E., F. Bäckhed, P. Turnbaugh, C. A. Lozupone, R. D. Knight and J. L. Gordon. 2005. Obesity alters gut microbial ecology. *Proc. Natl. Acad. Sci. USA* 102: 11070–11075.

Ley, R. E., D. A. Peterson and J. I. Gordon. 2006a. Ecological and evolutionary forces shaping microbial diversity in the human intestine. *Cell* 124: 837–848.

Ley, R. E., P. J. Turnbaugh, S. Klein and J. L. Gordon. 2006b. Human gut microbes associated with obesity. *Nature* 444: 1022–1023.

Lombardo, M. P. 2008. Access to mutualistic endosymbiotic microbes: An underappreciated benefit of group living. *Behav. Ecol. Sociobiol.* 62: 479–497.

Lorber, B. 2005. Infection and mental illness: Do bugs make us batty? *Anaerobe* 11: 303–307.

Ma, J. and 10 others. 2014. High-fat maternal diet during pregnancy persistently alters the offspring microbiome in a primate model. *Nature Commun.* 5: 3889.

MacDonald, T. T. and G. Monteleone. 2005. Immunity, inflammation, and allergy in the gut. *Science* 307: 1920–1925.

Makdad, A. H., J. S. Marks, D. F. Stroup and J. L. Gerberding. 2004. Actual causes of death in the United States, 2000. *J. Amer. Med. Assoc.* 291: 1238–1245.

Makino, H. and 11 others. 2013. Mother-to-infant transmission of intestinal bifidobacterial strains has an impact on the early development of vaginally delivered infant's microbiota. *PLoS ONE* 8(11): e78331.

Mandel, M. J., A. L. Schaefer, C. A. Brennan, E. A. Heath-Heckman, C. R. Deloney-Marino, M. J. McFall-Ngai and E. G. Ruby. 2012. Squid-derived chitin oligosaccharides are a chemotactic signal during colonization by *Vibrio fischeri*. *Appl. Environ. Microbiol.* 78: 4620–4626.

Manthey, C. F., C. A. Autran, L. Eckmann and L. Bode. 2014. Human milk oligosaccharides protect against enteropathogenic *Escherichia coli* attachment in vitro and EPEC colonization in suckling mice. *J. Pediatr. Gastroenterol. Nutr.* 58: 167–170.

Marcobal, A., A. M. Southwick, K. A. Earle and J. L. Sonnenburg. 2013. A refined palate: bacterial consumption of host glycans in the gut. *Glycobiology* 23: 1038–1046.

Margulis, L. 1971. *Origin of Eukaryotic Cells.* Yale University Press, New Haven, CT.

Margulis, L. 1981. *Symbiosis in Cell Evolution.* New York: W. H. Freeman

Margulis, L. 1994. Symbiogenesis and symbioticism. In *Symbiosis as a Source of Evolutionary Innovation*, L. Margulis and R. Fester (eds.). MIT Press, Cambridge, MA, pp. 1–14.

Margulis, L. and D. Sagan. 1986. *Origins of Sex : Three Billion Years of Genetic Recombination.* Yale University Press, New Haven, CT.

Martin, F., A. Kohler and S. Duplessis. 2007. Living in harmony in the wood underground: Ectomycorrhizal genomics. *Curr. Opin. Plant Biol.* 10: 204–210.

Mayer, E. A., R. Knight, S. K. Mazmanian, J. F. Cryan and K. Tillisch. 2014. Gut microbes and the brain: Paradigm shift in neuroscience. *J. Neurosci.* 34: 15490–15496.

Mazmanian, S. K., C. H. Liu, A. O. Tzianabos and D. L. Kasper. 2005. An immunomodulatory molecule of symbiotic bacteria directs maturation of the host immune system. *Cell* 122: 107–118.

Mazmanian, S. K., J. L. Round and D. L. Kasper. 2008. A microbial symbiosis factor prevents intestinal inflammatory disease. *Nature* 453: 620–625.

McFall-Ngai, M. J. 2002. Unseen forces: The influence of bacteria on animal development. *Dev. Biol.* 242: 1–14.

McFall-Ngai, M. 2008. Host-microbe symbiosis: The squid-*Vibrio* association, a naturally occurring, experimental model of animal-bacterial partnerships. *Adv. Exp. Med. Biol.* 635: 102–112.

McFall-Ngai, M. and 25 others. 2013. Animals in a bacterial world, a new imperative for the life sciences. *Proc. Natl. Acad. Sci. USA* 110: 3229–3236.

McLean, P. G., G. E. Bergonzelli, S. M. Collins and P. Bercik. 2012. Targeting the microbiota-gut-brain axis to modulate behavior:

Which bacterial strain will translate best to humans? *Proc. Natl. Acad. Sci. USA* 109: e174.

Messaoudi, M. and 10 others. 2011. Assessment of psychotropic-like properties of a probiotic formulation (*Lactobacillus helveticus* R0052 and *Bifidobacterium longum* R0175) in rats and human subjects. *Br. J. Nutr.* 105: 755–764.

Metchnikoff, E. 1907. Essais optimistes. Paris. (Trans. and edited by P. C. Mitchell as *The Prolongation of Life: Optimistic Studies.* Heinemann, London.)

Milius, S. 2006. They're all part fungus: Grass blades, coffee or cacao leaves … probably all plants. *Science News* 169: 231.

Millikan, D. S. and E. G. Ruby. 2001. Alterations in *Vibrio fischeri* motility correlate with a delay in symbiosis initiation and are associated with additional symbiotic colonization defects. *Appl. Env. Microbiol.* 68: 2519–2528.

Mills, N. E. and M. C. Barnhart. 1999. Effects of hypoxia on embryonic development in two *Ambystoma* and two *Rana* species. *Physiol. Biochem. Zool.* 72: 179–188.

Moran, N., P. H. Degnan, S. R. Santos, H. E. Dunbar and H. Ochman. 2005. The players in a mutualistic symbiosis: Insects, bacteria, viruses, and virulence genes. *Proc. Natl. Acad. Sci. USA* 102: 16919–16926.

Muscatine, L., P. G. Falkowski, W. Porter and Z. Dubinsky. 1984. Fate of photosynthetic fixed carbon in light- and shade-adapted colonies of the symbiotic coral *Stylophora pistillata*. *Proc. Roy. Soc. Lond. B Biol. Sci* 222: 181–202.

Nash, D. R., T. D. Als, R. Maile, S. R. Jones and J. J. Boomsma. 2008. A mosaic of chemical coevolution in a large blue butterfly. *Science* 319: 88–90.

Negri, I., A. Franchini, D. Daffonchio, P. J. Mazzoglio, M. Mandrioli and A. Alma. 2009. Unravelling the *Wolbachia* evolutionary role: The reprogramming of the host genomic imprinting. *Proc. Roy. Soc. Lond. B Biol. Sci* 276: 2485–2491.

Nicholson, J. K., E. Holmes, J. Kinross, R. Burcelin, G. Gibson, W. Jia and S. Pettersson. 2012. Host-gut microbiota metabolic interactions. *Science* 336: 1262–1267.

Niess, J. H., F. Leithäuser, G. Adler and J. Reimann. 2008. Commensal gut flora drives the expansion of proinflammatory CD4 T cells in the colonic lamina propria under normal and inflammatory conditions. *J. Immunol.* 180: 559–568.

Norsworthy, A. N. and K. L Visick. 2013. Gimme shelter: How *Vibrio fischeri* successfully navigates an animal's multiple environments. *Front Microbiol.* 4: 356.

Noverr, M. C. and G. B. Huffnagle. 2004. Does the microbiota regulate immune responses outside the gut? *Trends Microbiol.* 12: 562–568.

Nyholm, S. V. and M. J. McFall-Ngai. 2004. The winnowing: Establishing the squid-*Vibrio* symbiosis. *Nature Rev. Microbiol.* 2: 632–642.

Nyholm, S. V., E. V. Stabb, E. G. Ruby and M. J. McFall-Ngai. 2000. Establishment of an animal-bacterial association: Recruiting symbiotic *Vibrios* from the environment. *Proc. Natl. Acad. Sci. USA* 97: 10231–10235.

Ohlsson, C. and 8 others. 2014. Probiotics protect mice from ovariectomy-induced cortical bone loss. *PLoS ONE* 9(3): e92368.

Olivier, H. M. and B. R. Moon. 2010. The effects of atrazine on spotted salamander embryos and their symbiotic alga. *Ecotoxicology* 19: 654–661.

Olszak, T. and 10 others. 2012. Microbial exposure during early life has persistent effects on natural killer T cell function. *Science* 336: 489–493.

Pacheco, A. R., M. M. Curtis, J. M. Ritchie, D. Munera, M. K. Waldor, C. G. Moreira and V. Sperandio. 2012. Fucose sensing regulates bacterial intestinal colonization. *Nature* 492: 113–117.

Pannebakker, B. A., B. Loppin, C. P. H. Elemains, L. Humblot and F. Vavre. 2007. Parasitic inhibition of cell death facilitates symbiosis. *Proc. Natl. Acad. Sci. USA* 104: 213–215.

Pantoja-Feliciano, I. G., J. C. Clemente, E. K. Costello, M. E. Perez, M. J. Blaser, R. Knight and M. G. Dominguez-Bello. 2013. Biphasic assembly of the murine intestinal microbiota during early development. *ISME J.* 7: 1112–1115.

Petschow, B. and 17 others. 2013. Probiotics, prebiotics, and the host microbiome: The science of translation. *Ann. N. Y. Acad. Sci.* 1306: 1–17.

Peyer, S. M., M. S. Pankey, T. H. Oakley and M. J. McFall-Ngai. 2014. Eye-specification genes in the bacterial light organ of the bobtail squid *Euprymna scolopes*, and their expression in response to symbiont cues. *Mech. Dev.* 131: 111–126.

Philip, L., S. Simard and M. Jones. 2011. Pathways for belowground carbon transfer between paper birch and Douglas-fir seedlings. *Plant Ecol. Divers.* 3: 221–233.

Pinder, A. W. and S. C. Friet. 1994. Oxygen transport in egg masses of the amphibians *Rana sylvatica* and *Ambystoma maculatum*: Convection, diffusion, and oxygen production by algae. *J. Exp. Biol.* 197: 17–30.

Pollan, M. 2002. Power steer. *New York Times* March 31 2002. Available at www.michaelpollan.com/article.php?id=14.

Poulsen, M. and 23 others. 2014. Complementary symbiont contributions to plant decomposition in a fungus-farming termite. *Proc. Natl. Acad. Sci. USA* 2014 111: 14500–14505.

Rao, A. V., A. C. Bested, T. M. Beaulne, M. A. Katzman, C. Iorio, J. M. Berardi and A. C. Logan. A randomized, double-blind, placebo-controlled pilot study of a probiotic in emotional symptoms of chronic fatigue syndrome. *Gut Pathog.* 1: 6.

Rawls, J. F., M. A. Mahowald, R. E. Ley and J. I. Gordon. 2006. Reciprocal gut microbiota transplants from zebrafish and mice to germ-free recipients reveal host habitat selection. *Cell* 127: 423–433.

Rawls, J. F., B. S. Samuel and J. I. Gordon. 2004. Gnotobiotic zebrafish reveal evolutionarily conserved responses to the gut microbiota. *Proc. Natl. Acad. Sci. USA* 101: 4596–4601.

Rhee, K. J., P. Sethupathi, A. Driks, D. K. Lanning and K. L. Knight. 2004. Role of commensal bacteria in development of gut-associated lymphoid tissue and preimmune antibody repertoire. *J. Immunol.* 172: 1118–1124.

Roeselers, G. and I. L. Newton. 2012. On the evolutionary ecology of symbioses between chemosynthetic bacteria and bivalves. *Appl. Microbiol. Biotechnol.* 94: 1–10.

Rook, G. A. and J. L. Stanford. 1998. Give us this day our daily germs. *Immunol. Today* 19: 113–116.

Rosenberg, E., O. Koren, L. Reshef, R. Efrony and I. Zilber-Rosenberg. 2007. The role of microorganisms in coral health, disease, and evolution. *Nature Rev. Microbiol.* 5: 355–362.

Roth, M. S. 2014. The engine of the reef: Photobiology of the coral-algal symbiosis. *Front Microbiol.* 5: 422.

Sacchi, L. and 8 others. 1998. Some aspects of intracellular symbiosis during embryo development of *Mastotermes darwiniensis* (Isoptera: Mastotermitidae). *Parasitology* 40: 309–316.

Saikkonen, K., P. Lehtonen, M. Helender, J. Koricheva and S. H. Faeth. 2006. Model systems in ecology: Dissecting the endophyte-grass literature. *Trends Plant Sci.* 11: 428–433.

Sammarco, P. W. and K. B. Strychar. 2013. Responses to high seawater temperatures in zooxanthellate octocorals. *PLoS ONE* 8(2): e54989.

Samuel, B. S. and J. I. Gordon. 2006. Humanized gnotobiotic mouse model of host-archaeal-bacterial mutualism. *Proc. Natl. Acad. Sci. USA* 103: 10011–10016.

Sapp, J. 1994. *Evolution by Association: A History of Symbiosis*. Oxford University Press, New York.

Schiøtt, M., A. Rogowska-Wrzesinska, P. Roepstorff and J. J. Boomsma. 2010. Leaf-cutting ant fungi produce cell wall degrading pectinase complexes reminiscent of phytopathogenic fungi. *BMC Biol.* 8: 156.

Sela, D. A. and 8 others. 2011. An infant-associated bacterial commensal utilizes breast milk sialyloligosaccharides. *J. Biol. Chem.* 286: 11909–11918.

Selosse, M. A., F. Richard, X. He and S. W. Simard. 2006. Mycorrhizal networks: Les liaisons dangereuses? *Trends Ecol. Evol.* 21: 621–628.

Sharon, G., D. Segal, J. M. Ringo, A. Hefetz, I. Zilber-Rosenberg and E. Rosenberg. 2010. Commensal bacteria play a role in mating preference of *Drosophila melanogaster*. *Proc. Natl. Acad. Sci. USA* 107: 20051–20056.

Sheh, A., R. Chaturvedi, D. S. Merrell, P. Correa, K. T. Wilson and J. G. Fox. 2013. Phylogeographic origin of *Helicobacter pylori* determines host-adaptive responses upon coculture with gastric epithelial cells. *Infect. Immun.* 81: 2468–2477.

Shugart, J. 2014. Mother Lode: *Science News* Jan. 11, 2014. www.sciencenews.org/article/mother-lode?mode=magazine&context=4527

Simard, S. W., D. A. Perry, M. D. Jones, D. D. Myrold, D. M. Durall and R. Molina. 1997. Net transfer of carbon between ectomycorrhizal trees in the field. *Nature* 388: 579–582.

Simpson, S. 2010. Quoted in esciencenews.com/articles/2010/05/14/baby.corals.dance.their.way.home

Smil, V. 2001. *Enriching the Earth: Fritz Haber, Carl Boschm and the Transformation of World Food Production*. MIT Press, Cambridge, MA.

Smith, M. B., C. Kelly and E. J. Alm. 2014. Policy: How to regulate faecal transplants. *Nature* 506: 290–291.

Smith, M. I. and 18 others. 2013. Gut microbiomes of Malawian twin pairs discordant for kwashiorkor. *Science* 339: 548–654.

Smits, L. P., K. E. Bouter, W. M. de Vos, T. J. Borody and M. Nieuwdorp. 2013. Therapeutic potential of fecal microbiota transplantation. *Gastroenterology* 145: 946–953.

Song, Y. Y., M. Ye, C. Y. Li, R. L. Wang, X. C. Wei, S. M. Luo and R. S. Zeng. 2103. Priming of anti-herbivore defense in tomato by arbuscular mycorrhizal fungus and involvement of the jasmonate pathway. *J. Chem. Ecol.* 39: 1036–1044.

Song, Y.Y., R. S. Zeng, J. F. Xu, J. Li, X. Shen and W. G. Yihdego. 2010. Interplant communication of tomato plants through underground common mycorrhizal networks. *PLoS ONE* 5: e13324.

Sonnenburg, J. L., L. T. Angenent and J. I. Gordon. 2004. Getting a grip on things: How do communities of bacterial symbionts become established in our intestine? *Nature Immunol.* 5: 569–573.

Stappenbeck, T. S., L. V. Hooper and J. I. Gordon. 2002. Developmental regulation of intestinal angiogenesis by indigenous microbes via Paneth cells. *Proc. Natl. Acad. Sci. USA* 99: 15451–15455.

Steidler, L. 2001. Microbiological and immunological strategies for treatment of inflammatory bowel disease. *Microbes Infect.* 3: 1157–1166.

Stilling, R. M., S. R. Bordenstein, T. G. Dinan and J. F. Cryan. 2014. Friends with social benefits: Host-microbe interactions as a driver of brain evolution and development? *Front. Cell Infect. Microbiol.* 4: 147.

Stouthamer, R., J. A. J. Breeuwer and G. D. D. Hurst. 1999. *Wolbachia pipientis*: Microbial manipulator of arthropod reproduction. *Annu. Rev. Microbiol.* 53: 71–102.

Strathmann, R. R. and M. F. Strathmann. 1995. Oxygen supply and limits on aggregation of embryos. *J. Mar. Biol. Assoc.* UK 75: 413–428.

Suez, J. and 16 others. 2014. Artificial sweeteners induce glucose intolerance by altering the gut microbiota. *Nature* 514: 181–186.

Takahashi, K., Y. Sugi, K. Nakano, M. Tsuda, K. Kurihara, A. Hosono and S. Kaminogawa. 2011. Epigenetic control of the host gene by commensal bacteria in large intestinal epithelial cells. *J. Biol. Chem.* 286: 35755–35762.

Theis, K. R. and 7 others. 2013. Symbiotic bacteria appear to mediate hyena social odors. *Proc. Natl. Acad. Sci. USA* 110: 19832–19837.

Theis, K. R., T. M. Schmidt and K. E. Holekamp. 2012. Evidence for a bacterial mechanism for group-specific social odors among hyenas. *Science Rep.* 2: 615.

Thomas, J. A. 1995. The ecology and conservation of *Maculinea arion* and other European species of large blue butterfly. In A. S. Pullin (ed.), *Ecology and Conservation of Butterflies*. Chapman and Hall, New York.

Thomas, J. A. and 7 others 2002. Parasitoid secretions provoke ant warfare. *Nature* 417: 505.

Thomas, J. A. and G. W. Elmes. 1993. Specialized searching and the hostile use of allomones by a parasitoid whose host, the butterfly *Maculinea rebeli*, inhabits ant nests. *Anim. Behav.* 45: 593–602.

Thomas, L. 1974. *The Lives of a Cell: Notes of a Biology Watcher*. Viking, New York.

Tissier, H. 1900. Recherches sur la flore intestinale des nourrissons (e´tat normal et pathologique). Paris: G. Carre and C. Naud.

Tong, D., N. S. Rozas, T. H. Oakley, J. Mitchell, N. J. Colley and M. J. McFall-Ngai. 2009. Evidence for light perception in a bioluminescent organ. *Proc. Natl. Acad. Sci. USA* 106: 9836–9841.

Trasande, L., J. Blustein, M. Liu, E. Corwin, L. M. Cox and M. J. Blaser. 2013. Infant antibiotic exposures and early-life body mass. *Int. J. Obes.* 37: 16–23.

Trehan, I., H. S. Goldbach, L. N. LaGrone, G. J. Meuli, R. J. Wang, K. M. Maleta and M. J. Manary. 2013. Antibiotics as part of the management of severe acute malnutrition. *N. Engl. J. Med.* 368: 425–435.

Turnbaugh, P. J., F. Bäckhed, L. Fulton and J. I. Gordon. 2008. Diet-induced obesity is linked to marked but reversible alterations in the mouse distal gut microbiome. *Cell Host Microbiol.* 3: 213–223.

Turnbaugh, P. J., R. E. Ley, M. A. Mahowald, V. Magrini, E. R. Mardis and J. I. Gordon. 2006. An obesity-associated gut microbiome with increased capacity for energy harvest. *Nature* 444: 1027–1031.

Umesaki, Y. 1984. Immunohistochemical and biochemical demonstration of the change in glycolipid composition of the intestinal epithelial cell surface in mice in relation to epithelial cell differentiation and bacterial association. *J. Histochem. Cytochem.* 32: 299–304.

Underwood, M. A., K. M. Kalanetra, N. A. Bokulich, Z. T. Lewis, M. Mirmiran, D. J. Tancredi and D. A. Mills. 2013. A comparison of two probiotic strains of bifidobacteria in premature infants. *J. Pediatr.* 163: 1585–1591.

van der Heide, T. and 7 others. 2012. Ecosystem engineering by seagrasses interacts with grazing to shape an intertidal landscape. *PLoS ONE* 7: e42060.

van Nood, E. and 12 others. 2013. Duodenal infusion of donor feces for recurrent *Clostridium difficile*. *New Engl. J. Med.* 368: 407–415

Vandenkoorhuyse, P., S. L. Baldauf, C. Leyval, J. Straczek, and J. P. Young. 2002. Extensive fungal diversity in plant roots. *Science* 295: 2051.

Vermeij, M. J., K. L. Marhaver, C. M. Huijbers, I. Nagelkerken and S. D. Simpson. 2010. Coral larvae move toward reef sounds. *PLoS ONE* 5(5): e10660.

Visick, K. L., J. Foster, J. Donio, M. McFall-Ngai and E. G. Ruby. 2000. *Vibrio fischeri* lux genes play an important role in colonization and development of the host light organ. *J. Bacteriol.* 182: 4578–4586.

Walter, J., N. C. K. Heng, W. P. Hammes, D. M. Loach, G. W. Tannock and C. Hertel. 2003. Identification of *Lactobacillus reuteri* genes specifically induced in the mouse gastrointestinal tract. *Appl. Environ. Microbiol.* 69: 2044–2051.

Wang, S., A. K. Ghosh, N. Bongio, K. A. Stebbings, D. J. Lampe and M. Jacobs-Lorena. 2012. Fighting malaria with engineered symbiotic bacteria from vector mosquitoes. *Proc. Natl. Acad. Sci. USA* 109:12734–12739.

Wang, Y. and L. H. Kasper. 2014. The role of microbiome in central nervous system disorders. *Brain Behav. Immun.* 38: 1–12.

Watanabe, K. F. Yukuhiro, Y. Matsuura, T. Fukatsu, H. Noda 2014. Intrasperm vertical symbiont transmission. *Proc. Natl. Acad. Sci. USA* 111: 7433–7437.

Waterman, R. J. and M. I. Bidartando. 2008. Deception above, deception below: Linking pollination and mycorrhizal biology of orchids. *J. Exp. Botany* 59: 1085–1096.

Waycott, M. and 13 others. 2009. Accelerating loss of seagrasses across the globe threatens coastal ecosystems. *Proc. Natl. Acad. Sci. U.S.A.* 106: 12377–12381.

Werren, J. H. 1997. Biology of *Wolbachia*. *Annu. Rev. Entomol.* 42: 587–609.

Wesemann, D. R. and 9 others. Microbial colonization influences early B-lineage development in the gut lamina propria. *Nature* 501: 112–115.

Wolf, G. 2006. Gut microbiota: A factor in energy regulation. *Nutr. Rev.* 64: 47–50.

Woyke, T. and 17 others. 2006. Symbiosis insights through metagenomic analysis of a microbial consortium. *Nature* 443: 950–955.

Xu, J. and J. I. Gordon. 2003. Honor thy symbionts. *Proc. Natl. Acad. Sci. USA* 100: 10452–10459.

Yong, E. 2015. phenomena.nationalgeographic.com/2015/03/18/can-probiotic-bacteria-save-an-endangered-frog/

Zardus, J. D., B. T. Nedved, Y. Huang, C. Tran and M. G. Hadfield. 2008. Microbial biofilms facilitate adhesion in biofouling invertebrates. *Biol. Bull.* 214: 91–98.

Zchori-Fein, E. and 6 others. 2001. A newly discovered bacterium associated with parthenogenesis and a change in host selection behavior in parasitoid wasps. *Proc. Natl. Acad. Sci. USA* 98: 12555–12560.

Zivkovic, A. M., J. B. German, C. B. Lebrilla and D. A. Mills. 2011. Human milk glycobiome and its impact on the infant gastrointestinal microbiota. *Proc. Natl. Acad. Sci. USA* 108: 4653–4658.

Part 2

Ecological Developmental Biology and Disease States

Part 2 looks at the ways in which ecological agents can disrupt development. **Chapter 4** discusses the physiological mechanisms that embryos have evolved to expel and detoxify dangerous compounds. The end of **Chapter 4** details the ways by which global climate changes are stressing these systems. **Chapter 5** looks at the mechanisms through which certain toxic compounds such as ethanol and mercury create developmental abnormalities that are often seen at birth, and **Chapter 6** focuses on one set of these substances, the endocrine disruptors. These compounds perturb the hormones and disturb development in ways that might not be seen until long after birth. **Chapter 7** looks at the evidence that disease conditions can arise through the mismatch of environmental conditions experienced by the fetus and by the newborn, while **Chapter 8** looks at aging and cancer as diseases also involving the interactions between the developing organism and its environment.

Developmental Physiology for Survival in Changing Environments

Self-defense is Nature's eldest law.

John Dryden, 1681

Developmental biologists rightly focus on the mechanisms by which fertilized eggs become new and highly complex individuals. What is not as appreciated, however, is that there are *two* systems operating to ensure the forward movement of development: one system that ensures the fidelity of development, and another that ensures an embryo's chances of surviving stresses that it could encounter throughout its developmental stages. Fidelity is achieved by a buffering, or robustness, which ensures that small environmental or genetic perturbations do not alter development. Survival is achieved by a unique cellular physiology that allows the embryo to stay alive and continue its development in an environment teeming with such stresses as reactive oxygen, predatory animals, toxic compounds, and potentially lethal radiation (Figure 4.1) as well as hypoxia and climate change.

In addition, when we look at the diverse reproductive patterns in the animal kingdom, we find that the vast majority of organisms develop as **orphan embryos**, with little or no parental protection during their development (Figure 4.2). In many of these organisms, the adults can be viewed as players in a giant lottery, producing millions of embryos on the chance that one or two or a few will survive. In fact, most embryos share the fate of becoming food for other organisms. And the embryos and larvae that manage to escape predation must still cope with a constantly changing environment, including such challenges as variations in temperatures and humidity, microbial infection, and irradiation by solar wavelengths that can damage DNA.

Whatever the level of parental protection, the embryo's own protective mechanisms must be at the level of the cell, since the tissues and organs that provide protection for the mature individual haven't yet developed. And even at the cellular level, embryonic cells and adult cells do things very

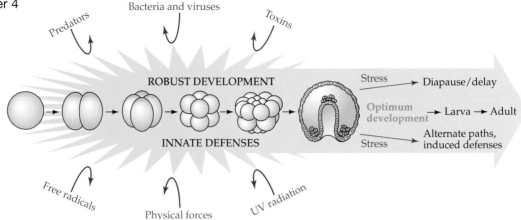

Figure 4.1 The embryo has not only a plan for development, but also a battery of defenses in order to survive in its anticipated environment. In early development of embryos with rapid early cleavages (e.g., most "orphan" embryos), the strategy is to have high levels of innate defenses already present when environmental stress is encountered. Later in development, the embryo can respond to the environment by inducing defenses, altering developmental pathways, or temporarily stopping development. These alternate paths can represent a "lottery approach," producing phenotypes that may or may not be adaptive. (After Hamdoun and Epel 2007.)

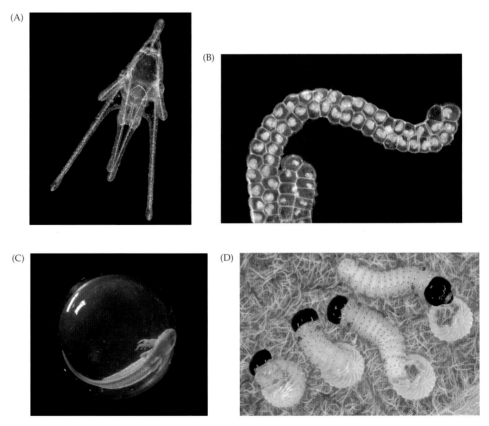

Figure 4.2 Most animals begin life as "orphan embryos" with no parental protection. (A) Pluteus-stage larva of a sea urchin. Sea urchins are prototypical of the myriad marine species whose embryos develop unaided in the ebbs and flows of the ocean currents. (B) Gastropod embryos such as these of the pond snail *Limnaea stagnalis* have been the source of much interesting data regarding the defense mechanisms of orphan embryos. (C) Most amphibian species lay fertilized eggs that develop and hatch unaided, as this salamander will do. (D) Larvae of the monarch butterfly (*Danaus plexippus*) get early nutrition from eating their egg cases immediately after hatching.

differently. Adult cells are sophisticated, with a strategy similar to that of wealthy nations: they maintain a large arsenal of defense mechanisms that are always available and can be quickly called up when danger is detected. These adult cells can respond by initiating synthesis of stereotypic defense proteins immediately upon infusion of some toxin from a recent meal, upon detecting a bacterial infection, or upon sensing denatured proteins from an increase in temperature.

By contrast, embryos of most organisms, especially the orphans, seem to be on a fast track, with seemingly no time or no capacity to divert their metabolism to defense. It appears that the gene activity of embryonic cells is taken up with instructing the cells in how to divide, diversify, and form the basic plan of the adult organism; there are not many resources left for them to spend on defense until after gastrulation.

Given these constraints, we may wonder how any embryo makes it to adulthood. But, in fact there are four strategies enhancing embryo survival. First is an innate robustness that buffers the embryo against minor environmental variations as well as the detrimental effects of minor mutations.

The second are defenses that are provided during oogenesis to handle the expected, anticipated, historical stresses that an embryo might encounter later during its development. Some of these are in the form of the familiar cell defenses seen in adult cells, such as the ability to repair DNA damage or protein-folding damage; paradoxically, these defenses are turned off during the cleavage stages in many embryos. There are also defenses that are unique to the embryo such as protective, embryo-specific antioxidants to protect against toxic free oxygen radicals, mechanisms to repair torn cell surfaces, sunscreens to protect from ultraviolet damage, and protective mechanisms thwarting pathogens and predators. These defenses are present in the egg and embryo and used if the expected stress occurs.

The third embryo survival strategy is the ability to mount specific epigenetic changes in response to environmental signals that can result in altered phenotype. This ability comes from alternate developmental programs that are available in the embryo for expected, anticipated historical stresses but that are only turned on if those stresses actually take place. Examples include responses to predators or seasonal changes, as described in Chapters 1 and 2. Another example, studied extensively in nematodes, annual fish and brine shrimp, is the set of changes that are turned on in response to food limitations and that result in cessation of development and entrance into a dormant stage.

Mammalian embryos make epigenetic changes in fundamental food metabolism if they sense food is limiting for the mother, thus altering their metabolism for a predicted world of little food, an alteration that is retained when those embryos become adults. As described in Chapter 7, these epigenetic changes can cause problems if the prediction is inaccurate; the embryo is adapted for the wrong environment, and serious health problems can arise in later life.

The fourth embryo survival strategy is parental protection, such as by the placenta in mammals or by the collective behavior of adult bees, ants, termites, and other social insects. Here, there is often a separate and sterile caste of workers that tend and care for the eggs and larvae. As we shall see, embryos use combinations of all the above strategies to survive the precarious journey from egg to adult.

In this chapter, we will first look at two characteristics of early development. One is a robustness that ensures normal development in the face of changing environments or the presence of mutations. The other characteristic is a paradoxical response to damage from environmental damage during the cleavage period; these early stages do not mount a stress response and do not remove severely injured cells by apoptosis (cell death). We will then look in great detail at mechanisms of embryo defenses. We end with a consideration of how embryos can cope—or not cope—with the consequences of changing environments resulting from global climate change.

Developmental Robustness: A Necessary but Paradoxical Defense

The developmental process is precise and, under the expected and anticipated conditions, proceeds in a predictable way. A *Drosophila melanogaster* egg fertilized in Los Angeles will develop identically if fertilized in New Delhi instead, even under a variety of different environmental conditions such as temperatures or humidity. Such **robustness**—also called **canalization**—is an essential property of developing systems (Waddington 1942; Nijhout 2002; Pujadas and Feinberg 2012).

On one level, canalization is the opposite of phenotypic plasticity, because it ensures that the same phenotype is produced regardless of environmental or genetic perturbations. At the molecular level, however, such robustness can be considered a product of plasticity, since the developmental interactions can adjust to compensate for genetic or environmental differences (see Gilbert 2002).

Two examples of developmental robustness

Major insights into the canalization of developmental pathways have come from studies of the fruit fly *Drosophila melanogaster*. These studies have focused on the stability of the insect's body segmentation pattern, which is determined in large part by the distribution of the Bicoid protein. Raising the embryo in different environments can alter Bicoid distribution but there is little or no effect on the segmentation pattern and subsequent development.

The mRNA for Bicoid is maternal (i.e., it is carried in the egg cytoplasm) and is localized to the anterior region of the future embryo during oogenesis. Fertilization and egg laying initiate mRNA translation, and the newly translated Bicoid protein diffuses from the anterior end to establish a concentration gradient across the embryo. In the fruit fly embryo, DNA synthesis and mitosis occur without cell division for the first 12 nuclear divisions such that thousands of nuclei are present in what is effectively a large, single-celled embryo. Bicoid protein extends through this cell, forming a gradient that is highest at the anterior and lowest at the posterior of the embryo. The resultant reading of this gradient establishes the precise distribution of the body segments (see Gilbert 2013).

Among its other functions, Bicoid can act as a transcription factor, inducing the expression of the *Hunchback* gene. This causes Hunchback protein to be produced in an anterior-to-posterior gradient in the embryo. Conversely, Nanos protein, which is synthesized in the posterior of the embryo, *inhibits* the translation of *Hunchback* mRNA, steepening the Hunchback protein gradient (Figure 4.3). Depending on the concentration of Hunchback present, a given region of the embryo is instructed to form the head (high concentration), the thorax (medium concentration), or the abdomen (low concentration). Together, they establish the boundaries of other segmentation genes.

Several investigators have shown that when *Drosophila* embryos are raised at different temperatures, the gradient of Bicoid expression is altered but the phenotype of the fly remains the same (see Houchmandzadeh et al. 2002; Lucchetta et al. 2005). This may be due to the fact that Bicoid can be both a transcription factor and a translational regulator, or it might be due to Hunchback's being synthesized both by maternal genes (during egg formation) and by zygotic genes (in the embryo). Several mathematical models have been formulated to explain this precision (see Gregor et al. 2007a,b; Jaeger et al. 2007; Lepzelter and Wang 2008). These models suggest that the stability of the body plan ensues just from interaction of many components of the system. The idea here is that the robustness is an inherent outcome of the complexity and integration of the gene networks through which development operates (Lander 2011; Staller et al. 2015).

Similarly, the formation of the vulva in the nematode *Caenorhabditis elegans*—a developmental process that has been extensively studied and is known to proceed usually from a specific and identifiable central cell—shows remarkable robustness in the presence of genetic and environmental variation. Interestingly, it is the underlying developmental plasticity that allows such robustness (Gleason et al. 2002; Chen and Grunwald 2004; Braendle and Felix 2009). In cases of starvation, for instance, the formation of the vulva will proceed from a different cell, indicating functional redundancy of vulval precursor cells. Furthermore, at least three signaling systems determine which differentiating cells become central to the structure and which are peripheral. If one of these signals fails to function, the other two will still persist.

(A)

(B)

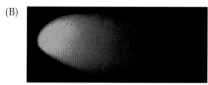

(C)

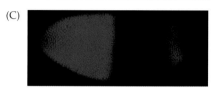

Figure 4.3 Bicoid and Hunchback proteins establish the *Drosophila* body axis. (A) The localization of *bicoid* mRNA (black) in the anterior pole of the *Drosophila* egg. (B) Later in development, the *bicoid* mRNA is translated into a Bicoid protein gradient, visualized here by a fluorescent green antibody. (C) Bicoid protein is a transcription factor that induces the expression of the Hunchback protein (blue fluorescent antibody; the posterior band of hunchback expression is due to its activation slightly later by a different set of factors). (A from Ephrussi and St Johnston 2004; B,C from Lopes et al. 2008.)

Mutational robustness

Canalization can also buffer against minor genetic perturbations, or even mutations, such that the same phenotype is always produced. As Waddington (1942) put it, " if wild animals of almost any species are collected, they will usually be found 'as like as peas in a pod.'" Researchers can experimentally "knock out" genes in embryos, and often there are dire consequences to such manipulations. But just as often there will be no phenotypic difference. It is as if the gene isn't even needed (Tautz 1992; Krakauer and Nowak 2014). This has led to the idea that robustness or buffering ensues from the presence of many genes controlling the same output, so removing one has no obvious effect on the outcome. It's as if you had ten people

Figure 4.4 Hsp90 suppresses the expression of mutant phenotypes. Fruit flies treated with geldanamycin, an Hsp90 inhibitor, exhibit a variety of developmental abnormalities, including (A) left egg has black facets, (B) part of wing is notched (From Rutherford and Lindquist 1998, photographs courtesy of S. Lindquist.)

(A)
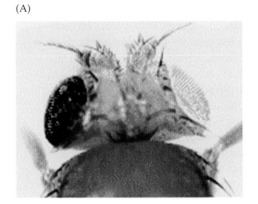

(B)

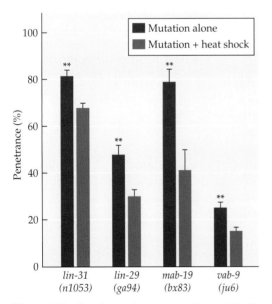

Figure 4.5 Penetrance (expression of a mutation) is reduced by mild stress—2 hours of high temperature. The red bars (controls) show the percent penetrance (frequency of expression) of four different mutants of *Caenorhabditis elegans*. Expression (penetrance) of the mutant phenotype is reduced in the heat-stressed larvae (blue). Asterisks indicate that the difference in expression is statistically significant. (After Casanueva et al. 2012.)

carrying a heavy box; removing one of them would not make a perceptible difference (see Cooke et al. 1997).

Another type of mutational robustness arises from the dampening or modulation of mutant gene activity. A surprising number of mutations are never expressed even though they are present in the genome. It seems that the amino acid substitutions in many mutant proteins cause folding problems that can be "cured" or mitigated by heat-shock proteins such that the mutant phenotype is not expressed. Hsp90, a major player in this silencing, helps aberrant proteins fold better by giving them "second chances" to fold properly and get their hydrophobic residues internalized. Susan Lindquist and her colleagues discovered the buffering capability of this chaperone in experiments where they manipulated the Hsp90 level in fruit flies (Rutherford and Lindquist, 1998; Sangster et al. 2004). One approach was to use drugs that prevent Hsp90 action; when they were applied to fly embryos, there was an appearance of a variety of developmental disorders (Figure 4.4) mimicking known mutations such as bristles and eyes appearing in the wrong places. These findings have been repeated in yeast and plants, indicating that a similar buffering of mutant protein activity exists in all organisms (Sangster et al. 2004; Jarosz et al. 2010).

Experiments with *C. elegans* embryos reveal that variation in heat-shock protein expression can regulate the extent/penetration/amount of mutant gene expression (Casanueva et al. 2012). Figure 4.5 shows the results of one of these experiments, where worms with several mutations with differing penetrance were subjected to a 2-hour period of 35°C and then brought back to a normal temperature of 20°C. The brief high temperature exposure resulted in an increase in Hsp.2, which like Hsp90 is a chaperone protein involved with protein folding. As shown in Figure 4.5, this increased amount of HSP resulted in more buffering of the mutant gene. This may also explain why some women with mutations in the *BRCA1* gene get breast cancer, while about 40% do not (Pasche and Yi 2010).

The paradox of robustness

The above section suggests that embryonic development can be insulated from environmental variation and even be buffered from expressing some mutations. Obviously, since we live in a constantly changing and evolving biosphere, there must be ways to overcome this robustness to produce altered phenotypes that can be selected for adapted properties. Recent work shows that environmental stress, acting on the HSPs, could be the key that unlocks these cryptic mutations.

Above we referred to Lindquist's work showing mutational buffering by inhibiting the action of the HSPs with drugs; inhibiting HSP prevented mutational buffering and resulted in expression of mutant phenotypes. Lindquist also found that subjecting the embryos to severe heat stresses resulted in a similar appearance of developmental mutants (Jarosz et al. 2010). This seems at odds with the above finding that decreasing Hsp90 activity also brings forth similar mutants. The explanation is that the high heat stress results in major protein unfolding in the cell, which disturbs the equilibrium of the bound and free Hsp90 such that the Hsp90 association with mutant proteins is decreased. The mutant phenotype is now expressed as the observed developmental anomalies.

These experiments, showing that stress can expose hidden mutations, have given rise to the hypothesis that the reservoir of cryptic mutations could provide variation for selection to act upon. This is discussed further in Chapter 11.

The Curious Absence of Stress and Cell Death Responses in Cleavage-Stages of Orphan Embryos

Stress responses differ in adults and embryos

The adult cell, when experiencing stresses such as starvation or temperature shock, initiates stress responses by up-regulating the synthesis of numerous protective **stress proteins** (molecular chaperones such as Hsp90) and initiating specific signaling cascades that protect the cell. In some cases, the response to temperature is so extreme that the cell shifts the bulk of its protein synthesis to the manufacture of these heat-shock proteins. In addition to heat, free radicals (which can damage DNA), toxic chemicals, and osmotic changes each induce their own specific set of stress proteins, as well as a general set of protective factors that can repair cellular damage or, in the worst-case scenario, direct the damaged cell to commit suicide (Feder and Hofmann 1999; Kultz 2005; Bukau et al. 2006).

Given these potent stress defenses in adults, it is paradoxical that the orphan embryos do not show any of these stress responses during the period between fertilization and gastrulation when their blastomeres are rapidly dividing (Heikkila et al. 1997; Lang et al. 1999). The rate of cell division is so rapid that the DNA is unavailable for gene transcription. Probably for this reason, the cleavage-stage cells are remarkably unresponsive to heat stress, and there is no induction of synthesis of the heat-shock proteins until the embryo enters the **midblastula transition**, a point in development when the tempo of cell division begins to slow down, the G phases of the cell cycle first appear, and the nucleus becomes active (Newport and Kirschner 1982). Therefore, when discussing the defenses of the zygote and the early embryo, we must remember that early embryonic cells of orphan embryos cannot call forth the arsenal of protective agents used by adult (or even later embryonic) cells. These cells are not helpless, however; as we have mentioned and will describe later in more detail, embryos are adapted to their historical, anticipated environment, have undergone the equivalent of a stress response to expected dangers during oogenesis, and can handle the expected challenges without mounting the classic stress response.

Cell death responses differ in adults and embryos

Should the proteins or DNA of an adult cell become badly damaged, the cell will sacrifice itself. The two main methods of cellular suicide are referred to as **apoptosis** (programmed cell death) and **autophagy** (the cell "eats" itself in a manner similar to the way in which it clears bacteria out of its cytoplasm). The two processes proceed by different mechanisms, but in both cases externally induced damage causes the cell to synthesize proteins that digest its own constituents. In the end, the cell is physically removed, with only limited consequences to the surrounding tissue.

Embryonic cells also undergo apoptosis when they are severely damaged, but the timing of such cell death is usually delayed until after gastrulation (Figure 4.6; Hensey and Gautier 1998; Ikegami et al. 1999; Thurber and Epel 2007). So if, during the cleavage stage, embryos are exposed to severe stresses such as radiation or DNA-damaging drugs, the cells continue to divide even though their DNA may be irreparably damaged. In many species, every one of the few cells present in an early embryo carries so much developmental information in its cytoplasm that the loss of even one cell can affect development. Embryonic cells thus appear to "postpone" apoptosis until around the time of the midblastula transition—the same point at which the heat-shock response is first seen.

The reason that orphan embryos do not use the stress response, which prolongs cell division (such as the mounting of transcriptional responses or delays in the cell cycle) is that every moment is a dangerous one for predation. Similarly, the loss of an early embryonic cell may mean the loss of an organ. So selection favors rapid development (Strathmann et al. 2002).

In mammalian embryos, development is much slower and the interactions more regulative. Strategies of cell death and slow stress responses (that can occur as early as the 2-cell stage) are therefore possible (Bensaude et al. 1983; Levy 2001; Artus et al. 2006).

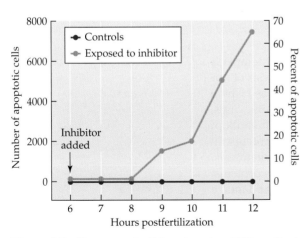

Figure 4.6   Exposure of a fish embryo to a DNA-synthesis inhibitor at 6 hours after fertilization does not induce apoptosis until the midblastula stage (at 9 hours). Controls not exposed to the inhibitor show no apoptosis. (After Ikegami et al. 1999.)

Strategies for Embryo Defense

The warnings to pregnant mothers not to smoke or to be wary of medications might suggest that the fetus is poorly protected and fragile. But embryos *are* well prepared for the stresses of the environment in which they normally develop.

The protective strategies found among animal embryos can be grouped into four major categories: induced polyphenism, parental protection, dormancy and diapause, and cellular stress defenses. These are described briefly below; the remainder of the chapter then details some of the mechanisms by which embryos implement these four strategies.

Strategy 1: Induced polyphenism

The first type of protective program, which was described in detail for a variety of organisms in Chapters 1 and 2 and will be detailed later for early embryonic cells, is induced polyphenism. Here, an embryo can search the environment for cues indicating potential dangers and then alter its developmental trajectory so that it can best survive these dangers. The developmental changes might be alterations in structure to foil predators, or changes in physiology and behavior in response to environmental conditions.

Strategy 2: Parental protection

There are many forms of parental protection. In some instances, especially in mammals, the embryo forms the placenta, which provides a partial barrier against pathogens and toxins. The placenta enables the mother to provide nutrients and oxygen to the embryo from her uterine capillaries. The mother's immune system also protects the placenta-embryo complex from some (although not all; see Chapter 5) environmental toxins, bacteria, and viruses (Glezen and Alpers 1999).

In most organisms, the oocyte or cells associated with the growing oocyte prepare a protective egg coat—variously named the shell, chorion, egg envelope, egg case, or sheath—that provides a physical barrier around the oocyte. Such barriers can comprise a large portion of the embryo mass and, as we describe later, provide protection against physical trauma as well as fungal, bacterial, and viral infections.

In many nonmammalian species, the mother prepares protective compounds that she transports into the egg. The hen, for instance, makes antibodies against common barnyard pathogens, and these antibodies get transported into the yolk of the egg (Ward 2004). In this way, the embryo and hatchling have a passively acquired immunity against common bacteria, viruses, and fungi before they are able to make their own antibodies. The hen also synthesizes several proteins that prevent bacterial growth and which are transported into the egg albumen. These include lysozyme (which prevents bacteria from making their cell walls) and avidin (which binds and inactivates biotin, a bacterial growth factor).

Vertebrates aren't the only animals whose mothers (and in some cases fathers) provide protective compounds for the egg and early embryo. The eggs of the arctiid moth *Utetheisa ornatrix* get a heavy dose of a foul-tasting alkaloid from both the mother and the father (Dussourd et al. 1988). The parents acquire these bitter alkaloids while still in the larval stage by eating the alkaloid-laden plants of the genus *Crotalaria*. These alkaloids do not harm the egg or the larvae, and they continue giving protection to the adult (Figure 4.7). The adult female puts these compounds into her eggs, and the father places the alkaloids into his semen. The female stores the sperm for later fertilization, and the alkaloids in the seminal fluid are transferred to the eggs. Both the male and female contributions to the egg inhibit the predation of the egg by beetles (Dussourd et al. 1988).

Another behavior critical to the protection of many insect larvae is **specific oviposition**, the behavior of the newly mated female to seek out only certain places to lay her eggs. This was discovered by the artist Marian Merian (1679), who noted that butterflies laid their eggs on plants that the *larvae* would eat, not on the food plants used by adults. In many cases, these plants contain poisons that the larva can tolerate but its predators cannot. The monarch butterfly, *Danaus plexippus*, for instance, lays its eggs on milkweed (*Asclepias*), and the larvae ingest a set of cardenolide lipids (similar to digitalis) that give the larva and adult a foul taste and possibly

(A)

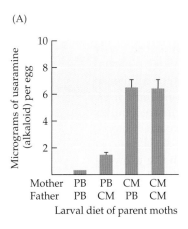

Figure 4.7 Eggs and embryos of the moth *Utetheisa ornatrix* acquire plant alkaloids such as usaramine from their parents. These chemicals appear to protect the embryos from being eaten by predators. (A) Alkaloid content of the eggs made by parents raised on diets containing alkaloids (CM) or no alkaloid (PB). (B) The presence of alkaloid is a major determinant of whether or not the eggs are eaten or not. (C) Adult *U. ornatrix*. (A,B after Dussourd et al. 1988.)

(B)

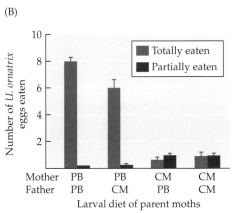

(C)

Figure 4.8   A naive blue jay has just eaten a monarch butterfly that contains cardiac glycosides stored from the milkweed plant that the caterpillar had eaten. About 12 minutes later, the jay vomits several times. Once a jay has had this negative experience, it avoids subsequent monarchs on sight alone. (Photographs courtesy of Lincoln Brower.)

induce nausea* (Parsons 1965; Rothschild et al. 1975; Figure 4.8). These foul-tasting compounds protect the larval and adult butterflies, since once a predator has experienced eating one, it does not do it again (Brower 1969; Brower and Moffit 1974). Moreover, the caterpillars "announce" themselves with bright colors.

The colorful heliconiid group of butterflies similarly get their unpalatability from related cardenolides in the plants they eat as larvae. They digest their poisons from the *Passiflora* vines of the New World and Asian forests (Murawski 1993; Moore 2009), and different species use different parts of the plant. The protection provided by cardenolide compounds is so successful that both the monarch and and heliconiid butterflies have mimics (non-noxious butterflies that look like them). This mimicry will be discussed in Chapter 9.

Strategy 3: Dormancy and diapause

When the going gets tough, some embryos sleep through it all. Like the seeds of many angiosperm plants, the embryos and larvae of many animals can curtail development, becoming metabolically inert and entering a state of **dormancy** when conditions are not favorable for growth and survival.

---

*The monarch has evolved insensitivity to these toxins (Petschenka et al. 2013). We will discuss this butterfly further in the Coda. It is becoming endangered due to the extermination of the larval food plant by new broad-spectrum herbicides.

They are able to restore normal metabolism and continue growing when favorable conditions return.

For example, in overly dry conditions, brine shrimp embryos and certain fly larvae get rid of all water. This anhydrous condition would normally destroy the cell's proteins, but in these larvae the proteins are kept intact and in a functional (although inert) state by the larval synthesis of high concentrations of the protective sugar trehalose (MacRae 2003). The dormancy ends when rainfall brings the larvae "back to life." Another well-studied example of dormancy is that of the nematode *C. elegans*, which in the absence of normal food supply can form a **dauer larva** after its second metamorphic molt (see box). This nonfeeding larval stage can be terminated by providing the larva with adequate food (Cassada and Russell 1975). Thus, one way for an embryo to protect itself is to form a state in which it can sleep through the bad times and wait for better days.

In many animal species, a variant of the dormancy strategy, called **diapause**, has evolved. Diapause is a genetically programmed suspension of development that can occur at the embryonic, larval, pupal, or adult stage, depending on the species. In most cases, diapause is not a physiological response brought about by harsh conditions. Rather, it is brought about by token stimuli that presage a change in the environment; thus diapause usually begins before the actual severe conditions arise. Some specific proteins are induced by the environmental cues, and many of these enable the survival of the animal (Guz et al. 2014; Sim et al. 2015).

In some species, a period of diapause has become an obligatory part of the life cycle. This is often seen in temperate zone insects that overwinter, where diapause is induced by changes in the photoperiod (the relative lengths of day and night). The point at which 50% of the population has entered diapause—an event that is usually quite sudden—is called the **critical day length** (Danilevskii 1965; Tauber et al. 1986).

Silkworm moths (*Bombyx mori*) overwinter as embryos, entering diapause just before segmentation. The gypsy moth (*Lymantia dispar*) initiates its diapause as a fully formed larva and is thus ready to hatch as soon as diapause ends. Hormones appear to be a determining factor in the stage at which an insect enters diapause. Diapause in the embryonic stage appears to be regulated in some insects by increased levels of PTTH (prothoracicotropic hormone), whereas in other insects larval diapause appears to be the result of inhibition of this hormone, which prevents the larvae from molting and entering pupation. In the laboratory, diapausing pupae can be reactivated by exposure to 20-hydroxyecdysone. However, under normal conditions, the brains of diapausing pupae (such as those of the moth *Hyalophora*) are activated by the exposure to cold weather for a particular duration. Moth pupae kept in warm conditions will remain in diapause until they die (see Nijhout 1994). The mechanism by which changes in temperature and day length regulate hormone production remains to be elucidated.

Diapause isn't only for invertebrates. More than a hundred mammalian species also have this ability. For example, some female mammals can delay the implantation of the embryo, keeping it suspended in the uterus. Further development does not proceed until the embryo has attached itself to the uterine lining. In this way, these mammals can mate in the winter and yet

The Dauer Larva of *C. elegans*

Studies on the nematode worm *Caenorhabditis elegans* have provided insights on how the availability of food governs whether, at the second molt stage, this animal continues its metamorphosis and gives rise to a reproductive adult or whether it becomes a dormant larval form known as the dauer larva (Wang et al. 2009). Genetic analysis reveals a complex of interacting sensory systems that determine this decision. The input to these systems comes from chemosensory neurons that detect the "state of the environment" by sensing the concentration of specific chemicals in the environment. One sensory system "measures" the density of the worm population, using a pheromone produced by other *C. elegans* individuals in the immediate environment. Another sensory system reports the availability of food, and a third is sensitive to high temperatures (Bargmann 2006). The worm responds to this sensory input with the activation of signaling cascades that result in the production of numerous neuropeptides, including an insulin-like growth factor and a paracrine factor called transforming growth factor-β (Hu 2007).

These neuropeptides can signal other cells to turn on the *DAF-9* gene, which encodes a P450-type cytochrome that converts cholesterol into a sterol-like compound called dafachronic acid (Yabe et al. 2005; Motola et al. 2006). Dafachronic acid is a ligand for the transcription factor DAF-12, and binding to DAF-12 initiates a cascade of events directing the larva to become a reproductive adult. If food is limiting, however, the neuropeptides are not made, *DAF-9* is not active, and in the absence of dafachronic acid the larva enters the dauer stage.

This dissection of the dauer response points to the common mechanisms used in environmental sensing during development. In the other well-studied cases, such as photoperiod or temperature control of phenotype in butterflies or the temperature control of gender in reptiles, the sensing of the environment is ultimately tied to production of a hormone, which then directs a suite of events leading to the new phenotype (see Chapter 2). The sensing mechanism, interestingly, is most likely also through neurons. This is obviously the case with photoperiod. The mechanism of temperature-sensitive hormone production is still not known, but the temperature-sensitive ion channels are an interesting candidate target (Rosenzweig et al. 2005; Voets et al. 2005).

"time" their pregnancies such that the young are born in the spring, when food is abundant (Renfree and Shaw 2000).

Strategy 4: Defense physiologies

A range of physiological strategies allows the embryo to cope with everyday stresses. These stresses include heat that can damage proteins, toxic chemicals and radiation that can damage DNA, and physical stresses that can damage cell membranes. These defense physiologies apparently do not interact with the developmental program, but they ensure that the embryo can survive while carrying out its development. The heat-shock proteins mentioned earlier, as well as molecular mechanisms for removing toxic substances from cells and the provision of "sunscreen" chemicals—both described in detail below—are among the prime examples of embryonic defenses against stress.

A general strategy: "Be prepared"

Many of the mechanisms that have evolved to protect the early embryo are versions of the Boy Scout injunction "Be prepared." Just as the scout is prepared to survive emergencies that might come up while camping or hiking, so has there been selection for adaptations that "prepare" the embryo to survive stressors it might encounter during its development (see Figure 4.1). Like the scout's preparedness, the embryo's preparedness is directed toward *potential* problems, not toward *all* problems. If historically the embryo develops in a cold environment such as the Antarctic, there will be adaptations (such as unique microtubular proteins that are functional at –2°C) to allow the embryo to survive and develop in a frigid habitat (Detrich et al. 2000). Cells of an embryo developing in the desert will have adaptations against heat (David et al. 2004). For the embryo developing in the full glare of sunlight on a beach (which would fry our DNA if we forgot sunscreen), embryonic cells have ways of avoiding or repairing UV damage. The protective and defensive evolutionary adaptations found in embryonic cells provide effective, and in many cases novel, means to survive in remarkably rigorous environments (see Hamdoun and Epel 2007).

Mechanisms of Embryo Defense

How do embryos defend themselves before the differentiation of the adult tissues and organs that normally provide protection? How does the early embryo handle toxic substances, whether natural or man-made? How does it keep its DNA intact in the face of such stresses as UV radiation or the free radicals produced by normal metabolism? How can it defend itself against physical damage? Is the membranous envelope that surrounds the embryo equivalent to skin or skeleton? What adaptations are there to ward off pathogens without an immune system, or predators without some sort of protective structure or behavior response? Finally, will embryos be able to adapt to the rapid climate changes resulting from increases in greenhouse gases, the increase in temperature, and the acidification of the oceans?

The answer to these questions, as you will see, is that there is a remarkable constellation of defenses that can work well at this few-celled level. Their presence indicates the environmental pressures that have operated over the eons and the remarkable selection that must have taken place to provide these defenses.

We start with how embryos thwart exposure to toxic substances while simultaneously carrying out the developmental program. Remember that these defenses have evolved to defend the embryo in its anticipated, historical milieu. But that milieu, that environment, is being changed rapidly by human activities. Can embryos evolve fast enough to withstand the effects of new chemicals and environmental conditions? Some chemicals will get through the embryos' defenses and affect development. These agents are called teratogens and will be discussed in Chapter 5. Some chemicals will affect developmental pathways in ways that won't become apparent until later in life; such agents will be considered in Chapter 6. And some conditions, such as climate change, will result from indirect activities of a rapidly growing and consuming human population. A question to consider, therefore, is whether these ancient embryonic defenses will prove adequate in this rapidly changing world (see Hamdoun and Epel 2007 and the last section of this chapter).

Protection against Toxic Substances

An embryo, whether protected by a parent or developing independently, can be exposed to toxic substances from its mother's diet or from the environment. Some of these toxic substances are from other organisms (such as defensive chemicals), some are from the abiotic environment (such as heavy metals resulting from volcanic eruptions or industrial spillage), and some are from chemicals added to consumer products (such as plasticizers, bisphenol A, and fire retardants).

The general plan: "Bouncers," "chemists," and "policemen"

Whereas the adult animal uses sophisticated enzyme pathways in the liver and kidney for inactivating, metabolizing, and removing poisons, the embryo must do this at the cellular level. Their general strategy is to use a three-tier scheme to rid themselves of poisons:

1. "Bouncers" that prevent the entry of toxic substances into the cell
2. A set of "chemists" that modify (detoxify) the dangerous molecules and then put some sort of chemical tag on the modified molecules
3. A set of "policemen" that then escort the tagged toxicants out of the cell

The "bouncer" and "policeman" functions are carried out by **efflux transporters**, which are members of the ATP-binding cassette (**ABC**) family of proteins, one of the largest protein families known. ABC efflux transporters use the energy of ATP hydrolysis to carry compounds across cell membranes, either transporting them into the cell or transporting foreign or endogenous compounds out of the cell (Figure 4.9).

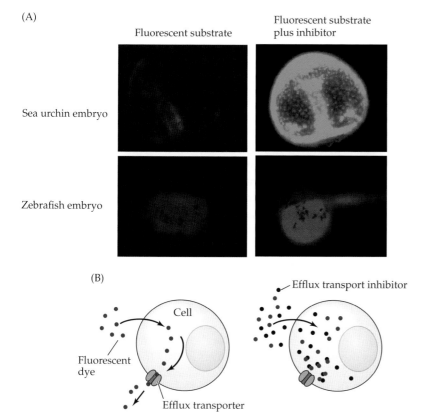

Figure 4.9 An efflux transporter in embryos of the sea urchin and zebrafish prevents entry of a fluorescent dye. (A) Micrographs on the left show effective exclusion of the fluorescent dye calcein AM. Micrographs on the right depict the consequences of inhibiting transport activity in the same material. As seen, there is little entry of the dye unless the transporter is inhibited. (B) The diagram shows what happens when transport activity is inhibited. (From Epel et al. 2008.)

The "bouncer" function, which is the most effective way of dealing with toxic compounds, is to not let the compounds get into the cell. This function is handled by members of the P-glycoprotein sub-family of ABC transporters. These glycoproteins interact with the toxicants within the plasma membrane and actively prevent their entry into the cell (see reviews by Higgins 1992; Cole and Deeley 1998; Epel et al. 2008).

The "chemist" role is carried out by suites of enzymes that modify the putative toxicant, usually by adding groups that make the toxic substance more water soluble. Typically this can involve oxidation by the enzymes of the cytochrome P450 system, followed by direct conjugation of a polar (hydrophilic) group to the oxidized toxicant. This polar group can be a glutathione (a tripeptide containing a cysteine), a sugar, or even a sulfate group. If the toxicant can be conjugated directly, the P450 step need not take place.

The next step is carried out by the "policeman." This step involves binding any toxicants that entered the cell and were modified and escorting them out of the cell. This is carried out by the multidrug-resistance proteins (**MRP**) sub-family of ABC transporters.

The properties of the efflux transporters in embryos have been best described in the sea urchin embryo (see Hamdoun et al. 2004; Goldstone et al. 2006). The unfertilized sea urchin egg has low ABC activity, but this activity increases markedly after fertilization. It appears that an ABC transporter belonging to the P-glycoprotein family (which primarily prevents toxicant entry into the cell) and a transporter belonging to the MRP family (which pumps toxicants out of cells after they have entered the cytoplasm and been modified) are already present in the egg in an inactive form. Fertilization initiates steps that move the transporter to the plasma membrane and tips of the microvilli (Figure 4.10), and this translocation is somehow tied to an almost 200-fold increase in activity that begins about 25 minutes after fertilization (Hamdoun et al. 2004; Gökirmak et al. 2014).

These transporters defend the embryo against toxic chemicals. As shown in Figure 4.11, cell division in the sea urchin embryo is inhibited by vinblastine, a microtubule-disrupting chemical. The concentration of vinblastine needed to inhibit 50% of the embryos from dividing is 2.5 μM. However, if the transporter activity is inhibited, the effective concentration drops to 0.1 μM. In other words, the transporter activity is so effective that 25 times more vinblastine is required to inhibit mitosis when the transporter is functioning than when it isn't. (Vinblastine is a protective alkaloid made by plants of the periwinkle family.)

The limited work on other embryos suggests drug transporters might be general defenses in all embryos. They are present in the embryos of several other marine organisms, such as the echiuroid worm *Urechis caupo* (Toomey and Epel 1993) and the mussel *Mytilus galloprovincialis* (McFadzen et al. 2000), and they are also found in embryos of the soil nematode *C. elegans* (Broeks et al. 1995). All of these organisms develop in

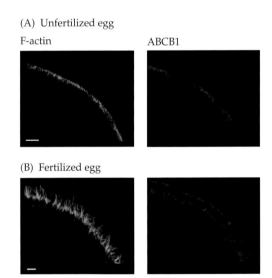

(A) Unfertilized egg

F-actin   ABCB1

(B) Fertilized egg

Figure 4.10   Movement of ABCB1 (P-glycoprotein) from plasma membrane to the microvilli after fertilization. (A) Unfertilized egg. The transporter protein (red) is distributed evenly in the sub-cortical region in the unfertilized egg. The actin (green) is present but not formed into microvilli. (B) Sixty minutes after fertilization. The actin (green) has polymerized and formed long, thin microvilli extending into the surrounding seawater. The transporter protein (red) has moved to the tips of microvilli where presumably they will interact with substrates to prevent them from entering the embryo cytoplasm. (From Whalen et al. 2012.)

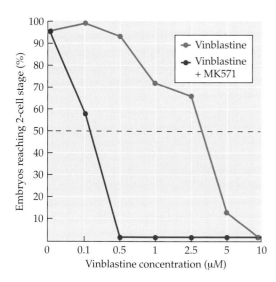

Figure 4.11 An efflux transporter protects sea urchin embryos from the effects of a cell division inhibitor. Incubation of sea urchin embryos in vinblastine destroys microtubules and inhibits mitosis. The 50% concentration necessary for inhibition (indicated by the dotted horizontal line) is about 2.5 μ*M* for vinblastine only. If the MRP transporter inhibitor MK571 is added, the 50% effective concentration (dotted line) goes down to 0.1 μ*M*. (After Hamdoun et al. 2004.)

environments that can be very toxic, and the titers of ABC proteins found in *U. caupo* (which lives in mudflats) are higher even than in drug-resistant cancer cells that overexpress ABC proteins (Toomey and Epel 1993). *C. elegans* uses an ABC transporter to export dafachronic acid, the signal regulating dauer formation (see p. 149). Among mammals, protective ABC transporters have been found in early mouse embryos (Ebling et al. 1993), as well as in the mouse placenta. Smit and colleagues (1999) showed that these placental transporters play important roles in preventing toxins from entering and accumulating in the mouse embryo; it is assumed that similar transporters serve a similar role in the human placenta.

The transporter defense, however, is of no use if the transporter protein does not recognize a dangerous chemical. One property of teratogens, those chemicals that can disrupt development, is that they can evade the ABC transporters as well as bypass the pathways for detoxification of the chemical should it get into the cell. One example (detailed in the next chapter) is thalidomide. This drug, given to pregnant women in the 1950s and early 60s, caused birth defects involving the limbs and ears. We now know that part of thalidomide's effectiveness as a teratogen was due to the drug's not being recognized by the ABC efflux transporters in the placenta (Zimmerman et al. 2006; Beghin et al. 2010); the drug thus had unfettered access to the embryo.

Toxic metals

Below we describe cellular mechanisms of defense against toxic metals such as mercury, lead, and cadmium. These defenses can be overwhelmed by high levels of these metals and, as detailed in Chapter 5, can result in horrific developmental abnormalities and birth defects. Their toxicity can ensue from (1) interfering with enzyme activity, as when mercury or cadmium binds to a protein (often to a critical cysteine residue on the protein); (2) preventing closure of a calcium ion channel, so toxic levels of calcium can enter a cell; or (3) generating reactive oxidative intermediates. Some metals can be toxic through all three mechanisms.

Whereas the protection against many poisonous organic compounds was through the "bouncers" (i.e., the family of efflux transporters that prohibit entry into the cell), the defenses against heavy metals are carried out by the "chemists" and "policemen," which in this case are enzymes along with specialized peptides and proteins that shackle the metals and then escort them out of the cell. These defenses are based on sequestering the metals through binding to sulfhydryl groups (–SH) on specific defense molecules.* The sulfhydryl groups bind avidly to heavy metals, and they often enclose them, thereby preventing the metals from doing damage. Sulfhydryl groups do double duty in cell protection, since they are also antioxidants (i.e., they reduce oxidant molecules, as discussed later in this chapter).

Another major defense molecule against heavy metals is the tripeptide glutathione, composed of glutamate-cysteine-glycine, which, with its sulfhydryl-containing cysteine group, binds with high affinity to heavy metals such as mercury and cadmium. Glutathione is very abundant in cells (in the millimolar range) and, as noted later in this chapter, also provides critical defenses against oxidant damage. The cells get rid of the metal-glutathione conjugate by pumping it out of the cytoplasm using members of the MRP family of efflux transporters, as described above (Cole and Deeley 1998).

Variants of the glutathione defense are the phytochelatins, which are polymers of glutathione and similarly bind to heavy metals. Phytochelatins, as the name implies, had been assumed to be plant-specific, but related phytochelatin synthase genes have also been found in animals, suggesting their use may be more widespread (Goldstone et al. 2006).

The third defense against heavy metal toxicity is a special class of metal-binding proteins, the metallothioneins (MTs). These proteins are cysteine-rich (at least 30% of their component amino acids are cysteines), and as in glutathione, the –SH group of the cysteines binds to the metals (Palmiter 1998). This class of protein also has antioxidant activity.

If the organism is exposed to heavy metals, its cells respond by increasing *MT* gene transcription and translation, which provides further protection. The genes encoding MTs (and other metal transporter proteins) have an enhancer region containing a metal-sensing regulatory element, or MRE. The MRE sequence binds the transcription factor MTF-1; MTF-1, however, will only induce *MT* gene expression if it has bound a heavy metal ion. The heavy metal ion appears to alter the shape of MTF-1 so that it can recruit a histone acetyltransferase capable of activating gene expression (Li et al. 2008). Thus, the genes encoding MT and other proteins that protect against heavy metals are activated by the heavy metals themselves. MTF-1 also regulates the transcription of antioxidant enzymes, thus providing multiple defenses against the consequences of metal exposure. Indeed, like the

---

*We have also learned to use sulfhydryl groups to combat heavy metals in medical practice. British anti-lewisite (BAL) is a 3-carbon backbone linked to two sulfhydryl groups. BAL was developed to treat exposure to the arsenic-containing compound lewisite (the "dew of death"), an agent of chemical warfare. BAL and its related sulfhydryl-containing compound, DMSA, have since become important in the treatment of people who have been exposed to heavy metals.

heat-shock protein scenario, a previous exposure of cells to low levels of a toxic metal will induce the synthesis of more protective molecules, in this case more MTs, thus providing protection against subsequent escalation of the metal concentration (Andrews 2000).

The defense against toxic metals during the cleavage stages of orphan embryos utilizes endogenous glutathione and MTs to sequester metals. If exposed to heavy metals during cleavage, these embryos do not initiate additional transcription of MTs. As in the situation with heat-shock proteins, the induction of new MT occurs only after gastrulation, meaning that the basal levels of protective molecules present in the egg are the only protection during the cleavage period against any heavy metal exposure (Scudiero et al. 1994). This level is presumably based on the anticipated exposure to heavy metals and will be effective as long as there is no deviation from the expected levels of metals in the embryos' environment. After gastrulation, when the pace of cell division slows, elective synthesis of MTs can occur (Nemer et al. 1991). Prior to this, the embryo depends on its "be prepared" strategy for the anticipated environment.

In the slower-developing (and non-orphan) mammalian embryos, there is a different scenario, most likely related to the long duration of the cell cycle in mammals, which then allows time for new transcription in response to environmental stresses. Here induction of MT by heavy metals can occur during cleavage, by the four-cell stage (Andrews et al. 1991). This behavior is similar for the induction of heat-shock proteins, which are also inducible in the mouse embryo during cleavage. Work on the mouse embryo, however, indicates that MT is not solely for protection from toxicants; MTs have other roles, protecting cells from oxidative damage and also acting as a metal buffer by providing metals needed for normal cell functioning, such as zinc (Andrews et al. 1991; Palmiter 1998).

Problems with metal detoxification

The trouble with metal-binding defenses is getting rid of the bound metal. In plants, the metal-loaded phytochelatins are secreted into the plant cell vacuole, which essentially isolates the toxicant from the cell cytoplasm (Cobbett 2000). A similar mechanism may exist in animal cells via transport of metal-loaded glutathione into lysosomal-type vesicles (Berry 1996). However, it is thought that the bulk of the metals bound to glutathione are ejected via the efflux transporters, which pump the metal conjugate into the extracellular medium; this method merely displaces the problem to adjacent cells or tissues (since the metal-loaded glutathione remains in the embryo or organism).

There is a similar problem with the metallothioneins. The metal-loaded MTs can stay in the cell a while but eventually are leaked or diffuse from the cell. In vertebrates—both larvae and adults—the metal-loaded MT travels to the kidneys, where hydrolysis of the MT leads to the release of toxic metals; if toxin levels are too high, there can be kidney damage (Klaassen and Liu 1997).

The defense provided by the MTs (and probably by glutathione as well) may actually be a sort of timing game. The immediate toxic effect of high metal levels is ameliorated by sequestration, but eventually the metal is released. The benefit may be that the subsequent release takes place over a longer time period, so the concentration is smaller and hence less toxic. This scheme works for limited exposure to metals but would not be effective if the exposure were chronic.

Protection against Physical Damage—The Egg Coat

Eggs are fragile, but as described below, animal eggs have adaptations that allow the embryos to survive in environments that could easily be expected to convert them into omelets.

One structural problem is size. The egg is often the largest cell in the organism and maintaining this large size requires a rigid cortex. This rigidity is provided by an actin cytoskeleton embedded in the cortical layer of the egg, which might be thought of as an actin "girdle" that maintains the egg's ovoid structure and rigidity (Elinson and Houliston 1990; Spudich 1992; Ryabova and Vassetzky 1997). If one depolymerizes the actin, as with the microfilament-destroying drug cytochalasin, the egg loses its spherical shape and begins to ooze or flow onto whatever surface it is on.

The rigid cortex is itself not an adequate protection from physical injury and the solution is some sort of extracellular envelope. In most eggs, the envelope is laid down before ovulation and fertilization so that when the egg is released into the environment, the coat is there to protect the newly released egg.

The presence of this coat also means that the fertilizing sperm has to pass through the egg coat in order to fertilize the egg. This is not a barrier to fertilization, however, but is a "fertilization filter," ensuring that only sperm of the same species can bind to the outer envelope or pass through it on the way to fusing with the egg plasma membrane (Vacquier 1998). Some eggs, such as bird eggs, have such an elaborate or thick shell that the sperm could never penetrate it. In birds, the sperm are stored in and fertilization occurs before the outer layers and shells are deposited. The avian egg also has another layer, the albumen, or "egg white" which serves as a shock absorber.

These outer egg coats are of course barriers to physical damage and also function to prevent bacterial and fungal infection (discussed in detail later in this chapter). They can also be a major structural barrier to viruses. Larva and adult frogs are highly susceptible to hantaviruses, but the embryo, which develops inside the chorion, is not affected by these pathogens. However, it can be infected if the virus is injected directly into the embryo (Haislip et al. 2011).

The early extracellular coat of mammalian eggs and embryos

The extracellular coat of the mammalian egg is an example of a protective coat that also acts as a gatekeeper for fertilizing sperm. The components of this thick extracellular matrix, called the **zona pellucida**, are synthesized by the developing oocyte. The zona pellucida surrounds the unfertilized egg and, as noted, is a physical barrier that challenges sperm on their way to the plasma membrane of the oocyte. The zona pellucida of the mouse egg has three main structural proteins: ZP1, ZP2, and ZP3. If a sperm is of the correct species, and if it is able to bind to ZP2, it can pass through the zona and reach the egg (Baibakov et al. 2012; Avella et al. 2014).

As the embryo undergoes cleavage inside the zona pellucida, the enclosing zona allows the embryo to reach the uterus without attaching to the Fallopian tubes. If such attachment does occur, it results in an ectopic, or "tubal," pregnancy that can kill the mother as well as the embryo. In addition, the zona apparently protects the embryo from viruses and bacteria that might have evaded the mother's immune system. Mateusen and colleagues (2004) found that if pig embryos were incubated with certain porcine viruses, none of the embryonic cells became infected as long as the zona pellucida was present. When the zona was removed, more than 50% of the cells were rapidly infected.

The sea urchin extracellular coat

The egg of the sea urchin elevates a protective fertilization envelope within a few seconds after sperm entry. This envelope arises from the conversion of a thin, ephemeral precursor layer adhering closely to the egg surface—the vitelline layer—into the substantial fertilization envelope (Figure 4.12). The vitelline layer is similar to the zona pellucida in that it also has sperm binding sites, and only sperm of the same species can attach to these sites (Vacquier 1998). The fertilizing sperm digests a hole through the vitelline layer, followed by fusion of the sperm with the egg plasma membrane (Vacquier 1998; Briggs and Wessel 2006).

The major role of the fertilization envelope is to protect the embryo from physical damage. Its efficacy was best demonstrated by a simple but elegant experiment by Miyake and McNeil (1998). They compared the resistance of unfertilized eggs (without fertilization envelopes) and zygotes (with fertilization envelopes) to the shear forces generated when an egg or embryo suspension is forced through a syringe and narrow needle. They found that the fertilized eggs were much more resistant to damage than were unfertilized eggs, supporting the idea that the elevation of the fertilization envelope provides physical protection against shear stresses that might be encountered in the environment. There may also be additional protection from an increase in polymerized actin in the cortex, which also occurs after fertilization (Spudich 1992).

The formation of the fertilization envelope results from a cascade of events that results in the exocytosis of thousands of protein-containing **cortical granules** embedded in the egg cortex. In sea urchins, this cortical granule exocytosis releases a large number of proteins and enzymes whose action simultaneously prevents further sperm-egg binding and causes the detachment and subsequent elevation of the vitelline layer from the egg surface.* This newly-elevated layer is initially quite soft, but within a few minutes it undergoes major structural changes to become a rigid physical barrier—the fertilization envelope—that protects the newly formed embryo (Wong and Wessel 2006, 2008).

---

*Similarly in the mouse, the bursting of the cortical granules releases enzymes that modify the ZP3 and ZP2 proteins mentioned earlier, preventing any other sperm from entering into the egg (Sun 2003; Wortzman-Show et al. 2007).

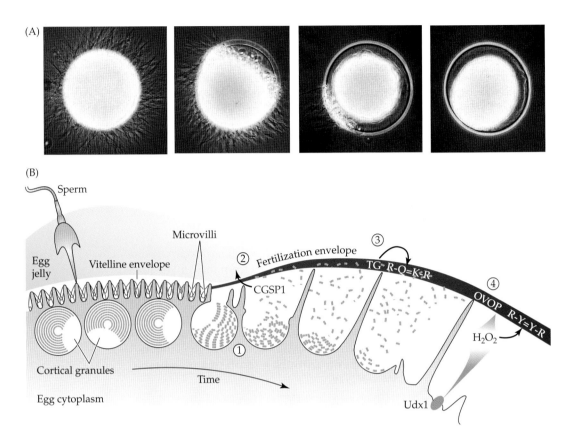

Figure 4.12 Formation of the sea urchin fertilization envelope. (A) Sequence of sperm binding, elevation of the fertilization envelope and concomitant detachment of the unsuccessful sperm from the egg surface. The first photo is 30 seconds after adding sperm; subsequent photos over the next 60 seconds. The fertilization envelope protects the early embryo from shear forces and also blocks other sperm from entering the egg (which would have lethal consequences). (B) Steps leading to the elevation and hardening of the fertilization envelope: (1) Signals arising from fusion of the sperm and egg cause the membranes of cortical granules to fuse with the egg cell membrane. This exocytosis releases the protein-digesting enzyme CGSP1, the ovoperoxidase (OVOP) enzyme, and several structural proteins targeted to the fertilization envelope. (2) CGSP1 cleaves the protein binding the vitelline envelope to the cell membrane, thereby freeing the vitelline envelope. Mucopolysaccharides from the cortical granules bring in water that separates the cell membrane from the vitelline envelope (which from this point on is called the fertilization envelope). (3) CGSP1 activates a transglutaminase enzyme, present on the tips of the egg's microvilli, which cross-links glutamines and lysines on adjacent proteins. (4) The peroxidase enzymes (Udx1 on the cell membrane and ovoperoxidase) make hydrogen peroxide, which cross-links adjacent tyrosine residues, hardening the envelope to become resistant to shear forces. (A from Vaquier and Payne 1973, photographs courtesy of V. D. Vaquier; B after Wong and Wessel 2008.)

A surprising aspect, best understood in the sea urchin, is the complexity of the assembly, elevation, and hardening of the vitelline layer as it is converted to fertilization envelope (see Figure 4.12). The constituents of the cortical granules cause the overlying vitelline envelope to rise and become a fragile sphere surrounding the egg; many of the proteins released from the granules are then incorporated into the nascent fertilization envelope. These released proteins, along with the vitelline envelope proteins, are linked together by the actions of a peroxidase enzyme (also released from the cortical granules) that cross-links tyrosine residues on adjacent proteins to convert the soft, membranous vitelline layer into the rigid fertilization envelope (Somers and Shapiro 1991; Wong and Wessel 2008).

Besides protecting the egg from physical stresses such as wave forces, this robust membrane could have many additional roles. This area is ripe for ecological analysis. For example, the fertilization envelope might also provide protection against predators, including organisms such as zooplankton that feed on embryos; or, like the zona pellucida, it could protect the embryo from pathogenic bacteria and fungi by forming a physical barrier to their entry.

Gastropod egg capsules

Forming a protective physical barrier is also the case with the elaborate egg capsules housing large numbers of gastropod embryos, including snails in the genus *Nucella*. These snails lay large egg capsules that are often attacked by intertidal isopods (Figure 4.13). The larvae develop completely inside these capsules, hatching out as fully formed small snails. The thickness of the *Nucella* capsule is variable, and Rawlings (1994, 1995) found that thin capsules were more vulnerable to predation from these isopods than thick capsules. He also found that the capsule wall protected against pathogens; if the wall was damaged, protozoa and bacteria entered, with larval death rapidly ensuing.

Figure 4.13   Two dogwinkle snails, *Nucella lapillus*, are seen among their oval yellow egg capsules.

The cost of protection

The protection provided by extracellular coatings comes at significant expense. Perron (1981) calculated that as much as 50% of the weight of the embryo mass in cone snails is in their large protective capsule. There are between 10,000 and 15,000 cortical granules in sea urchin eggs, comprising 5%–10% of the egg proteins (Vacquier 1975). That's a lot of energy for structures that are not used after development is complete. But the coatings persist until the larva hatches (about 12–24 hours in sea urchins and up to 3 weeks in cone snails), so its cost may be small relative to its protective value (see Strathmann 2007 and Ovarzun and Strathmann 2011 for a discussion of reproductive strategies).

Protection against Physical Damage—Protection of the Plasma Membrane

When a tire is punctured, compressed air leaks out, and usually the tire collapses. However, some tires have an extra layer inside, containing a sealant that wraps around the inserted object and then fills in the gap when the object is removed. The plasma membrane has such self-sealant properties and can quickly repair the membrane and prevent cell lysis.

Historically, such a mechanism could have been inferred from experiments that started in the late 1800s. At that time, embryologists trying to understand the organization of the egg were busily cutting regions out of eggs and embryos and transplanting other regions back in. In some instances, eggs were cut in half with glass needles, and they were found to heal. Later embryologists often injected dyes into eggs, and recently scientists began studying the electrical properties of eggs by inserting microelectrodes. In each case, the cell membrane was able to heal, to reseal, and the egg or embryo was capable of resuming development.

Although it was obvious that embryos could survive these assaults, few people paid attention to the question of how such physical insult could be tolerated. Unique were the 1930 studies of the cell physiologist Lewis Heilbrunn, who noted a requirement for calcium in order for the embryos to survive mechanical damage (Heilbrunn 1943). He suggested that calcium ions (Ca^{2+}) were required for "clotting" the cytoplasm, noting the similarity to the then recently discovered requirement of Ca^{2+} for blood clotting.* Heilbrunn's work, however, was largely forgotten until the question of how cells survive such damage was asked again in the 1990s.

Steinhardt and his colleagues (1994) revisited cell repair by studying how sea urchin eggs and embryos sealed themselves after insertion of a microneedle. They injected fluorescent dyes and determined how quickly the dyes diffused from the cell. If the cell was able to seal itself, the dye remained in the cell and there was no change in fluorescence. However, if the cell could not seal itself, the dye flowed out into the extracellular medium and was detected as a loss of cell fluorescence.

*Some embryologists took pains to get special water in which to do their dissections. Viktor Hamburger (1988, p. 22) tells of Hans Spemann's ordering water from Würzburg to be shipped to his laboratory because his dissected amphibians did better in that water than in any other. The secret ingredient of the "Würzburger Wasser"? Calcium ions.

These experiments confirmed Heilbrunn's findings of a calcium requirement for cell healing. But, surprisingly, they found that the calcium concentration required was quite high. Instead of being in the submicromolar range that is used for intracellular signaling events, the calcium requirement for sealing was between 1 mM and 10 mM. This level is more akin to the extracellular calcium concentration in the blood (as seen in mammals) or seawater (as for marine embryos) than to intracellular free calcium concentration.

These studies revealed that the repair was initiated by an influx of calcium ions from the outside media, into the cell, when it was wounded. This entry induced a calcium-mediated fusion of intracellular vesicles with each other and with the plasma membrane, as illustrated in Figure 4.14. The fusion-generated membrane material sealed together at the edges of the torn membrane (McNeil and Terasaki 2001). This sealing phenomenon appears to be a general property of cells; for example, a similar calcium-induced fusion of intracellular vesicles seems to operate in muscle cells that are torn during heavy exercise (McNeil and Steinhardt 2003).

This type of sealing, however, does not account for the ability of severely damaged eggs (such as those cut in half) to heal and to continue developing. This question was studied by Terasaki and colleagues (1997), who found that the mechanism for larger-scale wound repair also relied on a calcium-induced fusion of cell vesicles but that the mechanism was more similar to that of a tire patch being added to seal a hole in a pneumatic tire.

But why do embryos have a mechanism to protect them against the massive injuries inflicted upon them by embryologists or cell biologists? A reasonable hypothesis is that it may have originally evolved as an essential component of cytokinesis, exocytosis events (such as the bursting of cortical granules in eggs and synaptic vesicles in neurons), and fertilization (fusion of the sperm and egg membranes). The massive cell-sealing mechanism might also be an adaptation to protect embryos from predators

(A)   (B)

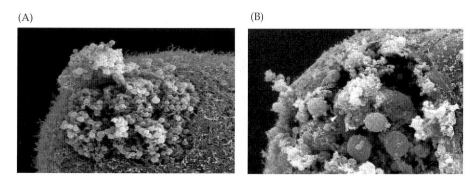

Figure 4.14   Calcium is required for cell vesicle fusion. In this experiment, the surfaces of sea urchin eggs were sheared to tear the plasma membrane and then immediately placed into a fixative, which stopped any further fusion. In these micrographs, the membrane has not had time to seal and is still open. (A) In cells sheared in the absence of Ca^{2+}, the intracellular vesicles are small and of uniform size. (B) In cells sheared in the presence of Ca^{2+}, the vesicles are larger and of varying sizes, suggesting that vesicle fusion took place. (From McNeil and Baker 2001, photographs courtesy of P. McNeil.)

who feed using small bites or from physical forces such as those applied by waves. Both types of actions could kill the early embryo if there were no mechanism to repair minor rips in the membrane.

Protection against Oxidative Damage

Reactive oxygen species (ROS), normal components of cell life, have their good and bad sides (Olsen et al. 2013). Oxidative damage is a major stressor of cells. It can be initiated by ROS, which are produced as a by-product of normal metabolism and respiration, but it can also arise from exposure to metal ions, ultraviolet radiation, and certain pollutants. The damage (as we will see in Chapter 8) is linked to mutagenesis, cancer, aging, and even Alzheimer's disease. But some types of ROS, such as nitric oxide, are essential signaling molecules for critical processes such as neural activity, regulating heart beat and blood vessel pressure, maintaining penile erection, and killing bacterial pathogens.

Embryo and adult cells have mechanisms to deal with the potential ROS damage, using antioxidant molecules such as glutathione, ascorbic acid, and the metallothioneins as well as enzymes that degrade ROS, such as catalase and superoxide dismutase. Embryos, moreover, have additional embryo-specific protective mechanisms that have probably evolved to mitigate the use of ROS as important signaling molecules in early development (Finkel 2003; Coffman and Denegre 2007), to protect embryos from environmental sources of ROS, and in some embryos to protect against the ROS used to harden extracellular protective coats, such as the fertilization envelopes of sea urchins (Wong et al. 2004; Wong and Wessel 2008).

Enormous amounts of ROS are produced during sea urchin fertilization. Here, hydrogen peroxide is produced for the peroxidase reactions that harden the fertilization envelope (Turner et al. 1987; Wong et al. 2004). Fertilization also results in the production of signaling molecules that are themselves free radicals or that produce free radicals as by-products of their formation. One of these molecules is nitric oxide, a free radical involved in regulating calcium levels after fertilization (Kuo et al. 2000; Leckie et al. 2003). Arachidonic acid derivatives, lipids that arise from the activation of a lipoxygenase enzyme at fertilization, can lead to reactive lipid intermediates that in turn can lead to reactive oxygen and damage to cell components (Perry and Epel 1981).*

So how does the embryo deal with this sudden bolus of ROS at fertilization? One coping mechanism is to use the standard defenses seen in other cells: molecules that scavenge ROS, such as ascorbate and glutathione, and the neutralization of ROS by such enzymes as superoxide dismutase and catalase (Dickinson and Forman 2002). But this is apparently not adequate. Instead, sea urchins use an embryo-unique sulfhydryl-containing antioxidant that is a modified histidine (Turner et al. 1987; Shapiro and Hopkins

*These events do not appear to be unique to sea urchin eggs, as such ROS-creating reactions have been detected in starfish eggs (Brash et al. 1991). Most eggs (including those of mammals) have not been studied for these reactions.

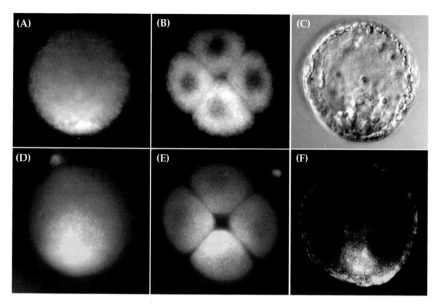

Figure 4.15 Oral-aboral axis specification in the sea urchin embryo is determined by the mitochondria and resultant redox state. The positions of the mitochondria were experimentally manipulated by centrifugation (A–C) or by microinjection of extra mitochondria (D–F). (A–C) Mitochondria were centrifuged to the bottom half of the egg (A); when the egg was fertilized, the future anus/archenteron developed in that area (B, C). (D–F) Mitochondria injected into the unfertilized egg are stained green (D). Following fertilization, the anus/archenteron similarly forms at the site of the extra mitochondria. (From Coffman et al. 2004.)

1991). This antioxidant is called ovothiol, acknowledging its uniqueness to eggs (*ovo*) and its sulfhydryl (*thiol*) group. Ovothiol concentration in sea urchin eggs is high, about 5 mM in the cytoplasm (similar to glutathione concentrations). It is also found in the eggs of several other invertebrate organisms (Shapiro and Hopkins 1991). Shapiro and colleagues, who first discovered ovothiol in sea urchins, found that it was even more effective than glutathione for the types of oxidants produced at fertilization (Turner et al. 1987).

This increased effectiveness may be important in sea urchin embryos, in which ROS are also used as signaling molecules in later development: mitochondrial gradients and a resultant gradient of ROS appear to have important roles in determining the animal's oral-aboral (mouth-posterior) axis (Figure 4.15; Coffman et al. 2004; see Coffman and Denegre 2007 and Coffman et al. 2014). The regulation of ROS therefore must be highly precise and must not allow high levels of ROS "noise" to interfere with the crucial signaling function of these free radicals.

Protection against Damage to DNA

We are constantly told that exposure to sunlight increases our risk of getting cancer and that we should put on sunscreens to avoid the DNA-damaging effects of ultraviolet (UV) light. But as we lie on the sunlit beach, embryos are floating in tide pools only a few feet away. How are they protected? It may seem as if their mothers put sunscreen on them, too.

Sunscreens prevent DNA damage

The problem for many aquatic embryos is that their physiology and habitat put them at the air-water interface. Fish and frog embryos, for example, have lipid reserves that can make them so buoyant that they float at the surface. Some tunicate embryos have extraembryonic cells attached to their chorions that are less dense than seawater, which also results in floating eggs and embryos. (This is made possible by the unique adaptation of substituting ammonium ion for sodium ion in the cytoplasm of the extraembryonic cells. Ammonium is lighter than sodium, so these cells act like life jackets that carry the eggs or embryos to the surface; see Lambert and Lambert 1978.)

A floating lifestyle can result in exposure to UV radiation and resultant DNA and protein damage. Damage from UV radiation can arise from the direct absorption of the UV by proteins and DNA, which have absorption peaks at 280 nm and at 260 nm, respectively.* UV at these wavelengths is blocked by the Earth's ozone layer, but enough light in the 300–320 nm range gets through to generate cross-linking between adjacent thymine nucleotides; this **dimer formation** prevents the proper replication of the DNA and results in mutation (Franklin and Haseltine 1986). Another form of damage, which is indirect, is the generation of ROS by radiation in the 320–400 nm range, which can generate oxidative damage to membranes, proteins, and even DNA (Heck et al. 2003).

One adaptation that prevents UV damage is spawning at night. This is the situation in Australia's Great Barrier Reef, where more than 100 species of coral spawn on the same night each year (Babcock et al. 1986). This mass spawning event is thought to maximize survival, since the embryos' predators rapidly become satiated by the large amount of food suddenly provided. But the nocturnal spawning also means that the embryos pass through most vulnerable cleavage stages before their first exposure to UV radiation at dawn.

In the waters of the Arctic and Antarctic, embryos face an even more severe problem. The embryos of the Antarctic sea urchin *Sterechinus neumayeri*, for example, experience extreme exposure to UV, since they develop during the polar summer and thus are exposed to 24 hours of sunlight a day. This exposure is compounded by the fact that in the frigid Antarctic water (–1.9°C), development is extremely slow—it takes 3 weeks for an *S. neumayeri* embryo just to reach the pluteus stage. Further problems could arise from increased UV exposure due to the thinning ozone layer in the Antarctic, which would allow even more damaging wavelengths to pass through the atmosphere. Karentz and colleagues (2004) showed that the strategy here is for the embryo to develop deep in the water column, where UV penetration is low. At these depths there is no UV damage to DNA, as estimated by measuring thymine dimers. However, if the embryos are placed in containers near the surface, there is a significant increase in dimer formation.

*The UV portion of the light spectrum is divided into UV-B, which is the light at wavelengths between 290 nm and 320 nm, and UV-A, with wavelengths between 320 nm and 400 nm.

Most aquatic embryos will be exposed to at least some UV radiation. Even if they don't float at the surface, their eggs contain enough lipid to provide moderate to extensive buoyancy, so they will be found near the water's surface. Even well below the surface, a small proportion of the UV wavelengths penetrate the water column. And because developmental times to a swimming larva are more than 12 hours for most species, some portion of development will inevitably take place in daylight, when UV radiation is present.

The solution most often used is a preventive one, sometimes whimsically referred to as the "Coppertone defense." The idea here is for the egg to accumulate chemical sunscreens that absorb UV radiation and prevent harmful wavelengths from penetrating deep into the cytoplasm. A potent sunscreen used in marine embryos (as well as in adult organisms) is provided by a group of chemicals called **mycosporines**, which have high absorption in the UV part of the spectrum. Mycosporine concentrations can reach the millimolar range, and their attenuation of UV can be especially potent at these high concentrations (Shick and Dunlap 2002).

Work by Adams and Shick (2001) provides the best evidence for the role of mycosporines in protecting embryos from UV radiation. Their studies took advantage of the finding that animals cannot synthesize these compounds, but must get them from plants that they eat. The researchers used this dietary requirement to manipulate the mycosporine content in sea urchin embryos by feeding the adult female urchins, during the period that they were making eggs, algae that varied in the amount of mycosporine pigments. Sure enough, the degree of UV protection was correlated with the amount of mycosporine present in the embryos (Figure 4.16).

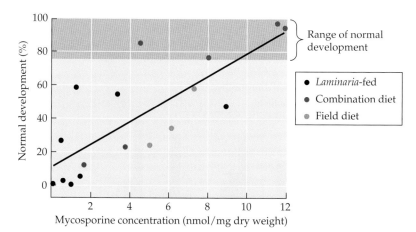

Figure 4.16 Mycosporines protect embryos from UV irradiation. Northern sea urchins (*Strongylocentrotus droebachiensis*) were fed algae that produce no sunscreen-containing mycosporines (*Laminaria*-fed); *Laminaria* plus *Chondrus crispus*, a green macroalga high in mycosporines (combination diet); and a control diet of algae collected in the field. Eggs laid by adults in these three experimental groups were then fertilized and exposed to UV. The extent of normal development (shaded bar at top of graph) achieved under conditions of adverse UV radiation was seen to be related to the mycosporine content in the parental diet. (After Adams and Shick 2001.)

An interesting variant of the mycosporine defense is seen in tunicate embryos, where Epel and colleagues (1999) found that the major attenuation of UV radiation is provided by a layer of extraembryonic "test cells" that surround the egg and embryo. These test cells have such a large amount of the mycosporine pigment that they can absorb 90% of the ambient UV radiation, with a peak at around 360 nm.

A problem with this defense is that the mycosporines are akin to vitamins; they are not synthesized by animals and depend on consumption of algae, bacteria, and fungi that make these molecules. Thus, there is a question of reliability: can the animals depend on being able to consume the algae that make these sunscreens? Studies on bacteria and algae show that mycosporine synthesis is induced in response to sunlight (Rastogi and Incharoensakdi 2014), which means that when UV radiation is high, the algal diet of animals will also have high levels of mycosporines. The animals do not seem to digest or degrade these chemicals and must have mechanisms to transport them to their ovaries and associated oocytes. These mechanisms remain to be discovered (see Mason et al. 1998).

Repairing damaged DNA

The above protective mechanisms provide some help to the embryos, but the final, critical defense is having a good DNA damage restorative system to repair any UV-induced dimers; if not repaired, the dimers would result in misreadings during replication and hence in mutations.

Cells have an extremely effective repair system carried out by a flavoprotein enzyme called photolyase. Photolyase absorbs light in the yellow range of the visible spectrum and uses the energy from this light absorption to reverse thymidine dimerization and restore the DNA to its original state (Sinha and Häder 2002; Häder and Sinha 2005). This is a very clever mechanism, since the "healing" visible light and the damaging UV part of the light spectrum are both present in sunlight.

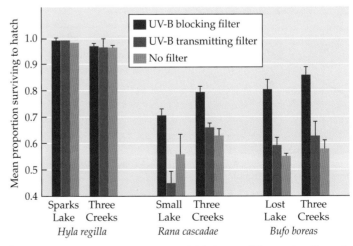

Figure 4.17 In studies at separate field sites at different elevations, the effect of UV-B radiation on tadpole hatching was noted for three amphibian species. Blocking UV-B radiation had no effect on *Hyla regilla* but significant effect on the other two species. (After Blaustein et al. 1994.)

TABLE 4.1 **Photolyase activity levels and UV exposure in the eggs of some amphibians**

Species	Photolyase activity (10^{11} CBPD/hr/µg)[a]	Egg-laying mode	UV exposure
Caudates (urodeles)			
Plethodon dunni	<0.1	Hidden	Not exposed
Taricha granulosa	0.2	Hidden	Limited exposure
Rhyacotriton variegatus	0.3	Hidden	Not exposed
Plethodon vehiculum	0.5	Hidden	Not exposed
Ambystoma macrodactylum	0.8	Open water	Limited exposure
Ambystoma gracile	1.0	Open, shallow water	Exposed
Anurans			
Xenopus laevis	0.1	Under vegatation	Limited exposure
Bufo boreas	1.3	Open, shallow water	Exposed
Rana cascadae	2.4	Open, shallow water	Highly exposed at high altitudes
Hyla regilla	7.5	Open, shallow water	Highly exposed

Source: Blaustein et al. 1994.

[a]CBPD, cyclobutane pyrimidine dimer. Refers to the rate at which the photolyase enzyme reverses thymidine dimerization (see text).

This seems like an efficient mechanism, but the few quantitative studies that have been done on UV radiation damage in embryos indicate that not all the dimers are returned to their original state. Also, the level of the photolyase might be linked to an anticipated level of UV radiation, and if that level is exceeded, repair will not be as efficient. Blaustein and colleagues (1994) looked at the levels of photolyase in different amphibian eggs and oocytes. Levels of photolyase varied 80-fold among the tested species, and the differences correlated with the site of egg laying: those species typically laying eggs in sites exposed to full sunlight had the highest levels of photolyase (Table 4.1).

In the same study, Blaustein and colleagues tested whether or not UV-B could be a factor in lowering the hatching rate of tadpoles. At two field sites, they divided the eggs of the three anuran species studied into three groups. The first group developed under a filter that blocked UV-B from reaching the eggs. The second group developed under a filter that allowed UV-B to pass through. The third group developed without any sun filter (Figure 4.17). For *Hyla regilla*, the filters had no effect, and hatching was excellent in all three conditions. For *Rana cascadae* and *Bufo boreas*, however, the UV-B–blocking filter raised the percentage of hatched tadpoles from about 60% to close to 80%, indicating that their innate defense against UV was not perfect.

Protection against Pathogens

Pathogens such as bacteria, fungi, viruses, and oomycetes (pathogenic algae, also called "water mold") are not a problem for embryos which develop in the mothers protective structures, such as the placenta in mammals (some reptiles and fish also have placental development!). However, pathogens pose a special problem for orphan embryos, as part of their development takes place before the adult immune system has differentiated. During this vulnerable period the embryos must persevere in a world teeming with potential pathogens. In the marine environment, for example, the surrounding seawater contains a million bacteria per milliliter, as well as 10 million viruses and 50 million bacteriophages (Munn 2012; Breitbart 2012). In the terrestrial environment, soil contains even more, some 40 million bacteria per gram, along with some 20 million viruses and many fungi (Whitman et al. 1998; Williamson et al. 2005). Recent research, described below, reveals a rich assortment of protective mechanisms that are unique to orphan embryos and appear to provide robust protection against pathogens.

Parental behavior

Defense against pathogens can also arise from parental behavior (Royle et al. 2012). Many animals brood their embryos in specialized brooding organs, and this clearly protects them from predators and possibly pathogens. Examples of pathogen protection are seen in brooding shrimp, where the mother has a specialized appendage for grooming the eggs; if this appendage is removed, there is increased embryo mortality from fungal infection (Fisher 1983; Förster and Baeza 2001). These shrimp have an additional antifungal protection provided by symbiotic bacteria surrounding the egg, as described later in this chapter.

A more elaborate behavioral protection is provided by fish, usually males, who guard their embryos after courtship and fertilization. This guarding includes removal of dead embryos that would foul adjacent eggs. Male stickleback fish go further, adding antibiotic glue to the nest. The male fish makes this proteinaceous glue to assemble the nest, and its antibacterial properties augment survival of embryos (Little et al. 2008).

Maternal provision of antimicrobial and antifungal defenses

We mentioned earlier that the mother hen transfers antibodies to the yolk of the egg. These antibodies are made in response to bacteria the mother has recently been exposed to, so these antibodies provide protection against ambient pathogens that could become problems to the developing embryo. Similar antibody transfer has been seen in fish (Hongmiao et al. 2012).

In some species, antibacterial and antifungal chemicals are associated with the egg surface. Kudo (2000) found that carp eggs contain antibacterial substances such as lysozyme in the cortical granules. These granules are discharged at fertilization and the lysozyme is transferred to the fertilization envelope, where it is hypothesized to prevent bacterial attachment to this envelope.* Cortical granules in frog eggs also contain lectins that become part of the egg surface after fertilization (Chang et al. 2004) and

---

*Lysozyme is also used in adult organisms. For example, it is found in human pores and is an important ingredient in human tears, where it helps keep the eye free of bacterial infections (see Root-Bernstein 1997).

could protect against pathogens by binding to the surface of bacteria and fungi (Van Die et al. 2004).

Another type of pathogen defense is through **defensins**, a group of small peptides with potent activity against bacteria. These are critical components of the innate defense system in adults and are also used in embryos. Such molecules have been found in shrimp eggs (Bachère et al. 2000) and embryos of the tobacco hornworm, *Manduca sexta* (Gorman et al. 2004).

A fascinating role for defensins is seen in the coelenterate *Hydra*, where defensins provided by the mother prevent infection during the early stages of development (Fraune et al. 2010). In its later stages, the embryo acquires specific symbiotic bacteria and related defensins govern the type of bacterial symbionts that are established. Similar defensin molecules regulate symbiotic populations of bacteria on our skin and in our intestines (Salzman 2010).

Fungi and oomycetes are major pathogens in embryos. Fungi grow slowly, and embryos on the "fast track" will be out of their egg coats before a pathogenic fungus or oomycete has been able to attach and pass through the chorion. Embryos with long developmental periods are at risk but possess two potent strategies to avoid fungal infection.

One is the provision of antifungal proteins in the gel of the egg capsules. An example is seen in the egg cases of the freshwater snail *Biomphalaria glabrata*, the infamous intermediate host for the parasite that causes schistosomiasis. Its embryos would seem to be prime targets for pathogens as they develop over a week-long period in ponds, marshes, irrigation channels, or open sewer drains.

Recent studies on these egg cases show that their embryos are highly protected from bacteria and fungi by a protein that is deposited by the mother in the egg capsule (Baron et al. 2013). This molecule, referred to as LBP/BPI, comprises 90% of the total protein of this egg matrix. The protein has antifungal and antibacterial activity, but its major role seems to be to protect the embryo against oomycetes. When Baron and colleagues (2013) prevented the mother from making the protein (by RNA-silencing technology), the eggs became susceptible to oomycete infection. As shown in Figure 4.18, the eggs without the protein were covered with hyphae, whereas the control eggs were unaffected.

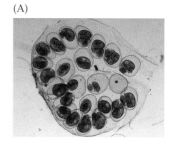

(A)

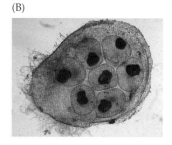

(B)

Figure 4.18 The major protein (LBP/BPI) of the egg sac of this snail protects the embryo from oomycete fungal-like infection. (A) Normal embryo sac. (B) Embryo sac lacking LBP/BPI. As seen, this egg sac is covered with hyphae from the oomycete, which will kill the embryos. (From Baron et al. 2013.)

Symbiosis and protection from fungi

The second antifungal defense uses an approach that will be familiar to readers from Chapter 3, the use of symbiotic bacteria. The shrimp *Palaemon macrodactylus* has a long developmental period and would be susceptible to fungal infection were it not for its symbionts, which live in the coats surrounding the crustacean embryos (Fisher 1983; Gil-Turnes et al. 1989). Culturing the bacterial population revealed that these symbionts produce a single chemical, 2,3-indolinedione, which is a fungicide. In experiments in which the symbiotic bacteria were removed by incubating the mothers in an antibiotic solution, the embryos were much more susceptible to fungal attack (Figure 4.19). Similarly, salmon eggs are prey to *Saprolegnia* oomycetes. However, certain bacteria (especially those of the Actinobacteria genus *Frondihabitans*) are bound by salmon eggs and inhibit the attachment of the parasitic *Saprolegnia* (Liu et al. 2014). Infection appears to be more

Figure 4.19 Symbiotic bacteria protect shrimp embryos from pathogenic fungi. Embryo masses were treated with antibiotics to kill the symbiotic bacteria, and after 12 days none of these embryos survived. However, if the antifungal agent produced by the bacteria was added back, the survival was not significantly different from that among the controls. These results indicate that the symbionts are providing protection against fungal infection. (After Gil-Turnes et al. 1989.)

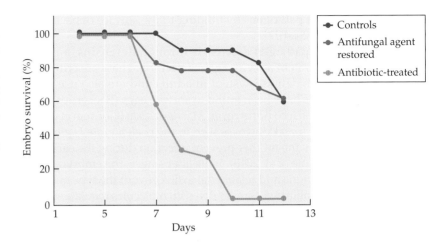

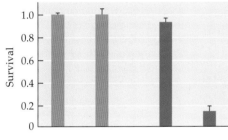

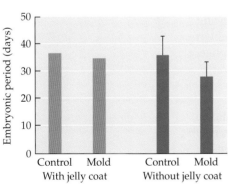

Figure 4.20 The jelly coat of the eggs of the salamander *Ambystoma maculatum* provides protection against pathogenic fungi. If the embryos are exposed to a water mold, the control and mold-exposed embryos survive and develop at similar rates. If the jelly coat is removed, survival is drastically reduced; however, those few eggs that do survive the mold are somehow induced to develop more rapidly. (After Gomez-Mestre et al. 2006.)

a problem in commercial aquaculture than in the wild, possibly because the natural populations have access to the symbionts.

The European beewolf wasp (*Philanthus triangulum*) uses a similar symbiotic strategy, but the outcome is orders of magnitude more potent than the shrimp-bacteria association (Kroiss et al. 2010). In the shrimp, the symbiont provides only one active chemical, and the effective concentration is in the millimolar range. In the beewolf wasp the symbiont(s) produces nine antifungal and antibacterial agents, and these are active at the micromolar level. This powerful defense is probably necessary since the wasp embryos develop in underground nests in soil teeming with potential pathogens.

Amphibian eggs, which are laid in water, are also susceptible to water molds and have developed several means of protection. In addition to having algal symbionts producing oxygen and nutrients (see Figure 3.6), the eggs of the spotted salamander (*Ambystoma maculatum*) are protected by a thick layer of jelly that the symbionts eat and that repels water molds (Figure 4.20). The wood frog (*Rana sylvatica*) is protected because the adult females lay eggs very early in the spring, while the pond water is still cold and mold growth is slow. The eggs of the American toad (*Bufo americanus*) are highly susceptible to water molds, having only a thin jelly coating and being laid after the pond water has warmed. However, if a *B. americanus* egg becomes infected, it will hatch more rapidly and develop immune responses that can repel the mold. Interestingly, *R. sylvatica* tadpoles (which hatch earlier in the season) can have a positive effect on *B. americanus* eggs—the feeding *Rana* tadpoles eat the mold off infected *Bufo* egg clutches (Gomez-Mestre et al. 2006).

Larva antipathogen strategies: A cellular immune system

The above examples focused on how embryos defend themselves before the adult immune system has developed. There is, however, an intermediate stage, the larva, the feeding and dispersive stage used by many embryos before metamorphosing into the adult. Larval forms are highly diverse, with some bearing no resemblance to the future adults. But do they have a functional immune system to protect them against pathogens? A perusal of the current literature reveals a paucity of work on this question, with most research focused on adult vertebrates.

Amazingly, it has been known for more than a century that larvae have cells that can recognize foreign invaders. In fact, the studies into larval defenses initiated the current science of immunology (Tauber and Chernyak 1991). The Russian scientist Elie Metchnikoff discovered this in 1883 when he utilized the transparent larvae of starfish to see what happened when he inserted various foreign objects into them. Figure 4.21A shows what happened to some plant tissue that had pierced a starfish larva. Metchnikoff saw that it was soon surrounded by cells that had migrated from all over the larval body, seemingly trying to eat or oust this intruder. Metchnikoff went on to look at this in other organisms, and his observations led him to identify the role of bacteria-eating cells (phagocytes) in protecting against bacteria throughout the animal kingdom (Metchnikoff 1891).

Metchnikoff's original observations on starfish embryos have been revisited by Furukawa and colleagues (2009). They found that bacteria injected into the blastocoel of starfish embryos are rapidly surrounded by and killed by macrophage-like cells that phagocytize (eat) bacteria. This clearing of bacteria (Figure 4.21B–E) is probably mediated by antimicrobial peptides. Li and colleagues (2014) found that the similar cells in the sea urchin embryo killed bacteria using an antimicrobial peptide named centrocin.

Protection from Predation

Another problem for embryos is that they often serve as food for other animals. Predatory *Homo sapiens*, for example, consume chicken eggs. Or go to a Japanese restaurant and you may find *uni*, the roe of sea urchins wrapped in rice and seaweed; or eat in an upscale restaurant and order caviar, the coveted roe of sturgeon. In many areas of the world, the nesting sites of endangered turtles have to be protected from the various primate and canine species who eat their eggs.

Leaving human predation aside, are there defenses against other predators? Some defenses are general, such as the oft-mentioned rapid development into a motile larva or small adult. Indeed, in most orphan embryos, the progression through cleavage and gastrulation to the point of larval motility is extremely rapid. This common feature

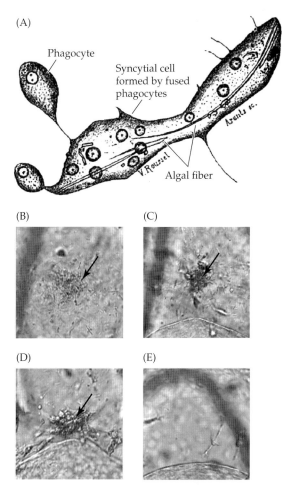

Figure 4.21 The initial discovery of phagocytosis by Metchnikoff in 1883 and the revisiting of that discovery by Furukawa et al. in 2009. (A) This 1883 drawing depicts the engulfment by phagocytic cells of a thin fiber of an algal cell that had penetrated the blastocoel of a starfish larva. The phagocytes have fused together into a giant syncytial cell and are attempting to isolate and digest this foreign object. (B–E) 2009 photographic series of what happens when bacteria are injected into the coelom of a starfish larva. (B) Arrow points to the clump of bacteria shortly after injection. (C,D) Two hours later, the intact bacteria are no longer visible, having been engulfed by the surrounding phagocytes. (E) Three days later, the area is cleared of bacteria (and phagocytes). (A from Metchnikoff 1891; B–E from Furukawa et al. 2009.)

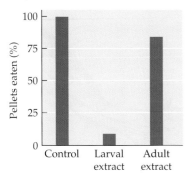

Figure 4.22 Embryos (but not adults) of *Bugula neritina* contain substances that make them unpalatable. Here artificial food pellets contain either larval extract or adult extract and are offered to a pinfish. Rejection of the pellet is seen only for the larval extract. (After Lopanik et al. 2001.)

suggests that the fast track has been selected in part because it allows the embryo to get off the ocean floor (or the base of a leaf, or the surface of a pond) and at least become a moving target instead of (so to speak) a sitting duck (Strathmann et al. 2002). Indeed, some orphan embryos become ciliated and motile while still in the early cleavage period.

Another approach is to make the embryos taste bad (Lindquist 2000). We earlier mentioned the alkaloids put into moth eggs by both the father and mother and the incorporation of noxious alkaloids into the larvae of certain butterflies. Bryozoans (also called ectoprocts) and tunicates are particularly good chemists, depositing all sorts of exotic protective compounds into their eggs. In the bryozoan *Bugula neritina*, symbiotic bacteria produce a toxic chemical called bryostatin. This chemical is not present in adult tissues; it is transferred to the eggs and embryos, which it renders unpalatable (Figure 4.22; Lopanik et al. 2004).

Recently, a new predator avoidance mechanism was discovered in sand dollars. When exposed to predator cues (the mucus from fish), sand dollar larvae cloned themselves by detaching buds from the region opposite the mouth. Slow-moving larvae have no way to escape predatory fish, and such cloning may provide a means for them to reproduce and become smaller than the fish can readily see (Figure 4.23; Vaughn and Strathmann 2009).

Global Climate Change

The preceding sections have emphasized that embryos are well protected against the anticipated environments they might experience occur during development. But what happens if there are sudden changes in the environment, ones that might be beyond the capacity of the embryo's physiology?

We are living in one of these periods, a time characterized by some of the most rapid climate changes in the world's history (Hansen et al. 2013), resulting from the combustion of fossil fuels and unprecedented rates of accumulation of atmospheric CO_2. One consequence will be in increases in the temperature of the Earth. Climate models from the National Research Council (2010) predict that by 2100, the world's average temperature will have increased by 1°C–6.4°C (2°F–11.5°F). The Intergovernmental Panel on Climate Change (2014), an international scientific organization sponsored by the United Nations, estimates that if global warming exceeds 1.6°C above the preindustrial temperatures, 9%–31% of species will become extinct. If temperatures rise 2.9°C, they estimate, 21%–52% of all species on Earth will be committed to extinction.

The Intergovernmental Panel on Climate Change (2014) also reports that oceanic pH has already decreased by about 0.1 pH unit since the beginning of the industrial revolution and that depending on the different scenarios for CO_2 production, the average ocean pH could drop from its current 8.1 to as low as 7.75 by 2100. The major consequences of a more acidic ocean are effects on the calcium carbonate skeletons of

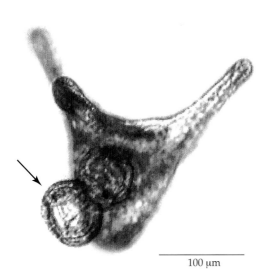

100 μm

Figure 4.23 Exposure of a sand dollar larva (*Dendraster*) to mucus of a fish predator induces the larva to form a clone of itself (arrow). The clone, which looks like a gastrula-stage embryo, develops into a smaller larva that can escape predatory fish. (From Vaughn and Strathmann 2008.)

many ocean plants and animals; these include algal coccolithophores, zooplankton such as foraminifera, and animals such as corals, mollusks, and echinoderms. A secondary effect of low pH could arise from the metabolic costs of compensating for the lowered extracellular pH.

Consequences of and adaptations to temperature change

TEMPERATURE AND LIFE CYCLE   Those organisms whose development is intimately connected with temperature will probably be affected earliest. First, many animals have a developmental stage that is very restrictive in temperature. Egg-laying species, especially tropical turtles, have embryonic stages that have much lower thermal tolerances than their adult stages (Pike 2014). Moreover, as the mother leaves the clutch after laying the eggs, the embryos cannot move to other environments. In many parts of their range, loggerhead turtles are at the upper limits of their thermotolerance, and these nesting sites are producing 30% fewer hatchlings per nest than more temperate sites. Similarly, increased temperatures threaten the survival of the leatherback turtle in the eastern Pacific Ocean (Santidrián Tomilo et al. 2012). Heat waves produce morphological abnormalities in pond turtles, and these female turtles have lesser fitness than the unstressed females (Telemeco et al. 2013b). Stenothermal larvae (i.e., larvae with narrow temperature tolerances), such as those of Antarctic sea urchin *Sterechinus neumayeri*, where a 2°C temperature increase alters the survivability and allometry of the larvae, are also at great risk in global climate change (Byrne et al. 2013).

Second, those animals with temperature-dependent sex determination are also expected to be affected early. Indeed, it is thought that the effects of increased temperature over the past 50 years are already being seen in many turtle species (Hawkes et al. 2009). As discussed in Chapter 1, temperature determines the sexual phenotype of some turtles. In many marine turtles, where females are produced at higher temperatures, the sex ratio is already skewed toward producing females, and the predicted increases in global temperature are expected to exacerbate this trend. In some rookeries of the endangered green turtle (*Chelonia mydas*, where females are produced at higher temperatures), the sex ratio is 95:5 female. Fortunately for the species, the males appear to be able to mate with multiple females and have a shorter mating period (Wright et al. 2012). However, some models (Telemeco et al. 2013a) predict that if current trends continue, the nesting sites of many species will be 100% female.

Third, temperature may disrupt developmental symbioses. Global climate change has already changed the phenology (timing of cyclic events in a plant's annual cycle) of numerous plants (Reyer et al. 2011; Cleland et al. 2007), with spring changes occurring 2.5 days earlier per decade since the 1970s. For example, the cherry trees on Mt. Takao in Tokyo are now flowering 5–6 days earlier than they did 25 years ago. Most cherry tree species flower 3–5 days earlier for each 1°C increase in temperature (and the February–March temperature at Mt. Takao has increased 1.8°C during that time), but some early-blooming varieties flower as much as 9 days earlier for each 1°C increase (Miller-Rushing et al. 2007). These altered rates in timing reflect the plasticity of the organism to environmental cues.

However, one can imagine that if insect pollinators of these trees had a different susceptibility to temperature than the plant (or if they were to cycle on a more stable cue, such as photoperiod), they would not eclose from their pupae at a time when the blossoms were open to them (see Stiling 1993; Rafferty et al. 2013). Pollinations in the early spring, which depends on the precise timing of insect maturation and bird migrations, may be the most susceptible to disruption (Bartomeus et al. 2013).

Experimental evidence shows that the effectiveness of a pollinator varies when flowering time is experimentally altered (see Rafferty and Ives 2012), and long-term observations of plant-pollinator interactions strongly suggest that changes in phenology can cause the extinction of specialist pollinators (such as those that can pollinate only a single species). Fortunately some studies suggest that plants and their pollinators often use similar environmental cues to time their emergence in the spring (Bartomeus et al. 2011; Forrest and Thomson 2011), and plant species with several pollinators appear to have "insurance" against extinction. Apples, for instance, that have several pollinators are able to survive their changes in phenology (Bartomeus et al. 2013).

PLASTICITY AND INTERSPECIES RELATIONSHIPS   There are two major ways for commensal symbioses to survive climate-induced asynchrony in their life cycles. One mechanism is to have genetic diversity in the population, such that there are different alleles conferring different timings. The second mechanism is to have plasticity such that at least one of the interacting species can accommodate the other. And if plasticity, itself, is an inherited trait, then the two mechanisms can be intimately linked. Moreover, as we've seen in Chapter 8, developmental plasticity can produce a phenotype that buys time for populations to evolve by mutation and recombination, while keeping fitness stable.

Many of the phenotypic changes seen in response to global warming involve phenotypic plasticity. Bradshaw and Holzapfel (2001) demonstrated that over the past 30 years, the pitcher plant mosquito (*Wyeomyia smithii*) has evolved an adaptive response to global warming by changing its photoperiodic response. With longer growing seasons now the norm in the northern part of their range, mosquitoes there have shifted toward using shorter day length cues to initiate larval dormancy. This is because winter is arriving later in the northern latitudes, and conditions remain favorable for reproductive activity longer. The northerly mosquitoes now use a photoperiod cue similar to that used by the conspecifics to the south; so what used to be a southerly behavior is now prevalent in the north as well. This change in behavior is the result of genetic alterations in the northerly populations.

The great tit (*Parus major*) has experienced selection for phenotypic plasticity in the timing of its reproduction. These birds feed caterpillars to their young. However, as spring has come progressively earlier to the Northern Hemisphere, the caterpillars have been maturing earlier, to the point that they now often pupate before the young birds hatch. The disappearance of their food source has led to a decline in the tit population. In both Great Britain and the Netherlands, researchers have documented that phenotypic plasticity has enabled some *Parus* individuals to adjust their egg-laying behavior to this changing environmental state of affairs.

There is genetic variation among individual females in how early they can lay eggs, and those birds that can lay their eggs the earliest have the best chance of having caterpillars available to their young. The birds that survive are the offspring of parents that are most plastic in their egg-laying response (i.e., those with the broadest norm of reaction) and whose egg-laying date thus can be advanced the most. Using a 47-year-old database, Charmantier and colleagues (2008) were able to show that the egg-laying date of female tits in Great Britain has advanced by 14 days over the past half century. In both this and a Dutch study (Nussey et al. 2005), selection has altered the boundaries of phenotypic plasticity for reproductive timing, allowing it to change with the ecological conditions. The broadest reaction norm has been selected such that the birds have a better chance of having their offspring survive.

In the Canadian Yukon, drier spring seasons have caused the white spruce to produce its pine cones earlier. These cones are a major food source for North American red squirrels. In response to this shift in the availability of this food, female squirrels have advanced their parturition dates by 18 days over a single decade, and much of this response has been due to plasticity (Reále et al. 2003; Reed et al. 2011).

In Chapter 1, we discussed the symbiosis between corals and their algal symbionts. When temperatures rise above certain levels, the algal symbionts die, and the "bleached" corals die as well. As the temperatures rise, the corals and symbionts attempt several strategies to survive, including transient changes in algal and host gene expression and exchanges of different clades of algal symbionts. These are transient changes, reverting back to the previous conditions if the temperature stress is relieved.

However, in some corals, adaptive changes in gene expression are made that are part of an acclimatization that remains in the coral if it is transplanted to a cooler site; the change is stable (Barshis et al. 2010; Palumbi et al. 2014). In a short time, a physiological response has become an inherited part of the coral's phenotype. "In less than two years, acclimatization achieves the same heat tolerance that we would expect from strong natural selection over many generations of these long-lived organisms" (Palumbi et al. 2014, p. 895).

These observations hold out some hope for endangered reef ecosystems, but they leave open the question of how this could happen. If it were an epigenetic effect, the stability would be an example of a stable adaptive change that would be of critical survival value to long-lived organisms such as corals.

Consequences of ocean acidification

Another important consequence of the rise in atmospheric CO_2, sometimes called "the other carbon dioxide problem," is a decrease in oceanic pH due to the transfer of part of the atmospheric CO_2 to the ocean. CO_2 forms H_2CO_3 (carbonic acid) in water and dissociates to H^+, HCO_3^-, and CO_3^{2-}; this chemical equilibrium provides the major pH buffering capacity of seawater. The increased CO_2 from fossil fuel combustion has already resulted in a more acidic ocean (lowered pH). There is no disagreement that added CO_2 can affect ocean pH: it is the only buffer.

The predicted effects of ocean acidification are (1) pH effects on metabolism, which might be seen in all organisms, and (2) pH effects on calcification, which would be seen only in organisms developing calcium carbonate skeletons (e.g., phytoplankton such as coccolithophores; plants such as corraline alga, and; animals such as corals, mollusks, and echinoderms).

Acidic pH is the natural enemy of calcium carbonate. Recall that caves are formed when water made acidic by CO_2 passes through a barrier of limestone (a type of calcium carbonate). The acidic water dissolves the limestone resulting in a network of tunnels, which, over time, erode into a cave.

Numerous studies have been done to examine the effects of lower pH on embryos. As with climate change studies, some embryos are negatively affected, some are positively affected, and in some there is no effect (Dupont and Portner 2013). However, a new study shows that even before an observable effect on skeletal growth is seen, altered pH affects the embryo. This study (Pan et al. 2015) on sea urchin embryos shows that an energy reallocation occurs at lower pH levels that could affect survival under stressful conditions. Earlier work had shown that the bulk of respiratory energy was used for just two processes: for maintenance of the sodium:potassium ratio by a Na/K ATPase and for protein synthesis. When Pan and colleagues (2015) exposed embryos to lower pH, they saw no visible effect on growth rate, morphology, or respiration. However when they looked at what happened to the two major ATP-consuming processes, they found remarkable changes. The ATP used for these two processes in the prefeeding stage went from 40% in the control to 62% in the low pH group. The ATP in feeding stages (the embryos had excess food available) went from 55% in the control to 84% in the low pH group. Stated another way, the reserve energy available for other processes in the prefeeding embryo dropped from 60% to 38%; the reserve energy in the feeding larva went from 45% to 16%.

Although we are not certain why this reallocation occurred (it could be diversion of energy to repair damaged proteins or from attempts to regulate internal pH), the consequences seem obvious. What the embryo has done is to deplete its savings account held in reserve for other tasks. This will not be a problem if there is enough food. But if food is limiting, embryo survival could be at risk. Life is dangerous as a planktonic larva, and the longer one spends in the plankton, the higher the chances of becoming someone else's meal.

Other work on mussel embryos shows that low pH affects calcium deposition and possibly causes a similar energy diversion (Gaylord et al. 2011). Low pH slightly lowered the thickness and strength of the embryonic calcium carbonate shell but there was a much larger effect on the body mass (Figure 4.24). The smaller mass could be a direct pH effect on metabolism, or it could result, as in the sea urchin, from a diversion of energy from growth to maintaining ion gradients or repairing damaged proteins. The thinner shell could make the embryos more vulnerable to predation.

Lower pH also affects digestion in sea urchin larvae (Stumpp et al. 2013). The pH of the larval intestine is quite alkaline, about 9.5. When these larvae were placed

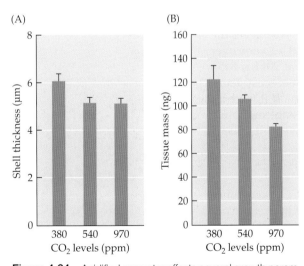

Figure 4.24 Acidified seawater affects several growth parameters of mussel larvae. (A) Shell thickness. (B) Tissue mass, an index of growth. The x-axis refers to various scenarios for CO_2 concentration in 2100: 380 ppm is the present-day level; 540 ppm is an optimistic level for 2100; 970 refers to a worst-case scenario. (After Gaylord et al. 2011.)

(A)

(B)

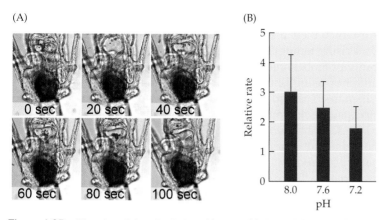

Figure 4.25 Digestion of algae by 8-day-old sea urchin larvae is impaired by lower seawater pHs that are predicted as outcomes of ocean acidification. (A) Time series of digestion of a naturally fluorescent alga over 100 seconds. (B) Measurement of rate of digestion at three different pH levels. A pH of 8.0 corresponds to the current pH. The other two pH levels represent different levels of continued acidification predicted for 2100. (From Stumpp et al. 2013.)

in seawater at the predicted 2100 pH of 7.7 or at an extreme pH of 7.4, the intestinal pH became more acidic (respective decrease of 0.3–0.5 pH units). Compared to the current ocean average pH levels of 8.0 (present day pH) to pH levels of 7.7 or 7.4 (predicted range of pHs in 2100), there was a significant slowing of the rate of digestion of algal food at the lower pHs (Figure 4.25). Again, the consequence is less energy available for growth. The common theme of the above examples is that there is an energy cost of acidification in addition to any effect on calcification. There may be no or little effect if food is in excess. But there would be a price in survival if food were limiting.

Looking Ahead

Development is precise and robust, with buffering and canalization of the embryonic programs to ensure normal development as well as embryonic defense mechanisms to ensure survival of the embryo while it develops. Such survival is challenging, especially since the embryos of most animal species develop as orphans, with no or little parental protection. This problem is further exacerbated by the fact that the embryo cannot use adult defense systems until differentiation of these systems has taken place.

This vulnerability is compensated for by a wide range of defenses operating at the cellular level and by epigenetic options available for adaptations that require alternative developmental pathways, such as seasonal changes or responses to predators (see Chapter 1). The defenses constitute a "be prepared" approach tailored to the evolutionary history of the embryo and, as a result, to those challenges the embryo might be expected or anticipated to face. These include ABC transporter proteins that prevent toxins from accumulating in cells; thiol-containing compounds that block heavy metal toxicity; extracellular coats to shield against physical damage and pathogens; symbioses with bacteria to handle fungi; and a series of enzymes, sunscreens, and ROS scavengers to protect proteins and nucleic acids from UV radiation and oxidative damage.

Problems with these defenses arise if there are rapid environmental changes for which there are no or limited defenses. Climate change is one of these, and rising temperatures and CO_2 levels are both concerns for the sustainability of embryonic development. Such change may become too rapid to adapt to, using gene mutations alone. It is possible that epigenetic effects will enable such adaptations to occur, and there are now studies pointing to unsuspected modes of rapid adaptation.

We briefly described work of Susan Lindquist (also discussed in Chapter 11) showing that extreme stress can unmask cryptic mutations that are normally buffered, held in check, by molecular chaperones such as heat-shock proteins. There also appear to be more "extreme" measures an organism might take. One of these is a "Hail Mary" strategy in which a developing embryo scrambles its gene expression patterns in an attempt to find something that works.

When exposed to high levels of a novel stress, some organisms will allow genes normally restricted to one tissue type to be expressed in new places and at new times, thus providing a mechanism for rapid change using already existing genes. In an ingenious experiment, Stern and his colleagues (2012) showed that when faced with a novel challenge, *Drosophila* embryos will try anything—turning on the expression of just about every gene, in various combinations, to try to get some combination that will allow survival. If some winning combination is found, that gene expression pattern can be stabilized by chromatin modulators and inherited.

To see whether stress would alter the location or time of expression, Stern and colleagues put a neomycin-resistance gene onto a promoter and enhancer of a gene that is normally only expressed in a small subset of cells during early embryonic development. They also linked a reporter gene, green fluorescent protein (GFP), onto the enhancer so they could see where the neomycin-resistance gene was being expressed. When the larvae were exposed to lethal doses of neomycin, the researchers saw that the GFP marker was turned on in numerous places throughout the body (Figure 4.26). Genes were being allowed to be expressed where and when they hadn't been expressed before (Stern et al. 2012, 2014). This did not happen when larvae were not exposed to the toxic levels of neomycin. In some instances, the expression of this neomycin-resistance gene was sufficient to rescue the larvae.

Remarkably, this tolerance to neomycin was inherited in the next two generations of flies. Polycomb proteins (those powerful suppressors of gene expression) now repress the regulatory proteins that normally turn the developmental gene off in the larvae.

Therefore, when facing a new challenge, embryos may scramble gene expression to randomly achieve a winning combination that will allow survival. And these new patterns of gene expression can be inherited epigenetically. The heritability is transient, but as Stern and colleagues note in their paper, "it may be stabilized by genetic assimilation, thus leading to stable incorporation of modified features into the developmental program." Development may be providing more means of coping with change than we had realized. Further work is needed to determine whether such strategies are widespread, how successful they might be in circumventing death, and what trade-offs might have to be made in order to survive.

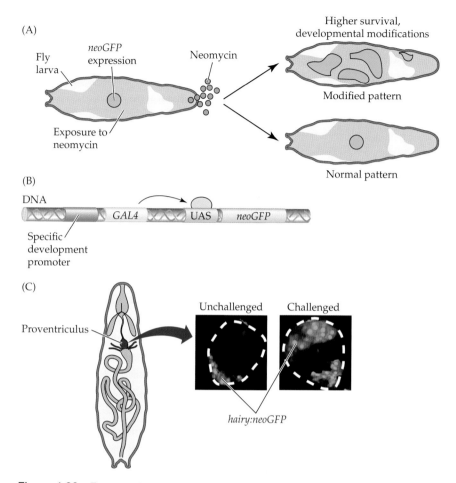

Figure 4.26 Extreme plasticity of gene expression in *Drosophila* embryos when confronted with a severe environmental challenge. A neomycin-resistance gene tagged with a fluorescent (GFP) marker is inserted into the genome of a *Drosophila* larva. The expression pattern is observed. The larva is then challenged with lethal doses of neomycin. The larva is tested to see if the neomycin resistance gene kept its normal expression pattern or if it changed. (A) Experimental protocol. (B) The marker system to see gene expression, where a developmentally specific enhancer regulates the expression of GFP. (C) Example of the expression of *hairy* in a section of the larval midgut. Its expression domain has enlarged significantly during the toxic challenge. (After Stern et al. 2012, photographs courtesy of Y. Soen.)

References

Adams, N. L. and J. M. Shick. 2001. Mycosporine-like amino acids prevent UVB-induced abnormalities during early development of the green sea urchin *Strongylocentrotus droebachiensis*. *Mar. Biol.* 138: 267–280.

Andrews, G. K. 2000. Regulation of metallothionein gene expression by oxidative stress and metal ions. *Biochem. Pharmacol.* 59: 95–104.

Andrews, G. K., Y. Huet-Hudson, B. C. Paria, M. T. McMaster and S. K. Dey. 1991. Metallothionein gene expression and metal regulation during preimplantation mouse embryo development (*MT* mRNA during early development). *Dev. Biol.* 145: 13–27.

Artus, J., C. Babinet and M. Cohen-Tannoudji. 2006. The cell cycle of early mammalian embryos: Lessons from genetic mouse models. *Cell Cycle* 5: 499–502.

Avella, M. A., B. Baibakov and J. Dean. 2014. A single domain of the ZP2 zona pellucida protein mediates gamete recognition in mice and humans. *J. Cell Biol.* 205: 801–809.

Babcock, R. C., G. D. Bull, P. L. Harrison, A. J. Heyward, J. K. Oliver, C. C. Wallace and B. L. Willis. 1986. Synchronous spawning of 105 scleractinian coral species on the Great Barrier Reef. *Mar. Biol.* 90: 379–394.

Bachère, E., D. Destoumieuxa and P. Bulet. 2000. Penaeidins, antimicrobial peptides of shrimp: A comparison with other effectors of innate immunity. *Aquaculture* 191: 71–88.

Baibakov, B., N. A. Boggs, B. Yauger, G. Baibakov and J. Dean. 2012. Human sperm bind to the N-terminal domain of ZP2 in humanized zonae pellucidae in transgenic mice. *J. Cell Biol.* 197: 897–905.

Bargmann, C. I. 2006. Chemosensitization in *C. elegans*. The *C. elegans* Research Community Wormbook. www.wormbook. org, doi/10.1895/wormbook.1.123.1.

Baron, L. B. and 7 others. 2013. Parental transfer of the antimicrobial protein LBP/BPI protects *Biomphalaria glabrata* eggs against oomycete infections. *PLoS Pathog.* 9(12): e1003792.

Barshis, D. J., J. H. Stillman, R. D. Gates, R. J. Toonen, L. W. Smith and C. Birkeland. 2010. Protein expression and genetic structure of the coral *Porites lobata* in an environmentally extreme Samoan back reef: Does host genotype limit phenotypic plasticity? *Mol. Ecol.* 19: 1705–1720.

Bartomeus, I., J. S. Ascher, D. Wagner, B. N. Danforth, S. Colla, S. Kornbluth and R. Winfree. 2011. Climate-associated phenological advances in bee pollinators and bee-pollinated plants. *Proc. Natl. Acad. Sci. USA* 108: 20645–20649.

Bartomeus, I., M. G. Park, J. Gibbs, B. N. Danforth, A. N. Lakso and R. Winfree. 2013. Biodiversity ensures plant-pollinator phenological synchrony against climate change. *Ecol. Lett.* 16: 1331–1338.

Beghin, D., J. L. Delongeas, N. Claude, R. Farinotti, F. Forestier and S. Gil. 2010. Comparative effects of drugs on P-glycoprotein expression and activity using rat and human trophoblast models. *Toxicol. In Vitro.* 24: 630–637.

Bensaude, O., C. Babinet, M. Morange and F. Jacob. 1983. Heat-shock proteins: First major products of zygotic gene activity in mouse embryo. *Nature* 305: 331–333.

Berry, J. P. 1996. The role of lysosomes in the selective concentration of mineral elements: A microanalytical study. *Cell. Mol. Biol. (Noisy-le-grand)* 42: 395–411.

Blaustein, A. R., P. D. Hoffman, D. G. Hokit, J. M. Kiesecker, S. C. Walls and J. B. Hays. 1994. UV repair and resistance to solar UV-B in amphibian eggs: A link to population declines? *Proc. Natl. Acad. Sci. USA* 91: 1791–1795.

Braendle, C. and M.-A. Felix. 2009. The other side of phenotypic plasticity: How a developmental system generates an invariant phenotype despite environmental variation. *J. Biosci.* 34: 543–551.

Bradshaw, W. E. and C. M. Holzapfel. 2001. Genetic shift in photoperiodic response correlated with global warming. *Proc. Natl. Acad. Sci. USA* 98: 14509–14511.

Brash, A. R., M. A. Hughes, D. J. Hawkins, W. E. Boeglin, W. C. Song and L. Meijer. 1991. Allene oxide and aldehyde biosynthesis in starfish oocytes. *J. Biol. Chem.* 266: 22926–22931.

Breitbart, M. 2012. Marine viruses: Truth or dare. *Ann. Rev. Mar. Sci.* 4: 425–448

Briggs, E. and G. M. Wessel. 2006. In the beginning: Animal fertilization and sea urchin development. *Dev. Biol.* 300: 15–26.

Broeks, A., H. W. Janssen, J. Calafat and R. H. Plasterk. 1995. A P-glycoprotein protects *Caenorhabditis elegans* against natural toxins. *EMBO J.* 14: 1858–1866.

Brower, L.P. 1969. Ecological chemistry. *Sci. Am.* 220: 22–29.

Brower, L. P. and C. M. Moffit. 1974. Palatability dynamics of cardenolides in the monarch butterfly. *Nature* 249: 280–283.

Bukau, B., J. Weissman and A. Horwich. 2006. Molecular chaperones and protein quality control. *Cell* 125: 443–451.

Byrne, M. and 7 others. 2013. Vulnerability of the calcifying larval stage of the Antarctic sea urchin *Sterechinus neumayeri* to near-future ocean acidification and warming. *Glob. Chang. Biol.* 19: 2264–2275.

Casanueva, M. O., A. Burga and B. Lehner 2012. Fitness trade-offs and environmentally induced mutation buffering in isogenic *C. elegans*. *Science* 335: 82–85

Cassada, R. C. and R. L. Russell. 1975. The dauerlarva, a postembryonic developmental variant of the nematode *Caenorhaditis elegans*. *Dev. Biol.* 46: 326–342.

Chang, B. Y., T. R. Peavy, N. J. Wardrip and J. L. Hedrick. 2004. The *Xenopus laevis* cortical granule lectin: cDNA cloning, developmental expression, and identification of the eglectin family of lectins. *Comp. Biochem. Physiol. A: Mol. Integr. Physiol.* 137: 115–129.

Charmantier, A., R. H. McCleery, L. R. Cole, C. Perrins, L. E. B. Kruuk and B. C. Sheldon. 2008. Adaptive phenotypic plasticity in response to climate change in a wild bird population. *Science* 320: 800–803.

Chen, N. and I. Grunwald. 2004. The lateral signal for LIN12/Notch in *C. elegans* vulval development comprises redundant secreted and transmembrane DSL proteins. *Dev. Cell* 6: 183–192.

Cleland, E. E., I. Chuine, A. Menzel, H. A. Mooney and M. D. Schwartz. Shifting plant phenology in response to global change. *Trends Ecol. Evol.* 22: 357–365.

Cobbett, C. S. 2000. Phytochelatins and their roles in heavy metal detoxification. *Plant Physiol.* 123: 825–832.

Coffman, J. A. and J. M. Denegre. 2007. Mitochondria, redox signaling, and axis specification in metazoan embryos. *Dev. Biol.* 308: 266–280.

Coffman, J. A., J. J. McCarthy, C. Dickey-Sims and A. J. Robertson. 2004. Oral–aboral axis specification in the sea urchin embryo: II. Mitochondrial distribution and redox state contribute to establishing polarity in *Strongylocentrotus purpuratus*. *Dev. Biol.* 273: 160–171.

Coffman, J. A., A. Wessels, C. DeSchiffart and K. Rydlizky. 2014. Oral–aboral axis specification in the sea urchin embryo, IV. Hypoxia radializes embryos by preventing the initial spatialization of nodal activity. *Dev. Biol.* 386: 302–307.

Cole, S. P. and R. G. Deeley. 1998. Multidrug resistance mediated by the ATP-binding cassette transporter protein MRP. *Bioessays* 20: 931–940.

Cooke, J., M. A. Nowak, M. Boerlijst and J. Maynard Smith. 1997. Evolutionary origins and maintenance of redundant gene expression during metazoan development. *Trends Genet.* 13: 360–364.

Danilevskii, A. S. 1965. *Photoperiodism and Seasonal Development of Insects*. Oliver and Boyd, Edinburgh.

David, J. R., R. Allemand, P. Capy, M. Chakir, P. Gibert, G. Pétavy and B. Moreteau. 2004. Comparative life histories and eco-physiology of *Drosophila melanogaster* and *D. simulans*. *Genetica* 120: 151–163.

Detrich III, H. W., S. K. Parker, R. C. Williams, Jr., E. Nogales and K. H. Downing. 2000. Cold adaptation of microtubule assembly and dynamics: Structural interpretation of primary sequence changes present in the alpha- and beta-tubulins of Antarctic fishes. *J. Biol. Chem.* 275: 37038–37047.

Dickinson, D. A. and H. J. Forman. 2002. Cellular glutathione and thiols metabolism. *Biochem. Pharmacol.* 64: 1019–1026.

Dupont, S. and H. Pörtner. 2013. Marine science: Get ready for ocean acidification. *Nature* 498: 429.

Dussourd, D. E., K. Ubik, C. Harvis, J. Resch, J. Meinwald and T. Eisner. 1988. Biparental defense endowment of eggs with acquired plant alkaloid in the moth *Utetheisa ornatrix*. *Proc. Natl. Acad. Sci. USA* 85: 5992–5996.

Ebling, L., W. Berger, A. Rehberger, T. Waldhor and M. Micksche. 1993. P-glycoprotein regulates chemosensitivity in early developmental stages of the mouse. *FASEB J.* 7: 1499–1506.

Elinson, R. P. and E. Houliston. 1990. Cytoskeleton in *Xenopus* oocytes and eggs. *Semin. Cell Biol.* 1: 349–357.

Epel, D., K. Hemela, M. Shick and C. Patton. 1999. Development in the floating world: Defenses of eggs and embryos against damage from UV radiation. *Amer. Zool.* 39: 271–278.

Epel, D., T. Luckenbach, C. N. Stevenson, L. A. MacManus-Spencer, A. Hamdoun and T. Smital. 2008. Efflux transporters: Newly appreciated roles in protection against pollutants. *Environ. Sci. Tech.* 42: 3914–3920.

Ephrussi, A. and D. St Johnston. 2004. Seeing is believing: The bicoid morphogen gradient matures. *Cell* 116: 143–152.

Feder, M. E. and G. E. Hofmann. 1999. Heat-shock proteins, molecular chaperones, and the stress response: Evolutionary and ecological physiology. *Annu. Rev. Physiol.* 61: 243–82.

Finkel, T. 2003. Oxidant signals and oxidative stress. *Curr. Opin. Cell Biol.* 15: 247–254.

Fisher, W. F. 1983. Eggs of *Palaemon macrodactylus*. III. Infection by the fungus *Lagenidium callinectes*. *Biol. Bull.* 164: 214–226.

Forrest, J. R. K. and J. D. Thomson. 2011. An examination of synchrony between insect emergence and flowering in Rocky Mountain meadows. *Ecol. Monogr.* 81: 469–491.

Förster, C. and J. A. Baeza. 2001. Active brood care in the anomuran crab *Petrolisthes violaceus* (Decapoda: Anomura: Porcellanidae): Grooming of brooded embryos by the fifth periopods. *J. Crust. Biol.* 21: 606–615.

Franklin, W. A. and W. A. Haseltine. 1986. The role of the (6–4) photoproduct in ultraviolet light-induced transition mutations in *E. coli*. *Mutat. Res.* 165: 1–7.

Fraune, S. and 7 others. 2010. In an early branching metazoan, bacteria colonization of the embryo is controlled by maternal antimicrobial peptides. *Proc. Natl. Acad. Sci. USA* 107: 18067–18072.

Furukawa, R., H. Funabashi, M. Matsumoto and H. Kaneko. 2012. Starfish ApDOCK protein essentially functions in larval defense system operated by mesenchyme cells. *Immunol. Cell Biol.* 90: 955–965.

Gaylord, B. and 7 others. 2011. Functional impacts of ocean acidification in an ecologically critical foundation species. *J. Exp. Biol.* 214: 2586–2594.

Gibson, G. and I. Dworkin. 2004. Uncovering cryptic genetic variation. *Nature Rev. Genet.* 5: 681–690.

Gilbert, S. F. 2002. Canalization. In M. Pagel (ed.), *Encyclopedia of Evolution*. Oxford University Press, New York, Volume 1, pp. 133–135.

Gilbert, S. F. 2013. *Developmental Biology*, 10th Ed. Sinauer Associates, Sunderland, MA.

Gil-Turnes, M. S., M. E. May and W. Fenical. 1989. Symbiotic marine bacteria chemically defend crustacean embryos from a pathogenic fungus. *Science* 246: 116–118.

Gleason, J. E., H. C. Koswagen and D. M. Eismann. 2002. Activation of Wnt signaling bypasses requirement for RTK/Ras signaling during *C. elegans* vulval induction. *Genes Dev.* 16: 1281–1290.

Glezen, W. P. and M. Alpers. 1999. Maternal immunization. *Clin. Infect. Dis.* 28: 219–224.

Gökirmak, T., L. E. Shipp, J. P. Campanale, S. C. Nicklisch and A. Hamdoun. 2014. Transport in technicolor: Mapping ATP-binding cassette transporters in sea urchin embryos. *Mol. Reprod. Dev.* 81: 778–793.

Goldstone, J. V. and 9 others. 2006. The chemical defensome: Environmental sensing and response genes in the *Strongylocentrotus purpuratus* genome. *Dev. Biol.* 300: 366–384.

Gomez-Mestre, I., J. C. Touchon and K. M. Warkentin. 2006. Amphibian embryo and parental defenses and a larval predator reduce egg mortality from water mold. *Ecology* 87: 2570–2581.

Gorman, M. J., P. Kankanala and M. R. Kanos. 2004. Bacterial challenge stimulates innate immune responses in extraembryonic tissues of tobacco hornworm eggs. *Insect Mol. Biol.* 13: 19–24.

Gregor, T., D. W. Tank, E. F. Wieschaus and W. Bialek. 2007a. Probing the limits to positional information. *Cell* 130: 153–164.

Gregor, T., E. F. Wieschaus, A. P. McGregor, W. Bialek and D. W. Tank. 2007b. Stability and nuclear dynamics of the bicoid morphogen gradient. *Cell* 130: 141–152.

Guz, N., U. Toprak, A. Dageri, M. O. Gurkan and D. L. Denlinger. 2014. Identification of a putative antifreeze protein gene that is highly expressed during preparation for winter in the sunn pest, *Eurygaster maura*. *J. Insect Physiol.* 68: 30–35.

Häder, D. P. and R. P. Sinha. 2005. Solar ultraviolet radiation-induced DNA damage in aquatic organisms: Potential environmental impact. *Mutat. Res.* 571: 221–233.

Hamburger, V. 1988. *The Heritage of Experimental Embryology: Hans Spemann and the Organizer*. Oxford University Press, Oxford.

Hamdoun, A., G. N. Cherr, T. A. Roepke, K. R. Foltz and D. Epel. 2004. Activation of multidrug efflux transporter activity at fertilization in sea urchin embryos (*Strongylocentrotus purpuratus*). *Dev. Biol.* 276: 413–423.

Hamdoun, A. and D. Epel. 2007. Embryo stability and vulnerability in an always changing world. *Proc. Natl. Acad. Sci. USA* 104: 745–750.

Hansen, J. and 17 others. 2013. Assessing "dangerous climate change": Required reduction of carbon emissions to protect young people, future generations and nature. *PLoS ONE* 8(12): e81648.

Haislip, N. A., M. J. Gray, J. T. Hoverman and D. L. Miller. 2011. Development and disease: How susceptibility to an emerging pathogen changes through anuran development. *PLoS ONE* 6(7): e22307.

Hawkes, L. A., A. C. Broderick, M. H. Godfrey and B. J. Godley. 2009. Climate change and marine turtles. *Endang. Sp. Res.* 7: 137–154.

Heck, D. E., A. M. Vetrano, T. M. Mariano and J. D. Laskin. 2003. Unexpected role for catalase. *Biol. Chem.* 278: 22432–22436.

Heikkila, J. J., N. Ohan, Y. Tam and A. Ali. 1997. Heat-shock protein expression during *Xenopus* development. *Cell Mol. Life Sci.* 53: 114–121.

Heilbrunn, L. V. 1943. *An Outline of General Physiology*, 2nd Ed. W. B. Saunders Co., Philadelphia.

Hensey, C. and J. Gautier. 1998. Programmed cell death during *Xenopus* development: A spatio-temporal analysis. *Dev. Biol.* 203: 36–48.

Higgins, C. F. 1992. ABC transporters: From microorganisms to man. *Annu. Rev. Cell Biol.* 8: 67–113.

Hongmiao, W., J. Dongrui, J. Shaob and S. Zhanga. 2012. Maternal transfer and protective role of antibodies in zebrafish *Danio rerio*. *Mol. Immunol.* 51: 332–336

Houchmandzadeh, B., E. Wieschaus and S. Leibler. 2002. Establishment of developmental precision and proportions in the early *Drosophila* embryo. *Nature* 415: 798–802.

Hu, P. J. 2007. Dauer. In The *C. elegans* Research Community Wormbook. www.wormbook.org/chapters/www_dauer/dauer.html

Ikegami, R., P. Hunter and T. D. Yager. 1999. Developmental activation of the capability to undergo checkpoint-induced apoptosis in the early zebrafish embryo. *Dev. Biol.* 209: 409–433.

Intergovernmental Panel on Climate Change (IPCC). 2014. Contribution of Working Groups I, II and III to the Fifth Assessment Report of the Intergovernmental Panel on Climate Change. In R.K. Pachauri and L.A. Meyer (eds.), *Climate Change 2014: Synthesis Report*. IPCC, Geneva, Switzerland. www.ipcc.ch/report/ar5/syr/

Ivarie, R. 2006. Competitive bioreactor hens on the horizon. *Trends Biotechnol.* 24: 99–101.

Jaeger, J., D. H. Sharp and J. Reinitz. 2007. Known maternal gradients are not sufficient for the establishment of gap domains in *Drosophila melanogaster*. *Mech. Dev.* 124: 108–128.

Jarosz, D. F., M. Taipale and S. Lindquist. 2010. Protein homeostasis and the phenotypic manifestation of genetic diversity: Principles and mechanisms. *Annu. Rev. Genet.* 44: 189–216.

Karentz, D., I. Bosch and D. M. Mitchell. 2004. Limited effects of Antarctic ozone depletion on sea urchin development. *Mar. Biol.* 145: 277–292.

Klaassen, C. D. and J. Liu. 1997. Role of metallothionein in cadmium-induced hepatotoxicity and nephrotoxicity. *Drug Metab. Rev.* 29: 79–102.

Krakauer, D. C. and M. A. Nowak. 2014. Evolutionary preservation of redundant duplicated genes. *Semin. Cell Dev. Biol.* 10: 555–559.

Kroiss, J. and 7 others. 2010. Symbiotic streptomycetes provide antibiotic combination prophylaxis for wasp offspring. *Nature Chem. Biol.* 6: 261–263.

Kudo, S. 2000. Enzymes responsible for the bactericidal effect in extracts of vitelline and fertilization envelopes of rainbow trout eggs. *Zygote* 8: 257–265.

Kultz, D. 2005. Molecular and evolutionary basis of the cellular stress response. *Annu. Rev. Physiol.* 67: 225–257.

Kuo, R., G. Baxter, S. H. Thompson, S. A. Stricker, C. Patton, J. Bonaventura and D. Epel. 2000. Nitric oxide is both necessary and sufficient for activation of the egg at fertilization. *Nature* 406: 633–636.

Lambert, C. C. and G. Lambert. 1978. Tunicate eggs utilize ammonium ions for flotation. *Science* 200: 64–65.

Lander, A. D. 2011. Pattern, growth, and control. *Cell* 144: 955–969.

Lang, L., D. Miskovic, P. Fernando and J. J. Heikkila. 1999. Spatial pattern of constitutive and heat-shock induced expression of the small heat-shock protein gene family, Hsp30, in *Xenopus laevis* tailbud embryos. *Dev. Genet.* 25: 365–374.

Leckie, C., R. Empson, A. Becchetti, J. Thomas, A. Galione and M. Whitaker. 2003. The NO pathway acts late during the fertilization response in sea urchin eggs. *J. Biol. Chem.* 278: 12247–12254.

Lepzelter, D. and J. Wang. 2008. Exact probabilistic solution of spatial-dependent stochastics and associated potential landscape for the bicoid protein. *Physical Rev. E* 77: 041917.

Levy, R. 2001. Genetic regulation of preimplantation embryo survival. *Int. Rev. Cytol.* 210: 1–37.

Li, C., H. M. Blencke, T. Haug, Ø. Jørgensen and K. Stensvåg. 2014. Expression of antimicrobial peptides in coelomocytes and embryos of the green sea urchin (*Strongylocentrotus droebachiensis*). *Dev. Comp. Immunol.* 43: 106–113.

Li, Y., R. W. Huyck, J. H. Laity and G. R. Andrews. 2008. Zinc-induced function of a co-activator complex containing the zinc-sensing transcription factor MTF-1, p300/CBP, and SP1. *Mol. Cell Biol.* 28: 4275–4284.

Little, T. J., M. Perutz, M. Palmer, C. Crossan and V. A. Braithwaite. 2008. Male three-spined sticklebacks *Gasterosteus aculeatus* make antibiotic nests: A novel form of parental protection? *J. Fish Biol.* 73: 2380–2389

Liu, Y. and 13 others. 2014. Deciphering microbial landscapes of fish eggs to mitigate emerging diseases. *Int. Soc. Microb. Ecol. J.* 8: 2002–2014.

Lindquist, N. 2000. Chemical defense of early life stages of benthic marine invertebrates. *J. Chem. Ecol.* 28: 1987–2000.

Lopanik, N., N. Lindquist and N. Targett. 2004. Potent cytotoxins produced by a microbial symbiont protect host larvae from predation. *Oecologia* 139: 131–139.

Lopes, F. J. P., F. M. C. Vieira, D. M. Holloway, P. M. Bisch and A. V. Spirov. 2008. Spatial bistability generates hunchback expression sharpness in the *Drosophila* embryo. *PLoS Comp. Biol.* 4(9): e1000184.

Lucchetta E. M., J. H. Lee, L. A. Fu, N. H. Patel and R. F. Ismagilov. 2005. Dynamics of *Drosophila* embryonic patterning network *per* turned in space and time using microfluidics. *Nature* 434: 1134–1138.

MacRae, T. H. 2003. Molecular chaperones, stress resistance and development in *Artemia franciscana*. *Semin. Cell Dev. Biol.* 14: 251–258.

Mason, D. S., F. Schafer, J. M. Shick and W. C. Dunlap. 1998. Ultraviolet radiation-absorbing mycosporine-like amino acids (MAAs) are acquired from their diet by medaka fish (*Oryzias latipes*) but not by SKH-1 hairless mice. *Comp. Biochem. Physiol. A* 120: 587–598.

Mateusen, B., R. E. Sanchez, A. Van Soom, P. Meerts, D. G. Maes and H. J. Nauwynck. 2004. Susceptibility of pig embryos to porcine circovirus type 2 infection. *Theriogenology* 61: 91–101.

McFadzen, I., N. Eufemia, C. Heath, D. Epel, M. Moore and D. Lowe. 2000. Multidrug resistance in the embryos and larvae of the mussel *Mytilus edulis*. *Mar. Environ. Res.* 50: 319–323.

McNeil, P. L. and M. M. Baker. 2001. Cell surface events during resealing visualized by scanning electron microscopy. *Cell Tiss. Res.* 304: 141–146.

McNeil, P. L. and R. A. Steinhardt. 2003. Plasma membrane disruption: Repair, prevention, adaptation. *Annu. Rev. Cell Dev. Biol.* 19: 697–731.

McNeil, P. L. and M. Terasaki. 2001. Coping with the inevitable: How cells repair a torn surface membrane. *Nature Cell Biol.* 3: E124–E129.

Merian, M. 1679. *Der Raupen wunderbare Verwandelung, und sonderbare Blumen-nahrung*. Nuremberg: Johann Andreas Graffen.

Metchnikoff, E. 1891. *Lectures on the Comparative Pathology of Inflammation*. Dover, NY.

Miller-Rushing, A. J., T. Katsuki, R. Primack, H. Higuchi, Y. Ishii, S. D. Lee and H. Higuchi. 2007. Impact of global warming on a group of related species and their hybrids: Cherry tree flowering at Mt. Takao, Japan. *Amer. J. Bot.* 94: 1470–1478.

Miyake, K. and P. L. McNeil. 1998. A little shell to live in: Evidence that the fertilization envelope can prevent mechanically induced damage of the developing sea urchin embryo. *Biol. Bull.* 195: 214–215.

Moore, P. D. 2009. *Tropical Forests*. InfoBase Publishing, New York, p. 147.

Motola, D. L. and 10 others. 2006. Identification of ligands for DAF-12 that govern dauer formation and reproduction in C. *elegans*. *Cell* 124: 1209–1223.

Munn, C. 2012. *Marine Microbiology: Ecology and Applications*. Garland Science, New York.

Murawski, D. A. 1993. A taste for poison. *Natl. Geog.* 184(6): 122–137.

Nemer, M., R. D. Thornton, E. W. Stuebing and P. Harlow. 1991. Structure, spatial, and temporal expression of two sea urchin metallothionein genes, SpMTB1 and SpMTA. *J. Biol. Chem.* 266: 6586–6593.

Newport, J. and M. Kirschner. 1982. A major developmental transition in early *Xenopus* embryos. I. Characterization and timing of cellular changes at the midblastula stage. *Cell* 30: 675–686.

Nijhout, H. F. 1994. *Insect Hormones*. Princeton University Press, Princeton, NJ.

Nijhout, H. F. 2002. The nature of robustness in development. *BioEssays* 24: 553–563.

NRC (National Research Council). 2010. *Advancing the Science of Climate Change*. The National Academies Press, Washington, DC.

Nussey, D. H., E. Postma, P. Gienapp and M. E. Visser. 2005. Selection on hereditable phenotypic plasticity in a wild bird population. *Science* 310: 304–306.

Olsen, L. F., O. G. Issinger and B. Guerra. 2013. The Yin and Yang of redox regulation. *Redox Rep.* 18: 245–252.

Oyarzun, X. and R R. Strathmann. 2011. Plasticity of hatching and the duration of planktonic development in marine invertebrates. *Integr. Comp. Biol.* 51: 81–90.

Palmiter, R. D. 1998. The elusive function of metallothioneins. *Proc. Natl. Acad. Sci. USA* 95: 8428–8430.

Palumbi, S. R., D. J. Barshis, N. Traylor-Knowles and R. A. Bay. 2014. Mechanisms of reef coral resistance to future climate change. *Science* 344: 895–898.

Pan, T.-C. F., S. L. Applebaum and D. T. Manahan. 2015. Experimental ocean acidification alters the allocation of metabolic energy. *Proc. Natl. Acad. Sci. USA* 112: 4696–4701.

Parsons, J. A. 1965. A digitallis-like toxin in the monarch butterfly *Danaus plexippus* L. *J. Physiol.* 178: 290–304.

Pasche, B. and N. Yi. 2010. Candidate gene association studies: Successes and failures *Curr. Opin. Genet. Dev.* 20: 257–261.

Perron, F. E. 1981. The partitioning of reproductive energy between ova and protective capsules in marine gastropods of the genus *Conus*. *Am. Nat.* 118: 110–118.

Perry, G. and D. Epel. 1981. Ca^{+2}-stimulated production of H_2O_2 from napthoquinone oxidation in *Arbacia* eggs. *Exp. Cell Res.* 114: 65–72.

Petschenka, G., S. Fandrich, N. Sander, V. Wagschal, M. Boppre and S. Dobler. 2013. Stepwise evolution of resistance to toxic cardenolides via genetic substitutions in the AT/K-ATPase of milkweed butterflies (Lepidoptera: Danaini). *Evolution* 69: 2753–2761.

Pike, D. A. 2014. Forecasting the viability of sea turtle eggs in a warming world. *Glob. Chang. Biol.* 20: 7–15.

Pujadas, E. and A. P. Feinberg. 2012. Regulated noise in the epigenetic landscape of development and disease. *Cell* 148: 1123–1131.

Rafferty, N. E., P. J. Caradonna, L. A. Burkle, A. M. Iler and J. L. Bronstein. 2013. Phenological overlap of interacting species in a changing climate: An assessment of available approaches. *Ecol. Evol.* 3: 3183–3193.

Rafferty, N. E. and A. R. Ives. 2012. Pollinator effectiveness varies with experimental shifts in flowering time. *Ecology* 93: 803–814.

Rastogi, R. P. and A. Incharoensakdi. 2014. Characterization of UV-screening compounds, mycosporine-like amino acids, and scytonemin in the cyanobacterium *Lyngbya* sp. CU2555. *FEMS Microbiol. Ecol.* 87: 244–256.

Rawlings, T. A. 1994. Encapsulation of eggs by marine gastropods: Effect of variation in capsule form on the vulnerability of embryos to predation. *Evolution* 48: 1301–1313.

Rawlings, T. A. 1995. Shields against ultraviolet radiation: An additional protective role for the egg capsules of benthic marine gastropods. *MEPS* 136: 81–95.

Reále, D., A. G. McAdam, S. Boutin and D. Berteaux. 2003. Genetic and plastic responses of a northern mammal to climate change. *Proc. Biol. Sci.* 270: 591–596.

Reed, T. E., D. E. Schindler and R. S. Waples. 2011. Interacting effects of phenotypic plasticity and evolution on population persistence in a changing climate. *Conserv. Biol.* 25: 56–63.

Renfree, M. B. and B. Shaw. 2000. Diapause. *Annu. Rev. Physiol.* 62: 353–375.

Reyer, C. P. and 20 others. 2011. A plant's perspective of extremes: Terrestrial plant responses to changing climatic variability. *Glob. Chang. Biol.* 19: 75–89.

Root-Bernstein, R. S. 1997. *Discovering*. Harvard University Press, Cambridge, MA.

Rosenzweig, M., K. M. Brennan, T. D. Tayler, P. O. Phelps, A. Patapoutian and P. A. Garrity. 2005. The *Drosophila* ortholog of vertebrate TRPA1 regulates thermotaxis. *Genes Dev.* 19: 419–424.

Rothschild, M., J. Von Euw, T. Reichstein, D. Smith and J. Pierre. 1975. Cardenolide storage in *Danaus chyrsippus* (L.) with additional notes on *D. plexippus* (L.). *Proc. R. Soc. Lond. B* 190: 1–31.

Royle, N. J., P. T. Smiseth, M. Kolliker, R. M. Kilner and C. A. Hinde. 2012. Parent–offspring conflict. In N. J. Royle, P. T. Smiseth and M. Kölliker (eds.), *The Evolution of Parental Care*. Oxford University Press, Oxford, pp. 119–132.

Rutherford, S. L. and S. Lindquist. 1998. Hsp90 as a capacitor for morphological evolution. *Nature* 396: 336–342.

Ryabova, L. V. and S. G. Vassetzky. 1997. A two-component cytoskeletal system of *Xenopus laevis* egg cortex. *Int. J. Dev. Biol.* 41: 843–851.

Salzman, N. H. 2010. Paneth cell defensins and the regulation of the microbiome: Détente at mucosal surfaces. *Gut Microbes* 1: 401–406.

Sangster, T. A., S. Lindquist and C. Queitsch. 2004. Under cover: Causes, effects and implications of Hsp90-mediated genetic capacitance. *BioEssays* 26: 348–362.

Santidrián Tomillo, P., V. S. Saba, G. S. Blanco, C. A. Stock, F. V. Paladino and J. R. Spotila. 2012. Climate driven egg and hatchling mortality threatens survival of eastern Pacific leatherback turtles. *PLoS ONE* 7(5): e37602.

Scudiero, R., C. Capasso, P. P. De Prisco, S. Capasso, S. Filosa and E. Parisi. 1994. Metal-binding proteins in eggs of various sea urchin species. *Cell Biol. Internatl.* 18: 47–53.

Shapiro, B. M. and P. B. Hopkins. 1991. Ovothiols: Biological and chemical perspectives. *Adv. Enzymol. Relat. Areas Mol. Biol.* 64: 291–316.

Shick, J. M. and W. C. Dunlap. 2002. Mycosporine-like amino acids and related gadusols: Biosynthesis, accumulation, and UV-protective functions in aquatic organisms. *Annu. Rev. Physiol.* 64: 223–262.

Sim, C., D. S. Kang, S. Kim, X. Bai and D. L. Denlinger. 2015. Identification of FOXO targets that generate diverse features of the diapause phenotype in the mosquito *Culex pipiens*. *Proc. Natl. Acad. Sci. USA* 112: 3811–3816.

Sinha, R. P. and D. P. Häder. 2002. UV-induced DNA damage and repair: A review. *Photochem. Photobiol. Sci.* 1: 225–236.

Sinha, R. P., M. Klisch, E. W. Helbling and D. P. Hader. 2001. Induction of mycosporine-like amino acids (MAAs) in cyanobacteria by solar ultraviolet-B radiation. *J. Photochem. Photobiol. B* 60: 129–135.

Smit, J. W., M. T. Huismann, O. V. Tellingen, H. Wiltshire and A. H. Schinkel. 1999. Absence or pharmacological blocking of placental P-glycoprotein profoundly increases fetal drug exposure. *J. Clin. Invest.* 104: 1441–1447.

Somers, C. E. and B. M. Shapiro. 1991. Functional domains of proteoliaisin, the adhesive protein that orchestrates fertilization envelope assembly. *J. Biol. Chem.* 266: 16870–16875.

Spudich, A. 1992. Actin organization in the sea urchin egg cortex. *Curr. Top. Dev. Biol.* 26: 9–21.

Staller, M. V., C. C. Fowlkes, M. D. Bragdon, Z. Wunderlich, J. Estrada and A. H. DePace. 2015. A gene expression atlas of a bicoid-depleted *Drosophila* embryo reveals early canalization of cell fate. *Development* 142: 587–596.

Steinhardt, R. A., G. Bi and J. M. Alderton. 1994. Cell membrane resealing by a vesicular mechanism similar to neurotransmitter release. *Science* 263: 390–393.

Stern, S., Y. Fridmann-Sirkis, E. Braun and Y. Soen. 2012. Epigenetically heritable alteration of fly development in response to toxic challenge. *Cell Rep.* 1: 528–542

Stern, S., O. Snir, E. Mizrachi, M. Galili, I. Zaltsman and Y. Soen. 2014. Reduction in maternal polycomb levels contributes to transgenerational inheritance of a response to toxic stress in flies. *J. Physiol.* 592: 2343–2355.

Stiling, P. 1993. Why do natural enemies fail in biological control campaigns? *Amer. Entomol.* 39: 31–37.

Strathmann, R. R. 2007. Three functionally distinct kinds of pelagic development. *Bull. Mar. Sci.* 81: 167–179.

Strathmann, R. S., J. M. Staver and J. R. Hoffman. 2002. Risk and the evolution of cell cycle durations of embryos. *Evolution* 56: 708–720.

Stumpp, M., M. Hu, I. Casties, R. Saborowski, M. Bleich, F. Melzner and S. Dupont. 2013. Digestion in sea urchin larvae impaired under ocean acidification. *Nature Clim. Chang.* 3: 1044–1049.

Sun, Q. Y. 2003. Cellular and molecular mechanisms leading to cortical reaction and polyspermy block in mammalian eggs. *Microsc. Res. Tech.* 61: 342–348.

Tauber, A. I. and L. Chernyak. 1991. *Metchnikoff and the Origins of Immunology: From Metaphor to Theory.* Oxford University Press, New York.

Tauber, M. J., C. A. Tauber and S. Masaki. 1986. *Seasonal Adaptations of Insects.* Oxford University Press, Oxford.

Tautz, F. 1992. Redundancies, development and the flow of information. *BioEssays* 14: 263–266.

Telemeco, R. S., K. C. Abbott and F. J. Janzen. 2013a. Modeling the effects of climate change-induced shifts in reproductive phenology on temperature-dependent traits. *Am. Nat.* 181: 637–648.

Telemeco, R. S., D. A. Warner, M. K. Reida and F. J. Janzen. 2013b. Extreme developmental temperatures result in morphological abnormalities in painted turtles (*Chrysemys picta*): A climate change perspective. *Integr. Zool.* 8: 197–208.

Terasaki, M., K. Miyake and P. L. McNeil. 1997. Large plasma membrane disruptions are rapidly resealed by Ca^{2+}-dependent vesicle-vesicle fusion events. *J. Cell Biol.* 139: 63–74.

Thurber, R. V. and D. Epel. 2007. Apoptosis in early development of the sea urchin, *Strongylocentrotus purpuratus. Dev. Biol.* 303: 336–346.

Toomey, B. H. and D. Epel. 1993. Multixenobiotic resistance in *Urechis caupo* embryos: Protection from environmental toxins. *Biol. Bull.* 185: 355–364.

Turner, E., R. Klevit, L. J. Hager and B. M. Shapiro. 1987. Ovothiols, a family of redox-active mercaptohistidine compounds from marine invertebrate eggs. *Biochemistry* 26: 4028–4036.

Vacquier, V. D. 1975. The isolation of intact cortical granules from sea urchin eggs: Calcium ions trigger granule discharge. *Dev. Biol.* 43: 62–74.

Vacquier, V. D. 1998. Evolution of gamete recognition proteins. *Science* 281: 1995–1998.

Vacquier, V. D. and J. E. Payne. 1973. Methods for quantitating sea urchin sperm in egg binding. *Exp. Cell Res.* 82: 227–235.

Van Die, I., A. Engering and Y. Van Kooyk. 2004. C-type lectins in innate immunity to pathogens. *Trends Glycotechnol.* 16: 265–279.

Vaughn, D. and R. R. Strathmann. 2009. Predators induce cloning in echinoderm larvae. *Science* 319: 1503.

Voets, T., K. Talavera, G. Owsianik and B. Nilius. 2005. Sensing with TRP channels. *Nature Chem. Biol.* 1: 85–92.

Waddington, C. H. 1942. Canalization of development and the inheritance of acquired characteristics. *Nature* 150: 563–565.

Wang, Y., A. N. Ezemaduka, Y. Tang and Z. Chang. 2009. Understanding the mechanism of the dormant dauer formation of *C. elegans*: From genetics to biochemistry. *IUBMB Life* 61: 607–612.

Ward, E. S. 2004. Acquiring maternal immunoglobulin: Different receptors, similar functions. *Immunity* 20: 507–508.

Whalen, K., A. M. Reitzel and A. Hamdoun. 2012. Actin polymerization controls the activation of multidrug efflux at fertilization by translocation and fine-scale positioning of ABCB1 on microvilli. *Mol. Biol. Cell* 23: 3663–3672.

Whitman, W. B., D. C. Coleman and W. J. Wiebe. 1998. Prokaryotes: The unseen majority. *Proc. Natl. Acad. Sci. USA* 95: 6578–6583

Williamson, K. E., M. Radosevich and E. K. Wommack. 2005. Abundance and diversity of viruses in six Delaware soils. *Appl. Environ. Microbiol.* 71: 3119–3125.

Wong, J. L., R. Créton and G. M. Wessel. 2004. The oxidative burst at fertilization is dependent upon activation of the dual oxidase Udx1. *Dev. Cell.* 7: 801–814.

Wong, J. L. and G. M. Wessel. 2006. Defending the zygote: Search for the ancestral animal block to polyspermy. *Curr. Top. Dev. Biol.* 72: 1–151.

Wong, J. L. and G. M. Wessel. 2008. Free-radical crosslinking of specific proteins alters the function of the egg extracellular matrix at fertilization. *Development* 135: 431–440.

Wortzman-Show, G. B., M. Kurokawa, R. A. Fissore and J. P. Evans. 2007. Calcium and sperm components in the establishment of the membrane block to polyspermy: Studies of ICSI and activation with sperm factor. *Mol. Hum. Reprod.* 13: 557–565.

Wright, L. I. and 7 others. 2012. Turtle mating patterns buffer against disruptive effects of climate change. *Proc. Biol. Sci.* 279: 2122–2127.

Yabe, T., N. Suzuki, T. Furukawa, T. Ishihara and I. Katsura. 2005. Multidrug resistance-associated protein MRP-1 regulates dauer diapause by its export activity in *C. elegans. Development* 132: 3197–3207.

Zimmermann, C., H. Gutmann and J. Drewe. 2006. Thalidomide does not interact with P-glycoprotein. *Cancer Chemother. Pharmacol.* 57: 599–606.

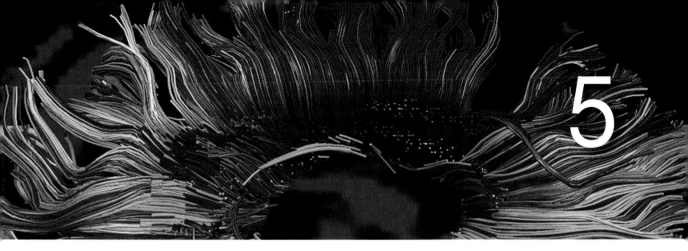

Teratogenesis

Environmental Assaults on Development

It is all of a piece, thalidomide and pesticides. They represent our willingness to rush ahead and use something without knowing what the results will be.

Rachel Carson, 1962

According to the Surgeon General, women should not drink alcohol during pregnancy because of the risk of birth defects.

U.S. Surgeon General's Office warning on alcoholic beverages

The environment is a source not only of instructive signals for normal development, but also of signals that can disrupt normal development. Between 2% and 5% of human infants are born with an observable anatomical abnormality (Thorogood 1997; Epstein 2008). These abnormalities may include missing limbs, missing or extra digits, cleft palate, eyes that lack certain parts, hearts that lack valves, and so forth. These anomalies can be caused by genetic influences, by environmental agents, or by the interactions between environmental and genetic factors.

Physicians need to know the causes of specific birth defects in order to counsel prospective parents, and experimental embryology originated in attempts to determine the causes of such birth defects in human populations (Oppenheimer 1968). By growing embryos at different temperatures or in different salt solutions, the early experimental embryologists attempted to reproduce in chicks, amphibians, and marine invertebrates anatomical abnormalities similar to those seen in anomalous human development. In addition, the study of birth defects can help us understand how the human body is normally formed. In the absence of experimental data on human embryos, we often must rely on nature's "experiments" to learn how the human body becomes organized.* In the eighteenth century, the physician Johann Friedrich Meckel was probably the first to realize that parts of the body that were affected together in malformation **syndromes** (Greek, "happening together") must have some common developmental origin or mechanism that was being affected (Opitz et al. 2006).

The modern medical term for "birth defect" is **congenital anomaly** (congenital, "at birth"; anomaly, "not normal"). These defects may be either structural or functional. Among the latter are brain-related defects that may involve intellectual (cognitive) and emotional changes or physiological deficits that may or may not have a readily identifiable structural basis. Structural birth defects are categorized as malformations, disruptions, deformations, or dysplasias. A *malformation* is a structural birth defect that results from failure of tissue to initially form properly. Anencephaly, for example, is a result of the failure of the anterior neural tube to properly form. It may have a genetic, environmental, or multifactorial basis. A *disruption* results from breakdown of tissue that has initially formed properly. The patterns of missing body parts that result from amniotic bands and limb strangulation (a condition in which torn amniotic tissue surrounds a portion of the body, often digits or extremities, resulting in deep grooves or amputations) are good examples of a disruption type of birth defect. A *deformation* is a result of extrinsic mechanical forces on otherwise normal tissue. This is illustrated in the characteristic pattern of abnormalities that includes the abnormal faces, small lungs, and limb contractures that result from a prolonged deficiency of amniotic fluid. Finally, if the primary defect is a lack of normal organization of cells into tissue, a *dysplasia* will result. This is best illustrated by the pattern of bony abnormalities found in achondroplasic dwarfism, where a defect in the gene encoding fibroblast growth factor receptor 3 results in abnormal cartilage formation.

This chapter is concerned with environmental agents that cause nonheritable birth defects; these are known as **teratogens** (Greek, "monster formers"), and the study of how these agents interfere with normal development to cause birth defects is called **teratology**. Emphasis in this chapter will be on teratogens that cause structural defects, though increasing research attention is being directed toward functional/behavioral teratology (Bushnell et al. 2010). The risk from anthropogenic chemical teratogens increases as more and more untested compounds enter the environment. Over 90,000 artificial chemicals are currently used in the United States, and each year the Environmental Protection Agency receives notification of the manufacture of 500–1000 (EPA 2008, 2014). Most industrial chemicals have not been screened for teratogenic effects. Standard screening protocols are expensive, take a long time, and are subject to the differences in metabolism between

---

*Our ability to learn about normal development from abnormal occurrences is seen in the language once used to describe birth defects. The word "monster," frequently encountered in medical textbooks prior to the mid-twentieth century to describe malformed infants, comes from the Latin *monstrare*, "to show" or "to point out." This is also the root of the English word "demonstrate."

humans and the test animals used in screenings. There is still no consensus on how to accurately test a substance's teratogenicity for human embryos.*

Medical Embryology and Teratology

At one time, doctors and biologists believed that the mother's body and the placenta afforded protection from the environment during prenatal life (see Dally 1998). But in 1941, Norman Gregg, an Australian ophthalmologist, documented that women who contracted rubella (German measles) during the first trimester of pregnancy had a 1 in 6 chance of giving birth to an infant with eye cataracts, heart malformations, and/or deafness. This study provided strong evidence that a pregnant mother could not fully protect the developing fetus from the outside environment. Twenty years later, the U.S. rubella epidemic of 1963–1965 resulted in over 10,000 fetal deaths and the births of some 20,000 infants with birth defects (CDC 2002). Children and most adults infected by the rubella virus show relatively mild symptoms (indeed, some are unaware of even being sick), but infants born to mothers who contract rubella during early pregnancy are likely, as Gregg found, to be born blind, deaf, or both. Many are also born with heart defects or mental deficiencies. The epidemic underscored the fact that prenatal development can be adversely affected by environmental factors.

Wilson's principles of teratology

In 1959, James Wilson put forth six principles that (with some modernization) are still applied to nearly all discussions of teratogenesis:

1. Susceptibility to the teratogenic effect of an agent depends on the genotype of the embryo, the genotype of the mother, and the ways in which their genotypes allow mother and fetus to interact with the adverse environmental factors.

2. There are critical periods of development when specific organ systems are most susceptible to being adversely affected by teratogenic agents.

3. Teratogenic agents act in specific ways on genes, cells, and tissues in the developing organism to interfere with normal developmental events.

4. Several factors affect the ability of a teratogen to interfere with normal development. These include the nature of the agent itself, the route and degree of maternal exposure, the ability of the mother to

*Only about 8,000 chemicals have been tested for their potential teratogenic effects. The European Union (EU) and the United States have different rules concerning testing. Following their respective legal traditions, chemicals in the United States (under the Toxic Substances Control Act) are considered benign until proven unsafe. Chemicals in Europe (regulated under REACH) are usually thought potentially dangerous until their safety is proven. The EU has also found that the evidence (e.g., Kimura-Kuroda et al. 2012) linking certain pesticides to memory dysfunction was sufficient to ban them, while the United States has not.

detoxify or block the agent, the rate of transfer through the placenta, the rate of absorption by the embryo or fetus, and the genotype of the mother and her conceptus.*

5. There are four major manifestations of abnormal development: death, malformation, growth retardation, and functional defects.

6. Manifestations of deviant development increase in frequency and degree as the teratogen dosage increases.[†]

Wilson also noted (1961, p. 191) that "an agent which is very damaging to the embryo may be relatively harmless to the mother."

Thalidomide and the window of susceptibility

In 1961, the subject of teratogens was brought to public attention in dramatic fashion. Two researchers working independently, Widukind Lenz and William McBride, accumulated evidence that the drug thalidomide, prescribed as a mild sedative to many pregnant women, caused an enormous increase in a previously rare syndrome of congenital anomalies. The most noticeable of these anomalies was phocomelia, a condition in which the long bones of the limbs are deficient or absent (Figure 5.1A). Worldwide, more than 7000 affected infants were born to women who took thalidomide, and a woman need only have taken one tablet to produce a child with malformations involving all four limbs (Lenz 1962, 1966; Toms 1962). Other abnormalities induced by maternal ingestion of this drug included ocular and heart defects, absence of the external ears, malformed intestines, and neurobehavioral changes (Miller and Strömland 1999; Vargesson 2009). About half the infants born to mothers who had taken thalidomide died shortly after birth. Thalidomide was withdrawn from the market in November 1961.[‡]

Nowack (1965) documented a **window of susceptibility** during which thalidomide caused malformations. The drug was found to be teratogenic only during days 20–36 after conception (34–50 days after the last menstruation). The developmental stage–related specificity of thalidomide-induced birth defects is shown in Figure 5.1B. From day 20 to day 24 of development, no limb abnormalities are seen. However, during this period, thalidomide can cause the absence or deficiency of ear components.

---

*The word "conceptus" means any product of the zygote, including the extraembryonic membranes. In general, this includes the embryo and fetus as well as the yolk sac, amnion, and portions of the placenta.

†As we will see in Chapter 6, this tenet must be modified when dealing with certain compounds that elicit responses to dose that are not linear. For instance, there are several compounds that interfere with development by impairing the endocrine system. These *endocrine disruptors* produce disease states at moderate concentrations, but (probably due to feedback regulation in the organism) not at higher concentrations.

‡The drug was marketed primarily in Germany, and it is estimated that 5000–7000 infants were born with thalidomide-induced anomalies. It was kept out of the United States largely by Frances Oldham Kelsey of the U.S. Food and Drug Administration. She did not want to rely solely on the industry-based investigations and required that further testing be done. When the drug company refused to do the testing, the FDA would not give license for the drug to be distributed in the United States.

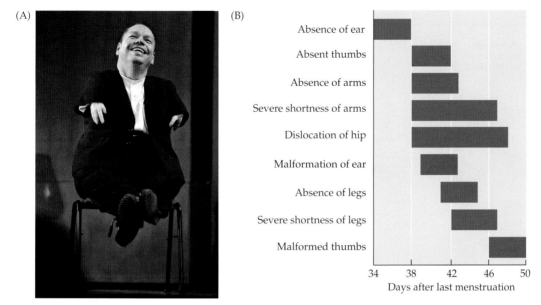

(A)

(B)

Absence of ear
Absent thumbs
Absence of arms
Severe shortness of arms
Dislocation of hip
Malformation of ear
Absence of legs
Severe shortness of legs
Malformed thumbs

34   38   42   46   50
Days after last menstruation

Figure 5.1   During the early 1960s, doctors (primarily in Europe) began prescribing the drug thalidomide to pregnant women as an effective mild sedative and remedy for morning sickness. The resulting epidemic of a specific syndrome of birth defects soon identified the drug as a teratogen. (A) German singer Thomas Quasthoff, Grammy-winning performer of classical and jazz music, was born with phocomelia (improper limb development). Phocomelia, in this case most pronounced in the arms, is the most visible of the birth defects resulting from the ingestion of thalidomide during pregnancy. (B) Thalidomide affects different structures at different times of human development. (Photograph © A2070 Rolf Haid/dpa/Corbis; B after Nowack 1965.)

Malformations of upper limbs are seen before those of the lower limbs, since the arms form slightly before the legs during development.

The concept of a window of susceptibility for teratogens is an important one; most teratogens produce structural defects only during certain critical periods of development. Human development is usually divided into two periods, the **embryonic period** (from conception to the end of week 8, or the first 2 months of gestation) and the **fetal period** (the remaining time in utero). It is during the embryonic period that most of the organ systems form; the fetal period is generally one of growth and modeling. Figure 5.2 indicates the time frames during which various organs are most susceptible to induction of structural malformations by teratogens.

Prior to week 3, exposure to teratogens does not usually produce congenital anomalies, because a teratogen encountered at this time typically either (1) damages most or all the cells of an embryo, resulting in its death, or (2) damages only a few cells, allowing the embryo to recover (since at this stage many cells retain the ability to generate several different cell types). The period of maximum susceptibility to teratogens is between weeks 3 and 8 after conception because that is when cell types are differentiating and most organs are formed. The nervous system, however, is forming continually throughout gestation (and, indeed, even after birth,

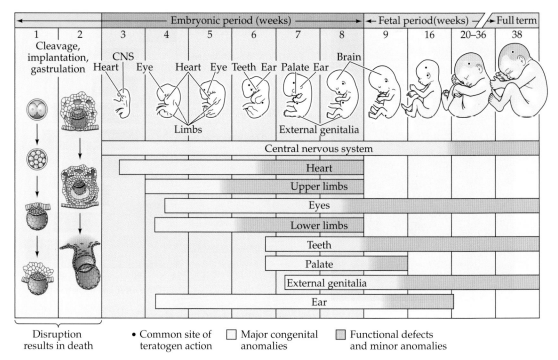

Figure 5.2 Periods (weeks of gestation) and degrees of sensitivity of embryonic organs to teratogens. The embryonic stage (weeks 3–8) is the period of maximum vulnerability. (After Moore and Persaud 1993.)

through adolescence) and thus remains susceptible to insult by environmental agents for this extended period.

Teratogenic Agents

The largest class of human teratogens includes drugs and chemicals (including heavy metals such as lead and mercury). Although teratogenic effects are usually associated with anthropogenic chemicals (i.e., those produced by humans), some chemicals found naturally in the environment can also cause birth defects. Quinine and alcohol, two common substances derived from plants, also cause developmental abnormalities. Quinine ingested by a pregnant mother can cause deafness, and (as we will describe shortly) alcohol can cause physical and mental abnormalities in her offspring.

Viruses, radiation, hyperthermia, and metabolic conditions in the mother can also act as teratogens. A partial list of agents that are teratogenic in human embryos and fetuses is provided in Table 5.1.

Thalidomide as a teratogenic agent

Thalidomide has been difficult to study because it is relatively ineffective in mice and rats, two predominant animals used for developmental toxicity testing. It is also difficult to study because thalidomide becomes metabolized into several products, some of which may be teratogenic and

TABLE 5.1 Some agents thought to disrupt human fetal development[a]

Drugs and chemicals	
Alcohol	Ionizing radiation (x-rays)
Aminoglycosides (e.g., gentamycin)	Hyperthermia (fever)
Aminopterin	Infectious microorganisms
Antithyroid agents (e.g., PTU)	Coxsackie virus
Bromine	Cytomegalovirus
Cortisone	Herpes simplex
Diethylstilbestrol (DES)	Parvovirus
Lead	Rubella (German measles)
Methylmercury	*Toxoplasma gondii* (toxoplasmosis)
Penicillamine	*Treponema pallidum* (syphilis)
Retinoic acid (isotretinoin, Accutane)	Metabolic conditions in the mother
Streptomycin	Autoimmune disease (including Rh incompatibility)
Tetracycline	Diabetes
Thalidomide	Dietary deficiencies, malnutrition
Trimethadione	Phenylketonuria
Valproic acid	
Warfarin	

Source: Opitz 1991.

[a]This list includes known and possible teratogenic agents and is not exhaustive.

others not. Moreover, it may have several ways of adversely affecting development (Vargesson 2009, 2014; Ito et al. 2011).

A major theory of thalidomide action is that it blocks angiogenesis (the formation of blood vessels) (D'Amato et al. 1994; Yabu et al. 2005). Using a teratogenic variant of thalidomide that could not be further metabolized, Therapontos and colleagues (2009) showed that thalidomide blocked the stabilization of new blood vessels and that the time sensitivity of thalidomide-induced limb defects correlated with the time that these immature blood vessels were forming. This form of thalidomide prevents the capillary-forming endothelial cells from migrating and forming the tubes that are critical in making blood vessels (Figure 5.3). Once the smooth muscles cover the tubes, thalidomide has no effect. It appears that blood vessel loss may be the trigger leading to

(A) Control

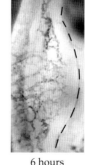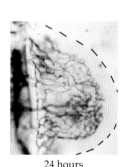

3 hours   6 hours   24 hours

(B) Thalidomide treated

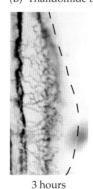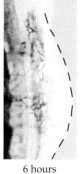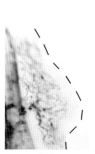

3 hours   6 hours   24 hours

Figure 5.3 Thalidomide causes limb malformations by interfering with blood vessel formation in limb buds. (A) Control chick embryos were injected with an inert substance at day 3, and the limb vasculature was subsequently stained with India ink at 3, 6, and 24 hours after injection. (B) Chick embryos injected with a teratogenic analogue of thalidomide show the failure of blood vessel formation at those times. (From Therapontos et al. 2009.)

increased cell death and misregulation of the paracrine factor pathways necessary for limb bud formation. Recent research by Siamwala and colleagues (2012) supports the premise that the antiangiogenic property of thalidomide is the basis for its teratogenesis. They demonstrated that nitric oxide, which promotes angiogenesis, provides protection from thalidomide-induced limb defects. Although it should not be used by pregnant women, thalidomide's antiangiogenic effect makes it a potential drug for the treatment of cancer and inflammatory diseases (Raje and Anderson 1999; Mahony et al. 2013).

Heavy metal teratogens: Industrial mercury and Minamata disease

As modern agriculture has become dependent on pesticides, herbicides, and fertilizers, and as chemical and mining industries have expanded their interests across the globe, the teratogenic effects of anthropogenic chemicals have traveled with them. The unregulated "production at all costs" approach of industrialized countries is associated with increasing birth defect rates. For example, in some regions of Kazakhstan, high concentrations of heavy metals in the drinking water, vegetables, and air have been found. In such locations, nearly half the people tested have extensive chromosome breakage, and in some areas the incidence of birth defects doubled between 1980 and 1994 (Edwards 1994). Before the 1960s, however, there were very few warnings that embryos might be endangered by the progress of human technologies. Other than Rachel Carson, whose book *Silent Spring* warned that our songbird population was being wiped out by DDT, few people thought that plastics, pesticides, or the chemicals that comprise them were dangerous. Perhaps the first disaster to raise public awareness of the danger from industrial chemicals occurred in Japan, where methylmercury was found to be the cause of neurological abnormalities in nearly 10% of the children born near the area of Minamata Bay.

In the years after World War II, the Minamata plant of Japan's Chisso Corporation manufactured numerous organic chemicals, including acetaldehyde, an intermediate in the manufacture of consumer products ranging from plastics and paints to perfumes. By 1951, the plant was producing more than 6000 tons of acetaldehyde a year. A by-product of the process (in which mercuric sulfate is used to oxidize acetylene into acetaldehyde) is mercury, which the company simply dumped into Minamata Bay. Microbes in the water of the bay converted the mercury into an organic form, methylmercury. Fish and shellfish—dietary staples for the townspeople in the area—consumed and concentrated this methylmercury.

As the 1950s progressed, cats in the bay area began dying amid fits of convulsions, crows fell from the sky, and fish floated dead on the bay's surface. Some villagers began having problems speaking and walking. Eventually, these same people had difficulty seeing, hearing, and swallowing. By October 1956, 40 such patients had been documented, of which 14 died of convulsions. A research team from Kumamoto University concluded that the symptoms were due, not to contagious microorganisms, but to a heavy metal (Figure 5.4). When mercury levels in the area were measured, the

(A)

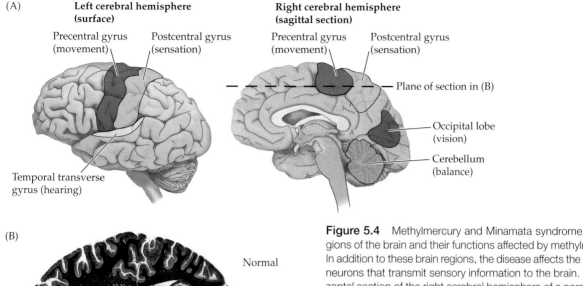

Left cerebral hemisphere (surface)

Precentral gyrus (movement)   Postcentral gyrus (sensation)

Temporal transverse gyrus (hearing)

Right cerebral hemisphere (sagittal section)

Precentral gyrus (movement)   Postcentral gyrus (sensation)

Plane of section in (B)

Occipital lobe (vision)

Cerebellum (balance)

(B)

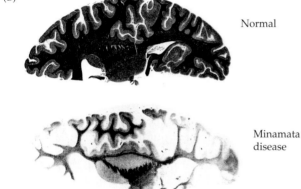

Normal

Minamata disease

Figure 5.4 Methylmercury and Minamata syndrome. (A) Regions of the brain and their functions affected by methylmercury. In addition to these brain regions, the disease affects the axons of neurons that transmit sensory information to the brain. (B) Horizontal section of the right cerebral hemisphere of a normal adult and that of an adult with Minamata disease. (After National Institute for Minamata Disease, www.nimd.go.jp/archives/english.)

amounts seen were remarkable.* The Chisso wastewater canal contained over 2 kg of mercury in every ton of sludge—so much that it was actually profitable to mine it!

One of the most important findings of Masazumi Harada and his team of physicians (see Harada 1972; Ui 1992) was that children born to unaffected mothers could be severely damaged by methylmercury. Simply by eating their normal diet, pregnant women were inadvertently exposing

*The Chisso management's response to the situation was to change the site where it dumped mercury (thereby killing the fish in the tributary Minamata River and spreading Minamata disease to new areas). The company prevented its own investigator from releasing his findings that water from the factory caused neurological problems in previously healthy cats, and instead it published its own report indicating that factors other than its wastes might be the cause of the disease. Moreover, government authorities—whose investigations revealed extremely large amounts of mercury in the bodies of the Minamata victims—failed to publish their findings (see Harada 1972; Ui 1992). Other companies in other countries have discharged heavy metal effluents into rivers where the townspeople have little power. Minamata syndrome has been documented in indigenous peoples in Ontario, Canada, and in Brazil's Amazon basin where new chemical plants have been established (Harada et al. 1976; Harada 1996; Gilbertson 2004).

Figure 5.5 The appearance of developmental anomalies such as two-headed trout hatchlings focused attention on heavy metal (selenium) dumping by mining companies into lakes of the western United States. Slag generated by mining operations reaches the lakes via stream runoff.

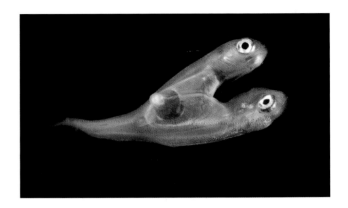

their embryos/fetuses to high doses of this compound. Mercury is selectively absorbed by regions of the developing cerebral cortex (Eto 2000; Kondo 2000; Eto et al. 2001), and when pregnant mice are given mercury on day 9 of gestation, nearly half the pups are born with small brains or small eyes (O'Hara et al. 2002). It had been thought that the placenta would protect fetuses against the outside environment. However, methylmercury appears to be concentrated by the placenta, and it can cross the placenta to cause brain and eye defects in the fetus. Methylmercury can also be concentrated in and transmitted through mother's milk.

This finding forced a shift in thinking about the protection of the conceptus from environmental chemicals. It also set in motion a change in attitude, to one that held corporations legally responsible for their actions and that held governments legally responsible for policing such corporations. Despite this, in the United States, industrial dumping of selenium, mercury, cadmium, and lead* and the lax enforcement of antipollution laws have created a situation where lakes throughout the country have warnings against eating fish caught therein. The International Joint Commission of the United States and Canada (2000) warns that "eating Great Lakes sport fish may lead to birth anomalies and serious health problems for children and women of childbearing age." In Idaho, recent photographs of two-headed trout (Figure 5.5) caused public recognition that selenium levels in lakes in the southern part of the state were much higher than government regulations supposedly permit (Kaufman 2012; USDI 2012). The selenium appears to be coming from mining tailings that flow down streams from the mines to the lakes.

Alcohol as a teratogen

In terms of the frequency of its effects and its cost to society, the most devastating human teratogen is undoubtedly ethanol (reviewed by Foltran et al. 2011; Warren and Li 2005). In 1968, Lemoine and colleagues noticed a

---

*Lead, another heavy metal, damages the developing brain in prenatal and childhood stages (Bellinger et al. 1987; Baghurst et al. 1992; Dietrich et al. 1993) and contributes to developmental delays and mental disabilities. Lead paint was banned in Europe in 1955, but lobbying efforts by the paint industry in the United States succeeded in keeping lead in American paints until 1977 (see Steingraber 2003). The usage of selenium as a teratogen is used as the basis for a Dick Francis mystery novel.

syndrome of birth defects in the children of alcoholic mothers. This **fetal alcohol syndrome**, or **FAS**, was confirmed by Jones and Smith (1973). Babies with FAS are physically characterized by a small head size, an indistinct philtrum (the pair of ridges that runs between the nose and mouth above the center of the upper lip), a narrow upper lip (vermillion) border, and a low nasal bridge. The brain of such a child may be dramatically smaller than normal and may show defects in neuronal and glial migration (Figure 5.6A,B; Clarren 1986). Moreover, in keeping with the occurrence of intellectual deficits and behavioral abnormalities in individuals with FAS (e.g.,

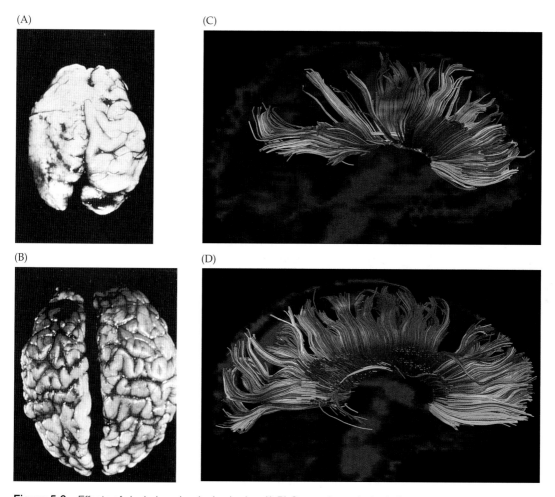

(A)

(C)

(B)

(D)

Figure 5.6 Effects of alcohol on developing brains. (A,B) Comparison of a brain from an infant with fetal alcohol syndrome (FAS) (A) with a brain from a normal infant of the same age (B). The brain from the infant with FAS is smaller, and the pattern of convolutions is obscured by glial cells that have migrated over the top of the brain. (C,D) Regionally specific abnormalities of the corpus callosum seen by diffusion tensor imaging (focusing at myelinated neurons). The brain of a child with fetal alcohol spectrum disorder (C) is compared with a same-age control child (D). The difference in fiber tracts suggests that there are significant abnormalities in neurons that would normally project through the posterior regions into the cortex of the parietal and temporal lobes. (A,B from Clarren 1986, courtesy of S. Clarren; C,D from Wozniak and Muetzel 2011, courtesy of the authors.)

Streissguth and LaDue 1987; Mattson et al. 2013) techniques that identify neural tracts in the brain have found subtle abnormalities that correlate with altered mental processing speed and executive functioning such as planning, memorizing, and retaining information (Wozniak and Muetzel 2011).

FAS represents an extreme portion of a range of defects caused by prenatal alcohol exposure.* The term **fetal alcohol spectrum disorder** (**FASD**) has been coined to encompass all of the alcohol-induced malformations and functional deficits that occur (Figure 5.6C,D). In many children who have defects within this spectrum, behavioral abnormalities exist in the absence of gross physical changes in head size or notable reductions in IQ (NCBDD 2009). FAS occurs in approximately 1 out of 750 live births in the United States, while FASD is much more prevalent, with current data suggesting that as many as 1 in every 100 live births are affected (see CDC 2015; Sampson et al. 1997). While FASD is 100% preventable if women refrain from drinking alcohol during pregnancy, sadly maternal alcohol use continues to be the leading known cause of congenital mental deficiency in the United States.

The severity and types of alcohol-induced anomalies depend on alcohol dosage and developmental stage at the time of prenatal exposure. As expected, the most severe outcomes result from high levels of alcohol consumption. Identification in humans of a potentially "universally" safe level of alcohol exposure is problematic because of confounding influences such as other concurrent teratogen exposures and because of differing genetic backgrounds among individuals. Illustrating the importance of the latter are studies employing animals that have specific gene mutations that have been shown to confer greater sensitivity to alcohol teratogenicity than is present in wild-type animals (Kietzman et al. 2014; McCarthy and Eberhart 2014). A recent epidemiological analysis (Flak et al. 2014) revealed that even moderate levels of alcohol are associated with childhood behavior problems, leading the researchers to conclude that there is no safe level of alcohol that can be consumed during pregnancy.

Animal model–based studies have shown that ethanol can cause permanent brain damage following exposure at virtually every stage of prenatal development, including times before most women realize that they are pregnant. Mice have proven to be particularly useful for identification of ethanol-sensitive stages and tissues. As illustrated in Figure 5.7, exposing mice to ethanol at the time of gastrulation induces concurrent abnormalities of the face and brain that vary in degrees of severity. The defects at the mild end of the spectrum appear consistent with those in FAS (Sulik et al. 1981) and involve deficiencies in the midline tissue of both the face and brain. Such mouse studies suggest that there may be several different mechanisms through which ethanol may induce its teratogenic effects (see Sulik 2014).

Among these mechanisms is interference with cell migration, in particular the migration of cranial neural crest cells. These neural crest cells usually migrate from the dorsal region of the neural tube (which will form

*For remarkable accounts of raising children with fetal alcohol syndrome, read Michael Dorris's *The Broken Cord* (1989) and Liz and Jodee Kulp's *The Best I Can Be* (2000). For an excellent account of the debates within the media and the medical profession about FAS, see Janet Golden's *Message in a Bottle* (2005).

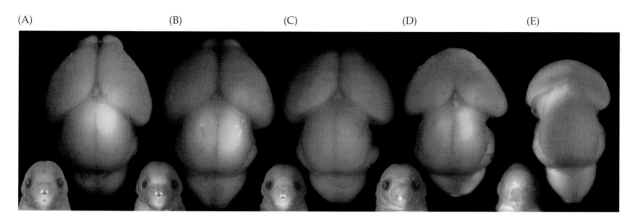

Figure 5.7 Alcohol-induced craniofacial and brain anomalies in mice. The facial anomalies, especially the reduction of the medial structures, correlate with medial forebrain deficiencies. The dorsal views of dissected brains are shown for a normal fetus (A) and for representative examples of increasingly abnormal facial characteristics (B–E). The medial facial deficiency (absence of central structures) was associated with the decrease of the cerebral cortices (B–E) and decrease in the growth (B,C) or absence of the olfactory bulbs (D,E) and incomplete division of the forebrain (D,E). (From Kietzman et al. 2014, courtesy of K. Sulik and R. J. Lipinski.)

the brain) to generate the bones of the face. Instead of migrating and dividing, ethanol-treated neural crest cells prematurely initiate their differentiation into facial bones (Hoffman and Kulyk 1999). Among the numerous genes that are misregulated following maternal alcohol exposure in mice are several involved in the cytoskeletal reorganization that enables cell movement (Green et al. 2007).

In addition, cell death appears to play a key role in alcohol's teratogenesis and is seen shortly after prenatal alcohol exposure (Figure 5.8).

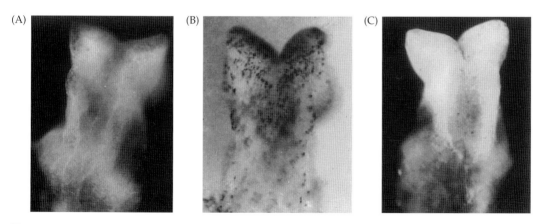

Figure 5.8 Cell death caused by ethanol-induced superoxide radicals is a possible mechanism producing fetal alcohol syndrome. Staining with Nile blue sulfate shows areas of cell death. (A) Control 9-day mouse embryo (head region). (B) Head region of ethanol-treated 9-day embryo. (C) Head region of embryo treated with both ethanol and superoxide dismutase, an inhibitor of superoxide radicals. (From Kotch et al. 1995, photographs courtesy of K. Sulik.)

When the time of alcohol exposure corresponds to the third and fourth weeks of human development, cells that should form the median portion of the forebrain, upper midface, and cranial nerves are killed by ethanol. The death of vulnerable cell populations, including neural crest cells, can be seen as early as 12 hours following the exposure. This has been confirmed in early chick embryos, where transient ethanol exposure at environmentally relevant doses (25 mM) decimates migrating cranial neural crest cells, causing cell death throughout the head region (Flentke et al. 2011).

Such ethanol-induced cell death might have multiple primary causes. Among these is that alcohol metabolism generates superoxide radicals that can oxidize cell membranes and lead to cytolysis (Davis et al. 1990; Kotch et al. 1995; Sulik 2005). In model systems, antioxidants have been shown to be effective in reducing both the cell death and the malformations caused by alcohol (Chen et al. 2004; Miller et al. 2013). Another basis for the induced cell death is the ethanol-induced down-regulation of *sonic hedgehog* expression and subsequent death of sonic hedgehog–dependent cells (Chrisman et al. 2004; Aoto et al. 2008). The sonic hedgehog protein is critical for normal formation of the brain and facial skeleton. This is the protein whose signaling is inhibited by cyclopamine, resulting in the formation of a single eye in the center of the head (see the box, p. 203). If sonic hedgehog–secreting cells are placed in the head mesenchyme at the sensitive time, they can prevent the ethanol-induced apoptosis of the cranial neural crest cells (Ahlgren et al. 2002).

Alcohol also seems to be able to lock neuroblasts into an undifferentiated state (Zhou et al. 2011a, b). It down-regulates those genes involved in neuron maturation (*Sox5* and *Ngn1*), growth, and cell cycle regulation. This may happen through ethanol's ability to induce significant increases in DNA methyltransferase activity (but not histone transacetylase activity) in certain regions of the brain (Perkins et al. 2013), and may be causally related to the anatomical changes mentioned above.

Alcohol's teratogenesis may also be related to its interference with the ability of the cell adhesion molecule L1 to hold cells together. In this case, ethanol is acting on the cell's proteins and not on genes that encode them. Ramanathan and colleagues (1996) have shown that, in vitro, ethanol can block the adhesive function of the L1 protein at levels as low as 7 mM—a concentration that can be produced in the blood or brain by a single drink (Figure 5.9). Moreover, addition of L1 prevents the alcohol-induced disruption of neurite growth and neural cell death (Bearer et al. 1999; Gubitosi-Klug et al. 2007). Mutations in the human gene for L1 cause a syndrome of mental retardation and malformations similar to those seen in severe cases of FAS. Interestingly, strains of mice having different

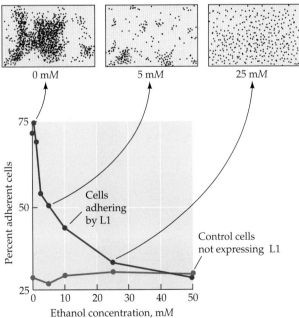

Figure 5.9   The inhibition of L1-mediated cell adhesion by ethanol is another possible factor in fetal alcohol syndrome. (From Ramanathan et al. 1996.)

strengths of signaling pathways enable alcohol to bind to L1 at different concentrations. This causes different predispositions to producing FASD and may explain how similar doses of alcohol can cause different severities of FASD in humans (Fransen et al. 1995; Dou et al. 2013).

Taken together, the above studies show that maternal alcohol use can result in a wide range of birth defects and that they may be the result of a variety of mechanisms through which alcohol can impact cells and tissues. Alcohol is a widely used and abused compound. Thus, the U.S. Surgeon General's office recommends that pregnant women (and those who are contemplating pregnancy) refrain from drinking alcoholic beverages.*

Retinoic acid

Even a compound normally involved in development can disrupt development if it is present in the wrong amounts and/or at the wrong times. Retinoic acid (RA) is important for many aspects of normal development, including the formation of the anterior-posterior axis of the mammalian embryo, as well as for proper heart and jaw formation. In normal development, RA is secreted from discrete cells and works in circumscribed areas of the embryo. However, if RA is present in large amounts, cells that normally would not receive high concentrations of this molecule are exposed to it and will respond to it.

Isotretinoin, also called 13-*cis*-retinoic acid and first sold under the trade name Accutane, has been useful in treating severe cystic acne and was made available for this purpose in 1982. While the deleterious effects of administering large amounts of retinoids to pregnant animals have been known since the 1950s (see Cohlan 1953; Giroud and Martinet 1959; Shenefelt 1972; Kochhar et al. 1984), a large number of women of childbearing age have taken isotretinoin, and some have taken it during pregnancy. Lammer and his coworkers (1985) studied a group of 59 women who were inadvertently exposed to isotretinoin and elected not to terminate the pregnancy. Of the offspring, 26 were born without any noticeable anomalies, 12 aborted spontaneously (miscarried), and 21 were born with obvious anomalies. The affected infants had a characteristic syndrome, including absent or defective ears, absent or small jaws, cleft palate, aortic arch abnormalities, thymic deficiencies, and abnormalities of the central nervous system.

This pattern of multiple congenital anomalies is similar to that seen in rat and mouse embryos whose pregnant mothers were given RA. Goulding and Pratt (1986) placed 8-day mouse embryos in a solution containing 13-*cis*-retinoic acid at a very low concentration (2×10^{-6} M). Even at this concentration, approximately one-third of the embryos developed a very

---

*Men should also be warned. Alcohol (like marijuana and tobacco) is believed to cause reduced sperm counts, and it increases the incidence of defective sperm. It also results in impotence and testosterone deficiency (Emanuele and Emanuele 1998; Battista et al. 2008). Interestingly, the fact that alcohol is bad for developing embryos appears to have been known for a long time, and then the knowledge was lost. Philip Pauly (1996) has documented that in the early twentieth century, the harmful effects of ethanol on an individual were seen as doing a public service, getting rid of weak embryos and allowing only the strong ones to survive. There is also evidence that ethanol consumption changes the methylation pattern of the sperm DNA (Govorko et al. 2012) and that this could alter behavior of the offspring, but these effects, though possibly important, remain hypothetical.

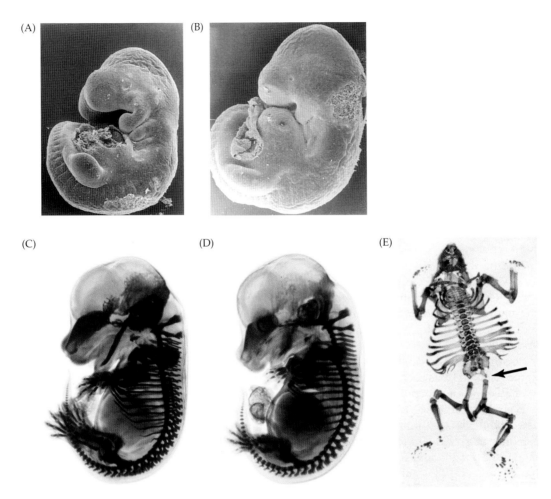

Figure 5.10 Effects of retinoic acid (RA) on mouse embryos. (A,B) Embryos cultured in control medium (A) or in medium containing RA (B), seen on day 10. The first pharyngeal (jaw) arch of the treated embryo is malformed and has fused with the second arch. The ossification of some of the skull has failed, and there are limb abnormalities. (C,D) Skeletal formation in a control embryo (C) and an embryo exposed to RA in utero (D), seen at day 17. Craniofacial malformations are seen; the neural crest-derived cartilage of the jaw and of the middle ear failed to form properly. (E) In some cases, exposure to RA results in severe loss of caudal vertebrae (arrow). (A–D courtesy of G. Morriss-Kay; C,D from Morriss-Kay 1993; E courtesy of M. Kessel.)

specific pattern of anomalies, including dramatic reduction in the size of the first and second pharyngeal arches (Figure 5.10). In normal mice, the first arch eventually forms the maxilla and mandible of the jaw and two ossicles of the middle ear, while the second arch forms the third ossicle of the middle ear, as well as other facial bones.

Like many other teratogens, RA probably disrupts development through several mechanisms. One of these mechanisms appears to be RA's ability to alter the expression of the Hox genes. Hox genes are critical in specifying which part of the embryo is anterior, which is posterior, and which region lies between them. Hox genes are also critical in specifying the fates of certain cells, and Hox proteins inhibit neural crest cells from forming cartilage and bone. By altering the expression of Hox genes, RA can re-specify portions of the anterior-posterior axis in a more posterior direction and can inhibit neural crest cells from migrating to form facial cartilage (Moroni et al. 1994; Studer et al. 1994). Radioactively labeled RA can be seen to bind to the cranial neural crest cells, and it arrests both their proliferation and their migration (Johnston et al. 1985; Goulding and Pratt 1986). This binding seems to be specific to the cranial neural crest cells, and the neural crest-related teratogenic effect of the drug is confined to a specific developmental period (days 8–10 in mice; days 20–35 in humans). Another mechanism is that excess RA will activate the negative feedback pathway that usually ensures the proper amount of this compound. Transient large increases in RA activate the synthesis of RA-degrading enzymes, causing a long-lasting *decrease* of RA. This deficiency in RA may explain why large amounts of RA produce phenotypes similar to RA deficiency (Lee et al. 2012).

Retinoic acid and public health

Retinoic acid exposure during pregnancy is a critical public health concern because there is significant overlap between the population using acne medicine and the population of women of childbearing age, and because it is estimated that half of the pregnancies occurring in the United States are unplanned (Finer and Zolna 2011). Vitamin A, which is transformed into RA in the body, is itself teratogenic in megadose amounts. Rothman and colleagues (1995) found that pregnant women who took more than 10,000 international units of vitamin A per day (in the form of vitamin supplements) had a 2% chance of having a baby born with malformations similar to those produced by RA. According to the rules of the U.S. Food and Drug Administration, every patient using isotretinoin, every physician prescribing it, and every pharmacy selling it must sign a registry. Moreover, women who use this drug are expected to take a pregnancy test within the 7 days before they fill their prescription and to agree to use two methods of birth control and adhere to pregnancy testing on a monthly basis.

Interference with RA signaling may be a public health concern for another reason, too. Glyphosate herbicides (such as Roundup) have been reported to up-regulate the activity of endogenous RA (Paganelli et al. 2010). When *Xenopus* (frog) embryos were incubated in solutions containing ecologically relevant concentrations of glyphosate herbicides, RA-responsive reporter gene activation was dramatically altered. Moreover, these embryos exhibited cranial neural crest defects and facial disorders similar to those seen in RA teratogenesis (Figure 5.11). Co-treatment with an RA inhibitor blocked the teratogenic effects of glyphosate. Scientists hired by the herbicide companies attempted to discredit this report.* The principal investigator (Carrasco 2011; Lopez et al. 2012) has criticized the chemical

---

*The chemical industry has denied these products cause human birth defects. For analysis of how the chemical industry's statement is at odds with the scientific data, see Antinou et al. 2012. In 2015, the International Agency for Research on Cancer, the major scientific advisory committee on the subject, reclassified glyphosate as "probably carcinogenic to humans." Monsanto, the makers of Roundup®, disagree (Guyton et al. 2015; Pollack 2015).

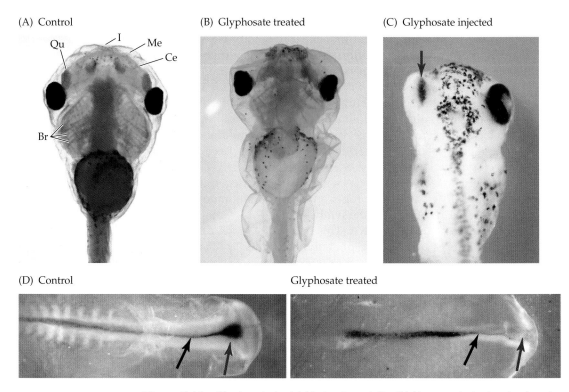

Figure 5.11 Glyphosate herbicide teratogenicity. (A) *Xenopus* tadpole raised under control conditions, stained with Alcian blue to show facial cartilages. Br, branchial; Ce, ceratohyal; I, infrarostral; Me, Meckel; Qu, quadrate. (B) *Xenopus* tadpole raised in environmentally relevant concentrations of glyphosate and similarly stained. Its branchial arches and midline facial cartilage (cranial neural crest derivatives) failed to develop properly. (C) If an embryo is injected such that only one side (arrow) is exposed to glyphosate, that side shows cranial neural crest anomalies. (D) Control chick embryos show sonic hedgehog gene expression in the notochord (black arrow) and prechordal mesoderm (red arrow). Chick embryos grown in glyphosate show a severe reduction of sonic hedgehog expression in the prechordal (craniofacial) mesoderm. (Photographs courtesy of A. Carrasco.)

industry statement and has pointed out that investigations on cultured cells have also shown the teratogenic potential of glyphosate herbicides, as have epidemiological studies on humans and wildlife studies. Relyea (2012) found that predator-induced developmental changes cause the tadpoles of some species to become even more sensitive to glyphosate herbicides (see the box, p. 210).

Glyphosate is the most widely used (and profitable) herbicide in America, where over 180 tons of it have been applied. It acts by blocking a plant enzyme that is critical for the synthesis of certain amino acids. One of the powerful abilities of genetic engineering has been to manufacture wide-spectrum herbicides, such as the glyphosate-based Roundup®, and then breed plants that are resistant to this herbicide. This means that if you spray a large area, all the weeds will get killed and the only plants that will remain standing are those that are glyphosate resistant. In 2010, 70% of the corn and 93% of the soybeans grown in the United States were from herbicide-resistant genetically modified seeds (Hamer 2010).

Natural Killers: Teratogens from Plants

Given the voracity of insects and their larvae, it's amazing that any plant survives. Plants can't run away from predators, so they must protect themselves in other ways. Poisoning one's enemies is a time-honored way to rid oneself of them, so perhaps it is not surprising that most of the poisons known to humans are substances that were originally found occurring naturally in plants. A somewhat more subtle way to get rid of a predator is to destroy its offspring, and several teratogenic compounds found among plants do just that.

Veratrum alkaloids

The plant *Veratrum californicum* (corn lily, or California false hellebore) produces several alkaloid substances, including veratramine, jervine, and cyclopamine, that can block the functions of the Hedgehog family of paracrine factors (Keeler and Binns 1968; Beachy et al. 1997; Chen et al. 2002. Figure A,B). One of the major functions of hedgehog protein in vertebrates is to stimulate the proliferation of cells in the ventral midline of the face and forebrain. Blocking the Sonic hedgehog protein signal leads to the failure of the optical field to separate and also prevents the pituitary gland from forming. If a pregnant ewe eats *Veratrum*, her lambs are likely to be born with a single eye (like a "cyclops") in the center of the head—hence the name "cyclopamine" for one of the compounds causing this effect (Figure C).

Cyclopia is an example of a birth defect that can result from either environmental or genetic insults, or a combination of these. In humans, mutations in several different genes can result in cyclopia or in milder defects that are within the same spectrum of median forebrain and facial deficiencies—defects in the "holoprosencephaly spectrum" (Muenke 1995; Cordero 2004). One of these genes encodes the enzyme sterol-Δ7 reductase, which is critical for cholesterol biosynthesis. Cholesterol is a key co-factor in allowing the proper function of Sonic hedgehog protein. Mutation of several different genes in the sonic hedghog pathway, including the gene that codes for the Sonic hedgehog protein itself,

(A)

(B)
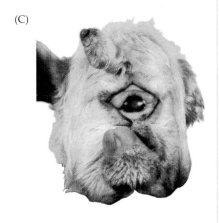

(C)

Cyclopia caused by *Veratrum* alkaloids. (A) Structures of cholesterol and two *Veratrum* alkaloids, cyclopamine and veratramine. (B) *Veratrum californicum*, California false hellebore. (C) Head of a cyclopic lamb stillborn to a ewe who grazed on *V. californicum* early in pregnancy. The cerebral hemispheres are lacking midline structures, and there is only one central eye and no pituitary gland. (B courtesy of D. Powell, USDA Forest Service, Bugwood.org; C courtesy of L. James, USDA ARS Poisonous Plant Research Laboratory.)

(Continued on next page)

Natural Killers: Teratogens from Plants (continued)

can yield this spectrum of defects (Roessler et al. 1996; Traiffort et al. 2004; Raam et al. 2011). Notably, related individuals who carry the *same gene mutation* can have *different phenotypes*. Such **phenotypic heterogeneity** (the ability of one mutant gene to produce a range of phenotypes) is thought to result from the interaction of the genotype with the environment. In cases of *Sonic hedgehog* mutations that result in low amounts of Sonic hedgehog protein, it is likely that this deficiency could be exacerbated by a limited amount of cholesterol (Dehart et al. 1997; Edison and Muenke 2003; Cordero et al. 2004). While the mechanism remains unknown, animals having *sonic hedgehog* mutations or mutations in genes important for *Sonic hedgehog* signaling have recently been shown to be more sensitive to the teratogenic effects of alcohol (Hong and Krauss 2012; Kietzman et al. 2014).

Plant juvenile hormones

One of the most fascinating examples of teratogenic control of animal pests occurs at the level of juvenile hormone. Juvenile hormone is made in the corpora allata of insect larvae, and its secretion prevents the larvae from undergoing metamorphosis at the next molt; instead, the next molt simply results in larger larvae. Inappropriate amounts of juvenile hormone at the wrong time can delay metamorphosis too long or precipitate it too soon, either one of which can have lethal outcomes for the developing insect.

The effects of delayed metamorphosis were seen when Karel Sláma came from Czechoslovakia to work in Carroll Williams's insect development laboratory at Harvard University. He brought his chief experimental animal with him—the European plant bug *Pyrrhocoris apterus*. To the consternation of the entire laboratory, his bugs failed to undergo normal metamorphosis, which for this species should take place at the end of the fifth instar.*

---

*"Instar" is the term used to describe the larva between molts. Thus, the first instar larva is that hatched from the egg. The first instar molts into the second instar, and so forth until the insect forms a pupa and then has a final molt into an adult.

Rather, they became something never before observed in nature or the laboratory: large, sixth-instar larvae that died before becoming adults. After testing many variables, which failed to provide an explanation, the papers lining the culture dishes were tested for their effect on the larvae. The results were as conclusive as they were surprising: larvae reared on European paper (including pages of the journal *Nature*) underwent metamorphosis as usual, whereas larvae reared on American paper (such as shredded copies of the journal *Science*) did not undergo metamorphosis. It was eventually determined that the source of the American paper was the balsam fir (*Abies balsamea*), a tree indigenous to the northern United States and Canada. *A. balsamea* synthesizes a biochemical compound that closely resembles juvenile hormone (Bowers et al. 1966; Sláma and Williams 1966; Williams 1970). This compound presumably protects the tree by interfering with the metamorphosis of its insect predators.

The teratogenic effect that some natural plant products have on insects has been studied and turned to human advantage in a number of pest-control products. The juvenile hormone mimic methoprene, for instance, is used to control ant and mosquito infestations and is a major ingredient in flea collars (EPA 2007).

Some plants have compounds that produce the same effect—the death of insect herbivores—but do so by eliciting *precocious* metamorphosis. Two compounds that have been isolated from herbaceous composites (the sunflower family, Compositae), including goatweed and floss flower (*Ageratum conyzoides* and *A. houstonianum*), have been found to cause the premature metamorphosis of certain insect larvae into sterile adults (Bowers et al. 1976). These compounds are called **precocenes**; their chemical structures are shown in Figure D. When the larvae or nymphs of these insects are dusted with either of these compounds, they undergo one more molt and then metamorphose into the adult form (Figure E).

Precocenes cause the selective death of the corpus allatum cells responsible for synthesizing juvenile hormone in the immature insect (Schooneveld 1979; Pratt

Other teratogenic agents

Cigarette smoke is less pervasive than it was a few decades ago but is still commonly encountered in regular day-to-day life. Marijuana smoke, on the other hand, may be becoming more pervasive. Most Americans take prescription drugs, and we are all affected by pathogens. Oil spills,

Natural Killers: Teratogens from Plants (*continued*)

et al. 1980). Without these cells to produce juvenile hormone, the larva commences its metamorphic and imaginal molts. Moreover, juvenile hormone is also responsible for the maturation of the insect egg. Without prolonged exposure to this hormone, females are sterile. Thus, precocenes are able to protect the plant by blocking the synthesis or actions of juvenile hormone, causing the larva to quickly metamorphose into sterile adults. Other recently discovered plant compounds accomplish the same goal by blocking the ability of juvenile hormone receptor complex to form (Lee et al. 2015). These compounds may become used as natural insecticides.

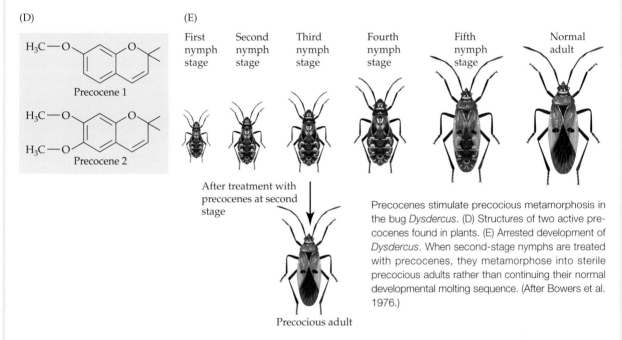

(D)

Precocene 1

Precocene 2

(E)

First nymph stage | Second nymph stage | Third nymph stage | Fourth nymph stage | Fifth nymph stage | Normal adult

After treatment with precocenes at second stage

Precocious adult

Precocenes stimulate precocious metamorphosis in the bug *Dysdercus*. (D) Structures of two active precocenes found in plants. (E) Arrested development of *Dysdercus*. When second-stage nymphs are treated with precocenes, they metamorphose into sterile precocious adults rather than continuing their normal developmental molting sequence. (After Bowers et al. 1976.)

dangerously high temperatures, and damaging sound frequencies are less ubiquitous but still affect some people's lives. They can all have teratogenic effects.

TOBACCO SMOKING   Cigarette smoke retards the growth of human fetuses and gives a heightened risk of fetal and newborn death. Indeed, compared with nonsmoking women having their first birth, women who smoked one or more packs of cigarettes a day have been shown to have a 56% greater risk of having the fetus or newborn infant die (Kleinman et al. 1988; Werler 1997). There is also evidence that nicotine, an important component of cigarette smoke, may damage the fetal brain and lungs during development. Dwyer and colleagues (2008) showed that nicotine induces abnormalities of synapse formation and cell survival in the developing brain. Maritz (2008) showed that nicotine can induce lung cells to have an altered metabolism that ages them more rapidly, and prenatal exposure to cigarette smoke is associated with increased risks of impaired lung functions later in life (Wang and Pinkerton 2008). Smoking also significantly lowers the number, quality, and motility of sperm in the semen of males who smoke at least four cigarettes a day (Kulikauskas et al. 1985; Mak et al. 2000; Shi et al. 2001).

Smoking may also alter the DNA methylation pattern of some important genes, including tumor suppressor genes, which can be very important for cancer formation. If these genes are not active, tumors can arise because the cells are not told to stop dividing (Belinsky et al. 2002). (As we will discuss in Chapter 8, many of the cancer-causing agents are actually factors that can alter DNA methylation.) Tobacco smoke appears to alter DNA methylation patterns not only in the smokers themselves but also in utero. Toledo-Rodriguez and colleagues (2010) found that teenagers who had been exposed to maternal smoking in utero had increased methylation in the gene encoding BDNF, a paracrine factor associated with learning ability and neural plasticity.

MARIJUANA   While the teratogenic potential of cannabinol, the active ingredient in marijuana, remains controversial, a recent surge in the use of highly potent synthetic cannabinoids argues strongly for more research in this arena. Chronic uptake of cannabinol has been shown to lead to the impairment of testes function in rats, to lower gonadotropin-stimulating hormone synthesis, and to cause impaired fertility (Banerjee et al. 2011.) Historically, many epidemiological studies failed to show that illicit drug use (cocaine or marijuana) leads to significant elevations in birth defects (Chiriboga 2003; van Gelder et al. 2009). However, animal research and studies of human behavior call attention to certain differences in brain anatomy that might be caused by maternal marijuana use. These would not be detected at birth and could cause behavioral problems later in life (Mereu et al. 2003; Saez et al. 2014). The endocannabinoid system, which includes the receptors that usually bind cannabinol-like substances, is important in regulating neural progenitor cell proliferation, specification, and the migration and differentiation of cortical neurons. Saez and colleagues (2014) showed that giving cannabinoids to pregnant rats changed the radial and tangential migration of the neuroblasts and induced a delay in neuronal differentiation in the embryonic rat cerebral cortex. This, they claim, might explain the cognitive and memory declines seen in adult rats that were exposed prenatally to cannabinoids and may also explain some of the epidemiological data suggesting that children exposed prenatally to marijuana have a significantly higher probability of cognitive impairment later in life. These studies are relatively new, but if they are confirmed, then marijuana may be a very subtle teratogen that alters brain development in ways that may become evident many years later, when the conceptus is an adult.

PRESCRIPTION DRUGS   Each year, hundreds of new, artificial chemical compounds come into general use in our industrial society, and many of these are found in pharmaceutical products. Some drugs used to control diseases in an adult may have deleterious effects on a fetus (see Table 5.1).

The teratogenic effects of the sedative thalidomide were described at the beginning of this chapter. Other such drugs include cortisone (used for rheumatoid arthritis), warfarin (used to prevent blood clots), tetracycline (an antibiotic), and valproic acid (an anticonvulsant used to treat epilepsy, whose teratogenicity was also previously considered).

As previously noted, prenatal valproic acid (VA) exposure is associated with both structural and functional birth defects. Finnell and colleagues (1997) showed that VA blocks folate (see below) from being absorbed by the embryo, thereby leading to neural tube defects. Barnes and colleagues (1996) have shown that VA decreases the level of *Pax1* transcription in chick somites. This decrease causes malformation of the somites and corresponding malformations of the vertebrae and ribs. Menegola and colleagues (2006) and Eikel and colleagues (2006) have shown that VA inhibits the activity of a class of enzymes called histone deacetylases, which are critical regulators of gene expression (see Chapter 2). About 20% of a cell's transcription is regulated by histone deacetylases, and VA causes a hyperacetylation of the genome within an hour of treatment, especially in the neural tube and somites. Many of these genes are also regulated by retinoic acid, leading to the possibility that VA prevents the deacetylation of the histones around those genes regulated by the retinoic acid receptors (Menegola et al. 2006).

Interestingly, VA's ability to inhibit histone deacetylation may cause autism-like developmental problems (Chomiak et al. 2013). Autism is thought to be caused by the accelerated early growth of those brain regions involved in emotional, social, and communicative functions (Schumann et al. 2010). Recent epidemiological studies—both a prospective study of pregnant women who were taking VA and a retrospective large-scale study of individuals with autism (Bromley et al. 2013; Christensen et al. 2013)—have linked prenatal VA exposure with autism spectrum disorders. Early exposure to VA (either prenatally or during the first days of postnatal life) can lead to overgrowth in certain regions of rat and mouse brains and produce anatomical, physiological, and behavioral phenotypes very similar to human autism spectrum disorders. Blocking histone deacetylases has been shown to increase synapse numbers and to cause neural cell proliferation (Balasubramaniyan et al. 2006; Akhtar et al. 2009; Almeida et al. 2014). While it is doubtful that VA causes most of the autism spectrum disorders, it is likely that it causes the early overgrowth of brain cortical neurons and is one of many agents that can do this.

Folic acid, or vitamin B9, is critical for neural tube closure. According to the U.S. Centers for Disease Control and Prevention, some 50% of neural tube defects can be prevented if pregnant women take folate supplements (CDC 1992). Folic acid has been added to foods as a supplement to ensure that pregnant women receive a sufficient amount of this compound. Although the folate supplementation of foods has probably caused a major decrease in the number of babies born with neural tube defects (see Chapter 10), recent studies suggest that aggressive folate supplementation during pregnancy may actually be teratogenic (Marean et al. 2011; Mikael et al. 2013; Vasquez et al. 2013).

Teratogens and Cognitive Function

Teratogens can cause brain damage in the absence of obvious malformations: damage that may be first detected by changes in normal cognitive or behavioral abilities. One possible cognitive alteration is lowered intelligence. Importantly, environmental teratogenesis is gaining prominence as a possible cause of at least some types of **autism**, a spectrum of behavioral and cognitive changes that includes repetitive behaviors and impaired communication and social interactions (see Autism Genome Consortium 2007; Hallmayer et al. 2011; Frazier et al. 2014). An important clue that autism might be caused by specific environmental agents came from studies of children with thalidomide syndrome. While autism is found in 0.1%–0.2% of the general population, it was identified in 5% of the Swedish children having the thalidomide syndrome described at the beginning of the chapter. Moreover, it was found specifically in those children whose mothers had taken thalidomide between days 21 and 24 of embryonic development (Miller et al. 2005).

Valproic acid (VA) is also known to disrupt normal embryonic development. Mice and humans exposed prenatally to VA often have neural tube closure defects as well as rib and facial malformations, and they may have cognitive disorders as well. The IQs of children whose mothers took VA during pregnancy are significantly (10–20 points) below those of their siblings (see Meador et al. 2006), and epidemiological studies suggest that as many as 11% of children born with valproic acid syndrome are autistic (Moore et al. 2000; Arndt et al. 2005).

Recent studies suggest that VA causes a comparable spectrum of neurological alterations in rodents and humans. In rats, reductions in the numbers of neurons in several cranial nerve ganglia, as well as markedly reduced numbers of Purkinje cells in the cerebellum, have been found (see Ingram et al. 2000; Arndt et al. 2005). Neurotransmitter abnormalities in these rats are similar to those seen in human patients (Narita et al. 2002). One recent study (Rinaldi et al. 2008) found that a single exposure to VA in utero caused changes in the rat neocortex, including a weaker excitatory synapse response, diminished contact between neuronal layers, and more autonomic neural activity. This is in keeping with models of human autism that predict greater activity of local neural modules at the expense of globally coordinated activity.

Some of the VA-induced anomalies found in rats may be caused by the misregulation of *Hoxa1* in the embryos. This gene is critically important in forming the cerebellum. In human embryos, exposure to thalidomide in the time period of 20–24 days after fertilization causes hindbrain anomalies that mimic loss-of-function *HOXA1* mutations (Miller and Strömland 1999; Tishfield et al. 2005). In mice, VA exposure in utero induces the overexpression of *Hoxa1*, as well as its expression at the wrong time during development (Stodgell et al. 2006). While *HOXA1* mutations are rarely found in humans and are probably not a major cause of autism, it appears that teratogen-induced misregulation of this gene can lead to altered neural patterning and perhaps to the cognitive anomalies seen in autistic children (Miller and Strömland 1999; Tischfield et al. 2005).

OIL SPILLS   Oil spills can affect larval and embryonic development in many ways. This was seen when over 200 million gallons of oil were released into the Gulf of Mexico in 2010. Water-soluble crude oil components from the Deepwater Horizon oil spill are teratogenic in zebrafish (Figure 5.12). They produce developmental anomalies in the size of head and gill cartilages derived from cranial neural crest cells (de Soysa et al. 2012.) In addition, polycyclic aromatic hydrocarbons in the non-water-soluble compartment of the oil spill appear to be largely responsible for abnormalities of heart development (Incardona et al. 2013). The naphthenic acids recovered as a by-product from the Canadian oil sands have been found to cause developmental anomalies in wood frog tadpoles at much lower concentrations than those found in the oil sands region's tailing ponds (Melvin and Trudeau 2012a,b).

(A) Control

(B) Treated

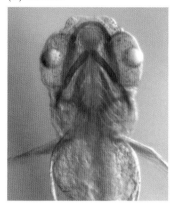
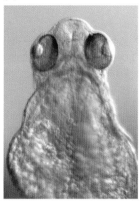

Figure 5.12 Water-soluble crude oil components from the Deepwater Horizon oil spill did not kill the fish directly but produced developmental anomalies—reduction in the size of head and gill cartilages (blue staining), which are cranial neural crest derivatives. (A) A normal zebrafish embryo 4 days postfertilization, showing the normal facial cartilage. (B) Those embryos developing in water containing water-soluble components of the oil spill had significant cartilage deficiencies and facial anomalies. (From de Soysa et al. 2012.)

PATHOGENS   Another class of teratogens includes viruses and other pathogens. At the beginning of this chapter, we described the effects of the rubella (German measles) epidemic of 1963–1965. The rubella virus is able to enter many types of cells, where it produces a protein that stops mitosis by blocking kinases that are necessary for the cell cycle to progress (Atreya et al. 2004). Thus, numerous organs are affected, and the earlier in pregnancy the rubella infection occurs, the greater the risk that the embryo will be malformed. The first 5 weeks of development appear to be most critical because that is when the heart, eyes, and ears begin to be formed (see Figure 5.2). Two other viruses, cytomegalovirus and the herpes simplex virus, are also known to be teratogenic. Cytomegalovirus infection of early embryos is nearly always fatal, and infection of later embryos can lead to blindness, deafness, cerebral palsy, and mental deficiency. Herpes virus is rarely transmitted across the placenta, and appears to cause skin and neural deficiencies when it is. Herpes is much more dangerous when the newborn receives the virus at birth from an infected mother. Here the exposed newborn is at a high risk for getting meningitis.

Bacteria and protists are rarely teratogenic, but at least two of them are known to damage human embryos. *Toxoplasma gondii*, a protist carried by rabbits and cats (and their feces), can cross the placenta and cause brain and eye defects. *Treponema pallidum*, the bacterium that causes syphilis, can kill early embryos and produce congenital deafness and facial damage in older ones.

HEAT   Hyperthermia, the elevation of core body temperature beyond a certain point, is dangerous to the embryo/fetus. An extended maternal temperature of 102°F (38.9°C) or higher during the first 6 weeks of

Deformed Frogs: A Teratological Enigma

One phenomenon connected with teratogenicity has received a great deal of attention in both the scientific and the national media. Throughout the United States and southern Canada, there has been a dramatic increase in the number of anomalous frogs and salamanders observed in what seem to be pristine woodland habitats. In some localized areas, it is estimated that as many as 60% of certain amphibian species have been documented to display visible malformations (Ouellet et al. 1997). These defects include extra or missing limbs (see figure), missing or misplaced eyes, abnormal jaws, and malformed hearts and guts. We do not know what is causing these abnormalities; David Stocum (2000) has called the situation "an 'eco-devo' riddle wrapped in multiple hypotheses by insufficient data." Numerous hypotheses have been put forward to explain the sudden prevalence of these anomalies (see Lannoo 2008), and in recent years two of these explanations have gained acceptance.

A combination of factors

The first hypothesis proposes a combination of causes. The proximate cause of many limb defects may be the infection of the larval amphibian limb buds by the larvae of trematode (flatworm) parasites (Johnson et al. 1999, 2002). Under most conditions, the tadpoles' immune system can destroy the trematode larvae. In some ponds, however, it appears that the tadpoles suffer from an acquired immune deficiency syndrome that leaves them vulnerable to the parasites (Christin et al. 2004).

A northern leopard frog (*Lithobates pipiens*) from a wild population, suffering from polymelia (an extra set of hindlimbs). (Photograph courtesy of U.S. Geological Survey.)

Kiesecker (2002) has shown that frogs living in habitats contaminated by certain pesticides are less able to resist trematode infection. He found that neither pesticide pollution alone nor trematodes alone was sufficient to cause deformities in wild populations of frogs, but that the combination of the two factors resulted in a significant proportion of the frog population showing limb deformities. Many endocrine disruptors (see Chapter 6), including pesticides and herbicides, can down-regulate the immune system.

At least in some species, then, trematode infestation has to be coupled with pollutants such as pesticides in order for significant cases of limb deformities to be observed. Similar multiagent causation, where one agent is a pesticide, has been shown by Davidson and Knapp (2007) to be responsible for the mortality of some frog populations.

Pesticides and herbicides

A second possibility is that pesticidal and herbicidal chemicals are the direct cause of physiological alterations. The spectrum of abnormalities seen in natural populations of malformed frogs resembles some of the malformations that result when tadpoles are exposed to retinoic acid (Crawford and Vincenti 1998), and it is possible that some herbicides and pesticides disrupt development after being converted naturally into chemicals that resemble RA. Experiments in the Gardiner laboratory (Gardiner and Hoppe 1999; Gardiner et al. 2003) have shown that RA can induce all the limb malformations seen in wild populations of some frogs. The research team has also purified a retinoid compound from the water of one lake that had a high incidence of malformed amphibians. Since no one is suspected of dumping retinoic acid into the lake, how did such a compound get there? One possibility involves the insecticidal chemical methoprene.

Methoprene is an insect juvenile hormone mimic that inhibits mosquito pupae from metamorphosing into adults. Because vertebrates do not have juvenile hormone, it was assumed that this pesticide would not harm fish or amphibians (or humans). Indeed, this is to some extent the case; methoprene itself has no teratogenic properties. However, on exposure to sunlight, methoprene breaks down into chemical products that have significant teratogenic activity in frogs. These compounds have a structure very similar to that of retinoic acid and will bind to the RA receptor (Harmon et al. 1995). When *Xenopus* eggs are incubated in water containing these

Deformed Frogs: A Teratological Enigma (*continued*)

compounds, the resulting tadpoles are often malformed and show a spectrum of defects similar to those found in the wild (La Claire et al. 1998).

Saving the frogs

Lannoo (2008) proposes that we probably know enough to save many of these species even though the specific mechanisms of developmental disruption remain unknown. He points out that nearly all the proposed causes for these anomalies can be prevented simply by controlling runoff from agricultural lands.

Lannoo notes that nutrients entering ponds from runoff contaminated by fertilizers encourage plant growth, which encourages snail population growth. Increased numbers of snails (the vectors of trematode larvae) means more parasitic trematodes to infest the frog larvae and disrupt their development. Moreover, pesticides and herbicides in the runoff may be destroying amphibian immune capabilities and/or directly inducing developmental anomalies.

In all these cases, the remedy—stopping erosion from farmlands—is the same. The idea gains support from recent studies by McCoy and colleagues (2008), who quantified the relationship between gonadal anomalies in male toads and the amount of agricultural activity at or near specific ponds. They found statistically significant increases in both gonadal abnormalities and reproductive failures as agricultural activity levels increased. However, as Davidson and Knapp (2007) have pointed out, erosive runoff is only one way chemicals get into the water. Pesticides and herbicides can be blown on the wind and still have severe consequences for the frogs in the ponds in which they finally settle. Still, controlling runoff would probably be a very good start.

---

pregnancy is considered detrimental to the closing of the embryonic neural tube (OTIS 2007). Such elevated body temperature can occur as a result of illness (fever), long exposure to unrelieved heat wave conditions, or extensive stays in a hot tub. Proper use of saunas does not appear to be as dangerous, because by causing one to sweat, saunas minimize changes in internal body temperature. A study by Saxén and colleagues (1982) showed that although 98.5% of pregnant Finnish women visited a sauna regularly, Finland had an extremely low incidence of neural tube defects.

NOISE Noise may also be able to interrupt normal development. Recall from earlier chapters that sonic vibrations could be used as signals for larval settlement. If these were disrupted such that they could not be heard or were coming from the wrong place, development would be halted. Low-frequency vibrations (5–100 Hz) in water can disrupt the development of fish and frog embryos, altering left-right axis formation and neural tube formation (Vandenberg et al. 2011, 2012). The assembly of tight junctions appears to be disrupted at 7 Hz. The human voice range is mostly in higher frequencies—between 75 and 1000 Hz—and the ultrasound used to visualize embryos is about 20,000 Hz, so few humans are exposed to disruptive vibrational frequencies in utero. The only women expected to experience such frequencies for prolonged durations would be those driving large vehicles or construction machines (McDonald et al. 1988).

Looking Ahead

Two major questions addressed by teratology studies are whether or not a substance is a teratogen and how the teratogenic agent acts to cause birth defects. Traditionally, the first question has been answered using epidemiological and experimental analyses. The experimental approach utilizes animal models with exposure to the agent in question at certain times and in certain dosages during development. Controls are then performed such that the only variable is the possible teratogen. These are "prospective studies," in that the animals are exposed to a specific agent and are then studied. The epidemiological approach is often "retrospective," wherein one studies a population that has already been exposed. Thus, epidemiological studies in humans compare control populations with populations that have been exposed to the substance in question, and then determine whether those exposed are at a statistically higher risk for birth defects. Both prospective and retrospective epidemiological studies have been employed in illustrating the human teratogensis of ethanol, retinoic acid, and valproic acid.

The question regarding how teratogens do their damage typically entails examination of effects on particular developmental, cellular, or genetic events or on physiological changes. For example, whether the substance modifies cells' adhesion, migration, differentiation, or proliferation or causes cells to die are commonly investigated. In this approach, only expected effects are studied.

Both epidemiological and experimental studies use large numbers of people or research animals. This causes the research to be extremely expensive. According to the European Union regulations for toxic substances (Prutner 2013), there are over 30,000 chemicals that need to be tested, and it has been estimated that testing these compounds would require over 7 million laboratory animals (Hofer et al. 2004; Lee et al. 2012). Therefore, finding alternatives to animal testing for teratogens is now a major priority.

One new way of studying the mechanism of teratogenesis has been provided by **systems biology** and **bioinformatics**. Here, the beginning of investigation is a computer database of what genes are influenced by the presence or absence of a particular chemical. The Comparative Toxigenomics Database (CTD), for instance, contains over 175,000 interactions between nearly 5000 chemicals and 16,000 genes in nearly 300 species. Using this database, Ahir and colleagues (2013) found that the heavy metals cadmium and arsenic caused several genes to be expressed at higher levels and several other genes to be expressed at much lower levels. Surprisingly, when the annotation of these genes was done (again by computer), the investigators found that many of these genes were in the glucocorticoid stress pathway. The administration of arsenic to chick embryos caused a suite of developmental anomalies, and these same anomalies could be induced by the addition of phenytoin, a teratogen that is known to function through the glucocorticoid pathway. Moreover, the arsenic-induced anomalies could be eliminated if the embryos were cotreated with arsenic and an inhibitor of the glucocorticoid pathway. Thus, a possible mechanism for teratogenicity was predicted by a computerized database and then confirmed by further experimentation. This may provide a cost-effective strategy for prioritizing which experiments should be performed and may provide a relatively efficient strategy for testing some of the numerous compounds being put into our environment each year.

Another new way of predicting teratogenicity involves testing cells and tissues in culture (rather than using pregnant animals). These techniques include **whole mouse embryo culture** (growing early embryos in incubators) and testing compounds on **embryonic stem cells** (Seiler et al. 2011; Lee et al. 2012; Kuske et al. 2012). Mouse embryonic stem cells are relatively easy to obtain, and even human embryonic stem cells can now be obtained by relatively simple means (using viruses to transform skin cell into pluripotent cells that can generate most types of human cell types; Awe et al 2014). By seeing whether potentially teratogenic compounds or treatments alter the differentiation of the stem cells into particular cell types, one can determine if the agent affects that pathway. Kuske and colleagues (2012) have shown that stem cell studies may be an effective way of studying agents that cause congenital bone and heart disease.

However, since it is difficult for such an *in vitro* assay to look at the normal or abnormal phenotype of an entire organ, it would be useful to have a marker, an easily recognized surrogate, for the phenotype. Here is where a new area of developmental biology may become incredibly important. Many developing cells have specific regulatory RNAs called microRNAs (miRNAs). These miRNAs help regulate protein synthesis, and they can block or permit networks of cellular events. Moreover, certain miRNAs are associated with certain specific developmental events (and could thus be used as a surrogate for that event). Recent evidence (Sathyan et al. 2007; Guo et al. 2011) have shown, for instance, that a relatively small subset of miRNAs in the neural stem cells are sensitive to ethanol. Moreover, some of these microRNAs are critical for certain developmental events of neuron formation (Miranda 2014). If miRNAs could be analyzed in cultured cells, this may enable relatively inexpensive and rapid testing of potentially teratogenic compounds. As one researcher (Miranda 2012) stated, "Clearly, research on miRNAs and teratology is in its infancy; but as outlined in this review, the possibilities for understanding and ultimately, for therapeutic intervention in teratology are enormous."

In addition to the established teratogens discussed in this chapter, a new and widespread class of teratogens has recently been uncovered. These teratogens work by disrupting the endocrine system, and they usually affect the physiological function of the organism more than they disrupt its anatomy. Because of this, their effects are often unnoticed until adulthood. These *endocrine disruptors* are the topic of the next chapter.

References

Ahir, B. K., A. P. Sanders, J. E. Rager and R. C. Fry. 2013. Systems biology and birth defects prevention: Blockade of the glucocorticoid receptor prevents arsenic-induced birth defects. *Environ. Health Perspect.* 121: 332–338.

Ahlgren, S. C., V. Thakur and M. Bronner-Fraser. 2002. Sonic hedgehog rescues cranial neural crest from cell death induced by ethanol. *Proc. Natl. Acad. Sci. USA* 99: 10476–10481.

Akhtar, M. W., J. Raingo, E. D. Nelson, R. L. Montgomery, E. N. Olson, E. T. Kavalali and L. M. Monteggia. 2009. Histone deacetylases 1 and 2 form a developmental switch that controls excitatory synapse maturation and function. *J. Neurosci.* 29: 8288–8297.

Almeida, L. E., C. D. Roby and B. K. Krueger. 2014. Increased BDNF expression in fetal brain in the valproic acid model of autism. *Mol. Cell Neurosci.* 59: 57–62.

Antoniou, M. and 7 others. 2012. Teratogenic effects of glyphosate-based herbicides: Divergence of regulatory decisions from scientific evidence. *J. Environ. Anal. Toxicol.* S4: 006.

Aoto, K., Y. Shikata, D. Higashiyama, K. Shiota and J. Motoyama. 2008. Fetal ethanol exposure activates protein kinase A and impairs *Shh* expression in prechordal mesendoderm cells in the pathogenesis of holoprosencephaly. *Birth Def. Res. A: Clin. Mol. Teratol.* 82: 224–231.

Ankley, G. T., S. A. Diamond, J. E. Tietge, G. W. Holcombe, K. M. Jensen, D. L. Defoe and R. Peterson. 2002. Assessment of risk of solar ultraviolet radiation to amphibians. I. Dose-dependent induction of hindlimb malformations in the northern leopard frog (*Rana pipiens*). *Environ. Sci. Tech.* 36: 2853–2858.

Arndt, T. L., C. J. Stodgell and P. M. Rodier. 2005. The teratology of autism. *Int. J. Devl. Neurosci.* 23: 189–199.

Atreya, C. D., K. V. Mohan and S. Kulkarni. 2004. Rubella virus and birth defects: Molecular insights into the viral teratogenesis at the cellular level. *Birth Def. Res. A: Clin. Mol. Teratol.* 70: 431–437.

Autism Genome Consortium. 2007. Mapping autism risk loci using genetic linkage and chromosomal rearrangements. *Nature Genet.* 39: 319–328.

Awe, J. P., A. Vega-Crespo and J. A. Byrne. 2014. Derivation and characterization of a transgene-free human induced pluripotent stem cell line and conversion into defined clinical-grade conditions. *J. Vis. Exp.* 93: e52158.

Baghurst, P. A. and 8 others. 1992. Environmental exposure to lead and children's intelligence at the age of 7 years: The Port Pirie Cohort Study. *New Engl. J. Med.* 327: 1279–1284.

Balasubramaniyan, V., E. Boddeke, R. Bakels, B. Küst, S. Kooistra, A. Veneman and S. Copray. 2006. Effects of histone deacetylation inhibition on neuronal differentiation of embryonic mouse neural stem cells. *Neuroscience.* 143: 939–951.

Banerjee, A., A. Singh, P. Srivastava, H. Turner and A. Krishna. 2011. Effects of chronic bhang (cannabis) administration on the reproductive system of male mice. *Birth Defects Res. B Dev. Reprod. Toxicol.* 92: 195–205.

Barnes, G. L., Jr., B. D. Mariani and R. S. Tuan. 1996. Valproic acid-induced somite teratogenesis in the chick embryo: Relationship with *Pax-1* gene expression. *Teratology* 54: 93–102.

Battista, N., C. Rapino, M. DiTomasso, M. Bari, N. Pasquaielo and M. Maccarone. 2008. Regulation of male fertility by the endocannabinoid system. *Mol. Cell Endocrinol.* 286(suppl.): S17–S23.

Beachy, P. A. and 7 others. 1997. Multiple roles of cholesterol in Hedgehog protein biogenesis and signaling. *Cold Spring Harb. Symp. Quant. Biol.* 62: 191–204.

Bearer, C. F., A. R. Swick, M. A. O'Riordan and G. Cheng. 1999. Ethanol inhibits L1-mediated neurite outgrowth in postnatal rat cerebellar granule cells. *J. Biol. Chem.* 274: 13264–13270.

Belinsky, S. A. and 13 others. 2002. Aberrant promoter methylation in bronchial epithelium and sputum from current and former smokers. *Cancer Res.* 62: 2370–2377.

Bellinger, D., A. Leviton, C. Waternaux, H. Needleman and M. Rabinowitz. 1987. Longitudinal analyses of prenatal and postnatal lead exposure and early cognitive development. *New Engl. J. Med.* 316: 1037–1043.

Bowers, W. S., H. M. Fales, M. J. Thompson and E. C. Uebel. 1966. Identification of an active compound from balsam fir. *Science* 154: 1020–1021.

Bowers, W. S., T. Ohta, J. S. Cleere and P. A. Marsella. 1976. Discovery of insect anti-juvenile hormones in plants. *Science* 193: 542–547.

Bromley, R. L. and 9 others. 2013. The prevalence of neurodevelopmental disorders in children prenatally exposed to antiepileptic drugs. *J. Neurol. Neurosurg. Psychiat.* 84: 637–643.

Bushnell, P. J., R. J. Kavlock, K. M. Crofton, B. Weiss and D. C. Rice. 2010. Behavioral toxicology in the 21st century: Challenges and opportunities for behavioral scientists. Summary of a symposium presented at the annual meeting presented at

the annual meeting of the neurobehavioral teratology society. *Neurotoxicol. Teratol.* 32: 313–328.

Carrasco, A. E. 2011. Reply to the letter to the editor regarding our article (Pagenelli et al., 2010). *Chem. Res. Toxicol.* 24: 610–613.

Caspers, K. M., P. A. Romitti, S. Lin, R. S. Olney, L. B. Holmes and M. M. Werler. 2013. National Birth Defects Prevention Study. Maternal periconceptional exposure to cigarette smoking and congenital limb deficiencies. *Paediatr. Perinat. Epidemiol.* 27: 509–520.

CDC (Centers for Disease Control). 1992. Recommendation for the use of folic acid to reduce the number of cases of spina bifida and other neural tube defects. *Morb. Mort. Wkly. Rpt.* 41: 1–7.

CDC (Centers for Disease Control). 2002. Rubella near elimination in the United States. Press release. www.cdc.gov/od/oc/media/pressrel/r020429.htm

CDC (Centers for Disease Control). 2015. Fetal alcohol spectrum disorders (FASDs). www.cdc.gov/ncbddd/fasd/data.html

Chen, S. Y., D. B. Dehart and K. K. Sulik. 2004. Protection from ethanol-induced limb malformations by the superoxide dismutase/catalase mimetic, EUK-134. *FASEB J.* 18: 1234–1236.

Chen, J. K., J. Taipale, M. K. Cooper and P. A. Beachy. 2002. Inhibition of hedgehog signaling by direct binding of cyclopamine to Smoothened. *Genes Dev.* 16: 2743–2748.

Chiriboga, C. A. 2003. Fetal alcohol and drug effects. *Neurologist* 9: 267–279.

Chomiak, T., N. Turner and B. Hu. 2013. What we have learned about autism dpectrum disorder from valproic acid. *Patholog. Res. Int.* 2013: 712758.

Chrisman, K., R. Kenney, J. Comin, T. Thal, L. Suchoki, Y. G. Yueh and D. F. Gardner. 2004. Gestational ethanol exposure disrupts the expression of *Fgf8* and *Sonic hedgehog* during limb patterning. *Birth Def. Res. A: Clin. Mol. Teratol.* 70: 163–171.

Christensen, J., T. K. Grønborg, M. J. Sørensen, D. Schendel, E. T. Parner, L. H. Pedersen and M. Vestergaard. 2013. Prenatal valproate exposure and risk of autism spectrum disorders and childhood autism. *JAMA* 309: 1696–1703.

Christin, M. S. and 7 others. 2004. Effects of agricultural pesticides on the immune system of *Xenopus laevis* and *Rana pipiens*. *Aquat. Toxicol.* 67: 33–43.

Clarren, S. K. 1986. Neuropathology in the fetal alcohol syndrome. In J. R. West (ed.), *Alcohol and Brain Development*. Oxford University Press, New York.

Cohlan, S. Q. 1953. Excessive intake of vitamin A as a cause of congenital anomalies in the rat. *Science* 117: 535–537.

Cordero, D., R. Marcucio, D. Hu, W. Gaffield, M. Tapadia and J. A. Helms. 2004. Temporal perturbations in sonic hedgehog signaling elicit the spectrum of holoprosencephaly phenotypes. *J. Clin. Invest.* 114: 485–494.

Crawford, K. and D. M. Vincenti. 1998. Retinoic acid and thyroid hormone may function through similar and competitive pathways in regenerating axolotls. *J. Exp. Zool.* 282: 724–738.

Dally, A. 1998. Thalidomide: Was the tragedy preventable? *Lancet* 351: 1197–1199.

D'Amato, R. J., M. S. Loughnan, E. Flynn and J. Folkman. 1994. Thalidomide is an inhibitor of angiogenesis. *Proc. Natl. Acad. Sci. USA* 91: 4082–4085.

Davidson, C. and R. A. Knapp. 2007. Multiple stressors and amphibian declines: Dual impacts of pesticides and fish on yellow-legged frogs. *Ecol. Appl.* 17: 587–597.

Davis, W. L., L. A. Crawford, O. Cooper, G. R. Farmer, D. Thomas and B. L. Freeman. 1990. Ethanol induces the generation of reactive free radicals by neural crest cells in culture. *J. Craniofac. Genet. Dev. Biol.* 10: 277–293.

Dehart, D. B., L. Lanoue, G. S. Tint and K. K. Sulik. 1997. Pathogenesis of malformations in a rodent model for Smith-Lemli-Opitz syndrome. *Am. J. Med. Genet.* 68: 328–337.

de Soysa, T. Y. and 13 others. 2012. Macondo crude oil from the Deepwater Horizon oil spill disrupts specific developmental processes during zebrafish embryogenesis. *BMC Biol.* 10: 40.

Dietrich, K. N., O. G. Berger, P. A. Succop, P. B. Hammond and R. L. Bornschein. 1993. The developmental consequences of low to moderate prenatal and postnatal lead exposure: Intellectual attainment in the Cincinnati Lead Study Cohort following school entry. *Neurotoxicol. Teratol.* 15: 37–44.

Dorris, M. 1989. *The Broken Cord.* Harper and Row, New York.

Dou, X., M. F. Wilkemeyer, C. E. Menkari, S. E. Parnell, K. K. Sulik and M. E. Charness. 2013. Mitogen-activated protein kinase modulates ethanol inhibition of cell adhesion mediated by the L1 neural cell adhesion molecule. *Proc. Natl. Acad. Sci. USA* 110: 5683–5688.

Dwyer, J. B., R. S. Broides and F. M. Leslie. 2008. Nicotine and brain development. *Birth Def. Res. C: Embryo Today* 84: 30–44.

Edison, R. and M. Muenke. 2003. The interplay of environmental and genetic factors in craniofacial morphogenesis: Holoprosencephaly and the role of cholesterol. *Congen. Anom. Kyoto* 43: 1–21.

Edwards, M. 1994. Pollution in the former Soviet Union: Lethal legacy. *Natl. Geog.* 186(2): 70–115.

Eikel, D., A. Lampen and H. Nau. 2006. Teratogenic effects mediated by inhibition of histone deacetylases: Evidence from quantitative structure activity relationships of 20 valproic acid derivatives. *Chem. Res. Toxicol.* 19: 272–278.

Emanuele, M. A. and N. V. Emanuele. 1998. Alcohol's effects on male reproduction. *Alcoh. Health Res. World* 22: 195–201.

EPA (Environmental Protection Agency). 2007. Insect growth regulator fact sheet. epa.gov/oppbppd1/biopesticides/ingredients/factsheets/factsheet-igr.htm

EPA (Environmental Protection Agency). 2014. TSCA chemical substance inventory. www.epa.gov/oppt/existingchemicals/pubs/tscainventory/basic.html

EPA-PCB. 2008. www.epa.gov/epawaste/hazard/tsd/pcbs/pubs/effects.htm

Epstein, C. 2008. Human malformations and their genetic basis. In C. Epstein, R. P. Erickson and A. Wynshaw-Boris (eds.), *Inborn Errors of Development: The Molecular Basis of Clinical Disorders of Morphogenesis,* 2nd Ed. Oxford University Press, New York, pp. 3–8

Eto, K. 2000. Minamata disease. *Neuropathology* 20(suppl.): S14–S19.

Eto, K. and 7 others. 2001. Methylmercury poisoning in common marmosets: A study of selective vulnerability within the cerebral cortex. *Toxicol. Pathol.* 29: 565–573.

Finer, L. B. and M. R. Zolna. 2011. Unintended pregnancy in the United States: Incidence and disparities, 2006. *Contraception* 84: 478–485.

Finnell, R. H., B. C. Wlodarczyk, J. C. Craig, J. A. Piedrahita and G. D. Bennett. 1997. Strain-dependent alterations in the expression of folate pathway genes following teratogenic exposure to valproic acid in a mouse model. *Am. J. Med. Genet.* 70: 303–311.

Flak, A. L., S. Su, J. Bertrand, C. H. Denny, U. S. Kesmodel and M. E. Cogswell. 2014. The association of mild, moderate, and binge prenatal alcohol exposure and child neuropsychological outcomes: A meta-analysis. *Alcohol Clin. Exp. Res.* 38: 214–226.

Flentke, G. R., A. Garic, E. Amberger, M. Hernandez and S. M. Smith. 2011. Calcium-mediated repression of b-catenin and its transcriptional signaling mediates neural crest cell death in an avian model of fetal alcohol syndrome. *Birth Defects Res. A: Clin. Mol. Teratol.* 91: 591–602.

Foltran, F., D. Gregori, L. Franchin, E. Verduci and M. Giovannini. 2011. Effect of alcohol consumption in prenatal life, childhood, and adolescence on child development. *Nutr. Rev.* 69: 642–659.

Franco, B. and 12 others. 1995. A cluster of sulfatase genes on Xp22.3: Mutations in chondrodysplasia punctata (*CDPX*) and implications for warfarin embryopathy. *Cell* 81: 15–21.

Fransen, E., V. Lemmon, G. Van Camp, L. Vits, P. Coucke and P. J. Willems. 1995. CRASH syndrome: Clinical spectrum of corpus callosum hypoplasia, retardation, adducted thumbs, spastic paraparesis and hydrocephalus due to mutations in one single gene, L1. *Eur. J. Hum. Genet.* 3(5): 273–284

Frazier, T. W., L. Thompson, E. A. Youngstrom, P. Law, A. Y. Hardan, C. Eng and N. Morris. 2014. A twin study of heritable and shared environmental contributions to autism. *J. Autism Dev. Disord.* 44: 2013–2025.

Govorko, D., R. A. Bekdash, C. Zhang and D. K. Sarkar. 2012. Male germline transmits fetal alcohol adverse effect on hypothalamic proopiomelanocortin gene across generations. *Biol. Psychiatry* 72: 378–388.

Gardiner, D. and D. M. Hoppe. 1999. Environmentally induced limb malformations in mink frogs (*Rana septentrionalis*). *J. Exp. Zool.* 284: 207–216.

Gardiner, D., A. Ndayibagira, F. Grün and B. Blumberg. 2003. Deformed frogs and environmental retinoids. *Pure Appl. Chem.* 75: 2263–2273.

Gilbertson, M. 2004. Male cerebral palsy hospitalization as a potential indicator of neurological effects of methylmercury in Great Lakes communities. *Environ. Res.* 95: 375–384.

Giroud, A. and M. Martinet. 1959. Teratogenese pur hypervitaminose A chez le rat, la souris, le cobaye, et le lapin. *Arch. Fr. Pediatr.* 16: 971–980.

GMWatch. 2011. Roundup and birth defects: Carrasco vs Monsanto. ww.gmwatch.org/index.php?option=com_content&view=article&id=13001

Golden, J. 2005. *Message in a Bottle: The Making of Fetal Alcohol Syndrome.* Harvard University Press, Cambridge, MA.

Goulding, E. H. and R. M. Pratt. 1986. Isotretinoin teratogenicity in mouse whole-embryo culture. *J. Craniofac. Genet. Dev. Biol.* 6: 99–112.

Green, M. L., A. V. Singh, Y. Zhang, K. A. Nemeth, K. K. Sulik and T. B. Knudson. 2007. Reprogramming of genetic networks during initiation of the fetal alcohol syndrome. *Dev. Dyn.* 236: 613–631.

Gregg, N. M. 1941. Congenital cataracts following German measles in the mother. *Trans. Opthalmol. Soc. Austral.* 3: 35.

Gubitosi-Klug, R., C. G. Larimer and C. F. Bearer. 2007. L1 cell adhesion molecule is neuroprotective of alcohol induced cell death. *Neurotoxicology* 28: 457–462.

Guo, Y., Y. Chen, S. Carreon and M. Qiang. 2011. Chronic intermittent ethanol exposure and its removal induce a different miRNA expression pattern in primary cortical neuronal cultures. *Alcohol. Clin. Exp. Res.* 36: 1058–1066

Guyton, K. Z. and 8 others. 2015. Carcinogenicity of tetrachlorvinphos, parathion, malathion, diazinon, and glyphosate. *Lancet Oncol.* pii: S1470-2045(15)70134-8.

Hallmayer, J. and 15 others. 2011. Genetic heritability and shared environmental factors among twin pairs with autism. *Arch. Gen. Psychiatry* 68: 1095–1102.

Hamer, H. 2010. Acreage. *National Agricultural Statistics Board Annual Report.* United States Department of Agriculture.

Harada, M. 1972. *Minamata Disease.* Kumamoto Nichinichi Shinbun Center, Iwanami Shoten, Tokyo. (Trans. 2004 by T. S. George and T. Sachie.)

Harada, M. 1996. Characteristics of industrial poisoning and environmental contamination in developing countries. *Environ. Sci.* 4(suppl.): S157–S169.

Harada, M., T. Fujino, S. Akagi and S. Nishigaki. 1976. Epidemiological and clinical study and historical background of mercury pollution on Indian reservations in northwestern Ontario, Canada. *Bull. Instit. Constitut. Med.* 26: 169–184.

Harmon, M. A., M. F. Boehm, R. A. Heyman and D. J. Mangelsdorf. 1995. Activation of mammalian retinoid-X receptors by the insect growth regulator methoprene. *Proc. Natl. Acad. Sci. USA* 92: 6157–6160.

Höfer, T. and 7 others. 2004. Animal testing and alternative approaches for the human health risk assessment under the proposed new European chemicals regulation. *Arch. Toxicol.* 78: 549–564.

Hoffman, L. M. and W. M. Kulyk. 1999. Alcohol promotes in vitro chondrogenesis in embryonic facial mesenchyme. *Internatl. J. Dev. Biol.* 43: 167–174.

Hong, M. and R. S. Krauss. 2012. Cdon mutation and fetal ethanol exposure synergize to produce midline signaling defects and holoprosencephaly spectrum disorders in mice. *PLoS Genet.* 8(10): e1002999.

Incardona, J. P. and 8 others. 2013. Exxon Valdez to Deepwater Horizon: Comparable toxicity of both crude oils to fish early life stages. *Aquat. Toxicol.* 142–143: 303–316.

Ingram, J. L., S. M. Peckham, B. Tisdale and P. M. Rodier. 2000. Prenatal exposure of rats to valproic acid reproduces the cerebellar anomalies associated with autism. *Neurotox. Teratol.* 22: 319–324.

International Joint Commission of the United States and Canada. 2000. *Tenth Biennial Report of Great Lakes Water Quality.* IJC, Ottawa.

Ito, T., H. Ando and H. Handa. 2011. Teratogenic effects of thalidomide: Molecular mechanisms. *Cell Mol. Life Sci.* 68: 1569–1579.

Johnson, P. T. J., K. B. Lunde, E. G. Ritchie and A. E. Launer. 1999. The effect of trematode infection on amphibian limb development and survivorship. *Science* 284: 802–804.

Johnson, P. T. J. and 9 others. 2002. Parasite (*Ribeiroia ondatrae*) infection linked to amphibian malformations in the western United States. *Ecol. Monogr.* 72: 151–168.

Johnston, M. C., K. K. Sulik, W. S. Webster and B. L. Jarvis. 1985. Isotretinoin embryopathy in a mouse model: Cranial neural crest involvement. *Teratology* 31: 26A.

Jones, K. L. and D. W. Smith. 1973. Recognition of the fetal alcohol syndrome. *Lancet* 2(7836): 999–1001.

Kaufman, L. 2012. Mutated trout raise new concerns around mine sites. *New York Times* Feb. 23, 2012.

Keeler, R. F. and W. Binns. 1968. Teratogenic compounds of *Veratrum californicum* (Durand). V. Comparison of cyclopian effects of steroidal alkaloids from the plant and structurally related compounds from other sources. *Teratology* 1: 5–10.

Kessel, M. 1992. Respecification of vertebral identities by retinoic acid. *Development* 115: 487–501.

Kiesecker, J. M. 2002. Synergism between trematode infection and pesticide exposure: A link to amphibian limb deformities in nature? *Proc. Natl. Acad. Sci. USA* 99: 9900–9904.

Kietzman, H. W., J. L. Everson, K. K. Sulik and R. J. Lipinski. 2014. The teratogenic effects of prenatal ethanol exposure are exacerbated by sonic hedgehog or Gli2 haploinsufficiency in the mouse. *PLoS ONE* 9(2): e89448.

Kimura-Kuroda, J., Y. Komuta, Y. Kuroda, M. Hayashi and H. Kawano. 2012. Nicotine-like effects of the neonicotinoid insecticides acetamiprid and imidacloprid on cerebellar neurons from neonatal rats. *PLoS ONE.* 7(2): e32432.

Kleinman, J. C., M. B. Pierre, Jr., J. H. Madans, G. H. Land and W. E. Schramm. 1988. The effects of maternal smoking on fetal and infant mortality. *J. Epidemiol.* 127: 274–282.

Kochhar, D. M., J. D. Penner and C. Tellone. 1984. Comparative teratogenic activities of two retinoids: Effects on palate and limb development. *Teratogen. Carcinogen. Mutagen.* 4: 377–387.

Kondo, K. 2000. Congenital Minamata disease: Warnings from Japan's experience. *J. Child Neurol.* 15: 458–464.

Kotch, L. E., S.-Y. Chen and K. K. Sulik. 1995. Ethanol-induced teratogenesis: Free radical damage as a possible mechanism. *Teratology* 52: 128–136.

Kulikauskas, V., A. B. Blaustein and R. J. Ablin. 1985. Cigarette smoking and its possible effects on sperm. *Fertil. Steril.* 44: 526–528.

Kulp, L. and J. Kulp. 2000. *The Best I Can Be: Living with Fetal Alcohol Syndrome-Effects.* Better Endings/New Beginnings Press.

Kuske, B., P. Y. Pulyanina and N. I. zur Nieden. 2012. Embryonic stem cell test: Stem cell use in predicting developmental cardiotoxicity and osteotoxicity. *Methods Mol. Biol.* 889: 147–179.

La Claire, J., J. A. Bantle and J. Dumont. 1998. Photoproducts and metabolites of a common insect growth regulator produce developmental deformities in *Xenopus. Environ. Sci. Technol.* 32: 1453–1461.

Lammer, E. J. and 11 others. 1985. Retinoic acid embryopathy. *New Engl. J. Med.* 313: 837–841.

Lannoo, M. 2008. *Malformed Frogs: The Collapse of Aquatic Ecosystems.* University of California Press, Berkeley.

Lee, H. Y., A. L. Inselman, J. Kanungo and D. K. Hansen. 2012. Alternative models in developmental toxicology. *Syst. Biol. Reprod. Med.* 58: 10–22.

Lee, L. M. Y. and 8 others. 2005. A paradoxical teratogenic mechanism for retinoic acid. *Proc. Natl. Acad. Sci. USA* 109: 13668–13673.

Lee, S. H. and 12 others. 2015. Identification of plant compounds that disrupt the insect juvenile hormone receptor complex. *Proc. Natl. Acad. Sci. USA* 112: 1733–1738.

Lemoine, E. M., J. P. Harousseau, J. P. Borteyru and J. C. Menuet. 1968. Les enfants de parents alcooliques: Anomalies observées. *Oest. Med.* 21: 476–482.

Lenkowski, J. R., J. M. Reed, L. Deninger and K. A. McLaughlin. 2008. Perturbation of organogenesis by the herbicide atrazine in the amphibian *Xenopus laevis. Environ. Health Persp.* 116: 223–230.

Lenz, W. 1962. Thalidomide and congenital abnormalities. *Lancet* 1: 45. (Reported in a symposium in 1961.)

Lenz, W. 1966. Malformations caused by drugs in pregnancy. *Amer. J. Dis. Child.* 112: 99–106.

López, S. L. and 8 others. 2012. Pesticides used in South American GMO-based agriculture: A review of their effects on humans and animal models. *Advances Molecular Teratology.* Elsevier, Oxford, UK, pp. 41–76.

Mahony, C., L. Erskine, J. Niven, N. H. Greig, W. D. Figg and N. Vargesson. 2013. Pomalidomide is nonteratogenic in chicken and zebrafish embryos and nonneurotoxic in vitro. *Proc. Natl. Acad. Sci. USA* 110: 12703–12708.

Mak, V., K. Jarvi, M. Buckspan, M. Freeman, S. Hechter and A. Zini. 2000. Smoking is associated with the retention of cytoplasm by human spermatozoa. *Urology* 56: 463–466.

Marean, A., A. Graf, Y. Zhang and L. Niswander. 2011. Folic acid supplementation can adversely affect murine neural tube closure and embryonic survival. *Hum. Mol. Genet.* 20: 3678–3683.

Maritz, G. S. 2008. Nicotine and lung development. *Birth Def. Res. C: Embryo Today* 84: 45–53.

Mattson, S. N. and 11 others, and The Collaborative Initiative on Fetal Alcohol Spectrum Disorders. 2013. Further development of a neurobehavioral profile of fetal alcohol spectrum disorders. *Alcohol Clin. Exp. Res.* 37: 517–528.

May, P. A. and J. P. Gossage. 2001. Estimating the prevalence of fetal alcohol syndrome: A summary. *Alch. Res. Health* 25: 159–167.

McCarthy, N. and J. K. Eberhart. 2014. Gene-ethanol interactions underlying fetal alcohol spectrum disorders. *Cell Mol. Life Sci.* 71: 2699–2706.

McCoy, K. A., L. J. Bortnick, C. M. Campbell, H. J. Hamlin, L. J. Guillette Jr. and C. M. St. Mary. 2008. Agriculture alters gonadal form and function in the toad *Bufo marinus*. *Environ. Health Persp.* 116: 1526–1532.

McDonald, A. D. and 7 others. 1988. Fetal death and work in pregnancy. *Br. J. Ind. Med.* 45: 148–157.

Meador, K. J. and 10 others. 2006. In utero antiepileptic drug exposure: Fetal death and malformations. *Neurology* 67: 407–412.

Melvin, S. D. and V. L. Trudeau. 2012a. Growth, development and incidence of deformities in amphibian larvae exposed as embryos to naphthenic acid concentrations detected in the Canadian oil sands region. *Environ. Pollut.* 167: 178–183.

Melvin, S. D. and V. L. Trudeau. 2012b. Toxicity of naphthenic acids to wood frog tadpoles (*Lithobates sylvaticus*). *J. Toxicol. Environ. Health A.* 75: 170–173.

Menegola, E., F. Di Renzo, M. L. Broccia, M. Prudenziati, S. Minucci, V. Massa and E. Giavini. 2006. Inhibition of histone deacetylase activity on specific embryonic tissues as a new mechanism for teratogenicity. *Birth Def. Res. B: Dev. Reprod. Toxicol.* 74: 392–398.

Mereu, G. and 9 others. 2003. Prenatal exposure to a cannabinoid agonist produces memory deficits linked to dysfunction in hippocampal long-term potentiation and glutamate release. *Proc. Natl. Acad. Sci. USA* 100: 4915–4920.

Meteyer, C. U. and 9 others. 2000. Hindlimb malformations in free-living northern leopard frogs (*Rana pipiens*) from Maine, Minnesota, and Vermont suggest multiple etiologies. *Teratology* 62: 151–171.

Mikael, L.G., L. Deng, L. Paul, J. Selhub and R. Rozen. 2013. Moderately high intake of folic acid has a negative impact on mouse embryonic development. *Birth Defects Res. A: Clin. Mol. Teratol.* 97: 47–52.

Miller, L., A. M. Shapiro and P. G. Wells. 2013. Embryonic catalase protects against ethanol-initiated DNA oxidation and teratogenesis in acatalasemic and transgenic human catalase-expressing mice. *Toxicol. Sci.* 134: 400–411.

Miller, M. T. and K. Strömland. 1999. Thalidomide: A review, with a focus on ocular findings and new potential uses. *Teratology* 60: 306–321.

Miller, M. T., K. Strömland, L. Ventura, M. Johansson, J. M. Bandim and C. Gillberg. 2005. Autism associated with conditions characterized by developmental errors in early embryogenesis: A minireview. *Internatl. J. Dev. Neurosci.* 23: 201–219.

Miranda, R. C. 2012. MicroRNAs and fetal brain development: Implications for ethanol teratology during the second trimester period of neurogenesis. *Front. Genet.* 3: 77.

Miranda, R. C. 2014. MicroRNAs and ethanol toxicity. *Int. Rev. Neurobiol.* 115: 245–284.

Moore, K. L. and T. N. N. Persaud. 1993. *Before We Are Born: Essentials of Embryology and Birth Defects*. W. B. Saunders, Philadelphia.

Moore, S. J., P. Turnpenny, A. Quinn, S. Glover, D. J. Lloyd, T. Montgomery and J. C. S. Dean. 2000. A clinical study of 57 children with fetal anticonvulsant syndrome. *J. Med. Genet.* 37: 489–497.

Moroni, M. C., M. A. Vigano and F. Mavilio. 1994. Regulation of human *Hoxd-4* gene by retinoids. *Mech. Dev.* 44: 139–154.

Morriss-Kay, G. M. 1993. Retinoic acid and craniofacial development: Molecules and morphogenesis. *Bioessays* 15: 9–15.

Morrow, E. M. and 24 others. 2008. Identifying autism loci and genes by tracing recent shared ancestry. *Science* 321: 218–223.

Muenke, M. 1995. Holoprosencephaly as a genetic model for normal craniofacial development. *Semin. Dev. Biol.* 5: 293–301.

NARCAM (North American Reporting Center for Amphibian Malformations). 2002. Northern Prairie Wildlife Research Center, Jamestown, ND. www.nprwc.usgs.gov/narcam.

Narita, N., M. Kato, M. Tazoe, M. Miyazake, M. Narita and N. Okado. 2002. Increased monoamine concentration in brain and blood of fetal thalidomide–and valproic acid-exposed rat: Putative animal models for autism. *Pediat. Res.* 52: 576–579.

Nowack, E. 1965. Die sensible Phase bei der Thalidomide-Embryopathie. *Humangenetik* 1: 516–536.

O'Hara, M. F., J. H. Charelap, R. C. Craig and T. B. Knudsen. 2002. Mitochondrial transduction of ocular teratogenesis during methylmercury exposure. *Teratology* 65: 131–144.

Oppenheimer, J. M. 1968. Some historical relationships between teratology and experimental embryology. *Bull. Hist. Med.* 42: 145–159.

Opitz, J., R. Schultka and L. Göbbel. 2006. Meckel on developmental pathology. *Am. J. Med. Genet. A* 140: 115–128.

OTIS (Organization of Teratology Information Specialists). 2007. otispregnancy.org/pdf/hyperthermia.pdf

Ouellet, M., J. Bonin, J. Rodriguez, J. L. DesGanges and S. Lair. 2007. Hindlimb deformities (ectromelia, ectrodactyly) in free-living anurans from agricultural habitats. *J. Wildlife Disord.* 33: 95–104.

Paganelli, A., V. Gnazzo, H. Acosta, S. L. López and A. E. Carrasco. 2010. Glyphosate-based herbicides produce teratogenic effects on vertebrates by impairing retinoic acid signaling. *Chem. Res. Toxicol.* 23: 1586–1595.

Pauly, P. J. 1996. How did the effects of alcohol on reproduction become scientifically uninteresting? *J. Hist. Biol.* 29: 1–28.

Perkins, A., C. Lehmann, R. C. Lawrence and S. J. Kelly. 2013. Alcohol exposure during development: Impact on the epigenome. *Int. J. Dev. Neurosci.* 31: 391–397.

Pollack, A. 2015. Weed killer, long cleared, is doubted. *New York Times*. 27 March 2015. B1.

Pratt, G. E., R. C. Jennings, A. F. Hammett and G. T. Brooks. 1980. Lethal metabolism of precocene-I to a reactive epoxide by locust corpora allata. *Nature* 284: 320–323.

Prutner, W. 2013. Hazard and risk assessment of teratogenic chemicals under REACH. *Methods Mol. Biol.* 947: 517–543.

Raam, M. S., B. D. Solomon and M. Muenke. 2011. Holoprosencephaly: A guide to diagnosis and clinical management. *Indian Pediatr.* 48: 457–466.

Raje, N. and K. Anderson. 1999. Thalidomide: A revival story. *New Engl. J. Med.* 341: 1606–1609.

Ramanathan, R., M. F. Wilkemeyer, B. Mittel, G. Perides and M. E. Charness. 1996. Alcohol inhibits cell-cell adhesion mediated by human L1. *J. Cell Biol.* 133: 381–390.

Relyea, R. A. 2012. New effects of Roundup on amphibians: Predators reduce herbicide mortality; herbicides induce anti-predator morphology. *Ecol. Appl.* 22: 634–647.

Rinaldi, T., G. Silverberg and H. Markram. 2008. Hyperconnectivity of local neocortical microcircuitry induced by prenatal exposure to valproic acid. *Cereb. Cort.* 18: 763–770.

Roessler, E., E. Belloni, K. Gaudenz, P. Jay, P. Berta, S. W. Scherer, L.-C. Tsui and M. Muenke. 1996. Mutations in the human *Sonic hedgehog* gene cause holoprosencephaly. *Nature Genet.* 14: 357–360.

Rothman, K. J., L. L. Moore, M. R. Singer, U. S. Nguyen, S. Mannino and A. Milunsky. 1995. Teratogenicity of high vitamin A intake. *New Engl. J. Med.* 333: 1369–1373.

Saez, T. M., M. P. Aronne, L. Caltana and A. H. Brusco. 2014. Prenatal exposure to the CB$_1$ and CB$_2$ cannabinoid receptor agonist WIN 55,212-2 alters migration of early-born glutamatergic neurons and GABAergic interneurons in the rat cerebral cortex. *J. Neurochem.* 10.1111/jnc.12636

Sampson, P. D. and 7 others. 1997. Incidence of fetal alcohol syndrome and prevalence of alcohol-related neurodevelopmental disorder. *Teratology* 56: 317–326.

Sass, J. B. and A. Colangelo. 2006. European Union bans atrazine, while the United States negotiates continued use. *Int. J. Occup. Environ. Health.* 12: 260–267.

Sathyan, P., H. B. Golden and R. C. Miranda. 2007. Competing interactions between micro-RNAs determine neural progenitor survival and proliferation after ethanol exposure: Evidence from an ex vivo model of the fetal cerebral cortical neuroepithelium. *J. Neurosci.* 27: 8546–8557.

Saxén, L., P. C. Holmberg, M. Nurminen and E. Kuosma. 1982. Sauna and congenital defects. *Teratology* 25: 309–313.

Schooneveld, H. 1979. Precocene-induced collapse and resorption of corpora allata in nymphs of *Locusta migratoria*. *Experientia* 35: 363–364.

Schumann, C. M. and 10 others. 2010. Longitudinal magnetic resonance imaging study of cortical development through early childhood in autism. *J. Neurosci.* 30: 4419–4427.

Seiler, A., M. Oelgeschläger, M. Liebsch, R. Pirow, C. Riebeling, T. Tralau and A. Luch. 2011. Developmental toxicity testing in the 21st century: The sword of Damocles shattered by embryonic stem cell assays? *Arch. Toxicol.* 85: 1361–1372.

Shi, Q., E. Ko, L. Barclay, T. Hoang, A. Rademaker and R. Martin. 2001. Cigarette smoking and aneuploidy in human sperm. *Mol. Reprod. Dev.* 59: 417–421.

Siamwala, J. H. and 8 others. 2012. Nitric oxide rescues thalidomide mediated teratogenicity. *Sci. Rep.* 2: 679.

Sláma, K. and C. M. Williams. 1966. The juvenile hormone. V. The sensitivity of the bug *Pyrrhocoris apterus* to a hormonally active factor in American paper-pulp. *Biol. Bull.* 130: 235–246.

Steingraber, S. 2003. *Having Faith*. Berkley Trade, New York.

Stocum, D. 2000. Frog limb deformities: An "eco-devo" riddle wrapped in multiple hypotheses surrounded by insufficient data. *Teratology* 62: 147–150.

Stodgell, C. J., J. L. Ingram, M. O'bara, B. K. Tisdale, H. Nau and P. M. Rodier. 2006. Induction of the homeotic gene *Hoxa1* through valproic acid's teratogenic mechanism of action. *Neurotoxicol. Teratol.* 28: 617–624.

Stopper, G. F., L. Hecker, R. A. Franssen and S. K. Sessions. 2002. How trematodes cause limb deformities in amphibians. *J. Exp. Zool.* 294: 252–263.

Streissguth, A. P. and R. A. LaDue. 1987. Fetal alcohol: Teratogenic causes of developmental disabilities. In S. R. Schroeder (ed.), *Toxic Substances and Mental Retardation.* American Association of Mental Deficiency, Washington, DC, pp. 1–32.

Studer, M., H. Pöpperl, H. Marshall, A. Kuroiwa and R. Krumlauf. 1994. Role of a conserved retinoic acid response element in rhombomere restriction of *Hoxb1. Science* 265: 1728–1732.

Sulik, K. K. 2005. Genesis of alcohol-induced craniofacial dysmorphism. *Exp. Biol. Med.* 230: 366–375.

Sulik, K. K. 2014. Fetal alcohol spectrum disorder: Parthogenesis and mechanisms. In E. V. Sullivan and A. Pfefferbaum (eds.), *Handbook of Clinical Neurology*, Vol. 125, *Alcohol and the Nervous System.* Elsevier B. B., Amsterdam, Netherlands.

Sulik, K. K., C. S. Cook and W. S. Webster. 1988. Teratogens and craniofacial malformations: Relationships to cell death. *Development* 103(suppl.): S213–S231.

Therapontos, C., L. Erskine, E. R. Gardner, W. D. Figg and N. Vargesson. 2009. Thalidomide induces limb defects by preventing angiogenic outgrowth during early limb formation. *Proc. Natl. Acad. Sci. USA* 106: 8573–8578.

Thorogood, P. 1997. The relationship between genotype and phenotype: Some basic concepts. In P. Thorogood (ed.), *Embryos, Genes, and Birth Defects.* Wiley, New York, pp. 1–16.

Tischfield, M. A. and 10 others. 2005. Homozygous *HOXA1* mutations disrupt human brainstem, inner ear, cardiovascular, and cognitive development. *Nature Genet* 37: 1035–1037.

Toledo-Rodriguez, M. and 7 others. 2010. Maternal smoking during pregnancy is associated with epigenetic modifications of the brain-derived neurotrophic factor-6 exon in adolescent offspring. *Am. J. Med. Genet. B Neuropsychiatr. Genet.* 153B(7): 1350–1354.

Toms, D. A. 1962. Thalidomide and congenital abnormalities. *Lancet* 2: 400.

Traiffort, E. and 7 others. 2004. Functional characterization of *SHH* mutations associated with holoprosencephaly. *J. Biol. Chem.* 279: 42889–42897.

Ui, J. 1992. *Industrial Pollution in Japan.* United Nations University Press, Tokyo. www.unu.edu/unupress/unupbooks/uu35ie/uu35ie00.htm#Contents

USDI (United States Department of the Interior). 2012. Technical Review: Smoky Canyon Mine Selenium Report (154 pages).

van Gelder, M. M., J. Reefhuis, A. R. Caton, M. M. Werler, C. M. Druschel and N. Roeleveld. National Birth Defects Prevention Study. 2009. Maternal periconceptional illicit drug use and the risk of congenital malformations. *Epidemiology* 20: 60–66.

Vandenberg, L. N., B. W. Pennarola and M. Levin. 2011. Low frequency vibrations disrupt left-right patterning in the *Xenopus* embryo. *PLoS ONE* 6(8): e23306.

Vandenberg, L. N., C. Stevenson and M. Levin. 2012. Low frequency vibrations induce malformations in two aquatic species in a frequency-, waveform-, and direction-specific manner. *PLoS ONE* 7(12): e51473.

Vargesson, N. 2009. Thalidomide-induced limb defects: Resolving a 50-year-old puzzle. *Bioessays* 31: 1327–1336.

Vargesson, N. 2014. Thanlidomide embryopathy: An enigmatic challenge. *ISRN Dev. Biol.* dx.doi.org/10.1155/2013/241016

Vasquez, K., S. Kuizon, M. Junaid and A. E. Idrissi. 2013. The effect of folic acid on GABA(A)-B 1 receptor subunit. *Adv. Exp. Med. Biol.* 775: 101–109.

Wang, L. and K. E. Pinkerton. 2008. Detrimental effects of tobacco smoke exposure during development on postnatal lung function and asthma. *Birth Def. Res. C: Embryo Today* 84: 54–60.

Warren, K. R. and T.-K. Li. 2005. Genetic polymorphisms: Impact on the risk of fetal alcohol spectrum disorders. *Birth Def. Res. A: Clin. Mol. Teratol.* 73: 195–203.

Werler, M. M. 1997. Teratogen update: Smoking and reproductive outcomes. *Teratology* 55: 382–388.

Williams, C. M. 1970. Hormonal interactions between plants and insects. In E. Sondheimer and J. B. Simeone (eds.), *Chemical Ecology*. Academic Press, New York, pp. 103–132.

Wilson, J. G. 1961. General principles in experimental teratology. In M. Fishbein (ed.), *Proceedings of the First International Conference on Congenital Malformations*. Lippincott, Philadelphia.

Wilson, J. G. (ed.). 1973. *Environment and Birth Defects*. Academic Press, London.

Wozniak, J. R. and R. L. Muetzel. 2011. What does diffusion tensor imaging reveal about the brain and cognition in fetal alcohol spectrum disorders? *Neuropsychol Rev.* 21: 133–147.

Xu, Y., P. Liu and Y. Li. 2005. Impaired development of mitochondria plays a role in the central nervous system defects of fetal alcohol syndrome. *Birth Def. Res. A: Clin. Mol. Teratol.* 73: 83–91.

Yabu, T., H. Tomimoto, Y. Taguchi, S. Yamaoka, Y. Igarishi and T. Okazaki. 2005. Thalidomide-induced anti-angiogenic action is mediated by ceramide through depletion of VEGF receptors, and is antagonized by sphingosine-1-phosphate. *Blood* 106: 125–134.

Yang, Y. and 10 others. Abnormal cortical thickness alterations in fetal alcohol spectrum disorders and their relationships with facial dysmorphology. *Cereb Cortex.* 22: 1170-1179.

Zhou, F. C., Y. Balaraman, M. Teng, Y. Liu, R. P. Singh and K. P. Nephew. 2011. Alcohol alters DNA methylation patterns and inhibits neural stem cell differentiation. *Alcohol Clin. Exp. Res.* 35: 735–746.

Zhou, F. C., Y. Chen and A. Love. 2013. Cellular DNA methylation program during neurulation and its alteration by alcohol exposure. *Birth Defects Res. A: Clin. Mol. Teratol.* 91: 703–715.

Endocrine Disruptors

And I brought you into a plentiful land, to eat the fruit thereof and the goodness thereof; but when ye entered, ye defiled my land, and made my heritage an abomination.

Jeremiah 2:7, quoted in Lannoo 2008

Chemicals that disrupt hormone messages have the power to rob us of the rich possibilities that have been the legacy of our species and, indeed, the essence of our humanity.

T. Colburn et al., 1996

One of the most active and controversial areas of teratology concerns the misregulation of the endocrine system during development. Such studies have led to the **endocrine disruptor hypothesis**, which states that hormonally active molecular compounds in the environment—endocrine disruptors—alter gene expression during early development in ways that have a significant impact on the health of human and wildlife populations. An endocrine disruptor has been defined (Zoeller et al. 2012) as "an exogenous chemical, or a mixture of chemicals, that interferes with any aspect of hormonal action." The changes produced by endocrine disruptors during development are not the obvious anatomical malformations classically produced by the teratogens described in Chapter 5; indeed, the anatomical alterations induced by endocrine disruptors are often internal and visible only with a microscope. Rather, the major changes are physiological, and in many cases the aberrant phenotypes are not seen until adulthood. These functional changes are more subtle than the visibly aberrant phenotypes produced by teratogens, but they nevertheless can be extremely important phenotypic alterations. Moreover, in some instances, the alterations can persist in offspring for generations after an organism's exposure to the disruptor.

Within the paradigms of teratology, it was once thought that there were only a few "bad" agents and that people were exposed to dangerous doses of these agents only inadvertently, by occupational exposure, or by accident. However, we now recognize that hormone-disrupting chemicals are everywhere in our technological society. Many ubiquitous products—including plastics, cosmetics, and pesticides—contain endocrine disruptors, and animal studies have shown that even low-dose in utero exposure to these chemicals can produce major disabilities later in life. While their manufacturers did not anticipate that these chemicals would affect organisms in such unexpected ways, the chemicals' rapid introduction and wide use have caused involuntary exposure of recent generations of embryos to a wide range of endocrine disruptors. Because these compounds are not as obviously toxic as some of the classic teratogens, it required detective work to establish that they could indeed disrupt the developing endocrine system. We now know that animals, including humans, are exposed to persistent endocrine-disrupting chemicals in utero and that this exposure continues through childhood. Endocrine disruptors line baby bottles, constitute soft toys, and leach from our brightly colored plastic containers. As we grow older, we are exposed to endocrine disruptors found in cosmetics such as sunblocks and hair rinses, chemical coatings that prevent clothing and bed linens from being highly flammable, steroids from the livestock and plants we eat, and even the substances responsible for air freshener fragrances and that "new car smell" (see Aitken et al. 2004; Schlumpf et al. 2004; Dodson et al. 2012).

The Nature of Endocrine Disruptors

The term "endocrine disruptor" was coined at the 1991 Wingspread conference (organized by Theo Colborn). The purpose of the conference was to determine whether observations of altered development in wildlife could be attributed to hormonally active chemicals and whether they portended a similar effect for humans. The conferees concluded that the developmental alterations observed in a diversity of wildlife species were due to exposure to multiple chemicals that, through different modes of action, disrupted the endocrine systems of the developing organisms. An example of one of these chemicals was found in the late 1980s. While investigating the estrogen sensitivity of human breast cells in culture conditions, the Soto-Sonnenschein lab accidentally found that a chemical with estrogenic activity, p-nonylphenol, leached out of their plastic centrifuge tubes (Soto et al. 1991). The finding of a "novel estrogen" suggested that the anomalies found in wildlife could not solely be attributed to already banned compounds such as DDT. The work of Frederick vom Saal, which showed that small physiological variations in hormone levels in utero produced effects that persisted in adult life, also suggested that exogenous exposures (i.e., those arising from outside the body) may cause altered development. The conclusions of this conference, The Wingspread Statement, brought together data from several fields and identified a common theme—endocrine disruption.

Since then, endocrine disruptors, which are also known by other names such as hormone mimics, environmental signal modulators, environmental estrogens, and hormonally active agents, have been found to affect hormonal functions in many ways:

- Endocrine disruptors can be **agonists**, mimicking the effects of natural hormones and binding to their receptors. An example of an agonist is the paradigmatic endocrine disruptor diethylstilbestrol (DES), which mimics the sex hormone estradiol (a common form of estrogen) and binds to the body's estrogen receptors.

- Endocrine disruptors can act as **antagonists**, either preventing the binding of hormones to their natural receptors or blocking the hormones' synthesis. DDE, a metabolic product of the insecticide DDT, can act as an anti-androgen (i.e., an antagonist of masculinizing hormones) by binding to the androgen receptor and preventing testosterone (which normally binds to this receptor) from functioning properly.

- Endocrine disruptors can increase hormone synthesis. Compounds such as the herbicide atrazine (see Chapter 2 and later in this chapter) act by elevating the synthesis of certain hormones (in this case, estrogen via induction of the aromatase enzyme).

- Endocrine disruptors can affect the elimination or transportation of a hormone within the body. One of the ways that the polychlorinated biphenyls (PCBs) disrupt the endocrine system is to interfere with the elimination and degradation of thyroid hormones. Several pesticides may act indirectly to affect hormone levels by inducing the synthesis of liver enzymes involved in the synthesis and degradation of certain hormones (see Guillette 2006).

- Endocrine disruptors can "prime" the organism to be more sensitive to hormones later in life. For instance, when a rat fetus is exposed to bisphenol A (BPA), the embryonic mammary gland cells are induced to make more progesterone receptors. These extra receptors alter mammary gland growth responses to natural hormones later in life, predisposing breast tissue to cancer formation (see Markey et al. 2005; Muñoz-de-Toro et al. 2005; Wadia et al. 2007; Soto et al. 2008).

The Endocrine Disruptor Hypothesis

The roots of the endocrine disruptor hypothesis can be found in Rachel Carson's 1962 book *Silent Spring*, which documented the effects of DDT and other insecticides on reproductive failures in birds and other wildlife. Within the next decade, endocrine disruption in human populations was documented when DES, a drug commonly prescribed during pregnancy, was found to cause reproductive abnormalities and also a rare form of carcinoma of the vagina. The drug's effects, however, were seen not in the mothers, but rather in their daughters, who were exposed to DES in utero when the mothers took the drug (see later in this chapter). The widespread occurrence of endocrine disruption was more fully appreciated when endocrinologist John A. McLachlan and wildlife biologist Theo Colborn synthesized the vast literature on the subject and published scientific and popular articles concerning reproductive failures and developmental anomalies (Colborn et al. 1996; Krimsky 2000).

According to Jerrold Heindel of the National Institute of Environmental Health Sciences, the endocrine disruptor hypothesis has had a difficult time being accepted in mainstream medical circles because its paradigms are so different from those of infectious disease, trauma relief, or toxicology—the usual Western approaches to medical care (Heindel 2006). The following insights point to the importance of thinking in terms of ecology and developmental biology in filling the gaps in our medical knowledge:

- Endocrine disruption is a functional change in a tissue that superficially appears normal. The pathology may be evident only upon microscopic examination, where one can see tissue-level anomalies in the sexually dimorphic regions of the brain or altered tissue formation in the mammary gland or reproductive tract. Anomalies may also manifest themselves as changes in gene expression that alter cell metabolism.

- The mechanism of pathology is not infection, injury, or toxicity. Rather, the disease is likely to result from altered gene expression caused by the environmental agent, resulting in altered morphogenesis leading to dysfunctional physiology.

- Sensitivity to disruptive agents will depend on the context of the exposure—it will depend on the stage of development, the dose of the agent, and even the sex of the exposed individual.

- The environmental insult may be additive or synergistic with nutritional influences and is influenced by the exposed organism's genetic background. As mentioned above, a tissue may be "primed" by exposure to an endocrine disruptor in utero and then "activated" years later, when the cell encounters the same or a similar endocrine disruptor.

- The effect can be transgenerational. That is to say, the agents may cause the changes in gene expression throughout the embryo, including the newly formed germ cells. If the DNA of the germ cells is altered (say, by methylation), the effect can be transmitted to the next generation. As we have seen in earlier chapters, transgenerational inheritance can also be mediated through maternal nursing behaviors.

This chapter focuses on a set of endocrine disruptors that interact with the reproductive hormone pathways: these include diethylstilbestrol (DES), bisphenol A (BPA), nonylphenol, atrazine, phthalates, and the polychlorinated biphenyls (PCBs). These are some of the most well-studied compounds for which there is evidence of endocrine disruption.

DDT: The start of it all

Silent Spring was one of the most influential books of the twentieth century. Its author, fisheries biologist Rachel Carson, warned that pesticides were destroying wildlife, that DDT in particular appeared to be exterminating shorebird populations, and that pesticides were becoming a staple of the American diet. For this she was reviled by the agricultural chemicals industry and called a fanatic, a Communist, and worse (see Lear 1998; Orlando 2002; Stouder 2012). But subsequent research (Frye 1995) bore out Carson's claims and revealed the first evidence of endocrine disruption due to exposure to environmental chemicals.

The chemical components of DDT (dichloro-diphenyl-trichloroethane; Figure 6.1A) cannot be broken down and eliminated by vertebrate organisms; DDT remains in their bodies and builds up, becoming especially concentrated in organisms that feed on the DDT-containing tissues of other animals. This persistence, along with its solubility in lipids, results in tremendous **bioaccumulation** such that even though DDT has not been legally used in the United States since 1972, most of us have this chemical in our bodies. Indeed, it is so resistant to metabolic degradation, and so much was manufactured, that it remains not only in humans and other terrestrial animals but also in fishes, marine mammals, and seabirds. It has an environmental half-life of about 15 years, which means it can take 100 years or more for the concentration of DDT in the soil to fall below active levels.

Bioaccumulation of DDT was especially pronounced in some birds of prey living at the top of the food chain. Peregrine falcons and bald eagles

(A)

(B)

Figure 6.1   The effects of pesticide bioaccumulation. (A) Chemical structures of DDT (dichloro-diphenyl-trichloroethane) and its metabolic by-product, DDE. DDT is an estrogenic compound, while DDE is an androgen inhibitor. (B) One notable effect of environmental bioaccumulation of these chemicals was the prevalence of thin, nonviable eggshells found among many bird species (particularly birds of prey), with subsequent severe population declines. This brown pelican egg cracked open long before the embryo inside was ready to hatch.

became endangered because of DDT-induced fragility of their eggshells (Cooke 1973). Even seabirds were affected when DDT in runoff accumulated in the fish they fed on. Fragile eggshells resulted in high mortality as the developing bird embryos became desiccated, were easily preyed upon, and often were not able to withstand even minor physical forces (Figure 6.1B).

Discovering why DDT contamination results in thin eggshells was a formidable research challenge. It turns out that DDT acts as an estrogenic compound (see the next section), while its chief metabolic product, DDE (which lacks one of DDT's chlorine atoms), inhibits androgens such as testosterone from binding to the androgen receptor (Davis et al. 1993; Kelce et al. 1995; Xu et al. 2006). Eggshell thinning is caused by several actions of DDT and DDE. Hens with high DDT levels have poorly developed shell glands, with capillaries deficient in carbonic anhydrase—an enzyme critical for the deposition of shell-strengthening calcium carbonate in the egg (Holm et al. 2006). High DDE levels in the shell gland also prevent calcium carbonate deposition by down-regulating the synthesis of prostaglandins (a group of fatty acid derivatives that regulate many vertebrate physiological processes). In birds, one of these prostaglandins is critical for the transport of calcium ions through the shell gland for use in shell formation (Lundholm 1997; Guillette 2006).

DDT use has been banned in the United States since 1972 and is also banned in most of Europe. The Stockholm Convention of 2001, which attempts a worldwide ban on the use of several dangerously persistent organic chemicals, limits the use of DDT to the control of disease-carrying insects (notably mosquitoes), and only in those countries where there is no affordable and/or comparably effective alternative for managing a serious public health threat (primarily malaria; see Chapter 8). As a result of the ban, levels of DDT in the birds diminished, and the bird population began recovering (Grier 1982; Newton and Wyllie 1992).

Despite these bans and limits, DDT is still having marked effects. It is still used extensively in tropical regions for growing exportable foods. With the globalization of our food supply, DDT continues to enter the U.S. food chain through imported fruits and vegetables and other products (Stutchbury 2008). Moreover, the spraying of DDT to grow exported foods in South America has been linked to the large decline in migratory songbirds (Stutchbury 2008, 2009).

Banning DDT turned out to be an excellent idea for reasons beyond eggshell fragility and wildlife endangerment. What wasn't known in 1972 was that women who had been exposed to DDT before age 14 would have a greater than fivefold increase in their incidence of breast cancers later in life (Figure 6.2). Using a remarkable database of American patients studied for

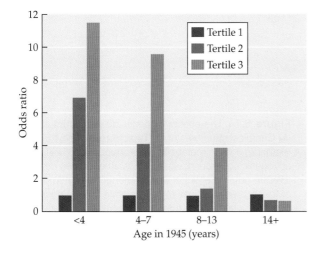

Figure 6.2 The odds ratio for breast cancer (how much more likely one is to have breast cancer diagnosed before age 50 than the average population) depends on how old a woman was when exposed to DDT. The tertiles (division into reference [tertile 1], middle third [tertile 2], and upper third [tertile 3]) reflect the concentrations of DDT in the women's blood samples taken in 1945. (After Cohn 2011.)

Establishing a Chain of Causation

DDT and DDE have been linked not only to fragile egg-shells in birds but also to other incidents of reproductive failure in wildlife. In the 1990s, researchers linked a pollutant spill in Florida's Lake Apopka (a discharge including DDT, DDE, and numerous PCBs) to a 90% decline in the birthrate of alligators and reduced penis size in young males (Guillette et al. 1994; Matter et al. 1998). Indeed, this work revealed how the interaction of many chemicals at once can result in unanticipated consequences on reproduction.

The Lake Apopka study is an example of how environmental toxicology research establishes **chains of causation**. Whether in law or in science, establishing such chains is a demanding and necessary task. In environmental toxicology, numerous end points must be checked, and many different levels of causation must be established (Crain and Guillette 1998; McNabb et al. 1999). For example, to establish whether the pollutant spill in Lake Apopka was responsible for the decline in the juvenile alligator population there, researchers first had to establish exactly *why* the population was declining, then follow through with various studies to show how the specific chemicals present in the spill could contribute to the observed reasons for the decline. The table shows the postulated chain of causation, from overall observations of the entire population through the molecular biochemistry.

As seen in the accompanying table, *population level* studies revealed that the Lake Apopka population decline was the result of a decrease in the number of alligators being born (the birthrate). At the *organism level*, researchers found unusually high levels of estrogens in juvenile female alligators and unusually low levels of testosterone in juvenile males. On the *tissue and organ level*, they observed elevated production of estrogens in juvenile male testes, along with penis and testis malformations, and changes in enzyme activity in the female gonads. On the *cellular level*, ovarian abnormalities in females correlated with elevated estrogen levels. These cellular changes could be explained at the *molecular level* by noting that many of the chemicals in the spill bind to alligator estrogen and progesterone receptors and that these chemicals are able to circumvent the cell's usual defenses against the overproduction of steroid hormones. The alligators exposed to these chemicals had altered expression patterns of those genes encoding proteins used in sex determination and steroid hormone production (Crain et al. 1998; Guillette et al. 2007; Kohno et al. 2008). Whereas newly hatched alligators from an uncontaminated lake (the controls) had sexually dimorphic expression of several of these genes, in newly hatched alligators from the contaminated lake, the expression of these genes was the same in both males and females (Milnes et al. 2008). Thus, the pollutant chemicals can be linked by a logical chain of events to reproductive anomalies that could explain the decreased birthrate among alligators in the lake.

In many instances, it is found that the "chain" forms a circle, in which endocrine disruptor chemicals from human activities enter the environment, the pollutant chemicals affect the ecosystem, and the environmental pollution returns to affect the human population. Such work shows that even though many of the candidate disruptors are compounds referred to as "pesticides," what constitutes a "pest" is more a value judgment than a scientific designation. We share enzymes and receptor molecules with every animal on the planet, and our mammalian hormones are especially similar to those of other vertebrates.

Chain of causation linking contaminant spill in Lake Apopka to endocrine disruption in juvenile alligators

Level	Evidence
Population	The juvenile alligator population in Lake Apopka has decreased.
Organism	Juvenile Apopka (JA) females have elevated circulating levels of 17β-estradiol.
	JA males have depressed circulating concentrations of testosterone.
Tissue/organ	JA females have altered gonad aromatase activity.
	JA males have poorly organized seminiferous tubules.
	JA males have reduced penis size.
	Testes from JA males have elevated estradiol (estrogen) production.
Cell	JA females have polyovular follicles that are characteristic of estrogen excess.
Molecule	Many contaminants bind the alligator estrogen receptors and progesterone receptor.
	Many of these contaminants do not bind to the alligator cytosol proteins that blockade excess hormones.

Source: Crain and Guillette 1998.

over 50 years, Cohn (2011) concluded that DDT exposure after adolescence might not lead to major health risks; but exposure to this endocrine disruptor in utero or at an early age can make a person susceptible to breast cancer later in life.* We will see this type of effect in our discussions of BPA later in the chapter.

Estrogen and Endocrine Disruption

Some of the best studied endocrine disruptors are those that mimic or block the actions of **estrogens**, a family of steroid hormones that regulate growth, differentiation, and function of reproductive tissues and of other organs such as the bones, brain, and cardiovascular organs (Brosens and Parker 2003). Although estrogen is popularly known as the female sex hormone, both sexes need estrogens for proper bone and connective tissue development.

We are continuously exposed to estrogen-like compounds in the environment. These compounds are often called **xenoestrogens** since they are foreign to the body (Greek, *xenos*, "foreigner"). The water we drink and the food we eat contain xenoestrogens. Although some of these compounds, such as the **phytoestrogens** (plant estrogens), are from natural sources, modern technological industries have manufactured an estrogenic environment. The plastics that store our water and beverages are slowly leaching these compounds into our drinks. Because the estrogen receptors in our bodies bind so many of the compounds used in plastics production, we now develop in and live in a world full of unanticipated estrogenic structures (Colborn et al. 1996; Figure 6.3).

When present early in development, xenoestrogens have consequences for health later in life. Indeed, Heindel (2007) has pointed to a "developmental estrogen syndrome" wherein exposure to estrogenic compounds early in development can cause both men and women to experience fertility problems, cancers, and obesity later in life. Fetal exposure to environmentally relevant doses of BPA, for example, has this effect (Sonnenschein et al. 2011).

The structure and function of estrogen receptors

The two major estrogen receptor proteins, **ERα** and **ERβ**, are nuclear transcription factors that are activated by binding ligands such as 17β-estradiol (the most common endogenous estrogen). Nuclear estrogen receptor proteins such as ERα and ERβ have three major domains. The first is the hormone-binding domain, which binds the estrogen. The second is a protein-interaction domain that allows the estrogen receptor to interact with other proteins (including other estrogen receptors). The third domain is the DNA-binding domain that binds to specific DNA sequences (the

---

*This idea that early exposure to an agent might predispose one to having cancers later in life also appears in the studies of breast cancer in the atomic bomb survivors of Hiroshima and Nagasaki. The part of that population most at risk comprised the women under the age of 20 at the time of the bombing (Ronckers et al. 2005).

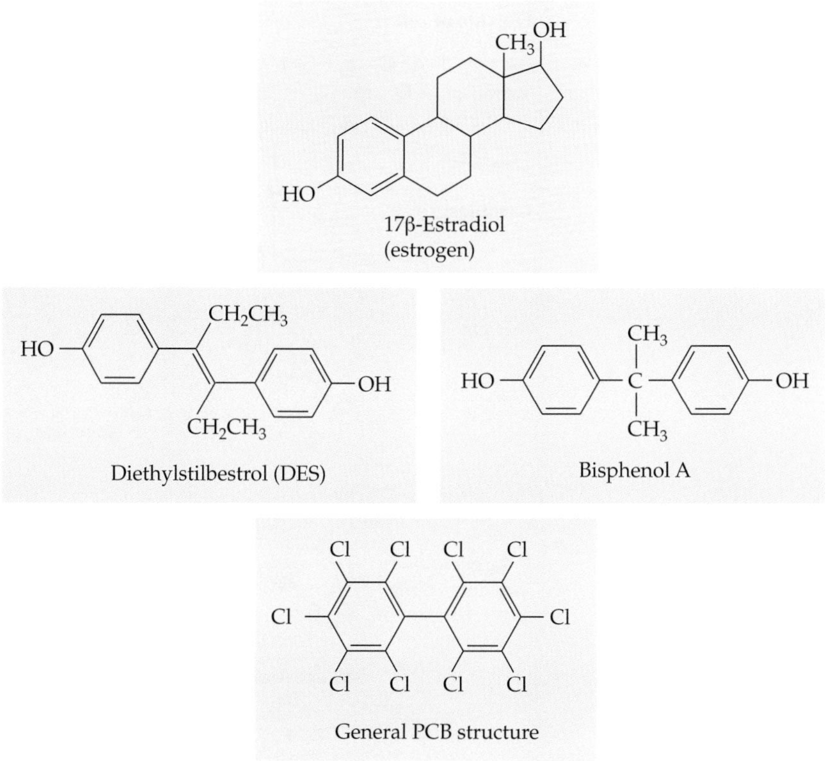

Figure 6.3   Chemical structures of estradiol (natural estrogen) and three of the endocrine disruptors discussed in this chapter.

estrogen-responsive elements, or EREs) and regulates the transcription of the genes adjacent to these regions. When an estrogen molecule binds to its receptor protein, the receptor protein breaks its connections to several other proteins and combines with another bound estrogen receptor (dimerizes). The paired and hormone-bound estrogen receptors bind to the EREs in the enhancers of specific genes, where they interact with other transcription factors to control the transcription of these genes and generate a cascade of cellular actions (Brosens and Parker 2003; Figure 6.4). When estrogen-bound estrogen receptors bind to their EREs on the DNA, they interact with histone-modification enzymes that can repress or promote transcription, depending on the other proteins there (Ali and Coombes 2000; Moggs and Orphanides 2001; Hervouet et al. 2013).

Some endocrine disruptors can similarly bind to estrogen receptors. This can occur because, unlike in most other receptors, the hormone-binding domain of the estrogen receptor can bind numerous differently shaped molecules (see Figure 6.3). Some endocrine disruptors are estrogen agonists and bind in a manner that mimics the effects of estrogen (which is not a good thing if it occurs in the embryo when or where estrogens are not normally produced). Other endocrine disruptors are estrogen antagonists and

Figure 6.4 Mechanism of the estrogen receptor within a general cell nucleus. The estrogen receptor is a protein that can bind endogenous steroidal estrogen (and numerous structurally diverse xenoestrogens). It is usually bound to numerous other proteins that prevent its binding to DNA. When the receptor is activated by hormonal ligands such as estradiol or an estrogen mimic, the inhibitory proteins leave, and the activated receptor then binds to a DNA sequence (estrogen-responsive element, or ERE). These EREs are in the enhancers of specific genes, where they control the transcription of target genes and generate a cascade of cellular actions.

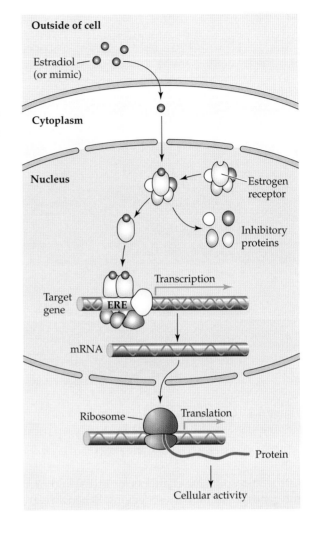

can enter the hormone-binding site of the receptor in a way that prevents it from functioning.

Recently, a third member of this receptor family has been discovered, estrogen-related receptor-γ (**ERR-γ**). This estrogen receptor appears to be important in regulating blood pressure during pregnancy, and it appears to mediate the effects of estrogen in certain breast and endometrial (uterine) tumors (Heckler et al. 2014; Sun et al. 2014). Interestingly, the human form of this receptor is strongly bound by the endocrine disruptor BPA, and in zebrafish ERR-γappears to be almost completely responsible for the ability of BPA to cause anomalies in the auditory system (Liu et al. 2014; Tohmé et al. 2014).

A fourth estrogen receptor, GPER, is located in the plasma membrane (Powell et al. 2001), where it binds to estrogens and also to xenoestrogens at very low (picomolar) concentrations. When this receptor binds estrogen, it becomes active and initiates several signaling cascades, such as those opening calcium channels and those generating active transcription factors (Wozniak et al. 2005; Watson et al. 2007). Recent studies also suggest that endocrine disruptors such BPA, nonylphenol, genistein, and DDE bind to both the nuclear and the membrane estrogen receptors (see Prossnitz and Barton 2014).

The fifth estrogen receptor is called the steroid and xenobiotic receptor (SXR, also called the pregnane X receptor). It is a nuclear receptor like ERα and ERβ, and like GPER it binds endogenous estrogens and xenoestrogens. It also binds pharmacological agents. Once it binds one of these compounds, the activated SXR enters the nucleus, where it induces the expression of those genes whose products detoxify and eliminate the compound. In other words, the SXR is part of our defense system and can be considered one of the strategies to eliminate potential poisons (Zhou et al. 2009). The SXR protein has a broad lipophilic binding domain, and it can be activated by endocrine disruptors as diverse as BPA, phthalates, pesticides, phthalic acid, and nonylphenol. SXR is predominantly found in the intestine and liver.

DES: The paradigmatic endocrine disruptor

One of the best studied of the xenoestrogens is the drug **diethylstilbestrol**, or **DES**. It provides a paradigmatic example of an endocrine disruptor (see Bell 1986 and Palmund 1996). Indeed, Sir Charles Dodds first synthesized the drug in 1938 in an attempt to obtain inexpensive synthetic estrogens that could be used to study how this hormone acts. It was later marketed as one of the many "miracle drugs" that became available shortly after World War II (Figure 6.5).

DES was prescribed for several reasons; it was used to suppress lactation, to balance hormones in menopausal women, and to rectify hormone imbalances that were thought to contribute to the premature termination of pregnancies (NRC 1999; Krimsky 2000). At the time DES was first marketed, the idea that a drug could damage the fetus, or that it could produce effects that would not show up until decades later, was not seriously considered.

It is estimated that in the United States, over a million fetuses were exposed to DES between 1947 and 1971, and this is a small fraction of exposures worldwide. In addition to DES administered to pregnant women, biologically relevant levels of DES were present in meat, since the drug was also used to accelerate livestock growth (Knight 1980). Even though research done in the 1950s showed that in fact DES had *no* effect on the maintenance of a pregnancy, the drug continued to be prescribed right up until the U.S. Food and Drug Administration banned its use in 1971. The ban was instituted when a specific type of tumor—clear-cell adenocarcinoma—was discovered in the reproductive tracts of women whose mothers had ingested DES prior to week 18 of pregnancy. This is a rare type of tumor, but epidemiology revealed a clear association of this cancer with early exposure to DES in utero. Although the frequency of this cancer in the women studied was in fact low (0.1%), these women were at high risk for numerous reproductive problems. Greater than 95% of females exposed

Figure 6.5   Aimed at obstetricians, this 1956 advertisement promoted the use of diethylstilbestrol (DES) for maintaining at-risk pregnancies. The U.S. Food and Drug Administration took DES off the market in 1971, when it was definitively linked to a specific type of cancer in the reproductive tracts of women whose mothers ingested the drug during early pregnancy.

to DES in utero had some abnormalities in the structure of their reproductive organs, and the risk of these women getting breast cancer after age 50 was increased threefold (Newbold 2004; Palmer et al. 2006; Hoover et al. 2011; Figure 6.6).

DES interferes with sexual development by causing cell type changes in the female reproductive tract (the derivatives of the Müllerian duct, which forms the upper portion of the vagina, the cervix, the uterus, and oviducts). In many cases, DES causes the boundary between the oviduct and the uterus (the uterotubal junction) to be lost, resulting in infertility or subfertility (Robboy et al. 1982; Newbold et al. 1983). Moreover, the distal Müllerian ducts (that form the cervix and uterus) often fail to come

Figure 6.6 Genital abnormalities can occur in women exposed to DES in utero. In these "DES daughters," the cervical tissue (red) is often displaced into the vagina. They may experience a T-shaped and constricted uterus, adenosis of cervix and vagina (where the lining differentiates into mucosal cells), precancerous cells (cervical dysplasia), ectopic pregnancies, adenocarcinoma, and other effects.

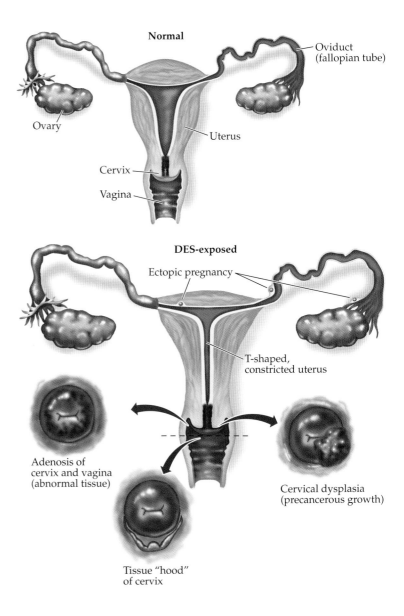

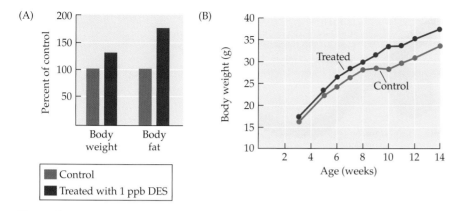

Figure 6.7   DES as an obesogen. (A) As little as 1 part per billion (ppb) DES can cause the accumulation of fat to nearly double that of control mice who were not exposed to DES. (B) Weight and fat gain occurs after "puberty" in mice. (After Newbold 2004.)

together to form a single cervical canal (see Figure 6.6). Symptoms similar to the human DES syndrome occur in mice exposed to DES in utero or shortly after birth.

Interestingly, DES is also an **obesogen**—a substance that stimulates the body both to produce fat cells and to accumulate fat within these cells. After studying the endocrine problems of the female mice born of mothers treated with DES, Newbold and her colleagues (2007) noticed that these mice seemed to be fatter than mice born of mothers who had not been treated with DES.* By examining the different groups of mice, the Newbold laboratory (2004, 2007) confirmed that mice treated shortly after birth with as little as 1 part per billion DES had a marked obesity syndrome later in life. The DES-exposed and control mice were the same until about 8 weeks after birth (roughly the time of "mouse puberty"). After that, the DES-exposed mice grew much fatter than the controls (Figure 6.7; see also Figure 3.22). DES had sensitized the mice early in life, and when the large concentrations of estrogen associated with sexual maturity began to be synthesized, the mice rapidly became obese.

Mechanisms of DES action

The particular regions of the female reproductive tract disrupted by exposure to DES are normally those specified by the expression of the *HOXA* genes.[†] Hox genes are physically linked together on human chromosome 7, and they are expressed in a nested fashion throughout the Müllerian

---

*Obesity may be a common effect of estrogenic compounds. For example, Rubin and colleagues (2001) made observations similar to Newbold's in rats treated with BPA.

[†]The term "Hox gene" has become common usage in biology, referring to a group of genes widely conserved across many families and phyla and involved in axis specification and pattern formation during animal development. The general term is not italicized, but the specific genes are. Following convention, human genes are named using capital letters, thus differentiating them from mouse genes, which are referred to with initial capital letters only.

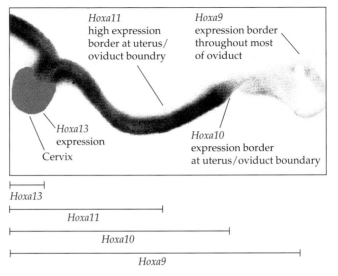

Figure 6.8 *Hoxa* gene expression in the reproductive system of a normal 16.5-day embryonic female mouse. *Hoxa9* expression extends from the cervix through the uterus to about halfway up the oviduct. *Hoxa10* expression has a sharp anterior border at the transition between the oviduct and the presumptive uterus. *Hoxa11* has the same anterior border as *Hoxa10*, but its expression diminishes closer to the cervix. *Hoxa13* is expressed only in the cervix and upper vagina. (After Ma et al. 1998.)

duct (see Ma 2009). *HOXA9* mRNA is detected throughout the uterus and continues to be found about halfway through the presumptive oviduct. *HOXA10* expression exhibits a sharp anterior boundary at the junction between the future uterus and the future oviduct. *HOXA11* has strong expression in the anterior regions where *HOXA10* is expressed, but its expression weakens in the posterior regions. *HOXA13* expression is seen only in the cervix (Figure 6.8).

Ma and colleagues (1998) showed that the effects of DES on the female mouse reproductive tract could be explained as the result of altered *Hoxa10* expression in the Müllerian duct. They showed that estrogen and progesterone are able to regulate certain genes of the HoxA cluster (*Hoxa9*, *Hoxa10*, *Hoxa11*, and *Hoxa13*). To determine whether DES changed Hox gene expression patterns, they injected DES under the skin of pregnant mice, and the fetuses were allowed to develop almost to birth. When the fetuses from the DES-injected mothers were compared with fetuses from mothers that had not received DES, it was seen that DES almost completely repressed the expression of *Hoxa10* in the Müllerian duct (Figure 6.9). This repression was most pronounced in the stroma (mesenchyme) of the duct, the place where experimental embryologists had localized the effect of DES (Boutin and Cunha 1997). The case for DES's acting through repression of *Hoxa10* is strengthened by the phenotype of the *Hoxa10* mutant mouse, in which this gene fails to function (Benson et al. 1996; Ma et al. 1998). In such *Hoxa10*-deficient mutants, there is a transformation of the proximal quarter of the uterus into oviduct tissue, and there are abnormalities at

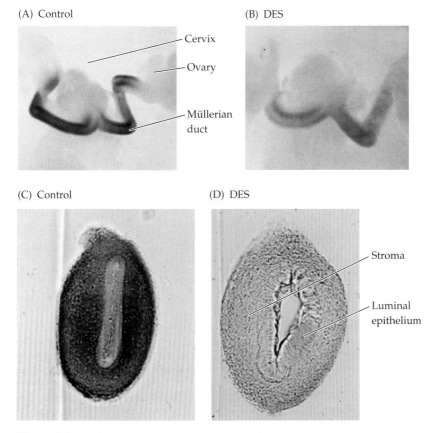

(A) Control

Cervix

Ovary

Müllerian duct

(B) DES

(C) Control

(D) DES

Stroma

Luminal epithelium

Figure 6.9   DES exposure represses *Hoxa10*. (A) Normal 16.5-day embryonic female mice show *Hoxa10* expression from the boundary of the cervix through the uterus primordium and most of the oviduct (compare with Figure 6.8). (B) In mice exposed prenatally to DES, this expression was severely repressed. (C) In control female mice at 5 days after birth (when the reproductive tissues are still forming), a section through the uterus shows abundant expression of the *Hoxa10* gene in the uterine stroma. (D) In female mice given high doses of DES 5 days after birth, *Hoxa10* gene expression in the mesenchyme is almost completely suppressed. (From Ma et al. 1998.)

the border of the uterus and oviduct. Similar changes are also seen in Hox gene expression in human cervical cells exposed to DES (Block et al. 2000).

The Hox gene expression in the stroma and epithelium of the female reproductive tract (i.e., the determination of whether a particular portion of the tract functions as a uterus, an oviduct, or a cervix) is linked to the Wnt proteins. Wnt proteins are associated with cell proliferation and protection against apoptosis (cell death), and the reproductive tracts of DES-exposed female mice also resemble those of mutant mice that lack *Wnt7a* genes. Miller and colleagues (1998) have shown that the Hox genes and the Wnt genes communicate with each other to keep each other activated during the specification and morphogenesis of the reproductive tissues (Figure 6.10). However, DES, acting through the estrogen receptor, represses the *Wnt7a* gene. This repression prevents the maintenance of the Hox gene expression

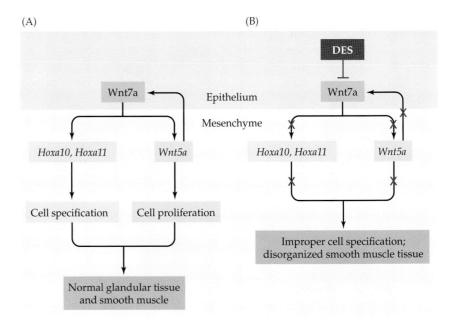

Figure 6.10   Misregulation of Müllerian duct morphogenesis due to DES exposure. (A) In normal female morphogenesis, the *Hoxa10* and *Hoxa11* genes in the mesenchyme are activated and maintained by the *Wnt7a* gene product. Wnt7a also induces *Wnt5a* gene expression; the *Wnt5a* gene product both maintains *Wnt7a* gene expression and stimulates cell proliferation and the normal morphogenesis of the uterus. (B) Acting through the estrogen receptor, DES blocks expression of *Wnt7a*. Proper activation of the Hox genes and *Wnt5a* thus does not occur (red Xs), resulting in a drastically altered female genital morphology. (After Gilbert 2013.)

pattern, and it also prevents the activation of another Wnt gene, *Wnt5a*, which encodes a protein necessary for cell proliferation.

The role of Wnt signaling in uterine morphology was confirmed by Carta and Sassoon (2004) when they observed that the *Wnt7a* gene product plays a key role as a cell death suppressor. The uteri of *Wnt7a* mutant mice, like those of DES-exposed mice, displayed high levels of cell death, while the wild-type uteri displayed almost no cell death. The expression pattern of ERα, Hox and Wnt genes, and other regulatory genes that guide uterine development reveals either abnormal regulation or atypical spatial distribution of transcripts in response to DES exposure (Carta and Sassoon 2004). This study demonstrates that *Wnt7a* coordinates multiple cell and developmental pathways in the morphogenetic field involved in uterine growth and hormonal responses, and disruption of these pathways by DES results in uterine cell death.*

Wnt signaling may also be important in regulating DNA methylation, a mechanism described in Chapter 2. DNA methylation regulates gene

---

*We will be discussing morphogenetic fields in Chapter 8. A morphogenetic field is that spatial region of the embryo that gives rise to a particular organ (the "limb field," the "eye field") or organ part ("the anterior heart field," "the retina field"). In this region, the roles of the paracrine factors can differ from their roles outside this region.

expression, and abnormalities of DNA methylation have been observed in many types of human tumors, including those of the uterus and cervix (see Chapter 7). DES causes the methylation of DNA and histones at particular sites in the genome, leading to abnormal gene expression. Moreover, the DES-induced chromatin methylation patterns at some of these sites persist through adulthood (Li et al. 2003; Jefferson et al. 2013). Li and colleagues (2003) have shown that exposure to a wide variety of endocrine disruptors during critical periods of mammalian development can interfere with normal DNA methylation patterns, leading to atypical gene expression. They demonstrated that exposure of mice to DES around the time of birth (when the mouse reproductive tract is being developed) changes the methylation patterns in the promoters of many estrogen-responsive genes associated with the development of reproductive organs.

The altered methylation hypothesis may also explain the transgenerational effects seen in DES syndrome. Although there are only sparse reports of reproductive anomalies and tumors in the grandchildren of mothers exposed to DES, the grandpups of mice who were exposed prenatally to DES have higher risks of reproductive tract anomalies and tumors later in life (Newbold et al. 1998, 2000, 2006). This will be discussed later in this chapter.

We noted earlier that a specific class of tumors was found in reproductive tissues in 0.1% of women who were exposed to DES in utero. These women were most likely to have been exposed to DES before week 13 of gestation, indicating a particular period of susceptibility. Studies by Cook and colleagues (2005) on rats provide some clues to the origin, and fortunate low frequency, of these tumors. They found that DES leaves a hormonal "imprint" on the uterine tissues, increasing the expression of estrogen-responsive genes prior to tumor formation. Moreover, they found that variations in a tumor-suppressor gene caused variations in the responses to DES. The lesson suggested by this study is that not all people will respond similarly to these endocrine disruptors, and that genetic polymorphism may be a critical factor in the types of responses seen. In the case of DES, almost all women exposed in utero experienced some sort of developmental modification of the uterus; the polymorphism leading to cancer upon DES exposure, however, was more limited.

Phytoestrogens and phthalates

More than 160 lipids from over 300 different plant species have been found to have estrogenic activity (Dixon 2004; Reynaud et al. 2005). These lipids are called phytoestrogens. In the human diet, phytoestrogens are especially abundant in soybeans, chickpeas, barley, flaxseed, and broccoli. Why should plants have estrogens? In a fascinating convergence, plants also use estrogenic compounds for signaling. In fact, they are the interspecies signals responsible for the formation of the nitrogen-fixing nodules of legumes, the symbiotic event responsible for maintaining life on Earth. The rhizobium-attracting flavonoids from the legume roots (mentioned in Chapter 3) are phytoestrogens. Moreover, each legume appears to use different phytoestrogens to attract different sets of bacteria.

Importantly, phytoestrogens bind differently to estrogen receptors in other plants and animals. The phytoestrogen of one species of legume can interfere with another legume's ability to attract bacteria. Not only that—phytoestrogens can bind to animals' estrogen receptors, and some of the endocrine disruptors that bind to human estrogen receptors (especially DDT, pentachlorophenol, and BPA) also inhibit normal phytoestrogen signaling (Fox et al. 2004, 2007; Figure 6.11). Thus, human-made substances

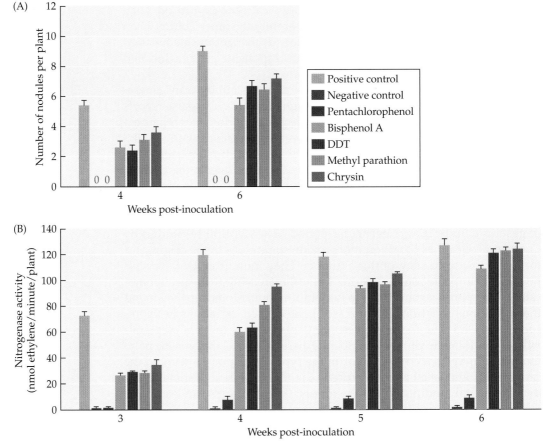

Figure 6.11 Pesticides interfere with legume-rhizobium symbiosis and nitrogen fixation. (A) Pesticides and other known endocrine disruptors inhibit recruitment of bacteria to host plants and delay nodulation. Alfalfa seeds were inoculated with *Sinorhizobium meliloti* bacteria and treated with one of five different environmental chemicals (at 5×10^{-6} *M*) at day 0. Alfalfa was harvested at 2, 4, and 6 weeks after inoculation, and the number of root nodules per plant was determined for each treatment group. At 2 weeks, none of the plants had nodules. At 4 and 6 weeks post-inoculation, all treatment groups had significantly fewer nodules compared with the positive control (bacteria without disruptor). 0 indicates no nodules present for this treatment group. (B) Nitrogenase activity was measured by using an acetylene reduction assay. Five replicates per treatment per time point were assayed, and average nanomolar ethylene produced per minute per plant was calculated for each treatment group. At weeks 3, 4, and 5 after inoculation, nitrogenase activity was significantly reduced in all treatment groups compared with the positive control. At 6 weeks after inoculation, nitrogenase activity of groups exposed to bisphenol A and pentachlorophenol was significantly reduced. (After Fox et al. 2007.)

(such as DDT) can possibly disrupt those critically important symbiotic relationships that lead to the fixation of atmospheric nitrogen.

So one important thing about phytoestrogens is that they are targets of endocrine disruption. But phytoestrogens are also the causes of endocrine disruption. Phytoestrogens can bind to mammalian estrogen receptors (especially ERβ) and change reproduction, fertility, and behavior (Kuiper et al. 1998; Patisaul et al. 2004, 2006; Wasserman et al. 2013). One particularly dramatic example of this ability of phytoestrogens to mimic normal estrogen was documented in Western Australia, where phytoestrogens in the clover (*Trifolium subterraneum*) consumed by domesticated sheep caused widespread infertility in the flock, leading to a reduced number of lambs (Bennetts and Underwood 1951; Cornwell et al. 2004).

Like the artificial estrogens found in birth control pills, phytoestrogens can bind to the hypothalamic estrogen receptors, where they act to inhibit the hormones responsible for ovulation. Therefore, ingesting phytoestrogens may actually lead to a decrease of the body's estrogen production, longer estrous cycles, and smaller ovaries (Kouki et al. 2003). Doses are important, and different species react differently to different estrogen mimics (Trisomboon et al. 2005; Wasserman et al. 2013).

Genistein, a phytoestrogen found in soybeans (and soy products such as tofu) binds to estrogen receptors ERα and ERβ. Moreover, it seems to have both beneficial effects (such as cancer prevention; see Cortina et al. 2013) and detrimental effects (endocrine disruption; Newbold et al. 2001; Wiszniewski et al. 2005; Jefferson et al. 2007; Cederroth et al. 2010), depending on both its dosage and the developmental stage of the embryo experiencing it (Allred et al. 2001; Cotroneo et al. 2002). Studies of cultured human placental cells indicate that genistein decreases the production of placentally made human chorionic gonadotropin (hCG) and progesterone, both of which are needed to sustain pregnancy (Jeschke et al. 2005; Richter et al. 2009), and Chan (2009) found that genistein slowed the rate of mouse oocyte maturation and increased the number of spontaneous abortions (where the embryos are resorbed by the uterus). Newborn mice treated with genistein have reproductive deficiencies and aberrant mammary gland development (Padilla-Banks et al. 2006; Jefferson et al. 2007), and exposure to this phytoestrogen can delay breast development in girls (Cheng et al. 2010; Wolff et al. 2010; Mervish et al. 2013). Moreover, Chavarro and colleagues (2008) found that men with extremely high dietary soy intake had significantly lower sperm counts than men who did not eat large amounts of soy. Genistein may be detrimental to wildlife, as well. Frog tadpoles treated with genistein fail to undergo normal thyroid-dependent metamorphosis (Ji et al. 2007).

Phthalates are multifunctional synthetic chemicals used in plastics, polyvinyl chloride products, cosmetics, hair spray, and children's toys. They are organic esters that are added to plastics to enhance their flexibility, durability, or transparency. Phthalates bind to estrogen receptors and several other types of receptors, and they have been linked to thyroid anomalies and changes in male fertility. Phthalates may have important anti-androgenic effects. Swann and her colleagues (2005) reported that phthalate exposure decreased the anogenital distance in boys (a measure of endocrine disruption in mammals). In a subsequent study, Swann's laboratory (Swann et al. 2009) confirmed those studies, concluding that phthalates may also make the brain less sexually dimorphic. In humans, phthalates have been shown to inhibit testosterone production, alter testes morphology, and change the anatomy of the genital region (Duty et al. 2003; Swan et al. 2005; Desdoit-Lethimonier et al. 2012). Albert and Jegou (2014) concluded that animal experiments, cultured human organ experiments, and epidemiological studies converge on phthalates as candidates for altering human male anatomy and behaviors. This has led to discussions that phthalates may play a significant role in a syndrome of testicular dysgenesis.

Declining Sperm Counts and Testicular Dysgenesis Syndrome

Females are not the only ones whose reproductive success may be affected by the estrogenic compounds being released into the environment. During the past two decades, there has been an increase in testicular cancers and a decrease in sperm count in men throughout the industrialized world (Carlsen et al. 1992; Aitken et al. 2004). Sperm count (number of sperm per milliliter of semen) has dropped precipitously throughout much of Europe and the Americas. When Skakkebaek and his team at the University of Copenhagen reviewed 61 international studies involving 14,947 men between 1938 and 1992, they found that the average sperm count had fallen from 113 million per milliliter in 1940 to 66 million in 1990 (Carlsen et al. 1992). In addition, the number of "normal" sperm fell from 60 million per milliliter to 20 million in the same period.

Subsequent studies have confirmed and extended Skakkebaek's findings. A survey of 1350 sperm donors in Paris found a decline in sperm counts by about 2% each year over the past 23 years, with younger men having the poorest quality semen (Auger et al. 1995). This age group had an average sperm count of 90 million sperm per milliliter in 1973, but only 60 million in 1992. A study from Finland (Pajarinen et al. 1997) showed a similar decline. Using testicular tissue from 528 middle-aged Finnish men who died suddenly, it showed that among the men who died in 1981, 56.4% had normal sperm production, but by 1991 this figure had dropped to 26.9%. Moreover, the average weight of the men's testes also decreased over the decade, while the proportion of useless fibrous testicular tissue increased at the expense of the sperm-producing seminiferous tubules. A Scottish study (Irvine et al. 1996) showed that men born in the 1970s were producing some 24% fewer motile sperm in their ejaculate than men who were born in the 1950s. In addition to the drop in sperm count, there was an increase in testicular cancers during this time (Figure 6.12).

Recent data from Jørgensen and colleagues (2012) have brought both good and bad news. The good news is that in the past 15 years (to 2012), there has been a slight increase in both the number of sperm ejaculated and the average sperm concentration. The bad news is that of the 4867 men studied, only 23% had "normal" sperm numbers. Moreover, over 90% of the sperm had abnormal morphology and only 69% were motile. The authors suggest that this paucity of functional sperm is responsible for the

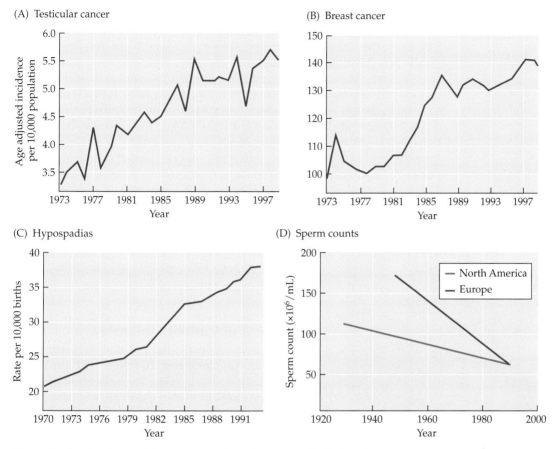

Figure 6.12 Developmental estrogen syndrome is manifest in climbing breast cancer rates and testicular dysgenesis. The rise of testicular cancers (A) parallels the rise of breast cancers (B) and abnormalities of penis development such as hypospadias (the failure to completely close the penis) (C). Sperm counts among North American males, moreover, have declined nearly 50% within the past century (D); the decline has been even steeper among European men. (After Sharpe and Irvine 2004.)

prolonged wait to have a baby in one-quarter of Danish families, as well as the increased use of fertility clinics.

Skakkebaek (2004) has analyzed these data and has hypothesized that there exists a **testicular dysgenesis syndrome** (Greek, *dys*, "bad" + *genesis*, "formation") characterized by disorganized testis development, testicular germ cell tumors, and low sperm count. Moreover, Sharpe (1994) has suggested that this syndrome may be due, at least in large part, to endocrine disruptors. While the chain of causation has not been completely established, subsequent studies have provided further evidence that the components of this syndrome can be caused by environmentally relevant concentrations of endocrine disruptors (Juul et al. 2014).

Many endocrine disruptors, including dioxins, nonylphenol, BPA, acrylamide, and certain pesticides and herbicides, can adversely affect testes morphology and sperm production (see Aitken et al. 2004). The sunscreen

4-MBC, a camphor derivative (approved in Europe and Canada, but not in the United States or Japan), binds to ERβ and has also been found to decrease the size of testes and prostate glands, and it can delay puberty in male rats (Schlumpf et al. 2004).

Indeed, many aspects of testicular dysgenesis syndrome (including all of the developmental anomalies, but not the testicular tumors) can be induced in the laboratory by administering phthalate derivatives to pregnant rats (Fisher et al. 2003; Mahood et al. 2007). Male rodents exposed in utero to dibutyl phthalate had an extremely high rate (>60%) of cryptorchidism (undescended testes), hypospadias (improperly formed penis), low sperm counts, Leydig cells trapped in the seminiferous tubules, and other testis abnormalities—conditions very similar to human testicular dysgenesis syndrome. Other evidence, obtained from newborn human males, suggests that those babies exposed in utero to relatively high phthalate levels had morphological changes in their testes (Swan et al. 2005; Huang et al. 2008). In response to these types of findings, there is increasing regulation to limit exposure to the phthalate chemicals.

Pesticides and Male Infertility

The link between pesticides and male infertility has been known for a long time (Carlsen et al. 1992; Colborn et al. 1996). Data from Swann and colleagues (2003, 2006) showed lower sperm counts in men living in agricultural regions, where pesticide and herbicide use is prevalent, than in men living in urban areas. Epidemiological evidence also associates the consumption of fruits and vegetables having high levels of pesticide residues with lower sperm count and a lower percentage of morphologically normal sperm (Chiu et al 2015). A recent quantitative study of toad populations throughout Florida demonstrated that the frequency of testicular abnormalities and intersex gonads increased linearly with the amount of agricultural activity (McCoy et al. 2008). In highly agricultural sites, testosterone production in male toads had declined so significantly that the sexual dimorphism in skin color had vanished, with males looking the same as the females.

The fungicide vinclozolin, widely used on vineyard grapes, is an androgen antagonist, inserting itself into the androgen (i.e., testosterone) receptor and preventing testosterone from binding there (Grey et al. 1999; Monosson et al. 1999). Rat embryos experimentally exposed to high concentrations of this chemical during the period of sex determination have malformations of the penis, absent sex accessory glands, and very low (around 20% of normal) sperm production (Figure 6.13;

(A)

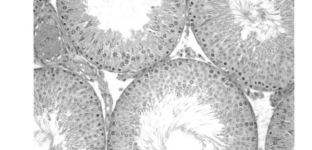

(B)

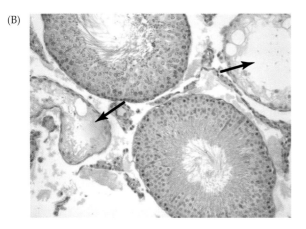

Figure 6.13 Cross section of seminiferous tubules from the testes of a control rat and a rat whose grandfather was born from a mother who had been injected with vinclozolin. (A) Normal seminiferous tubules are present in the control rat. (B) In the vinclozolin-affected rat, many seminiferous tubules (two are indicated by arrows) lack sperm-producing cells. (Photographs courtesy of Michael Skinner.)

Anway et al. 2005, 2006). This exposure not only impaired the fertility of males in the generation exposed in utero, but also impaired fertility in males for at least three generations afterward. We will discuss this transgenerational effect in detail later in the chapter. But to end our discussion of agriculturally prominent estrogenics, we return to the discussion of the controversial herbicide atrazine.

Atrazine, again

Discussed in regard to sex determination in Chapter 2, the estrogenic herbicide atrazine is the second-largest-selling weed killer in the world (Miller et al. 2000; Capel and Larson 2001; Sass and Colangelo 2006)—this in spite of the fact that its use is banned in Europe (including in Switzerland, the headquarters of the company that manufactures it). Most sales of atrazine are in the United States, where nearly 80 million pounds are applied to the soil annually (USGS 2014). Atrazine is the world's most common herbicidal pollutant of groundwater and surface water, and it is extremely stable.

REPRODUCTIVE FAILURES   As mentioned in Chapter 2, atrazine induces the enzyme aromatase, which converts testosterone into estrogen (estradiol). In several classes of vertebrates—fishes, amphibians, and laboratory mammals—the decrease in testosterone results in "chemical castration" in the form of low sperm count, reduced fertility, and smaller male sex organs. In those animals whose sex determination can be affected by hormones, males are known to have been converted into females or hermaphrodites (organisms having both sexes in the same body) by exposure

---

Sensitivity to Disruption: A Genetic Component

Some scientists have argued that claims about endocrine disruptors affecting reproduction are exaggerated, citing tests on mice that indicate that litter size, sperm count, and gonadal development are not affected by *environmentally relevant* concentrations of environmental estrogens. In fact, it is difficult to determine the concentrations of endocrine disruptors that would cause testicular dysgenesis in a majority of human males. One reason is the genetic heterogeneity of human populations.

Mice are inbred, and each genetic strain responds differently. Investigations by Spearow and colleagues (1999) have shown a remarkable genetic difference in sensitivity to estrogen among these different strains of mice. The CD-1 strain of laboratory mice used for testing environmental estrogens is at least 16 times more resistant to endocrine disruption than are the more sensitive mouse strains, such as B6. When estrogen-containing pellets were implanted beneath the skin of young male CD-1

mice, very little happened. However, when the same pellets were placed beneath the skin of B6 mice, their testes shrank and the number of sperm seen in the seminiferous tubules dropped dramatically (see figure). This widespread range of sensitivities, which can vary for different tissues in the same strain of mice (see Wadia et al. 2007), has important consequences for determining safety limits for humans. Safe levels could also be related to human gene polymorphisms, meaning that some men will be highly resistant to these chemicals whereas others could be sensitive. This genetic heterogeneity of human populations was confirmed by studies of human foreskin fibroblast cells. These cells (which are responsive to steroid hormones and are relatively easy to acquire) differ in their response to BPA, depending on the individual (Qin et al. 2012).

(Continued on next page)

Sensitivity to Disruption: A Genetic Component *(continued)*

These considerations confirm several important principles of ecological developmental biology. The first is that genetic differences can be critical in how a developing organism responds to environmental cues. The second is that these differences exist (as Darwin said they would) among individuals of the same species. The third is that testing for endocrine disruption is challenging, since different species may have evolved different means for coping with or using environmental cues, and the choice of an experimental model is probably critical in the detective work needed to see if disruption is occurring.

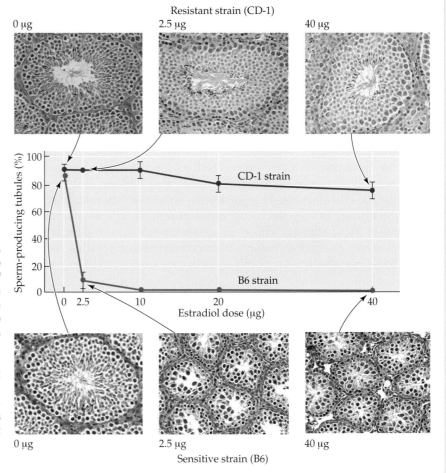

Varied effects of estrogen implants on different strains of mice. The graph shows the percentage of seminiferous tubules containing elongated spermatozoa. (The mean ± standard error is for an average of six individuals.) The micrographs show cross sections of the testicles (all at the same magnification). Although 40 mg of estradiol did not affect spermatogenesis in the CD-1 mice, as little as 2.5 mg almost completely abolished spermatogenesis in the B6 strain. (After Spearow et al. 1999; photographs courtesy of J. L. Spearow.)

to estrogenic compounds. Hayes and colleagues (2003) have documented that atrazine causes the formation of oocytes within what had been the testes (see p. 57), and they have established a probable chain of causation for the reproductive problems of frog populations exposed to atrazine (Figure 6.14). Like the story of DDT and many other widely used endocrine-disrupting chemicals, this chain of causation reflects the difficult issue of how human political and social needs and agendas interact with geological and biological agents.

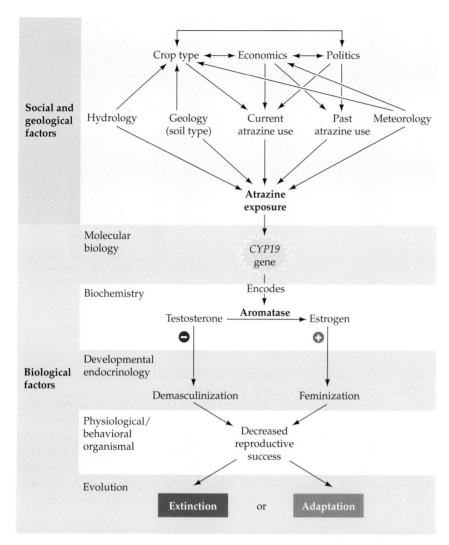

Figure 6.14 Possible chain of causation leading to the demasculinization of male frogs and the decline of frog populations in regions where atrazine use is prevalent. Both social and biological agents contribute to the chain. Sanderson and colleagues (2000) have shown that transcription of the human *CYP19* gene that encodes aromatase is induced by herbicides such as atrazine. (After Hayes 2005.)

IMMUNOSUPPRESSION   Ecologically, one of the most impressive dangers of atrazine involves its immunosuppressive effects. The chemical has been shown to produce an immunodeficiency syndrome in amphibians. Many amphibians exposed to atrazine (which they absorb through their skin) lack the ability to fight off common pathogens such as bacteria, fungi, parasites, and viruses. Just as cancer chemotherapy or HIV infection prevents humans from defeating the normal opportunistic pathogens in the environment, so atrazine-induced immunosuppression prevents amphibians from fighting off waterborne opportunistic pathogens (Kiesecker 2002; Christin et al.

Figure 6.15 Atrazine-induced immunodeficiency in frogs. (A) Meningitis in a leopard frog due to immune system breakdown following a "cocktail" of pesticides, including atrazine. The bacteria that cause this disease are always present, but a normally functioning immune system can exclude them. (B) Exposure of frogs to 21 parts per billion atrazine (or to pH 5.5 water, a known immune disruptor in frogs) resulted in reduced immune stimulation of frog white blood cells. (A courtesy of T. Hayes; B after Brodkin et al. 2007.)

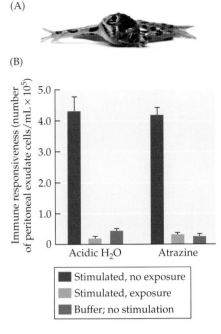

2003; Gendron et al. 2003; Forson and Storfer 2006a,b; Hayes et al. 2006; Brodkin et al. 2007; Paetow et al. 2013; Figure 6.15).

Evidence is accumulating that exposure to atrazine and other endocrine disruptors causes immune deficiencies in mammals as well (Filipov et al. 2005; Karrow et al. 2005; Rowe et al. 2006). Here, however, the spectrum of disease includes not only immunosuppression but also hyperpotentiation (allergies). When Rowe and colleagues (2006) exposed mice to atrazine for the last part of their gestation and suckling, their adult immune function included hypersensitivity to certain common environmental bacteria.

Regulating Fat Stem Cells: The Organic Tin Model

We mentioned above that DES, in addition to causing reproductive abnormalities, is also an obesogen, a compound that predisposes one toward becoming fat. The obesogenic effect has been seen with several other estrogenic endocrine disruptors* (Rubin et al. 2001; vom Saal et al. 2012), including tributyltin (TBT). This organic tin compound had been used as an antifouling agent for the hulls of ships, until it was found to be an endocrine disruptor, converting testosterone into estrogen and changing the sex of mollusks living near shipyards (Oberdörster and McClellan-

---

*Although several genetic loci have become associated with obesity, their total contribution to the average person's body weight is less than 5% (Loos 2009; Loos and Yeo 2014). Environment appears to be much more important than genes. Obesogens active during prenatal development may be critically important in causing weight gain later in life, and these compounds include DES, BPA, phthalates, and possibly high fructose corn syrup (Goran et al. 2013).

Green 2002). TBT is still used in fungicides, wood preservatives, and heat stabilizers in plastics. When ingested by a pregnant mouse, TBT makes her pups obese. It does this by activating a transcription factor, PPARγ (Evans et al. 2004; Janesick and Blumberg 2011, 2012). When activated in the liver, PPARγ activates the genes involved in fat synthesis and storage. TBT not only activates the transcription factor here, but also appears to demethylate the region of DNA regulating the *PPARγ* gene, making PPARγ even more abundant in the cell (Grün et al. 2006; Kirschner et al. 2010).

But the main developmental effect of TBT's activation of *PPARγ* is its effect in mesenchymal stem cells. When it is activated there, PPARγ instructs the descendants of these cells to become adipose (fat) cells instead of bone or cartilage cells, the two other main derivatives of the mesenchymal stem cell (Figure 6.16). By altering the commitment of mesenchymal cells toward the adipose fate, TBT causes the embryo to make a larger number of fat cells, and by activating the fat synthesis and storage genes, it makes them functional adipocytes. Estrogen mimics such as DES appear to act in a similar manner, activating the synthesis of PPARγ in mesenchymal stem cells (Hao et al. 2012). If the number of fat cells at birth is a major factor of adult obesity (see Janesick and Blumberg 2011, 2012), then endocrine disruptors may be a significant part of the current epidemic of obesity in the industrialized world.

Plastics and Plasticity

All human fetuses and adults have substantial levels of estrogenic endocrine disruptors in their circulatory systems. Many of these chemicals probably come from the contamination of food products with material

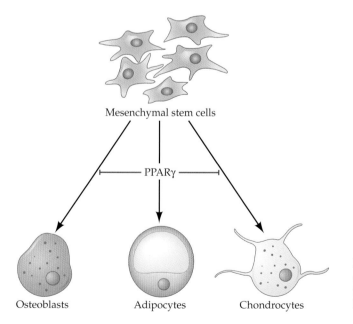

Figure 6.16 Postulated mechanism for obesity due to the activation of PPARγ, which biases mesenchymal stem cells to become adipocytes and activates the fat-storing enzymes in these cells. (After Kirschner et al. 2010.)

used in the stabilizing or hardening of the plastics in which our foods are stored, cooked, and served.

The discovery of the estrogenic effect of plastic stabilizers was made in a particularly alarming way. Investigators at Tufts University School of Medicine had been studying estrogen-responsive tumor cells. These cells are inhibited from proliferating when exposed to serum proteins; estrogen overcomes this inhibition. The studies were going well until 1987, when the experiments suddenly went awry. The control cells began to show high growth rates suggesting stimulation comparable to that of the estrogen-treated cells. It was as if someone had contaminated the control culture medium by adding estrogen to it. But where could such contamination possibly be coming from? After spending 4 months testing all the components of their experimental system, the investigators found that the source of the estrogenic factor was the plastic containers that held the water and serum. The company that made the containers refused to describe its new process for stabilizing the polystyrene plastic, so the researchers had to discover it for themselves.

The culprit turned out to be *p*-nonylphenol, a compound that is used to harden the plastic of the pipes that bring us water and to stabilize the polystyrene plastics that hold water, milk, orange juice, and other common liquid food products (Soto et al. 1991; Colborn et al. 1996). This compound is also a degradation product of detergents, household cleaners, and contraceptive creams. Nonylphenol has been shown to alter reproductive physiology in female mice and to disrupt sperm function. It is also correlated with developmental anomalies in wildlife (Fairchild et al. 1999; Kim et al. 2002; Adeoya-Osiguwa et al. 2003; Smith and Hill 2006; Kurihara et al. 2007). Unfortunately, nonylphenol is only one of many components of plastics that can be an estrogenic endocrine disruptor.

Bisphenol A

One of the most ubiquitous plasticizing compounds is **bisphenol A** (**BPA**). In the early years of hormone research, when the actual steroid hormones were difficult to isolate, chemists looked for synthetic analogues that would accomplish the same tasks. The pioneering British chemist Sir Charles Dodds discovered that BPA was estrogenic in 1936. Later, polymer chemists realized that BPA could be used in plastic production, and today it is one of the most abundant chemicals in production worldwide. (Interestingly, BPA is therefore a chemical whose endocrine effects were known *before* its use in plastics.) Four corporations in the United States make almost *2 billion pounds* of it each year for use in the resin lining in food cans, in the polycarbonate plastic used in bottles and children's toys, and in dental sealant. Its modified form, tetrabromobisphenol A, is the major flame retardant coating the world's fabrics.

BPA ingested by pregnant rats appears to pass readily into the fetus and is not hindered by steroid-binding hormones. Within an hour of ingestion, BPA is found in the fetus at the same doses that it had been in the mother (Miyakoda et al. 1999). Human exposure comes primarily from

BPA that has leached from food containers (von Goetz et al. 2010). Babies and infants acquire BPA through polycarbonate bottles, while teenagers and adults get most of their BPA through the consumption of canned food that had been stored in containers lined with BPA-containing resins. BPA crosses the human placenta and accumulates in concentrations that can alter development in laboratory animals (Ikezuki et al. 2002; Schönfelder et al. 2002). Moreover, since 93% of urine samples taken from people in the United States and Japan have had measurable BPA levels (Calafat et al. 2005; Nelson et al. 2012), public health concerns have been raised over the roles BPA might play in causing reproductive failure, cancer, and behavioral anomalies (vom Saal and Hughes 2005; Figure 6.17). Most critically, children have higher concentrations of BPA in their blood than do adolescents or adults.

BPA AND REPRODUCTIVE HEALTH   We have mentioned the ubiquitous presence of BPA in plastics. However, studies by several laboratories have demonstrated BPA is not fixed in plastic forever (Krishnan et al. 1993; vom Saal 2000; Howdeshell et al. 2003). If water sits in an old polycarbonate rat cage at room temperature for a week, the water can acquire levels of BPA that

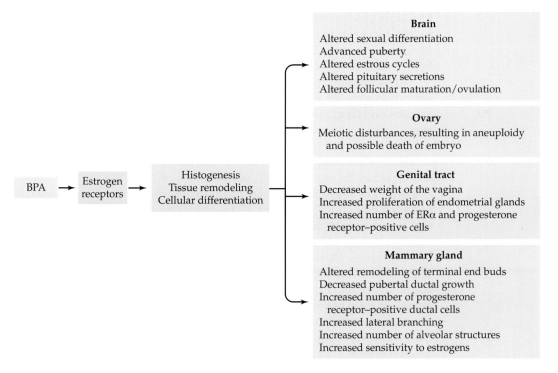

Figure 6.17   Some effects of perinatal BPA exposure. Schematic representation of the effects of BPA on adult mice. BPA works through the estrogen receptors, thereby activating an assortment of genes. The products of these genes can disrupt cell-cell and cell-matrix associations, leading to altered organ construction and cell division in the various organs. (After Murray et al. 2006.)

(A)

(B)

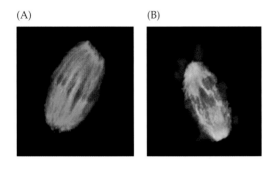

Figure 6.18 Bisphenol A causes meiotic defects in maturing oocytes. (A) During the first meiotic metaphase, chromosomes (stained red) normally line up in a paired fashion at the center (equator) of the spindle. (B) Brief exposure to BPA causes chromosomes to align randomly on the spindle. Different numbers of chromosomes then enter the egg and polar body, resulting in chromosomal aneuploidy (incorrect number of chromosomes in the daughter cells of meiosis) and infertility. (From Hunt et al. 2003, photographs courtesy of P. Hunt.)

will cause weight changes in the uteruses of young mice. This leaching from plastic can cause chromosome anomalies. When a laboratory technician in Patricia Hunt's laboratory mistakenly rinsed some polycarbonate cages in an alkaline detergent, there was significant leaching of BPA. The female mice housed in these cages showed meiotic abnormalities in 40% of their oocytes (the normal level of such anomalies is about 1.5%). When BPA was administered to pregnant mice under more controlled circumstances, Hunt and colleagues (2003) showed that a short, low-dose exposure was sufficient to cause similar meiotic defects in maturing mouse oocytes (Figure 6.18). This production of meiotic anomalies was also seen in primates. Exposure of fetal female monkeys to low-dose BPA (at a level comparable to that found in human serum) caused several disturbances of ovarian function, including abnormal meiotic chromosome behavior and aberrant follicle formation (Hunt et al. 2012).

Not surprisingly, then, female mice exposed to low doses of BPA in utero had reduced fertility and fecundity as adults (Cabaton et al. 2011). Studies on laboratory animals indicate that BPA at environmentally relevant concentrations can cause disruptions in the morphology of the fetal sex organs, as well as low sperm counts (vom Saal et al. 1998; Palanza et al. 2002). Moreover, BPA also appears to affect the sexual biology of the rodent brain. In rats and mice, certain areas of the brain are different in males and females, and there are behavioral differences that are more stereotyped than those in humans. Kubo and colleagues (2003) demonstrated that low-dose exposure to BPA in utero can destroy the sex differences in the developing rat brain and elicit behavioral changes when these fetuses become adults. Moreover, BPA (and other endocrine disruptors) are found to prevent the sex-specific maturation of those parts of the mouse brain regulating ovulation (Narita et al. 2006; Rubin et al. 2006; Gore et al. 2011). Female mice exposed in utero to low doses of BPA (2000 times lower than the dosage considered "safe" by the U.S. government) had alterations in the organization of their uterus, vagina, breast tissue, and ovaries, as well as altered estrous cycles as adults (Howdeshell et al. 1999, 2000; Markey et al. 2003). Such experiments cannot be done on humans; but women exposed to high levels of BPA during pregnancy had an 83% higher rate of miscarriages than women who had not been so heavily exposed (Lathi et al 2014).

BPA AND CANCER SUSCEPTIBILITY   Female mouse embryos exposed to BPA in utero have shown a number of developmental and physiological changes: altered mammary development that manifested during the period of exposure as morphological alterations of the mesenchymal stroma and epithelial ducts (Vandenberg et al. 2007); alterations of the gene expression profiles of both compartments (Wadia et al. 2013) at puberty; and alterations in the organization of the mammary tissue and ovaries and altered estrous cyclicity as adults (Markey et al. 2003). Each mammary gland of the prenatally exposed mice also produced more terminal buds and was more sensitive to estrogen. Murray and colleagues (2007) showed that fetal exposure to BPA caused the development of carcinoma in situ (early-stage cancer) in the mammary glands of 33% of the mice exposed in utero to low levels of BPA. None of the control rats developed such tumors. Moreover, fetal exposure to BPA increased the number of "preneoplastic lesions" (areas of rapid cell growth within the ducts) three- to fourfold. Furthermore, gestational exposure to as little as 25 milligrams BPA (per kilogram each day), followed at puberty by a "subcarcinogenic dose" of a chemical mutagen, resulted in the formation of tumors only in the animals exposed to the BPA (Durando et al. 2007; Vandenberg et al. 2007). Studies on fetal and adolescent rats showed that perinatal exposure to environmentally relevant doses of BPA causes changes in anatomical patterns of mammary gland development at puberty (Muñoz-de-Toro et al. 2005; Vandenberg et al. 2007). They also found that gestational exposure to BPA induces conditions that can lead to tumors when a second exposure of estrogenic hormones or carcinogens is experienced later in life (see Figure 6.17).

In the above experiments, BPA was shown to be a factor that predisposed the rats to develop a cancer when they encountered estrogenic chemicals later in life. However, new studies with a different strain of rats have shown that when rat embryos are exposed to relatively small doses of BPA (considered safe by the EPA), they can develop palpable tumors 50 days after birth (not needing a second dose later in life). Acevedo and colleagues (2013) conclude, "BPA may act as a complete mammary gland carcinogen."

Especially noteworthy is the fact that in all of the above studies, the effective dosage of BPA was extremely low—as much as *2000 times lower* than the dosage set as safe by the U.S. government. When BPA was given to pregnant rhesus monkeys to achieve the levels of BPA found currently in American women, similar changes in the density of mammary gland buds were found in the fetal female monkey mammary tissue (Tharp et al. 2012; Figure 6.19).*

---

*This may have implications for environmental justice. Most hazardous waste landfills and incinerators are located in communities with large minority populations, and such groups "experience higher than average exposures to selected air pollutants, hazardous waste facilities, contaminated fish, and agricultural pesticides in the workplace" (GAO 1983; EPA 1992, 2011; quotation from EPA, 1992). Landigran and colleagues (2010) note, "The concept of environmental injustice has been further elaborated in studies examining ethnic disparities in exposures to automotive exhaust and ambient air pollution; in studies in New York City documenting that virtually all diesel bus depots, places at which buses may idle for hours while emitting pollutants are located in minority, mostly disadvantaged neighborhoods; in studies examining disparities in housing quality; and in studies of residential proximity to polluting industrial facilities. Environmental injustice has been well documented in occupational settings. It has served as an operational concept to guide pollution prevention programs."

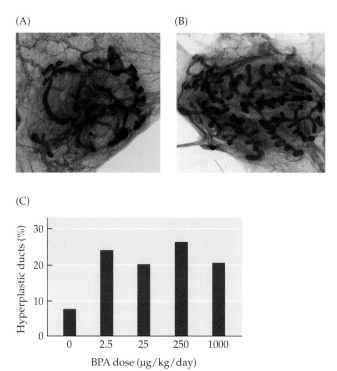

(A)

(B)

(C)

Figure 6.19 Bisphenol A induces altered mammary gland development. (A,B) Whole-mount stained preparation of mammary glands from newborn female rhesus monkeys. (A) Control mammary gland. (B) Mammary gland from fetus exposed in utero to BPA. Twice as many buds (incipient branches) are seen. (C) The percentage of mouse mammary glands showing intraductal hyperplasia (indicating a cancer-prone state) is significantly increased at postnatal day 50 in BPA-treated animals. (A,B from Tharp et al. 2012; C after Murray et al. 2007.)

Females aren't the only sex affected by BPA. The chemical has also been implicated (along with other estrogenic compounds) in the decline in sperm quality and density. Several of these investigations also suggest that BPA may result in an increased incidence of prostate enlargement in men; in the United States, 65% of men in their 60s have enlarged prostate glands, and 1 out of every 6 American men develop prostate cancer sometime in their lives. Studying the mouse prostate gland (whose size is sensitive to estrogens), vom Saal and colleagues (1997) found that when they gave pregnant mice 2 parts per billion BPA—that is, 2 *nanograms* per gram of body weight—for 7 days at the end of pregnancy (equivalent to the period when human reproductive organs are developing), the male progeny showed an increase in prostate size of about 30%. The sperm count of these mice was also lowered. Increased exposure to BPA is associated with increasing risks of early-onset prostate cancer in humans. BPA increases the rate of mitosis and chromosome anomalies in human prostate cells (Wetherill et al. 2002; Timms et al. 2005; Tarapore et al. 2014).

BPA'S EFFECTS ON THE NERVOUS SYSTEM   Prenatal exposure of rhesus monkeys to relatively low levels of BPA during the final 2 months of gestation produced abnormalities in the fetal ventral mesencephalon and hippocampus (Leranth et al. 2008; Elsworth et al. 2013, 2015). Specifically, BPA delivered to the fetus at a dosage considered safe by the U.S. government induced a decrease in the number of dopaminergic neurons in the midbrain of these BPA-exposed fetuses, as well as a reduction in the number and efficiency of dendritic connections in the hippocampus. Because remodeling of these synapses is thought to play a critical role in cognition and mood, the ability of BPA to interfere with synapse formation has profound implications.

Indeed, BPA has been associated with behavioral changes in zebrafish and mammals. It is not surprising that BPA, as it was synthesized as a synthetic estrogen, affects behavior. In zebrafish, the binding of BPA to the estrogen receptor activates the genes producing aromatase, the enzyme that converts testosterone into estrogen. Male fish exposed to synthetic estrogens become intersexual and cannot reproduce (Kidd et al. 2007). BPA appears to activate aromatase gene expression in the same set of neurons as estrogen (Chung et al. 2011).

Exposure of rodents to BPA around the time of birth (while the brain is developing) modifies sex differences in the brain (Patisaul et al. 2006; Rubin et al. 2006). In the offspring of monkeys exposed in utero or perinatally to BPA, males displayed fewer social behaviors (Nakagami et al. 2009). In humans, BPA exposure in the uterus has been associated with hyperactivity and aggression in 2-year-old children (Braun et al. 2009) and with anxiety and depression in older children (Braun et al. 2011). As we will see later, such BPA-induced behaviors in mice may be transmitted from generation to generation.

The dose-response curve of BPA action

Many of BPA's effects are observed at extremely low concentrations, and higher concentrations often do not exert the same effects. This strange dose-response relationship is probably a function of the nature of endocrine disruptors, which have a different dose-response curve from that of classic teratogens such as alcohol, thalidomide, and heavy metals. Classical teratology, based on the principles of toxicology, operates on the principle that "the dosage makes the poison" and that there is a linear relationship between the dosage of the toxic substance and the severity or the incidence of the abnormality. Toxicity is traditionally reported as an LD50 ("lethal dose," the concentration at which 50% of the animals die) or ED50 ("effective dose," at which 50% of the animals show an effect).

Such linear effects are not seen with most endocrine disruptors. Instead, there is usually an increasing effect (typically the developmental anomalies described above) from low to medium concentrations and then the gradual *disappearance* of this effect at higher concentrations (Figure 6.20A). That is to say, the dosage relationship is not linear, and the abnormalities might not even occur at high concentrations of the chemical. This sort of response is seen in the effects of estrogen on cultured human breast cancer cells (Welshons et al. 2003), on the induction of terminal end buds in the

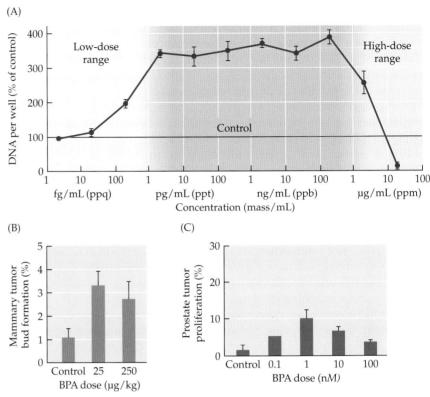

Figure 6.20 Endocrine disruptors often have a non-monotonic dose-response curve. (A) The inverted U shape of a typical non-monotonic curve indicates that medium-level dosage has the greatest effect and that neither low nor high doses of the compound produce as striking an effect. (B,C) This type of dose-response curve is revealed in histograms showing the effects of BPA on mammary terminal end bud formation (B) and prostate tumor growth (C). (A after Wellshons et al. 2003; B after Markey et al. 2001; C after Wetherill et al. 2002.)

mammary gland (Figure 6.20B; Markey et al. 2001), and in prostate tumor proliferation (Figure 6.20C; Wetherill et al. 2002). The curve seen in Figure 6.20A is often referred to as an **inverted U response**. Such a response is expected when there are negative feedback loops, such that high levels of compound can actually inhibit the effect, or when there are two or more different effects, such that the effects at the higher concentrations inhibit each other. This inverted U response has been seen for numerous endocrine disruptors, including BPA, atrazine, vinclozolin, and the phthalate DEHP, as well as for the estrogen receptor–binding drug tamoxifen (Fagin 2012; Vandenberg et al. 2012, 2013). The inverted U response is very important in testing the safety of any compound, since testing at high levels (standard in toxicology and teratology) does not necessarily reveal the compound's effect on development (see vom Saal and Welshons 2006).

The molecular biology of the BPA effect

The effects of BPA may be mediated by different estrogen receptors in different tissues. In some tissues, the effects of BPA seem to be mediated by

ERβ. BPA may cause its effects on meiosis by interfering with the normal functioning of ERβ. We noted earlier the work of Hunt's lab showing that BPA exposure could lead to a high degree of aneuploidy (an abnormal number of chromosomes in the daughter cells produced by meiosis) in mouse oocytes (see Figure 6.18). Hunt's group has recently shown that a knockout of ERβ gives the same phenotype of aneuploidy as does adding BPA (Susiarjo et al. 2007). Moreover, BPA does not increase aneuploidy in this ERβ knockout, indicating that BPA causes its effects by somehow inactivating ERβ in wild-type mice. This inactivation could be produced by interference with estradiol binding, or perhaps by binding to a yet un-described, meiosis-specific protein that acts through ERβ.

The importance of ERβ for mediating the effects of BPA is also seen in the human islets of Langerhans, which include the insulin-secreting cells of the pancreas. BPA has been seen to alter the function of these pancreatic cells at a low dose (1 nM, less than the doses producing the same effect in mice). Interestingly, the effect is so rapid that the receptor may be binding BPA on the cell membrane and not working its affect through gene action (Soriano et al. 2012).

The induction of gene expression by BPA can be seen in living fish. There had been controversy over whether BPA would be able to activate estrogen-responsive genes at environmental concentrations. While the appearance of certain phenotypes in mice (obesity, cancers) indicated that it could, strong visual evidence was also obtained with transgenic zebrafish wherein estrogen-responsive enhancers were linked to a gene for green fluorescent protein (Gorelick and Halpern 2011; Lee et al. 2012). The enhancer would be activated wherever an estrogenic compound bound to its receptor. The enhancer was activated by BPA as well as by water samples from rivers from the Mid-Atlantic United States (Gorelick et al. 2014; Figure 6.21). As mentioned earlier in the chapter, ERRγ appears to be responsible for the otic (auditory) malformations caused in zebrafish by BPA.

Some of the targets of BPA's activation of estrogen receptors have been identified. The DNA methylation patterns of numerous genes, including those encoding several transcription factors, are altered by fetal exposure to BPA (Dolinoy et al. 2007; Susiarjo et al. 2013). One of the targets is *Hoxa10*, one of the genes involved

(A) DES exposed

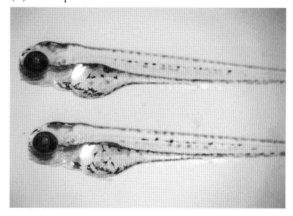

(B) BPA exposed

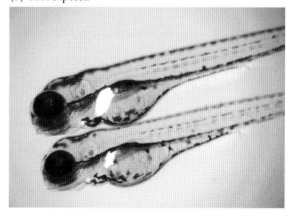

Figure 6.21 Activation of the estrogen receptor enhancer by DES and BPA. A gene construct fusing the enhancer regions of zebrafish estrogen receptors to the gene for green fluorescent protein causes the gene to be expressed when exposed to (A) 100 nM DES or (B) 2.3 mg/L BPA. The very bright area is the liver (which is activated by the ERα receptor), and the smaller glowing spots beneath it are heart valves (which are activated by the ERβ receptor). The control embryos (that are not exposed to either compound) lack any bright spots. (After Gorelick and Halpern 2011.)

in uterine development. Here (in an effect opposite to that of DES), BPA down-regulates methylation and increases the amount of Hoxa10 being produced (Bromer et al. 2010). Another one of the targets of BPA's activation of estrogen receptors is a long noncoding RNA called HOTAIR. This RNA recruits the enzymes (such as those of the polycomb repressive complex) required for gene silencing to the promoters of target genes, thereby leading to specific gene silencing. HOTAIR is expressed in many types of tumor cells, and it is necessary for the survival of breast cancer cells. It is transcriptionally activated by estradiol. However, recent studies (Bhan et al. 2014) have shown that HOTAIR is also up-regulated by low concentrations of DES and by BPA in the mouse mammary gland. Like estradiol, DES and BPA alter the histone acetylation and methylation status at the HOTAIR promoter.

The transcriptome of the mammary gland also changes, and this appears to be mediated primarily by ERα. This estrogen receptor is predominantly in the mesenchymal stroma of the mammary gland, and BPA up-regulates several stromal genes (such as one for a paracrine factor that induces epithelial cells to become mesenchymal cells) that may cause changes in the overlying epithelial cells (Wadia et al. 2013). Thus, as we will see later, in the chapter on cancer, carcinogens can elicit cancers by damaging the interactions between the mesenchyme and epithelial compartments of a tissue.

BPA exposure also alters the chromatin of genes that are known to be important in nerve transmission and in prostate cancer susceptibility. For instance, BPA alters the chromatin acetylation of the gene encoding the potassium chloride transmitter that directs sexually dimorphic nerve transmission (Yeo et al. 2013). This may be critical in the ability of BPA to alter sexual behaviors. Also, there appears to be a pathway connecting perinatal BPA exposure to the development of prostate cancer or breast cancer later in life (Ho et al. 2006; Doherty et al. 2010). BPA-exposed rats exposed to elevated testosterone and estrogen during adult life were more likely than unexposed animals to develop a type of cancerous lesion in the prostate. This rat tumor corresponds to an early stage of prostate cancer in human males. In this case, BPA (or estrogen) has been seen to reduce the methylation of specific regions of the *PDE4D4* gene. *PDE4D4* encodes a phosphodiesterase that degrades cyclic AMP, a crucial regulatory molecule for normal cell growth and differentiation. It has been proposed that decreased methylation of *PDE4D4* allows for the production of greater than normal amounts of phosphodiesterase in prostate cells, which results in more cAMP degradation and thus less cAMP, with consequent abnormal cell proliferation and development (Ho et al. 2006).

Polychlorinated biphenyls

Polychlorinated biphenyls (PCBs) are a family of about 200 related chemicals that have been used since 1928 in the manufacture of transformers, capacitors, flame retardants, hydraulic fluids, insulating coolants, and liquid seals (Colborn et al. 1996). Studies with mammals have shown PCBs to be associated with reproductive abnormalities, neurological and cognitive deficits, and thyroid deficits, as well as cancers and immune dysfunctions

(Brouwer et al. 1999). PCBs have also been associated with abnormalities of neural development in animals, and abnormalities of neural and psychological development were seen in Taiwanese children born to mothers accidentally exposed to high levels of PCBs in cooking oil (Schantz et al. 2003; Guo et al. 2004). Although banned in the United States and most European countries since the 1980s, PCBs still contaminate landfills and waterways (Colborn et al. 1996; Domingo and Bocio 2007).

PCB exposure appears to affect reproductive development and has different impacts on males and females. In male mammals, studies have found that perinatal PCB exposure can cause cryptorchidism (failure of testes to descend), hypospadias (abnormalities of penis formation), testicular cancers, and low sperm count (Facemire et al. 1995; Den Hond and Schoeters 2006). In females, PCB exposure has led to a decrease in the age of menarche (onset of menstruation) and an increased incidence of endometriosis (Den Hond and Schoeters 2006).

The highest concentration of PCBs in humans is not found in city dwellers, but in the indigenous populations of the Arctic. Here, the food chain lacks primary producers (i.e., plants) and the diet is high in animal fat (AMAP 2002). Similar concentrations are seen in nonhuman animals high on the food chain. PCB concentrations in polar bears have been linked to malformations of the reproductive and genital organs (Christian et al. 2006). Studies in mammals also show decreased immune response after consumption of PCB-laden fish oils. In seals, decreased size of the thymus has been documented, which results in a reduced immune response (Yoshimura 2003; Ross 2004). PCB exposure affects thyroid function by influencing the hypothalamus-pituitary-thyroid axis through various mechanisms and leads to a decrease in thymus size in rhesus monkeys and other animals. Wang and colleagues (2005) have shown that in utero exposure to PCBs and dioxins causes feedback alterations in the hypothalamus, causing an overproduction of thyroid hormones. This in turn can cause the PCB-associated neurological disturbances that include deficits in visual recognition, short-term memory, and learning, as well as interference with sexual differentiation of the brain (Ribas-Fito et al. 2001; EPA-PCB 2008).

Possible mechanisms for the effects of PCBs

PCBs can affect the estrogen, androgen, thyroid, retinoid, corticosteroid, and other endocrine pathways (Brouwer et al. 1999). Some PCBs interrupt estrogen metabolism through the cytochrome P450 enzyme system, inducing aromatase and thus increasing production of estradiol from testosterone (Brouwer et al. 1999). These same cytochrome P450 enzymes convert adrenal progesterone into cortisol and aldosterone, two steroids that can induce hypertension (Lin et al. 2006). Several PCB metabolites appear to bind to both the estrogen receptor and thyroid hormone receptors (Brouwer et al. 1999). PCBs have a weaker affinity for sex hormone binding globulins (SHBGs), which are proteins that bind excess steroids, than does estrogen. This property leaves PCBs free to bind with estrogen receptors when endogenous estrogen is bound up by SHGB.

Unbound ("free") PCBs often interfere with the production or elimination of naturally occurring compounds, and this can cause damage in several ways. One effect is the decrease in follicle-stimulating hormone (FSH) levels, thereby inhibiting the production of male Sertoli cells and consequently the production of sperm (Mol et al. 2002). Another effect is induction of the liver enzymes uridine-5-diphosphate-glucuronyltransferases (UGTs), which promote the elimination and excretion of essential thyroid hormones. Exposure to PCBs is known to result in reduced plasma thyroid hormone (Brouwer et al. 1999). This can cause thyroid dysplasia and tumors (due to the lack of thyroid hormone inhibition on TSH, the hormone that instructs the thyroid follicle cells to divide).

PCBs also may directly affect the brain. Royland and colleagues (2008) showed that in utero exposure of rats to one particular PCB caused dramatic changes in the gene expression pattern in the hippocampus. The genes altered were predominantly those involved in intracellular signaling (such as the genes responsible for normal calcium signaling) and axonal guidance. These changes are consistent with the hypothesis that PCBs change neuronal growth patterns and that existing neurons do not function properly (Royland and Kodavanti 2008; Lilienthal et al. 2013).

Other endocrine disruptors affecting thyroid function

The thyroid gland is thought to be involved in regulating metabolism and maintaining energy balance, as well as maintaining the functions of the nervous, cardiovascular, skeletal, pulmonary, auditory, and reproductive systems.* Fetal mammalian neurogenesis depends on thyroid hormone, and there is evidence that endocrine disruptors are suppressing this important phenomenon (Demeneix 2014; Préau et al. 2015). In amphibians, thyroid hormones are largely responsible for the metamorphosis from tadpole to frog, where every organ system is modified. As expected for an agent that inhibits thyroid function, exposure of tree frog tadpoles to the contact antibiotic triclosan severely inhibits thyroxine-dependent metamorphosis (Marlatt et al. 2013).

Several endocrine disruptors can alter thyroid hormone production and function. In addition to triclosan, these include perchlorate, perfluorooctanoic acid (PFOA; used in the nonstick surfaces of frying pans and carpets), and parabens (like triclosan, widely used in antiseptic soaps and toothpastes). Exposing pregnant rats to triclosan lowers the maternal, fetal, and early postnatal concentrations of thyroid hormones, possibly by elevating the removal rate of thyroid hormone by the liver (Paul et al. 2012). For parabens, epidemiological research using the National Health and Nutrition Examination Survey found evidence of inverse relationships between parabens and thyroid hormone levels in adult women, but not in men. No

---

*As with estrogen, the major way thyroxine works is to bind to a receptor protein in the cytoplasm, converting it into an active transcription factor that activates or represses thyroxine-sensitive genes. And as with estrogen, there seems to be a membrane-associated pathway, as well. One of the major thyroxine receptors (TRβ) is tethered to the cell membrane. Here, it blocks the activation of Akt. When TRβ binds thyroid hormone, it leaves the membrane, allowing Akt to be active. Akt regulates the mTOR pathway (discussed in Chapter 8), which accelerates metabolism and growth (Martin et al. 2014).

correlations were seen with triclosan (Koeppe et al. 2013.) However, there was no assessment of prenatal exposure and adult disease.

Thyroid deficiencies inhibit the maturation of lymphocytes, the cells responsible for fighting infections. Immunodeficiencies associated with thyroid-depressing endocrine disruptors have been seen in animal populations and have been suggested for humans (Clayton et al. 2011; Jobling et al. 2013). As mentioned earlier, amphibian populations are threatened with decline, largely by parasitic infections, and these have been associated with immune dysfunction throughout the world. It is possible that low levels of endocrine disruptors in ponds are giving frogs an immunodeficiency syndrome that prevents them from defeating parasites (Gilbertson et al. 2003; Albert et al. 2007; Jobling et al. 2013). Indeed, PFOA causes immune dysfunction; it has been found to reduce the humoral immune response to routine childhood inoculations against diphtheria and tetanus and can prevent the immunizations that should result from inoculations (Grandjean et al. 2012).

Fracking: Hydraulic Fracturing and the Poisoning of the Wells

Hydraulic fracturing ("fracking") is the set of procedures used to extract natural gas (mostly methane) from shale. It involves the high-pressure injection of millions of gallons of water into the shale, along with over 750 chemicals and suspended solids. Of these chemicals, over 100 are suspected to be endocrine disruptors, and about one-quarter of them are known to cause tumors (Colburn et al. 2011). It is estimated that about 50% of the fracking fluid returns to the surface (DOE 2009). Under President Bush and Vice President Cheney, fracking was made exempt from the Safe Drinking Water Act, the Clean Water Act, and the Clean Air Act in the United States (Wiseman 2009). Moreover, the materials in the fluids used in fracking are not allowed to be classified as pollutants. These rules are currently being challenged.

There have been numerous reports of changes in ground and surface water near fracking regions, and the U.S. Environmental Protection Agency concluded that the domestic water supplies in certain towns near fracking sites have been severely contaminated. Independent analyses (Osborn et al. 2011; Jackson et al. 2013) found that gas contamination from fracking was common in the drinking water of those living within a kilometer of fracking sites. Moreover, water samples taken from the standing water and groundwater of fracking sites (Kassotis et al. 2014; Webb et al. 2014) revealed that they contained compounds that are estrogenic, antiestrogenic, and antiandrogenic (antitestosterone). There were compounds in the water that activated the enhancers of estrogen-responsive genes and that prevented the activation of testosterone-responsive genes (Figure 6.22). One of the sites tested had been used as a ranch prior to drilling and fracking, but ranching had to be discontinued because the animals were no longer producing offspring. Numerous studies strongly suggest that fracking operations can elevate levels of endocrine-disrupting compounds in nearby groundwater and surface water, and a recent study concerning fracking in rural Colorado documented an increased incidence of congenital heart

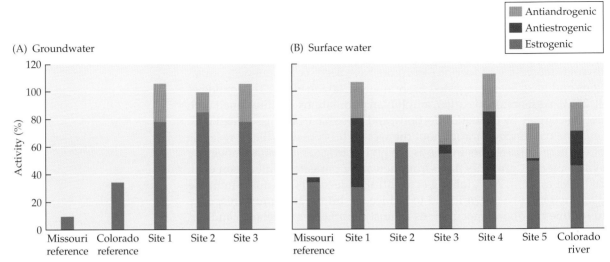

Figure 6.22 Combined estrogen and androgen receptor stimulation by groundwater just below the surface (A) and surface water (B) in creeks, lakes, and streams at various sites close to fracking activities in Colorado. The graphs represent the combined means of estrogenic (blue), anti-estrogenic (red), and anti-androgenic (green) activities at each sampling site. Estrogenic activities are measured as a percentage of the activation by 10 pM 17β-estradiol at 40% concentration; anti-estrogenic and anti-androgenic activities are based on the ability of the water sample to inhibit 17β-estradiol- and testosterone-mediated gene activation. The reference sites are not near fracking operations. (After Kassotis et al. 2014.)

disease in children born in families residing close to the fracking wells (McKenzie et al. 2014).

Transgenerational Effects of Endocrine Disruptors

One of the surprising lessons of the DES affair was that the effects of the endocrine disruptor lasted for at least another generation. Not only were the daughters of the women who took DES at risk for reproductive tract tumors and uterine abnormalities, *their daughters* were as well (Newbold et al. 1998, 2000.) In mice, both the female and the male grandpups of females exposed in utero to DES also had reproductive tract anomalies and tumors (Turusov et al. 1992; Walker and Kerth 1995; Walker and Haven 1997).

The idea that a deformity caused in one generation could be transferred to another generation through sexual reproduction goes against one of the most important principles of genetics—that acquired traits cannot be transmitted (see Appendix D). This "Weismannian block" to the transmission of acquired traits is a genetic block. A lifetime of chopping wood will not give your offspring bulging biceps; nor would the loss of one's arms in an accident cause one's offspring to have a propensity for limblessness. This is because the environmental agent does not cause mutations in the DNA. And mutations, if they are to be transmitted, must not only be somatic, they must enter the germline. So mutations acquired in the skin by being in sunlight will not be transmitted.

However, DNA methylation seems to be a mechanism that can circumvent that block. Certain agents can cause the same alterations of DNA methylation in many cell types throughout the body, and these alterations in methylation can be transmitted by the sperm and egg. Jablonka and Raz (2009) have documented dozens of such cases wherein different "epialleles," DNA containing different methylation patterns, can be stably transmitted from generation to generation. In mammals, epiallelic inheritance was first documented by studies of the endocrine disruptor vinclozolin, a fungicide used widely on grapes. When injected into pregnant rats during particular days of gestation, vinclozolin will cause testicular dysgenesis in the male offspring. The testes will start forming normally, but as the rat gets older, its testes will degenerate and no more sperm will be made. What's more disturbing is that the male rats fathered by rats that get testicular dysgenesis will also get testicular dysgenesis. So do their male offspring and the subsequent generation's male offspring (see Figure 6.13; Anway et al. 2005, 2006; Guerrero-Bosagna et al. 2010). Thus, when a pregnant mouse or rat is given vinclozolin, her "great-grandsons" will still be affected (Figure 6.23).

How does DNA methylation work this way? The promoters of over 100 genes in the Sertoli cells have their methylation pattern changed by vinclozolin, and altered promoter methylation can be seen in the sperm DNA for at least three subsequent generations (Guerrero-Bosagna et al. 2010; Stouder and Paolini-Giacobino 2010). These genes include those whose products are necessary for cell proliferation, G proteins, ion channels, and receptors. It is important to note that by the third (F_3) generation, there can have been no direct exposure to vinclozolin: The fetus (F_1 generation) is inside the treated dam; the fetus has germ cells (of the F_2 generation) inside itself. The offspring of the F_3 and F_4 generations have never been exposed to vinclozolin, but their phenotype has been changed by the initial injection to their great-granddam.

Since intrauterine vinclozolin exposure had been linked to several adult-onset disorders of numerous organs (testes, prostate, kidneys, nervous system, and immune system), it was hypothesized that the epialleles of the sperm genome influenced the development and function of many bodily organs. This has been found to be the case. When third-generation animals are compared with control animals whose lineage has not been exposed to vinclozolin, the DNA methylation patterns are seen to differ in all tissues studied. Moreover the transcriptomes (the types of mRNA) of the exposed and unexposed animals differ in each organ (Skinner et al. 2012).

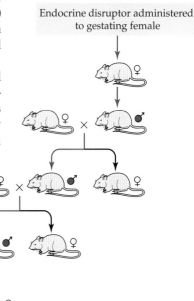

Endocrine disruptor administered to gestating female

Figure 6.23 Epigenetic transmission of endocrine disruption. Transmission of testicular dysgenesis syndrome (red circles) is shown through four generations of mice. The only mice exposed in utero were in the F_1 generation. (After Anway and Skinner 2006.)

Thus, it appears that the methylation pattern in the sperm DNA is being retained by the different cells of the body.

This conclusion was recently confirmed by a large study involving a mixture of three plastics-derived endocrine disruptors (BPA and the phthalates DEHP and DBP). Menikkam and colleagues (2013) showed that this mixture promoted transgenerational inheritance of testicular dysgenesis, polycystic ovaries, puberty anomalies, and obesity. In the F_3 generation (which had no exposure to the mix), altered DNA methylation was seen in the sperm DNA (Menikkam et al. 2013). The affected genes encoded proteins that could be linked in pathways that are involved in both sperm production and obesity.

Similar studies have indicated that BPA and other endocrine disruptors—DES and TBT—have transgenerational effects (Skinner et al. 2010; Walker and Gore 2011; Chamorro-Garcia et al. 2013), and these effects can literally be seen—as syndromes of familial obesity. Indeed, the administration of TBT to pregnant mice increased the amount of white adipose tissue, adipocyte size, and adipocyte number in the offspring, and it reprogrammed the mesodermal stem cells toward the adipocyte lineage at the expense of bone for at least three generations. The embryonic TBT exposure up-regulated the hepatic expression of genes involved in lipid storage/ transport, lipogenesis, and lipolysis, causing a syndrome resembling non-alcoholic fatty liver disease for three generations of mice (Chamorro-Garcia et al. 2013; Janesick et al. 2014).

Indeed, the behavioral changes induced by BPA in mice may last at least four generations. Wolstenholme and colleagues (2012) gave pregnant mice food containing BPA and measured levels of BPA in their blood that were within the range of those found in humans. The offspring were significantly less social than control mice (using metrics used to assess some aspects of autism in children). Figure 6.24 shows that two measures of behavior, side-by-side sitting (Figure 6.24A) and exploratory sniffing (Figure 6.24B), were markedly affected by prenatal BPA exposure. BPA appeared to have these effects by interfering with the way that the transcription of oxytocin and vasopressin occurs in the brain (Figure 6.24C,D). These two hormones are involved in mediating social behaviors, especially trust and intimacy. Moreover, the effects of exposure on behaviors and gene expression in the brain could be seen three generations later. The public health ramifications of this type of inheritance are just beginning to be explored.

Looking Ahead

The environment can act in many ways during an organism's development. It can act in its normal role to produce the expected phenotype (as in mammalian-bacterial developmental symbiosis). It can act instructively to enable an organism to develop an alternate phenotype predicted to be especially fit for a given environment (as in polyphenisms). It can act in an overtly deleterious manner, disrupting normal development (as in teratogenesis). And it can act disruptively but covertly, as when endocrine disruptors modify gene expression in a manner that does not become apparent until much later in life.

(A)

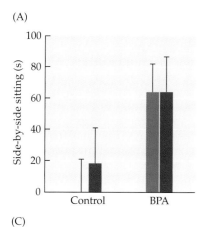

(B)

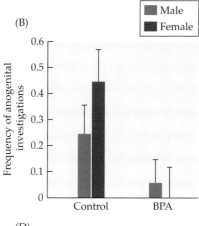

Male
Female

Figure 6.24 Effects of gestational BPA exposure on behaviors and neuroendocrine gene expression in mice. Pregnant mice were fed food containing BPA such that the serum BPA concentrations approximated those found in adult humans. Gene expression was measured at the last day of gestation (to prevent effects from nursing), and behaviors were measured in juveniles. Juvenile mice born with prenatal exposure to BPA had greater side-by-side sitting behaviors (A), but less exploratory sniffing (B), than their control counterparts. BPA caused decreases in vasopressin (C) and oxytocin (D) mRNAs. In all experiments, the differences between control and BPA-exposed animals were statistically significant ($p < 0.05$). (After Wolstenholme et al. 2012.)

(C)

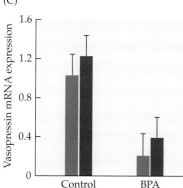

(D)

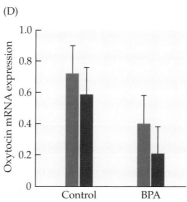

Regulatory and Policy Decisions on BPA and Other Endocrine Disruptors

Numerous plastic compounds have estrogenic activity. Indeed, when robotic testing was done to see which compounds bind to estrogen receptors, some of the plastics labeled "BPA-free" had higher estrogenic activity than BPA (Yang et al. 2011; Bittner et al. 2014). (The same testing has identified some plastics that can be made cheaply from high-molecular-weight subunits that resist degradation and do not have estrogenic activity.) Plastics containing BPA, however, have gotten most of the recent attention.

The Food and Drug Administration of the United States holds that BPA is safe and does not recommend that consumers avoid BPA-containing products (FDA 2008, 2014). Similarly, the National Institute of Environmental Health Sciences (NIEHS 2008), while expressing "some concern" over the possible effects of BPA on

the developing brain, has not recommended legislation against it. However, in 2007, a consensus statement of 38 researchers in the field called for an immediate ban on BPA, stating that currently prevalent levels of this chemical in humans are already higher than those shown to cause developmental damage to laboratory animals (vom Saal et al. 2007).

In the United States, a chemical is assumed to be safe until proven dangerous, and government agencies have based their policy decisions on the ambiguity surrounding the assessments of BPA's safety, and on the fact that no scientific experiments have yet shown that the chemical is harmful to human fetuses. However,

(Continued on next page)

Regulatory and Policy Decisions on BPA and Other Endocrine Disruptors *(continued)*

the studies claiming to show no low-dose effects of BPA and other endocrine disruptors were performed by scientists paid by the chemical industry, and experiments exposing human fetuses to specific concentrations of BPA cannot be done. So this ambiguity, according to independent scientists, has been manufactured by the chemical industry. These independent researchers have claimed that, whereas all 14 of the studies showing that BPA is safe were funded by the chemical industry, of the 204 studies funded through government grants, 93% concluded that BPA is harmful (vom Saal 2008). Similarly, every study finding that atrazine is safe was funded by Syngenta, the makers of atrazine. In contrast, across several species and test sites, every government, private, and independent laboratory looking at this issue found that atrazine is an endocrine disruptor that demasculin-izes frogs (Hayes 2004). As such a dichotomy points out, there are numerous problems, both scientific and politi-cal, in assessing the effects of BPA or any other endo-crine disruptor on human individuals and populations.

In 2013, a group of industry scientists put forth a declaration that the European ban on BPA was based on "scientifically unfounded precaution." Reaction to this declaration, largely by independent scientists, was that it ignored the scientifically demonstrated idea that endo-crine disruptors could instruct developmental changes, and that it also ignored the inverted U activity charts that demonstrated the endocrine disruptors may have more pronounced effects at low concentrations than at higher ones. While the spokespeople for the chemical industry contend that the studies on animals have little relevance for human beings, there is no other choice for these studies. Animal model systems are the only ethically al-lowable way to get data on the mechanism of hormone action, and the use of several animal models is critically important.

The chemical industry has also attempted to under-mine the reputations of scientists who have linked certain compounds to endocrine disruption. One such targeted researcher is Tyrone Hayes. He had been hired by the makers of atrazine to show whether atrazine caused cancer. He showed that it didn't; but he demonstrated that it caused sexual reversals in male frogs. The mak-ers of atrazine (Syngenta) did not allow him to publish his results. So Hayes started anew, got new funding, and reproduced and extended his findings. The Syn-genta company sent people to check his taxes, harass him at meetings, and discredit him with his employers. The strategies used against him are reported in *The*

Chronical of Higher Education (Blumenstyk 2003) and in the *New Yorker* (Aviv 2014). The court forced the release of Syngenta's memos and emails concerning atrazine and Hayes, and these can be found on Hayes's website (www.atrazinelovers.com).

Similarly, the makers of plastics have tried to ignore and have attempted to prevent the publication of findings that the materials in their products cause damage to the reproductive and immune systems of people and other animals. An interview with Dr. Fred vom Saal (1998), who first found evidence linking BPA and cancer, shows some of the tactics, as does a more recent report in *Nature* (Borrell 2010).

Scientific issues

As mentioned earlier, one set of problems in assessing the effects of endocrine disruptors on humans is scientific. An overarching issue, as noted several times in this chapter, is that the effects of BPA and other endocrine disruptors are not anatomically visible and are often not functionally apparent until adult life. This temporal disconnect makes it difficult to be certain that problems observed in adults are the result of prenatal exposure to a given substance.

In virtually all biomedical research, there are issues based on the differences between laboratory animals and humans. This is especially relevant to endocrine disruptor research, because animals differ widely in their responses to hormone disruptors. (This problem was evident in the case of teratogen studies described in Chapter 5, in which laboratory mice and rats were not susceptible to thalidomide, whereas rabbits were.) Genetic differences among individuals affect sensitivity to these substances (see the box, p. 243). And, whereas laboratory animals are highly inbred, humans are not; thus, among humans one would expect to find greater genetic differences in regulatory and homeostatic mechanisms between individuals, meaning that some humans will be more sus-ceptible than others to different chemicals. This means that the data on humans will probably be epidemiologi-cal and not as clear-cut as the evidence from rodents. Such epidemiological data will be difficult to document properly, however, since there is no control group—the entire human population is exposed to these ubiquitous compounds.

Another confounding factor in endocrine disruptor research is the interaction of environmental estrogens. Silva and colleagues (2002) have shown that if cells are exposed to a mixture of environmental estrogens, each of which is at a concentration that induces an

Regulatory and Policy Decisions on BPA and Other Endocrine Disruptors *(continued)*

estrogen-responsive gene only very weakly, the mixture can induce a response that is much more than the additive responses of the individual components (Figure A). Moreover, Rajapakse and colleagues (2001, 2002) have shown that when normal estrogen is supplemented with amounts of environmental xenoestrogens so small that they have inconsequential effects alone, the small amounts of xenoestrogen potentiate the effect of normal estrogen and makes it much greater. This may make the mixture of endocrine disruptors we receive daily even more problematic for us and our offspring. It also makes blaming a particular compound for a specific effect very difficult.

The good news is that we can get rid of the endocrine-disrupting effects of these mixtures in a many-at-a-time fashion. One example is the cleanup of Boulder Creek in Colorado. In 2005, this creek, which carries the processed sewage of the city of Boulder, contained high enough concentrations of estrogenic compounds that adult male fathead minnows were demasculinized when they were held in creek water for 2 weeks. They had elevated levels of vitellogenin (egg protein), decreased sperm production, and reduced expression of male secondary sex traits. When the city of Boulder upgraded their wastewater treatment facility to include activated sludge (which gets rid of many endocrine-disrupting compounds, as well as heavy metals), the concentrations of estrogenic compounds in the stream plummeted, and the fish population responded. The males made normal sperm, had normal gonadal development, and no longer expressed the vitellogenin gene (Figure B–E; Barber et al. 2012; Vadja et al. 2011). The cost of the project was $50 million, around the same cost as building a baseball stadium (and one-third the price of building Yankee Stadium in New York), which brings us to government priorities.

Political issues

In the United States, rigorous scientific proof is needed to claim that a product is harmful. As if in a homicide case, the product is assumed to be innocent until proven guilty beyond a reasonable doubt. In past instances of environmental concern (about lead, mercury, asbestos, tobacco, global warming, and ozone depletion), the affected industries have financed scientific studies that have been able to raise such doubt (see Markowitz and

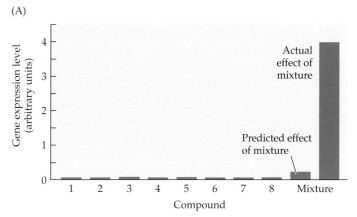

(A) A mixture of estrogenic compounds at low concentrations activates an estrogen-responsive gene. Plasmids containing an estrogen response element attached to a *lacZ* gene were added to cells, and each of eight environmental estrogens was added at a concentration that only weakly induced the expression of the *lacZ* gene. However, when a mixture of these eight compounds was added, the response was far greater than the additive values of the separate responses.

Rossner 2000; Steingraber 2001; Michaels 2005, 2008). In 2008, when researchers at Yale University (Leranth et al. 2008) showed that BPA disrupted brain development in monkeys (at concentrations lower than what the EPA considers safe), the American Chemical Council, an industry trade group, replied that "there is no direct evidence that exposure to BPA adversely affects human reproduction or development" (see Layton 2008). But the catch-22 exists in that such "direct evidence" (either for or against) may be impossible to obtain, given that we cannot test drugs on human fetuses.

Another series of problems comes from the way chemicals are patented and licensed. As an example, heavily brominated chemicals such as polybrominated diphenyl ethers (PBDEs) are a family of chemicals used as flame retardants in fabrics, furniture, and plastics. These substances, which are related to PCBs, were discovered to bioaccumulate in humans and wildlife and to be hazardous to human health. Some of them have been banned in the United States and Europe, some are banned only in Europe, and some are banned in a few particular U.S. states. When one of these compounds is banned, however, the industry can easily manufacture

(Continued on next page)

Regulatory and Policy Decisions on BPA and Other Endocrine Disruptors *(continued)*

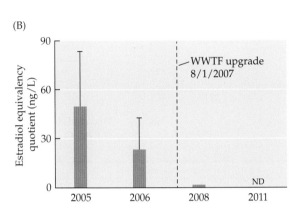

(B)

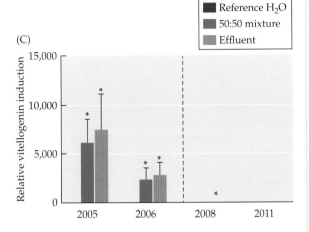

(C)

Reference H_2O
50:50 mixture
Effluent

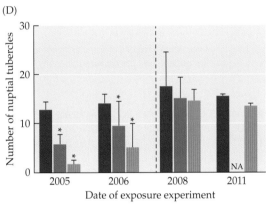

(D)

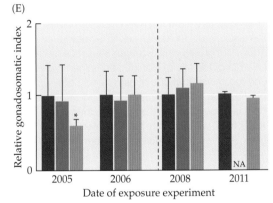

(E)

(B) Summary of estrogenic effects in the Boulder wastewater treatment facility (WWTF) effluent before the upgrade (2005 and 2006) and after the upgrade (2008 and 2011). The results are compared with upstream Boulder Creek reference water (Reference H_2O) and a 50:50 mixture of the two. The upstream waters (before reaching Boulder) show no estrogenic components, and the development of male fish is normal. Average WWTF effluent estradiol equivalency quotient as a function of exposure experiment based on weekly measurements of multiple endocrine-disrupting chemicals. (C) Relative plasma vitellogenin concentrations (normalized to the mean reference concentration) in adult male fathead minnows exposed to the test waters for 28 days. (D) Number of nuptial tubercles (head structures that are a measure of male reproductive capability) in adult male fathead minnows exposed to the waters for 28 days. (E) Gonadosomatic index (gonadal mass as a proportion of total body weight) normalized to the mean reference concentration in adult male fathead minnows exposed to the waters for 28 days. Bars with asterisks indicate a significant difference from the upstream Boulder Creek reference water exposure for the same year. (A after Silva et al. 2002; B–E after Barber et al. 2012.)

another that is not much different from the original. Most of these alternative retardants still contain bromine and similarly bioaccumulate.* In 2007 Donald Kennedy, then

*A similar strategy is being questioned with regard to bisphenol F and bisphenol S, two molecules structurally related to bisphenol A. Both have been found to cause endocrine disruption (Eldak et al. 2015; Goldinger et al. 2015).

editor of the journal *Science,* compared the regulation of PBDEs to the game of Whac-A-Mole, in which whacking down one problem only leads to another one popping up (Kennedy 2007).

Under current regulations, a new chemical cannot be distributed in the United States until the manufacturer has submitted data to the Environmental Protection Agency detailing the properties of the chemical

Regulatory and Policy Decisions on BPA and Other Endocrine Disruptors *(continued)*

and probability of any harmful effects. The EPA then has 90 days to respond. But the required data do not currently include assessment of endocrine disruptor or teratogenic effects, and so far lobbyists have prevented implementation of changes intended to legislate such testing (Michaels 2008; Kollipara et al. 2015). These tests, they note, would be extremely expensive, which would retard innovation and make products more costly to consumers.

International perspectives

In contrast to the "harmless unless proven otherwise" approach common in the United States, some Western European nations are apt to invoke the "precautionary principle" when approaching issues of public health and safety (see Harremoes et al. 2002). This principle asserts that when an action or a policy could seriously or irreversibly damage the public welfare or the environment, in the lack of full scientific certainty that harm will not take place, the burden of proof falls on those who advocate the policy or action (Raffensberg and Tickner 1999). This, however, calls for a substantial judgment concerning what constitutes irreversible or serious damage, how much scientific information is needed for such a case to be made, and how one balances the risk of bad outcomes with the potential gains from the innovation (Sunstein 2005).

Currently the European Union (EU) is attempting to develop a new screening approach for chemicals, referred to as REACH (Registration, Evaluation, Authorisation and Restriction of Chemicals). This program aims to develop tests to assess the toxicity of thousands of chemicals, including endocrine disruptor and reproductive and developmental toxicity assays. The size of the EU market may result in most international chemical companies conforming to the EU guidelines (see Schapiro 2007).

Right now, public relations and consumer confidence seem to be the major forces driving the regulation of potential endocrine disruptors. Canada, for instance, has banned the use of BPA in the manufacture of baby bottles, while the United States, the EU, and Japan still allow it. Commercial manufacturers, however, have independently moved to discontinue BPA-containing products. Wal-Mart has said they would stop selling BPA-containing baby bottles in Canada and the United States. Similarly, Nalgene is discontinuing the use of BPA in the production of their brightly colored reusable water bottles (Layton and Lee 2008; Nalgene 2015). Many supermarket chains are phasing out BPA in their cans and cash register receipts.

Because endocrine disruptors recognize no political boundaries and persist through many administrations and regimes, international agencies can be critically important. In May 2001, the Stockholm Convention of the United Nations had been signed by delegates from 122 nations (including the United States). This treaty, which regulates certain industrial pollutants, incorporates the precautionary principle (Kaiser and Enserink 2000; UNEP 2008). It went into effect in 2004 after being ratified by the necessary 50 nations, but as of 2014 the U.S. Congress has not endorsed it.

The crucial question, as the authors of *Our Stolen Future* (Colborn et al. 1996) ask, is "Are we threatening our fertility, intelligence, and survival?" And how much is at risk if the answer should turn out to be "yes"?

Endocrine disruption has opened up a new chapter in the study of environmental toxicology. Usually, compounds that change development have been studied as teratogens, but endocrine disruptors have changed the rules. First, they can have effects at low doses, rather than having to be present at high doses. Second, they are ubiquitous, rather than confined to a small exposed population. Third, by operating through the endocrine system, they establish an inverted U dose curve, wherein high amounts may actually have less of an effect than slightly lower amounts. Fourth, several endocrine disruptors are usually present in the same environment (and in the same person), and each endocrine disruptor can have complex interactions with other endocrine disruptors.

Moreover, unlike most teratogens, the effects of some endocrine disruptors can be transmitted by DNA methylation differences or other epigenetic alterations and thus are transgenerational. Endocrine disruptors can alter DNA methylation in all cells (including the precursors of the germline cells), and if these methylation patterns take place in the germ cells, the change can become stably inherited (see Barresi et al. 2013 for a video review of some endocrine disruptor research). As we will see in the next chapter, other types of stress during fetal development, such as low-protein diets, can act very similarly and modify gene expression patterns to generate phenotypes that are not apparent at birth but that show up later in the adult.

The nature of endocrine disruptors opens up a new set of questions, many of which concern how these chemicals might cause or predispose people and other animals to certain diseases. Since wild animals and humans are genetically diverse (unlike laboratory mammals), the integration of epidemiology, ecology, developmental biology, and molecular genetics is needed to ascertain how these compounds affect individual health and who is at risk. In this way, the study of endocrine disruptors will include both a public and a private component. This latter component is similar to "personalized medicine," only in the opposite direction.

The ability of endocrine disruptors to alter behavior in laboratory animals causes concern about the extent to which these compounds alter brain development (Crews et al. 2014). This will no doubt be a major research program during the next decade.

But as we've seen in Chapter 3, animals are holobionts, and holobionts contain bacteria that are critically important in how animals respond to environmental factors. Indeed, the metabolism of endocrine disrupting compounds can depend on the microbes in the gut, and how the endocrine disruptor is digested can give an organism different phenotypes (Frankenfeld et al. 2014). The interaction of holobiont and endocrine disruptor is a new and potentially very important area. Another, related, area is the use microbes and other organisms to perform bioremediation on endocrine disruptors. Here, bacteria, fungi, or plants would digest the endocrine disruptors and transform them into harmless (or even useful) chemicals (see Gattullo et al. 2013; Macellaro et al. 2014; Gavrilescu et al. 2015).

Certainly one of the most critical and challenging questions will be the means by which endocrine disruptive compounds can cause disease and how the effects of endocrine disruption can be inherited. If BPA is causing behavioral alterations in mice, then we need to know how these changes are accomplished and if humans are responding in similar ways and at environmentally relevant doses of BPA. If vinclozolin is causing defects in sperm production, then we need to know how this is done. Moreover, if both these compounds are altering chromatin structure and DNA methylation pattern in a manner that gets transmitted from one generation to the next, then we need to understand how this altered chromatin state is made and how it can be propagated. We also need to know if the relatively new epidemics of obesity, asthma, and behavioral disorders may be due to endocrine disruptors and their interaction with the holobiont organism, both the host and the symbionts. These results can affect policy decisions that may have enormous economic and medical impact.

References

Acevedo, N., B. Davis, C. M. Schaeberle, C. Sonnenschein and A. M. Soto. 2013. Perinatally administered bisphenol A as a potential mammary gland carcinogen in rats. *Environ. Health Perspect.* 121: 1040–1046.

Adeoya-Osiguwa, S. A., S. Markoulaki, V. Pocock, S. R. Milligan and L. R. Fraser. 2003. 17β Estradiol and environmental estrogens significantly affect mammalian sperm function. *Human Reprod.* 18: 100–107.

Aitken, R. J., P. Koopman and S. E. Lewis. 2004. Seeds of concern. *Nature* 432: 48–52.

Albert, A., K. Drouillard, G. D. Haffner and B. Dixon. 2007. Dietary exposure to low pesticide doses causes long-term immunosuppression in the leopard frog (*Rana pipiens*). *Environ. Toxicol. Chem.* 26: 1179–1185.

Albert, O. and B. Jegou. 2014. A critical assessment of the endocrine susceptibility of the human testis to phthalates from fetal life to adulthood. *Hum. Reprod. Update* 20: 231–249.

Ali, S. and R. C. Coombes. 2000. Estrogen receptor alpha in human breast cancer: Occurrence and significance. *J. Mammary Gland Biol. Neoplasia* 5: 271–281.

Allred, C. D., Y. H. Ju, K. F. Allred, J. Chang and W. G. Helferich. 2001. Dietary genistin stimulates growth of estrogen-dependent breast cancer tumors similar to that observed with genistein. *Carcinogenesis* 22: 1667–1673.

AMAP (Arctic Monitoring and Assessment Programme). 2002. *Arctic Pollution 2002.* Oslo, Norway.

Anway, M. D., A. S. Cupp, M. Uzumcu and M. K. Skinner. 2005. Epigenetic transgenerational actions of endocrine disruptors and male fertility. *Science* 308: 1466–1469.

Anway M. D., C. Leathers and M. K. Skinner. 2006. Endocrine disruptor vinclozolin induced epigenetic transgenerational adult onset disease. *Endocrinology* 147: 5515–5523.

Anway, M. D. and M. K. Skinner. 2006. Epigenetic transgenerational actions of endocrine disruptors. *Endocrinology* 147(suppl.): S43–S49.

Auger, J., J. M. Kunstmann, F. Czyglick and P. Jouannet. 1995. Decline in semen quality among fertile men in Paris during the past 20 years. *New Engl. J. Med.* 332: 281–285.

Aviv, R. 2014. A valuable reputation: After Tyrone Hayes said that a chemical was harmful, its maker pursued him. *New Yorker.* Feb. 10, 2014. www.newyorker.com/magazine/2014/02/10/a-valuable-reputation

Barber, L. B., A. M. Vajda, C. Douville, D. O. Norris and J. H. Writer. 2012. Fish endocrine disruption responses to a major wastewater treatment facility upgrade. *Environ. Sci. Technol.* 46: 2121–231

Bell, S. E. 1986. A new model of medical technology development: A case study of DES. *Sociol. Health Care* 4: 1–32.

Bell, S. E. 1995. Gendered medical science: Producing a drug for women. *Feminist Studies* 21: 469–500.

Bennetts, H. W. and E. J. Underwood. 1951. The oestrogenic effects of subterranean clover (trifolium subterraneum): Uterine maintenance in the ovariectomised ewe on clover grazing. *Aust. J. Exp. Biol. Med. Sci.* 29: 249–253.

Benson, G. V., H. Lim, B. C. Paria, I. Satokata, S. K. Dey and R. Maas. 1996. Mechanisms of female infertility in *Hoxa-10* mutant mice: Uterine homeosis versus loss of maternal *Hoxa-10* expression. *Development* 122: 2687–2696.

Bhan, A., I. Hussain, K. I. Ansari, S. A. Bobzean, L. I. Perrotti and S. S. Mandal. 2014. Bisphenol-A and diethylstilbestrol exposure induces the expression of breast cancer associated long noncoding RNA HOTAIR in vitro and in vivo. *J. Steroid Biochem. Mol. Biol.* 141: 160–170.

Bittner, G. D., C. Z. Yang and M. A. Stoner. 2014. Estrogenic chemicals often leach from BPA-free plastic products that are replacements for BPA-containing polycarbonate products. *Environ. Health* 13: 41.

Block, K., A. Kardanga, P. Igarashi and H. S. Taylor. 2000. In utero DES exposure alters Hox gene expression in the developing Müllerian system. *FASEB J.* 14: 1101–1108.

Blumenstyk, G. 2003. The story of Syngenta and Tyrone Hayes at UC Berkeley: A Berkeley scientist says a corporate sponsor tried to bury his unwelcome findings and then buy his silence. *Chron. Higher Ed.* 50 (10) October. www.mindfully.org/Pesticide/2003/Syngenta-Tyrone-Hayes31oct03.htm

Borrell, B. 2010. Toxicology: The big test for bisphenol A. *Nature* 464: 1122–1124.

Boutin, E. L. and G. R. Cunha. 1997. Estrogen-induced proliferation and cornification are uncoupled in sinus vaginal epithelium associated with uterine stroma. *Differentiation* 62: 171–178.

Braun, J. M., A. E. Kalkbrenner, A. M. Calafat, K. Yolton, X. Ye, K. N. Dietrich and B. P. Lanphear. 2011. Impact of early-life bisphenolA exposure on behavior and executive function in children. *Pediatrics* 128: 873–882.

Braun, J. M., K. Yolton, K. N. Dietrich, R. Hornung, X. Ye, A. M. Calafat and B. P. Lanphear. 2009. Prenatal bisphenol A exposure and early childhood behavior. *Environ. Health Perspect.* 117: 1945–1952.

Brodkin, M. A., H. Madhoun, M. Rameswaran and I. Vatnick. 2007. Atrazine is an immune disruptor in adult northern leopard frogs (*Rana pipiens*). *Environ. Toxicol. Chem.* 26: 80–84.

Bromer, J. G., Y. Zhou, M. B. Taylor, L. Doherty and H. S. Taylor. 2010. Bisphenol-A exposure in utero leads to epigenetic alterations in the developmental programming of uterine estrogen response. *FASEB J.* 24: 2273–2280.

Brosens, J. J. and M. G. Parker. 2003. Oestrogen receptor hijacked. *Nature* 423: 487–488.

Brouwer, A. and 7 others. 1999. Characterization of potential endocrine-related health effects at low-dose levels of exposure to PCBs. *Environ. Health Persp.* 107(suppl. 4): 639–649.

Cabaton, N. J. and 10 others. Perinatal exposure to environmentally relevant levels of bisphenol A decreases fertility and fecundity in CD-1 mice. *Environ. Health Perspect.* 119: 547–552.

Calafat, A. M., Z. Kuklenyik, J. A. Reidy, S. P. Caudill, J. Ekong and L. L. Needham. 2005. Urinary concentrations of bisphenol A and 4-nonylphenol in a human reference population. *Env. Health Persp.* 113: 391–395.

Capel, P. and S. Larson. 2001. Effect of scale on the behavior of atrazine in surface waters. *Environ. Sci. Tech.* 35: 648–657.

Carlsen, E., A. Giwercman, N. Keiding and N. Skakkebæk. 1992. Evidence for decreasing quality of semen during past 50 years. *Brit. Med. J.* 305: 609–613.

Carson, R. 1962. *Silent Spring.* Houghton Mifflin, New York.

Carta, L. and D. Sassoon. 2004. Wnt7a is a suppressor of cell death in the female reproductive tract and is required for postnatal and estrogen-mediated growth. *Biol. Reprod.* 71: 444–454.

Cederroth, C. R. and 9 others. 2007. A phytoestrogen-rich diet increases energy expenditure and decreases adiposity in mice. *Env. Health Persp.* 115: 1467–1473.

Cederroth, C. R. and 9 others. 2010. Potential detrimental effects of a phytoestrogen-rich diet on male fertility in mice. *Mol. Cell Endocrinol.* 321: 152–160.

Chamorro-García, R., M. Sahu, R. J. Abbey, J. Laude, N. Pham and B. Blumberg. 2013. Transgenerational inheritance of increased fat depot size, stem cell reprogramming, and hepatic steatosis elicited by prenatal exposure to the obesogen tributyltin in mice. *Environ. Health Perspect.* 121: 359–366.

Chan, W-H. 2009. Impact of genistein on maturation of mouse oocytes, fertilization, and fetal development. *Reprod. Toxicol.* 28: 52–58.

Chavarro, J. E., T. L. Toth, S. M. Sadio and R. Hauser. 2008. Soy food and isoflavone intake in relation to semen quality parameters among men from an infertility clinic. *Human Reprod.* 23: 2584–2590.

Cheng, G., T. Remer, R. Prinz-Langenohl, M. Blaszkewicz, G. H. Degen and A. E. Buyken. 2010. Relation of isoflavones and fiber intake in childhood to the timing of puberty. *Am. J. Clin. Nutr.* 92: 556–564.

Chiu, Y. H. and 7 others. 2015. Fruit and vegetable intake and their pesticide residues in relation to semen quality among men from an infertility clinic. *Hum. Reprod.* doi: 10.1093/humrep/dev064.

Christian, S. and 8 others. 2006. Xenoendocrine pollutants can reduce size of sexual organs in East Greenland polar bears (*Ursus maritimus*). *Environ. Sci. Tech.* 40: 5668–5674.

Christin, M.-S. and 7 others. 2003. Effects of agricultural pesticides on the immune system of *Rana pipiens* and on its resistance to parasitic infection. *Environ. Toxicol. Chem.* 22: 1127–1133.

Chung, E., M. C. Genco, L. Megrelis and J. V. Ruderman. 2011. Effects of bisphenol A and triclocarban on brain-specific expression of aromatase in early zebrafish embryos. *Proc. Natl. Acad. Sci. USA* 108: 17732–17737.

Clayton, E. M., M. Todd, J. B. Dowd and A. E. Aiello. 2011. The impact of bisphenol A and triclosan on immune parameters in the U.S. population, NHANES 2003–2006. *Environ. Health Perspect.* 119: 390–396.

Cohn, B. A. 2011. Developmental and environmental origins of breast cancer: DDT as a case study. *Reprod. Toxicol.* 31: 302–311.

Colborn, T., D. Dumanoski and J. P. Myers. 1996. *Our Stolen Future.* Dutton, New York.

Colburn, T., C. Kwiatkowski, K. Schultz and M. Bachran. 2011. Natural gas operations fro a public health perspective. *Human Ecol. Risk Assess.* 17: 1039–1056.

Colborn, T., F. S. vom Saal and A. M. Soto. 1993. Developmental effects of endocrine-disrupting chemicals in wildlife and humans. *Environ. Health Persp.* 101: 378–384.

Cook, J. D., B. J. Davis, S. L. Cai, J. C. Barrett, C. J. Conti and C. L. Walker. 2005. Interaction between genetic susceptibility and early-life environmental exposure determines tumor-suppressor gene penetrance. *Proc. Natl. Acad. Sci. USA* 102: 8644–8649.

Cooke, A. S. 1973. Shell thinning in avian eggs by environmental pollutants. *Environ. Pollut.* 4: 85–152.

Cornwell, T., W. Cohick and I. Raskin. 2004. Dietary phytoestrogens and health. *Phytochemistry* 65: 995–1016.

Cortina, B. and 9 others. 2013. Improvement of the circulatory function partially accounts for the neuroprotective action of the phytoestrogen genistein in experimental ischemic stroke. *Eur. J. Pharmacol.* 708: 88–94.

Cotroneo, M. S., J. Wang, W. A. Fritz, I. E. Eltoum, and C. A. Lamartiniere. 2002. Genistein action in the prepubertal mammary gland in a chemoprevention model. *Carcinogenesis* 23: 1467–1474.

Crain, D. A. and L. L. Guillette, Jr. 1998. Reptiles as models of contaminant-induced endocrine disruption. *Anim. Reprod. Sci.* 53: 77–86.

Crain, D. A., A. Rooney, E. F. Orlando and L. J. Guillette. 1998. Endocrine-disrupting contaminants and hormone dynamics: Lessons from wildlife. In L. L. Guillette, Jr., and D. A. Crain (eds.), *Environmental Endocrine Disruptors: An Evolutionary Perspective.* Taylor & Francis, New York, pp. 1–21.

Crews, D., R. Gillette, I. Miller-Crews, A. C. Gore and M. K. Skinner. 2014. Nature, nurture, and epigenetics. *Mol. Cell Endocrin.* 398: 45–52.

Davis, D. L., H. L. Bradlow, M. Wolff, T. Woodruff, D. G. Hoel and H. Anton-Culver. 1993. Xenoestrogens as preventable causes of breast cancer. *Environ. Health Persp.* 101: 372–377.

Demeneix, B. 2014. *Losing Our Minds: How Environmental Pollution Impairs Human Intelligence and Mental Heath.* Oxford University Press, Oxford.

Den Hond, E. and G. Shoeters. 2006. Endocrine disrupters and human puberty. *Internatl. J. Androl.* 29: 264–271.

Desdoits-Lethimonier, C. and 9 others. 2012. Human testis steroidogenesis is inhibited by phthalates. *Hum. Reprod.* 27: 1451–1459.

Dixon, R. A. 2004. Phytoestrogens. *Annu. Rev. Plant Biol.* 55: 225–261.

Dodson, E. E., M. Nishioka, L. J. Standley, L. J. Perovich, J. G. Brody and R. A. Rudel. 2012. Endocrine disruptors and asthma-associated chemicals in consumer products. *Environ. Health Perspect.* 120: 935–943

DOE (U.S. Department of Energy). 2009. State Oil and Natural Gas Regulations Designed to Protect Water Resources, May 2009. www.gwpc.org/sites/default/files/state_oil_and_gas_regulations_designed_to_protect_water_resources_0.pdf

Doherty, L. F., J. G. Bromer, Y. Zhou, T. S. Aldad and H. S. Taylor. 2010. In utero exposure to diethylstilbestrol (DES) or bisphenol-A (BPA) increases EZH2 expression in the mammary gland: An epigenetic mechanism linking endocrine disruptors to breast cancer. *Horm. Cancer* 1: 146–155.

Dolinoy, D. C., D. Huang and R. L. Jirtle. 2007. Maternal nutrient supplementation counteracts bisphenol A–induced DNA hypomethylation in early development. *Proc. Natl. Acad. Sci. USA* 104: 13056–13061.

Domingo, J. L. and A. Bocio. 2007. Levels of PDCC/PCDFs and PCBs in edible marine species and human uptake: A literature review. *Environ. Internatl.* 33: 397–405.

Durando, M., L. Kass, J. Piva, C. Sonnenschein, A. M. Soto, E. H. Luque and M. Muñoz-de-Toro. 2007. Prenatal bisphenol A exposure induces preneoplastic lesions in the mammary gland in Wistar rats. *Environ. Health Persp.* 115: 80–86.

Duty, S. M. and 8 others. 2003. Phthalate exposure and human semen parameters. *Epidemiology* 14: 269–277.

Eladak, S. and 9 others. 2015. A new chapter in the bisphenol A story: Bisphenol S and bisphenol F are not safe alternatives to this compound. *Fertil. Steril.* 103: 11–21.

Elsworth, J. D., J. D. Jentsch, S. M. Groman, R. H. Roth, D. E. Redmond and C. Leranth. 2015. Low circulating levels of Bisphenol-A induce cognitive deficits and loss of asymmetric spine synapses in dorsolateral prefrontal cortex and hippocampus of adult male monkeys. *J. Comp. Neurol.* doi: 10.1002/cne.23735

Elsworth, J. D., J. D. Jentsch, C. A. Vandevoort, R. H. Roth, D. E. Jr. and C. Leranth. 2013. Prenatal exposure to bisphenol A impacts midbrain dopamine neurons and hippocampal spine synapses in non-human primates. *Neurotoxicology* 35: 113–120.

EPA (Environmental Protection Agency [Internet]). 1992. Washington (DC): EPA. Press release, Environmental equity: reducing risks for all communities; 1992 Jul 22 [cited 2011 Apr 10]. Available from: www.epa.gov/history/topics/justice/01.htm

EPA (Environmental Protection Agency). 2011. Environmental equity: reducing risks for all communities [Internet]. Washington (DC): EPA; [cited 2011 Apr 10]. Available from: www.epa.gov/compliance/ej/resources/reports/annual-project-reports/reducing_risk_com_vol1.pdf

EPA-PCB. 2008. Polychlorinated biphenyls (PCBs). www.epa.gov/pcb/pubs/effects/html

Evans, R. M., G. D. Barish and Y. X. Wang. 2004. PPARS and the complex journey to obesity. *Natature Med.* 10: 355–361.

Facemire, C. F., T. S. Gross and L. J. Guillette, Jr. 1995. Reproductive impairment in the Florida panther: Nature or nurture, *Environ. Health Persp.* 103(suppl. 4): 79–86.

Fagin, D. 2012. The learning curve. *Nature* 490: 462–465.

Fairchild, W. L., E. O. Swansburg, J. T. Arsenault and S. B. Brown. 1999. Does an association between pesticide use and subsequent declines in catch of Atlantic salmon (*Salmo salar*) represent a case of endocrine disruption? *Environ. Health Perspect.* 107: 349–358.

FDA. 2008. Draft assessment of Bisphenol A for use in food contact applications. www.fda.gov/ohrms/dockets/ac/08/briefing/2008-0038b1_01_02_FDA%20BPA%20Draft%20Assessment.pdf

FDA. 2014. Bisphenol A (BPA). www.fda.gov/Food/FoodborneIllnessContaminants/ChemicalContaminants/ucm166145.htm

Filipov, N. M., L. M. Pinchuk, B. L. Boyd and P. L. Crittenden. 2005. Immunotoxic effects of short-term atrazine exposure in young male C57BL/6 mice. *Toxicol. Sci.* 86: 324–32.

Fisher, J. S., S. Macpherson, N. Marchetti and R. M. Sharpe. 2003. Human testicular dysgenesis syndrome: A possible model using in utero exposure of the rat to dibutyl phthalate. *Human Reprod.* 18: 1383–1394.

Forson, D. and A. Storfer. 2006a. Effects of atrazine and iridovirus infection on survival and life history traits of the long-toed salamander (*Ambystoma macrodactylum*). *Environ. Toxicol. Chem.* 25: 168–173.

Forson, D. and A. Storfer. 2006b. Atrazine increases Ranavirus susceptibility in the tiger salamander, *Ambystoma tigrinum*. *Ecol. Applic.* 16: 2325–2332.

Fox, J. E., J. Gulledge, E. Engelhaupt, M. E. Burow and J. A. McLachlan. 2007. Pesticides reduce symbiotic efficiency of nitrogen-fixing rhizobia and host plants. *Proc. Natl. Acad. Sci. USA* 104: 10282–10287.

Fox, J. E., M. Starcevic, P. E. Jones, M. E. Burow and J. A. McLachlan. 2004. Phytoestrogen signaling and symbiotic gene activation are disrupted by endocrine-disrupting chemicals. *Environ. Health Perspect.* 112: 672–677.

Frankenfeld, C. L., C. Atkinson, K. Wähälä and J. W. Lampe. 2014. Obesity prevalence in relation to gut microbial environments capable of producing equol or O-desmethylangolensin from the isoflavone daidzein. *Eur. J. Clin. Nutr.* 68: 526–530.

Frye, D. M. 1995. Reproductive effects in birds exposed to pesticides and industrial chemicals. *Env. Health Persp.* 103: 165–171.

GAO (Government Accountability Office). 1983. Siting of hazardous waste landfills and their correlation with racial and economic status of surrounding communities. Washington (DC): GAO;1983. Report No.: RCED-83-168.

Gattullo, C. E., B. B. Cunha, A. H. Rosa and E. Loffredo. 2013. Removal of a combination of endocrine disruptors from aqueous systems by seedlings of radish and ryegrass. *Environ. Technol.* 34: 3129–3136.

Gavrilescu, M., K. Demnerová, J. Aamand, S. Agathos and F. Fava. 2015. Emerging pollutants in the environment: Present and future challenges in biomonitoring, ecological risks and bioremediation. *New Biotechnol.* 32: 147–156.

Gendron, A. and 7 others. 2003. Exposure of leopard frogs to a pesticide mixture affects life history characteristics of the lungworm *Rhabdias ranae*. *Oecologia* 135: 469–476.

Gilbert, S. F. 2013. *Developmental Biology*, 10th Ed. Sinauer Associates, Sunderland, MA.

Gilbertson, M. K., G. D. Haffner, K. G. Drouillard, A. Albert and B. Dixon. 2003. Immunosuppression in the northern leopard frog (*Rana pipiens*) induced by pesticide exposure. *Environ. Toxicol. Chem.* 22: 101–110.

Goldinger, D. M. and 7 others. 2015. Endocrine activity of alternatives to BPA found in thermal paper in Switzerland. *Regul Toxicol Pharmacol.* 2015 S0273–2300. doi: 10.1016/j.yrtph.2015.01.002

Goran, M. I., K. Dumke, S. G. Bouret, B. Kayser, R. W. Walker and B. Blumberg. 2013. The obesogenic effect of high fructose exposure during early development. *Nature Rev. Endocrinol.* 9: 494–500.

Gore, A. C., D. M. Walker, A. M. Zama, A. E. Armenti and M. Uzumcu. 2011. Early life exposure to endocrine-disrupting chemicals causes lifelong molecular reprogramming of the hypothalamus and premature reproductive aging. *Mol. Endocrinol.* 25: 2157–2168.

Gorelick, D. A. and M. E. Halpern. 2011. Visualization of estrogen receptor transcriptional activation in zebrafish. *Endocrinology* 152: 2690–2703.

Gorelick, D. A., L. R. Iwanowicz, A. L. Hung, V. S. Blazer and M. E. Halpern. 2014. Transgenic zebrafish reveal tissue specific differences in estrogen signaling in response to environmental water samples. *Environ. Health Perspect.* 122: 356–362.

Grandjean, P., E. W. Andersen, E. Budtz-Jørgensen, F. Nielsen, K. Mølbak, P. Weihe and C. Heilmann. 2012. Serum vaccine antibody concentrations in children exposed to perfluorinated compounds. *JAMA* 307: 391–397.

Grey, L. E., Jr., J. Ostby, R. L. Cooper and W. R. Kelce. 1999. The estrogenic and antiandrogenic pesticide methoxychlor alters the reproductive tract and behavior without affecting pituitary size or LH and prolactin secretion in male rats. *Toxicol. Indust. Health.* 15: 37–47.

Grier, J. W. 1982. Ban of DDT and subsequent recovery of reproduction in bald eagles. *Science* 218: 1232–1235.

Grün F. and 9 others. 2006. Endocrine-disrupting organotin compounds are potent inducers of adipogenesis in vertebrates. *Mol. Endocrinol.* 20: 2141–2155.

Guerrero-Bosagna, C., M. Settles, B. Lucker and M. K. Skinner. 2010. Epigenetic transgenerational actions of vinclozolin on promoter regions of the sperm epigenome. *PLoS ONE* 5(9): e13100.

Guillette, L. J., Jr. 2006. Endocrine disrupting contaminants: Beyond the dogma. *Environ. Health Persp.* 114(suppl 1): 9–12.

Guillette, L. J., Jr., T. M. Edwards and B. C. Moore. 2007. Alligators, contaminants and steroid hormones. *Environ. Sci.* 14: 331–347.

Guillette, L. J., Jr., T. S. Gross, G. R. Masson, J. M. Matter, H. F. Percival and A. R. Woodward. 1994. Developmental abnormalities of the gonad and abnormal sex hormone concentrations in juvenile alligators from contaminated and control lakes in Florida. *Environ. Health Persp.* 102: 680–688.

Guo, Y. L., G. H. Lambert, C. C. Hsu and M. M. Hsu. 2004. Yucheng: Health effects of prenatal exposure to polychlorinated biphenyls and dibenzofurans. *Internatl. Arch. Occup. Environ. Health* 77: 153–158.

Hao, C. J., X. J. Cheng, H. F. Xia and X. Ma. 2012. The endocrine disruptor diethylstilbestrol induces adipocyte differentiation and promotes obesity in mice. *Toxicol. Appl. Pharmacol.* 263: 102–110.

Harremoes, P., D. Gee, M. MacGarvin, A. Stirling, J. Keys, B. Wynne and S. G. Vaz (eds.). 2002. *The Precautionary Principle in the 20th Century: Late Lessons from Early Warnings*. Earthscan Publications Ltd., London.

Hayes, T. B. 2004. There's no denying this: Defusing the confusion about atrazine. *BioScience* 54: 1138–1149.

Hayes, T. B. 2005. Welcome to the revolution. Integrative biology and assessing the impact of endocrine disruptors on environmental and public health. *Integr. Comp. Biol.* 45: 321–329.

Hayes, T. B., K. Haston, M. Tsui, A. Hoang, C. Haeffele and A. Vonk. 2003. Atrazine-induced hermaphroditism at 0.1 ppb in American leopard frogs (*Rana pipiens*): Laboratory and field evidence. *Environ. Health Persp.* 111: 568–575.

Hayes, T. B. and 10 others. 2006. Pesticide mixtures, endocrine disruption and amphibian declines: Are we underestimating the impact? *Environ. Health Persp.* 114(suppl 1): 40–50.

Heckler, M. M., H. Thakor, C. C. Schafer and R. B. Riggins. 2014. ERK/MAPK regulates ERRγ expression, transcriptional activity and receptor-mediated tamoxifen resistance in ER+ breast cancer. *FEBS J.* 281: 2431–2442.

Heindel, J. 2006. Role of exposure to environmental chemicals in the developmental basis of reproductive disease and dysfunction. *Semin. Reprod. Med.* 24: 168–177.

Heindel, J. 2007. *Endocrine disruption: Status and outlook*. Talk presented at the symposium Endocrine Disruptors: Relevance to Humans, Animals and Ecosystems in Macolin, Switzerland.

Hervouet, E., P. F. Cartron, M. Jouvenot and R. Delage-Mourroux. 2013. Epigenetic regulation of estrogen signaling in breast cancer. *Epigenetics* 8: 237–245.

Ho, S.-M., W.-Y. Tang, J. Belmonte de Frausto and G. S. Prins. 2006. Developmental exposure to estradiol and bisphenol A increases susceptibility to prostate carcinogenesis and epigenetically regulates phosphodiesterase type 4 variant 4. *Cancer Res.* 66: 5624–5632.

Holm, L., A. Blomqvist, I. Brandt, B. Brunstrom, Y. Ridderstrale and C. Berg. 2006. Embryonic exposure to *o,p_*-DDT causes eggshell thinning and altered shell gland carbonic anhydrase expression in the domestic hen. *Environ. Toxicol. Chem.* 25: 2787–2793.

Hoover, R. N. and 18 others. 2011. Adverse health outcomes in women exposed in utero to diethylstilbestrol. *N. Engl. J. Med.* 365: 1304–1314.

Howdeshell, K. L., A. K. Hotchkiss, K. A. Thayer, J. G. Vandenbergh and F. S. vom Saal. 1999. Plastic bisphenol A speeds growth and puberty. *Nature* 401: 762–764.

Howdeshell, K. L. and F. S. vom Saal. 2000. Developmental exposure to bisphenol A: Interaction with endogenous estradiol during pregnancy in mice. *Amer. Zool.* 40: 429–437.

Howdeshell, K. L. and 7 others. 2003. Bisphenol A is released from used polycarbonate animal cages into water at room temperature. *Environ. Health Persp.* 111: 1180–1187.

Huang, P. C., P. L. Kuo, Y. Y. Chou, S. J. Lin and C. C. Lee. 2008. Association between prenatal exposure to phthalates and the health of newborns. *Environ. Internatl.* doi: 10.1016/jenvint.2008.05.012

Hunt, P. A. and 8 others. 2003. Bisphenol A exposure causes meiotic aneuploidy in the female mouse. *Curr. Biol.* 13: 546–553.

Hunt, P. A. and 7 others. 2012. Bisphenol A alters early oogenesis and follicle formation in the fetal ovary of the rhesus monkey. *Proc. Natl. Acad. Sci. USA* 109: 17525–17530.

Ikezuki, Y., O. Tsutsumi, Y. Tahai, Y. Kamzi and Y. Taketa. 2002. Determination of bisphenol A concentrations in human biological fluids reveals significant prenatal exposure. *Human Reprod.* 17: 2839–2841.

Irvine, S., E. Cawood, D. Richardson, E. MacDonald and J. Aitken. 1996. Evidence of deteriorating semen quality in the United Kingdom: Birth cohort study in 577 men in Scotland over 11 years. *Brit. Med. J.* 312: 467–471.

Jackson, R. B. and 8 others. 2013. Increased stray gas abundance in a subset of drinking water wells near Marcellus shale gas extraction. *Proc. Natl. Acad. Sci. USA* 110: 11250–11255.

Janesick, A. and B. Blumberg. 2011. Minireview: PPARγ as the target of obesogens. *J. Steroid Biochem. Mol. Biol.* 127: 4–8.

Janesick, A. and B. Blumberg. 2012. Obesogens, stem cells and the developmental programming of obesity. *Int. J. Androl.* 35: 437–448.

Janesick, A., T. Shioda and B. Blumberg. 2014. Transgenerational inheritance of prenatal obesogen exposure. *Mol. Cell Endocrin.* 398: 31–35.

Jefferson, W. N., E. Padilla-Banks and R. R. Newbold. 2007. Disruption of the developing female reproductive system by phytoestrogens: Genistein as an example. *Mol. Nutr. Food Res.* 51: 832–844.

Jefferson, W. N. and 8 others. 2013. Persistently altered epigenetic marks in the mouse uterus after neonatal estrogen exposure. *Mol. Endocrinol.* 27: 1666–1677.

Jeschke, U. and 7 others. 2005. Effects of phytoestrogens genistein and daidzein on production of human chorionic gonadotropin in term trophoblast cells in vitro. *Gynecol. Endocrinol.* 21: 180–184.

Ji, L., R. C. Domanski, R. C. Skirrow and C. C. Helbing. 2007. Genistein prevents thyroid hormone-dependent tail regression of *Rana catesbeiana* tadpoles by targeting protein kinase C and thyroid hormone receptor α. *Devel. Dynam.* 236: 777–790.

Jobling, S., A. Baynes and Y. W. J. Garner. 2013. Frogs at risk and possible implications for humans. www.CHEM-Trust-Frogs-Immune-Report-FINAL.pdf

Jørgensen N. and 11 others. 2012. Human semen quality in the new millennium: A prospective cross-sectional population based study of 4867 men. *BMJ Open* 2: e000990.

Juul, A. and 8 others. 2014. Possible fetal determinates of male infertility. *Nature Rev. Endocrinol.* 10: 553–562.

Kaiser, J. and M. Enserink. 2000. Treaty takes a POP at the dirty dozen. *Science* 290: 2053.

Karrow, N. A., J. A. McCay, R. D. Brown, D. L. Musgrove, T. L. Guo, D. R. Germolec and K. L. White, Jr. 2005. Oral exposure to atrazine modulates cell-mediated immune function and decreases host resistance to the B16F10 tumor model in female B6C3F1 mice. *Toxicology* 209: 15–28.

Kassotis, C. D., D. E. Tillitt, J. W. Davis, A. M. Hormann and S. C. Nagel. 2014. Estrogen and androgen receptor activities of hydraulic fracturing chemicals and surface and ground water in a drilling-dense region. *Endocrinology* 155: 897–907.

Kelce, W. R., C. R. Stone, S. C. Laws, L. E. Gray, J. A. Kemppainen and E. M. Wilson. 1995. Persistent DDT metabolite *p,p_*-DDE is a potent androgen receptor antagonist. *Nature* 375: 581–585.

Kennedy, D. 2007. Toxic dilemmas. *Science* 318: 1217.

Kidd, K. A., P. J. Blanchfield, K. H. Mills, V. P. Palace, R. E. Evans, J. M. Lazorchak and R. W. Flick. 2007. Collapse of a fish population after exposure to a synthetic estrogen. *Proc. Natl. Acad. Sci. USA* 104: 8897–8901.

Kiesecker, J. M. 2002. Synergism between trematode infection and pesticide exposure: A link to amphibian limb deformities in nature? *Proc. Natl. Acad. Sci.* 99: 9900–9904.

Kirchner, S., T. Kieu, C. Chow, S. Casey and B. Blumberg. 2010. Prenatal exposure to the environmental obesogen tributyltin predisposes multipotent stem cells to become adipocytes. *Mol. Endocrinol.* 24: 526–539.

Kim, H. S. and 8 others. 2002. Comparative estrogenic effects of *p*-nonylphenol by 3-day uterotrophic assay and female pubertal onset assay. *Reprod. Toxicol.* 16: 259–268.

Knight, W. M. 1980. Estrogens administered to food-producing animals: Environmental considerations. In J. McLachlan (ed.), *Estrogens in the Environment*. Elsevier, Amsterdam, pp. 391–401.

Koeppe, E. S., K. K. Ferguson, J. A. Colacino and J. D. Meeker. 2013. Relationship between urinary triclosan and paraben concentrations and serum thyroid measures in NHANES 2007–2008. *Sci. Total Environ.* 445–446: 299–305.

Kohno, S., D. S. Bermudez, Y. Katsu, T. Iguchi and L. B. Guillette, Jr. 2008. Gene expression patterns in juvenile American alligators (*Alligator mississippiensis*) exposed to environmental contaminants. *Aquatic Toxicol.* 88: 95–101.

Kollipara, P. 2015. Reform of toxics law is contentious. *Science* 347: 1403–1404.

Kouki, T., M. Kishitake, M. Okamoto, I. Oosuka, M. Takebe and K. Yamanouchi. 2003. Effects of neonatal treatment with phytoestrogens, genistein and daidzein, on sex difference in female rat brain function: estrous cycle and lordosis. *Horm. Behav.* 44: 140–145.

Krimsky, S. 2000. *Hormonal Chaos: The Scientific and Social Origins of the Environmental Endocrine Hypothesis*. Johns Hopkins University Press, Baltimore.

Krishnan, A. V., P. Starhis, S. F. Perlmuth, I. Tokes and D. Feldman. 1993. Bisphenol A: An estrogenic substance is released from polycarbonate flasks during autoclaving. *Endocrinology* 132: 2279–2286.

Kubo, K., O. Arai, M. Omura, R. Watanabe, R. Ogata and S. Aou. 2003. Low-dose effects of bisphenol A on sexual differentiation of the brain and behavior in rats. *Neurosci. Res.* 45: 345–356.

Kuiper, G. G. and 7 others. 1998. Interaction of estrogenic chemicals and phytoestrogens with estrogen receptor beta. *Endocrinology* 139: 4252–4263.

Kurihara, R., E. Watanabe, Y. Ueda, A. Kakuno, K. Fujiki, F. Shiraishi and S. Hishimoto. 2007. Estrogenic activity in sediments contaminated by nonylphenol in Tokyo Bay (Japan) evaluated by vitellogenin induction in male mummichogs (*Fundulus heteroclitus*). *Mar. Pollut. Bull.* 54: 1315–1320.

Landrigan, P. J., V. A. Rauh and M. P. Galvez. 2010. Environmental justice and the health of children. *Mt. Sinai J. Med.* 77: 178–187.

Lannoo, M. 2008. *Malformed Frogs: The Collapse of Aquatic Ecosystems*. University of California Press, Berkeley.

Lathi, R. B., C. A. Liebert, K. F. Brookfield, J. A. Taylor, F. S. vom Saal, V. Y. Fujimoto and V. L. Baker. Conjugated bisphenol A in maternal serum in relation to miscarriage risk. *Fertil. Steril.* 102: 123–128.

Layton, L. 2008. BPA linked to primate health issues. *Washington Post*, September 4, 2008, p. A2.

Layton, L. and C. Lee. 2008. Canada bans BPA from baby bottles. *Washington Post*, April 19, 2008, p. A3.

Lear, L. 1998. *Rachel Carson: Witness for Nature*. Holt, New York.

Lee, O., A. Takesono, M. Tada, C. R. Tyler and T. Kudoh. 2012. Biosensor zebrafish provide new insights into potential health effects of environmental estrogens. *Environ. Health Perspect.* 120: 990–996.

Leranth, C., T. Hajszan, K. Szigeti-Buck, J. Bober and N. J. Maclusky. 2008. Bisphenol A prevents the synaptogenic response to estradiol in hippocampus and prefrontal cortex of ovariectomized nonhuman primates. *Proc. Natl. Acad. Sci. USA* 105: 14187–14191.

Li, S., S. D. Hursting, B. J. Davis, J. A. McLachlan and J. C. Barrett. 2003. Environmental exposure, DNA methylation, and gene regulation: Lessons from diethylstilbestrol-induced cancers. *Ann. N.Y. Acad. Sci.* 983: 161–169.

Lilienthal, H., P. Heikkinen, P. L. Andersson, L. T. van der Ven and M. Viluksela. 2013. Dopamine-dependent behavior in adult rats after perinatal exposure to purity-controlled polychlorinated biphenyl congeners (PCB52 and PCB180). *Toxicol. Lett.* 224: 32–39.

Lin, T.-C. E., S.-C. Chien, P.-C. Hsu and L.-A. Li. 2006. Mechanistic study of polychlorinated bisphenyl 126-induced CYP11B1 and CYP11B2 upregulation. *Endocrinology* 147: 1536–1544.

Liu, X., A. Matsushima, M. Shimohigashi and Y. Shimohigashi. 2014. A characteristic back support structure in the bisphenol A-binding pocket in the human nuclear receptor ERRγ. *PLoS ONE* 9(6): e101252.

Loos, R. J. 2009. Recent progress in the genetics of common obesity. *Br. J. Clin. Pharmacol.* 68: 811–829.

Loos, R. J. and G. S. Yeo. 2014. The bigger picture of FTO: The first GWAS-identified obesity gene. *Nat. Rev. Endocrinol.* 10: 51–61.

Lundholm, C. D. 1997. DDE-induced eggshell thinning in birds: Effects of p,p_-DDE on the calcium and prostaglandin metabolism of the eggshell gland. *Comp. Biochem. Physiol. C: Pharmacol. Toxicol. Endocrinol.* 118: 113–128.

Ma, L. 2009. Endocrine disruptors in female reproductive tract development and carcinogenesis. *Trends Endocrinol. Metab.* 20: 357–363.

Ma, L., G. V. Benson, H. Lim, S. K. Dey and R. Maas. 1998. *Abdominal B (AbdB) Hoxa* genes: Regulation in adult uterus by estrogen and progesterone and repression in Mullerian duct by the synthetic estrogen diethylstilbestrol (DES). *Dev. Biol.* 197: 141–154.

Macellaro, G., C. Pezzella, P. Cicatiello, G. Sannia and A. Piscitelli. 2014. Fungal laccases degradation of endocrine disrupting compounds. *Biomed. Res. Int.* 2014: 614038. doi: 10.1155/2014/614038

Mahood, I. K., H. M. Scott, R. Brown, N. Hallmark, M. Walker and R. M. Sharpe. 2007. In utero exposure to di(n-butyl)phthalate and testicular dysgenesis: Comparison of fetal and adult endpoints and their dose sensitivity. *Environ. Health Persp.* 115(Supp 1): 55–61.

Markey, C. M., M. A. Coombs, C. Sonnenschein and A. M. Soto. 2003. Mammalian development in a changing environment: Exposure to endocrine disruptors reveals the developmental plasticity of steroid hormone target organs. *Evol. Dev.* 5: 67–75.

Markey, C. M., E. H. Luque, M. Muñoz-de-Toro, C. Sonnenschein and A. M. Soto. 2001. In utero exposure to bisphenol A alters the development and tissue organization of the mouse mammary gland. *Biol. Reprod.* 65: 1215–1223.

Markey, C. M., P. R. Wadia, B. S. Rubin, C. Sonnenschein and A. M. Soto. 2005. Long-term effects of fetal exposure to low doses of the xenoestrogen bisphenol-A in the female mouse genital tract. *Biol. Reprod.* 72: 1344–1351.

Markowitz, G. and D. Rosner. 2000. "Cater to the children": The role of the lead industry in a public health tragedy, 1900–1955. *Am. J. Publ. Health* 90: 36–46.

Marlatt, V. L. and 11 others. 2013. Triclosan exposure alters postembryonic development in a Pacific tree frog (*Pseudacris regilla*) Amphibian Metamorphosis Assay (TREEMA). *Aquat. Toxicol.* 126: 85–94.

Martin, N. P. and 9 others. 2014. A rapid cytoplasmic mechanism for PI3 kinase regulation by the nuclear thyroid hormone receptor, TRβ, and genetic evidence for its role in the maturation of mouse hippocampal synapses in vivo. *Endocrinology* 155: 3713–3724.

Matter, J. M. and 8 others. 1998. Effects of endocrine-disrupting contaminants in reptiles: Alligators. In R. J. Kendall, R. L. Dickerson, J. P. Geisy and W. A. Suk (eds.), *Principles and Processes for Evaluating Endocrine Disruptions in Wildlife.* SETAC Press, Pensacola, FL, pp. 267–289.

McCoy, K. A., L. J. Bortnick, C. M. Campbell, H. J. Hamlin, L. J. Guillette, Jr., and C. M. St. Mary. 2008. Agriculture alters gonadal form and function in the toad *Bufo marinus. Env. Health Persp.* 116: 1526–1532.

McKenzie, L. M., R. Guo, R. Z. Witter, D. A. Savitz, L. S. Newman and J. L. Adgate. 2014. Birth outcomes and maternal residential proximity to natural gas development in rural Colorado. *Environ. Health Perspect.* 122: 412–417.

McNabb, A. and 7 others. 1999. Basic physiology. In R. T. Di Giulio and D. E. Tillitt (eds.), *Reproductive and Developmental Effects of Contaminants in Oviparous Vertebrates.* SETAC Press, Pensacola, FL, pp. 113–223.

Menikkam, M., R. Tracey, C. Guerreri-Bosagnia and M. K. Skinner. 2013. Plastics derived endocrine disruptors (BPA, DEHP and DBP) induce epigenetic transgenerational inheritance, of obesity, reproductive disease, and sperm epimutations. *PLoS ONE* 8(1): e55387.

Mervish, N. A. and 10 others. 2013. Dietary flavonol intake is associated with age of puberty in a longitudinal cohort of girls. *Nutr. Res.* 33: 534–542.

Michaels, D. 2005. Doubt is their product. *Sci. Am.* 292(6): 96–101.

Michaels, D. 2008. *Doubt Is Their Product: How Industry's Assault on Science Threatens Your Health.* Oxford University Press, New York.

Miller, C., K. Degenhardt and D. A. Sassoon. 1998. Fetal exposure to DES results in de-regulation of *Wnt7a* during uterine morphogenesis. *Nature Genet.* 20: 228–230.

Miller, S. M., C. W. Sweet, J. V. DePinto and K. C. Hornbuckle. 2000. Atrazine and nutrients in precipitation: Results from the Lake Michigan mass balance study. *Environ. Sci. Tech.* 34: 55–61.

Milnes, M. R., T. A. Bryan, Y. Katsu, S. Kohno, B. C. Moore, T. Iguchi and L. J. Guillette, Jr. 2008. Increased post-hatching mortality and loss of sexually dimorphic gene expression in alligators (*Alligator mississippiensis*) from a contaminated environment. *Biol. Reprod.* 78: 932–938.

Miyakoda, H., M. Tabata, S. Onodera and K. Takeda. 1999. Passage of bisphenol A into the fetus of the pregnant rat. *J. Health Sci.* 45: 318–323.

Moggs, J. G. and G. Orphanides. 2001. Estrogen receptors: Orchestrators of pleiotropic cellular responses. *EMBO Rep.* 2: 775–781.

Mol, N. M. and 7 others. 2002. Spermaturia and serum hormone concentrations at the age of puberty in boys prenatally exposed to polychlorinated biphenyls. *Eur. J. Endocrinol.* 146: 357–363.

Monosson, E., W. R. Kelce, C. Lambrigh, J. Ostby and L. E. Grey, Jr. 1999. Peripubertal exposure to the antiandrogenic fungicide, vinclozolin, delays puberty, inhibits the development of androgen-dependent tissues and alters androgen receptor function in the male rat. *Toxicol. Indust. Health* 15: 65–79.

Muñoz-de-Toro, M., C. Markey, P. R. Wadia, E. H. Luque, B. S. Rubin, C. Sonnenschein and A. M. Soto. 2005. Perinatal exposure to bisphenol A alters peripubertal mammary gland development in mice. *Endocrinology* 146: 4138–4147.

Murray, T. J., M. V. Maffini, A. A. Ucci, C. Sonenschein and A. M. Soto. 2007. Induction of mammary gland ductal hyperplasias and carcinoma in situ following bisphenol-A exposure. *Reprod. Toxicol.* 23: 383–390.

Nakagami, A. and 8 others. 2009. Alterations in male infant behaviors towards its mother by prenatal exposure to bisphenol A in cynomolgus monkeys (*Macaca fascicularis*) during early suckling period. *Psychoneuroendocrinology* 34:1189–1197.

Nalgene. 2015. Anatomy of a Nalgene. Nalgene.com/anatomy

Narita, M., K. Miyagawa, K. Mizui, T. Yoshida and T. Suzuki. 2006. Prenatal and neonatal exposure to low-dose of bisphenol A enhance the morphone-induced hyperlocomotion and rewarding effect. *Neurosci. Lett.* 402: 249–252.

Nelson, J. W., M. K. Scammell, E. E. Hatch and T. F. Webster. 2012. Social disparities in exposures to bisphenol A and polyfluoroalkyl chemicals: A cross-sectional study within NHANES 2003–2006. *Env. Health* 11: 10.

Newbold, R. R. 2004. Lessons learned from prenatal exposure to diethylstilbestrol. *Toxicol. Appl. Pharmacol.* 199: 142–150.

Newbold, R. R., E. P. Banks and W. N. Jefferson. 2006. Adverse effects of the model environmental estrogen diethylstilbestrol are transmitted to subsequent generations. *Endocrinology* 147(suppl): S11–S17.

Newbold, R. R., E. P. Banks, R. J. Snyder and W. N. Jefferson. 2007. Perinatal exposure to environmental estrogens and the development of obesity. *Mol. Nutr. Food Res.* 51: 912–917.

Newbold, R. R., R. B. Hanson, W. N. Jefferson, B. C. Bullock, J. Haseman and J. A. McLachlan. 1998. Increased tumors but uncompromised fertility in the female descendants of mice exposed developmentally to diethylstilbestrol. *Carcinogenesis* 19: 1655–1663.

Newbold, R. R., R. B. Hanson, W. N. Jefferson, B. C. Bullock, J. Haseman and J. A. McLachlan. 2000. Proliferative lesions and reproductive tract tumors in male descendants of mice exposed developmentally to diethylstilbestrol. *Carcinogenesis* 21: 1355–1363.

Newbold, R. R., E. Padilla-Banks, B. C. Bullock and W. N. Jefferson. 2001. Uterine adenocarcinoma in mice treated neonatally with genistein. *Cancer Res.* 61: 4325–4328.

Newbold, R. R., S. Tyrey, A. F. Haney and J. A. MacLachlan. 1983. Developmentally arrested oviduct: A structural and functional defect in mice following prenatal exposure to diethylstilbestrol. *Teratology* 27: 417–426.

Newton, I. and I. Wyllie. 1992. Recovery of a sparrowhawk population in relation to declining pesticide contamination. *J. Applied Ecol.* 29: 476–484.

NIEHS (National Institute of Environmental Health Sciences). 2008. www.niehs.nih.gov/news/media/questions/sya-bpa.cfm#4

NRC (National Research Council). 1999. *Hormonally Active Agents in the Environment.* National Academies Press, Washington, DC.

Oberdörster, E. and P. McClellan-Green. 2002. Mechanisms of imposex induction in the mud snail, *Ilyanassa obsoleta*: TBT as a neurotoxin and aromatase inhibitor. *Mar. Environ. Res.* 54: 715–718.

Orlando, L. 2002. Industry attacks on dissent: From Rachel Carson to Oprah. www.dollarsandsense.org/archives/2002/0302orlando.html

Osborn, S. G., A. Vengosh, N. R. Warner and R. B. Jackson. 2011. Methane contamination of drinking water accompanying gas-well drilling and hydraulic fracturing. *Proc. Natl. Acad. Sci. USA* 108: 8172–8176.

Padilla-Banks, E., W. N. Jefferson and R. R. Newbold. 2006. Neonatal exposure to the phytoestrogen genistein alters mammary gland growth and developmental programming of hormone receptor levels. *Endocrinology* 147: 4871–4882.

Paetow, L. J., J. D. McLaughlin, B. D. Pauli and D. J. Marcogliese. 2013. Mortality of American bullfrog tadpoles *Lithobates catesbeianus* infected by *Gyrodactylos jennyae* and experimentally exposed to *Batrachochytrium dendrobatidis*. *J. Aquat. Anim. Health* 25: 15–26.

Pajarinen, J., P. Laippala, A. Penttila and P. J. Karhunen. 1997. Incidence of disorders of spermatogenesis in middle-aged Finnish men 1981–1991: Two necropsy series. *Brit. Med. J.* 314: 13–18.

Palanza, P., K. L. Howdeshell, S. Parmigiani and F. S. vom Saal. 2002. Exposure to a low dose of bisphenol A during fetal life or in adulthood alters maternal behavior in mice. *Environ. Health Persp.* 110(suppl.): 415–422.

Palmer, J. R. and 10 others. 2006. Prenatal DES exposure and risks of breast cancer. *Cancer Epidem. Biomarkers Prevent.* 15: 1509–1514.

Palmund, I. 1996. Exposure to a xenoestrogen before birth: The diethylstilbestrol experience. *J. Psychosom. Obstet. Gynecol.* 17: 71–84.

Patisaul, H. B., A. E. Fortino and E. K. Polston. 2006 Neonatal genistein or bisphenol-A exposure alters sexual differentiation of the AVPV. *Neurotoxicol. Teratol.* 28: 111–118.

Patisaul, H. B., J. R. Luskin and M. E. Wilson. 2004. A soy supplement and tamoxifen inhibit sexual behavior in female rats. *Horm. Behav.* 45: 270–277.

Paul, K. B., J. M. Hedge, R. Bansal, R. T. Zoeller, R. Peter, M. J. De-Vito and K. M. Crofton. 2012. Developmental triclosan exposure decreases maternal, fetal, and early neonatal thyroxine: A dynamic and kinetic evaluation of a putative mode-of-action. *Toxicology* 300: 31–45.

Powell, C. E., A. M. Soto and C. Sonnenschein. 2001. Identification and characterization of membrane estrogen receptor form MCF7 estrogen-target cells. *J. Steroid Biochem. Mol. Biol.* 77: 97–108.

Préau, L., J. B. Fini, G. Morvan-Dubois and B. Demeneix. 2015. Thyroid hormone signaling during early neurogenesis and its significance as a vulnerable window for endocrine disruption. *Biochim. Biophys. Acta.* 1849: 112–121.

Prossnitz, E. R. and M. Barton. 2014. Estrogen biology: New insights into GPER function and clinical opportunities. *Mol. Cell Endocrinol.* 389: 71–83.

Qin, X. Y. and 15 others. 2012. Individual variation of the genetic response to bisphenol a in human foreskin fibroblast cells derived from cryptorchidism and hypospadias patients. *PLoS ONE* 7(12): e52756.

Raffensberger, C. and J. Tickner (eds.). 1999. *Protecting Public Health and the Environment: Implementing the Precautionary Principle.* Island Press, Washington, D.C.

Rajapakse, N., D. Ong and A. Kortenkamp. 2001. Defining the impact of weakly estrogenic chemicals on the action of steroidal estrogens. *Toxicol. Sci.* 60: 296–304.

Rajapakse, N., E. Silva and A. Kotenkamp. 2002. Combining xenoestrogens at levels below individual no-observed-effect concentrations dramatically enhances steroid hormone action. *Environ. Health Persp.* 110: 917–921.

Reynaud, J., D. Guilet, R. Terreux, M. Lussignol and N. Walchshofer. 2005. Isoflavonoids in non-leguminous families: An update. *Nat. Prod. Rep.* 22: 504–515.

Ribas-Fito, N., M. Sala, M. Kogevinas and J. Sunyer. 2001. Polychlorinated biphenyls (PCBs) and neurological development in children: A systematic review. *J. Epid. Comm. Health* 55: 537–546.

Richter, D. U., I. Mylonas, B. Toth, C. Scholz, V. Briese, K. Friese and U. Jeschke. 2009. Effects of phytoestrogens genistein and daidzein on progesterone and estrogen (estradiol) production of human term trophoblast cells in vitro. *Gynecol. Endocrinol.* 25: 32–38.

Robboy, S. J., R. H. Young and A. L. Herbst. 1982. Female genital tract changes related to prenatal diethylstilbestrol exposure. In A. Blaustein (ed.), *Pathology of the Female Genital Tract*, 2nd Ed. Springer-Verlag, New York, pp. 99–118.

Roberge, M., H. Hakk and G. Larsen. 2004. Atrazine is a competitive inhibitor of phosphodiesterase but does not affect the estrogen receptor. *Toxicol. Lett.* 154: 61–68.

Ronckers, C. M., C. A. Erdmann and C. E. Land. 2005. Radiation and breast cancer: A review of current evidence. *Breast Cancer Res.* 7: 21–32.

Ross, G. 2004. The public health implications of polychlorinated biphenyls (PCBs) in the environment. *Ecotoxicol. Environ. Safety* 59: 275–291.

Rowe, A., K. M. Brundage, R. Schafer and J. B. Barnett. 2006. Immunomodulatory effects of maternal atrazine exposure on male Balb/c mice. *Toxicol. Appl. Pharmacol.* 214: 69–77.

Royland, J. E. and P. R. S. Kodavanti. 2008. Gene expression profiles following exposure to a developmental neurotoxicant, Aroclor 1254: Pathway analysis for possible mode(s) of action. *Toxicol. Appl. Pharmacol.* 231: 179–196.

Royland, J. E., J. Wu, N. H. Zawia and P. R. S. Kodaventi. 2008. Gene expression profiles in the cerebellum and hippocampus following exposure to a neurotoxicant, Aroclor 1254: Developmental effects. *Toxicol. Appl. Pharmacol.* 231: 165–178.

Rubin, B. S., J. R. Lenkowski, C. M. Schaeberle, L. N. Vandenberg, P. M. Ronscheim and A. M. Soto. 2006. Evidence of altered brain sexual differentiation in mice exposed perinatally to low, environmentally relevant levels of bisphenol A. *Endocrinology* 147: 3681–3691.

Rubin, B. S., M. K. Murray, D. A. Damassa, D. A. King and A. M. Soto. 2001. Perinatal exposure to low doses of bisphenol A affects body weight, patterns of estrous cyclicity and plasma LH levels. *Environ. Health Persp.* 109: 675–680.

Sanderson, J. T., W. Seinen, J. P. Giesy and M. van den Berg. 2000. 2-Chloro-s-triazine herbicides induce aromatase (*CYP19*) activity in H295 human adrenocortical carcinoma cells: A novel mechanism of estrogenicity? *Toxicol. Sci.* 54: 121–127.

Sass, J. B. and A. Colangelo. 2006. European Union bans atrazine, while the United States negotiates continued use. *Int. J. Occup. Environ. Health.* 12: 260–267.

Schantz, S. L., J. J. Widholm and D. C. Rice. 2003. Effects of PCB exposure on neuropsychological function in children. *Environ. Health Persp.* 111: 357–376.

Schapiro, M. 2007. *Exposed: The Toxic Chemistry of Everyday Products and What's at Stake for American Power.* Chelsea Green Publishing, White River Junction, VT.

Schlumpf, M. and 14 others. 2004. Endocrine activity and developmental toxicity of cosmetic UV filters: An update. *Toxicology* 205: 113–122.

Schönfelder, G., W. Wittfoht, H. Hopp, C. E. Talsness, M. Paul and I. Chahoud. 2002. Parent bisphenol A accumulation in the human maternal-fetal-placental unit. *Environ. Health Persp.* 110: A703–A707.

Sharpe, R. M. 1994. Could environmental oestrogenic chemical be responsible for some disorders of human male reproductive development? *Curr. Opin. Urol.* 4: 295–301.

Sharpe, R. M. and D. S. Irvine. 2004. How strong is the evidence of a link between environmental chemicals and adverse effects on human reproductive health? *Brit. Med. J.* 328: 447–451.

Silva, E., N. Rajapakse and A. Kortenkamp. 2002. Something from "nothing": Eight weak estrogenic chemicals combined at concentration below NOECs produce significant mixture effects. *Environ. Sci. Technol.* 36: 1751–1756.

Skakkebaeck, N. E. 2004. Testicular dysgenesis syndrome: New epidemiological evidence. *Internatl. J. Andrology* 27: 189–191.

Skinner, M. K., M. Manikkam and C. Guerrero-Bosagna. 2010. Epigenetic transgenerational actions of endocrine disruptors. *Reprod. Toxicol.* 31: 337–343.

Skinner, M. K., M. Manikkam, M. M. Haque, B. Zhang and M. I. Savenkova. 2012. Epigenetic transgenerational inheritance of somatic transcriptomes and epigenetic control regions. *Genome Biol.* 13: R91.

Smith, M. D. and E. M. Hill. 2006. Profiles of short-chain oligomers in roach (*Rutilus rutilus*) exposed to waterborne polyethoxylated nonulphenols. *Sci. Total Environ.* 356: 100–111.

Sonnenschein, C., P. R. Wadia, P. R. Rubin and P. R. Soto. 2011. Cancer as development gone awry: The case for bisphenol-A as a carcinogen. *J. Dev. Origins Health Dis.* 2: 9–16.

Soriano, S. and 8 others. 2012. Rapid insulinotropic action of low doses of bisphenol-A on mouse and human islets of Langerhans: Role of estrogen receptor β. *PLoS ONE* 7(2): e31109.

Soto, A. M., H. Justicia, J. Wray and C. Sonnenschein. 1991. *p*-Nonylphenol: An estrogenic xenobiotic released from "modified" polystyrene. *Environ. Health Persp.* 92: 167–173.

Soto, A. M., M. U. Maffini and C. Sonnenschein. 2008. Neoplasia as development gone awry: The role of endocrine disruptors. *Internatl. J. Androbgy* 31: 280–293.

Spearow, J. L., P. Doemeny, R. Sera, R. Leffler and M. Barkley. 1999. Genetic variation in susceptibility to endocrine disruption by estrogen in mice. *Science* 285: 1259–1261.

Steingraber, S. 2001. *Having Faith: An Ecologist's Journey into Motherhood.* Berkley, New York.

Stouder, C. and A. Paoloni-Giacobino. 2010. Transgenerational effects of the endocrine disruptor vinclozolin on the methylation pattern of imprinted genes in the mouse sperm. *Reproduction* 139: 373–379.

Stouder, W. 2012. *On a Farther Shore: The Life and Legacy of Rachel Carson.* Broadway Books, New York.

Stutchbury, B. 2008 Did your shopping list kill a songbird? *New York Times.* www.nytimes.com/2008/03/30/opinion/30stutchbury.html?_r=0

Stutchbury, B. 2009. *The Silence of the Songbirds.* Walker, New York.

Sun, Y., C. Wang, H. Yang, and X. Ma. 2014. The effect of estrogen on the proliferation of endometrial cancer cells is mediated by ERRγ through AKT and ERK1/2. *Eur. J. Cancer Prev.* 23: 418–424.

Sunstein, C. 2005. *Laws of Fear: Beyond the Precautionary Principle* (The Seeley Lectures). Cambridge University Press, Cambridge.

Susiarjo, M., T. J. Hassold, E. Freeman and P. Hunt. 2007. Bisphenol A exposure in utero disrupts early oogenesis in the mouse. *PLoS Genet.* 3: 63–70.

Susiarjo, M., I. Sasson, C. Mesaros, and M. S. Bartolomei. 2013. Bisphenol a exposure disrupts genomic imprinting in the mouse. *PLoS Genet.* 9: e1003401.

Swan, S. H. 2006. Semen quality in fertile U.S. men in relation to geographical area and pesticide exposure. *Internatl. J. Androl.* 29: 62–68.

Swan, S. H. and 9 others. 2003. Semen quality in relation to biomarkers of pesticide exposure. *Environ. Health Perspect.* 111: 1478–1484.

Swan, S. H. and 11 others. 2005. Decrease in anogenital distance among male infants with prenatal phthalate exposure. *Env. Health Persp.* 113: 1056–1063.

Tarapore, P., J. Ying, B. Ouyang, B. Burke, B. Bracken and S. M. Ho. 2014. Exposure to bisphenol A correlates with early-onset prostate cancer and promotes centrosome amplification and anchorage-independent growth in vitro. *PLoS ONE* 9: e90332.

Tharp A.P., M. V. Maffini, P. A. Hunt, C. A. VandeVoort, C. Sonnenschein and A. M. Soto. 2012. Bisphenol A alters the development of the rhesus monkey mammary gland. *Proc. Natl. Acad. Sci. USA* 109: 8190–8195.

Timms, B. G., K. Howdeshell, L. Barton, S. Bradley, C. A. Richter and F. S. vom Saal. 2005. Estrogenic chemicals in plastic and oral contraceptives disrupt development of the fetal mouse prostate and urethra. *Proc. Natl. Acad. Sci. USA* 102: 7014–7019.

Tohmé, M. and 9 others. 2014. Estrogen-related receptor γ is an in vivo receptor of bisphenol A. *FASEB J.* 28: 3124–3133.

Trisomboon, H., S. Malaivijitnond, G. Watanabe and K. Taya. 2005. Ovulation block by *Pueraria mirifica*: A study of its endocrinological effect in female monkeys. *Endocrine* 26: 33–39.

Turusov, V. S., L. S. Trukhanova, Y. D. Parfenov and L. Tomatis. 1992. Occurrence of tumours in the descendants of CBA male mice prenatally treated with diethylstilbestrol. *Internatl. J. Cancer* 50: 131–135.

UNEP. 2008. Status of ratifications. chm.pops.int/TheConvention/Overview/History/Overview/tabid/3549/Default.aspx; chm.pops.int/Countries/StatusofRatifications/PartiesandSignatories/tabid/252/Default.aspx

U.S.G.S. (United States Geological Survey). 2014. pubs.usgs.gov/bat/fig1.gif.

Vajda, A. M., L. B. Barber, J. L. Gray, E. M. Lopez, A. M. Bolden, H. L. Schoenfuss and D. O. Norris. 2011. Demasculinization of male fish by wastewater treatment plant effluent. *Aquat. Toxicol.* 103: 213–321.

Vandenberg, L. N., P. A. Hunt, J. P. Myers and F. S. vom Saal. 2013. Human exposures to bisphenol A: Mismatches between data and assumptions. *Rev. Environ. Health* 28: 37–58.

Vandenberg, L. N., M. V. Maffini, P. R. Wadia, C. Sonnenschein, B. S. Rubin and A. M. Soto. 2007. Exposure to environmentally relevant doses of the xenoestrogen bisphenol A alters development of the fetal mouse mammary gland. *Endocrinology* 148: 116–127.

Vandenberg, L. N. and 11 others. 2012. Hormones and endocrine-disrupting chemicals: Low-dose effects and nonmonotonic dose responses. *Endocr. Rev.* 33: 378–455.

vom Saal, F. S. 1998. Frontline Interview. www.pbs.org/wgbh/pages/frontline/shows/nature/interviews/vomsaal.html

vom Saal, F. S. 2000. Very low doses of bisphenol A and other estrogenic chemicals alter development in mice. Endocrine Disruptors and Pharmaceutical Active Compounds in Drinking Water Workshop.

vom Saal, F. S. 2008. Bisphenol A references. endocrinedisruptors. missouri.edu/vomsaal/vomsaal.html

vom Saal, F. S. and C. Hughes. 2005. An extensive new literature concerning low-dose effects of bisphenol A shows the need for a new risk assessment. *Environ. Health Persp.* 113: 926–933.

vom Saal, F. S., S. C. Nagel, B. L. Coe, B. M. Angle and J. A. Taylor. 2012. The estrogenic endocrine disrupting chemical bisphenol A (BPA) and obesity. *Mol. Cell Endocrinol.* 354: 74–84.

vom Saal, F. S. and W. Welshons. 2006. Large effects from small exposures. II. The importance of positive controls in low-dose research on bisphenol A. *Environ. Res.* 100: 50–76.

vom Saal, F. S. and 7 others. 1998. A physiologically based approach to the study of bisphenol A and other estrogenic chemicals on the size of reproductive organs, daily sperm production, and behavior. *Toxicol. Indust. Health* 14: 239–260.

vom Saal, F. S. and 9 others. 1997. Prostate enlargement in mice due to fetal exposure to low doses of estradiol or diethylstilbestrol and opposite effects at high doses. *Proc. Natl. Acad. Sci. USA* 94: 2056–2061.

vom Saal, F. S. and 37 others. 2007. Chapel Hill bisphenol A expert panel consensus statement: integration of mechanisms, effects in animals and potential to impact human health at current levels of exposure. *Reprod Toxicol.* 24: 131–138.

von Goetz, N., M. Wormuth, M. Scheringer and K. Hungerbühler. 2010. Bisphenol A: How the most relevant exposure sources contribute to total consumer exposure. *Risk Anal.* 30: 473–487.

Wadia, P. R., N. J. Cabaton, M. D. Borrero, B. S. Rubin, C. Sonnenschein, T. Shioda and A. M. Soto. 2013. Low-dose BPA exposure alters the mesenchymal and epithelial transcriptomes of the mouse fetal mammary gland. *PLoS ONE* 8(5): e63902.

Wadia, P. R., L. N. Vandenberg, C. M. Schaeberle, B. S. Rubin, C. Sonnenschein and A. M. Soto. 2007. Perinatal bisphenol A exposure increases estrogen sensitivity of the mammary gland in diverse mouse strains. *Environ. Health Persp.* 115: 592–598.

Walker, B. C. and M. I. Haven. 1997. Intensity of multigenerational carcinogenesis from diethylstilbestrol in mice. *Carcinogenesis* 18: 791–793.

Walker, B. C. and L. A. Kerth. 1995 Multi-generational carcinogenesis from diethylstilbestrol investigated by blastocyst transfers in mice. *Internatl. J. Cancer* 61: 249–252.

Walker, D. M. and A. C. Gore. 2011. Transgenerational neuroendocrine disruption of reproduction. *Nature Rev. Endocrinol.* 7: 197–207.

Wang, S. L., P. H. Su, S. B. Jong, Y. L. Guo, W. L. Chou and O. Päpke. 2005. In utero exposure to dioxins and polychlorinated biphenyls and its relations to thyroid function and growth hormone in newborns. *Environ. Health Persp.* 113: 1645–1650.

Wasserman, M. D., K. Milton and C. A. Chapman. 2013. The roles of phytoestrogens in primate ecology and evolution. *Int. J. Primatol.* 34: 861–878.

Watson, C. S., R. A. Alyea, Y. J. Jeng and M. Y. Kochukov. 2007. Nongenomic actions of low concentration estrogens and xenoestrogens on multiple tissues. *Mol. Cell Endocrinol.* 274: 1–7.

Webb, E., S. Bushkin-Bedient, A. Cheng, C. D. Kassotis, V. Balise and S. C. Nagel. 2014. Developmental and reproductive effects of chemicals associated with unconventional oil and natural gas operations. *Rev. Environ. Health* 29: 307–318.

Welshons, W. V., K. A. Thayer, B. M. Judy, J. A. Taylor, E. M. Curran and F. S. vom Saal. 2003. Large effects from small exposures. I. Mechanisms for endocrine disrupting chemicals with estrogenic activity. *Environ. Health Persp.* 111: 994–1006.

Wetherill, Y. B., C. E. Petse, K. R. Monk, A. Puga and K. E. Knudson. 2002. Xenoestrogen bisphenol A induces inappropriate androgen receptor activation and mitogenesis in prostatic adenocarcinoma cells. *Mol. Cancer Ther.* 1: 515–524.

Wiseman, H. J. 2009. Untested Waters: The rise of fracturing in oil and gas production and the need to revisit regulation. *Fordham Envir. Law Rev.* 115: 142–192.

Wiszniewski, A. B., A. Cernetich, J. P. Gearhart and S. L. Klein. 2005. Perinatal exposure to genistein alters reproductive development and aggressive behavior in male mice. *Physiol. Behav.* 84: 327–334.

Wolff, M. S. and 11 others. 2010. Investigation of relationships between urinary biomarkers of phytoestrogens, phthalates, and phenols and pubertal stages in girls. *Environ. Health Perspect.* 118: 1039–1046.

Wolstenholme, J. T., M. Edwards, S. R. Shetty, J. D. Gatewood, J. A. Taylor, E. F. Rissman and J. J. Connelly. 2012. Gestational exposure to bisphenol A produces transgenerational changes in behaviors and gene expression. *Endocrinology.* 153: 3828–3838.

Wozniak, A., N. Bulayeva and C. S. Watson. 2005. Xenoestrogens at picomolar to nanomolar concentrations trigger membrane estrogen receptor-a-mediated Ca^{2+} fluxes and prolactin release in GH3/B6 pituitary tumor cells. *Environ. Health Persp.* 113: 431–439.

Xu, L. C., H. Sun, J. F. Chen, Q. Bian, L. Song and X. R. Wang. 2006. Androgen receptor activities of p,p'DDE, fenvalerate, and phoxim detected by androgen receptor reporter assay. *Toxicol. Lett.* 160: 151–157.

Yang, C. Z., S. I. Yaniger, V. C. Jordan, D. J. Klein and G. D. Bittner. 2011. Most plastic products release estrogenic chemicals: A potential health problem that can be solved. *Environ. Health Perspect.* 119: 989–996.

Yeo, M. and 10 others. 2013. Bisphenol A delays the perinatal chloride shift in cortical neurons by epigenetic effects on the Kcc2 promoter. *Proc. Natl. Acad. Sci. USA* 110: 4315–4320.

Yoshimura, T. 2003. Yusho in Japan. *Ind. Health* 41: 139–148.

Zhou, C., S. Verma and B. Blumberg. 2009. The steroid and xenobiotic receptor SXR, beyond xenobiotic metabolism. *Nucl. Recept. Signal.* 7: e001.

Zoeller, R. T. and 7 others. 2012. Endocrine-disrupting chemicals and public health protection: A statement of principles from The Endocrine Society. *Endocrinology* 153: 4097–4110.

The Developmental Origin of Adult Diseases

Time present and time past
Are both perhaps present in time future
And time future contained in time past.

T. S. Eliot, 1935

Now there are more overweight people in America than average-weight people.
So overweight people are now average. Which means you've met your New
Year's resolution.

Jay Leno, 2006

During the past 50 years, adult diseases have usually been sectioned into three large areas: (1) infectious diseases, which were caused by bad microbes; (2) intrinsic diseases, which were caused by bad genes; and (3) traumatic diseases, which were caused by bad luck. This chapter deals with intrinsic diseases such as hypertension (high blood pressure), diabetes, metabolic syndrome, and excessive obesity. Usually, whether one had them or not was considered the result of certain genes or of certain genes in certain environments. But the causes of these disease states turn out to be more interesting and more developmental.

"Originally we were told that common genetic variants would explain common diseases, and that's turned out largely not the case," says Andrew Feinberg (2014). "The environment is extremely important, and we pay almost no attention to it in biology." That may be beginning to change. New data have shown that the prenatal environment may be critical for adult health, and developmental plasticity may play an enormous role in our responses to the prenatal environment. Moreover, epigenetic changes, such as DNA methylation, may actually be able to transmit disease susceptibility from one generation to the next.

Organisms respond to environmental changes on three timescales. First, physiological homeostatic mechanisms can immediately respond to adverse conditions. On a longer timescale, developmental plasticity can result in adaptive changes in the organism by enabling the emergence of a phenotype appropriate for the environment it is expected to encounter. And, on the longest timescale, as changes in phenotypes interact with the environment, the genetic composition of populations can change as a result of natural selection.

Medicine deals predominantly with the first category, when short-term physiological responses are inadequate and lead to sickness and disease. However, as we document in this chapter, some medical conditions are the result of mismatches generated on the second timescale. Developmental plasticity is key to a strategy that uses cues present in early development to prepare an organism for its future life. The success of such a strategy assumes (1) that the cues are accurate and (2) that the early environment is a good predictor of the adult environment. If either assumption is not met, the survival and/or health of the resulting individual may be compromised.

But developmental plasticity must remain within the boundaries of the grand scale of evolution. Adaptive biological responses evolve to maximize fitness, not necessarily health (see Gluckman and Hanson 2007). Since evolutionary fitness is measured by the relative number of fertile progeny an individual produces, our biological responses have evolved to get us to the point where we can successfully reproduce and rear our offspring. "No sooner than we have a child," writes the melancholy Jean Rostand (1962, p. 9), we become "lateral excrescences on the tree of life"; evolution cares nought for us. However, public health programs, improved sanitation, and advances in medicine have enabled much of the population to live well beyond the years of prime reproductive life that marked the extent of average life expectancy throughout most of history. As a result, late-onset diseases such as hypertension, type 2 (adult onset) diabetes, and cancer have become major medical and public health concerns.

One of the most exciting new areas of public health and medicine concerns the developmental origins of adult disease. The previous chapter discussed the health implications of endocrine disruption. In this chapter, we will discuss the fetal origin of diseases such as hypertension, obesity, and heart failure. Here, the diet experienced in utero appears to play a role in making the individual more or less susceptible to developing these conditions much later in life.

The Developmental Origins of Health and Disease

Probably the most obvious instance of co-development is mammalian pregnancy. Entire volumes have been written on "maternal-fetal interactions," and "maternal-fetal medicine" is a recognized subspecialty of obstetrics and gynecology. The fetus and its extraembryonic layer (the trophoblast, which forms the portion of the placenta derived from the fetus) induce changes in the structure and physiology of the uterus. The fetus and placenta produce hormones such as progesterone and human chorionic gonadotropin (hCG) that maintain the lining of the uterus. These hormones enable the uterine stromal cells to proliferate rather than be sloughed off during menstruation; they also cause blood vessels to enter into this uterine tissue. (This set of changes to the uterus is called the *decidual reaction*.)

The fetus and its placental tissues also produce substances that block the maternal immune responses against the fetus, a critical action in maintaining the pregnancy. Also critical to the maintenance of pregnancy, the uterus induces proteins on the extraembryonic trophoblast that allow it to adhere to and ingress into the uterine lining (see Dey et al. 2004). Later, of course, the uterus provides the developing mammal with oxygen, nutrients, and a temperature-regulated residence for the duration of the pregnancy, as well as structures for removing its waste products. And, toward the end of pregnancy, trophoblast cells are involved in the production of relaxin, a hormone that softens the maternal cervix and pelvic ligaments in preparation for childbirth.

Almost all the research on maternal-fetal interaction concerns these permissive aspects of development. The presence of a fetus induces the decidual reaction in the mother, but this reaction has nothing to do with specifying the phenotype of the fetus. Rather, it allows the fetus to survive. Similarly, the transport of nutrients and gases into and out of the fetus is also permissive. It is critical for fetal growth and survival, but it is not usually discussed in terms of providing information to the fetus on how to develop its phenotype. Here we will explore evidence that the mother is indeed providing the fetus with instructions about how to develop phenotypes that will survive in the expected environment.

Instructing the Fetus

Mammals developing within the uterus can fine-tune their phenotypes to suit an expected future environment. The mother's diet and hormonal milieu provide information about the environment the offspring will be born into. However, when environmental conditions change greatly between conception and adulthood (as has been the case for many modern human populations), the potential exists that a phenotype developed for one situation may not be particularly healthy in the new situation.

Epidemiological, physiological, and molecular evidence are now converging to show that numerous human diseases—diabetes, obesity, and hypertension among them—can be attributed to the mismatch between (1) the phenotype developed from nutritional cues sensed by the fetus in utero and (2) the nutritional environment in which that individual finds itself as an adult (Gluckman and Hanson 2006a,b, 2007). The terms "fetal programming" and the more accurate and less deterministic "fetal plasticity" refer to the processes by which a stimulus given during an intrauterine period of development can have a lasting or lifelong effect (Lucas 1991; Barker 1998).*

Maternal effects

It is known that the maternal environment contains cues that can override genetic information. One place where this phenomenon is readily seen is in

---

*Mammals are not the only animals for whom signals received during embryogenesis can lead to disease or other phenotypic changes in adult life. In the marine environment, "orphan embryos" (see Chapter 4) that experience short-term starvation, salinity stress, or exposure to sublethal concentrations of pollutants may, after metamorphosis, experience altered growth and fecundity rates and heightened mortality rates. The mechanisms of these "latent effects" remain largely unexplored (Pechenik 2006).

the size of newborn mammals. Among healthy humans, an infant's birth weight is determined much more by its mother than by its father. When the same woman has borne children with different men, the birth weights of the babies are usually similar. However, when the same man fathers children with different women, the birth weights are often very different (Morton 1955; Robson 1955; Gluckman and Hanson 2005). This effect is probably what historically has allowed small women to bear children fathered by large men. The baby must be able to remain inside the uterus before birth and must be small enough to pass through the mother's birth canal (Hanson and Godfrey 2008). Thus, genes alone do not determine the size of the newborn. Maternal constraints play a major role here.

In controlled studies, researchers crossed large Shire horses with small Shetland ponies (Figure 7.1). If a Shire stallion was crossed (by artificial insemination) with a Shetland mare, the size of the resulting foal at birth was closer to the size of a typical Shetland foal. If, however, a Shire mare was crossed with a Shetland stallion, the newborn foal was much larger, closer in size to that of a typical Shire foal. Even though in both cases the foal genomes were 50% Shetland and 50% Shire, the size of the newborn was typically that of the mother's strain. The Shetland mare in some manner limited the growth of a fetus whose genotype, if fully expressed, would make it too large to pass through her birth canal. Uninhibited by such constraints, the same genotype could grow to be much larger, as shown by its growth in the Shire mother. While we don't know the exact mechanism regulating this growth constraint, it probably has to do with the nutrient supply that can be delivered by the smaller placenta. Moreover, the offspring of the mothers with smaller uteruses and placentas developed into smaller adults than did the analogous hybrids whose mothers were of the larger breed (Walton and Hammond 1938; Tischner 1987; Allen et al. 2004).

The importance of maternal effects on behavioral phenotypes was highlighted in a study (Francis et al. 2003) that separated the genetics of the mouse from its intrauterine environment and from its perinatal environment (immediately after birth). B6 mice and BALB/c mice are inbred mouse strains, where each member of the respective strain is genetically identical to every other member (the only difference being whether a mouse possesses a Y chromosome). Moreover, B6 mice and BALB/c mice have measurable and statistically significant differences in their behaviors. The questions are (1) whether the behavioral differences are due primarily to genes within the newborn and (2) if the behavioral differences are due to the environment, is the critical environmental component the prenatal intrauterine environment, the postnatal nursing environment, or both.

To answer these questions, the investigators obtained B6 embryos at a young stage and then transplanted them into the uteri of either B6 or BALB/c females. After the mice were born, the resulting offspring were given to new parents, either B6 mothers or BALB/c mothers. There, they experienced strain-specific maternal rearing behaviors for 3 weeks. After reaching adulthood, the mice were tested for behavioral traits (anxiety,

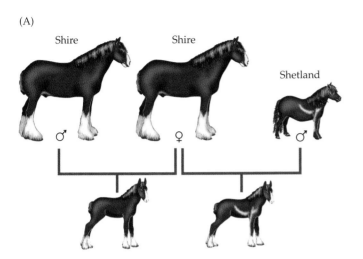

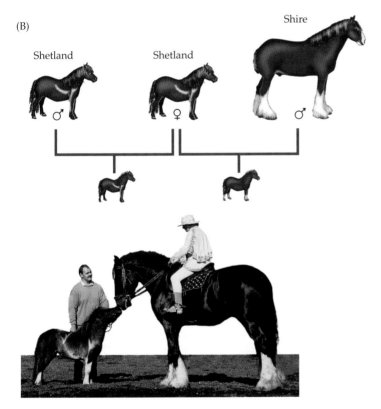

Figure 7.1 Foal size in outcrosses between Shire and Shetland horse breeds is determined by the mother. (A) A Shire mare crossed with a Shetland stallion produces a foal about the size of a purebred Shire. (B) When a Shetland mare is crossed with a Shire stallion, the foal is closer to the size of a purebred Shetland pony. (After Gluckman and Hanson 2005.)

learning, and sensory processing) that differed between the two strains (Figure 7.2).

Those B6 mice that experienced B6 uteri and B6-specific nursing (as would occur normally in the laboratory) had B6-like behaviors on all tests. Similarly, as expected, those BALB/c mice that experienced both BALB/c

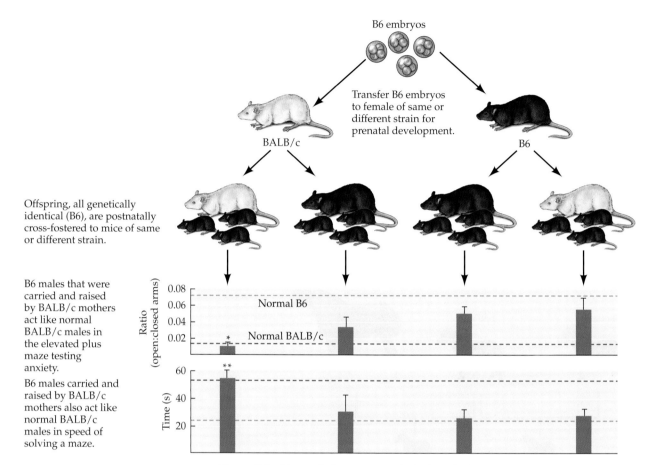

B6 embryos

Transfer B6 embryos to female of same or different strain for prenatal development.

BALB/c

B6

Offspring, all genetically identical (B6), are postnatally cross-fostered to mice of same or different strain.

B6 males that were carried and raised by BALB/c mothers act like normal BALB/c males in the elevated plus maze testing anxiety.

B6 males carried and raised by BALB/c mothers also act like normal BALB/c males in speed of solving a maze.

Ratio (open:closed arms)

Normal B6

Normal BALB/c

Time (s)

Figure 7.2 Maternal effect on mouse behaviors. A simplified representation of the experiments by Francis and colleagues. B6 mouse offspring showed behaviors typical of BALB/c mice (in three of four tests) only when they had been transplanted as embryos into BALB/c mice and foster-mothered by BALB/c mice. The litters were postnatally cross-fostered, and each litter contained pups prenatally reared by either B6 or BALB/c mothers. The results of two of the tests where the behaviors of B6 mice were transformed by maternal effects into the behavior of BALB/c mice are shown here. The ratio of time spent in open versus closed arms of an elevated plus maze is a measure of anxiety. Time to solve a maze is a measure of learning ability. In these cases, the B6 mice behaved like BALB/c mice only if they experienced BALB/c mothers both prenatally and postnatally. Asterisk (*) marks differences that are statistically significant to $p < 0.05$. Double-asterisks (**) indicate statistical differences of $p < 0.01$. (After Francis et al. 2003.)

uteri and nursing developed BALB/c behaviors. If a B6 mouse experienced either prenatal or postnatal BALB/c maternal environments, it still developed a B6 behavioral phenotype. However, if the B6 embryo experienced *both* a BALB/c intrauterine environment and a BALB/c postnatal environment, it was behaviorally like BALB/c mice on three out of the four tests. Thus, the maternal effect involves both prenatal and neonatal environments.

A potentially important maternal effect has been demonstrated in humans, and it concerns the length of the chromosomal telomeres. Telomeres

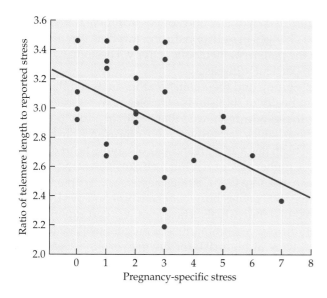

Figure 7.3 Pregnancy-specific stress and the average telomere length in the newborn cord blood. This scatterplot represents cord blood telomere length (adjusted for age, sex, and weight of fetus) as declining with pregnancy-related stress. The stress was based on a standard questionaire used for self-assessment of anxieties felt about being pregnant, the health of the fetus, and labor problems. (After Entringer et al. 2013.)

are the termini of chromosomes, the "tassels" of the DNA "shoelaces." Telomeres stop the DNA from unwinding prematurely, and they play a critical role in stabilizing the DNA. At each cell division, however, the telomeres get shorter because DNA polymerase is unable to completely replicate the 3′ end of the helix. Eventually, the telomeres reach a critically short length, at which point the cells undergo senescence or death. Shortened telomeres are thought to contribute to onset of age-associated diseases such as heart failure, hypertension, atherosclerosis, and diabetes. In humans, maternal psychological stress (as measured by both self-reporting and hormone levels) is associated with a decline in the size of telomeres at birth (Figure 7.3; Entringer et al. 2013.) Between the high- and low-stress groups, the difference was about 500 base pairs of DNA. As can be seen from the figure, this is a statistical correlation, and it is not determinative.*

Fetal plasticity in humans

D. J. P. Barker and colleagues cite evidence that, in humans, certain adult-onset diseases may result largely from conditions in the uterus prior to birth (Barker 1994; Barker and Osmond 1986). Based on epidemiological evidence, they hypothesized that there are critical periods of development during which insults or stimuli cause specific changes in the body's physiology. Barker's concept was that certain anatomical and physiological

*The expression "Stress will kill you" has received some physiological support (see Sapolsky 2004; Cossins 2015). Two provocative and insightful papers (Epel et al. 2004; Chae et al. 2014) show that in two highly stressed populations (women caring for family members with severe disabilities and African-American men), the perceived and hormonally measured stress levels were correlated with an accelerated decline in telomere length. Moreover, the effect was linear: the greater the stress, the smaller the telomere length. In this regard, one should note that African-American men have a disproportionately high level of chronic age-related diseases, and the life expectancy of American black men is 6 years less than that of white men (Chae et al. 2014).

parameters become specified by the actions of environmental agents of fetal plasticity, and that changes in nutrition during this time can produce permanent changes in the pattern of metabolic activity. These changes can predispose the adult to particular diseases.

Specifically, Barker's group and their collaborators analyzed the birth records and subsequent health of thousands of babies born between 1911 and 1948. Remarkably, these records showed that health in adulthood was associated with how these babies had grown in the uterus. A British baby weighing 2.7 kg (6 lb) at birth has a 25% greater risk of having heart disease in adulthood and a 30% higher risk of having a stroke, compared with a similar British baby born at 4.1 kg (9 lb). Similar results soon came from a Finnish cohort of some 20,000 babies born between 1924 and 1944. These and further data showed that individuals whose mothers had experienced nutritional deprivation during certain months of pregnancy (due to war, famine, or migrations) were at high risk for having certain diseases as adults (Barker 1997; Barker et al. 2002; Figure 7.4). Undernutrition during

(A)

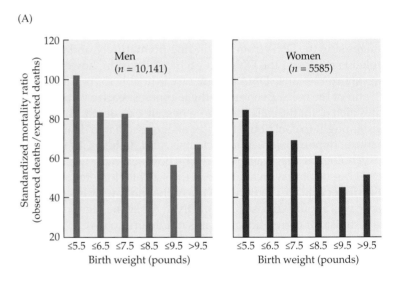

(B)

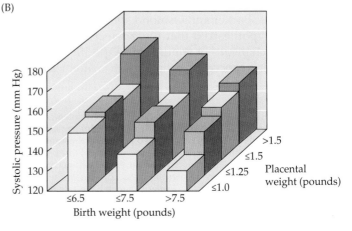

Figure 7.4 Low birth weight and adult-onset coronary disease. (A) Epidemiological data demonstrating the association of adult-onset coronary disease with low birth weight in both men and women. (B) The association of mean systolic blood pressure (mm Hg) of 50-year-old men and women is associated with both low birth weight and a large placenta. Those especially at risk would be those whose low birth weight could not be explained by a small placenta. (After Godfrey 2006.)

the first trimester led to a relatively high risk of having hypertension and stroke as an adult, while those fetuses experiencing undernutrition during the second trimester had a relatively high risk of developing heart disease and diabetes as adults.

Several studies have tried to determine whether there are molecular or anatomical reasons for these correlations (Gluckman and Hanson 2004, 2005; Lau and Rogers 2004). Anatomically, undernutrition can change the number of cells produced during critical periods of organ formation. When pregnant rats are fed low-protein diets at certain times during pregnancy, the resulting offspring are at high risk for hypertension as adults. The poor diet appears to cause low nephron numbers in the adult kidney (see Moritz et al. 2003). Nephrons are the filtering units of the kidneys, and they synthesize proteins that regulate blood pressure. In humans, men with hypertension were shown to have only about half the number of nephrons found in men without hypertension (Figure 7.5; Keller et al. 2003; Hoppe et al. 2007). It is possible that malnutrition activates a glucocorticoid response, which causes cell death among the developing kidney cells (Moritz et al. 2003; Welham et al. 2005).

The ability to sacrifice nephrons so that the limited nutrition available can go to the brain, heart, and other essential places seems like good evolutionary sense. One can survive well and reproduce with only one kidney, and the accumulated effects leading to hypertension usually are seen only after one's fortieth birthday. Therefore, when the life expectancy of the average human was less than 40 years, that is, during most of human history, one could function efficiently with half the number of nephrons throughout one's life. It is only since life expectancy in the industrialized world has reached the seventh decade that the detrimental effects of nephron deficiency have become apparent.

Developmentally plastic changes in anatomy have also been reported for the pancreas, heart, and liver. Poor nutrition during fetal development reduces the number of insulin-secreting cells in the pancreas, causing the person to synthesize smaller amounts of

(A)

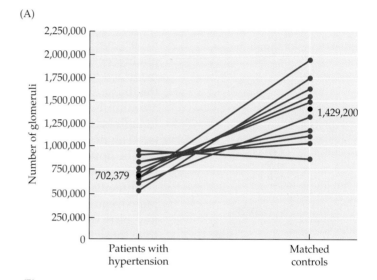

(B)

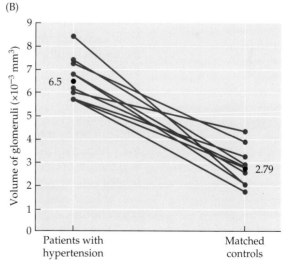

Figure 7.5 Anatomical changes associated with hypertension. (A) In age-matched individuals, the kidneys of men with hypertension had about half the number of nephrons as did the kidneys of men with normal blood pressure. (B) The glomeruli (filtering units of the nephron) of hypertensive kidneys were much larger than the glomeruli of control subjects. Average values for subjects in both categories are indicated in black. (After Keller et al. 2003.)

insulin (Hales et al. 1991; Hales and Barker 1992; Petrik et al. 1999). This insulin deficiency predisposes these individuals to type 2 diabetes and metabolic syndrome (an energy utilization disease of high glucose levels and obesity). In rats, undernutrition before (but not after) birth changes the histological architecture in the liver. Liver cells (hepatocytes) are divided into roughly two types, depending upon their location in the liver lobule: an upstream "periportal" population that maintains gluconeogenesis (glucose production), and a downstream "perivenous" population that is involved in glycolysis (glucose utilization). A low-protein diet during gestation appears to increase the number of periportal cells that produce the glucose-synthesizing enzyme phosphoenolpyruvate carboxykinase and to decrease the number of perivenous cells that synthesize the glucose-degrading enzyme glucokinase (Desai et al. 1995). As a result, more glucose is made and less is degraded. As in the case of kidney nephrons, it is thought that these changes may be coordinated by glucocorticoid hormones that are stimulated by malnutrition and that act to conserve resources, even though such actions might make the person prone to disease later in life (see Fowden and Forhead 2004).

The opposite side of this coin—that of offspring who experience high-protein diets in utero and perform less well in calorie-impoverished environments than do their siblings who experience low-protein diets in utero—has not been as well studied. However, during famines in Ethiopia, low–birth weight children appear to have had less chance of getting rickets than those born with high birth weights (Chali et al. 1998), and Bateson (2007) relates anecdotal evidence that in World War II concentration camps, large individuals appeared to die more readily than smaller inmates.*

Gene methylation in utero and the adult phenotype

How can environmental conditions experienced only while the fetus is in the uterus result in anatomical and physiological states that are maintained throughout the individual's adulthood? As we will see throughout this book, there are many levels at which changes to developmental parameters can alter the phenotype. Many teratogens, for instance, can affect protein function (e.g., ethanol blocks the L1 adhesion protein), and the endocrine disruptor DES can act by interfering with the Runx2 transcription factor. But one centrally important place where environmental agents affect future development is at the site of DNA methylation. As we saw in Chapter 2, the chromatin we inherit can be modified by enzymes that alter DNA or histones. Methylation of DNA and some histone residues can compress DNA and prevent its transcription. A gene is activated when the enzyme RNA polymerase binds to a DNA sequence (the promoter) in front of the actual gene sequence. Once bound to the promoter, RNA polymerase can transcribe the gene sequence into messenger RNA, which in turn enters the cell cytoplasm, where it is translated into a specific sequence of amino acids that forms a protein. RNA polymerase is stabilized on the promoter by other DNA sequences called enhancers (see Figure 2.5). The addition of methyl groups to the DNA of

---

*Later in this chapter, we will discuss the pathological effects of fetal overnutrition.

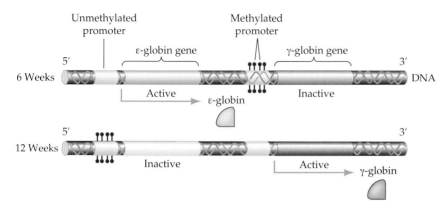

Figure 7.6 Methylation of globin genes in human embryonic blood cells. The activity of these human globin genes correlates inversely with the methylation of their promoters. (After Gilbert 2006.)

enhancer or promoter regions usually prevents the RNA polymerase from binding to the promoter and, as a result, blocks gene activity, whereas de-methylation (removing existing methyl groups) from these regions activates gene expression (Figure 7.6; see also Figure 2.4).

Studies have shown that rats born of mothers given a low-protein diet during gestation have a different pattern of liver gene methylation than do the offspring of mothers fed a diet with a normal amount of protein, and that these differences in methylation change the metabolic profile of a rat's liver. As mentioned in Chapter 2, these different forms of the gene due to differential methylation (or any other persistent chromatin change that does not alter the genetic code) are called **epialleles**.* For instance, the methylation of the promoter region of the *PPARα* gene (whose product is critical in the regulation of carbohydrate and lipid metabolism) is 20% lower in rats fed protein-restricted diets in utero, and consequently its activity is 10 times greater (Lillycrop et al. 2005; Figure 7.7). In addition, the promoter region of the glucocorticoid receptor gene (critical in blood pressure regulation) was 22% less methylated and three times more tran-scriptionally active in the pups born of mothers given a protein-restricted diet (Burdge et al. 2006). These methylation patterns persisted after the dietary restrictions ceased, thereby showing stable modifications in gene expression due to nutritional influences (Lillycrop et al. 2008). DNA meth-ylation thereby provides a mechanism allowing changes in fetal chromatin to persist throughout life.

Interestingly, the difference in methylation patterns could be abolished by including folate in the protein-restricted diet. Folic acid is a methyl-group donor. Thus, the difference in methylation is a result of changes in folate metabolism caused by the limited amount of protein available to the

*Alleles are genetic differences at a locus. Epialleles (sometimes called "epimutations") are epigenetic differences at a locus. "Allele" comes from the Greek *allo*, meaning "different." If a locus is a parking place, and an allele is what type of car is in the parking place, then an epiallele might be the color of the car.

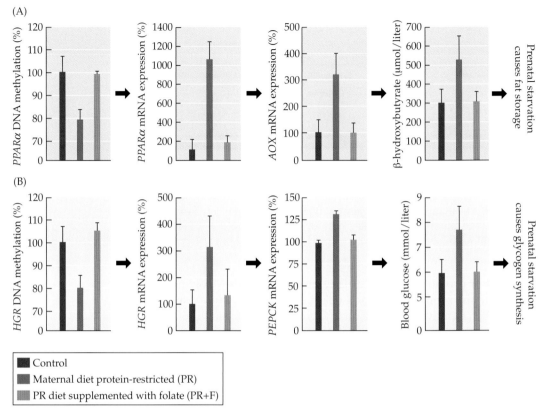

Figure 7.7   The offspring of rats fed a protein-restricted diet (PR) showed higher fat synthesis and breakdown of carbohydrates into glucose. The mechanism for this may involve reduced promoter methylation and increased mRNA expression of at least two major genes. (A) Reduced promoter methylation and increased expression of the gene for the transcription factor peroxisomal proliferators-activated receptor alpha (PPARα) in the liver, as compared with those of control rats fed a normal diet. Increased *PPARα* expression is associated with increased expression of acyl-CoA oxidase (AOX), a key enzyme in fatty acid oxidation, as well as increased circulating concentrations of the ketone β-hydroxybutyrate. These metabolic effects can be prevented by supplementing the protein-restrictive maternal diet with folate (PR+F). (B) Reduced promoter methylation and increased gene expression for hepatic glucocorticoid receptor (HGR) in rats whose mothers were protein-restricted during pregnancy. Increased *HGR* expression is associated with increased expression of the gene for gluconeogenic enzyme phospho-enolpyruvate carboxykinase (PEPCK) and consequent increased blood glucose levels. Again, the effects are not seen in offspring of PR rats whose diets were supplemented with folate. The *p* values for comparisons of the PR+F group with the control group were >0.05 in all cases, indicating no significant difference between these two groups. Data are means; error bars indicate standard error. (After Gluckman et al. 2008.)

fetus. Folic acid (also called folate and vitamin B9) is often added to cereal grain–based foods or taken as a supplement because of its importance for reducing neural-tube closure birth defects. However, as discussed in Chapter 5, increased folate may activate other genes and produce deleterious effects.

New research substantiates this and has shown that environmental dietary conditions can make epialleles that become transmitted between generations. Exposure of mouse fetuses to general undernutrition (50% normal caloric intake) causes intrauterine growth restriction in the fetuses as well as a predisposition to obesity and adult-onset diabetes when they become adults (Martínez et al. 2014). Such in utero malnutrition of male fetuses reduces the expression of several important genes involved in lipid production. This is due primarily to the reduction of *Lxra* transcription. *Lxra* encodes a transcription factor that activates several of the genes encoding lipid biosynthetic enzymes (Figure 7.8). The reduction of *Lxra* transcription

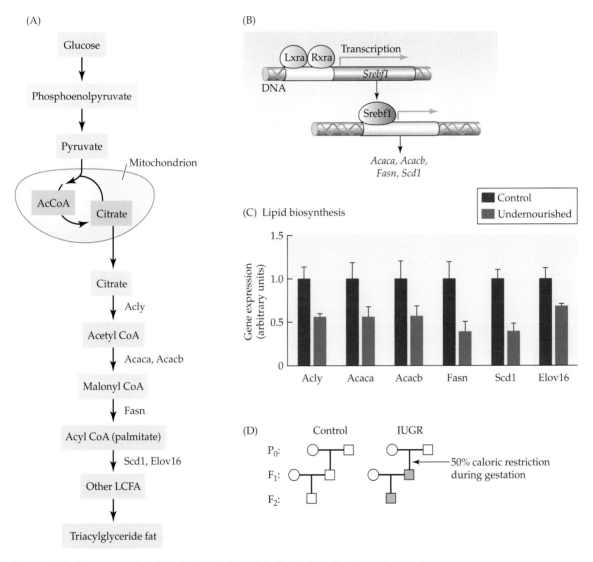

Figure 7.8   Transgenerational regulation of lipid metabolism induced by the environment. (A) Pathway for making triacylglyceride fats. The relevant enzymes are to the sides of the arrows. (B) Regulation of the pathway by transcription factors Lxra and Rxra. The combination activates the *Srebf1* gene, whose transcription factor product activates the genes encoding much of the lipid biosynthetic pathway. (C) Depression of mRNA for the major lipid biosynthetic enzymes. In each case, the gene expression in the undernourished mice was statistically lower than that of mice that received normal nourishment (controls). (D) Breeding strategy to generate prenatally challenged first generation (F_1) and unchallenged seond generation (F_2). (After Martínez et al. 2014.)

has been shown to be caused by an altered DNA methylation pattern that places repressive methyl modifications (H3K9me2, H3K27me3) on the histones of the *Lxra* promoter. These chromatin modifications have been found in the livers of F_1 males who experienced malnutrition in utero. They have *also* been found in the *sperm* of the adult F_1 males and in the *livers* of adult F_2 males that are their offspring. In other words, a set of chromatin modifications caused by malnutrition is found throughout the male offspring and changes their metabolism. These modifications are also made in the mouse's germline DNA, alter the sperm chromatin, and are thereby inherited by the offspring that did not experience malnutrition. Thus, the effects of dietary restriction can be inherited as an epiallele. Therefore, the environment can make a heritable change that can be transmitted and change the phenotype in further generations.

High-fat diets have also been shown to have epigenetic effects. The daughters of rats fed high-fat diets have different liver metabolism than do control females, even when they become adults. This is discussed in Appendix B.

Predictive Adaptive Responses

Data such as those shown in Figures 7.7 and 7.8 may have important consequences. How individuals utilize their nutrition may have to do with an interaction between genes, tissues, and environment that played out while they were in the uterus.

Hales and Barker (1992, 2001) proposed the **thrifty phenotype hypothesis**, which holds that malnourished fetuses are "programmed" during their plastic stages to expect an energy-deficient environment and, in response, set their biochemical parameters to conserve energy and store fat. When these individuals develop into adults, if they do indeed find themselves in the expected nutrient-poor environment, they are ready for it and can survive better than if their metabolism had been "set" to utilize energy less efficiently. However, if they find themselves living in a calorie- and protein-rich environment, their energy-efficient metabolism means that their cells simply store the abundance as more and more fat. In addition, because their hearts and kidneys developed to face more stringent conditions, their altered phenotype puts such persons at risk for diseases such as diabetes (Figure 7.9).

The thrifty phenotype hypothesis has since been expanded to link prenatal influences with many other types of adult illness, including lung disease, cognitive development abnormalities, breast cancer, prostate cancer, and leukemia (Godfrey and Barker 2000; De Boo and Harding 2006). This broadened version of Barker's hypothesis, encompassing a large range of prenatally influenced illnesses, has been referred to as the **developmental origins of health and disease** (**DOHaD**) hypothesis. The shift to the DOHaD nomenclature was proposed for two reasons. First, research has demonstrated the plasticity of development in both the prenatal and early postnatal periods (see the case of sweat gland activation, in the next section). Second, the change in terminology emphasizes not only the causes

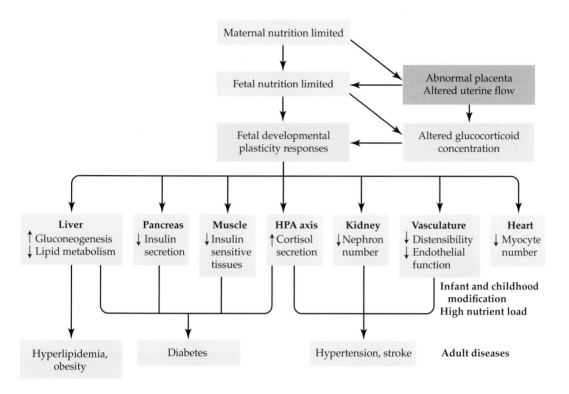

Figure 7.9   The thrifty phenotype hypothesis. Altered maternal nutrition leads to a cascade of consequences mediated by the phenotypic plasticity of the fetal organs. The combination of biochemical and anatomical changes in these organs produces a phenotype that stores excess energy as fats and sacrifices kidney growth for brain growth. This phenotype is beneficial under environments of low nutrition, but high nutrition during infancy and/or childhood can result in adult obesity, diabetes, and hypertension. (After De Boo and Harding 2006.)

of disease, but also disease prevention and adult health promotion (Gluckman and Hanson 2006a,b). The broader terminology reduces temptation to adopt the fatalistic idea that prenatal events can completely determine adult health. *Both* pre- and postnatal environments are influential in development.

Why is it advantageous for fetal development to pivot on environmental cues? The benefits of fetal developmental plasticity can be realized in two ways (Figure 7.10). First, there can be **immediate adaptive responses** that alter development in response to a particular challenge that is already occurring in the embryo. Thus, malnutrition or oxygen deprivation can cause changes in blood flow so that the brain continues to develop at the expense of less critical tissues (Bateson et al. 2004).

A second type of plasticity is the **predictive adaptive response**, or **PAR**. The PAR model states that early environmental cues can shift the developmental pathway to modify the phenotype in expectation of the later environment (Gluckman et al. 2005; Gluckman and Hanson 2005). As

Figure 7.10 Developmental responses and fitness. Mutations or severe environmental disturbances can disrupt development and cause *immediate responses* to the injury or defect. Less extreme disturbances can lead to *predictive adaptive responses* (PARs), including the induction of thrifty phenotypes. Within the normal range of variation, an individual's developmental trajectory will be conferred by the actions of PARs. (After Gluckman and Hanson 2006b.)

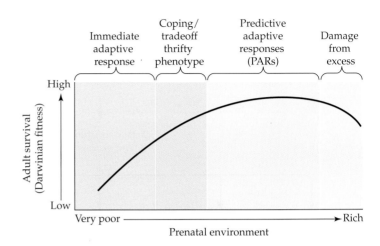

opposed to immediate adaptive responses, PARs benefit the organism later in life—as long as the predicted and actual postnatal environments match. The properties of PARs include the following characteristics:

- They are induced by early environmental factors and act via developmental plasticity to modify the phenotype to match an anticipated environment.
- They permanently change the physiology and/or anatomy of the organism.
- They respond to a range of developmental environments, not just to extreme conditions.

PARs can give the organism a survival advantage if the response matches the predicted environment; however, if the environment is not the one predicted by the PAR, the mismatched condition can lead to disease.

The essential elements of the PAR model are not new to biologists. Indeed, the plasticity seen in the wing color polyphenisms in butterflies (see Figure 1.5) is essentially a predictive adaptive response, since the color made within the imaginal discs of the larva will be manifest only in the adult. Similarly, many animals have a cyclic development of their fur. In some arctic animals whose coats are brown in summer but white in winter (Figure 7.11), the decreased sunlight as summer ends initiates increased melatonin secretion from the pineal gland. Melatonin secretion triggers a cascade of hormones that ultimately induces a molt followed by the regrowth of hair, this time in the absence of melanin production (i.e., the fur grows in white; Feder 1990). Again, the processes were initiated by a signal (decreasing photoperiod) long before the first snows of winter fell.

The environmental mismatch hypothesis

The PAR model can be seen as extending the thrifty phenotype hypothesis from the clinical and epidemiological realm into evolutionary biology. The evolutionary advantages of the PAR strategy depend on the accuracy of

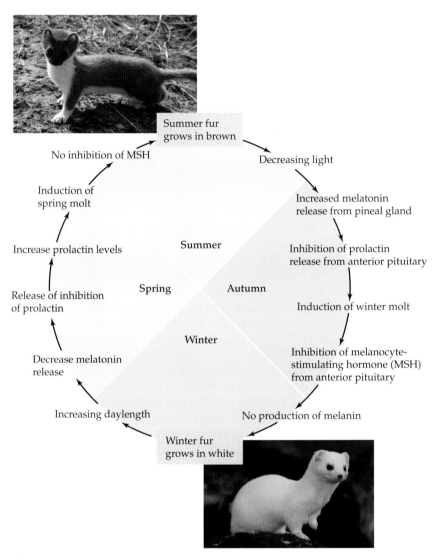

Figure 7.11   A well-known predictive adaptive response (PAR). Levels of the light-stimulated hormone melatonin are the agents of change in pelt pigmentation for the short-tailed weasel (*Mustela erminea*). The cycle of molts and color changes is environmentally mediated by altered day length as the seasons change. The white winter coat is the source of the fur known as ermine. (After Feder 1990.)

the cues, the intrinsic costs of plasticity, and the frequency of mismatches (Gluckman et al. 2005; Bateson et al. 2014). If the predicted and future environments do not match, health risks or other disadvantages may arise. Certain ideas concerning the inaccurate predictions of PARs have come to be known as the **environmental mismatch hypothesis**. One readily available example of such environmental mismatch would be the white pelt of an arctic fox or short-tailed weasel appearing in a winter with no snow (as is happening with global climate change; see Scott Mills et al. 2013).

One of the first recorded human cases of PARs making the "wrong" decisions came from the records of Japanese soldiers during World War II. In this case, the thermal environment triggered a minimal number of active sweat glands in human infants, and this condition became maladaptive when in later life the adults needed to survive in a hot and humid environment (Gluckman et al. 2005). The Japanese invasion of southeast Asia placed soldiers into environments that were new to them. Some soldiers got heatstroke, whereas other soldiers adapted more readily to the hotter climate. Most of the soldiers who got heatstroke came from the northern provinces of Japan. Their propensity to heatstroke was found to be due to an inability to sweat efficiently, and this inability was not genetic. Rather, it was due to a developmental decision made during the early years of life (see Diamond 1991; Barker 1998). The researchers found that each person is born with nearly the same number of sweat glands but that initially none of these sweat glands function. The glands are activated during the first 3 years after birth, and the percentage of sweat glands that becomes functional depends on the temperatures the child experiences during those years: the hotter the conditions, the greater the percentage of sweat glands that function. This is due to the innervation of the sweat glands by interaction with axons of the sympathetic nervous system; a sweat gland does not become functional until that interaction (Stevens and Landis 1988).

This high degree of plasticity allows the adjustment of human sweat glands to new thermal environments within a single generation. Indeed, Gluckman and Hanson (2005) suggest that this ability to adapt to new climates may have been crucial in allowing prehistoric human migrations to occur as rapidly as they did. We see in the example of the Japanese soldiers that an adult human phenotype can develop in a manner consistent with the environment experienced early in life. However, once the phenotype is induced, it remains set for the remainder of the person's life; for sweat gland function, the period of plasticity ends after 3 years. Should a person move to a new environment, the predictive response may not have been appropriate, and disease (such as heatstroke) may ensue.

One of the most important environmental mismatches involves fetal malnutrition followed by excess nutrition in the adult. This combination of circumstances has become more common as nutrition improves among people born in impoverished societies, as when some of these individuals migrate from regions of poor nutrition into regions of relative affluence. For instance, many people in the world, particularly pregnant women, live in a state of relatively poor nutrition. Fetuses in utero are therefore likely to sense a low-protein condition and actively adapt to alter their patterns of development to survive in an environment where food is scarce. In other words, the fetus predicts that getting adequate food will be difficult and sets its metabolism for a "thrifty phenotype" that will use every calorie efficiently and conserve any excess food as fat to guard against times of famine. Until recently, such a prediction was likely to be a successful one, enabling the individual to survive to reproductive age while staying small and thin. But rising affluence in some traditionally poorer countries has led to increasingly plentiful and available food supplies within a single generation. In this case, the fetus makes the same "thrifty" PAR (based either on maternal size constraints or on maternal diet), but the adult suddenly lives in a world full of food. The embryo develops resistance to insulin, fewer blood vessels feeding the tissues, and enzyme levels set to convert any unused food into fat. With more than enough food, the adult with this phenotype tends to be obese and diabetic and to have a high risk of heart disease. Such a syndrome is becoming commonplace throughout the developing world (Gluckman and Hanson 2007).

The problematic effects of an inaccurate predictive response were seen on a smaller scale following the Dutch Hunger Winter of 1944–1945. In reprisal for the Dutch resistance in 1944, Nazi authorities imposed severe food rationing in the western Netherlands. Almost immediately, the typical 1800 calorie per day diet was reduced to about 600 calories per day (Stein 1975; Hart 1993; Figure 7.12). This food restriction lasted 7 months (until Allied forces liberated the Netherlands), after which food intake returned immediately to the original level. Amazingly, despite the horrors of the war, doctors and midwives continued to deliver babies and to keep careful records of the health of these newborns. When analyzed decades later, these records showed that those adults who were undernourished in utero tended to develop insulin resistance and obesity.

In other short-lived famines that have been studied, similar results have been found. Those individuals exposed to maternal malnutrition in

Figure 7.12 Near-starvation conditions were imposed on portions of the Dutch population during the Nazi occupation. Those fetuses in utero during the winter of 1944–1945 experienced harsh protein deprivation. (Photograph courtesy of Nationaal Bevrijdingsmuseum Groesbeek/Beeldbank WO2, Netherlands.)

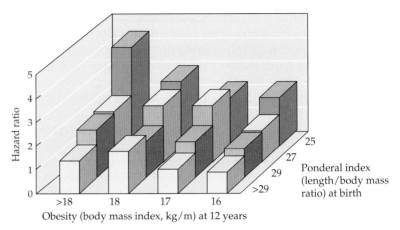

Figure 7.13 Data showing that the risk of coronary heart disease is increased by both small birth size and greater fatness in childhood. The graph shows that these conditions interact such that the highest risk of heart disease is in those people who were born at low birth weight (i.e., in conditions that may have caused a thrifty phenotype) but subsequently experienced ample nutrition. (After Eriksson et al. 2001.)

utero have had higher rates of hypertension, obesity, and impaired glucose tolerance than those who did not experience the famine gestationally (see Ganu et al. 2012).

Humans come in many shapes and sizes, and these different shapes and sizes mean different risks for diabetes and heart problems. The incidence of diabetes and cardiovascular disease is much higher in those people who are born small and thin and later become fat. Data from one study of a Finnish population (Eriksson et al. 2001) showed that the risk of coronary heart disease is increased by the presence of two factors: small birth size and then becoming relatively fat in childhood (Figure 7.13). What is important from these data is that high risk does not have to be associated with the extremes of malnutrition (as in the Dutch Hunger Winter). Even normal variations in nutrition can result in relatively high risks for heart disease and diabetes. It appears the fetus is making developmental decisions in response to low nutrient intake that anticipate its living after birth in a nutritionally deprived environment. Accordingly, it "expects" to stay small and thin and programs its development accordingly. However, if it is born into a nutritionally abundant environment, the thrifty phenotype that the fetus selected stores the extra calories and causes the person to become obese and at risk for cardiovascular problems.

Animal experiments have also provided data consistent with the environmental mismatch hypothesis (see Gluckman and Hanson 2005; Taylor and Poston 2007). As we have mentioned, poor maternal nutrition can lead to low birth weight, which has been correlated (depending on various factors, including the nature of the nutritional deprivation) with heightened risk of heart attack, obesity, and type 2 diabetes. For instance, rats that had been undernourished in utero have a higher probability of having hypertension and insulin resistance as adults (Figure 7.14; Vickers et al.

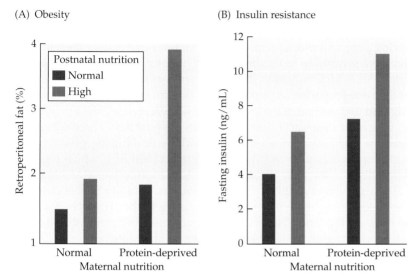

Figure 7.14 Experimental evidence for the fetal origin of adult disease in rats. Rat fetuses were exposed to either normal or protein-deficient maternal diets. After birth, the pups were given either normal or high-protein nutrition. They were then measured for (A) obesity (percent fatty tissue in the abdominal area between the peritoneal cavity and the skin) and (B) insulin resistance (as measured by the amount of insulin in their blood after fasting). Obesity and insulin resistance were greater in the pups whose mothers were fed a protein-deficient diet. High nutrition after birth heightened these differences. (After Vickers et al. 2000.)

2000). If, after weaning, these rats are exposed to a high-calorie diet, the level of hypertension and insulin resistance is even greater. As Gluckman and Hanson (2005) concluded:

> The full spectrum of the "couch potato" syndrome could be explained by a combination of an antenatal event coupled with postnatal amplification. (The only part of the syndrome that escaped us was that we could not demonstrate that the rats had a preference for a TV remote control.)

The likelihood that these diseases are caused by mismatched PARs was suggested by experiments in which the newborn rats were given injections of leptin. When rats are injected with leptin, it tells their bodies that they are fat. When newborn rats from protein-starved mothers were given leptin, they set the metabolic phenotype of their livers to levels appropriate to a high-nutrition diet (Gluckman and Hanson 2007; Gluckman et al. 2007).

Environment-genotype interactions in diabetes and fatty liver disease

It seems that obesity often begets obesity, and animal models, especially those of nonhuman primates, give us insights into how this happens. Exposure of rat and monkey embryos to high-fat diets in utero makes the offspring susceptible as adults to obesity, type 2 diabetes, and hypertension (see Dunn and Bale 2009; Ganu et al. 2012).*

Giving pregnant macaque monkeys a high-fat diet causes their infants to develop a syndrome replicating human nonalcoholic fatty liver disease (Figure 7.15A,B; McCurdy et al. 2009). This appears to be a teratogenic effect, rather than one preparing an organism to cope with a postnatal environment. The high-fat diet decreases the level of histone deacetylase SIRT1 in the fetal livers, which increases acetylated H3K14 in the fetal liver chromatin. This changes the expression of several genes. Many of the gene products are transcription factors such as PPARγ, PPARα, and SREBF1, which regulate fat metabolism in the liver (Figure 7.15C; Aagard-Tillery et al. 2008; Suter et al. 2012a; O'Neil et al. 2013). Thus, in mothers, "obesity begets obesity" does not necessarily involve a preexisting obesity. Rather, it may be the food intake during pregnancy. However, in fathers, pre-conception obesity may be an important factor in the obesity of their children (see box).

Another group of agents through which "obesity begets obesity" are endocrine disruptors such as diethylstilbestrol, bisphenol A, and tributyl tin (see Chapter 6). One of the most prevalent of these endocrine-disrupting obesogens is DDT. In rats, 50% of the pups of the F_3 generation (the great-grandoffspring of a F_0 pregnant rat exposed to DDT) were obese (Skinner et al. 2013). In these F_3 rats, there were specifically methylated regions that differed between the obese and non-obese littermates, and these epialleles were in genes associated with obesity. Both the male and female germline transmitted this condition, and the doses of DDT used in these studies were those within the range of human exposures in the United States during the 1950s. It is possible, then, that the DDT sprayed on our crops over a half century ago is causing obesity and illness today, when 35% of the adults and 17% of the children are obese (Ogden et al. 2014).

One of the generalizations of ecological developmental biology is that you can silence a gene as readily by chromatin modification as by muta-tion. There are several cases wherein either mutational or environmental repression of the same gene can cause disease. For instance, the *Hnf4α* gene encodes a transcription factor that is important for insulin production in the pancreas. Mutations of that gene can predispose people or mice to diabetes (see Arya et al. 2014). Mice given poor nutrition in utero have *epigenetic* repression of the *Hnf4α* genes in their pancreatic cells, and they are prone to adult-onset diabetes (Sandovici et al. 2011). Interestingly, as mice age, the promoter region of this gene is prone to methylation and epigenetic silencing, causing diabetic symptoms as the mice grow older. As we have

---

*The high-fat diet during pregnancy does many other things. Another gene induced by the high-fat diet in the pregnant macaque (and not induced by maternal obesity, per se) is *Npa2* (Suter et al. 2011). This gene encodes a transcription factor that is also involved in regulating the metabolic clock, through which circadian rhythms in metabolism are maintained. Disrup-tions in adult circadian rhythms are also correlated with obesity. Also, the maternal high-fat diet induces changes in thyroid gene expression (Suter et al. 2012b). The thyroid gland is critical for regulating growth and weight gain. And interestingly, the high-fat maternal diet appears to promote the growth of different microbes in the mother's gut, enabling different bacteria types to colonize the newborn macaque. Thus, the high-fat maternal diet appears to persistently change the microbiome of the newborn monkey (Ma et al. 2014).

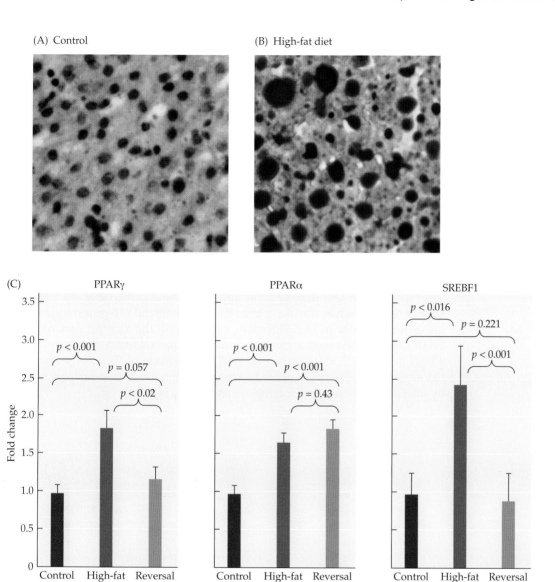

Figure 7.15 Macaque monkeys fed a high-fat diet during gestation developed fatty accumulations in their livers. Livers of fetal macaques fed (A) a control diet (14% fat) are compared with those of fetal macaques fed (B) a high-fat diet (32% fat). The sections are stained with Sudan O, which stains fat accumulations red. Fat accumulations with the high-fat diet resemble those seen in human nonalcoholic fatty liver disease. (C) Transcription factors regulated by elevated SIRT1 levels in the liver include PPARγ, PPARα, and SREBF1, and they all increased in the fetal macaque liver. When mothers were fed control diets after high-fat diets (reversal), the PPARγ and SREBF1 returned to normal levels. PPARα, however, remained elevated throughout the macaque's life. The y-axis refers to the amount of change (2-fold equaling 200%). (A,B from McCurdy et al. 2009, courtesy of J. Friedman; C after Suter et al. 2012a.)

seen throughout these chapters, genes do not work in isolation from their environment.

Another example of a gene that is regulated by the environment but that can have the same effects when it mutates is one of the "thrifty genes," *PPARγ2*. This gene encodes a transcription factor protein that activates particular genes after it combines with the retinoid receptor RXR. The PPARγ2 protein promotes the formation and storage of fat, induces the formation of more fat cells, increases insulin sensitivity, and reduces hypertension (by suppressing angiotensin II). Consequently, this gene can be critical in susceptibility to diabetes. Humans have several versions of *PPARγ2*. One allele, which encodes alanine as its twelfth amino acid, is found in about 10% of the Caucasian and Chinese populations and is less functional than the other, more common allele, which has proline at that position.

Research by Lindi and colleagues (2002) showed that the alanine-containing allele was associated with a predisposition to diabetes, especially in those having low birth weight (Figure 7.16). However, the association of low birth weight and adult-onset diabetes was not seen in people having the more predominant form of *PPARγ2*. In the presence of a genetic situation that made insulin resistance more likely, the connection between low birth weight and diabetes was apparent. If the genes were such that insulin resistance was not likely, then the effects of fetal programming had lower impact later in life. Moreover, people with the alanine variant of *PPARγ2* responded well to exercise and other lifestyle intervention strategies (Lindi et al. 2002; Rittig et al. 2007). The effects of the environment have to be mediated through the metabolic parameters that are established by interactions between the environment and genes.

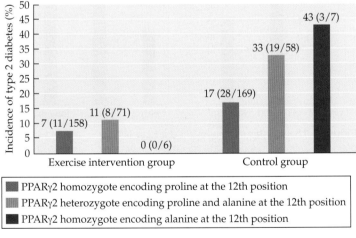

Figure 7.16 Genetics, predisposition, and behavioral intervention affect the incidence of type 2 diabetes in adults. Three-year weight change in control groups differs from that in exercise intervention groups with polymorphism of the *PPARγ2* gene. In the control group, those with the less common allele (which encodes alanine rather than proline at position 12 of the gene) had a high incidence of the disease. Individuals in an exercise program had the lowest incidence of type 2 diabetes regardless of which allele they carried. (After Lindi et al. 2002.)

Paternal Epigenetic Effects

Maternal effects in mammals can be readily accounted for by the interactions between the mother and the fetus. Paternal effects, however, have not been nearly as well studied (see Friedler 1985, 1996; Hansen 2008; Soubry et al. 2014). It was assumed that teratogenic effects would be passed on through the mother's exposure to chemicals, and that the father's germline, since it didn't transmit much but a genome, was relatively secure from environmental effects. For example, in 1995, a report from the Institute of Medicine (1995, p. 44) claimed that exposure of human males to environmental chemicals "is generally unrelated to developmental endpoints such as miscarriage, birth defects, growth retardation, and cancer." But as Richardson and colleagues (2014) remind us, "Don't blame the mothers. Careless discussion of epigenetic research on how early life affects health across generations could harm women." Indeed, we now know that environmental agents *can* effect epigenetic changes in the male germline and that these alterations can be passed from one generation to the next.

We saw (p. 240) the inheritance of testicular dysgenesis through the male germline. Here, there was abnormal methylation of the enhancers of certain genes whose products were necessary for testis morphogenesis and sperm production. This sperm-mediated inheritance of testicular dysgenesis has been seen in the exposure of male fetuses to the fungicide vinclozolin, and also in exposures to the herbicide simazine and the plasticizers/endocrine disruptors bisphenol A and dibutyl phthalate (Park and Bae 2012; Manikkam et al. 2013; Guerrero-Bosagna et al. 2014).

Moreover, if epigenetic modification can be inherited through the germline, then the male's germline can be as important as the woman's. And this means that the father's lifestyle and exposure could be as significant as the mother's. This is not usually the case in teratology, where the mother's lifestyle and exposure are considered much more significant. Diet may be critical here, too. Although much attention has been given to the mother's diet and weight, the father's body weight may also matter. Paternal obesity may trigger epigenetic changes in the sperm chromosomes, making the offspring more susceptible for obesity and metabolic disease. The Newborn Epigenetics Study (Soubry et al. 2013a,b) showed that paternal obesity is associated with the hypomethylation of the human insulin-like growth factor-2 (*IGF2*) gene and other genes that are usually expressed from the male-derived chromosomes. Insulin-like growth factor-2 is a growth hormone that is made during embryonic development. It is one of the few genes that is usually active only on sperm-derived chromatin (i.e., an "imprinted gene"), and higher levels of circulating IGF2 are associated with obesity.

These epidemiological data have been supported by laboratory experiments. Male mice that had been fed a low-protein diet from weaning produced offspring with increased methylation at the promoter of the gene encoding PPARα, a key lipid regulator (Carone et al. 2010). When the males were given high-fat diets, their female offspring had impaired glucose metabolism and pancreatic transcriptomes that differed from the norm (Ng et al. 2010, 2014; Wei et al. 2014).

While it is possible that paternal obesity may cause methylation changes resulting in obesity in his offspring, maternal undernourishment may also cause methylation differences in sperm. And these changes may initiate alterations in the metabolism of the male's offspring. Mouse models have shown that when a pregnant mouse is undernourished during the time her male embryos' germ cells are undergoing normal methylation changes during spermatogenesis, certain genes within the sperm become hypomethylated. Later, when these sperm fertilize eggs, the resulting offspring have metabolic disease (Radford et al. 2014).

Interestingly, epidemiological studies indicate that sperm might be affected by the nutritional environment of a grandparent (Pembrey et al. 2006). Analysis of medical records from an isolated community in northern Sweden showed that poor nutrition of the paternal grandfather during middle childhood was linked to a greater risk of early deaths in grandsons (but not granddaughters; Figure A). Conversely, the *paternal grandmother*'s food supply during middle childhood was reflected in the higher mortality risk of her granddaughters (and not her grandsons; Figure B,C). This effect was not observed in the maternal grandmother or maternal grandfather, indicating that this influence was coming from the children's father. The report argues that exposure of a male to environmental influences can affect the development and the health of his sons and grandsons.

Transgeneration inheritance of early trauma through sperm

DNA methylation and chromatin acetylation may not be the only means whereby the environmental effects can be inherited. The sperm contain numerous small noncoding mRNAs (sncRNAs) that can regulate transcription

(Continued on next page)

Paternal Epigenetic Effects (*continued*)

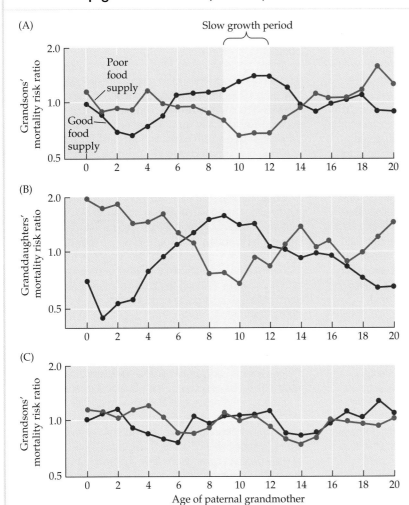

The effects of varying food supply of paternal grandparents on the mortality rates of their grandchildren. The age of the paternal grandparent refers to the beginning of a year or more of exposure to a poor duration in a "good" food supply, or exposure to a "good" duration in a poor food supply. (A) Mean mortality risk ratio of grandsons is plotted against age at which paternal grandparents' food supply changed. The curves show increased mortality risk in grandsons of those whose "good" food supply was interrupted during a crucial growth period. (B) A similar graph for granddaughters and paternal grandmothers. (C) A graph of grandsons' mortality ratio against paternal grandmothers shows no such correlation. A comparison of granddaughters' mortality risk against paternal grandfathers, not shown here, similarly showed no correlation. (After Pembrey et al. 2006.)

and translation. In mammals, sperm-borne microRNAs are required for the zygote to divide (Rassoulzadegan et al. 2006; Liu et al. 2012). Gapp and colleagues (2014) have proposed that behavioral phenotypes induced in male mice by stress are transmitted to the next generation by sncRNAs. These researchers gave newborn male pups an unstable and stressful environment. When raised in this unpredictable environment, the males showed depressive behaviors, differed significantly from control male mice in several behavioral traits, and had dramatically depressed glucose levels in their blood. Remarkably, their offspring showed a depressive phenotype and abnormally low sugar levels, even if these mice had not experienced the stressful environment (Figure D).

The ability of males to transmit such behavioral phenotype caused the researchers to focus on the sperm. Gapp and colleagues found that several sncRNAs were significantly more abundant in the sperm of the stressed males. These were also present in the sperm of their male offspring. To demonstrate that the sncRNAs were causally involved in producing the depressed phenotype, the investigators injected these RNAs into newly fertilized eggs from unstressed males and females. The resulting mice acted like the traumatized males and their offspring, showing the same depressed behavior and altered sugar metabolism (Figure E,F). The mechanisms for the

Paternal Epigenetic Effects (*continued*)

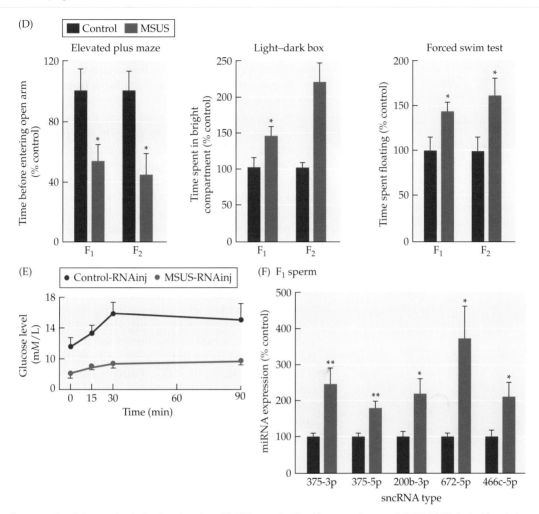

Paternally transmitted depressive behaviors in mice. (D) Differences between control male mice and male mice who had experienced experimentally induced trauma as newborns (MSUS). Traumatized mice spent more time in the light, entered the open arm of a maze more rapidly, and spent more time floating when forced to swim. Their serum glucose levels were much lower. Moreover, these effects were seen in the F_2 generation (the sons of the F_1 generation that experienced the trauma). (E) When sperm sncRNAs from traumatized mice were injected into mouse oocytes fertilized by control sperm (MSUS-RNAinj), the blood glucose levels of the offspring were severely depressed compared to injections from untraumatized mice. (F) Moreover, the expression of particular sncRNAs (mi375-3p, mi375-5p, etc.) in the sperm differed between the traumatized and control mice. Asterisk (*) marks differences that are statistically significant to $p < 0.05$. Double-asterisks (**) indicate statistical differences of $p < 0.01$. (After Gapp et al. 2014.)

production of these sncRNAs is not yet known.

It has long been thought that the male germline (and for that matter, males) exist as the means whereby diversity is brought into the lineage. Soubry (2015) speculates that the sperm may be "one of nature's tools to capture messages from our continually changing environment and to transfer this information to subsequent generations."

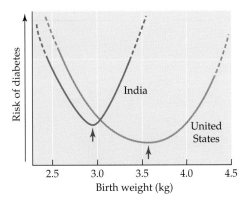

Figure 7.17 The risk of adult-onset diabetes is related to birth weight in both India and the United States. The U-shaped curve indicates that the risk for this disease increases at both very high and very low birth weights. However, the optimum birth weight is different for the two populations (arrows), suggesting that developmental plasticity may have resulted in adaptation to distinct adult nutritional milieus. (After Gluckman and Hansen 2005.)

Developmental Plasticity and Public Health

Understanding developmental plasticity has major implications for public health policy (see Law and Baird 2006; Gluckmann et al. 2009). Most public health programs have at their core an interest in survival. This is seen, for instance, in the United Nations Millennium Development Goals for substantially reducing infant mortality. But it is important that such remediation programs not shift the burden of illness from the early years of life to the later ones. We have seen that developmental plasticity needs to become a central part of the fields of developmental biology and conservation biology. It also needs to become a central part of public health policy making.

First, interventions designed to improve adult health need to begin early in life or even in utero. Lifestyle interventions aimed at obesity-related problems in adults, for example, may be less effective if the person's gene expression pattern is already "set" to a phenotype that efficiently conserves energy and stores it as fat (Wells 2007). Second, interventions designed for infant survival must take into account the effects of such changes. For instance, babies born weighing 2.5 kg in rural India may be well adapted both to their maternal milieu and to the nutritional environment where they will live as adults (Figure 7.17). Food supplementation programs aimed at increasing the birth weight of such babies or providing food in schools may increase the chance that those individuals will get diabetes later in life (Bhargava 2004; Yajnik 2005). Third, it must be recognized that different evolutionary forces have acted on different populations. A "universal standard" of human growth erroneously assumes that life history strategies are the same worldwide and does not take into account the plasticity that may have evolved in different locations (Adams and White 2005; Wells 2007). This new area demands research from many disciplines.

Looking Ahead

Although the epigenetic origin of human disease remains a hypothesis, we have seen that many of the most critical diseases of adult humans—diabetes, hypertension, heart attack, and obesity—have major epigenetic components. The susceptibility to such diseases, whether because of maternal diet, the environment, or lifestyle, is regulated both by genetic alleles and by epigenetic changes in the genome. These new insights can lead us to better preventive measures in public health and better treatments through drug design. While the epigenetic origin of adult diseases is a relatively new hypothesis and not as well established as others (such as the germ theory of infectious disease), it is providing explanations for many medical phenomena that had until now resisted our understanding.

One of the critical questions is "What takes the epigenetic marks off?" If environmental factors put them on, then you might expect that after 100 generations, a species would be saturated with methylated DNA and repressive histones. So are these marks taken off during the routine demethylations of mammalian germ cells? But if some get through these screens, how do they do it, and what gets rid of them?

One of the most important areas of research will be the effects of intrauterine and neonatal stress on adult behavioral and physiological phenotypes. Childhood stress (including intrauterine stress) is a major factor contributing to adult-onset age-related diseases (see Cohen et al. 2007; Entringer et al. 2010). Indeed, some of the most interesting data of medical sociology strongly suggest that one's *early* life socioeconomic status (SES) determines the probability of infectious, respiratory, and cardiac diseases later in life (Cohen et al. 2004; Galobardes et al. 2004, 2006; Thomas et al. 2008). These disease susceptibilities are largely independent of one's *adult* SES. For instance, one study (Kittleson et al. 2006) found that among highly educated and wealthy members of a medical school class, those who came from low SES backgrounds (independent of race) had over twice the risk of developing coronary disease by age 50. It has been suggested (Zhang et al. 2006; Miller et al. 2009) that repeated adversity in early life gives one a "defensive phenotype" characterized by heightened glucocorticoid and cytokine (inflammatory) reactions to stressful environments. This chapter has presented evidence that telomere shortening can be accelerated by stress, and researchers are looking to see whether differential chromatin modification might be also responsible (Hackman et al. 2010; Lam et al. 2012; Essex et al. 2013).

Another area of stress research concerns its inheritance. As we have seen in this chapter, certain stress-induced behaviors in mice can be transmitted by sperm. Whether this is possible in humans is not yet known. Offspring of parents with post-traumatic stress disorder (i.e., survivors of war trauma, childhood sexual abuse, torture, Holocaust experiences) have a higher probability than others of developing debilitating anxiety and depression (Kellermann 2013). Is the "transgenerational transmission of trauma" (see Kellerman 2009) a biological reality? Our ideas about disease and inheritance are changing. They may have to change even more to deal with the possible epigenetic causation and transmission of psychological conditions.

Another area to study is the possibility that in vitro fertilization may be changing the epigenetic constitution of the embryo's chromatin and that these changes may cause health effects later in life. Mice conceived by in vitro fertilization have, statistically, altered insulin signaling and glucose tolerance, compared with their naturally conceived controls. Even if the embryos are transferred into healthy mice at the two-cell stage, adverse health effects persist (Rexhaj et al. 2013; Feuer et al. 2014a,b). These mice also have alterations of gene methylation (see Lane et al. 2014). Mice are physiologically very different from humans, and their short lifespans and large litter sizes may make them poor proxies for humans. So extrapolation is a risky game. But given that thousands of children are conceived each year through in vitro fertilization, we might be doing an enormous medical experiment on human populations. Studies need to be done to see if being conceived by IVF in any way causes or predisposes people to particular diseases later in life.

In his poem on the glories of nature, Wordsworth (1802) claims, "The Child is father to the Man" and that the emotional experiences of youth condition those of the adult. But what are the biochemical agents whereby prenatal and childhood experiences would be allowed to influence adult health and cognition? The answers will be thought provoking and perhaps policy changing. They also will most likely be statistical answers, analogue not digital. As Jablonka (2013) has written (concerning the possibility of transmitting behavioral traits epigenetically), "What we know about epigenetic marks is that they can dispose one towards developing some behaviors, but the specific behavior depends on specific inputs the person gets in its own lifetime." Epigenetics is not a deterministic science; rather it is one of interpretations, and context is critical.

References

Aagaard-Tillery, K. M., K. Grove, J. Bishop, X. Ke, Q. Fu, R. McKnight and R. H. Lane. 2008. Developmental origins of disease and determinants of chromatin structure: Maternal diet modifies the primate fetal epigenome. *J. Mol. Endocrinol.* 41: 91–102.

Adams, J. and M. White. 2005. When the population approach to prevention puts the health of individuals at risk. *Int. J. Epidem.* 34: 40–43.

Allen, W. R., S. Wilsher, C. Tiplady, and R. M. Butterfield. 2004. The influence of maternal size on pre- and postnatal growth in the horse. III. Postnatal growth. *Reproduction* 127: 67–77.

Arya, V. B., S. Rahman, S. Senniappan, S. E. Flanagan, S. Ellard, and K. Hussain. 2014. HNF4A mutation: Switch from hyperinsulinaemic hypoglycaemia to maturity-onset diabetes of the young, and incretin response. *Diabetic Med.* 31: e11–15.

Barker, D. J. P. 1994. *Mothers, Babies and Disease in Later Life.* Churchill Livingstone, London.

Barker, D. J. P. 1997. Fetal nutrition and cardiovascular disease in later life. *Br. Med. Bull.* 53: 96–108.

Barker, D. J. P. 1998. In utero programming of chronic disease. *Clin. Sci.* 95: 115–128.

Barker, D. J. P. and C. Osmond. 1986. Infant mortality, childhood nutrition, and ischaemic heart disease in England and Wales. *Lancet* 1: 1077–1081.

Barker, D. J. P., J. G. Eriksson, T. Forsén and C. Osmond. 2002. Fetal origins of adult disease: Strength of effects and biological basis. *Int. J. Epidemiol.* 31: 1235–1239.

Bateson, P. 2007. Developmental plasticity and evolutionary biology. *J. Nutr.* 137: 1060–1062.

Bateson, P., P. Gluckman and M. Hanson. 2014. The biology of developmental plasticity and the predictive adaptive response hypothesis. *J. Physiol.* 592: 2357–2368.

Bateson, P. and 14 others. 2004. Developmental plasticity and human health. *Nature* 430: 419–421.

Bhargava, S. K. 2004. Relation of serial changes in childhood body mass index to impaired glucose tolerance in young adulthood. *New Engl. J. Med.* 350: 865–875.

Blaze, J. and T. L. Roth. 2015. Evidence from clinical and animal model studies of the long-term and transgenerational impact of stress on DNA methylation. *Semin. Cell Dev. Biol.* pii: S1084-9521(15)00079-8.

Burdge, G. C., J. Slater-Jefferies, C. Torrens, E. S. Phillips, M. A. Hanson and K. A. Lillycrop. 2006. Dietary protein restriction of pregnant rats in the F_0 generation induces altered methylation of hepatic gene promoters in the adult male offspring in the F_1 and F_2 generations. *Br. J. Nutr.* 97: 435–439.

Carone, B. R. and 14 others. 2010. Paternally induced transgenerational environmental reprogramming of metabolic gene expression in mammals. *Cell* 143: 1084–1096.

Chae, D. H., A. M. Nuru-Jeter, N. E. Adler, G. H. Brody, J. Lin, E. H. Blackburn and E. S. Epel. 2014. Discrimination, racial bias, and telomere length in African-American men. *Am. J. Prev. Med.* 46: 103–111.

Chali, D., E. Enquselassie and M. Gesese. 1998. A case-control study of determinants of rickets. *Ethiop. Med. J.* 36: 227–234. (Quoted in Bateson 2007.)

Cohen, S., W. J. Doyle, R. B. Turner, C. M. Alper and D. P. Skoner. 2004. Childhood socioeconomic status and host resistance to infectious illness in adulthood. *Psychosom. Med.* 66: 553–558.

Cohen, S., D. Janicki-Deverts and G. E. Miller. 2007. Psychological stress and disease. *J. Am. Med. Assoc.* 298: 1685–1687.

Cossins, D. 2015. Stress fractures. *The Scientist* Jan: 33–38.

De Boo, H. A. and J. E. Harding. 2006. The developmental origins of adult disease (Barker) hypothesis. *Aust. N. Z. J. Obstet. Gynaecol.* 46: 4–14.

Desai, M., N. Crowther, A. Lucas and C. M. Hales. 1995. Adult glucose and lipid metabolism may be programmed during fetal life. *Biochem. Soc. Trans.* 23: 331–335.

Dey, S. K., H. Lim, S. K. Das, J. Reese, B. C. Paria, T. Daikoku and H. Wang. 2004. Molecular cues to implantation. *Endocr. Rev.* 25: 341–373.

Diamond, J. 1991. Pearl Harbor and the emperor's physiologists. *Nat. Hist.* 1991(12): 2–7.

Dunn, G. A., and T. L. Bale. 2009. Maternal high-fat diet promotes body length increases and insulin insensitivity in second-generation mice. *Endocrinology* 150: 4999–5009.

Entringer, S., C. Buss and P. D. Wadhwa. 2010. Prenatal stress and developmental programming of human health and disease risk: Concepts and interpretation of empirical findings. *Curr. Opin. Endocrinol. Diabetes Obes.* 17: 507–516.

Entringer, S. and 7 others. 2013. Maternal psychosocial stress during pregnancy is associated with newborn leukocyte telomere length. *Am. J. Obstet. Gynecol.* 208: e1–7.

Epel, E. S., E. H. Blackburn, J. Lin, F. S. Dhabhar, N. E. Adler, J. D. Morrow and R. M. Cawthon. 2004. Accelerated telomere shortening in response to life stress. *Proc. Natl. Acad. Sci. USA* 101: 17312–17315.

Eriksson, J. G., T. Forsen, J. Tuomilehto, C. Osmond and D. J. Barker. 2001. Early growth and coronary heart disease in later life: Longitudinal study. *Br. Med. J.* 322: 949–953.

Essex, M. J., W. T. Boyce, C. Hertzman, L. L. Lam, J. M. Armstrong, S. M. Neumann and M. S. Kobor. 2013. Epigenetic vestiges of early developmental adversity: Childhood stress exposure and DNA methylation in adolescence. *Child Dev.* 84: 58–75.

Feder, S. 1990. Environmental determinants of seasonal coat color change in weasels (*Mustela erminea*) from two populations. M.S. Thesis, University of Alaska, Fairbanks.

Feinberg, A. 2014. Quoted in E. Callaway. Epigenetics starts to make its mark. *Nature* 508: 22.

Feuer, S. K. and 10 others. 2014a. Use of a mouse in vitro fertilization model to understand the developmental origins of health and disease hypothesis. *Endocrinology* 155: 1956–1969.

Feuer, S. K., A. Donjacour, P. K. Simbulan, W. Lin, X. Liu, E. Maltepe and P. F. Rinaudo. 2014b. Sexually dimorphic effect of in vitro fertilization (IVF) on adult mouse fat and liver metabolomes. *Endocrinology* 155: 4554–4567.

Fowden, A. L. and A. J. Forhead. 2004. Endocrine mechanisms of intrauterine programming. *Reproduction* 127: 515–526.

Francis, D. D., K. Szegda, G. Campbell, W. D. Martin and T. R. Insel. 2003. Epigenetic sources of behavioral differences in mice. *Nature Neurosci.* 6: 445–446.

Friedler, G. 1985. Effects of limited paternal exposure to xenobiotic agents on the development of progeny. *Neurobehav. Toxicol. Teratol.* 7: 739–743.

Friedler, G. 1996. Paternal exposures: Impact on reproductive and developmental outcome—An overview. *Pharmacol. Biochem. Behav.* 55: 691–700.

Galobardes, B., J. W. Lynch and G. Davey Smith. 2004. Childhood socioeconomic circumstances and cause-specific mortality in adulthood: Systematic review and interpretation. *Epidemiol. Rev.* 26: 7–21.

Galobardes, B., G. D. Smith and J. W. Lynch. 2006. Systematic review of the influence of childhood socioeconomic circumstances on risk for cardiovascular disease in adulthood. *Ann. Epidemiol.* 16: 91–104.

Ganu, R. S., R. A. Harris, K. Collins and K. M. Aagaard. 2012. Early origins of adult disease: Approaches for investigating the programmable epigenome in humans, nonhuman primates, and rodents. *ILAR J.* 53: 306–321.

Gapp, K. A. and 8 others. 2014. Implication of sperm RNAs in transgenerational inheritance of the effects of early trauma in mice. *Nat. Neurosci.* 17: 667–669.

Gilbert, S. F. 2006. *Developmental Biology*, 8th Ed. Sinauer Associates, Sunderland, MA, pp. 677–681.

Gluckman, P. D. and M. A. Hanson. 2004. Living with the past: Evolution, development, and patterns of disease. *Science* 305: 1733–1736.

Gluckman, P. D. and M. A. Hanson. 2005. *The Fetal Matrix: Evolution, Development, and Disease*. Cambridge University Press, Cambridge.

Gluckman, P. D. and M. A. Hanson. 2006a. The developmental origins of health and disease: An overview. In P. D. Gluckman and M. A. Hanson (eds.), *Developmental Origins of Health and Disease*. Cambridge University Press, Cambridge, pp. 1–5.

Gluckman, P. D. and M. A. Hanson. 2006b. The conceptual basis for the developmental origins of health and disease. In P. D. Gluckman and M. A. Hanson (eds.), *Developmental Origins of*

Health and Disease. Cambridge University Press, Cambridge, pp. 33–50.

Gluckman, P. D. and M. A. Hanson. 2007. *Mismatch: Why Our World No Longer Fits Our Bodies*. Oxford University Press, Oxford.

Gluckman, P. D., M. A. Hanson and H. G. Spencer. 2005. Predictive adaptive responses and human evolution. *Trends Ecol. Evol.* 20: 527–533.

Gluckman, P. D. A. S. Beedle, M. A. Hanson and M. H. Vickers. 2007. Leptin reversal of the metabolic phenotype: Evidence for the role of developmental plasticity in the development of the metabolic syndrome. *Horm. Res.* 67(Suppl. 1): 115–120.

Gluckman, P. D., M. A. Hanson, C. Cooper and K. L. Thornburg. 2008. Effect of the in utero and early-life conditions on adult health and disease. *New Engl. J. Med.* 359: 61–73.

Gluckman, P. D. and 19 others. 2009. Towards a new developmental synthesis: Developmental plasticity and human disease. *Lancet* 373: 1654–1657.

Godfrey, K. M. 2006. The "developmental origins" hypothesis. In P. D. Gluckman and M. A. Hanson (eds.), *Developmental Origins of Health and Disease*. Cambridge University Press, Cambridge, pp. 6–32.

Godfrey, K. M. and D. Barker. 2000. Fetal nutrition and adult diseases. *Am. J. Clin. Nutr.* 71(Suppl.): S1344–S1355.

Guerrero-Bosagna, C., S. Weeks and M. K. Skinner. 2014. Identification of genomic features in environmentally induced epigenetic transgenerational inherited sperm epimutations. *PLoS ONE* 9: e110194.

Hackman, D. A., M. J. Farah and M. J. Meaney. 2010. Socioeconomic status and the brain: Mechanistic insights from human and animal research. *Nat. Rev. Neurosci.* 11: 651–659.

Hales, C. N. and D. J. Barker. 1992. Type 2 (non-insulin-dependent) diabetes mellitus: The thrifty phenotype hypothesis. *Diabetologia* 35: 595–601.

Hales, C. N. and D. J. Barker. 2001. The thrifty phenotype hypothesis. *Br. Med. Bull.* 60: 5–20.

Hales, C. N. and 6 others. 1991. Fetal and infant growth and impaired glucose tolerance at age 64. *Br. Med. J.* 303: 1019–1022.

Hansen, D. A. 2008. Paternal environmental exposures and gene expression during spermatogenesis: Research review to research framework. *Birth Defects Res., Part C* 84: 155–163.

Hanson, M. A. and K. M. Godfrey. 2008. Commentary: Maternal constraint is a pre-eminent regulator of fetal growth. *Int. J. Epidemiol.* 37: 252–254.

Hart, N. 1993. Famine, maternal nutrition and infant mortality: A re-examination of the Dutch Hunger Winter. *Popul. Stud. (Camb.)* 47: 27–46.

Hoppe, C. C., R. G. Evans, J. F. Bertram and K. M. Moritz. 2007. Effects of dietary protein restriction on nephron number in the mouse. *Am. J. Physiol.: Regul. Integr. Comp. Physiol.* 292: R1768–1774.

Institute of Medicine. 1995. *Adverse Reproductive Outcomes in Families of Atomic Veterans: The Feasibility of Epidemiologic Studies*. National Academy Press, Washington, D.C.

Jablonka, E. 2013. Quoted in Kellermann 2013 Op. cit., p. 35.

Keller, G., G. Zimmer, G. Mall, E. Ritz and K. Amann. 2003. Nephron number in patients with primary hypertension. *New Engl. J. Med.* 348: 101–108.

Kellerman, N. P. F. 2009. *Holocaust Trauma: Psychological Effects and Treatment*. iUniverse, NY.

Kellermann, N. P. F. 2013. Epigenetic transmission of Holocaust trauma: Can nightmares be inherited? *Isr. J. Psychiatry Relat. Sci.* 50: 33–39.

Kittleson, M. M., L. A. Meoni, N. Y. Wang, A. Y. Chu, D. E. Ford and M. J. Klag. 2006. Association of childhood socioeconomic status with subsequent coronary heart disease in physicians. *Arch. Intern. Med.* 166: 2356–2361.

Lam, L. L., E. Emberly, H. B. Fraser, S. M. Neumann, E. Chen, G. E. Miller M. S. Kobor. 2011. Factors underlying variable DNA methylation in a human community cohort. *Proc. Natl. Acad. Sci. USA* 109(suppl 2): 17253–17260.

Lane, M., R. L. Robker and S. A. Robertson. 2014. Parenting from before conception. *Science* 345: 756–760.

Lau, C. and J. M. Rogers. 2004. Embryonic and fetal programming of physiological disorders in adulthood. *Birth Defects Res., Part C* 72: 300–312.

Law, C. and J. Baird. 2006. Developmental origins of health and disease: Public-health perspectives. In P. D. Gluckman and M. A. Hanson (eds.), *Developmental Origins of Health and Disease.* Cambridge University Press, Cambridge, pp. 446–455.

Lillycrop, K. A., E. S. Phillips, A. A. Jackson, M. A. Hanson and G. C. Burdge. 2005. Dietary protein restriction of pregnant rats induces and folic acid supplementation prevents epigenetic modification of hepatic gene expression in the offspring. *J. Nutr.* 135: 1382–1386.

Lillycrop, K. A., E. S. Phillips, C. Torrens, M. A. Hanson, A. A. Jackson and G. C. Burdge. 2008. Feeding pregnant rats a protein-restricted diet persistently alters the methylation of specific cytosines in the hepatic PPARα promoter of the offspring. *Br. J. Nutr.* 100: 278–282.

Lindi, V. I. and 11 others. 2002. Association of the Pro12Ala polymorphism in the *PPAR-γ2* gene with 3-year incidence of type 2 diabetes and body weight change in the Finnish diabetes prevention study. *Diabetes* 51: 2581–2586.

Liu, W. M., R. T. Pang, P. C. Chiu, B. P. Wong, K. Lao, K. F. Lee and W. S. Yeung. 2012. Sperm-borne microRNA-34c is required for the first cleavage division in mouse. *Proc. Natl. Acad. Sci. USA* 109: 490–494.

Lucas, A. 1991. Programming by early nutrition in man. In G. R. Bock and J. Whelan (eds.), *The Childhood Environment and Adult Disease.* Wiley, Chichester, UK, pp. 38–55.

Ma, J. and 10 others. 2014. High-fat maternal diet during pregnancy persistently alters the offspring microbiome in a primate model. *Nature Commun.* 5: 3889.

Manikkam, M., R. Tracey, C. Guerrero-Bosagna and M. K. Skinner. 2013. Plastics derived endocrine disruptors (BPA, DEHP and DBP) induce epigenetic transgenerational inheritance of obesity, reproductive disease and sperm epimutations. *PLoS ONE* 8(1): e55387.

Martínez, D. and 12 others. 2014. In utero undernutrition in male mice programs liver lipid metabolism in the second-generation offspring involving altered *Lxra* DNA methylation. *Cell Metab.* 19: 941–951.

McCurdy, C. E., J. M. Bishop, S. M. Williams, B. E. Grayson, M. S. Smith, J. E. Friedman and K. L. Grove. 2009. Maternal high-fat diet triggers lipotoxicity in the fetal livers of nonhuman primates. *J. Clin. Invest.* 119: 323–335.

Miller, G. E. and 7 others. 2009. Low early-life social class leaves a biological residue manifested by decreased glucocorticoid and increased proinflammatory signaling. *Proc. Natl. Acad. Sci. USA* 106: 14716–14721.

Moritz, K. M., M. Dodic and E. M. Wintour. 2003. Kidney development and the fetal programming of adult disease. *BioEssays* 25: 212–220.

Morton, N. E. 1955. The inheritance of human birth weight. *Ann. Hum. Genet.* 20: 125–134.

Ng, S. F., R. C. Lin, D. R. Laybutt, R. Barres, J. A. Owens and M. J. Morris. 2010. Chronic high-fat diet in fathers programs b-cell dysfunction in female rat offspring. *Nature* 467: 963–966.

Ng, S. F., R. C. Lin, C. A. Maloney, N. A. Youngson, J. A. Owens and M. J. Morris. 2014. Paternal high-fat diet consumption induces common changes in the transcriptomes of retroperitoneal adipose and pancreatic islet tissues in female rat offspring. *FASEB J.* 28: 1830–1841.

Nilsson, E. and 15 others. 2014. Altered DNA methylation and differential expression of genes influencing metabolism and inflammation in adipose tissue from subjects with type 2 diabetes. *Diabetes* 63: 2962–2976.

Ogden, C. L., M. D. Carroll, B. K. Kit and K. M. Flegal. 2014. Prevalence of childhood and adult obesity in the United States, 2011–2012. *J. Amer. Med. Assoc.* 311: 806–814.

O'Neil, D., H. Mendez-Figueroa, T. A. Mistretta, C. Su, R. H. Lane and K. M. Aagaard. 2013. Dysregulation of Npas2 leads to altered metabolic pathways in a murine knockout model. *Mol. Genet. Metab.* 110: 378–387.

Park, H. O. and J. Bae. 2012. Disturbed relaxin signaling pathway and testicular dysfunction in mouse offspring upon maternal exposure to simazine. *PLoS ONE* 7(9): e44856.

Pechenik, J. A. 2006. Larval experience and latent effects: Metamorphosis is not a new beginning. *Integr. Comp. Biol.* 46: 323–333.

Pembrey, M. E. and 7 others. 2006. Sex-specific, male-line transgenerational responses in humans. *Eur. J. Human Genet.* 14: 159–166.

Petrik, J., B. Reusens, E. Arany, C. Remacle, J. J. Hoet and D. J. Hill. 1999. A low protein diet alters the balance of islet cell replication and apoptosis in the fetal and neonatal rat and is associated with reduced pancreatic expression of insulin-like growth factor II. *Endocrinology* 140: 4861–4873.

Radford, E. J. and 13 others. 2014. In utero undernourishment perturbs the adult sperm methylome and intergenerational metabolism. *Science* 345: 785.

Rassoulzadegan, M., V. Grandjean, P. Gounon, S. Vincent, I. Gillot and F. Cuzin. 2006. RNA-mediated non-mendelian inheritance of an epigenetic change in the mouse. *Nature* 441: 469–474.

Rexhaj, E. and 9 others. 2013. Mice generated by in vitro fertilization exhibit vascular dysfunction and shortened life span. *J. Clin. Invest.* 123: 5052–5060.

Richardson, S. S., C. R. Daniels, M. W. Gillman, J. Golden, R. Kulka, C. Kuzawa and J. Rich-Edwards. 2014. Don't blame the mothers. *Nature* 512: 131–132.

Rittig, S. and 7 others. 2007. The Pro12Ala polymorphism in *PPARγ2* increases the effectiveness of primary prevention of cardiovascular disease by a lifestyle intervention. *Diabetologia* 50: 1345–1347.

Robson, E. B. 1955. Birth weight in cousins. *Ann. Hum. Genet.* 19: 262–268.

Rostand, J. 1962. *The Substance of Man.* I. Brandeis (trans.). Doubleday, New York.

Sapolsky, R. M. 2004. Organismal stress and telomere aging: An unexpected connection. *Proc. Natl. Acad. Sci. USA* 101: 17323–17325.

Scott Mills, L., M. Zimova, J. Oyler, S. Running, J. T. Abatzoglou and P. M. Lukacs. 2013. Camouflage mismatch in seasonal coat color due to decreased snow duration. *Proc. Natl. Acad. Sci. USA* 110: 7360–7365.

Skinner, M. K., M. Manikkam, R. Tracey, C. Guerrero-Bosagna, M. Haque and E. E. Nilsson. 2013. Ancestral dichlorodiphenyltrichloroethane (DDT) exposure promotes epigenetic transgenerational inheritance of obesity. *BMC Med.* 11: 228.

Soubry, A. 2015. Epigenetic inheritance and evolution: A paternal perspective on dietary influences. *Prog. Biophys. Mol. Biol.* doi: 101016/1.pbiomemolbio.2015.02.008

Soubry, A. and 9 others. 2013a. Paternal obesity is associated with *IGF2* hypomethylation in newborns: Results from a Newborn Epigenetics Study (NEST) cohort. *BMC Med.* 11: 29.

Soubry, A. and 10 others. 2013b. Newborns of obese parents have altered DNA methylation patterns at imprinted genes. *Int. J. Obes.* 39: 650–657.

Soubry, A., C. Hoyo, R. L. Jirtle and S. K. Murphy. 2014. A paternal environmental legacy: Evidence for epigenetic inheritance through the male germ line. *Bioessays* 36: 359–371.

Stein, Z. 1975. *Famine and Human Development: The Dutch Hunger Winter of 1944–1945.* Oxford University Press, New York.

Stevens, L. M. and S. C. Landis. 1988. Developmental interactions between sweat glands and the sympathetic neurons which innervate them: Effects of delayed innervation on neurotransmitter plasticity and gland maturation. *Dev. Biol.* 130: 703–720.

Suter, M. A. and 8 others. 2011. Epigenomics: Maternal high-fat diet exposure in utero disrupts peripheral circadian gene expression in nonhuman primates. *FASEB J.* 25: 714–726.

Suter, M. A. and 9 others. 2012a. A maternal high-fat diet modulates fetal SIRT1 histone and protein deacetylase activity in nonhuman primates. *FASEB J.* 26: 5106–5114.

Suter, M. A. and 10 others. 2012b. Maternal high-fat diet modulates the fetal thyroid axis and thyroid gene expression in a nonhuman primate model. *Mol. Endocrinol.* 26: 2071–2080.

Sweatt, J. D. 2013. The emerging field of neuroepigenetics. *Neuron* 80: 624–632.

Taylor, P. D. and L. Poston. 2007. Developmental programming of obesity in mammals. *Exp. Physiol.* 92: 287–298.

Thomas, C., E. Hyppönen and C. Power. 2008. Obesity and type 2 diabetes risk in midadult life: The role of childhood adversity. *Pediatrics* 121: e1240–e1249.

Tischner, M. 1987. Development of Polish-pony foals born after embryo transfer to large mares. *J. Reprod. Fertil.* 35(Suppl.): 705–709.

Vickers, M. H., B. H. Breier, W. S. Cutfield, P. I. Hofmann and P. D. Gluckman. 2000. Fetal origins of hyperphagia, obesity, and hypertension and its postnatal amplification by caloric nutrition. *Am. J. Physiol.* 279: E83–87.

Walton, A. J. and J. Hammond. 1938. The maternal effects on growth and conformation in Shire horses–Shetland pony crosses. *Proc. R. Soc. London, Ser. B* 125: 311–335.

Welham, S. J. M., P. R. Riley, A. Wade, M. Hubank and A. S. Woolf. 2005. Maternal diet programs embryonic kidney gene expression. *Physiol. Genet.* 22: 48–56.

Wells, J. C. K. 2007. Commentary: Why are South Asians susceptible to central obesity? The El Niño hypothesis. *Int. J. Epidemiol.* 36: 226–227.

Whitelaw, E. 2006. Sins of the fathers and their fathers. *Eur. J. Hum. Genet.* 14: 131–132.

Wordsworth, W. 1802. "My heart leaps up when I behold." www.bartleby.com/145/ww194.html

Yajnik, C. S. 2005. Size and body composition at birth and risk of type-2 diabetes: A critical evaluation of "fetal origins" hypothesis for developing countries. In G. Hornstra, R. Uauy and X. Yang (eds.), *The Impact of Maternal Nutrition on the Offspring.* Karger AG, Basel, pp. 169–183.

Zhang, T. Y. and 9 others. 2006. Maternal programming of defensive responses through sustained effects on gene expression. *Biol. Psychol.* 73: 72–89.

Developmental Models
of Cancer and Aging

In those days people knew (or at least they sensed) that they had death inside them, as the fruit has its seed. Children had a small death inside them, adults a large one. The women had it in their womb and the men in their chest. They had it there, and that gave them a peculiar dignity and quiet pride.

R. M. Rilke, 1910

In later life, the individuation field splits into smaller separate fields such as leg fields, head fields, etc. These are the agents from which the cancerous growth has escaped.

C. H. Waddington, 1935

Aging is the leading risk factor for those chronic illnesses that most limit survival and well-being. These diseases include atherosclerosis, stroke, dementias, diabetes, and cancer. During the past 50 years, narratives of aging and cancer have focused on the genetic causation of these conditions—allelic differences between individuals, cells, and species. Recently, however, developmental explanations of aging and cancer have become more important, as studies find that stem cells and their niches appear to regulate longevity and cancer initiation, and that aging and cancer can be regulated by epigenetic phenomena such as DNA methylation. Moreover, studies of cell senescence are beginning to link cancer and aging in ways that have their bases in normal development. To understand aging and cancer, one has to understand the phenomena of normal development. In many ways, cancers are to development what mutations are to genetics.

Aging and cancer depend on both chance and inheritance. Inheritance can be seen in the different age limitations for different species. Mice can live up to 3 years; humans can live past 100. Chance can be seen at many levels. Certain genes may predispose a person toward certain cancers or even toward certain aging phenotypes. Indeed, it is estimated that only about a third of a cell's variation in cancer risk is due to genetic predispositions or defined environmental causes (Tomasetti and Vogelstein 2015). A majority is due to random genetic or epigenetic events occurring as stem cells divide. We will first look at chance epigenetic events in aging, and we will then discuss age-related diseases as being caused by developmental processes that had earlier played helpful roles in eliminating redundant tissues and in modeling organs within the embryo. Last, we will look at one of these aging conditions, cancer, as a product of such altered development.

Developmental Models of Aging

Recent consideration of the causes of aging has broadened the field of contending theories to include: (1) Environmental factors that cause changes in the insulin signaling pathway that affect life spans in many species; (2) Epigenetic modifications such as histone acetylation and DNA methylation that have been seen to cause differences in disease development and that may be responsible for our deterioration as we age; and (3) At the tissue level, the cell senescence that was crucially important during development and which may cause chronic inflammation and frailty at the end of the life cycle.

Aging and the insulin signaling cascade

The idea that aging is a normal genetically controlled developmental process has been criticized: One might ask, "How could evolution have selected for regulating aging?" After all, once an organism has passed reproductive age and raised its offspring to sexual maturity, evolutionarily it becomes, as quoted in Chapter 7, an "excrescence" on the tree of life (Rostand 1962). Natural selection presumably cannot act on traits that affect an organism only after it has reproduced. But "How can evolution select for a way to degenerate?" may be the wrong question. Evolution probably can't select for such traits. However, there is often a trade-off between reproduction and maintenance, and in many species reproduction and senescence are closely linked. So the right question may be "How can evolution select for phenotypes that postpone reproduction or sexual maturity?"

Research on mice, *Caenorhabditis elegans*, and *Drosophila* suggest that there is a conserved pathway that regulates aging, and that it can indeed be selected for. This pathway involves the response to insulin and insulin-like growth factors, and it can be regulated, in part, by the environment. In *C. elegans*, a larva proceeds through four larval stages, after which it becomes an adult. If the nematodes are overcrowded or if there is insufficient food (an environmental factor), the larvae can enter a metabolically dormant dauer larva stage, a nonfeeding state of diapause, which is a condition in which development and aging are suspended (see Chapter 4). The nematodes can remain in the dauer stage for up to 6 months, and in this state they have increased resistance to oxygen radicals that can cross-link proteins and destroy DNA. The pathway that regulates both dauer larva formation and longevity has been identified as the **insulin signaling pathway** (Kimura et al. 1997; Guarente and Kenyon 2000; Gerisch et al. 2001; Pierce et al. 2001). In many species throughout the animal kingdom, this signaling pathway integrates the environment with development, reproduction, and aging (Antebi 2013; Schaedel et al. 2012).

In *C. elegans*, favorable environments signal activation of the insulin receptor homologue DAF-2, and this receptor stimulates the onset of adulthood (Figure 8.1A). Poor environments fail to activate the DAF-2 receptor, and dauer formation ensues. While severe loss-of-function alleles in the insulin signaling pathway cause the formation of dauer larvae in any environment, weak mutations enable the animals to reach adulthood and live four times longer than wild-type animals.

Down-regulation of the insulin signaling pathway has several other effects as well. First, it appears to influence metabolism, decreasing mitochondrial electron transport. Second, the lack of insulin signaling decreases fertility (Gems et al. 1998). Third, when the DAF-2 receptor is not active, cells increase the production of enzymes that prevent oxidative damage and of enzymes that repair DNA (Honda and Honda 1999; Tran et al. 2002). The increase in these two types of enzymes is due to the Foxo/DAF-16 transcription factor (see Figure 8.1). This forkhead-family transcription factor is inhibited by the insulin receptor (DAF-2) signal. When that signal is absent, Foxo/DAF-16 can function, and it promotes longevity in ways not yet deciphered. It is possible that Foxo/DAF-16 activates the expression of genes involved in producing antistress proteins within the cell as

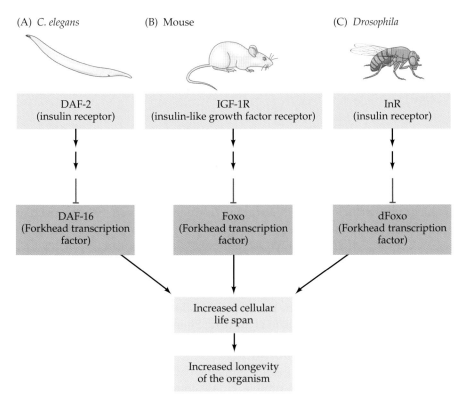

Figure 8.1   A possible pathway for regulating longevity. In each case, the insulin signaling pathway inhibits the synthesis of the Foxo transcription factor proteins that would otherwise increase cellular longevity.

well as lipid signals that help extend life to those cells nearby (Zhang et al. 2013). The Foxo transcription factor has been associated with longevity throughout the animal kingdom. Indeed, it has recently been shown to be one of the major drivers of stem cell renewal in immortal hydras (Boehm et al. 2012).

It is possible that this system also operates in mammals, but the mammalian insulin and insulin-like growth factor pathways are so integrated with embryonic development and adult metabolism that mutations often have numerous deleterious effects (such as diabetes or Donohue syndrome). However, there is some evidence that the insulin signaling pathway does affect life span in mammals (Figure 8.1B). Dog breeds with low levels of insulin-like growth factor-1 (IGF-1) live longer than breeds with higher levels of this factor. Also, mice with loss-of-function mutations of the insulin signaling pathway live longer than wild-type mice (see Partridge and Gems 2002; Blüher et al. 2003; Kurosu et al. 2005). Holzenberger and colleagues (2003) found that mice heterozygous for the IGF-1 receptor (IGF-1R) not only lived about 30% longer than wild types, but they also had greater resistance to oxidative stress. In addition, mice lacking one copy of their *IGF-1R* gene lived about 25% longer than wild-type mice.

The insulin signaling pathway does appear to regulate life span in *Drosophila* (Figure 8.1C). Flies with weak loss-of-function mutations of the insulin receptor gene or genes in the insulin signaling pathway live nearly 85% longer than wild-type flies (Clancy et al. 2001; Tatar et al. 2001). These long-lived mutants are sterile, and their metabolism resembles that of flies that are in diapause (Kenyon 2001). The insulin receptor in *Drosophila* is thought to regulate a forkhead transcription factor (dFoxo) similar to the Foxo/DAF-16 protein of *C. elegans*. When the *Drosophila dFoxo* gene is activated in the fat body, it can lengthen the fly's life span (Giannakou et al. 2004; Hwangbo et al. 2004).

From an evolutionary point of view, the insulin pathway may mediate a trade-off between reproduction and survival/maintenance. Many (although not all) of the long-lived mutants have reduced fertility. Thus, it is interesting that another longevity signal originates in the gonad. When the germline cells are removed from *C. elegans*, the nematodes live longer. The germline stem cells produce a substance that blocks the effects of a longevity-inducing steroid hormone (Hsin and Kenyon 1999; Gerisch et al. 2001; Shen et al. 2012).

Caloric restriction may reduce levels of IGF-1 and of circulating insulin, although other mechanisms are also being explored (e.g., Selman et al. 2009). However, studies in primates have not concluded that it extends longevity (see Mattison et al. 2012; Lorenzini 2014), although a low caloric intake appears to retard the age-associated decline of heartbeat variability (Colman et al. 2009; Mattison et al. 2012; Stein et al. 2012).

Aging and the TORC pathway

One of the main ways the insulin signaling pathway might lower longevity is by activating TORC1, a protein kinase complex that promotes translation in response to nutrients and hormones (Lamming et al. 2012; Hasty et al. 2013; Johnson et al. 2013). The insulin signaling pathway depresses Foxo in mice and at the same time activates mTORC1. Dietary restriction reduces mTORC1 activity, and mice with *reduced* mTORC1 levels have longer lives, better protection against age-related cognitive dysfunction, and more functional stem cells than control mice (Chen et al. 2009; Harrison et al. 2009; Halloran et al. 2012; Majumder et al. 2012; Yilmaz et al. 2012). Reducing mTORC1 also increases the amount of **autophagy**, the removal and replacement of damaged organelles and senescent cells. As we will see below, many of the maladies associated with old age appear to be the result of failed autophagy and replacement (Baker et al. 2011). The mechanisms by which reduced mTORC1 improves cell maintenance are still unknown, and this pathway is an area of active study.

Epigenetic regulation of aging: Evidence from identical twins

Another hypothesis relating development and aging emphasizes random epigenetic events. First proposed by Boris Vanyushin in 1973, this view contends that inappropriate methylation not only causes metabolic diseases (as detailed in the previous chapter) but is also the critical factor in aging and cancers. Some of the earliest evidence for the roles of epigenetic methylation in aging and diseases came from studies of identical twins. Human monozygotic ("identical") twins account for 1 in every 250 live births. Monozygotic twins, as the name indicates, arise from a single zygote and therefore have exactly the same DNA. However, such twins can have a relatively high rate of discordance in many characteristics,* including disease susceptibility—that is, identical twins often develop different diseases. These diseases include conditions that are seen early, such as juvenile diabetes and autism, as well as conditions that manifest later in life, such as ulcerative colitis and various cancers. There is no correlation between the age at a disease's onset and its concordance between twins (Figure 8.2; Petronis 2006). The cause of twin-pair discordance is not known, but recent evidence suggests that differences in DNA methylation may be involved: DNA methylation patterns can differ between twins, even though their DNA is identical.

If a gene's promoter becomes methylated when it should not be, it will usually become inactive, just as if the DNA had been mutated. The gene's function is lost either way, and in fact, methylation is a much easier way to lose function. Similarly, anomalies arise if a gene's promoter becomes unmethylated and thus activated in the "wrong" cells. Such methylation differences appear to be at the root of a case in which one of a pair of monozygotic twin girls had a severe anomaly—a duplicated portion of the spine in the posterior portion of her body. Her phenotype reminded clinicians of a similar phenotype seen in mice when the *Axin1* gene is mutated. Axin protein is an inhibitor of the Wnt pathway in development,

---

*Stephen Jay Gould liked to point out that conjoined twins Eng and Chang Bunker (who inspired the term "Siamese twins") were two very different people (see Gould 1997). Eng was a quiet, content vegetarian who didn't touch alcoholic beverages, while Chang was more aggressive fellow who liked strong drinks. This was despite the fact that the twins shared the same genome *and* were subject to exactly the same environments throughout their lives. So something else must have contributed to their phenotypes. Random epigenetic effects may provide one answer. (Eng and Chang, by the way, worked out a scheme whereby they adhered to the lifestyle of the twin in whose house they were living that week.)

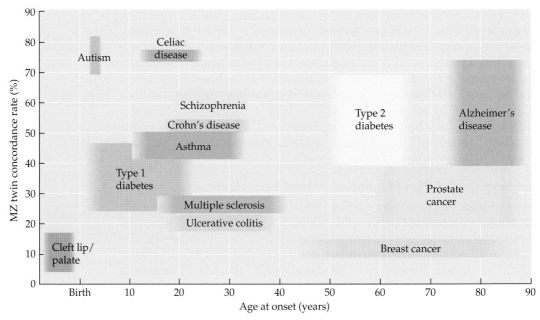

Figure 8.2 Age of disease onset and concordance of occurrence among monozygotic (MZ) twins for a number of pathologies. The concordance rate reflects the probability that both twins will be affected; a low rate indicates that often one twin has the disease and the other does not. Note that there is no correlation between age of onset and the degree of concordance. (After Petronis 2006.)

and in mice, mutations of the *Axin1* gene cause duplications of the caudal axis, resulting in extra spines and bifurcated tails. Research has revealed that methylation of the *Axin1* promoter represses *Axin1*, preventing it from functioning and giving an abnormal tail phenotype like that produced by the mutation. Blood samples from the twin girls showed that although both girls had the same allele for *AXIN1* (the human analogue of the mouse gene), there was significantly more methylation at the locus of the promoter in the affected twin than in the unaffected twin, or in either parent (Oates et al. 2006). The regions around the gene showed no significant differences in methylation.

Epigenetic marks, such as DNA methylation, appear randomly and accumulate with age. Most "identical" twins start life with very few differences in appearance or behaviors but accumulate these differences with age (Figure 8.3). Experience counts, and both random events and lifestyles may be reflected in phenotypes. Monozygotic twin pairs that were nearly indistinguishable in DNA methylation and histone acetylation patterns when young have exhibited very different patterns of both these epigenetic markers when older (Fraga et al. 2005; Bell et al. 2012). This affected their gene expression patterns, such that older twin pairs had different patterns of gene expression, while younger twin pairs had very similar expression patterns. Nilsson and coworkers (2014) have shown that identical twins discordant for diabetes have differently methylated DNA in several genes that have been implicated for this disease.

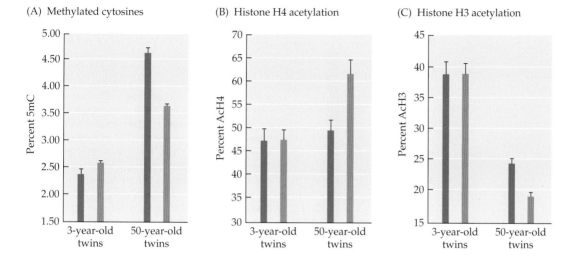

Figure 8.3 Monozygotic twins start off with identical amounts of (A) methylated cytosines, (B) histone H4 acetylation, and (C) histone H3 acetylation. However, in older pairs, the patterns of methylation and acetylation have become discordant. (After Fraga et al. 2005.)

Note that it is not only the amounts of methylation and acetylation that change as twins age, but also the patterns. The differences can be seen at specific gene sequences when the DNA is cut with enzymes sensitive to methyl groups on the cytosines (Cs). There are some enzymes that will cleave DNA at a sequence containing a C residue but will not cut that sequence if the cytosine is methylated. Figure 8.4 shows that as a twin pair ages, there is an increase in the discrepancies between their DNAs. Indeed, more recent studies that look at methylation differences in individuals at different times in their lives show that DNA methylation patterns change as a person ages. Moreover, these same studies suggest that methylation maintenance might be genetically controlled, since different families have shown different patterns of methylation changes (Bjornsson et al. 2008). It certainly seems likely that methylation differences will turn out to be critical in explaining both phenotypic divergence and discordant susceptibility for different diseases.

Aging and random epigenetic drift

The idea that **random epigenetic drift** inactivates important genes without any particular environmental cue gives rise to an entirely new hypothesis of aging. Instead of randomly accumulated mutations—which might be due to specific mutagens—we are at the mercy of chance accumulations of errors made by the DNA-methylating and demethylating enzymes. Indeed, unlike the DNA polymerases, our DNA-methylating enzymes are prone to errors. At each round of DNA replication, DNA methyltransferases must methylate the appropriate Cs and leave the other Cs unmethylated, and they are not the most fastidious of enzymes, making errors at the rate of about 4% (see Appendix B). Within certain genetic parameters (which may affect the speed at which methylation changes occur and which may differ

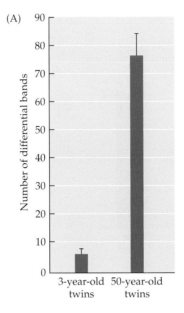

Figure 8.4 Differential DNA methylation patterns in aging twins. (A) In bisulfite sequence mapping, regions of DNA that are unmethylated will be cut by restriction enzymes (because bisulfite converts unmethylated cytosine to uracil) but methylated sites will not. The histogram summarizes the number of differences in the resulting restriction maps of 3-year-old twins and 50-year-old twins. (B) A more recent technique for revealing methylation differences and similarities between twins is to mark the DNA from one twin with a red dye and that from the other with a green dye. One can then collect only the nonmethylated DNA and bind it to metaphase chromosomes. If the bands are red or green, it means that the DNA from one twin bound but the DNA from the other twin did not. If the region is yellow, it means that the red and green DNAs bound equally. (After Fraga et al. 2005, photographs courtesy of M. Esteller.)

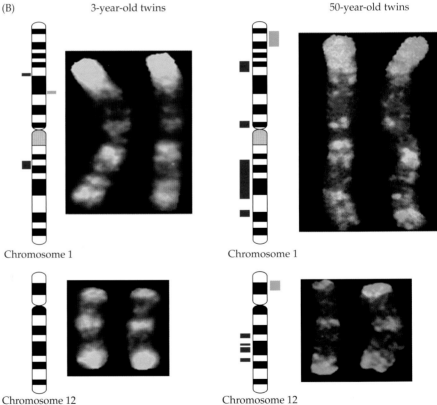

between species and between individuals), our cells may be accumulating errors of gene expression throughout our lives.

Random epigenetic drift may have profound effects on our physiology. For instance, methylation of the promoter regions of the α and β estrogen

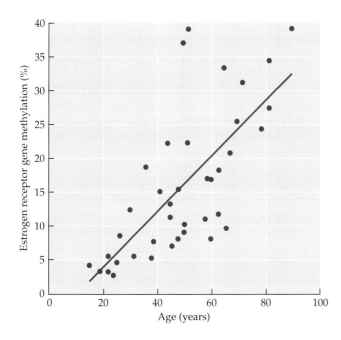

Figure 8.5 Methylation of an estrogen receptor gene occurs as a function of normal physiological aging. (After Issa et al. 1994.)

receptors is known to increase linearly with age (Figure 8.5; Issa et al. 1994), and such methylation is thought to bring about the inactivation of these genes in the smooth muscle cells of blood vessels. This decline in estrogen receptors would prevent estrogen from maintaining the elasticity of these muscles and would thereby lead to "hardening of the arteries." Increased methylation of the estrogen receptor genes is even more prominent in the atherosclerotic plaques that occlude the blood vessels (Figure 8.6); these plaques show more methylation of estrogen receptor genes than do the surrounding tissues (Post et al. 1999; Kim et al. 2007). Thus, methylation-associated inactivation of the estrogen receptor genes in these cells may play a role in the age-related deterioration of the vascular system. This

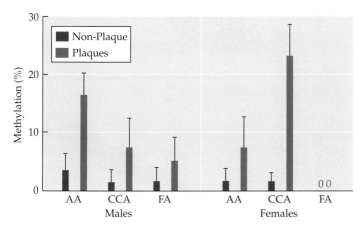

Figure 8.6 Methylation of the β estrogen receptor gene in atherosclerotic plaques and adjacent nonplaque blood vessel tissue in the ascending aorta (AA), common carotid artery (CCA), and femoral artery (FA). (After Kim et al. 2007.)

potentially reversible defect may provide a new target for intervention in heart disease. Several neurological diseases, including bipolar disorder, depression, and stress responses, have been linked to DNA methylation and/or histone modifications (Sweatt 2013). Recent studies of Alzheimer's disease (and its mouse model) showed epigenetic signals (chromatin methylation and acetylation) indicating a loss of synaptic plasticity functions and a gain of immune function in the hippocampus. These strongly implicate the immune system (and inflammation) in predisposing people to Alzheimer's dementia (Gjoneska et al. 2015).

Horvath (2013) extended and refined the epigenetic aging clock by wide-scale genomic investigations of over 50 healthy individuals at different ages. He was able to analyze over 350 sites of possible DNA methylation, showing that as people age, these sites become progressively more methylated. Cells removed from early embryos have hardly any methylation at these sites, while cells taken from centenarians are heavily methylated. Using cells from a person's saliva, Horvath's analysis of DNA methylation allows prediction of a person's age to within 2 years (Figure 8.7). Moreover, Hannum and colleagues (2013) showed that tumors of breast, kidney, and lung tissues had more heavily methylated DNA than surrounding nontumor tissues, causing them to appear about 40% "older" than the patients from whom they were removed. This could be due to the methylation of a gene involved with the chromatin remodeling processes.

So in this new hypothesis for aging, there appears to be random epigenetic drift that is not determined by the type of allele or any specific environmental factor. Random epigenetic drift may be the cause of the various phenotypes associated with aging as different genes randomly

(A) Saliva

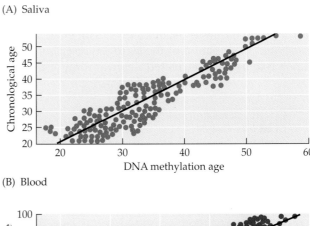

(B) Blood

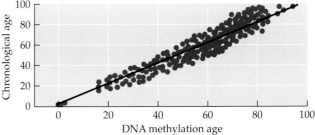

Figure 8.7 Chronological age (*y*-axis) of a person versus his or her DNA methylation "age" (*x*-axis). Each point corresponds to a DNA methylation sample taken from cells in the saliva (A) or cells in the blood (B). (After Horvath 2013.)

get repressed or ectopically activated. Mistakes in the DNA methylation process accumulate with age and may be responsible for the deterioration of our physiology and anatomy. If this is so, some genes may be more important targets than others. (The above-mentioned estrogen receptors, for instance, are critical not only in the vascular system but also for skeletal and muscular health.) It is not known what mechanisms set the rate of random DNA methylation, but these may be critically important for regulating aging both within and between species.

Aging and cellular senescence

So far, we have looked at aging by two processes. One process (involving the insulin signaling pathway) concerns a general pathway that limits life span. The other (involving random epigenetic drift) concerns random disruptions that may be repaired or not, depending on the cell type and species. Both are at the gene level. At another level, however, one has to ask how ectopic gene silencing or activation would cause an aging phenotype. What processes might be being activated or suppressed? In recent years, several lines of evidence have pointed to **chronic sterile inflammation** (an inflammatory response not initiated by bacteria) as a major common factor of the age-related diseases. Chronic sterile inflammation can occur in just about any type of tissue. The hallmarks of inflammation include elevated cytokine levels that attract lymphocytes and macrophages and digestive enzymes. These have been found in tissues where diabetes, dementias, atherosclerosis, and cancer originate (see Tchkonia et al. 2013; Muñoz-Espin and Serrano 2014). The lymphocytes and macrophages are thought to kill or cripple the cells and prevent the organ from functioning optimally. Inflammation is a major factor producing the age-related frailty syndrome, wherein the body becomes susceptible to dying from stresses it could otherwise tolerate.

So, then, what stimulates chronic sterile inflammation? The evidence is pointing to the persistence of **senescent cells** in a tissue as the cause of the inflammatory reaction. Senescent cells are cells that no longer divide. They are usually the mesenchymal stromal cells that underlie the epithelial cells (that perform the functions of the organ). Senescent cells are not necessarily bad cells to have around. In fact, like apoptosis, senescence seems to be a normal and important part of organogenesis. For instance, after the axes of the fetal limbs have been established, the apical ectodermal ridges that formed those axes become senescent and are removed. Likewise, the developing kidney and inner ear each use cellular senescence to change the shape of the organ and allow new cells entry into new places (Muñoz-Espin et al. 2013; Storer et al. 2013). Cell senescence also prevents the proliferation of disabled cells, and it thereby may stop cells from initiating tumors (Campisi 2001; Krtolika et al. 2001; Prieur and Peeper 2008). During development, cell senescence appears to organize a local immune response (i.e., an inflammation), attracting macrophages that ingest the senescent cells and enabling tissue modeling. Chronic sterile inflammation may be the result of having the senescent cells, helpful during development, accumulate in adult tissue, where they are no longer helpful.

Inflammation is good and normative in small amounts and small durations at specific places in the embryo, and it may also be beneficial when produced locally and for small durations in the adult to get rid of senescent cells and replace them (through stem cell division) with functional cells. But chronic (long-term) inflammation may be deleterious and unhealthy in adults. In this case, what is normal in the embryo becomes abnormal in the adult. This pattern will be seen again in our discussions of cancer.

The senescent cells appear to be inducing the inflammation by producing a cocktail of secreted factors called the **senescence-associated secretory phenotype** (**SASP**). SASP includes paracrine factors (proteins that can change the behaviors of nearby cells), cytokines (proteins that attract and activate lymphocytes and macrophages), and proteases (enzymes that digest proteins). Senescent cells and SASP increase as we grow older, primarily because senescence is a defense against cells with DNA damage. In small doses and short durations, SASP can remove the cells with DNA damage and replace them with healthy cells. This is thought to protect against cancers. However, if the senescent cells persist and are not removed, then the tissue is compromised by having such nonfunctional cells, which are also producing a chronic inflammatory response that might alter tissue structure and function (Tchkonia et al. 2013).

So, then, if the phenotypes of aging are caused by chronic inflammations, and chronic inflammations are caused by the accumulation of senescent cells, then what causes cellular senescence? In the mammalian embryo, senescence is mediated by paracrine factors produced by the mesenchymal stromal cells. In the embryonic ear these factors include TGF-β family proteins, while in the limb they are probably members of the Fgf family. These paracrine factors activate transcription factors that elevate levels of the p21 protein that inhibits the cell cycle and initiates the senescent phenotype (Figure 8.8; Banito and Lowe 2013; Campisi 2014; Muñoz-Espin and Serrano 2014). In adult cells, "oncogenic stress"—mainly damage to DNA from mutations, telomere shortening, and reactive oxygen species—can activate the p53 protein, which is often seen as the "guardian of the genome." Among p53's many powers is the ability to activate p21. In this way, both the normal developmental mechanisms of senescence and the normal anticancer mechanism of senescence feed into the same pathway. However, if the adult senescent cells accumulate, chronic inflammation and functional decline may be the result.

We therefore have a model wherein cellular senescence normally coordinates developmental tissue modeling by inducing a stable arrest in proliferation followed by a secretory phenotype (SASP) that recruits

Figure 8.8 The hypothesized senescence pathway leading to the aging syndrome. The central node is the protein p21, which inhibits cyclin-dependent kinases, blocking cells in the G1 phase of the cell cycle, through the function of the retinoblastoma protein. This leads to the senescent phenotype, characterized by specific secretion products (SASP). These proteins recruit immune system macrophages and lymphocytes, leading to inflammation. If the inflammation is transient, the senescent cells are removed, the tissue can be remodeled, and stem cells can restore more functional cells to the tissue. If the senescent cells are not removed, the inflammation can become chronic, in which case tissue degradation and the symptoms of aging can occur. In adults, this is thought to arise from oncogenic stress, and it is used as a way of removing senescent cells. In the embryo, senescence of cells happens normally in development of organs, such as in the placenta, ear, and limbs. (After Muñoz-Espin et al. 2013; Muñoz-Espin and Serrano 2014; Tchkonia et al. 2014.)

macrophages and other immune cells to eliminate the senescent tissue. This prevents the further functioning of this tissue. In some cases, the removal of senescent cells may allow nearby progenitor cells to repopulate the tissue with nonsenescent, more functional cells. In other cases, the removal of senescent cells may enable nearby motile cells to enter into the area. In adults, senescence can enable the removal of nonfunctional tissues and their

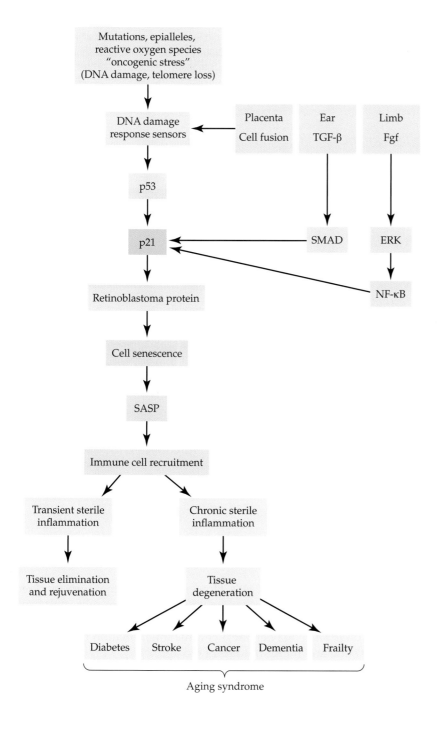

Stem Cells and Aging

One of the hallmarks of aging is the declining ability of stem cells and progenitor cells to restore damaged or nonfunctioning tissues. A decline in muscle progenitor (satellite) cell activity is seen when Notch signaling is lost and this results in a significant decrease in the ability to maintain muscle function. Similarly, an age-dependent decline in liver progenitor cell division impairs liver regeneration due to a decline in transcription factor cEBPα. And the age-associated graying of mammalian hair appears to be due to the apoptosis of melanocyte stem cells in the hair bulge niche (Nishimura et al. 2005; Robinson and Fisher 2009). One of the questions, then, becomes: Is this part of the aging syndrome caused by the declining function of stem cells or by a declining ability of the stem cell niche to support them?

One way to test this is by "fusing" an old mouse to a young mouse. This can be done by a technique called **parabiosis**, wherein the animals' circulatory systems are surgically joined so that the two mice share one blood supply. If an aged and a young mouse are parabiosed, the stem cells of the old mouse are exposed to factors in young blood serum (and vice versa). This *heterochronic parabiosis* has been seen to restore the activity of old stem cells. Notch signaling of the muscle stem cells regained its youthful levels, and muscle cell regeneration was restored. Similarly, liver progenitor cells regained "young" levels of cEBPα—and their ability to regenerate (Conboy et al. 2005; Conboy and Rando 2012). Young blood promoted the repair of aged spinal cords, reversed the thickening (hypertrophy) of the heart walls, and caused the formation of new neurons in aged mice (Ruckh et al. 2012; Loffredo et al. 2013; Villeda et al. 2011, 2014).

Loffredo and colleagues (2013) identified the "rejuvenating" blood-borne agent as paracrine factor GDF11, an extracellular signaling protein. GDF11 can circulate through the blood in young mice, and its levels decline with age. Moreover when GDF11 is transfused into older mice, the youthful levels reversed age-related hypertrophy. GDF11 also acted in the brain to counteract some

of the deterioration of aging. Injecting GDF11 into older mice increased brain capillary production, neuron formation, and olfactory discrimination (see figure; Katsimpardi et al. 2014).

Since the stem cells are not transfused from one animal to the other, it appears that GDF11 helps the function of the stem cell niche. It is not yet known whether GDF11 (or young blood plasma, in general) extends the lifespan or improves the health of mice or humans (see Scudellari 2015). There is also the danger that when working with stem cells, there is a fine balance between underproliferation (leading to aging) and overproliferation (leading to cancer).

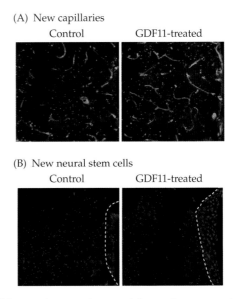

(A) New capillaries

Control   GDF11-treated

(B) New neural stem cells

Control   GDF11-treated

GDF11 promotes vascular remodeling and neurogenesis in the mouse brain. Region of the dentate gyrus of the brain stained for new capillaries (A) or neural stem cells (B). The GDF11-treated brains were from 22-month old (elderly) mice injected with GDF11 for 4 weeks. (From Katsimpardi et al. 2014, courtesy of L. Rubin.)

replacement by more functional cells. However, as an animal ages, random events (epigenetic and genetic changes) may make more cells senescent and put strains on the immune system to clear them away. If the senescent cells persist, their SASP may create a chronic inflammation that leads to tissue malfunction and the symptoms of old age.

Cancers as Diseases of Development

Cancers take many forms, but they are increasingly being seen as diseases of development. A *New York Times* headline summarized it well: "A Tumor, the Embryo's Evil Twin." Although mutations play roles in the formation and maintenance of hereditary tumors, most carcinogens (cancer-causing agents) are not mutagens (mutation-causing agents). The endocrine disruptors (such as bisphenol A and DES) mentioned in Chapter 6 often can cause cancer, but they do not cause gene mutations. Rather, carcinogenesis (cancer formation) can be viewed as aberrations of the very processes that underlie morphogenesis (Stevens 1953; Pierce et al. 1978). It had previously been thought that carcinogenesis (the formation of tumors) and metastasis (the spread of tumor cells to other regions of the body) were caused by the proliferation of a cell that had acquired mutations enabling it to become "autonomous" (Hanahan and Weinberg 2000; Vogelstein and Kinzler 2004; Stratton et al. 2009). Cancer-causing agents (carcinogens) were thought to be agents that caused mutations, and cancer was therefore defined by the intracellular mechanisms that enabled a cell to become independent of its environment. This **somatic mutation hypothesis** turns out to be only a part (and maybe a very small part) of the explanation (see Sonnenschein et al. 2014). Cancer cells harbor numerous mutated genes (Garaway and Lander 2013; Vogelstein et al. 2013), but these might not be responsible for causing the cancer.* Indeed, in 2014, Robert Weinberg, one of the foremost proponents of the somatic mutation hypothesis, stated that although the somatic mutation theory assumed that mutations were the major cause of cancers, "most human carcinogens are actually not mutagenic."

If cancer is not a disease caused by mutant genes, then what is it? What are carcinogens doing? For the answer, we may have to turn to development. An alternative set of hypotheses emphasizes that cancer is caused by disruptions of tissue interactions. According to the **tissue organization field theory** (**TOFT**; Sonnenschein and Soto 1999; Soto and Sonnenschein 2011; Sonnenschein et al. 2014), cancers are not diseases of cells, but diseases of tissues: cancer arises from disruptions in tissue organization. In this hypothesis, mutations are not required for a cell to become cancerous. Rather, what is required for cancer is a breakdown of the normal paracrine and juxtacrine interactions that determine the structure of a tissue by imposing constraints on its component cells. These constraints include those that limit the ability of the tissue's cells to proliferate and move. Proliferation with variation and motility are considered a cell's "default" states, and the cells are prevented from such expansion through their interactions with other cells. When there is disruption of this intercellular regulation, the cells can proliferate, move, and organize into pathological structures such

---

*Many cancer biologists have likened cancer to evolution, where genetic variants allow the survival of the fittest. Cancer might indeed be evolutionary, but more like the type of evolution depicted in Chapters 10 and 11. Here, there is a developmental event (usually involving cell interactions) that initiates a new phenotype, and the allelic changes allowing survival of the fittest fine-tune the phenotype for environmental changes.

Normal tissue developmental field

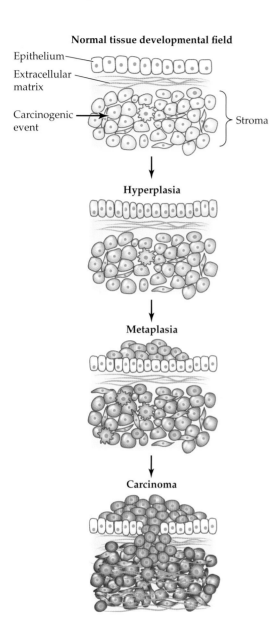

Figure 8.9 The tissue organization field theory of cancer. According to TOFT, epithelial cell cancers (carcinomas) originate when some event interferes with the biochemical or biophysical communication between the epithelial cells (parenchyma) and the mesenchymal cells (stroma) that support them. The stroma normally restricts the motility and the proliferation of the epithelial cells; so the epithelial cells now undergo proliferation (hyperplasia) and lose their ordered patterning (metaplasia). Meanwhile, the stroma also becomes disordered, as more inflammatory cells are allowed to enter. Eventually, the epithelial cells become neoplastic cells (tumors) and destroy the extracellular matrix and invade the stroma. (After Soto and Sonnenschein 2011.)

as hyperplasia (a mass of proliferating cells), dysplasia (a mass of cells with aberrant morphology), metaplasia (a mass where some cells have differentiated into other cell types), and neoplasia (a tumor), to eventually become "malignant tumors" that can spread rapidly (Figure 8.9).

The TOFT ideas come from embryology, specifically from the notion of "morphogenetic fields." These fields designate areas of developmental constraint that limit the organs formed there. The component cells of these fields create a web of interactions such that any cell is defined by its position within its respective field. For instance, there are the "cardiac fields" that produce the heart, and there are limb fields that produce our appendages (see Weiss 1939; Guss et al. 2001; Gilbert 2013; Levin 2012). These fields are physical entities, such that if such a field is removed from an embryo, that structure won't form, and if a field is transplanted to another organism, it will form a new structure. In different fields, paracrine factors do different things. For instance the paracrine factor BMP4 causes cell death in the late limb field, but it specifies the differentiation of the aorta and pulmonary artery in the late heart field.

These same interactions that are used to generate organs are also used in adults to maintain organ integrity and proper growth. Moreover, adult cells are able to respond to paracrine factors (Cunha et al. 1983; Kaplan et al. 2006). In fact, when tumors are injected into mice, the different growth rates of the tumor cells, from anterior to posterior, "may reflect the existence of morphogenetic gradients reminiscent of those invoked to explain patterns of differentiation during ontogeny" (Auerbach et al. 1978; Auerbach and Auerbach 1982; quotation from Auerbach et al., p. 697). The discovery of adult stem cells led to the discovery of stem cell niches, morphogenetic fields where paracrine and juxtacrine factors in adults carefully regulate cell division and differentiation (see Gilbert 2013).

In adult tissues, there are usually two major components (other than the blood vessels and nerves). First, there is the epithelial component, often called the "parenchyma" in medical texts. This is the highly differentiated tissue that performs the "business" of the organ, and it is tightly held together by adhesion molecules. The second component consists of the mesenchymal connective tissue. These cells are also called the "stroma." These are the cells that produce factors supporting the epithelial cells. It had been thought that they were merely supporting cells, but new evidence suggests they play major roles in tissue organization. The TOFT model is recasting cancer as a "stepwise progression" of conditions that depends on reciprocal interactions between the parenchyma and the stroma. Cancers appear to proceed by recapitulating steps of normal development, including the formation of a niche in which to grow (see Gilbert 2013).

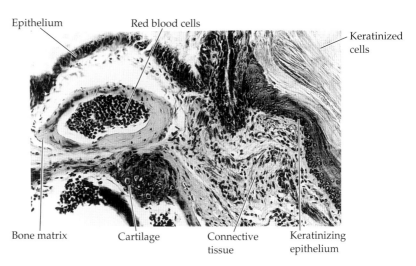

Figure 8.10   Micrograph of a section through a teratocarcinoma shows numerous differentiated cell types. (From Gardner 1982; photograph by C. Graham, courtesy of R. L. Gardner.)

Context-dependent cancer cells

The morphogenetic field model of cancer suggests new mechanisms of treatment (besides the destruction of cancer cells). According to TOFT, cells isolated from tumors can revert back to normal function and proliferation patterns if placed back into a supportive environment such as the normal tissue of origin, where they would reacquire their proper anatomy and function. In this model, cancer is not produced by "bad cells." Rather, it is made by dysfunctional environments. Changing the environment can turn cancer cells into normal and functioning members of a tissue.

Tumor cells, especially in their early stages of growth, can have normal genomes (see Pierce 1974; Mack et al. 2014), and whether or not these tumors are malignant depends on their environment. One of the most remarkable of these cases is the **teratocarcinoma**, which is a tumor of germ cells or stem cells (Illmensee and Mintz 1976; Stewart and Mintz 1981; Figure 8.10). Teratocarcinomas are malignant growths of cells that resemble the inner cell mass of the mammalian blastocyst, and they can kill the organism. However, if a mouse teratocarcinoma cell is placed on the inner cell mass of a mouse blastocyst, it will integrate into the early embryo, lose its malignancy, and divide normally. Its cellular progeny can become part of numerous embryonic organs. Should its progeny form part of the germline, sperm or egg cells formed from the tumor cell will transmit *the tumor genome* to the next generation. Thus, whether the cell becomes a tumor or part of the embryo can depend on its surrounding cells.*

The mechanism by which the stem cell environment can suppress tumor formation may be due to its secretion of inhibitors of the paracrine

---

*And it works both ways. The major way of inducing a teratocarcinoma to form in rodents is to take a normal mouse or rat embryo of a certain early stage and place it into an ectopic environment, such as the kidney or testis. In this new environment, the cells become teratocarcinomas instead of embryos (Solter et al. 1970; Stevens 1970; Damjanov 1993).

pathways. For instance, many tumor cells, such as melanomas, secrete the paracrine factor Nodal. This aids their proliferation and also helps supply them with blood vessels. When placed in an environment of embryonic stem cells (which secrete Nodal inhibitors), aggressive melanoma tumor cells become normal pigment cells (Hendrix et al. 2007; Postovit et al. 2008). Remarkably, such malignant melanoma cells (which are derived from neural crest cells), when transplanted into early chick embryos, down-regulate their Nodal expression and migrate nonmalignantly along the neural crest cell migration pathways (Figure 8.11; Kasemeier-Kulesa et al. 2008).

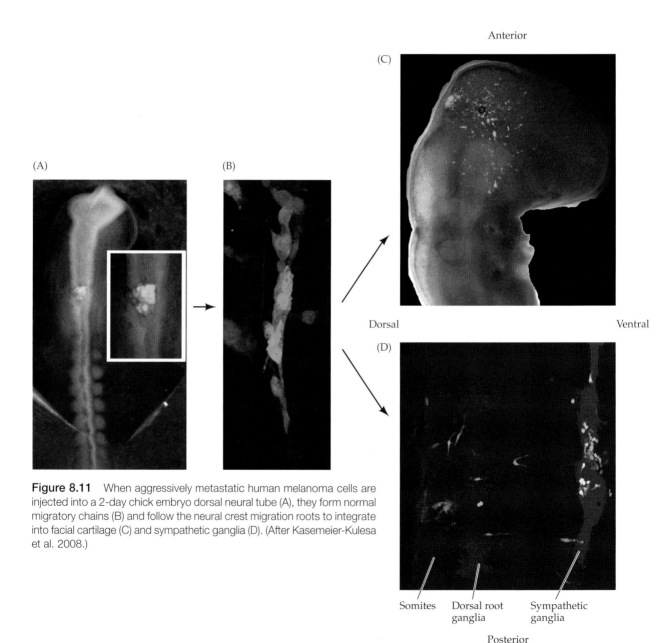

Figure 8.11 When aggressively metastatic human melanoma cells are injected into a 2-day chick embryo dorsal neural tube (A), they form normal migratory chains (B) and follow the neural crest migration roots to integrate into facial cartilage (C) and sympathetic ganglia (D). (After Kasemeier-Kulesa et al. 2008.)

This brings us to the idea that cancer can be caused by miscommunication between cells. In many cases, tissue interactions are required to prevent cells from dividing. Thus, tumors can arise through defects in tissue architecture, and the surroundings of a cell are critical in determining malignancy (Sonnenschein and Soto 1999, 2000; Bissell et al. 2002). Studies have shown that tumors can be caused by altering the structure of the tissue, and that these tumors can be suppressed by restoring an appropriate tissue environment (Weaver et al. 1997; Booth et al. 2010). In particular, whereas 80% of human tumors are from epithelial cells, these cells do not always appear to be the site of the cancer-causing lesion. Rather, epithelial cell cancers are often caused by defects in the *mesenchymal* stromal cells that surround and sustain the epithelia. When Maffini and colleagues (2004; Figure 8.12) separated, treated, and then recombined normal and carcinogen-treated

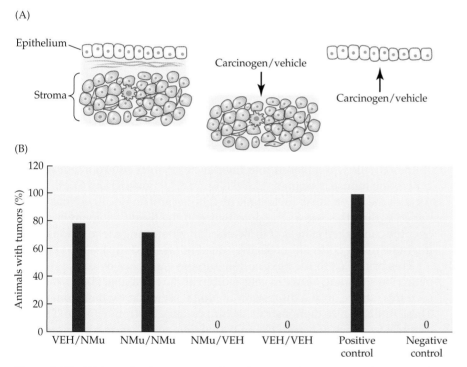

Figure 8.12 Evidence that the stroma regulates the production of epithelial (parenchymal) tumors. (A) Schematic drawing of the experimental protocol. The mammary gland tissue contains both epithelium and stroma (mesenchymal cells). The two groups of cells can be isolated and then recombined. One can add a cancer-forming substance (carcinogen) to the epithelium and not the stroma, or to the stroma but not the epithelium. Then one can combine them so that the cancer-causing substance has been experienced by the epithelium (but not the stroma), by the stroma (but not the epithelium), by both stroma and epithelium, or by neither. (B) Results when the cancer-causing mutagen N-methyl-nitrosourea (NMu) or just the control vehicle (VEH) was applied to either the stroma or epithelium and transplanted back into the rat mammary gland. On the horizontal axis, the upper numerator refers to the epithelium and the lower denominator refers to the stroma. Only animals whose stroma was treated with NMu developed tumors, regardless of whether the epithelium was exposed to NMu or not. NMu-treated, intact animals (positive controls) developed tumors, and none of the rats receiving control solutions (negative controls) had tumors. (After Maffini et al. 2004.)

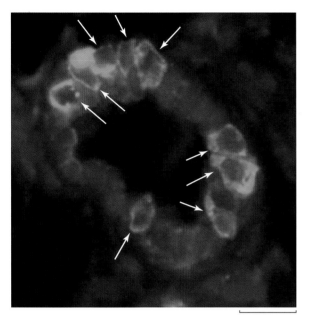

10 μm

Figure 8.13 Human mammary tumor cells become part of the normal milk ducts of regenerating mouse mammary glands. Human mammary tumor cells (green) are found incorporated into the regenerating mouse mammary ducts (blue). (From Bussard and Smith 2012.)

epithelia and mesenchyme in rat mammary glands, tumorous growth of mammary epithelial cells occurred not in carcinogen-treated epithelia, but only in epithelia combined with carcinogen-treated mesenchyme. Thus, the carcinogen caused defects in the mesenchymal stroma of the mammary gland, and the carcinogen-treated stroma promoted the formation of epithelial ducts prone to form cancers. It seems that the carcinogen-exposed stroma could no longer provide the complete set of instructions needed to form normal epithelial structures. In turn, these abnormal structures exhibited a loose control of cell proliferation.

Moreover, multipotent human teratocarcinoma cells will form normal mammary ducts when injected into the stroma of normal adult rats (Bussard et al. 2010), indicating that the stromal tissue is critical in regulating the cell proliferation and differentiation of the epithelial cells. Human breast cancer cells can be redirected to become normal mammary epithelial cells when placed into the developmental environment of a regenerating mouse mammary gland (Figure 8.13; Bussard and Smith 2012). As with the reversion of metastatic melanoma and teratocarcinomas, mentioned earlier, this shows the dominance of the tissue microenvironment over the cancer cell state and indicates that whatever genetic lesions there might be in the epithelial cells, they need not necessarily lead to cancer.

This brings us to the next notion: tumors can occur by disruptions of paracrine signaling between cells (Rubin and de Sauvage 2006). We have seen this, above, in the discussion of Nodal secretion by melanoma cells. These findings also demonstrate the importance of stromal tissue. There is a critical feedback, whereby the epithelial cells and their underlying stroma (mesenchymal cells) regulate each other's growth and differentiation. If the extracellular matrix or paracrine factors involved in this regulation are disrupted, tumors can occur (Figure 8.14). Many tumors, for instance, secrete the paracrine factor sonic hedgehog (Shh), which can act in either of two ways. First, it can act in an autocrine fashion, stimulating the producing cell to grow. Shh is normally required for the maintenance of granule neuron progenitor cells and hematopoietic stem cells, and inhibitors of the Shh pathway can reverse certain medulloblastomas and leukemias (Rubin and De Sauvage et al. 2006; Zhao et al. 2009).

Second, the Shh produced by the tumor cells can act not only on the tumor cells themselves but on the stromal cells, causing the stromal cells to produce factors (such as insulin-like growth factor) that support the growth of tumor cells. If the Shh pathway is blocked, the tumor regresses (Yauch et al. 2008, 2009; Tian et al. 2009). Indeed, cyclopamine, a teratogen

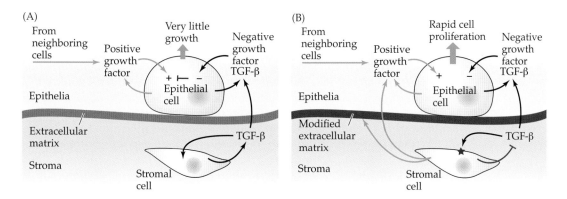

Figure 8.14 Regulation of the proliferation and differentiation of both the stroma and epithelia by the stromal cells of an organ. (A) Stromal cells secrete factors such as TGF-β that signal both stromal and epithelial cells to stop proliferation. Each epithelial cell, however, secretes and receives (from itself and its neighbors) positive growth factors that promote proliferation (green arrows). But as long as the stromal cells counter these positive signals, proliferation is tightly regulated. (B) Epigenetic or mutational changes in stromal cells block their inhibitory signaling (red star). The stroma can then start secreting paracrine factors (e.g., hepatocyte growth factor; green arrows) that modify the extracellular matrix and promote the proliferation and transformation of the epithelial cells. (After Bhowmick et al. 2004.)

that blocks Shh signaling (see Chapter 5), can prevent certain of these tumors from growing (Berman et al. 2002, 2003; Thayer et al. 2003; Song et al. 2011).

The observation that the most aggressive human pancreatic cancers were characterized by a depletion of stromal cells led researchers to see what would happen if the pancreatic stromal cells were removed from less invasive tumors. When these stromal mesenchyme cells were genetically ablated from noninvasive pancreatic cancer precursor tissue, those cells became highly invasive and malignant tumors. In this case, the stromal cells were restraining the epithelial cells from proliferating. Moreover, this function, too, was found to be dependent on Shh. Shh was made by the pancreatic epithelial cells, and it also promoted the stromal cells to produce factors that inhibited the epithelial cells' proliferation. If the stromal cells were removed or if the Shh synthesis was blocked, the epithelial cells divided unchecked (Özdemir et al. 2014; Rhim et al. 2014).*

The idea that cancer is a developmental problem and that the microenvironment of the cell plays a critical role is a new paradigm for cancer research. There is now a growing recognition that cancer-causing agents (including ionizing radiation) are causing epigenetic changes and that they are altering the normal relationships between cells that are essential for

*Although the compound made by the stromal cells to impede epithelial cell division is not yet known, the extracellular matrix glycoprotein hyaluronan may be important. Naked mole rats do not get tumors. Even when their cells are put into culture and given chemicals that make mouse cells malignant, the mole rats' cells appear to stop growing when they touch one another. The mole rat cells secrete enormous amounts of hyaluronan, and this appears to stop the cells from proliferating (Tian et al. 2013).

holding tissues together (Bizzarri and Cucina 2014; Dotto 2014; Sonnen-schein et al. 2014).

Knowing which signaling pathways activate cancer stem cells can be critically important for blocking the spread of cancers. Indeed, the same chemicals that can cause teratogenesis by blocking a pathway in embryonic development may be very useful in blocking the activation of cancer stem cells. Cyclopamine and other antagonists of the hedgehog pathway, for instance, can cause malformation in embryos, but they also appear to be useful in preventing the generation and proliferation of medulloblastoma stem cells (Berman et al. 2002; De Smaele et al. 2010). Even the classic teratogen thalidomide is being "rehabilitated" to fight cancer. Thalidomide and its derivatives have been found to inhibit the Wnt and FGF pathways and prevent new blood vessels that are critical for tumor growth (Hansen et al. 2002; Knobloch et al. 2007; see Chapter 5). These compounds have recently been approved for the treatment of certain myelomas and leukemias (List et al. 2005; Aragon-Ching et al. 2007). As mentioned at the beginning of this chapter and shown by several examples, to understand cancer, one has to understand development.

Thus, in developmental models of cancer formation such as TOFT, cancer is not so much a cell gone bad as it is a relationship gone awry. If altered paracrine factor signaling may cause cancer, then correcting such signaling may cure it. This is called **differentiation therapy**. Pierce and his colleagues (1978) noted that cancer cells were in many ways reversions to embryonic cells, and they hypothesized that cancer cells should revert to normalcy if they were made to differentiate. Treatment of acute promyelocytic leukemia patients with all-*trans* retinoic acid results in remission in more than 90% of cases because the additional retinoic acid is able to effect the differentiation of the leukemic cells into normal neutrophils (Hansen et al. 2000; Fontana and Rishi 2002). Differentiation therapy (with retinoic acid, histone deacetylases, and other agents) is currently being attempted on numerous types of cancers.

Embryonic Development and Cancer Growth

There are many ways that cancer cells reactivate the genes that enabled embryonic organ formation. Some of these genes are those promoting epithelial-mesenchymal transformation, enabling the cells to lose their connections with one another in an epithelium and **metastasize** (travel to new locations) in the body. Another set of genes are those that prevent the death of migrating cells. And another set of genes are those responsible for getting the tumor a blood supply. Here, cancers use the same strategies—and molecules—of migrating embryonic cells.

Like any other proliferating mass of cells, a cancer needs oxygen and nutrients, and it obtains these from blood vessel capillaries. Most clusters of tumor cells do not get these blood vessels and they die. Indeed,

autopsies have shown that every person over 50 years old has microscopic tumors in their thyroid glands, although less than 1 in 1000 persons have thyroid cancer (Folkman and Kalluri 2004). Folkman suggested that cells capable of forming tumors develop at a certain frequency, but that they most never form observable tumors. The reason is that a solid tumor, like any other rapidly dividing tissue, needs oxygen and nutrients to survive. Without a blood supply, potential tumors either die or remain as dormant "microtumors"; stable cell populations wherein dying cells are replaced by new cells.

The critical point at which a node of cancerous cells becomes a rapidly growing tumor occurs when the node becomes vascularized. A microtumor can expand

Embryonic Development and Cancer Growth (*continued*)

to 16,000 times its original volume in the two weeks following vascularization (Folkman 1974; Ausprunk and Folkman 1977). To achieve vascularization, the micro-tumor secretes substances called tumor angiogenesis factors, which often include the same factors that engender blood vessel growth in the embryo—VEGF, Fgf2, placenta-like growth factor, and others. Tumor angiogenesis factors stimulate mitosis in endothelial cells and direct their differentiation into blood vessels in the direction of the tumor.

Tumor angiogenesis can be demonstrated by implanting a piece of tumor tissue within the layers of a rabbit or mouse cornea. The cornea itself is not vascularized, but it is surrounded by a vascular border, or limbus. The tumor tissue induces blood vessels to form and grow toward the tumor (see figure; Muthukkaruppan and Auerbach 1979). Once the blood vessels enter the tumor, the tumor cells undergo explosive growth, eventually bursting the eye. Other adult solid tissues do not induce blood vessels to form. It might therefore be possible to block

tumor development by blocking angiogenesis. Numerous chemicals are being tested as natural and artificial angiogenesis inhibitors. These compounds act by preventing endothelial cells from responding to the angiogenetic signal of the tumor (Tammela et al. 2008).

However, therapies acting against tumor-derived angiogenesis factors are possible, but so far, the tumors have found ways that circumvent the molecules that block angiogenesis. This is probably because in the absence of blood vessels, the tumor becomes hypoxic (stressed by low oxygen levels), and one of the strategies against hypoxia is to trigger the expression of angiogenesis factors (Bellou et al. 2013; Giuliano and Pagès 2013). Just as in embryonic development, hypoxia induces the formation of blood vessels, and when one system is blocked, other systems can be brought in to compensate. Anti-angiogenesis is an excellent way of preventing tumor growth; but in order to prevent angiogenesis, we have to learn more about how it occurs during normal development (Wang et al. 2015).

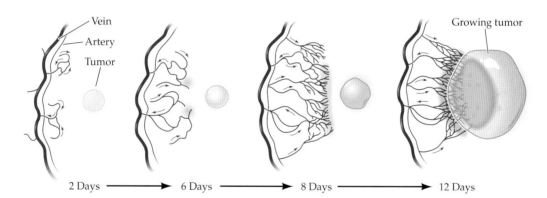

2 Days → 6 Days → 8 Days → 12 Days

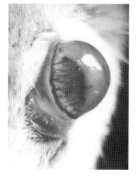

New blood vessels are stimulated by tumor angiogenesis factors to grow toward the tumor and supply it with oxygen and nutrients. Using the same molecules as embryonic tissues, a tumor placed into the cornea of a mouse eye will induce the blood vessels surrounding the cornea to make new capillaries. When they enter the tumor the tumor undergoes explosive growth. Moreover, the blood vessels become the routes by which the tumor spreads to new sites. (From Muthukkaruppan and Auerbach 1979, photograph courtesy of R. Auerbach.)

Cancer and epigenetic gene regulation

There is also a hypothesis that cancer is a disease of epigenetic methylation. In the **epigenetic progenitor model** (Feinberg et al. 2006), alterations of DNA methylation can prevent the normal functioning of DNA repair genes. Without the DNA repair proteins, mutations accumulate. If these mutations prevent the normal functioning of growth regulatory genes, cancers may develop.

Earlier in this chapter we saw evidence that the methylation patterns of mammalian genes change with age. We specifically looked at genes that might cause elements of the aging phenotype. But what would happen if the random, age-dependent patterns of gene methylation altered the transcription of the genes regulating cell division and cell signaling? One might expect that certain genes necessary for the usual constraints on cell division or signal reception might be inappropriately expressed or repressed.

One of these genes is the estrogen receptor that regulates cell division. In some cells (such as those in estrogen-dependent breast cancers), estrogen activates cell proliferation. In the colon, however, estrogen stops the proliferation of cells. Issa and colleagues (1994) showed that in addition to the age-associated methylation of estrogen receptors, there was a much higher level of DNA methylation in the estrogen receptor genes in colon cancers. Even the smallest colon cancers had nearly 100% methylation of the cytosines in the promoter of the estrogen receptor gene.

The epigenetic causation of cancer does not exclude a genetic cause. Indeed, several studies indicate that these mechanisms augment one another. For example, changes in tissue architecture have been found to change patterns of gene expression (Spencer et al. 2007; Vidi et al. 2012), meaning that the chromatin of the genome has been remodeled. And these changes in gene expression may cause further alterations in tissue structure (such as the loosening of the epithelium and the growth of cells). In many cancers, there are numerous somatic mutations (Sjöblom et al. 2006). But it is possible that these mutations may have accumulated because of DNA methylation. In some cancer cells, the genes encoding DNA repair enzymes appear to be susceptible to inactivation by methylation. Once DNA repair enzymes have been down-regulated, the number of mutations increases (Jacinto and Esteller 2007). This might explain (1) why there are so many somatic mutations in cancer cells, (2) why mutations in cancer cells differ from one another, and (3) how cancer cells can become immune to certain chemotherapy agents.

This interaction between genetic mutation and epigenetic changes has been seen in human tumors (see Esteller 2008). In breast cancers, for instance, the tumor suppressor genes encoding BRCA1 (which restricts cell division, and which predisposes women to breast cancer if mutated) and E-cadherin (which tightens the junctions between cells) become heavily methylated, while the genes for growth-promoting factors such as cyclooxygenase-2 are demethylated (Ma et al. 2004). In acute myeloid leukemia, several patients were found to have mutations in the DNA methylase gene *DNMT3*. This mutation may cause methylation defects that alter the epigenetic condition of tumor suppressor genes and lead to cancer stem cells (Shlush et al. 2014). In gliomas (brain tumors), the gene encoding the DNA repair enzyme MGMT is hypermethylated, as is the gene encoding thrombospondin, a tumor-suppressor protein needed for cell-cell and cell-matrix adhesion.

Aberrant methylation and other epigenetic processes appear to play major roles in cancer initiation and progression (Hattori and Ushijima 2014; Hovestadt et al. 2014). The methylation of tumor-suppressor genes (those genes encoding proteins involved in the networks supressing cell division) may explain the increased prevalence of sporadic tumors with age (Fraga and Esteller 2007). In the colon, where estrogen receptors function as tumor suppressors, there is a linear increase of estrogen receptor gene methylation with age (see Figure 8.7). In addition, Issa and colleagues (1994) showed extremely high levels of DNA methylation in the estrogen receptor genes of colon cancer cells, even the smallest of which had nearly 100% methylation of the estrogen receptor promoter region. Moreover, microarrays (which detect differences in mRNA populations) have found several genes whose methylation patterns changed dramatically in colon cancer cells (Figure 8.15). In normal human breast tissue, there is an age-dependent increase in the promoter methylation of another tumor-suppressor gene, *RASSF1A*, and this methylation is highly correlated with breast cancer risk (Euhus et al. 2008).

Epigenetic changes help explain the causes of tumor cell formation and malignancy. Sequencing the entire genomes of 47 ependymomas (tumors of the brain lining), Mack and colleagues (2014) found no mutational variants. Rather, they found widespread methylation on the genes encoding

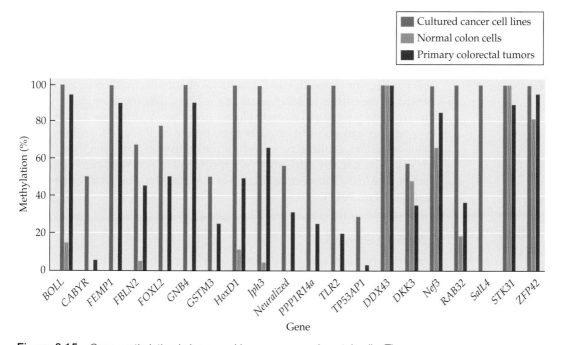

Figure 8.15   Gene methylation is increased in cancerous colorectal cells. The genes were selected from a list of genes analyzed by microarrays. Green bars represent samples from normal colon cells; note that several genes show essentially no methylation in normal tissue. The red bars represent genes sampled from primary colon tumors, while the blue bars represent samples from colorectal cancer cell cultures. (After Schuebel et al. 2007.)

polycomb repressive complex 2. This protein complex represses the transcription of differentiation genes by methylating histone 3. In Chapter 3, we mentioned that the bacteria *Helicobacter pylori* may cause cancer by altering the methylation patterns of certain genes in the stomach epithelium. The promoter of the tumor-suppressor gene encoding E-cadherin is hypermethylated by secretions of *H. pylori* (Chan et al. 2003, 2006; Miyazaki et al. 2007), while the gene encoding the growth-promoting enzyme cyclooxygenase-2 has reduced methylation (Akhtar et al. 2001). In lung cancer, the malignancy of a given tumor has been correlated with the promoter methylation of four tumor-suppressor genes, including a gene that regulates cell division (*p16*), one that increases cell adhesion (*H-cadherin*), one that causes the death of dividing cells (*APC*), and one that is necessary for stabilizing epithelial cell structure (*RASSF1A*).

New data are suggesting that ionizing radiation—the classic mutagen—can also cause epigenetic effects. Not only does ionizing radiation generate reactive oxygen species that can damage nuclear DNA, but it also damages mitochondrial DNA, which is more susceptible because it is the source of oxygen radicals. Mitochondrial damage decreases the activity of DNA methyltransferases, causing a global hypomethylation of nuclear DNA (Antwih et al. 2013; Szumiel 2014).

DNA methylation can also explain several lifestyle-related tumors. For instance, tobacco smoke is known to initiate lung cancers even though it is not a very powerful mutagen. Russo and colleagues (2005) and Liu and colleagues (2006) found that in lung cancers attributed to cigarette smoke, several tumor-suppressor genes in the early cancerous cells were heavily methylated. These included genes encoding cell adhesion proteins, apoptosis accelerators, and mitosis inhibitors. In addition, Liu and colleagues (2007) showed that the gene for synuclein-γ, which is not normally expressed in lung tissues, is activated by cigarette smoke. This gene appears to become demethylated, and its subsequent activation promotes the spread of the tumor to other parts of the body. The mechanism by which the synuclein-γ gene is demethylated gives us an important clue about how tumors can occur: it appears to be by the methylation, and therefore *down-regulation*, of the gene encoding the methyltransferase DNMT3B, an enzyme that usually adds methyl groups to DNA. Thus, carcinogenesis may be due to the slow accumulation of hypermethylated promoter regions (Landan et al. 2012).

Looking Ahead

Remarkably, both aging and cancer may be phenomena initiated by the activation in the adult of processes involved with embryonic development. And since aging is a major risk factor for most cancers (only a small number of cancers are hereditary and occur in children), the question of how aging and cancer are related is becoming a fascinating area of research.

As mentioned above, cancer appears to be one of those diseases incited by persistent senescent cells. Persistent DNA damage to cells triggers the senescence-associated secretory phenotype (Rodier et al. 2009). The secreted mixtures contain inflammatory cytokines (such as interleukin-6), paracrine growth factors (such as TGF-β), and proteases that degrade the extracellular matrices. These factors, in addition to providing cues for the origins of atherosclerosis, osteoarthritis, atherosclerosis, and Alzheimer's disease, also promote tumor development (see Davalos et al. 2010). We know that when fibroblasts (which are the major type of stromal cell) become senescent, they perturb the mammary epithelial microenvironment and promote epithelial growth and tumor formation in the mammary gland (Tsai et al. 2005; Barcellos-Hoff and Ravani 2000). This ability of senescent fibroblasts and their ability to promote cell migration have also been seen in the skin and prostate. Senescent cells may even induce the epithelial-to-mesenchymal transitions necessary for metastasis (Coppe et al. 2008, 2010).

So the senescent cell notion of aging fits remarkably well with the TOFT theory of carcinogenesis. Dotto (2014; Goruppi and Dotto 2013) emphasizes that in a majority of cases, epithelial cancerous lesions, having many of the genetic changes associated with invasive and metastatic tumors, do not become malignant tumors unless the stroma is altered. He and his colleagues view cancer as a disease of tissue or organ fields, where the mesenchymal stroma can permit or repress the cell proliferation and metastasis characteristic of cancer. Moreover, in this view, the initiator of cancer may be senescent stromal cells that produce an inflammatory response. This would explain the occurrence of multifocal and recurrent epithelial tumors—several tumors from the same area and those tumors continuing to be formed from the same tissue—that are often preceded by (and associated with) local areas of inflammation (Figure 8.16). In this view, though, there must be not only a stromal alteration, but also a potentially dangerous epithelial alteration. Not every epithelial cell becomes a tumor, and the precancerous cell must have some bias toward becoming a tumor. This can be through mutations of tumor-suppressor genes (such as *p53*) or the increased methylation of the promoters of such genes. (The ability of *Helicobacter pylori* bacterial infections to induce gastric cancer may be due not only to inflammation but to the epigenetic silencing of certain tumor suppressor genes by the microbes [Ushijima 2007; Niwa and Ushijima 2010].) The TOFT synthesis generally holds that rapid proliferation is the default state of all cells and that a loosening of stromal controls over cell proliferation is all that is needed for rapid cell multiplication. The "field cancerization" view (Dotto 2014) is a modification of TOFT, in that it also supports the ideas that cancer is caused by changes in organizational fields and that the mesenchymal stroma is critical for cancer initiation. However, it views the epithelium as also having to be hit by a change. This change is not usually manifest; but it can become critical when the stromal suppression of epithelial growth is minimized.

The view that cancer is initiated by changes in tissue fields will also have to be united with the view that only a few of the cells of a tumor are actually malignant cells. That is to say, while most cells of a tumor are not rapidly growing and cannot initiate a tumor in a new site, there is a small population of **cancer stem cells** that generate most of the other cells and that can initiate a new tumor when transplanted. These cancer stem cells may be the cells that separate from the epithelium in a process called **epithelial-mesenchymal transformation**, the same process used by cells of the embryo to detach and migrate (Mani et al. 2008; Morel et al. 2008, 2012; Singh and Settleman 2010). It is not known whether each cell of the

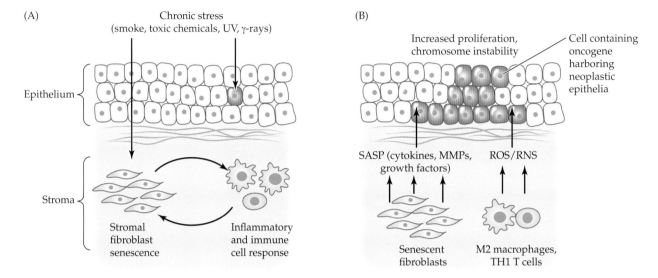

Figure 8.16 Interactions of stromal and epithelial cells within a field. (A) Epithelial cells acquire a potentially deleterious alteration. This could be due to environmental effects (such as mutations due to irradiation and carcinogens, epigenetic drift). However, these changes cannot initiate cancer. Cancer is initiated when the stroma is also affected, causing the cells to become senescent. The senescent cells recruit the cells that initiate a chronic inflammatory response, and these positively reinforce each other. (B) The stromal cells produce the moelcules defining a senescence-associated secretory phenotype (SASP), while the inflammatory cells release reactive oxygen species (ROS). Collectively, these factors promote mitosis and DNA rearrangements in the epithelium, allowing the damaged epithelial cells to proliferate and mutate. (After Goruppi and Dotto 2013.)

epithelium has an equal chance of becoming a cancer stem cell or whether such a cell has to been preconditioned (say, by a mutation) to become a cancer stem cell. However, there is evidence that the generation of cancer stem cells is under the regulation of the stroma (Cuiffo et al. 2014; Rajaram et al. 2015).

The relationships of cancer to aging, stem cells, and the roles of the stroma and senescent fibroblasts suggest new treatments for cancer. A focus on what causes cell senescence would be a critical way of providing a different type of cancer therapy (Hasty et al. 2013; Tchkonia et al. 2013), and perhaps some preventative actions, as well.

References

Akhtar, M., Y. Cheng, R. M. Magno, H. Ashktorab, D. T. Smoot, S. J. Melter and K. T. Wilson. 2001. Promoter methylation regulates *Helicobacter pylori*-stimulated cyclooxygenase-2 expression in gastric epithelial cells. *Cancer Res.* 61: 2399–2403.

Antebi, A. 2013. Steroid regulation of *C. elegans* diapause, developmental timing, and longevity. *Curr. Top. Dev. Biol.* 105: 181–212.

Antwih, D. A., K. M. Gabbara, W. D. Lancaster, D. M. Ruden and S. P. Zielske. 2013. Radiation-induced epigenetic DNA methylation modification of radiation-response pathways. *Epigenetics* 8: 839–848.

Aragon-Ching, J. B., H. Li, E. R. Gardner and W. D. Figg. 2007. Thalidomide analogues as anticancer drugs. *Res. Pat. Anticancer Drug Discov.* 2: 167–174.

Auerbach, R. and W. Auerbach. 1982. Regional differences in the growth of normal and neoplastic cells. *Science* 215: 127–134.

Auerbach, R., L. W. Morrissey and Y. A. Sidky. 1978. Gradients in tumour growth. *Nature* 274: 697–699.

Baker, D. J. and 7 others. 2011. Clearance of p16Ink4a-positive senescent cells delays ageing-associated disorders. *Nature* 479: 232–236.

Banito, A. and S. W. Lowe. 2013. A new development in senescence. *Cell* 155: 977–978.

Barcellos-Hoff, M. H. and S. A. Ravani. 2000. Irradiated mammary gland stroma promotes the expression of tumorigenic potential by unirradiated epithelial cells. *Cancer Res.* 60: 1254–1260.

Bell, J. T. and T. D. Spector. 2012. DNA methylation studies using twins: What are they telling us? *Genome Biol.* 13: 172.

Bellou, S., G. Pentheroudakis, C. Murphy and T. Fotsis. 2013. Anti-angiogenesis in cancer therapy: Hercules and hydra. *Cancer Lett.* 338: 219–228.

Berman, D. M. and 10 others. 2002. Medulloblastoma growth inhibition by hedgehog pathway blockade. *Science* 297: 1559–1561.

Berman, D. M. and 10 others. 2003. Widespread requirement for Hedgehog ligand stimulation in growth of digestive tract tumors. *Nature* 425: 846–851.

Bhowmick, N. A., E. G. Neilson and H. L. Moses. 2004. Stromal fibroblasts in cancer initiation and progression. *Nature* 432: 332–337.

Bissell, M. J., D. C. Radisky, A. Rizki, V. M. Weaver and O. W. Petersen. 2002. The organizing principle: Microenvironmental influences in the normal and malignant breast. *Differentiation* 70: 537–546.

Bizzari, M. and A. Cucina. 2014. Tumor and the microenvironment: A chance to reframe the paradigm of carcinogenesis? *BioMed Res. Int.* doi: 10.1155/2014/934038.

Bjornsson, H. T. and 15 others. 2008. Intraindividual change over time in DNA methylation with familial clustering. *J. Am. Med. Assoc.* 299: 2877–2883.

Blüher, M., B. B. Kahn and C. R. Kahn. 2003. Extended longevity in mice lacking the insulin receptor in adipose tissue. *Science* 299: 572–574.

Boehm, A. M. and 10 others. 2012. FoxO is a critical regulator of stem cell maintenance in immortal *Hydra*. *Proc. Natl. Acad. Sci. USA* 109: 19697–19702

Booth, B. W., C. A. Boulanger, L. H. Anderson and G. H. Smith. 2011. The normal mammary microenvironment suppresses the tumorigenic phenotype of mouse mammary tumor virus-neu-transformed mammary tumor cells. *Oncogene* 30: 679–689.

Bussard, K. M., C. A. Boulanger, B. W. Booth, R. D. Bruno and G. H. Smith. 2010. Reprogramming human cancer cells in the mouse mammary gland. *Cancer Res.* 70: 6336–6343.

Bussard, K. M. and G. H. Smith. 2012. Human breast cancer cells are redirected to mammary epithelial cells upon interaction with the regenerating mammary gland microenvironment in-vivo. *PLoS ONE* 7(11): e49221.

Campisi, J. 2001. Cellular senescence as a tumor-suppressor mechanism. *Trends Cell Biol.* 11: S27–31.

Campisi, J. 2014. Cell biology: The beginning of the end. *Nature* 505: 35–36.

Chan, A. O. and 8 others. 2003. Promoter methylation of *E-cadherin* gene in gastric mucosa associated with *Helicobacter pylori* infection and in gastric cancer. *Gut* 52: 502–506.

Chan, A. O. and 9 others. 2006. Eradication of *Helicobacter pylori* infection reverses E-cadherin promoter hypermethylation. *Gut* 55: 463–468.

Chen, C., Y. Lui and P. Zheng. 2009. mTOR regulation and therapeutic rejuvenation of aging hematopoietic stem cells. *Sci. Signaling* 2(98): ra75.

Clancy, D. J. and 7 others. 2001. Extension of life-span by loss of Chico, a *Drosophila* insulin receptor substrate protein. *Science* 292: 104–106.

Colman, R. J. and 10 others. 2009. Caloric restriction delays disease onset and mortality in rhesus monkeys. *Science* 325: 201–204.

Conboy, I. M., M. J. Conboy, A. J. Wagers, E. R. Girma, I. L. Weissman and T. A. Rando. 2005. Rejuvenation of aged progenitor cells by exposure to a young systemic environment. *Nature* 433: 760–764.

Conboy, I. M. and T. A. Rando. 2012. Heterochronic parabiosis for the study of the effects of aging on stem cells and their niches. *Cell Cycle* 11: 2260–2267.

Coppé, J. P., P. Y. Desprez, A. Krtolica and J. Campisi. 2010. The senescence-associated secretory phenotype: The dark side of tumor suppression. *Annu. Rev. Pathol.* 5: 99–118.

Coppé, J. P. and 8 others. 2008. Senescence-associated secretory phenotypes reveal cell-nonautonomous functions of oncogenic RAS and the p53 tumor suppressor. *PLoS Biol.* 6: 2853–2868.

Cuiffo, B. G. and 18 others. 2014. MSC-regulated microRNAs converge on the transcription factor FOXP2 and promote breast cancer metastasis. *Cell Stem Cell* 15: 762–774.

Cunha, G. R., H. Fujii, B. L. Neubauer, J. M. Shannon, L. Sawyer and B. A. Reese. 1983. Epithelial-mesenchymal interactions in prostatic development. I. Morphological observations of prostatic induction by urogenital sinus mesenchyme in epithelium of the adult rodent urinary bladder. *J. Cell Biol.* 96: 1662–1670.

Damjanov, I. 1993. Teratocarcinoma: Neoplastic lessons about normal embryogenesis. *Int. J. Dev. Biol.* 37: 39–46.

Davalos, A. R., J. P. Coppe, J. Campisi and P. Y. Desprez. 2010. Senescent cells as a source of inflammatory factors for tumor progression. *Cancer Metastasis Rev.* 29: 273–283.

De Smaele, E., E. Ferretti and A. Gulino. 2010. Vismodegib, a small molecule inhibitor of the hedgehog pathway for the treatment of advanced cancers. *Curr. Opin. Investig. Drugs* 11: 707–718.

Dotto, G. P. 2014. Multifocal epithelial tumors and field cancerization: Stroma as a primary determinant. *J. Clin. Invest.* 124: 1446–1452.

Esteller, M. 2008. Epigenetics in cancer. *New Engl. J. Med.* 358: 1148–1159.

Euhus, D. M., D. Bu, S. Milchgrub, X.-J. Xie, A. Bian, A. M. Leitch and C. M. Lewis. 2008. DNA methylation in benign breast epithelium in relation to age and breast cancer risk. *Cancer Epidemiol. Biomarkers Prev.* 17: 1051–1059.

Feinberg, A. P., R. Ohlsson and S. Henikoff. 2006. The epigenetic progenitor origin of human cancer. *Nature Rev. Genet.* 7: 21–33.

Folkman, J. 1974. Tumor angiogenesis. *Adv. Cancer Res.* 19: 331–358.

Folkman, J. and R. Kalluri. 2004. Cancer without disease. *Nature* 427: 787.

Fontana, J. A. and A. K. Rishi. 2002. Classical and novel retinoids: Their targets in cancer therapy. *Leukemia* 16: 463–472.

Fraga, M. F. and M. Esteller. 2007. Epigenetics and aging: The targets and the marks. *Trends Genet.* 23: 413–418.

Fraga, M. F. and 20 others. 2005. Epigenetic differences arise during the lifetime of monozygotic twins. *Proc. Nat. Acad. Sci USA* 102: 10604–10609.

Gardner, R. L. 1982. Manipulation of development. In C. R. Austin and R. V. Short (eds.), *Embryonic and Fetal Development*. Cambridge University Press, Cambridge, pp. 159–180.

Garraway, L. A. and E. S. Lander. 2013. Lessons from the cancer genome. *Cell* 153: 17–37.

Gems, D. and 7 others. 1998. Two pleiotropic classes of *daf-2* mutation affect larval arrest, adult behavior, reproduction and longevity in *Caenorhabditis elegans*. *Genetics* 150: 129–155.

Gerisch, B., C. Weitzel, C. Kober-Eisermann, V. Rottiers and A. Antebi. 2001. A hormonal signaling pathway influencing *C. elegans* metabolism, reproductive development, and life span. *Dev. Cell* 1: 841–851.

Giannakou, M. E., M. Goss, M. A. Junger, E. Hafen, S. J. Leevers and L. Partridge. 2004. Long-lived *Drosophila* with overexpressed dFoxo in adult fat body. *Science* 305: 361.

Gilbert, S. F. 2013. *Developmental Biology*, 10th Ed. Sinauer Associates, Sunderland, MA.

Gjoneska, E., A. R. Pfenning, H. Mathys, G. Quon, A. Kundaje, L.-H. Tsai and M. Kellis. 2015. Conserved epigenomic signals in mice and humans reveal immune basis of Alzheimer's disease. *Nature* 518: 365–369.

Goruppi, S. and G. P. Dotto. 2013. Mesenchymal stroma: Primary determinant and therapeutic target for epithelial cancer. *Trends Cell Biol.* 23: 593–602.

Gould, S. J. 1997. Dolly's fashion, Louis' passion. *Nat. Hist.* 106(5): 18–23.

Guarente, L. and C. Kenyon. 2000. Genetic pathways that regulate ageing in model organisms. *Nature* 408: 255–262.

Giuliano, S. and G. Pagès. 2013. Mechanisms of resistance to anti-angiogenesis therapies. *Biochimie.* 95: 1110–1119.

Guss, K. A., C. E. Nelson, A. Hudson, M. E. Kraus and S. B. Carroll. 2001. Control of a genetic regulatory network by a selector gene. *Science* 292: 1164–1167.

Halloran, J. and 9 others. 2012. Chronic inhibition of mammalian target of rapamycin by rapamycin modulates cognitive and non-cognitive components of behavior throughout lifespan in mice. *Neuroscience* 223: 102–113.

Hanahan, D. and R. A. Weinberg. 2000. The hallmarks of cancer. *Cell* 100: 57–70.

Hannum, G. and 15 others. 2013. Genome-wide methylation profiles reveal quantitative views of human aging rates. *Mol. Cell* 49: 359–367.

Hansen, J. M., S. G. Gong, M. Philbert and C. Harris. 2002. Misregulation of gene expression in the redox-sensitive NF-κ-β-dependent limb outgrowth pathway by thalidomide. *Dev. Dyn.* 225: 186–194.

Hansen, L. A., C. C. Sigman, F. Andreola, S. A. Ross, G. J. Kelloff and L. M. De Luca. 2000. Retinoids in chemoprevention and differentiation therapy. *Carcinogenesis* 21: 1271–1279.

Harrison, D. E. and 13 others. 2009. Rapamycin fed late in life extends lifespan in genetically heterogeneous mice. *Nature* 460: 392–395.

Hasty, P., Z. D. Sharp, T. J. Curiel and J. Campisi. 2013. mTORC1 and p53: Clash of the gods? *Cell Cycle* 12: 20–25.

Hattori, N. and T. Ushijima. 2014. Compendium of aberrant DNA methylation and histone modifications in cancer. *Biochem. Biophys. Res. Commun.* 455(1-2): 3–9.

Hendrix, M. J., E. A. Seftor, R. E. Seftor, J. Kasemeier-Kulesa, P. M. Kulesa and L. M. Postovit. 2007. Reprogramming metastatic tumour cells with embryonic microenvironments. *Nature Rev. Cancer.* 7: 246–255.

Holzenberger, M. and 7 others. 2003. IGF-1 receptor regulates lifespan and resistance to oxidative stress in mice. *Nature* 421: 182–186.

Honda, Y. and S. Honda. 1999. The *daf-2* gene network for longevity regulates oxidative stress resistance and Mn-superoxide dismutase gene expression in *Caenorhabditis elegans*. *FASEB J.* 13: 1385–1393.

Horvath, S. 2013. DNA methylation age of human tissues and cell types. *Genome Biol.* 14(10): R115.

Hovestadt, V. and 44 others. 2014. Decoding the regulatory landscape of medulloblastoma using DNA methylation sequencing. *Nature* 510: 537–541.

Hsin, H. and C. Kenyon. 1999. Signals from the reproductive system regulate the lifespan of *C. elegans*. *Nature* 399: 362–365.

Hwangbo, D. S., B. Gersham, M.-P. Tu, M. Palmer and M. Tatar. 2004. *Drosophila* dFOXO controls lifespan and regulates insulin signaling in brain and fat body. *Nature* 429: 562–566.

Illmensee, K. and B. Mintz. 1976. Totipotency and normal differentiation of single teratocarcinoma cells cloned by injection into blastocysts. *Proc. Natl. Acad. Sci. USA* 173: 549–553.

Issa, J.-P., Y. L. Ottaviano, P. Celano, S. R. Hamilton, N. E. Davidson and S. B. Baylin. 1994. Methylation of the oestrogen receptor CpG island links aging and neoplasia in human colon. *Nature Genet.* 7: 536–540.

Jacinto, F. V. and M. Esteller. 2007. Mutator pathways unleashed by epigenetic silencing in human cancer. *Mutagenesis* 22: 247–253.

Johnson, S. C., P. S. Rabinovitch and M. Kaeberlein. 2013. mTOR is a key modulator of ageing and age-related disease. *Nature* 493: 338–345.

Kaplan, F. S., J. Fiori, L. S. de la Peña, J. Ahn, P. C. Billings and E. M. Shore. 2006. Dysregulation of the BMP-4 signaling pathway in fibrodysplasia ossificans progressiva. *Ann. NY Acad. Sci.* 1068: 54–65.

Kasemeier-Kulesa, J. C., J. M. Teddy, L.-M. Postovit, E. A. Seftor, R. E. B. Seftor, M. J. C. Hendrix and P. M. Kulesa. 2008. Reprogramming multipotent tumor cells with the embryonic neural crest microenvironment. *Dev. Dyn.* 237: 2657–2666.

Katsimpardi, L. and 9 others. 2014. Vascular and neurogenic rejuvenation of the aging mouse brain by young systemic factors. *Science* 344: 630- 634.

Kenyon, C. J. 2001. A conserved regulatory system for aging. *Cell* 105: 165–168.

Kenyon, C. J. 2010. The genetics of ageing. *Nature* 464: 504–512.

Kim, J. and 8 others. 2007. Epigenetic changes in estrogen receptor beta gene in atherosclerotic cardiovascular tissues and in vitro vascular senescence. *Biochim. Biophys. Acta* 1772: 72–80.

Kimura, K. D., H. A. Tissenbaum, Y. Liu and G. Ruvkun. 1997. Daf-2, an insulin receptor-like gene that regulates longevity and diapause *in Caenorhabditis elegans*. *Science* 277: 942–946.

Knobloch, J., J. D. Shaughnessy, Jr. and U. Ruther. 2007. Thalidomide induces limb deformities by perturbing the Bmp/Dkk1/Wnt signaling pathway. *FASEB J.* 21: 1410–1421.

Krtolica, A., S. Parrinello, S. Lockett, P. Y. Desprez and J. Campisi. 2001. Senescent fibroblasts promote epithelial cell growth and tumorigenesis: A link between cancer and aging. *Proc. Natl. Acad. Sci. USA* 98: 12072–12077.

Kurosu, H. and 15 others. 2005. Suppression of aging in mice by the hormone Klotho. *Science* 309: 1829–1833.

Lamming, D. W. and 12 others. 2012. Rapamycin-induced insulin resistance is mediated by mTORC2 loss and uncoupled from longevity. *Science* 335: 1638–1643.

Landan, G. and 12 others. 2012. Epigenetic polymorphism and the stochastic formation of differentially methylated regions in normal and cancerous tissues. *Nature Genet.* 44: 1207–1214.

Levin, M. 2012. Morphogenetic fields in embryogenesis, regeneration, and cancer: Non-local control of complex patterning. *Biosystems.* 109: 243–261.

List, A. and 9 others. 2005. Efficacy of lenalidomide in myelodysplastic syndromes. *New Engl. J. Med.* 352: 549–557.

Liu, H., Q. Lan, J. M. Siegfried, J. D. Luketich and P. Keohavong. 2006. Aberrant promoter methylation of *p16* and *MGMT* genes in lung tumors from smoking and never-smoking patients. *Neoplasia* 8: 46–51.

Liu, H., Y. Zhou, S. E. Boggs, S. A. Belinsky and J. Liu. 2007. Cigarette smoke induces demethylation of prometastatic oncogene synuclein-γ in lung cancer cells by downregulation of DNMT3B. *Oncogene* 26: 5900–5910.

Loffredo, F. S. and 19 others. 2013. Growth differentiation factor 11 is a circulating factor that reverses age-related cardiac hypertrophy. *Cell* 153: 828–839.

Lorenzini, A. 2014. How much should we weigh for a long and healthy life span? The need to reconcile caloric restriction versus longevity with body mass index versus mortality data. *Front. Endocrinol.* 5: 121.

Ma, X., Q. Yang, K. T. Wilson, N. Kundu, S. J. Meltzer and A. M. Fulton. 2004. Promoter methylation regulates cyclooxygenase expression in breast cancer. *Breast Canc. Res.* 6: R316–R321.

Mack, S. C. and 88 others. 2014. Epigenomic alterations define lethal CIMP-positive ependymomas of infancy. *Nature* 506: 445–450.

Maffini, M. V., A. M. Soto, J. M. Calabro, A. A. Ucci and C. Sonenschein. 2004. The stroma as a crucial target in rat mammary gland carcinogenesis. *J. Cell Sci.* 117: 1495–1502.

Majumder, S. and 8 others. 2012. Lifelong rapamycin administration ameliorates age-dependent cognitive deficits by reducing IL-1β and enhancing NMDA signaling. *Aging Cell* 11: 326–335.

Mani, S. A. and 14 others. 2008. The epithelial-mesenchymal transition generates cells with properties of stem cells. *Cell* 133: 704–715.

Mattison, J. A. and 14 others. 2012. Impact of caloric restriction on health and survival in rhesus monkeys from the NIA study. *Nature* 489: 318–321.

Miyazaki, T. and 8 others. 2007. E-cadherin promoter methylation in *H. pylori*-induced enlarged fold gastritis. *Helicobacter* 12: 523–531.

Morel, A.-P. and 16 others. 2012. EMT inducers catalyze malignant transformation of mammary epithelial cells and drive tumorigenesis towards claudin-low tumors in transgenic mice. *PLoS Genet.* 8: e1002723.

Morel, A.-P., M. Lièvre, C. Thomas, G. Hinkal, S. Ansieau and A. Puisieux. 2008. Generation of breast cancer stem cells through epithelial-mesenchymal transition. *PLoS ONE* 3: e2888.

Muñoz-Espín, D. and M. Serrano. 2014. Cellular senescence: From physiology to pathology. *Nat. Rev. Mol. Cell Biol.* 15: 482–496.

Muñoz-Espín, D. and 10 others. 2013. Programmed cell senescence during mammalian embryonic development. *Cell* 155: 1104–1118.

Muthukkaruppan, V. R. and R. Auerbach. 1979. Angiogenesis in the mouse cornea. *Science* 205: 1416–1418.

Nilsson, E. and 15 others. 2014. Altered DNA methylation and differential expression of genes influencing metabolism and inflammation in adipose tissue from subjects with type 2 diabetes. *Diabetes* 63: 2962–2976.

Nishimura, E. K., S. R. Granter and D. E. Fisher. 2005. Mechanisms of hair graying: Incomplete melanocyte stem cell maintenance in the niche. *Science* 307: 720–724.

Niwa, T. and T. Ushijima. 2010. Induction of epigenetic alterations by chronic inflammation and its significance on carcinogenesis. *Adv. Genet.* 71: 41–56.

Oates, N. A. and 9 others. 2006. Increased DNA methylation at the *AXIN1* gene in a monozygotic twin from a pair discordant for a caudal duplication anomaly. *Am. J. Hum. Genet.* 79: 155–162.

Özdemir, B. C. and 20 others. 2014. Depletion of carcinoma-associated fibroblasts and fibrosis induces immunosuppression and accelerates pancreas cancer with reduced survival. *Cancer Cell* 25: 719–734.

Partridge, L. and D. Gems. 2002. Mechanisms of ageing: Public or private? *Nature Rev. Genet.* 3: 165–175.

Petronis, A. 2006. Epigenetics and twins: Three variations on a theme. *Trends Genet.* 22: 347–350.

Pierce, G. B. 1974. Neoplasms, differentiations, and mutations. *Am. J. Pathol.* 77: 103–118.

Pierce, G. B., R. Shikes and L. M. Fink. 1978. *Cancer: A Problem of Developmental Biology*. Prentice-Hall, Englewood Cliffs, NJ.

Pierce, S. B. and 12 others. 2001. Regulation of DAF-2 receptor signaling by human insulin and ins-1, a member of the unusually large and diverse *C. elegans* insulin gene family. *Genes Dev.* 15: 672–686.

Post, W. S. and 8 others. 1999. Methylation of the estrogen receptor gene is associated with aging and atherosclerosis in the cardiovascular system. *Cardiovasc. Res.* 43: 985–991.

Postovit, L. M. and 8 others. 2008. Human embryonic stem cell microenvironment suppresses the tumorigenic phenotype of aggressive cancer cells. *Proc. Natl. Acad. Sci. USA* 105: 4329–4334.

Prieur, A. and D. S. Peeper. 2008. Cellular senescence in vivo: A barrier to tumorigenesis. *Curr. Opin. Cell Biol.* 20: 150–155.

Rajaram, R. D. and 8 others. 2015. Progesterone and Wnt4 control mammary stem cells via myoepithelial crosstalk. *EMBO J.* 34: 641–652.

Rhim, A. D. and 16 others. 2014. Stromal elements act to restrain, rather than support, pancreatic ductal adenocarcinoma. *Cancer Cell* 25: 735–747.

Rilke, R. M. 1910. "He died a hard death" from *The Notebooks of Malte Laurids Brigge*, Penguin, New York.

Robinson, K. C. and D. E. Fisher. 2009. Specification and loss of melanocyte stem cells. *Semin. Cell Dev. Biol.* 20: 111–116.

Rodier, F. and 9 others. 2009. Persistent DNA damage signalling triggers senescence-associated inflammatory cytokine secretion. *Nature Cell Biol.* 11: 973–979.

Rostand, J. 1962. *The Substance of Man*. Doubleday, Garden City, NJ, p. 9.

Rubin, L. L. and F. J. de Sauvage. 2006. Targeting the Hedgehog pathway in cancer. *Nature Rev. Drug Discov.* 5: 1026–1033.

Ruckh, J. M., J. W. Zhao, J. L. Shadrach, P. van Wijngaarden, T. N. Rao, A. J. Wagers and R. J. Franklin. 2012. Rejuvenation of regeneration in the aging central nervous system. *Cell Stem Cell* 10: 96–103.

Russo, A. L. and 9 others. 2005. Differential DNA hypermethylation of critical genes mediates the stage-specific tobacco smoke–induced neoplastic progression of lung cancer. *Clin. Cancer Res.* 11: 2466–2470.

Schaedel, O. N., B. Gerisch, A. Antebi and P. W. Sternberg. 2012. Hormonal signal amplification mediates environmental conditions during development and controls an irreversible commitment to adulthood. *PLoS Biol.* 10(4): e1001306.

Schuebel, K. E. and 18 others. 2007. Comparing the DNA hypermethylome with gene mutations in human colonorectal cancer. *PLoS Genet.* 3(9): e157.

Scudellari, M. 2015. Blood to blood. *Nature* 517: 426–429.

Selman, C. and 20 others. 2009. Ribosomal protein S6 kinase 1 signaling regulates mammalian lifespan. *Science* 326: 140–144.

Shen, Y., J. Wollam, D. Magner, O. Karalay and A. Antebi. 2012. A steroid receptor-microRNA switch regulates life span in response to signals from the gonad. *Science* 338: 1472–1476.

Shlush, L. I. and 25 others. 2014. Identification of pre-leukaemic haematopoietic stem cells in acute leukaemia. *Nature* 506: 328–333.

Singh, A. and J. Settleman. 2010. EMT, cancer stem cells, and drug resistance: An emerging axis of evil in the war on cancer. *Oncogene* 29: 4741–4751.

Sjöblom, T. and 28 others. 2006. The consensus coding sequences of human breast and colorectal cancers. *Science* 314: 268–274.

Solter, D., N. Skreb and I. Damjanov. 1970. Extrauterine growth of mouse egg cylinders results in malignant teratoma. *Nature* 227: 503–504.

Song, Z. and 9 others. 2011. Sonic hedgehog pathway is essential for maintenance of cancer stem-like cells in human gastric cancer. *PLoS ONE* 6: e17687.

Sonnenschein, C. and A. M. Soto. 1999. *A Society of Cells: Cancer and Control of Cell Proliferation*. Oxford University Press, Oxford.

Sonnenschein, C., A. M. Soto, A. Rangarajan and P. Kulkarni. 2014. Competing views on cancer. *J. Biosci.* 39: 281–302.

Soto, A. M. and C. Sonnenschein. 2011. The tissue organization field theory of cancer: A testable replacement for the somatic mutation theory. *Bioessays* 33: 332–340.

Spencer, V. A., R. Xu and M. J. Bissell. 2007. Extracellular matrix, nuclear and chromatin structure, and gene expression in normal tissues and malignant tumors: A work in progress. *Adv. Cancer Res.* 97: 275–294.

Stein, P. K., A. Soare, T. E. Meyer, R. Cangemi, J. O. Holloszy and L. Fontana. 2012. Caloric restriction may reverse age-related autonomic decline in humans. *Aging Cell* 11: 644–650.

Stevens, L. C. 1970. The deyelopment of transplantable teratocarcinomas from intratesticular grafts of pre- and post-implantation mouse embryos. *Dev. Biol.* 21: 364–382.

Stevens, L. C. and C. C. Little. 1953. Spontaneous testicular teratomas in an inbred strain of mice. *Proc. Natl. Acad. Sci. USA* 40: 1080–1087.

Stewart, T. A. and B. Mintz. 1981. Successive generations of mice produced from an established culture line of euploid teratocarcinoma cells. *Proc. Natl. Acad. Sci. USA* 78: 6314–6318.

Storer, M. and 10 others. 2013. Senescence is a developmental mechanism that contributes to embryonic growth and patterning. *Cell* 155: 1119–1130.

Stratton, M. R., P. J. Campbell and P. A. Futreal. 2009. The cancer genome. *Nature* 458: 719–724.

Sweatt, J. D. 2013. The emerging field of neuroepigenetics. *Neuron* 80: 624–632.

Szumiel, I. 2014. Ionising radiation-induced oxidative stress, epigenetic changes and genomic instability: The pivotal role of mitochondria. *Int. J. Radiat. Biol.* 17: 1–55.

Tammela, T. and 20 others. 2008. Blocking VEGFR-3 suppresses angiogenic sprouting and vascular network formation. *Nature* 454: 656–660.

Tatar, M., A. Kopelman, D. Epstein, M. P. Tu, C. M. Yin and R. S. Garofalo. 2001. A mutant *Drosophila* insulin receptor homolog that extends life-span and impairs neuroendocrine function. *Science* 292: 107–110.

Tchkonia, T., Y. Zhu, J. van Deursen, J. Campisi and J. L. Kirkland. 2013. Cellular senescence and the senescent secretory phenotype: Therapeutic opportunities. *J. Clin. Invest.* 123: 966–972.

Thayer, S. P. and 13 others. 2003. Hedgehog is an early and late mediator of pancreatic cancer tumorigenesis. *Nature* 425: 851–856.

Tian, H. and 6 others. 2009. Hedgehog signaling is restricted to the stromal compartment during pancreatic carcinogenesis. *Proc. Natl. Acad. Sci. USA* 106: 4254–4259.

Tian, X. and 9 others. 2013. High-molecular-mass hyaluronan mediates the cancer resistance of the naked mole rat. *Nature* 499: 346–349.

Tlsty, T. D. and L. M. Coussens. 2006. Tumor stroma and regulation of cancer development. *Annu. Rev. Pathol.: Mech. Dis.* 1: 119–150.

Tomasetti, C. and B. Vogelstein. 2015. Cancer etiology. Variation in cancer risk among tissues can be explained by the number of stem cell divisions. *Science* 347: 78–81.

Tran, H. and 7 others. 2002. DNA repair pathways stimulated by the forkhead transcription factor FOXO3a through the Gadd45 protein. *Science* 296: 530–534.

Tsai, K. K., E. Y. Chuang, J. B. Little and Z. M. Yuan. 2005. Cellular mechanisms for low-dose ionizing radiation-induced perturbation of the breast tissue microenvironment. *Cancer Res.* 65: 6734–6744.

Ushijima, T. 2007. Epigenetic field for cancerization. *J. Biochem. Mol. Biol.* 40: 142–150.

Vanyushin, B. L., L. E. Nemirovsky, V. V. Klimenko, V. K. Vasilev and A. N. Belozersky. 1973. The 5-methylcytosine in DNA of rats: Tissue and age specificity and the changes induced by hydrocortisone and other agents. *Gerontologia* 19: 138–152.

Vidi, P. A. and 9 others. 2012. Interconnected contribution of tissue morphogenesis and the nuclear protein NuMA to the DNA damage response. *J. Cell Sci.* 125(Pt 2): 350–361.

Villeda, S. A. and 21 others. 2011. The ageing systemic milieu negatively regulates neurogenesis and cognitive function. *Nature* 477: 90–94.

Villeda, S. A. and 17 others. 2014. Young blood reverses age-related impairments in cognitive function and synaptic plasticity in mice. *Nature Med.* 20: 659–663.

Vogelstein, B. and K. W. Kinzler. 2004. Cancer genes and the pathways they control. *Nature Med.* 10: 789–799.

Vogelstein, B., N. Papadopoulos, V. E. Velculescu, S. Zhou, L. A. Diaz Jr. and K. W. Kinzler. 2013. Cancer genome landscapes. *Science* 339: 1546–1558.

Waddington, C. 1935. Cancer and the theory of organisers. *Nature* 135: 6006–6008.

Wang, Z. and 31 others. 2015. Broad targeting of angiogenesis for cancer prevention and therapy. *Semin. Cancer Biol.* pii: S1044-579X(15)00002-4.

Weaver, V. M., O. W. Petersen, F. Wang, C. A. Larabell, P. Briand, C. Damsky and M. J. Bissell. 1997. Reversion of the malignant phenotype of human breast cells in three-dimensional culture and in vivo using integrin blocking antibodies. *J. Cell Biol.* 137: 231–246.

Weinberg, R. A. 2014. Coming full circle-from endless complexity to simplicity and back again. *Cell* 157: 267–271.

Whisenant, J. and E. Bergsland. 2005. Anti-angiogenic strategies in gastrointestinal malignancies. *Curr. Treat. Options Oncol.* 6: 411–421.

Weiss, P. 1939. *Principles of Development*. Holt, New York.

Yauch, R. L. and 14 others. 2008. A paracrine requirement for Hedgehog signalling in cancer. *Nature* 455: 406–410.

Yauch, R. L. and 17 others. 2009. Smoothened mutation confers resistance to a Hedgehog pathway inhibitor in medulloblastoma. *Science* 326: 572–574.

Yilmaz, Ö. H. and 15 others. 2012. mTORC1 in the Paneth cell niche couples intestinal stem-cell function to calorie intake. *Nature* 486: 490–495.

Zhang, P., M. Judy, S. J. Lee and C. Kenyon. 2013. Direct and indirect gene regulation by a life-extending FOXO protein in *C. elegans*: Roles for GATA factors and lipid gene regulators. *Cell Metab.* 17: 85–100.

Zhao, C. and 13 others. 2009. Hedgehog signalling is essential for maintenance of cancer stem cells in myeloid leukaemia. *Nature* 458: 776–779.

Part 3

Toward a Developmental Evolutionary Synthesis

The relationships among evolution, development, and environment have been contentious themes in evolutionary biology. **Chapter 9** reviews the ideas of the Modern Synthesis, wherein Darwin's and Wallace's ideas concerning descent with modification were fused with genetic mechanisms for the origin and propagation of variation. **Chapter 10** concerns evolutionary developmental biology (evo-devo), focusing on the ideas that evolution can be viewed as changes in developmental trajectories and that heritable variation comes from changes in the regulatory genes active during embryonic and larval development. **Chapter 11** integrates ecology into this view of evolution, showing the environment (the symbionts, agents of plasticity, and epialleles of Part 1) as providing instruction for the production of new phenotypes. This points to a new synthesis of evolutionary, ecological, and developmental biology that can more fully explain the production and maintenance of biodiversity.

The Modern Synthesis
Natural Selection of Allelic Variation

Biology points out the individuality of every being, and at the same time reminds us of the brotherhood of all.

Jean Rostand, 1962

Our own genomes carry the story of evolution, written in DNA, the language of molecular genetics, and the narrative is unmistakable.

Kenneth R. Miller, 2009

During the past 25 years, a new interpretation of evolutionary biology has matured. This approach to evolution, called evolutionary developmental biology or "evo-devo," sees evolutionary change as heritable alterations in development. If, as ecological developmental biology claims, development encompasses the use of environmental signals in producing phenotypes, then this has profound implications for evolutionary biology.

Evolutionary biology is the scientific attempt to explain the origins and patterns of biodiversity and the processes that maintain it. The field has itself evolved through a series of syntheses. The idea that one form of life could give rise to another form had been mooted by biologists such as Jean-Baptiste Lamarck and Robert Chambers, as well as by Charles Darwin's grandfather, Erasmus Darwin. These schemes were often referred to as "transformationist theories." However, it was Charles Darwin's *On the Origin of Species* (1859) that synthesized evidence from embryology, biogeography, paleontology, agricultural breeding, and anatomy to show how species originated from other species and that life's diversity was not the result of myriad independent acts of creation. Moreover, Darwin proposed a mechanism for evolution—natural selection—that could be demonstrated on a smaller scale in the changing of the phenotype that characterized a species.*

Beginning in the 1920s, a second synthesis emerged that fused the science of genetics to Darwin's theory of evolution. This **Modern Synthesis** (sometimes called **neo-Darwinism**) emphasized gene mutations and rearrangements as the mechanisms producing the variation that is the grist for natural selection. The Modern Synthesis has been extremely successful in showing how environmental pressures favor the survival and propagation of a subset of variations within a species. We believe that we are currently seeing the third evolutionary synthesis, the "Developmental Synthesis" (sometimes called the "Extended Evolutionary Synthesis"), which emphasizes the idea that evolution consists of inherited changes in the patterns of development and focuses on the mechanisms whereby such changes are effected. This third synthesis focuses on the embryonic origin of adult variations, and it attempts to explain evolution above the species level through alterations in the use of regulatory genes.

This chapter will discuss both Darwin's ideas and what Haldane (1953) called "the current instar of the evolutionary theory," the Modern Synthesis. It will concern evolution within species and show that such evolution can be explained by natural selection (the link between differential reproductive success and traits) and that natural selection works by changing the frequency of alleles[†] within the population. The subsequent two chapters will document an expanded evolutionary theory that brings development and ecology into the synthesis.

Charles Darwin's Synthesis

In the nineteenth century, debates over the classification of species pitted two ways of viewing nature against each other. One view, championed by Georges Cuvier and Charles Bell, focused on the *differences* that allowed each distinct species to adapt to its environment. Thus, the hand of the human, the flipper of the seal, and the wing of the bird were seen as marvelous contrivances, each fashioned by the Creator to allow these animals to adapt to their "conditions of existence." The other view, championed by Étienne Geoffroy Saint-Hilaire and Richard Owen, was that the "unity of type" (the similarities among organisms, which Owen called "homologies") was critical. The human hand, the seal's flipper, and the bird's wing were all modifications of the same basic plan (Figure 9.1). In discovering that plan,

---

*One of the reasons we recognize Darwin as the major proponent of evolutionary biology rather than his contemporary, A. R. Wallace, is that *Origin of Species* provided a synthesis that encompassed nearly all of the biology of its time. Wallace's original paper (read, as was Darwin's abstract, at an 1858 meeting of the Linnean Society in London) shared Darwin's views on natural selection and biogeography, but it was Darwin's volume that showed how descent with modification would explain many of the biological enigmas that confounded scientists in the late-nineteenth century. For other reasons why Darwin gets the lion's share of the credit for the theory of evolution, see the excellent short essay by Berry and Browne (2008).

[†]A gene *locus* is a place. Thus, one can talk about the human β-globin gene as a locus on chromosome 11. An *allele* is the specific DNA sequence at that locus—for instance, the sickle-cell allele of the β-globin gene. Thus, the locus can be viewed as a parking spot, the allele is the car in that spot, and the allele might be a Toyota, Ford, or BMW.

Figure 9.1   Homologies of structure among a human arm, a seal forelimb, a bird wing, and a bat wing. (Homologous structures are shown in the same color.) All four are homologous as forelimbs because they derive from a common tetrapod ancestor. The adaptations of bird and bat forelimbs for flight, however, evolved independently of each other, after the two lineages diverged from a common ancestor. Therefore they are homologous as forelimbs but analogous as wings.

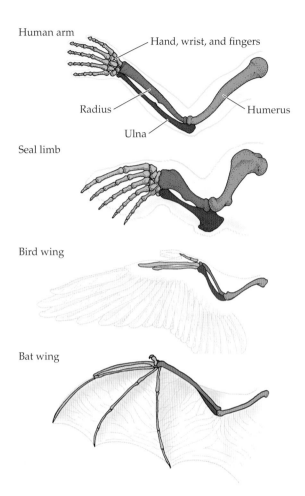

one could find the form upon which the Creator designed these animals; the distinctive adaptations were secondary.

Darwin acknowledged his debt to both sides of this debate when he wrote in *Origin*, "It is generally acknowledged that all organic beings have been formed on two great laws—Unity of Type, and Conditions of Existence." His theory, he stated, would explain "unity of type" as descent from a common ancestor. The changes creating the marvelous adaptations to the "conditions of existence" would be explained by natural selection. New adaptations were modifications of old parts. Darwin called this concept **descent with modification**.

Darwin noted that the homologies between the embryonic and larval structures of different phyla provided excellent evidence for descent with modification. He also argued that adaptations that depart from the "type" and allow an organism to survive in the "conditions" of its particular environment develop later in the embryo. Thus, one could emphasize *common descent* in the embryonic homologies between two or more groups of animals, or one could emphasize the *modifications* by showing how development had changed over generations to produce diverse adaptive structures (Gilbert 2003).

Classical Darwinism: Natural selection

Classical Darwinian emphasis on natural selection can be summarized in a few sentences:

1. There is variation among the individual organisms that make up a population of a species.
2. There is an enormous amount of death, and most individuals will not survive to reproduce.
3. Death is selective. Those individuals that best fit into the environment they encounter are more likely to survive; those that do not fit the environment well are usually eliminated.
4. When those individuals that survive reproduce, their progeny have a high likelihood of inheriting the variations that allowed their parents to survive. If individuals who carry those variations continue to be favored (selected), over time this natural selection will alter the overall characteristics of the population.

(A)

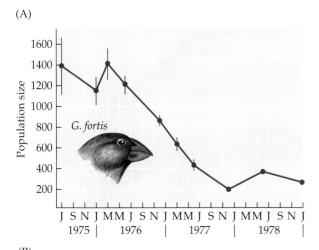

(B)

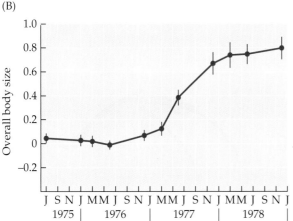

Figure 9.2   A contemporary example of Darwinian natural selection. (A) In 1976–1977, during a severe drought in the Galápagos Islands, the ground finch *Geospiza fortis* populations underwent a large drop in numbers. (B) Those that survived were the larger birds that could crack the larger, hard-shelled seeds—the only food available during the drought. (After Grant 1986.)

5. When populations of a species become reproductively isolated (i.e., separated in such a way that members of one population cannot mate with members of another*), each population can randomly acquire a distinct and separate suite of variations. If the environmental conditions faced by the isolated populations are different, different variations will be selected. Anatomical and physiological differences can result from the accumulation of even small distinctions, eventually causing the two populations to be recognized as different species.

There are many contemporary examples of classical Darwinism at work, some of the best known of which come from research done in a place that inspired Darwin himself. Peter and Rosemary Grant headed a team that observed evolution in action on the finches of the Galápagos Islands (see Grant 1986). In one of their several studies, drought conditions in 1976–1977 caused a dearth of nuts (especially the softer-shelled varieties), the main food for the finch *Geospiza fortis*. As the *G. fortis* population declined (Figure 9.2A), the individuals that survived were those with larger bodies (Figure 9.2B), whose stronger beaks could crack large, hard-shelled nuts, thus allowing them access to food that was unavailable to their smaller conspecifics.

GLOBAL CLIMATE CHANGE AS A SELECTIVE AGENT   Today we are seeing Darwinian selection in action as a result of the ubiquitous environmental stress of global climate change (see Chapter 4; Stern 2006). Global climate change (sometimes called "global warming" or "global climate disruption") threatens to change our planet into a different place, one that will almost certainly not be favorable to the diversity of life as we currently know it (Figure 9.3). Both the abundance of species at different locations and the ranges over which particular species can survive have been shifting. In the Northern Hemisphere, for instance, the ranges where certain trees and other plants grow and over which certain insects are found have extended north as winter temperatures get milder and springlike temperatures occur earlier. Conversely,

*While geographic isolation, as Darwin believed happened to finch and turtle populations on the Galápagos Islands, could explain reproductive isolation, Darwin believed that behavioral and ecological differences could also cause such isolation (Darwin 1844, 1859 p. 339). In his second edition of *Origin*, Darwin faced a major battle with Moritz Wagner, who thought geographic isolation the critical feature in species production.

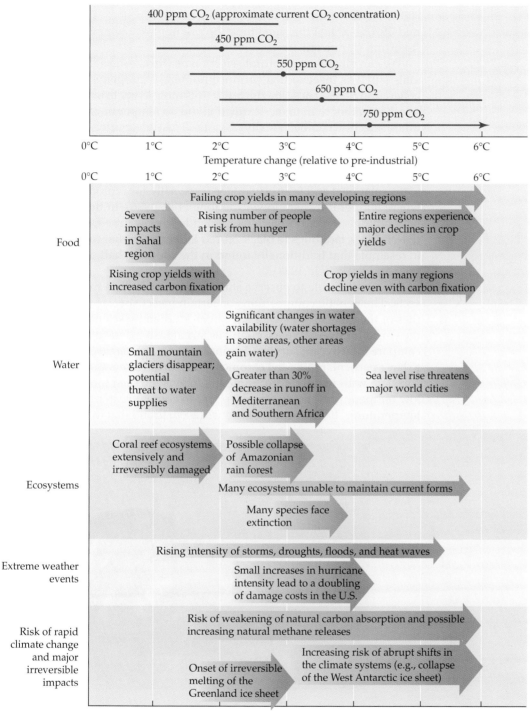

Figure 9.3 Summary of global changes that could be brought about by increased levels of carbon dioxide (CO_2, the major anthropogenic greenhouse gas) and the elevated temperatures these CO_2 levels would be expected to generate. A total environmental overhaul is to be expected if the average global temperature increases by 4°C, which is certainly a possible scenario. We are currently seeing the effects of an increase of about 0.8°C, and the amount of CO_2 already in the atmosphere may have irreversibly committed us to a 2°C increase. (After Stern 2006.)

the range of hundreds of arctic species is narrowing (Parmesan and Yohe 2003). Furthermore, the rate of these changes became significant in the mid-1970s—at just the same time the rise in global temperatures became steep and significant.

We can already see natural selection at work, especially in those animals (such as insects) that produce numerous progeny and have short generation times. These are the organisms most likely to evolve via natural selection rather than to face extinction due to temperature change. Sometimes selection can be visually obvious. The two-spotted ladybug beetle (*Adalia bipunctata*) has two major forms (Figure 9.4). The black-spotted red form is found predominantly in warmer climes, while a red-spotted black form is more common in the colder regions. The black (melanic) color absorbs solar heat and is an advantage (especially for reproductive activity) where temperatures are lower (de Jong et al. 1996). In the past 20 years, temperatures in the northern Netherlands have warmed, and the ratio of red-on-black ladybugs to black-on-red ladybugs in that region has begun to resemble that traditionally found in the southern half of the nation (de Jong and Brakefield 1998).

We are also able to observe shifts in gene and allele frequencies in insects. The genetic compositions of certain insect species (especially fruit flies of the genus *Drosophila*) have been studied by scientists throughout the world for over 50 years, giving us an extensive database with which to compare more recent surveys. For instance, *Drosophila subobscura* is known to have particular inherited chromosomal rearrangements (inversions) that correlate with temperature. Some inversions are seen more frequently at low temperatures, while others appear to confer greater fitness at higher temperatures. Balanyá and colleagues (2006) found that in North America,

Figure 9.4 Two genetically variant polymorphisms—black-on-red and red-on-black—of the spotted ladybug *Adalia bipunctata*. The ratio of these polymorphisms at any given location depends on temperature, with the red-on-black morph (which, being darker, absorbs sunlight better) found in colder latitudes.

South America, and Europe, the inversion frequencies have shifted away from the equator by about 1° of latitude from their historical norm. In other words, the inversion frequencies once found around Naples are now being seen as far north as Rome.

In Australia, the shift may be much greater. The climate along the eastern coast of Australia is becoming warmer and drier, and recent evidence demonstrates that in the past 50 years, the average temperature has increased dramatically as a result of human activities (see Williams 2008). The variants of the alcohol dehydrogenase gene in *Drosophila* correlate well with temperature, and the allele frequencies of this gene show a shift of 4° of latitude from the period 1978–1981 to the period 2000–2003 (Umina et al. 2005). Laboratory studies suggest that temperature (and, to some extent, humidity) selects for particular alleles and that this shift in alleles is probably due to natural selection rather than migration.

In many species, the timing of developmental events such as flowering and mating is also changing because of phenotypic plasticity (see Chapter 4). In Mediterranean ecosystems, the leaves of most deciduous plants emerge 2 weeks earlier and fall about 2 weeks later than they did 50 years ago, and in western Canada, the quaking aspen tree (*Populus tremuloides*) blooms almost a full month earlier than it did during the 1950s (Peñuelas and Filella 2001). Butterflies in northern Spain are eclosing 11 days earlier than they did in 1952, and the mating calls of some frog species in upstate New York are now heard an average of 10 days earlier than was common a century ago. All of these traits are developmentally plastic over a norm of reaction (see Chapter 1), and the populations mentioned above have not shown changes in gene frequency. However, other plants and animals that are not so developmentally plastic are perishing. Since Henry David Thoreau's detailed observations of the Walden Pond ecosystem in the 1850s, the average temperature there has risen 2°C. About 20% of the more than 400 plant species observed by Thoreau can no longer be found there, and other species that have not shifted their flowering times (especially irises, orchids, bladderworts, and lilies) are declining (Miller-Rushing and Primack 2008).

However, many species will not be able to evolve to counter climate change. Although extrapolation is always risky, it is possible that 50%–95% of existing species will perish by 2100, including half the existing number of bird species (Thomas et al. 2004; Kolbert 2014; UN Convention 2014). As of 2000, one-quarter of all existing mammal species in the world are threatened with extinction. Some species are being lost because of developmental inabilities to deal with rising temperatures. Others are lost secondarily when temperature impacts cause the loss of key species that support environments. Loss of coral reefs is due to the heat-induced elimination of symbionts from coral cells, for instance, and this negatively affects the fitness of numerous other species (as well as the livelihoods of millions of people). Changes in habitat due to thermal change can eliminate those species that have adapted to their original grasslands, forests, or prairies. One recent report (Pimm et al. 2014) claims that the current rate of animal extinction (due to all sources) is about 1000 times the usual extinction rate.

HUMAN BEINGS AND SELECTION PRESSURE   In addition to climate changes, human beings and their technology have significantly altered the natural world in many other ways. Moreover, in so doing, we have introduced selection pressures on a wide range of organisms. One of the first such cases to be documented was that of the peppered moth (*Biston betularia*) in the United Kingdom. During the nineteenth and early-twentieth centuries, there was a dramatic rise in the frequency of an allele (the *carbonaria* allele) that produces a black (melanic) phenotype along with a corresponding drop in the wild-type allele (the *typica* allele) that causes the lightly peppered phenotype (Figure 9.5A).

The first melanic *Biston* specimens appear in collections from the mid-1800s. By the 1890s, the proportion of melanic moths reached 90% in several urban centers, and a staggering 98% in Manchester (Grant et al. 1996;

(A)

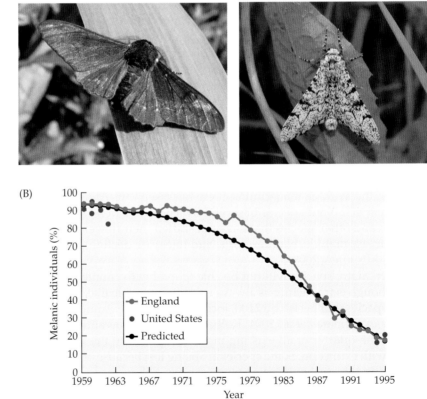

(B)

Figure 9.5 Industrial melanism. (A) Melanic (left) and typical (right) forms of the moth *Biston betularia*. (B) Frequency of the melanic form from 1959 to 1995 in industrialized areas in England (Caldy Common; blue circles) and the United States (George Reserve, Michigan; red circles). Sampling was continuous in Britain, while the data for the American region only compares samples taken at the beginning and end of the time period. The green plot represents mathematical predictions based on a constant selection coefficient against the melanic allele. (B after Grant et al. 1996.)

Grant and Wiseman 2002). By contrast, the black phenotype remained rare in the countryside. This "industrial melanism" has been well correlated with the darkening of trees and buildings by industrially generated soot and with the corresponding greater conspicuousness of the wild-type form of the moth. It is widely believed that the change in allele frequency was the result of selection pressure from bird predation, wherein the more conspicuous moths were preferentially eaten (Kettlewell 1973; Majerus 1998; Grant 1999).

But the story doesn't end there. If the prevalence of the dark phenotype was caused by airborne soot, one might expect (if the same predators were present) that selection would revert to favor the *typica* allele once air quality improved. In fact, this is precisely what happened. In both Britain and America, the frequency of melanic moths dropped precipitously as clean air legislation lessened the amount of soot that darkened urban areas (Figure 9.5B; Grant et al. 1996; Grant and Wiseman 2002).

Other human innovations have had dramatic effects on the evolution of certain populations. When insecticides were first used to kill malaria-carrying mosquitoes, most of the insects died. However, some few individuals survived because they carried preexisting mutations that either prevented their systems from absorbing the pesticide chemicals, or else enabled them to degrade those chemicals (Hemingway and Karunaratne 1998; Martinez-Torres et al. 1998). Some of the surviving individuals, for example, had duplications of the esterase gene whose product allows the cells to destroy DDT. In other individuals, a mutation in the potassium ion channel prevented certain insecticides from binding to their targets. Because only individuals with such mutations survived to mate with one another, the resistant genotypes quickly became the norm in many insect populations, with the result that it takes much higher concentrations of insecticides to kill them; indeed, some strains have become completely immune to the effects of many pesticides. A similar trajectory of human intervention explains the evolution of antibiotic resistance in many strains of bacteria and fungi (Nikaido 2009; Ford et al. 2015).

Several recent studies have shown that human predation practices have caused the evolution of different growth rates among prey species. Intensive selection pressure from human fishing has caused Norwegian graylings and North Atlantic cod to reach sexual maturity at smaller sizes and, in some cases, a year earlier than they did a century ago (Haugen and Vøllestad 2001; Andersen et al. 2007). This is probably because fish that reproduce at younger ages have a better chance of leaving offspring. Sport hunting of bighorn trophy rams has caused phenotype-selective mortality, wherein those rams with big horns are disproportionately killed. This practice inadvertently selects for rams with smaller horns, and indeed today's male bighorn sheep display a slower rate of horn growth than they did 30 years ago* (Coltman et al. 2003).

---

*Olivia Judson (2008) has noted that all of these growth changes were predicted by evolutionary theory.

Embryology and Darwin's synthesis

Darwin recognized that embryonic resemblances were a very strong argument in favor of the common ancestry of different animal groups (see Oppenheimer 1959; Ospovat 1981). "Community of embryonic structure reveals community of descent," Darwin said in *Origin of Species*. Thus, he looked to embryonic and larval stages for homologies that would be obscured in the adult. In *Origin*, Darwin celebrated the case of the barnacle, whose larvae showed it was a shrimplike crustacean rather than a clam-like mollusk, and in *The Descent of Man* (1874), he gloried in Alexander Kowalevsky's (1866) discovery that the tunicate—hitherto also classified as a mollusk—was actually a chordate (Figure 9.6). Tunicate larvae have a notochord and pharyngeal slits arising from the same cell layers that give rise to those structures in fishes and chicks. Thus, the "great divide" between the invertebrates and the vertebrates was bridged by the discovery of larval homologies.

Comparative embryology became evolutionary morphology as new information about the homologies of the germ layers in various animals became paramount in answering questions about phylogeny (e.g., Lankester 1877; Balfour 1880–1881; Oppenheimer 1940). Müller (1869) united

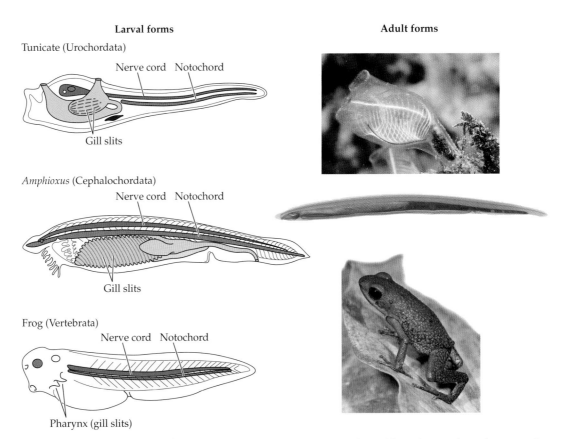

Figure 9.6 Tunicates, lancelets (*Amphioxus*), and frogs have a homologous notochord and pharyngeal gill slits and look similar as larvae but have very dissimilar adult morphologies.

the Crustacea (including in this phylum not only the barnacles but a whole range of parasites) through their possession of a larval type called the nauplius, and E. B. Wilson (1898) proposed a superphyletic taxon that included all those animals with spiral cleavage and possessing a particular mesoderm-forming cell. A century later, Wilson's proposed taxon was recognized by molecular and cladistic techniques and named the Lophotrochozoa (see Halanych et al. 1995; Halanych 2004).

To many in Darwin's generation, the embryo's development, or unfolding, could be seen as the motor driving evolution. Ernst Haeckel (1866) and others envisioned development as producing new structures that were added onto earlier stages of development. Individual development (ontogeny) would recapitulate (follow) the stages of previous generations of ancestors and then add something new. Haeckel proposed a causal parallelism between an animal's embryological development (ontogeny) and its evolutionary history (phylogeny). His so-called biogenetic law—that "ontogeny recapitulates phylogeny"—was based on the idea that the successive (and, to Haeckel, progressive) origin of new species was based on the same laws as the successive and progressive origin of new embryonic structures. Natural selection would prune the tree of life by eliminating the earlier (and hence less fit) species. To Haeckel, development produced novelties, and natural selection would get rid of entire species that had become obsolete. While this view of development and evolution was not tenable, the next generation of embryologists worked with and modified Haeckel's notions and showed how new morphological structures can arise via the hereditary modification of embryonic development.

One example of such modification, studied by Frank Lillie in 1898, is brought about by an alteration in the typical pattern of spiral cleavage in unionid clams. Unlike most clams, *Unio* and its relatives live in swiftly flowing streams. Streams create a problem for the dispersal of free-swimming larvae because such larvae are always carried downstream by the current. *Unio*, however, has adapted to the fast-flowing environment through modification of development. This modification begins with an alteration in embryonic cleavage (Figure 9.7). In typical molluscan cleavage, either all the macromeres (larger set of blastomeres) are equal in size or the 2D blastomere is the largest cell. However, cell division in *Unio* is such that the 2d blastomere gets the largest amount of cytoplasm. The 2d blastomere then divides to produce most of the larval structures, including a gland capable of producing a large shell on the resulting larva, which is called a glochidium. The glochidium resembles a tiny

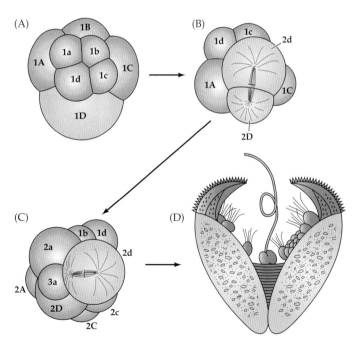

Figure 9.7 Formation of the glochidium ("bear trap") larva by modification of spiral cleavage. After the 8-cell embryo is formed (A), the placement of the mitotic spindle causes most of the 1D cytoplasm to enter the 2d (B). This enlarged 2d blastomere divides (C), eventually giving rise to the "bear trap" shell of the larva (D). This adaptation allows the larva to attach to the gills of passing fish. (After Raff and Kaufman 1983.)

bear trap; sensitive hairs cause the valves of the shell to snap shut when the larva encounters the gills or fins of a fish. The larva thus attaches itself to the fish and "hitchhikes" until it is ready to drop onto the substrate and metamorphose into an adult clam. In this manner, clam larvae can spread upstream as well as downstream and are less likely to be carried outside their favorable habitat. The marine biologist Walter Garstang used such evidence (and his own studies of snail larvae) to show that the evolution of new features was based on changes in developmental stages, not in adult stages. Garstang reversed Haeckel's relationship between ontogeny and phylogeny. In his address to the Linnean Society, Garstang (1922, p. 724) remarked that "ontogeny does not recapitulate phylogeny: it creates it."

The unionid clams continue to provide excellent examples of evolution through the action of natural selection on changes in development. In some unionid species, glochidia are released from the female's brood pouch and then wait passively for a fish to swim by. Some other species, such as *Lampsilis altilis*, have increased the chances of their larvae finding a fish by yet another series of developmental modifications (Welsh 1969). Many clams develop a thin mantle that flaps around the shell and surrounds the brood pouch. In some unionids, the shape of the brood pouch (marsupium) and the undulations of the mantle mimic the shape and swimming behavior of a minnow. To make the deception even better, the clams develop a black "eyespot" on one end and a flaring "tail" on the other. The "fish" in Figure 9.8 is not a fish at all, but the brood pouch and mantle of the female clam beneath it. When a predatory fish is lured within range of this "prey," the clam discharges the glochidia from the brood pouch. Thus, the modification

Figure 9.8 Phony fish atop the unionid clam *Lampsilis altilis*. The "fish" is actually the brood pouch and mantle of the clam. The "eyes" and flaring "tail" attract predatory fish, and the glochidium larvae attach to the fish's gills. (Photograph courtesy of Wendell R. Haag/USDA Forest Service.)

of existing developmental patterns to produce new phenotypes has permitted unionid clams to survive in challenging environments.

The failure of developmental morphology to explain evolution

While embryology played an important role in providing evidence for evolution, there were no mechanisms of development that allowed embryology to be cited as a causal agent for evolution. In the early twentieth century, embryology was predominantly a descriptive science, and any experimentation tended to be concerned with how embryonic development might be keyed to its environment (more about that in the next chapter). There were arguments over which structures were homologous (similar because of common ancestry, as in the chicken wing and the human hand), which were analogous (achieving the same function but arising from independent sources, as do the limbs of insects and vertebrates), and which were parallel (arising from two related but independent sources that diverged in the same way, as in the curved beaks of the diverse species of nectar-feeding birds, which evolved independently from different straight-beaked ancestral birds).

For example, to pose a broad question, are the segmented body plans that are seen in arthropods, annelid worms, and vertebrates (but not in other animals) homologous (arising from the same segmented ancestor), analogous (arising separately and independently because segmentation is a strongly favored adaptation that enhances movement), or parallel (arising not from a single common ancestor, but from several separate but related ancestors that all possessed the genetic basis for achieving segmentation)? No one knew. Fin de siècle evolutionary embryology had no independent method for ascertaining homology, no independent method of classification, and no knowledge of the cellular interactions that constituted embryogenesis. Thus, the field became mired in speculation (see Bowler 1996; Gilbert 1998).

Wilhelm Roux, a student of Haeckel, suggested that embryological ideas about evolution were premature and would remain so until we learned something about the specific ways in which animal bodies were constructed. According to Roux (1894), embryology had to leave the seashore and forest and go into the laboratory. Rather than being handmaiden to evolution, embryology should emulate physiology and look for the proximate causes of alterations in development. Roux predicted that embryology would someday return to prominence in evolutionary biology, bringing with it new knowledge of how animals were generated and how evolutionary changes might occur. But in the twentieth century, embryology "turned inward," away from evolution, and focused on the fascinating questions of how a fertilized egg gives rise to all the cells, tissues, and organs of the body.

The Modern Synthesis

The position vacated by embryology was soon filled by genetics. The fusion of Darwinian ideas of natural selection with the revelation of the gene as the agent of heredity resulted in the Modern Synthesis, or neo-Darwinism. This crucial evolutionary synthesis provided the theoretical and mathematical framework for evolutionary biology that embryology could not provide.

The origins of the Modern Synthesis

The existence of genes and their role as the source of hereditary information were not realized until the last years of the nineteenth century. Until that time, genetics and embryology were effectively joined in the science of "heredity," and genetics began largely as a subsection of that field (see Gilbert 1998). The work of the renowned geneticist Thomas Hunt Morgan in the early twentieth century was in large part responsible for separating genetics from embryology and establishing the former as a strong and independent field. The mathematical models of population genetics pioneered by Morgan and elaborated by his successors offered sound and demonstrable mechanical explanations for natural selection. And because population genetics can only study interbreeding populations and the barriers that prevent interbreeding, the *explanandum* (that which is to be explained) of evolutionary biology changed from the determination of phylogeny among different species (which was the goal of evolutionary morphology) to the quantification of natural selection within the same species.*

But if there were a "modern" synthesis, then there had to have been some "unmodern" synthesis; to the geneticists, that outdated synthesis was the one in which embryology had been united with evolution. Two embryologists-turned-geneticists were particularly important in making a case that evolutionary morphology had come to an end and that genetics was to take over. The first of these men, William Bateson, stated that "the embryological method has failed" when it came to determining the mechanisms of evolution (Bateson 1894). Bateson had been an evolutionary morphologist looking at developmental changes that could produce variation without selection. However, in 1900, he confessed that despite all that embryology and comparative anatomy had done, we were ignorant of the mechanism by which transmitted traits produce new species:

> Let us recognize at the outset that as to the essential nature of these phenomena we still know absolutely nothing. We have no glimmering of an idea as to what constitutes the essential process by which the likeness of the parent is transmitted to the offspring. We can study the process of fertilization and development in the finest detail which the microscope manifests to us, and we may fairly say that we now have a thorough grasp of the visible phenomena; but of the physical basis of heredity we have no conception at all…. Not only is our ignorance complete, but no one has the remotest idea how to set to work on that part of the problem.

Bateson admitted to a "crisis of faith" and was at a loss to provide a mechanism for evolution. However, he told how he was then "converted" to

---

*To be sure, this is an oversimplification. The early evolutionary biologists came from different subdisciplines within biology, and some of them professed more than one opinion during the course of their careers. Still, the ascendancy of the population geneticists can be seen in the frustration of those who felt that there was more to evolution than the mathematics of allele frequencies and "bean bag genetics" (Mayr 1959; Smocovitis 1996).

Mendelism and how he came to believe the work of Mendel could explain all the things that embryology couldn't. In his 1922 essay "Evolutionary Faith and Modern Doubts," Bateson announced the birth of a new science out of the decay of the old, proclaiming that "the geneticist is the successor of the morphologist."

The second great champion of genetics, Thomas Hunt Morgan, renounced embryology and wrote an article entitled "The Rise of Genetics" (1932a) and a book entitled *The Scientific Basis of Evolution* (1932b) in which he contrasted the new genetics with the old-fashioned embryology. Genetics, and only genetics, he wrote, constituted the "scientific basis of evolution." Embryology was one of the "metaphysical" and unscientific approaches that led nowhere. Moreover, Morgan asserted that the proper unit for study for evolutionary biology should be the population (the unit that evolves), not the individual (the unit that natural selection acts on).

The Fisher-Wright synthesis

It should be clear from the above discussion that the origins of the Modern Synthesis took place in a very anti-embryological environment, and it must be emphasized how complete the replacement of embryology by genetics has been. Darwin had shown that evolution occurred, and he used embryology to support such a view. However, whereas continental evolutionary biologists (notably in Germany) had seen in embryology possible mechanisms for evolutionary change, the British and American evolutionists sought this mechanism in Darwin's own idea of natural selection.

Many embryologists (beginning with Thomas Huxley; see Huxley 1893) accepted natural selection but did not think natural selection alone was sufficient to generate new types of organisms (see Lyons 1999). However, many of the early geneticists, starting with Morgan, did. But the early geneticists did not have the mathematical reasoning to show that this was the case; that task fell to three remarkable men, R. A. Fisher, J. B. S. Haldane, and Sewall Wright. In particular, Fisher's book *The Genetical Theory of Natural Selection* (1930) showed how the interplay of selection and mutation could bring about changes in the populations. Fisher's extensive equations and analyses were expanded, challenged, and restructured by Haldane and Wright (see Provine 1986; Edwards 2001) in what Mayr (2004) called the "Fisherian synthesis." This early version of the Modern Synthesis added a new genetic principle to evolutionary biology: that the inherited factors ("units of heredity') that would make an organism more or less fit for its environment were in fact the genes it received from its parents.

Another important concept for the Modern Synthesis was Sewall Wright's coefficients of relatedness and kinship (Wright 1922). The coefficient of kinship is the numerical probability that the alleles present at a particular locus chosen at random from two related individuals are identical by descent; the **coefficient of relatedness** (r) is twice that value. In diploid species, the r between parent and offspring is 0.5 (1/2), since half the offspring's genome comes from each parent. Full siblings also have an r coefficient of 0.5, indicating that half their genes are expected to be the same. Uncles and aunts have an r of 0.25 in relation to their nieces and nephews; half-sibs also have an r of 0.25. First cousins have an r of 0.125 (1/8).

Figure 9.9 Wendy Darling (Peter Pan's sometime girlfriend) shields her two brothers from Captain Hook, thereby illustrating Haldane's take on Wright's coefficient of relatedness, *r*. In this view, sacrificing oneself for two brothers (or eight first cousins) would offset the genes lost in one's own demise.

The kinship coefficient became a very important idea in the Modern Synthesis, leading to the concepts of **inclusive fitness** and **kin selection** (Hamilton 1967; Axelrod and Hamilton 1981), wherein individuals would be expected to favor the reproductive success of those relatives who carried the greatest proportion of the same genes. Thus, when asked if he would risk his life to save his brother, Haldane is said to have replied, "No, but I would do so to save two brothers—or eight cousins" (McElreath and Boyd 2007; Figure 9.9).

The Dobzhansky-Mayr synthesis

Evolutionary syntheses can appear to encapsulate one another like Russian dolls, each synthesis enfolding the previous one while explaining further phenomena (Wilkins 2011). Even with Wright's embellishments, the Fisherian synthesis was incomplete. It was based predominantly on research done on experimental organisms in the laboratory and thus seemed more relevant to artificial selection than to natural selection. Moreover, the mathematical underpinnings of this synthesis were not accessible to most biologists.

To show that natural selection actually worked in nature, Sergei Chetverikov demonstrated natural selection in natural populations of *Drosophila*. Chetverikov's program was expanded by Ukrainian-born Theodosius

Sturtevant on Snails

During the early days of the evolutionary synthesis, there was a huge debate between geneticists and embryologists over whether genetics could explain development or evolution. One thing embryologists could argue was that there were no mutations that were known to affect early development, and that the inherited traits that were known to affect early development did not seem to be inherited in the expected Mendelian manner. One of these traits was the sinistral (left-handed) shell-coiling variant in the snail *Limnaea*.

Lillie, Wilson, Garstang, and others (including Piaget, as we will see in Chapter 11) used snail embryos to link embryology to evolution, and molluscan development was well characterized. This coiling variant was widely discussed in embryology texts, but when Boycott and Diver (1923) did a study of its inheritance, they could not explain their observed ratios of right- to left-coiling offspring by Mendelian genetics.

It was Morgan's student Alfred Sturtevant who in 1923 explained the phenomenon, in a paper that covered only two columns in the journal *Science*. But in these two columns, Sturtevant told Boycott and Diver that their data were wrong, that their ratios were merely "fortuitous," and that they would have gotten different results if they had done the experiments properly. He then told them what the results *should* have been and explained why (see figure).

Sturtevant hypothesized that the case is a simple Mendelian one, with the dextral character dominant, but that the phenotype of the individual is determined not by its own genetic constitution, but by that of its mother. He concluded, "It seems likely that we shall have a model case of the Mendelian inheritance of an extremely 'fundamental' character, and a character that is impressed on the egg by the mother." We now call such inheritance (which includes the famous cases of the *Bicoid* and *Nanos* genes that specify the anterior-posterior axis in *Drosophila* embryos) "maternal effects."

Eventually, Boycott, working with senior embryologist S. L. Garstang, got exactly the results that Sturtevant had predicted he would. Boycott and colleagues (1930) called Sturtevant's analysis "his inspired guess." However, it was not so much an inspired guess as a beautiful example of predictive theory. Sturtevant did not have to do the laborious crosses to "know" the result. The geneticists claimed to have superseded the embryologists, and this round ended in their favor.

In an interesting "twist" on this story, Hoso and colleagues (2010) have shown that this single gene, working

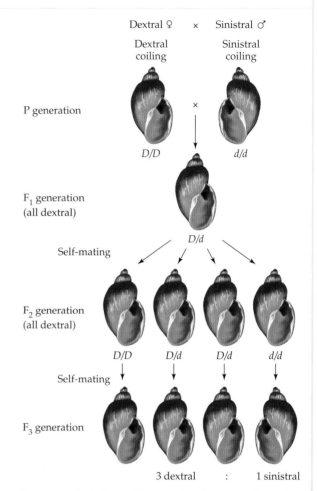

Coiling genetics in the snail *Limnaea*. The factor encoded by the *D* gene is needed in the egg for directing the embryo to coil in a dextral (rightward) direction. Without the factor encoded by *D*, the embryo coils in a sinistral (leftward) direction. Therefore, whether the snail coils to the right or to the left depends on the mother's genotype (i.e., whether or not she put the D factor into her eggs) and not on the genotype of the embryo. In the first cross here, since the mother expresses *D*, all her offspring coil to the right, and all are heterozygous. However, if the male parent expresses the D factor and the female does not, all the offspring will be similarly heterozygous, but they will all coil to the left (since the mother lacked the D factor). When the F₁ offspring are bred to one another, *dd* females are produced. These homozygous females will themselves coil to the right if their mother produced factor D; however, all the *progeny* of these *dd* females will coil to the left. (After Morgan 1927.)

(Continued on next page)

Sturtevant on Snails *(continued)*

early in development, can lead to the speciation of snails. The right-coiling snails and the left-coiling snails have difficulty mating with each other, and so a new mutation in the right-coiling species produces a left-coiling group that is reproductively isolated and has the potential to form a new species immediately. The rareness of the variant should preclude its reproducing, though, and speciation resulting from such single-gene mutations in chirality (right-left coiling) should be very rare. However, researchers have found several cases of left-handed snail species that have split from the right-handed coilers. This is because of a selective advantage that the lefties have—snakes can't eat them well. The Pareatidae family of snakes contain major predators of snails.

Since most snails coil to the right, these snakes have developed asymmetric jaws whose teeth are optimized for extracting snails from right-coiling shells. Their jaws don't hold left-coiling snails. In the laboratory, feeding experiments have supported the idea that this ability of the left-coiling snail to escape snake predation and the positive frequency-dependent mating of these snails after a certain threshold population is reached could explain the rapid speciation of the left-coiling snails into genetically separate groups soon after the mutation occurred. Therefore, a developmental mutation can provide the material for reproductive isolation and the rapid formation of a new species.

Dobzhansky, who brought this idea to New York, where he worked with Thomas Hunt Morgan and Sewall Wright. In 1937, Dobzhansky wrote *Genetics and the Origin of Species*, which in many ways translated Wright and Fisher into accessible English and added insights gained from studying natural populations, thus allowing the framework of population genetics to be expanded into the rest of biology. In his book, Dobzhansky redefined evolution as changes in allelic frequency.

But Dobzhansky's synthesis was also incomplete. For all the hopes that mathematical models could explain the origin of new species, these analyses really showed only how mutation and recombination could change the predominant phenotype of a population from one form to another. It did not show how speciation occurred. To approach this latter question, Ernst Mayr (1942) provided conceptual tools that defined a species as a real or potentially interbreeding population and showed that speciation could take place if this population were geographically divided or otherwise prevented from interbreeding.

By the 1940s, the major tenets of the Modern Synthesis had been established. These included the premises that all evolutionary phenomena could be explained by the known genetic mechanisms, that natural selection was the overwhelmingly important mechanism of evolutionary change, and that the large anatomical changes seen over the course of evolution were the result of accumulations of smaller changes. George Gaylord Simpson had shown that neo-Darwinism was compatible with paleontology (a premise that many other paleontologists, especially those in Germany, did not agree with), and G. Ledyard Stebbins had brought plant biology into the synthesis. By the time the third edition of Dobzhansky's book was published in 1951, he could state that "the study of mechanisms of evolution falls within the province of population genetics." Macroevolution (evolution at the level of the species or higher taxa) could be explained by microevolution (the evolution of populations within a species). Population genetics had become the core of evolutionary biology, and genetics was seen as "Darwin's missing evidence" (Kettlewell 1959).

During the 1970s, studies showed that sexual selection and polymorphic inheritance (i.e., when different environmental conditions within a species' range select for different phenotypes of that species) also can be explained by the mathematics of the Modern Synthesis. Research on kin selection then brought a mathematical analysis of behavioral biology into the fold as well. Evolutionary biology went from being a phenotype-based science of genealogical relationships to being the epiphenomenon of the gene pools of populations. The Modern Synthesis fully justified Dobzhansky's 1964 declaration that "nothing in biology makes sense except in the light of evolution."

The Triumph of the Modern Synthesis: The Globin Paradigm

But would the Modern Synthesis withstand a new challenge from molecular biology? In the 1960s, groundbreaking discoveries in the field of molecular biology began revolutionizing biology in much the way evolutionary biology had done a century earlier, this time from the "bottom up" rather than the "top down." Like evolution, the field of molecular biology was based on heredity; like evolution, it promised a common language for all biological disciplines; and like evolution, it claimed to be the explanation for life. But while evolution was primarily about ultimate causes (the "why?" questions), molecular biology was about proximate causes (the "how?" questions). Would the two approaches be compatible, or would molecular biology supersede evolution as an explanatory mechanism?

As it turned out, molecular biology was able to integrate the "how" into the "why," and the population genetic theory of allele frequencies was able to integrate the findings of molecular biology. In this reciprocal confirmation, molecular biology provided remarkable evidence for the Modern Synthesis, and evolution provided a new framework for molecular biology.* One of the best examples of this integration was the elucidation of the evolution of malarial resistance in human populations. It is somewhat ironic that, in the modern analysis, some of the best evidence for evolution by natural selection has been found in *Homo sapiens*—a species that fascinated Darwin, but the one he tried to ignore in *Origin*. However, these molecular studies became paradigmatic examples showing how genetic variation and natural selection produce phenotypes adaptive for a particular environment.

---

*One of the areas where the Modern Synthesis was *not* confirmed by molecular biology was the idea that all genetic variation was selected. Molecular biology revealed that individuals and populations harbor enormous amounts of genetic variation that does not appear to be acted on by selection (Lewontin and Hubby 1966; Kimura 1983; Kreitman 1983). At first, neo-Darwinians fought this idea, but the neutral mutation theory became a major part of the evolutionary canon (see Nei 2005) and has been important in documenting phylogenetic relationships and the timing of evolutionary changes (Bromham and Penny 2003). In fact, Darwin had written in *Origin* that some variations would not be affected by natural selection.

In the following examples, the agent of natural selection is malaria. The eukaryotic malarial parasites of the genus *Plasmodium* are some of the most successful predators of human populations (see the next box).

Hemoglobin S and sickle-cell disease

The course of malarial infection has resulted in natural selection favoring certain people whose red blood cells are better able to either resist the entry of *Plasmodium* or block the parasite's reproductive cycle. One of the most important genetic mechanisms for combating *Plasmodium* involves a mutation of the gene for the β-globin chain of human hemoglobin. The prevalence of this mutation in certain populations was a puzzle at first because it can result in mutant "hemoglobin S," deformed red blood cells, and sickle-cell disease, a condition that is more deadly than malaria.

THE GENETIC BASIS OF SICKLE-CELL DISEASE   The story of the human globins, hemoglobin S, malaria, and sickle-cell disease unites evolutionary biology, classical genetics, population genetics, medical genetics, and molecular biology. Sickle-cell disease is characterized by the sickling and loss of flexibility of the normally round and supple red blood cells. Their sickled shape and diminished flexibility cause these cells to obstruct capillaries, preventing blood flow to the areas supplied by the capillaries (Figure 9.10). This causes severe pain to the person and damage to organs. The spleen and the brain, with their narrow capillaries and large flow of blood, are among the organs most severely affected, and the condition is eventually lethal.

Sickle-cell disease was known to be an autosomal recessive trait, so only a person with two copies of the mutant gene (i.e., a homozygote) expresses the disease. In 1949, Linus Pauling discovered that the hemoglobin of patients with sickle-cell disease was abnormal, and shortly thereafter Itano and Neel (1950) showed that this hemoglobin abnormality was inherited as predicted by Mendelian genetics for an autosomal recessive allele. Red blood cells from wild-type individuals had normal hemoglobin (HbA); red blood cells from patients with sickle-cell disease had sickle hemoglobin

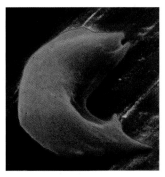
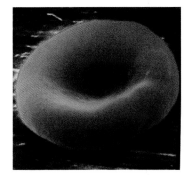

Sickle-cell phenotype   Normal phenotype

Figure 9.10 Sickled and normal red blood cells. The misshapen red blood cell on the left is the result of a mutation that changes the amino acid at a crucial position in the β subunit protein of hemoglobin.

(HbS); and the *parents* of sickle-cell patients had both HbA and HbS hemoglobin in their red blood cells.

In 1957, Vernon Ingram mapped the amino acids that compose hemoglobin and showed that glutamate, the amino acid normally found at position 6 in the β-globin chain, was replaced by valine in sickle hemoglobin.* When information about amino acids and the genetic code became known, Ingram was able to predict the genetic mutation in sickle-cell disease. This was later confirmed by DNA sequencing, making sickle-cell disease the first genetic disorder whose molecular basis was known. The gene defect is the mutation of a single nucleotide from A to T. This mutation in the β-globin gene on chromosome 11 changes a GAG codon to GTG, resulting in the replacement of glutamate by valine at position 6. At low oxygen concentrations (such as in the capillaries), the valine at position 6 forms a hydrophobic bond with a similar region on an adjacent HbS protein, causing hemoglobin S molecules to aggregate, come out of solution, and form fibers that stretch across the red blood cell.

SELECTION FOR AN UNFAVORABLE GENE   In 1933, Diggs and colleagues estimated that about 7.5% of the African population had at least one copy of the *HbS* allele. How could an allele so detrimental be present at such high frequencies? One would expect natural selection to have reduced the frequency of the sickle-cell allele to extremely low levels. Two scientists, J. B. S. Haldane and Anthony Allison, came up with a possible answer: to explain the high frequency of an allele that is deadly when homozygous, the *heterozygous* individuals must have a fitness advantage not only over those with the lethal homozygous condition, but also over "normal" homozygotes. In other words, *HbA/HbS* heterozygotes must be more fit than either *HbA/HbA* or *HbS/HbS* homozygotes.

Haldane, a pioneer of evolutionary biochemistry, found evidence that heterozygosity might indeed be favorable under certain conditions: "The corpuscles of the anaerobic heterozygote are smaller than normal, and more resistant to hypotonic solution. It is at least conceivable that they are also more resistant to attacks by the sporozoa that cause malaria" (Haldane 1949). This was remarkably good thinking. Once a method for culturing the malarial parasite had been perfected, it was found that *P. falciparum* infection increased the rate of sickling and dehydration of blood cells containing a mixture of HbS and HbA proteins. This led to the destruction of the blood cell—and of the parasite within it (Pasvol and Wilson 1980; Lew and Bookchin 2005). Thus it seems that the red blood cells of heterozygotes are not as good incubators for *Plasmodium* larvae as are normal erythrocytes. This would mean that, while *HbS/HbS* individuals are afflicted with sickle-cell disease and *HbA/HbA* "normals" are susceptible to malaria, *HbA/HbS* individuals do not have the disease and are less susceptible to malarial infections.

---

*Hemoglobin is made up of four globin protein subunits, two α-globins and two β-globins. Each of the subunits is complexed to a heme group that is capable of binding oxygen. Sickle-cell disease does not affect the fetus or newborn because while in utero we have γ-globin instead of β-globin. Recent studies (Cyrklaff et al. 2011) have shown that hemoglobin S also interferes with the remodeling of actin by the *Plasmodium* parasites.

While Haldane was looking at the biochemistry of the red blood cells, Allison, doing research in the months between university and medical school, was looking at African populations around Mt. Kenya. He found that different tribes had very different frequencies of the *HbS* allele. In some tribes, as much as 40% of the population was heterozygous for the *HbS* allele, while in others almost no one carried *HbS*. As he later wrote (Allison 2002):

> These observations raised questions; first, if there is strong selection against the sickle-cell homozygote, why is the frequency of the heterozygote high? Second, why is it high in some areas but not others? I formulated an exciting hypothesis; the heterozygotes have a selective advantage, because they are relatively resistant to malaria. This would operate only in areas of intense transmission of *Plasmodium falciparum*, and the selective advantage of the heterozygote would maintain a stable polymorphism. Testing the hypothesis had to wait until I had completed medical studies and received training in parasitology.

And in fact, once he had completed his studies, Allison (1954a) was able to show that heterozygous children were more resistant to induced malarial infections* than children with homozygous wild-type hemoglobin, and that these heterozygous individuals were common only in the regions where *P. falciparum* malaria was prevalent. Indeed, further studies showed that the only places where a high percentage of the population carried the *HbS* allele were those places where malaria was prevalent (Figure 9.11). Such areas included not only most of Africa, but also Greece and southern India.

Population genetics showed that, under normal selection, the *HbS* allele should be eliminated from the gene pool but that widespread presence of the

*Infecting a person with malaria may seem unethical, but at that time inducing malarial infections was a normal treatment for syphilis (see Allison 2002).

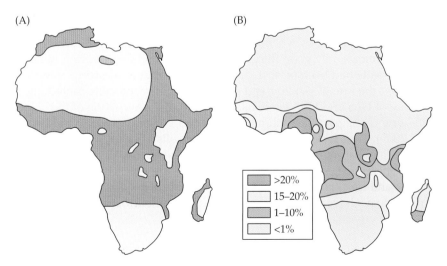

Figure 9.11   The sickle-cell allele (*HbS*) in regions of high malarial distribution. (A) The distribution of the malarial parasite *Plasmodium falciparum* in Africa before mosquito control was introduced. (B) Frequencies of *HbA/HbS* heterozygotes in different parts of Africa. (After Allison 2002.)

(A)

(B)

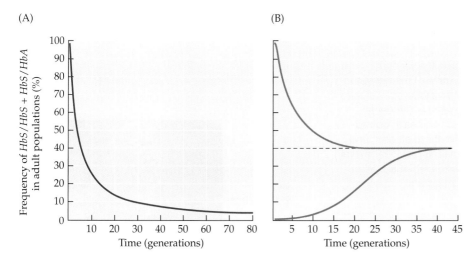

Figure 9.12 Importance of selection in maintaining the *HbS* allele in human populations. (A) The change in sickle-cell heterozygote + homozygote frequency that would occur in the absence of malaria, assuming that the fitness of the sickle-cell homozygote is 0.25 and that of sickle-cell heterozygote and normal homozygote are 1.0. (B) The change in sickle-cell heterozygote + homozygote frequency that would occur from a high or a low level when the fitness of the normal homozygote is 0.95, the fitness of the sickle-cell heterozygote is 1.19, and that of the sickle-cell homozygote is 0.30. The interrupted horizontal line represents an equilibrium frequency of 40% heterozygotes, the highest observed. (After Allison 1954b, 2002.)

malarial parasite favored selection for heterozygous individuals, who were more resistant to the parasite and thus more likely to survive to reproductive age (Figure 9.12; Allison 1954b; Vandepitte et al. 1955; Cavelli-Sforza and Bodmer 1971). Moreover, when the hemoglobin gene and the DNA adjacent to it were sequenced, it was determined that the *HbS* allele has evolved at least five separate times! There are separate GAG → GTG substitutions in the β-globin genes in the Bantu, Senegal, Benin, and Cameroon regions of Africa, and a fifth instance of the allele appears in populations from India. Thus the same mutation appears to have arisen randomly and been selected in at least five instances (Labie et al. 1989; Lapoumeroulie et al. 1992).

Favism

There is another mutation, this time in the gene for an enzyme, that seems to render red blood cells inhospitable to *Plasmodium*. Interestingly, this mutation can also be the cause of a disease, called favism, that manifests in certain environmental circumstances.

The ancient Greek mathematician Pythagoras, best known for his triangle theorem, was also a mystic and cult master who told his disciples that, although they must adhere to a vegetarian diet, they should avoid eating beans. By "beans," Pythagoras almost certainly meant fava beans (*Vicia faba*, sometimes called broad beans), as they were the only cultivated bean in the ancient Mediterranean. What could be the reason behind such a prohibition? There have been many speculations about this question (one of which involved the vegetable's supposed resemblance to the human

Malaria and Evolution

Allelic variation is critical for providing the variation needed for natural selection, and some of the best-studied examples (outside of the extensive laboratory studies of *Drosophila*, *Caenorhabditis elegans*, and rodents) come from human populations. Human health considerations are the primary impetus for these studies, and the widespread tropical disease malaria provides several examples of natural selection by environmental agents. Malaria is endemic in over 91 countries and kills some 2

Life cycle of *Plasmodium falciparum*. At the point marked "START," gametocytes enter the mosquito through infected human red blood cells. They fuse into zygotes and encyst themselves in the mosquito gut, where they develop into sporozoites. Mature sporozoites travel to the insect's salivary gland, from which they are transferred to the human bloodstream, infecting the liver and red blood cells. The micrograph shows encysted *Plasmodium* zygotes (artificially stained blue) covering the stomach wall of an *Anopheles* mosquito.

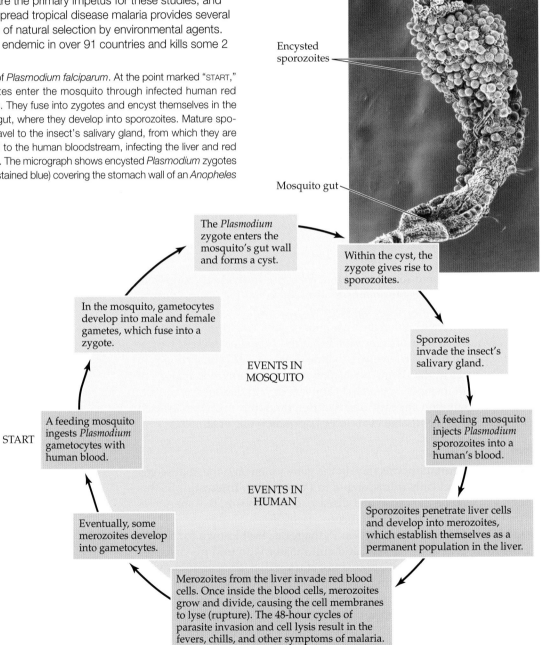

Encysted sporozoites

Mosquito gut

The *Plasmodium* zygote enters the mosquito's gut wall and forms a cyst.

Within the cyst, the zygote gives rise to sporozoites.

In the mosquito, gametocytes develop into male and female gametes, which fuse into a zygote.

EVENTS IN MOSQUITO

Sporozoites invade the insect's salivary gland.

A feeding mosquito ingests *Plasmodium* gametocytes with human blood.

START

A feeding mosquito injects *Plasmodium* sporozoites into a human's blood.

EVENTS IN HUMAN

Eventually, some merozoites develop into gametocytes.

Sporozoites penetrate liver cells and develop into merozoites, which establish themselves as a permanent population in the liver.

Merozoites from the liver invade red blood cells. Once inside the blood cells, merozoites grow and divide, causing the cell membranes to lyse (rupture). The 48-hour cycles of parasite invasion and cell lysis result in the fevers, chills, and other symptoms of malaria.

Malaria and Evolution (*continued*)

million people each year. It remains a powerful selective force on many human populations.*

Malaria in humans is the result of infection by a eukaryotic parasitic protist of the genus *Plasmodium* (notably *P. falciparum*). Plasmodial parasites have a complex life cycle, part of which is spent in the body of an insect vector (i.e., intermediate host), usually a mosquito of the genus *Anopheles*. As seen in the figure, *Plasmodium*

*Malaria has probably been a major factor in the constitution of the southern European gene pool. In 410, the Goths captured Rome, but their victory was thwarted when the conquering army was ravaged by malaria. In 452, when Attila and his Huns pillaged their way to Rome, they were met by Pope Leo I and malaria. While Church historians assure us that the Pope's eloquence convinced Attila not to sack the city, other historians note that malaria may have presented a more convincing argument to the Scourge of God. The Romans paid a high price for this protection. Analysis of bones and DNA from a children's cemetery dating from about 450 shows that falciparum malaria was rampant, and some historians suggest that malaria may have been the most critical factor in why the Roman army there at that time was unable to repel the barbarian invaders (see Sallares 2002).

invades the mosquito's salivary glands and can enter the human bloodstream via the mosquito's bite. There the plasmodia develop into parasites that infect and destroy red blood cells. Eventually some of the parasites in the bloodstream form gametocytes, which mosquitoes ingest when they bite an infected person, and the cycle starts again.

Malaria has not only affected the human gene pool, but has also affected the evolution of the vector (i.e., the mosquito) and the plasmodial parasite. Humans have used vast quantities of pesticides, including DDT (see Chapter 6), in our attempt to rid the world of malaria-infested mosquitoes. (Indeed, today DDT use is legal only in those tropical countries where the public health threat of malaria is greater and more immediate than the threat of the pesticide's toxicity.) But in doing so, we have applied a selective pressure that has resulted in pesticide-resistant mosquito populations. In addition, the drug chloroquine has been effective in destroying *Plasmodium*, but many parasite populations are now resistant to this treatment. It seems that in these populations, mutations have been selected for in a protein that transports the drug (Tran and Saier 2004; Yang et al. 2007).

---

testicle), but it is just possible that the ancient philosophers were aware that these beans could occasionally be lethal.

Fava beans contain high levels of oxidants such as divicine and isouramil that are capable of destroying the membranes of red blood cells. In most people, these compounds are harmless because the blood cells produce ample amounts of the enzyme G6PD (glucose-6-phosphate dehydrogenase). This enzyme is important in making glutathione, a reducing agent that combats environmental oxidants such as those found in fava beans (Jollow and McMillan 2001; Ho et al. 2007). The gene for G6PD is carried on the X chromosome. Men with a mutation in this gene synthesize a mutant enzyme that does not function well. Less glutathione is synthesized, and their red blood cells can break apart when exposed to the compounds released by digested fava beans. (Women have two X chromosomes, and it is very rare that both copies will have mutations.)

This disease of red blood cell destruction is called favism, but favism only manifests as a disease when the diet interacts with the genome. Otherwise, a person with the mutant *G6PD* allele is normal. G6PD deficiency is actually rather common; in some areas of the Mediterranean, about 30% of the male population do not make the G6PD enzyme. Why should such a large percentage of the population contain a gene that could become lethal under commonplace dietary circumstances? Again, natural selection appears to be responsible and, again, malaria appears to be the agent of selection.

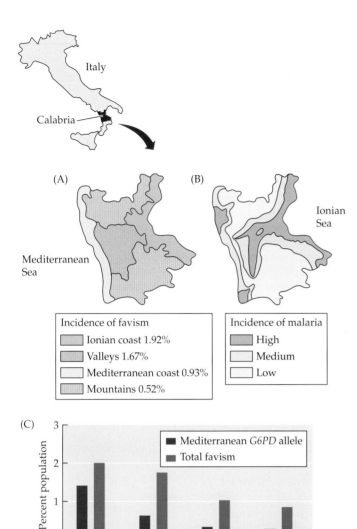

There is an advantage to being G6PD-deficient if one lives in a region where malaria is endemic. Figure 9.13 shows that the prevalence of the Mediterranean allele of *G6PD* (and hence favism) coincides with the areas where malaria has been endemic in a particular area of Italy. The relative lack of the G6PD enzyme and glutathione apparently make it more difficult for *Plasmodium* to reproduce in the red blood cells (see box). Thus, males with low *G6PD* gene activity appear to have an advantage in that they are less likely to die from a severe case of malaria. Even a lessening of severity of the disease could be an important factor in maintaining this mutant allele in the population (Mason et al. 2007).

Sexual Selection

Sexual selection is a subset of natural selection wherein some individuals are better at obtaining mates than others. In other words, the selection is not done directly by competition for resources, but by mate preference. As Darwin noted in 1874, there are two major types of sexual selection: "in the one it is between the individuals of the same sex, generally the males, in order to drive away or kill their rivals, the females remaining passive; while in the other, the struggle is likewise between the individuals of the same sex, in order to excite or charm those of the opposite sex, generally the females, which no longer remain passive, but select the more agreeable partners." The first type is called intrasexual (within-sex) competition; the second type is called intersexual (between-sex) competition.

One example of intersexual competition noted by Darwin was that of the blue-footed booby (*Sula nebouxii*), a bird whose males hunt in coordinated groups but compete individually for females through an elaborate courtship dance where they display their blue-colored webbed feet. The female gets to choose which male to pair with by the quality of his dance and feet.

But the females are extremely choosy, and they know what to seek. They want males whose feet are not dark blue, but bright aquamarine. The only natural way to get aquamarine feet is to be a good hunter. The yellow-orange carotenoid pigment from the fish they eat (which also aids the bird's immune system) is transferred to the feet, turning them aquamarine. Aquamarine feet become, then, an "honest indicator" for the male's

Figure 9.13 Correlation between the areas of favism (A) and malaria (B) in the district of Cosenza in Calabria, Italy. (C) Frequency gradients of favism and of its more frequent gene, Mediterranean *G6PD*. (After Tagarelli et al. 1991.)

being a good hunter.* Females seem to care about who gets to father their broods, because they do not copulate much or lay many eggs with males whose feet suggest they won't be good providers.

Scientists, however, can cause color changes in the male boobies with their own manipulations (Velando et al. 2006). They can give the male boobies food without carotenoid pigments and thus keep their webbing dark blue. They can even take the good hunters and paint their feet with dark blue mascara, obscuring the sign of a good provider (Figure 9.14).

*This notion of the beauty of an organism being an "honest signal" was not something that Darwin proposed. This was Wallace's idea. Darwin felt that the female had a "taste for the beautiful" and made the choice through her "aesthetic capacity." Alfred Wallace (1895), the co-originator of the theory of evolution by natural selection, criticized this. Having spent much of his adult life among the beautiful fauna of the tropics, he felt that beauty was an advertisement that indicated good health and prowess. It was difficult to keep up appearances, and having brightly colored feet or feathers would indicate a healthy body that could afford to produce such extravagant (and otherwise unneeded) phenotypes (Cronin 1991; Prum 2012). The debate continues today.

(A)

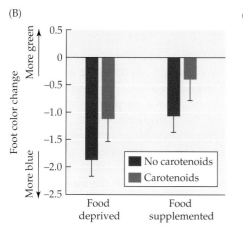

(B)

(C)

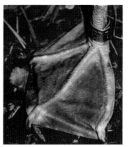

Figure 9.14 Sexual selection in the blue-footed booby. (A) Part of the courtship dance of the male bird, where he prominently displays his feet to a potential partner. (B) Food deprivation causes the foot to have a bluer color than if the bird is well fed, and feeding artificial carotenoids can give a poorly fed bird a greener foot. (C) The feet of a poorly fed male booby (right) compared with one that is well fed (left) give a striking signal to the female. (B,C after Velando et al. 2006.)

The females prefer not to pair with them. The females are so tied to this sign that when the males' feet were colored dark blue later in the breeding season, the females stayed with those males, but laid smaller eggs.

The webbing of the male booby's feet may actually integrate several environmental cues. If the immune system is very active (signaling that the bird is infected), the aquamarine feet of a good hunter turn dark blue again (Torres and Velando 2007). As sperm quality diminishes, apparently with age, so does the attractiveness of the foot webbing (Velando et al. 2011). So the blue color of the foot appears to integrate several components of fitness—hunting prowess, age, and health—into a single critical signal for the female bird. Here we see, too, that the environment (e.g., the amount of colorful lipids from the fish or the load of pathogens in the habitat) is able to change the phenotype of the male such that he becomes more or less attractive to the female. The environment can be literally incorporated into the organism to alter its phenotype, making it more attractive to a mate.

Looking Ahead

The Modern Synthesis linking genetics and evolutionary biology demonstrated (1) that evolution within the species can be modeled mathematically, (2) that genes under selection were the units whose frequencies changed under Darwinian selection, (3) that those organisms possessing genes that allowed them to become fit would produce more fit offspring, and (4) that these offspring had a high likelihood of inheriting the genes that made their parents and grandparents fit. These genes were the material bases of Darwin's inherited traits. Moreover, the change in allelic frequencies predicted by the mathematics of population genetics was confirmed by DNA sequencing, making the Modern Synthesis one of the most successful and important explanatory theories in science.

Evolutionary biology is being strengthened and revised by the revolution in genomics. This has enabled the revision of some of the phylogenetic trees that had been based primarily on morphology and the fossil record. The ability to sequence entire genomes and to see how the DNA of different groups has evolved is giving us a new appreciation of life's inflorescence. For instance, a new group of animals, the Afrotherians, emerged as a result of genomic sequencing. This group includes elephants, golden moles, sea cows, tenrecs, sengis, and aardvarks. While these animals share very few morphological features in common, genomic research, starting in the 1990s, showed that they were related (Stanhope et al. 1998; Springer et al. 2004; Murphy et al. 2007; Nikolaev et al. 2007). Similarly, genomic analyses have finally found the phylogenetic position of turtles—as the sister group to the Archosaurs, the dinosaurs, birds, and crocodilians (see Crawford et al. 2015).

Furthermore, genomic studies have shown surprising and nonintuitive relationships at the "root" of the evolutionary tree. The placement of the lineages at the base of the animal kingdom may finally be resolved (Figure 9.15). The analysis of the genome of ctenophores and other invertebrates strongly suggests that the ctenophore clade (and not the sponges) are the sister group to all other animals (Ryan et al. 2013; Moroz et al. 2014). The sponges have genes to produce nervous systems, even though they currently lack them, indicating that the sponges have

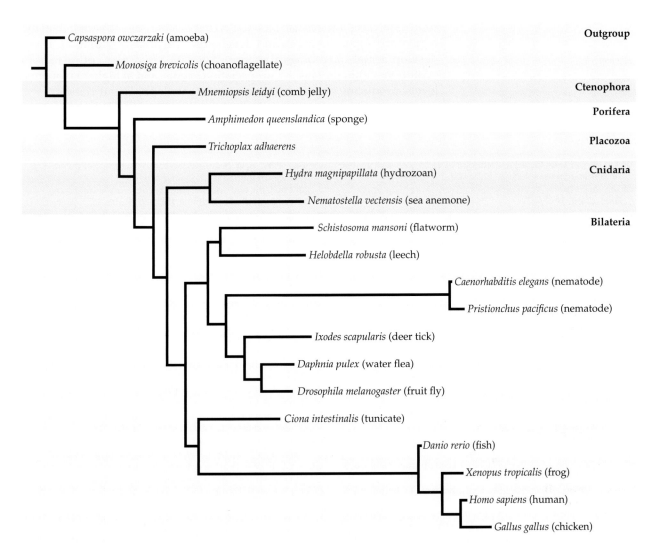

Figure 9.15 A new proposal for the tree of animal life. The maximum-likelihood analysis compared the DNA sequences from numerous organisms and calculated the probabilities of which set was derived from the other sets. In this analysis, the ctenophores are the group that is the sister clade to (i.e., which branched off earliest from) the remainder of the animals. (After Ryan et al. 2013.)

lost their nervous systems (rather than never having evolved them). As more genomes become sequenced (as the price of sequencing genomes has plummeted), one can expect a much fuller knowledge of the ramified branches of the tree of life.

However, the Modern Synthesis is mostly a set of theories about adults competing for reproductive success (i.e., who leaves the most offspring, relative to others). There remains the notion that there is more to evolution than the frequency of alleles within a species. Natural selection can only work on existing variation, and the next two chapters will document that in addition to allelic variation in the structural genes, there are two other sources of variation that can be acted on by selection: allelic variation in the regulatory regions of genes and developmentally plastic variation. These developmental and ecological sources of variation may play critical roles in evolution within the species, in speciation events, and at higher taxonomic levels, producing the diversity we associate with phyla and classes.

References

Allison, A. C. 1954a. Protection afforded by sickle-cell trait against subtertian malaria infection. *Br. Med. J.* 1: 290–294.

Allison, A. C. 1954b. Notes on sickle-cell polymorphism. *Ann. Hum. Genet.* 19: 39–57.

Allison, A. C. 2002. The discovery of the resistance to malaria in sickle-cell heterozygotes. *Biochem. Mol. Biol. Educ.* 30: 279–287.

Andersen, K. H., K. D. Farnsworth, U. H. Thyesen and J. E. Beyer. 2007. The evolutionary pressure from fishing on size at maturation of Baltic cod. *Ecol. Model.* 204: 246–252.

Axelrod R. and W. D. Hamilton. 1981. The evolution of cooperation. *Science* 211: 1390–1396.

Balanyá, J., J. M. Oller, R. B. Huey, G. W. Gilchrist and L. Serra. 2006. Global genetic change tracks global climate warming in *Drosophila subobscura*. *Science* 313: 1773–1775.

Balfour, F. M. 1880–1881. *A Treatise on Comparative Embryology*. (2 vols.) Macmillan, London.

Bateson, W. 1894. *Materials for the Study of Variation: Treated with Especial Regard to Discontinuity in the Origin of Species*. Macmillan, London. (Reprinted 1992, Johns Hopkins University Press, Baltimore.)

Bateson, W. 1900. Problems of heredity as a subject for horticultural investigation. *J. R. Hort. Soc.* 25: 1–8.

Bateson, W. 1922. Evolutionary faith and modern doubts. *Science* 55: 1412.

Berry, A. and J. Browne. 2008. The other beetle-hunter. *Nature* 453: 1188–1190.

Bowler, P. 1996. *Life's Splendid Drama*. University of Chicago Press, Chicago.

Boycott, A. E. and C. Diver. 1923. On the inheritance of sinistrality in *Limnaea peregra*. *Proc. R. Soc. Lond.* B 95: 207–213.

Boycott, A. E., C. Diver, S. L. Garstang and F. M. Turner. 1930. The inheritance of sinistrality in *Limnaea peregra* (Mollusca: Pulmonata). *Philos. Trans. R. Soc. Lond.* B 219: 51–131.

Bromham, L. and D. Penny. 2003. The modern molecular clock. *Nature Rev. Genet.* 4: 216–224.

Cavalli-Sforza, L. L. and W. Bodmer. 1971. *The Genetics of Human Populations*. W. H. Freeman, San Francisco.

Coltman, D. W., P. O'Donoghue, J. T. Jorgensen, J. T. Hogg, C. Strobeck and M. Festa-Blanchet. 2003. Undesirable evolutionary consequences of trophy hunting. *Nature* 426: 655–658.

Crawford, N. G. and 8 others. 2015. A phylogenomic analysis of turtles. *Mol. Phyl. Evol.* 83: 250–257.

Cronin, H. 1991. *The Ant and the Peacock*. Cambridge University Press, Cambridge, UK.

Cyrklaff, M. and 6 others. 2011. Hemoglobin S and C interfere with actin remodeling in *Plasmodium farciparum*–infected erythrocytes. *Science* 334: 1283–1286.

Darwin, C. 1844. *Life and Letters* II: 28.

Darwin, C. 1859. *On the Origin of Species*. John Murray, London.

Darwin, C. 1874. *The Descent of Man, and Selection in Relation to Sex*, 2nd Ed. (2 vols.) John Murray, London.

de Jong, P. W., S. W. S. Gussekloo and P. M. Brakefield. 1996. Differences in thermal balance and activity between non-melanic and melanic two-spot ladybird beetles (*Adalia bipunctata*) under controlled conditions. *J. Exp. Biol.* 199: 2655–2666.

de Jong, P. W. and P. M. Brakefield. 1998. Climate and change in clines for melanism in the two-spot ladybird, *Adalia bipunctata* (Coleoptera: Coccinellidae). *Proc. R. Soc. Lond.* B 265: 39–43.

Diggs, L. W., G. F. Ahmann and S. Bibb. 1933. The incidence and significance of the sickle cell trait. *Ann. Intern. Med.* 7: 769–778.

Dobzhansky, Th. 1937. *Genetics and the Origin of Species*. Columbia University Press, New York.

Dobzhansky, Th. 1951. *Genetics and the Origin of Species*, 3rd Ed. Columbia University Press, New York.

Dobzhansky, Th. 1964. Biology, molecular and organismic. *Am. Zool.* 4: 443–452.

Edwards, A. W. F. 2001. Darwin and Mendel united: The contributions of Fisher, Haldane and Wright up to 1932. In E. C. R. Reeve (ed.), *Encyclopedia of Genetics*. Fitzroy Dearborn, London.

Fisher, R. A. 1930. *The Genetical Theory of Natural Selection*. Clarendon Press, Oxford.

Ford, C. B. and 13 others. 2015. The evolution of drug resistance in clinical isolates of *Candida albicans*. *eLife* 4: e00662.

Garstang, W. 1922. The theory of recapitulation: A critical restatement of the biogenetic law. *Proc. Linn. Soc. Lond.* 35: 81–101.

Gilbert, S. F. 1998. Bearing crosses: The historiography of genetics and embryology. *Am. J. Med. Genet.* 76: 168–182.

Gilbert S. F. 2003. *Developmental Biology*, 7th Ed. Sinauer Associates, Sunderland, MA.

Grant, P. 1986. *The Ecology and Evolution of Darwin's Finches*. Princeton University Press, Princeton, NJ.

Grant, B. S. 1999. Fine tuning the peppered moth paradigm. *Evolution* 53: 980–984.

Grant, B. S., D. F. Owen and C. A. Clarke 1996. Parallel rise and fall of melanic peppered moths in America and Britain. *J. Hered.* 87: 351–357.

Grant, B. S. and L. L. Wiseman. 2002. Recent history of melanism in American peppered moths. *J. Hered.* 93: 86–90.

Haeckel, E. 1866. *Generelle Morphologie der Organismen: Allegeneine Grundzüge der organischen Formen-Wissenschaft, mechanisch begründendet durch die von Charles Darwin reformite Descendenz-Theorie*. Vol. 1. G. Reimer, Berlin, pp. 43–60.

Halanych, K. M. 2004. The new animal phylogeny. *Annu. Rev. Ecol. Evol. Syst.* 35: 229–256.

Halanych, K. M., J. D. Bachelier, A. M. Aguinaldo, S. M. Liva, D. M. Hillis and J. A. Lake. 1995. Evidence from 18S ribosomal DNA that the lophophorates are protostome animals. *Science* 267: 1641–1643.

Haldane, J. B. S. 1949. Disease and evolution. *La Ricerca Scientifica*, Suppl. A 19: 68–76.

Haldane, J. B. S. 1953. Foreword. In R. Brown and J. F. Danielli (eds.), *Evolution: Society of Experimental Biology Symposium 7*. Cambridge University Press, Cambridge, pp. ix–xix.

Hamilton, W. D. 1967. Extraordinary sex ratios. *Science* 156: 477–488.

Haugen, T. O. and L. A. Vøllestad. 2001. A century of life-history evolution in grayling. *Genetica* 112: 475–491.

Hemingway, J. and S. H. Karunaratne. 1998. Mosquito carboxylesterases: A review of the molecular biology and biochemistry of a major insecticide resistance mechanism. *Med. Vet. Entomol.* 12: 1–12.

Ho, H. Y., M. L. Cheng, P. F. Cheng and D. T. Chu. 2007. Low oxygen tension alleviates oxidative damage and delays cellular senescence in G6PD-deficient cells. *Free Radical Res.* 41: 571–579.

Hoso, M., Y. Kameda, S. P. Wu, T. Asami, M. Kato and M. Hori. 2010. A speciation gene for left-right reversal in snails results in anti-predator adaptation. *Nature Commun.* 1: 133.

Huxley, T. H. 1893. *Darwiniana: Collected Essays*, Vol. II. Macmillan, London.

Ingram, V. M. 1957. Gene mutations in human haemoglobins: The chemical difference between normal and sickle-cell haemoglobin. *Nature* 180: 326.

Itano, H. A. and J. V. Neel. 1950. A new inherited abnormality of human hemoglobin. *Proc. Nat. Acad. Sci. USA* 36: 613–617.

Jollow, D. J. and D. C. McMillan. 2001. Oxidative stress, glucose-6-phosphate dehydrogenase, and the red cell. *Adv. Exp. Med. Biol.* 500: 595–605.

Judson, O. 2008. Optimism in evolution. *New York Times* 12 Aug. 2008. www.nytimes.com/2008/08/13/opinion/13judson.html

Kettlewell, H. B. D. 1959. Darwin's missing evidence. *Sci. Am.* 200(3): 48–53.

Kettlewell, H. B. D. 1973. *The Evolution of Melanism*. Clarendon Press, Oxford.

Kimura, M. 1983. *The Neutral Theory of Molecular Evolution*. Cambridge University Press, New York.

Kolbert, E. 2014. *The Sixth Extinction: An Unnatural History*. Henry Holt, New York.

Kowalevsky, A. 1866. Entwickelungsgeschichte der einfachen Ascidien. *Mémoires de L'Académie Impériale des Sciences de St.-Pétersbourg*, VII Série. Tome X: 1–22.

Kreitman, M. 1983. Nucleotide polymorphism at the alcohol dehydrogenase locus of *Drosophila melanogaster*. *Nature* 304: 412–417.

Labie, D. and 11 others. 1989. Haplotypes in tribal Indians bearing the sickle gene: Evidence for the unicentric origin of the β^S mutation and the unicentric origin of the tribal populations in India. *Hum. Biol.* 61: 479–491.

Lapoumeroulie, C. and 9 others. 1992. A novel sickle-cell mutation of yet another origin in Africa: The Cameroon type. *Hum. Genet.* 89: 333–337.

Lankester, E. R. 1877. Notes on the embryology and classification of the animal kingdom: comprising a revision of speculations relative to the origin and significance of the germ layers. *Quart. Rev. Micr. Sci.* 2/17: 399–454.

Lew, V. L. and R. M. Bookchin. 2005. Ion transport pathology in the mechanism of sickle cell dehydration. *Physiol. Rev.* 85: 179–200.

Lewontin, R. C. and J. L. Hubby. 1966. A molecular approach to the study of genic heterozygosity in natural populations. II. Amount of variation and degree of heterozygosity in natural populations of *Drosophila pseudoobscura*. *Genetics* 54: 595–609.

Lillie, F. R. 1898. Adaptation in cleavage. In *Biological Lectures from the Marine Biological Laboratory, Woods Hole, Massachusetts*. Ginn, Boston, pp. 43–67.

Lyons, S. 1999. *Thomas Henry Huxley: The Evolution of a Scientist*. Prometheus Books, Amherst, New York.

Majerus, M. E. N. 1998. *Melanism: Evolution in Action*. Oxford University Press, Oxford.

Martinez-Torres, D. and 8 others. 1998. Molecular characterization of pyrethroid knockdown resistance (*kdr*) in the major malaria vector *Anopheles gambiae* s.s. *Insect Mol. Biol.* 7: 179–184.

Mason, P. J., J. M. Bautista and F. Gilsanz. 2007. G6PD deficiency: The genotype-phenotype association. *Blood Rev.* 21: 267–283.

Mayr, E. 1942. *Systematics and the Origin of Species*. Columbia University Press, New York.

Mayr, E. 1959. Where are we? *Cold Spring Harb. Symp. Quant. Biol.* 24: 1–14.

Mayr, E. 2004. *What Makes Biology Unique? Considerations of the Autonomy of a Scientific Discipline*. Cambridge University Press, Cambridge.

McElreath, R. and R. Boyd. 2007. *Mathematical Models of Social Evolution: A Guide for the Perplexed*. University of Chicago Press, Chicago, p. 82.

Miller, K. R. 2009. Seals, evolution, and the real "missing link." *The Guardian* www.theguardian.com/commentisfree/belief/2009/apr/29/religion-charles-darwin

Miller-Rushing, A. J. and R. B. Primack. 2008. Global warming and flowering times in Thoreau's Concord: A community perspective. *Ecology* 89: 332–341.

Morgan, T. H. 1927. *Experimental Embryology*. Columbia University Press, New York.

Morgan, T. H. 1932a. The rise of genetics. *Science* 76: 261–267.

Morgan, T. H. 1932b. *The Scientific Basis of Evolution*. W.W. Norton, New York.

Moroz, L. L. and 35 others. 2014. The ctenophore genome and the evolutionary origins of the neural system. *Nature* 510: 109–114.

Müller F. 1869. *For Darwin*. (W. S. Dallas, transl.) John Murray, London.

Murphy, W. J., T. H. Pringle, T. A. Crider, M. S. Springer and W. Miller. 2007. Using genomic data to unravel the root of the placental mammal phylogeny. *Genome Res.* 17: 413–421.

Nei, M. 2005. Selectionism and neutralism in molecular evolution. *Mol. Biol. Evol.* 22: 2318–2342.

Nikaido, H. 2009. Multidrug resistance in bacteria. *Annu. Rev. Biochem.* 78: 119–146.

Nikolaev, S. Y., J. I. Montoya-Burgos, E. H. Margulies, NISC Comparative Sequencing Program, J. Rougemont, B. Nyffeler and E. S. Antonarakis. 2007. Early History of Mammals Is Elucidated with the ENCODE Multiple Species Sequencing Data. *PLoS Genet.* 3(1): e2.

Oppenheimer, J. M. 1940. The nonspecificity of the germ-layers. *Q. Rev. Biol.* 15: 1–27.

Oppenheimer, J. M. 1959. An embryological enigma in *The Origin of Species*. In B. Glass, O. Temkin and W. L. Straus, Jr. (eds.), *Forerunners of Darwin 1745–1859*. Johns Hopkins University Press, Baltimore, pp. 292–322.

Ospovat, D. 1981. *The Development of Darwin's Theory: Natural History, Natural Theology, and Natural Selection, 1838–1859*. Cambridge University Press, Cambridge.

Parmesan, C. and G. Yohe. 2003. A globally coherent fingerprint of climate change impacts across natural systems. *Nature* 421: 37–42.

Pasvol, G. and R. J. Wilson. 1980. The interaction between sickle-cell haemoglobin and the malarial parasite *Plasmodium falciparum*. *Trans. R. Soc. Trop. Med. Hyg.* 74: 701–705.

Pauling, L., H. Itano, S. J. Singer and I. C. Wells. 1949. Sickle-cell disease: A molecular disease. *Science* 110: 543.

Peñuelas, J. and I. Filella. 2001. Responses to a warming world. *Science* 294: 793–795.

Pimm, S. L. and 8 others. 2014. The biodiversity of species and their rates of extinction, distribution, and protection. *Science* 344: 1246752.

Provine, W. B. 1986. *Sewall Wright and Evolutionary Biology.* University of Chicago Press, Chicago.

Prum, R. O. 2012. Aesthetic evolution by mate choice: Darwin's really dangerous idea. *Philos. Trans. R. Soc. Lond. B Biol. Sci.* 367: 2253–2265.

Raff, R. A. and T. C. Kaufman. 1983. *Embryos, Genes, and Evolution: The Developmental-Genetic Basis of Evolutionary Change.* Macmillan, New York.

Rostand, J. 1962. *The Substance of Man.* Doubleday, Garden City, New York, p. 12.

Roux, W. 1894. The problems, methods, and scope of developmental mechanics. In *Biological Lectures of the Marine Biology Laboratory, Woods Hole, Massachusetts.* Ginn, Boston, pp. 149–190.

Ryan, J. F. and 17 others. 2013. The genome of the ctenophore *Mnemiopsis leidyi* and its implications for cell type evolution. *Science* 342:1242592.

Sallares, R. 2002. *Malaria and Rome: A History of Malaria in Ancient Italy.* Oxford University Press, Oxford.

Smocovitis, V. B. 1996. *Unifying Biology: The Evolutionary Synthesis and Evolutionary Biology.* Princeton University Press, Princeton, NJ.

Springer, M. S., M. J. Stanhope, O. Madsen and W. W. de Jong. 2004. Molecules consolidate the placental mammal tree. *Trends Ecol. Evol.* 19: 430–438.

Stanhope, M. J. and 7 others. 1998. Molecular evidence for multiple origins of Insectivora and for a new order of endemic African insectivore mammals. *Proc. Nat. Acad. Sci. USA* 95: 9967–9972.

Stern, N. 2006. *Stern Review on the Economics of Climate Change.* Cambridge University Press, Cambridge.

Sturtevant, A. 1923. Inheritance of direction of coiling in *Limnaea. Science* 58: 269–270.

Tagarelli, A., L. Bastone, R. Cittadella, V. Calabro, M. Bria and C. Brancati. 1991. Glucose-6-Phosphate dehydrogenase (G6PD) deficiency in Southern Italy: A study on the population of the Cosenza province. *Gene Geog.* 5: 141–150.

Thomas, C. D. and 9 others. 2004. Extinction risk from climate change. *Nature* 427: 145–148.

Torres, R. and A. Velando. 2007. Male reproductive senescence: The price of immune-induced oxidative damage on sexual attractiveness in the blue-footed booby. *J. Anim. Ecol.* 76: 1161–1168.

Tran, C. V. and M. H. Saier, Jr. 2004. The principle chloroquine resistance protein of *Plasmodium falciparum* is a member of a drug/metabolite transporter superfamily. *Microbiology* 150: 1–3.

Umina, P. A., A. R. Weeks, M. R. Keraney, S. W. McKechnie and A. A. Hoffmann. 2005. A rapid shift in a classic clinal pattern in *Drosophila* reflecting climate change. *Science* 308: 691–693.

UN Convention (United Nations Convention of Biodiversity). 2014. *Global Biodiversity Outlook.* www.cbd.int/gbo1/

Vandepitte, J. M., W. W. Zuelzer, J. V. Neel and J. Colaert. 1955. Evidence on the inadequacy of mutation as an explanation of the frequency of the sickle-cell gene in the Belgian Congo. *Blood* 10: 341.

Velando, A., R. Beamonte-Barrientos and R. Torres. 2006. Pigment-based skin colour in the blue-footed booby: An honest signal of current condition used by females to adjust reproductive investment. *Oecologia* 149: 535–542.

Velando, A., J. C. Noguera, H. Drummond and R. Torres. 2011. Senescent males carry premutagenic lesions in sperm. *J. Evol. Biol.* 24: 693–697.

Wallace, A. R. 1895. *Natural Selection and Tropical Nature,* 2nd ed. Macmillan, New York.

Welsh, J. H. 1969. Mussels on the move. *Nat. Hist.* 78: 56–59.

Wilkins, A. S. 2011. Why did the Modern Synthesis give short shrift to "soft inheritance"? In Gissis S. B. and E. Jablonka (eds), *Transformations of Lamarckism from Subtle Fluids to Molecular Biology,* MIT Press, Cambridge, MA, pp. 127–132.

Williams, N. 2008. Australia fears over climate change gloom. *Curr. Biol.* 18: R633–R634.

Wilson, E. B. 1898. Cell lineage and ancestral reminiscence. In *Biological Lectures from the Marine Biological Laboratories, Woods Hole, Massachusetts.* Ginn, Boston, pp. 21–42.

Wright, S. 1922. Coefficients of inbreeding and relationship. *Am. Natur.* 56: 330–338.

Yang, Z. and 8 others. 2007. Molecular analysis of chloroquine resistance in *Plasmodium falciparum* in Yunnan Province, China. *Trop. Med. Internatl. Health* 12: 1051–1060.

Evolution through Developmental Regulatory Genes

Nature interests me because it's beautiful, complex, and robust. Evolutionary theory interests me because it explains why nature is beautiful, complex and robust.

David Quammen, 2007

A study of the effects of genes during development is as essential for an understanding of evolution as are the study of mutation and that of selection.

Julian Huxley, 1942

The Modern Synthesis explained the genetic mechanisms of natural selection, but it did not provide an adequate explanation for the origin of the anatomical variations that natural selection selects. To explain the mechanisms of how, and under what circumstances, complex phenotypes come into existence, the Modern Synthesis needs to be supplemented with a theory of how the construction of bodies can change (Amundson 2005; Moczek et al. 2015). This is the next layer of evolutionary theory. Indeed, without the integration of developmental genetics and developmental plasticity, evolutionary biology has no complete theory of variation. In 1964, Theodosius Dobzhansky proclaimed, "Nothing in biology makes sense except in the light of evolution." In 2006, evolutionary developmental biologist Jukka Jernvall paraphrased that famous statement, saying, "Nothing about variation makes sense except in the light of development."

For instance, when confronted with the question of how arthropod body plans arose, Hughes and Kaufman (2002) stated:

> To answer this question by invoking natural selection is correct—but insufficient. The fangs of a centipede... and the claws of a lobster accord these organisms a fitness advantage. However, the crux of the mystery is this: From what developmental genetic changes did these novelties arise in the first place?

In other words, while the Modern Synthesis could explain the *survival* of the fittest, it could not explain the *arrival* of the fittest.* For that, one needed a theory of body construction and its possible changes, a theory of developmental change.

This chapter will discuss the developmental genetic sources of variation and how they function in evolution; Chapter 11 will focus on the evolutionary importance of the variation derived from developmental plasticity, epialleles, and developmental symbiosis (i.e., the subjects of Chapters 1–3). Without an account of development, there can be no complete theory of variation; and without an adequate theory of variation, there can be no adequate explanation of evolution that embraces changing genes and changing bodies. This account of evolution, seeing evolution as the product of heritable changes in development is called **evolutionary developmental biology**.

The Origins of Evolutionary Developmental Biology

How does one bridge a theory of changing genes with a theory of evolving bodies? Perhaps the first stage is the realization that one needs to do so. This realization was proclaimed in a well-publicized 1975 research paper by Mary-Claire King and Alan Wilson. Entitled "Evolution at Two Levels in Humans and Chimpanzees," this study showed that despite the obvious anatomical differences between chimps and humans, their DNA was almost identical. Indeed, some scientists would opine that the paucity of differences places us as the third species of chimpanzee (Goodman 1999; Diamond 2002). In other words, the theory of evolving bodies and the theory of genetic change were not lining up. The proposed solution was elegant and harkened back to those developmental biologists such as Conrad Waddington (see Appendix C), who felt that evolutionary change was predicated upon specific changes in gene regulation. Specifically, the King and Wilson paper states:

> The organismal differences between chimpanzees and humans would... result chiefly from genetic changes in a few regulatory systems, while amino acid substitutions in general would rarely be a key factor in major adaptive shifts.

That is to say, the allelic substitutions of the genes that encode protein sequences—which seem to be pretty much the same for chimps and

---

*The term "arrival of the fittest" is very apt, since it was one of the weakest points of the Modern Synthesis. Darwin alludes to this problem in the *Origin of Species*, noting that "characters may have originated from quite secondary sources, independently from natural selection." He discusses this problem at length in his book on variation, noting that natural selection must have something from which to select and that the origin of variation appears to be internal to the organism and not a consequence of environmental factors. The term "arrival of the fittest" has been coined several times, but one of its earliest uses was by Arthur Harris, who in 1904 wrote that "natural selection may explain the survival of the fittest, but it cannot explain the arrival of the fittest."

humans—are not what is important. The important differences are where, when, and how much the genes are activated.

In 1977, four publications further paved the way for evolutionary developmental biology. These publications were Stephen J. Gould's *Ontogeny and Phylogeny*, François Jacob's "Evolution and Tinkering," and two papers concerning the techniques for DNA sequencing. In *Ontogeny and Phylogeny*, Gould demonstrated how the German biologist Ernst Haeckel had overstepped his data and had built a developmental theory of evolution that could not stand up to scrutiny. Indeed, the first half of Gould's book exorcised Haeckel's ghost so that some other model of evolution and development could be substituted for Haeckel's famous (some would say infamous) biogenetic law wherein "ontogeny recapitulates phylogeny." Jacob's paper suggested a new model, generalizing the work of King and Wilson and focusing attention on how the same genes can create new types of body plans by altering their expression pattern during development. This could become a testable theory if the genetic sequences could be read, and papers by Maxam and Gilbert (1977) and by the Sanger laboratory (Sanger et al. 1977) made this possible. Over the next 30 years, it became possible to study gene expression and gene sequence together. Recombinant DNA technology, polymerase chain reaction, in situ hybridization, and high-throughput RNA analysis enable us to look at gene sequences and to compare their expression between species. This ability has revolutionized the way we look at evolutionary processes. The new science of **evolutionary developmental biology** ("**evo-devo**") is looking at the mechanisms by which changes in gene regulation cause changes in anatomy. It looks at the mechanisms of tinkering.

Evolutionary developmental biology attempts to explain biodiversity in terms of developmental trajectories. It seeks to discern how changes in development create evolutionary novelties, how development might constrain certain phenotypes from arising, and how developmental mechanisms themselves evolve (Raff 1996; Hall 1999; Gilbert and Burian 2003; Carroll et al. 2005; Müller 2007; Moczek et al. 2015). As we will see in the next chapter, it also looks at the roles that environments can play in regulating developmental processes and how organisms may have evolved to incorporate environmental signaling.*

It used to be thought that changes in development would be detrimental to the organism because they would throw a finely honed harmonious system out of synchrony. Evolutionary developmental biologists have shown that there are three conditions that circumvent this problem and allow evolution to occur through changes in development: The first two, molecular parsimony and modularity, will be discussed here. The third condition, robustness and the adaptability of developmental processes,

---

*There are many variants of evolutionary developmental biology. The one that we are focusing on in this chapter is one where developmental genetics predominates. This is being emphasized because it links the genetics of the Modern Synthesis on the one hand with physiological ecology on the other. (For other concerns and foci of evolutionary developmental biology, see Odling-Smee et al. 2003; Kirschner and Gerhart 2005; Arthur 2004; Laubichler 2013; Wagner 2014; Chapter 11.)

was discussed in Chapter 4 and will be revisited in Chapter 11. Together, these processes allow the generation of diversity through changes in gene regulation.

Combining Evo-Devo and the Modern Synthesis: Natural Selection Working on Developmental Variation

As mentioned in the previous chapter, biologists have identified "microevolution" with the allelic differences within a species, whereas "macroevolution" concerns the origin and maintenance of biodiversity: how bats got their wings, how turtles got their shells, how limbs emerged from fish fins, and so forth. To explain macroevolution, one needs to understand changes in development. As one paper (Gilbert et al. 1996) tried to explain it:

Functional biology = anatomy, physiology, gene expression

Development = δ (anatomy, physiology, gene expression)$/\delta t$

Evolution = δ (development)$/\delta t$

In other words, development forms the needed transition between functional biology and evolutionary biology (i.e., to go from functional biology to evolution without development is like going from displacement to acceleration without velocity).

If this is so, one should be able to find examples *within* a species where changes in development cause selectable phenotypes (Shbailat and Abouheif 2013; Nunes et al. 2013). These should be discoverable using the new techniques of population genetics, developmental genetics, and genome-wide association studies (GWAS). Population genetics can identify which regions of the genome are associated with the presence or absence of a particular trait, GWAS can scan for specific nucleotide differences at loci associated with the trait, and developmental genetics can validate whether that candidate gene is a causative factor for producing that phenotype.

This combination of DNA sequencing and developmental genetics has uncovered one of the most important principles of evolution: **enhancer modularity**. Indeed, enhancer modularity makes evolution possible through genetic changes in developmental processes.

Enhancer modularity

One of the crucial insights of evolutionary developmental biology is that not only are the *anatomical* units modular (such that one part of the body can develop differently than the others), but the DNA regions that form the *enhancers* of genes are modular (see Davidson 2006; Gilbert 2013). The modularity of development is determined at least in part by the modularity of the gene enhancers. Each enhancer element allows the gene to be expressed in a different tissue or at a different time. As described in Chapter 2, the *Pax6* gene is expressed in the pancreas, the neural tube, the retina, and the cornea and lens. For each of these tissues, there is an enhancer element in the regulatory region of the *Pax6* gene (Figure 10.1A), so *Pax6*

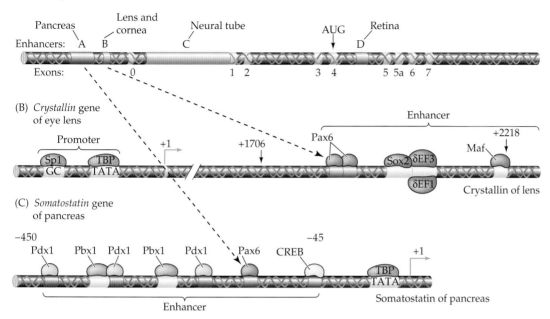

Figure 10.1 Modularity of enhancer regions. (A) The *Pax6* gene is expressed in the pancreas, lens/cornea, neural tube, and retina. There are specific regions of DNA that bind transcription factors in each of these primordia that act to allow the transcription of *Pax6* gene in these tissues. This is the "OR" condition. Pax6 is, itself, a transcription factor, and it binds to modular enhancers in genes that are expressed in the pancreas and eye. (B) In the eye Pax6 works primarily with Maf and Sox2 to initiate the transcription of crystallin in the lens. (δEF1 is an inhibitory transcription factor that can be blocked by δEF3. Sp1 and TBP are ubiquitous activators of transcription.) (C) In the pancreas it works with Pbx1 and Pdx1 transcription factors to activate numerous pancreatic genes. (After Gilbert 2006a.)

can be independently activated in each separate tissue type (the Boolean "OR" condition).* Within each of these enhancer elements are DNA binding sites that are recognized by transcription factor proteins. Often, several transcription factors are needed to bind in order to activate a gene (the Boolean "AND" function). For instance, the *Pax6* gene encodes a transcription factor. The Pax6 transcription factor that is produced in the lens can interact with other transcription factors (such as Maf and Sox2) to activate the transcription of the *crystallin* gene that encodes the transparent protein

*The Boolean logic terms used here can be understood conventionally, in that the "AND" cases demand that transcription factors A, B, and C all must be present for the gene to be activated. This is what happens within an enhancer unit. In the "OR" conditions, either A or B or C will function to activate the enhancer. Usually, within an enhancer element (say, for expressing the gene in the cornea), A, B, and C have to be present. Moreover, the same gene may be able to be expressed in the neural tube. Here, the enhancer element is activated by the "AND" conditions of R, S, and T. However, the "OR" condition applies *between* these enhancer elements; thus the gene can be expressed in those tissues having the A, B, AND C transcription factors (i.e., in the cornea) OR the R, S, AND T transcription factors (in the neural tube).

of the lens (Figure 10.1B). The Maf transcription factor is expressed in the presumptive lens cells only after these cells receive Fgf8 (a paracrine factor) from the neighboring optic cup. In this way, lens development is induced only in the head ectoderm (expressing the *Pax6* gene) that is also in contact with the optic cup (expressing *Fgf8*). Similarly, in the pancreas, Pax6 protein acts with other transcription factors (Pdx1, Pbx1) to activate the *somatostatin* gene of the islet cells (Figure 10.1C).

The modularity of enhancer elements allows particular sets of genes to be activated together and permits the same gene to become expressed in several discrete places. Thus, if a particular gene loses or gains a modular enhancer element, the organism containing that particular allele will express that gene in different places or at different times than those organisms retaining the original allele. Thus there are "regulatory alleles" of various genes. This has two critically important consequences for evolution. First, this mutability can result in the development of different anatomical and physiological morphologies (Sucena and Stern 2000; Shapiro et al. 2004; McGregor et al. 2007). Second, this means that a mutation affecting a particular protein does not have to affect it in all the places it's expressed. One can lose a particular protein's expression in the limbs, but not in the gut, for instance. Whereas a structural loss-of-function mutation (in the protein-encoding sequence) would knock out this protein's function everywhere it was expressed, a regulatory loss-of-function mutation could affect only a particular organ, leaving the others alone. We will now look at some examples.

LE FIN DU FIN   The evolutionary importance of such enhancer modularity has been dramatically demonstrated by the molecular analysis of evolution in three-spined stickleback fish (*Gasterosteus aculeatus*). Freshwater sticklebacks evolved from marine sticklebacks about 12,000 years ago, when marine populations colonized the newly formed freshwater lakes at the end of the last ice age. Marine sticklebacks have a pelvic spine that serves as protection against predation, lacerating the mouths of those predatory fish who try to eat the sticklebacks (Figure 10.2A). Freshwater sticklebacks, however, do not have pelvic spines (Figure 10.2B). This may be because freshwater species lack the piscine predators that the marine fish face but must deal instead with invertebrate predators that can easily capture them by grasping onto such spines. Thus, a pelvis *without* lacerating spines evolved in the freshwater populations of this species.

To determine which genes might be involved in this difference, researchers mated individuals from marine (with spines) and freshwater (without spines) populations. The resulting offspring were bred to each other and produced numerous progeny, some of which had pelvic spines and some of which didn't. Using molecular markers to identify specific regions of the parental chromosomes, Shapiro and coworkers (2004) found that the primary gene for pelvic spine development mapped to the distal end of chromosome 7. That is to say, nearly all the fish with pelvic spines inherited this "hindlimb-encoding" chromosomal region from the marine parent, while fish lacking pelvic spines inherited this region from the freshwater parent. Shapiro and coworkers then tested numerous candidate genes (e.g., genes known to be present in the hindlimb structures of mice) and found that the gene encoding transcription factor Pitx1 was located on this region of chromosome 7.

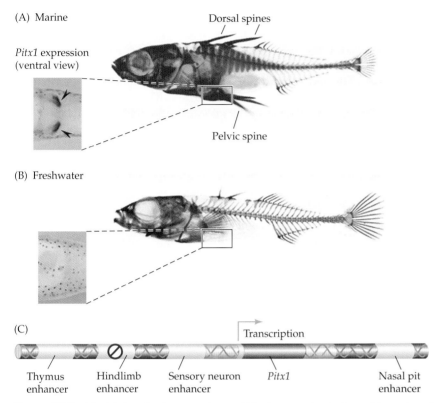

(A) Marine

Dorsal spines

Pitx1 expression
(ventral view)

Pelvic spine

(B) Freshwater

(C)

Transcription

| Thymus enhancer | Hindlimb enhancer | Sensory neuron enhancer | *Pitx1* | Nasal pit enhancer |

Figure 10.2 Modularity of enhancers. Loss of *Pitx1* expression in the pelvic region of freshwater three-spined sticklebacks. Bony plates cover more of the marine three-spined sticklebacks (A) than freshwater sticklebacks (B), and the marine forms have a prominent pelvic spine. The pelvic regions of the marine and freshwater populations show differences in *Pitx1* expression that can be readily observed at higher magnifications (from the area enclosed by the dashed lines). (C) Model for the evolution of pelvic spine loss in the freshwater three-spined stickleback. Four enhancers are postulated to reside near the coding region of the *Pitx1* gene. The enhancers direct the expression of this gene in the thymus, pelvic spine, sensory neurons, and nose, respectively. In the freshwater populations of this species, the pelvic spine (hindlimb) enhancer module has been mutated so that it fails to function. (Photographs courtesy of D. M. Kingsley.)

When they compared the amino acid sequences of the Pitx1 protein between marine and freshwater sticklebacks, there were no differences. However, there was a critically important difference when they compared the *expression patterns* of *Pitx1*. In both species, *Pitx1* was expressed in the precursors of the thymus, nose, and sensory neurons. However, in the marine species, *Pitx1* was also expressed in the pelvic region, whereas pelvic expression of *Pitx1* was absent or severely reduced in the freshwater populations (Figure 10.2C). Since the coding region of *Pitx1* was not mutated (and since the gene involved in the pelvic spine differences maps to the site of the *Pitx1* gene, and the difference between the freshwater and marine species involves the expression of this gene at a particular site), one reasonable hypothesis is that the *enhancer region* allowing *Pitx1* to be expressed in the pelvic area no longer functions in the freshwater species. The pelvic spine enhancers have been identified and sequenced, and they differ between the freshwater and marine forms, enabling the pelvic spines to be made in the marine fish (Chan et al. 2010).

Interestingly, the loss of the pelvic spines in other stickleback species appears to have been the result of independent losses of this *Pitx1* expression domain. The loss of hindlimbs in manatees may also be accounted for by the lack of pelvic *Pitx1* (Shapiro et al. 2006). This finding suggests that if the loss of *Pitx1* expression in the pelvis occurs, this trait can be readily selected (Colosimo et al. 2004). Thus we see that by combining population genetics approaches with developmental genetics approaches, one begins to understand the mechanisms by which evolution can occur.

Comparative developmental studies of the insect eye (Oakley and Cunningham 2002), stickleback fish armor plates and spines (Colosimo et al. 2004, 2005), and avian and *Drosophila* pigment patterns (Gompel et al. 2005; Mundy 2005) reinforce the fact that parallel evolution can result from the independent recruitment of similar developmental pathways by different organisms. In addition to selective pressures playing a role in such parallel evolution (see Schluter 2000; Meehan and Martin 2003), we can now identify intrinsic developmental factors for producing such parallel variation. Such parallel evolution was once the justification for the "creativity" of natural selection. Now we can see that development is what is creative (Gilbert 2006).

MALARIA, AGAIN   Enhancer modularity can be seen in human evolution as well. In Chapter 9 we discussed at length the case of malaria and how a mutation in the gene that encodes the β-globin chain of hemoglobin provides resistance against the malarial agent *Plasmodium falciparum*. But not all malaria comes from this species of *Plasmodium*. A relative, *P. vivax*, causes about 75 million cases of malaria each year. This form of the disease is not as lethal as falciparum malaria, but it can be incapacitating, with severe pain, diarrhea, and fever. Some human populations that have in constant exposure to *P. vivax* have evolved immunity to this protist. This ability to resist *P. vivax* infection is due to their red blood cells' lacking a protein, the Duffy glycoprotein, that *P. vivax* needs to attach itself to the host's red blood cells. The Duffy glycoprotein is probably one of several receptors for interleukin-8 (IL-8, a paracrine factor involved in the circulatory system), and it is found on the cells of the Purkinje neurons, veins, and red blood cells. People who lack Duffy glycoprotein on their red blood cells still have it on their veins and Purkinje neurons. So if one asks, "Why don't these people have Duffy glycoprotein on their red blood cells?" the ultimate answer is probably that the lack of the Duffy glycoprotein was selected for in these populations because it gives these people resistance to vivax malaria, given that *Plasmodium* uses the red blood cell protein to infect the cell. The proximate answer is that the lack of the Duffy glycoprotein is caused by a mutation in the enhancer, a C → G substitution at position −36, which prevents the binding of the GATA1 transcription factor present in red blood cell precursors (Tournamille et al. 1995). Thus, the mutation can be selected because it blocks only one of the enhancers (the one for its expression in red blood cells), while permitting the enhancers that allow the gene's expression in veins and Purkinje neurons to work.

This looking at microevolution through the eyes of developmental regulation has been remarkably successful, especially now that genomic sequencing can be accomplished relatively easily and cheaply. We will next look at two sets of examples, phenotypes that have long fascinated biologists: how hair colors and patterns are formed and selected, and how one kind of animal can mimic another animal's phenotype.

HAIR COLOR AND PATTERNING   A proper theory of evolution must explain both the origin and the maintenance of variation. That is to say, the arrival of the fittest and the survival of the fittest must both be accounted for. One of the functions that developmental biology can perform is to show how the genes selected by natural selection work to generate the selectable variants within a population. Now that genes and genomes can be sequenced, we have the privilege of seeing evolution at work at the molecular level, thus complementing earlier observations of evolution occurring at the morphological level. One of the most complete series of studies (begun by Francis Sumner in the 1920s and currently being continued by Hopi Hoekstra) involves the pigmentation colors and patterns of the oldfield mouse (*Peromyscus polionotus*). This field mouse inhabits the dark loamy soil of Alabama farmlands and the white sandy beaches of Florida's Gulf coast. Using genetic analysis, these mice were shown to be part of one large, interbreeding population (Mullen and Hoekstra 2008). And as would be expected, the mice living on the beaches tended to be much lighter in pigmentation than their relatives living by the plowed fields. Vignieri and colleagues (2010) tested the hypothesis that their color afforded the mice protection in each of these environments. They placed colored clay models in the various environments and then recorded the number of predation events (the number of models that showed signs of attack by a predator). Predators found the discordant models (sandy-colored mouse model on brown soil; brown-colored model on the beach) much more readily than they found those having colors that matched their backgrounds. The results support the hypothesis that color afforded these mice protection against predators (Figure 10.3), and documented that natural selection has a strong role in defining the color and patterns of pigmentation.

A gradual change from one phenotype to another in different parts of the geographic range of a species is called a **cline** (see Haldane 1948). Typically, clines are the results of genetic heterogeneity within the range of the species. Hoeckstra's group showed that the cline in oldfield mouse pigmentation arose, in large part, from differences in the allele frequencies of the *Agouti* gene. Two *Agouti* alleles, both in enhancer regions of the gene, not in the region for the structural protein, showed clinal variation along with the phenotype. There was a "light" allele that was usually homozygous (found in both copies) in individuals of the beach populations, there was a "dark" allele that was usually homozygous in the inland populations, and there was a region between these extreme situations where many of the individual mice were heterozygotes (having one of each allele).

The *Agouti* gene (which is also involved in other organs to regulate fat deposition), is a major gene controlling the color and pattern of mammalian pigmentation. One of its functions is to inhibit the migration of melanocytes

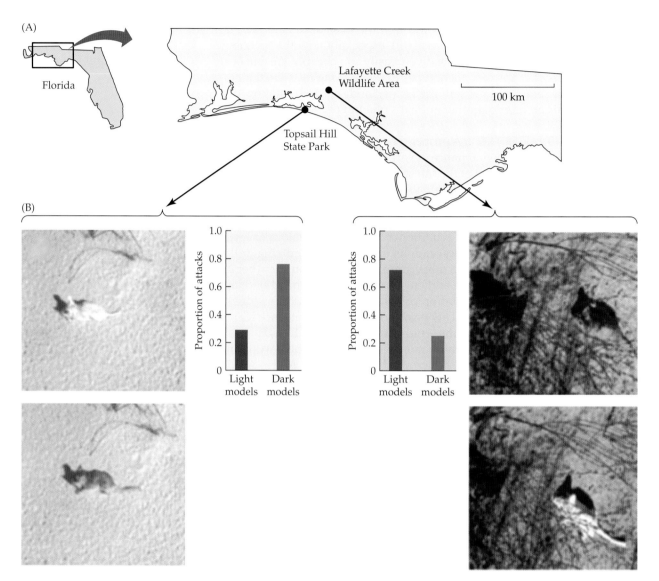

Figure 10.3 Predation on dark and light clay models of *Peromyscus polionotus* on dark and light soils. (A) Location of experiments in dark inland habitat and light beach habitat in Florida. (B) "Bird's eye views" of light and dark models on soil at dark and light locations. The graphs show the corresponding levels of predation events occurring in each category. The light color gave protection on light soil; the dark color gave protection on dark soil. (After Vignieri et al. 2010.)

(pigment cells) from the skin's dermis into the epidermis and hair shafts.* As a result, those regions expressing *Agouti* have white fur, because the pigment cells do not reach the hair. In those mice living in the beach, the

---

*The Agouti protein helps control the color of the hair shaft, as well as its pattern. There are "light" and "dark" mutations in other enhancers of this gene that contribute to color variations. In a related species of *Peromyscus*, the color mutations were also found to be selected by visually hunting predators (Linnen et al. 2013).

ventral region of *Agouti* expression has expanded dorsally, leaving only a band of pigmentation in the dorsal region (Figure 10.4A), while the darker mice (Figure 10.4B) have a smaller dorsal region of *Agouti* expression, enabling the dorsal pigmentation to extend far down the trunk and face of the mouse (Manceau et al. 2011). Thus, the oldfield mice have evolved their protective coloration by establishing different allele frequencies of a gene that will allow their ventral fur to be dark or light.

The hypothesis that the alleles of the *Agouti* gene establish the pigmentation pattern during development could be tested further by looking at other

(A) Beach

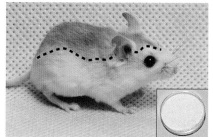

Adult and soil color

Melanocyte migration

Dorsal

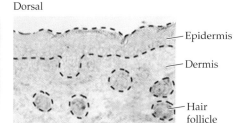

Epidermis

Dermis

Hair follicle

Ventrolateral

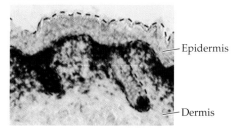

Epidermis

Dermis

Agouti expression

(B) Mainland

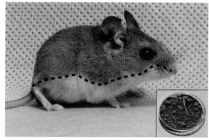

Adult and soil color

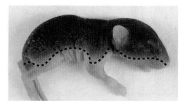

Melanocyte migration

Dorsal

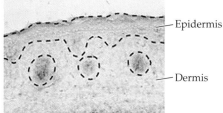

Epidermis

Dermis

Ventrolateral

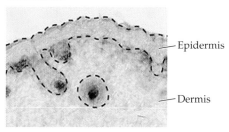

Epidermis

Dermis

Figure 10.4 Beach mice and mainland *Peromyscus polionotus* differ in coat color pattern, which provides camouflage in their respective habitats. (A) In the mice inhabiting beach areas (the soil can be seen in the insert), the nonpigmented region expands dorsally, making the mice whiter. This can be seen in dorsal expression of the *Agouti* gene in the newborn pup (center). The expression of the *Agouti* gene is seen in the ventral lateral dermis of the late embryo. (B) In the mice occurring in the darker soils (insert), the dark hair extends more ventrally and there is significantly less *Agouti* gene expression in the ventrolateral dermis. (After Manceau et al. 2011, photographs courtesy of H. Hoekstra.)

examples of this gene in this species. For example, there exists a naturally occuring strain of this species, where the *Agouti* gene has been lost. As a result, there is no patterning, and the entire mouse is black (like some laboratory mice). This observation confirms that the *Agouti* gene is necessary for establishing the dorsal and ventral pigment pattern. Moreover, if *Agouti* genes on viral promoters (i.e., active *Agouti* genes) are infected into the early embryos of the farmland *Peromyscus*, the melanocytes don't enter the hair shafts and the mice have light pigmentation. Thus, natural selection works on the variants of the *Agouti* gene already in the population. It allows those that have the regulatory alleles that give them protection to survive predators in their particular environment. Here, we also see how changes in gene expression during *development* can cause selectable changes in the adult phenotype and lead to evolution. And we see that there are regulatory alleles—differences in the enhancer regions of the gene—that confer such differences.

In humans, hair-patterning genes (explaining why we don't have hair patterns like apes, for instance) have not yet been discovered. However, evolutionary developmental biology has given us some fascinating clues about hair color. By combining genome-wide association databases, transcription factor interaction databases, and experimental analysis of recombinant mice, the molecular basis for the classic blond hair of northern Europeans has been elucidated. This blond hair is due to a mutation in the enhancer of the *KIT ligand* (also called the "steel factor") gene. KIT ligand is a paracrine factor necessary for the continued growth of melanocytes and the production of melanin pigment (Figure 10.5; Guenther et al. 2014). In blonds, there is an A → G mutation in the enhancer that binds the Lef1 transcription factor. This change decreases the binding and reduces, but does not eliminate, KIT ligand production, leading to lower amounts of pigment in the hair follicle. Importantly, this enhancer is only for the hair

(A)

(B)

Figure 10.5 *KIT ligand* gene. (A) College students expressing the wild-type and blond variants of the *Kit ligand* gene. (B) Mice carrying a regulatory mutation upstream of the *KIT ligand* gene show reduction in melanin. The homozygous wild type is at left, the homozygous recessive is at right, and a heterozygote (with slightly lighter pigmentation than the wild type) is in the center. (B from Gunther et al. 2014, courtesy of C. Guenther.)

pigmentation, so light-haired people with brown eyes and dark-haired people with blue eyes are not uncommon.

Mimicry: The convergence of phenotypes

Some of the most important breakthroughs in evolution immediately after Darwin's *Origin* were made by field naturalists for whom the evolutionary theory solved many mysteries (Punnett 1915; Carroll 2009). In the Amazon River rain forest, Wallace's traveling companion, Henry Bates (1862), found evidence for what would later be called "Batesian mimicry." In his words: "Members of a palatable species (the 'mimic') gain a degree of protection from predators by resembling an unpalatable species." For instance, bird predators would eat an awful-tasting butterfly, and, not wanting to ever eat anything like it again, would avoid butterflies that looked like the noxious one. If a tasty butterfly were to evolve a wing color pattern similar to that of the noxious butterfly, it would gain a great protective advantage. (The trick for the mimic is not to be too numerous, or else it ruins the game for everyone.) In other words, there would be a convergence of phenotypes—the phenotype of the mimic would evolve toward that of a noxious model.

The mechanisms through which these phenotypes converged were unknown then, and there were other puzzling observations that needed to be explained. For instance, in most species with Batesian mimicry, only the female forms converged, but not the male. Indeed, the male form stayed the same over a wide range, but female wing pattern was free to converge on the wing patterns of different poisonous species in different parts of their range. Figure 10.6 shows the polymorphism of wing pattern in the tropical swallowtail *Papilio polytes*. The male form resembles the non-mimetic female form (called the *cyrus* form), while the mimetic female forms look like distantly related, toxic butterflies of

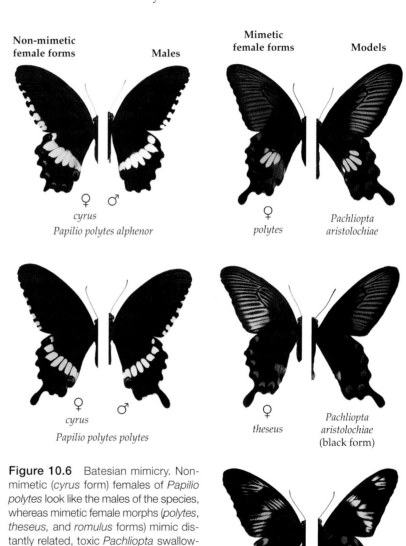

Figure 10.6 Batesian mimicry. Non-mimetic (*cyrus* form) females of *Papilio polytes* look like the males of the species, whereas mimetic female morphs (*polytes*, *theseus*, and *romulus* forms) mimic distantly related, toxic *Pachliopta* swallowtails. (After Kunte et al. 2014, courtesy of M. Kronforst.)

the genus *Pachliopta*. It wasn't until 2014 that these observations began to be explained. Then it was shown (by sequencing numerous entire genomes of genetically backcrossed lines) that the inheritance of the wing pattern was due to the expression of the *Doublesex* gene in these wings. *Doublesex* is a gene involved in sex determination, and in females its expression distinguishes those female-specific traits as being different from those found in males. In females of *Papilio polytes*, the expression of *Doublesex* maps out the contours of the mimetic color patterns (Kunte et al. 2014). Different regulatory alleles of *Doublesex* give different patterns of pigment expression.

Another form of mimicry is named after its discoverer, Fritz Müller, a brilliant and iconoclastic physician who fled Germany following the political upheaval of 1848 to become a field naturalist in Brazil (West 2003). He discovered a type of mutualistic mimicry (Batesian mimicry is essentially parasitic) where different toxic species would all converge on the same phenotype. A predator having a bad experience with one member of any species would avoid the others, as well. As Figure 10.7 shows, the convergences between species can be remarkable, fooling not only predators, but

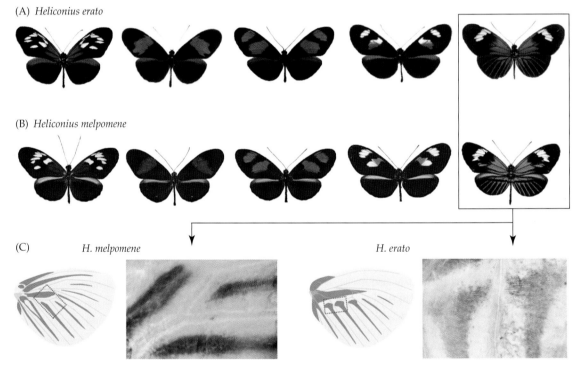

Figure 10.7 Müllerian mimicry. The wing patterns of two unpalatable butterfly species, *Heliconius erato* (A) and *Heliconius melpomene* (B), have evolved to have remarkably similar wing color patterns and shapes in different parts of their ranges. DNA studies suggest that these species generated similar patterns independently, showing convergent evolution; several species share the same color-producing genes. (C) The hindwing imaginal discs of *H. erato erato* (right) and *H. melpomene maletti* (left) show the expression of *optix* gene expression (for red pigmentation) in the two species. (After Reed et al. 2011, photographs courtesy of A. Martin and R. Reed.)

unwary humans as well. *Heliconius erato* and *H. melpomene* have undergone parallel changes throughout their habitats. When one has changed, so has the other.

The molecular bases for these convergences can be mapped to enhancer differences in the genes controlling the placement of wing pigments. The enhancer for the WntA paracrine factor causes the expression of this gene in the areas that are going to become black, and the enhancer for the *optix* gene controls the regions that will be expressing the red pigment. Remarkably, the convergent evolution of color patterns in different species is regulated by changes in the enhancers of the same genes (Reed et al. 2011; Martin et al. 2012, 2014). Thus convergent phenotypes can be produced by natural selection working on developmental regulatory genes.

From Microevolution to Macroevolution: "Toolkit Genes"

It would be difficult to understand macroevolution if each phylum developed in a completely different manner than any other phylum and used a completely different set of genes. However, just as there is descent with modification among living organisms, there is descent with modification among the genes that construct the organism. The genes that are involved in human development are also used in the development of hydroids, flatworms, butterflies, and squids. Indeed, one of the surprising concepts that came out of evolutionary developmental biology is the notion of molecular parsimony, sometimes called the "small toolkit." In other words, although development differs enormously from lineage to lineage, development within all lineages uses the same types of molecules. The transcription factors, paracrine factors, adhesion molecules, and signal transduction cascades are remarkably similar from one phylum to another. Whereas early evolutionary morphologists discussed homologies between bones and tissues, molecular biologists can now discuss an even deeper homology—the homologies of genes.* The Hox genes and the genes for certain transcription factors such as Pax6 are found in all animal phyla, from sponges and cnidarians to insects and primates. In fact, there are "toolkit genes" that appear to play the same roles in all animal lineages.

Evolutionary developmental biologists such as Walter Gehring (1998, 2005) demonstrated that there is a developmental genetic pathway specifying photoreceptor development and that this is common to all animals with eyes. The basic genetic pattern for eye development is specified by homologues of the *Pax6* gene. This pathway probably evolved only once, in early metazoans, and every eye on the planet comes from a modification of this central scheme. Indeed, experimenters have taken the mouse *Pax6* gene and expressed it in flies, specifically in those cells that should give rise

---

*The "deepest" homologies are that all organisms use DNA as their genetic material and a genetic code that is so common that bacterial ribosomes can translate human and fly mRNAs. This indicates a common ancestry for all existing life-forms. The use of animals in medical research is based on their developmental and physiological homologies to our bodies.

(A)

(B)

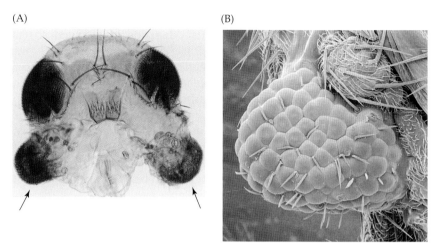

Figure 10.8 Evolutionary similarity of *Pax6* expression in vertebrates and insects. (A) Expression of a fly's *Pax6* gene in a non-eye (jaw) region of the fly produces ectopic fly eyes (arrows). The eye-specifying expression of the *Pax6* gene can interact with other components of the developing fly to induce the compound eye in new locations. (B) Expression of mouse *Pax6* in the leg region of a developing fly also causes an eye—an insect eye—to form there. Despite the enormous morphological differences in insect and vertebrate eyes, the mouse *Pax6* gene specifies insect eye development when placed in a fly embryo. (Photographs courtesy of W. Gehring and G. Halder.)

to the fly's jaw; the jaw then develops into a fly eye (Figure 10.8). If the fly *Pax6* gene is expressed in frog skin, the resultant fly Pax6 protein initiates eye development, causing the cells that should be the frog's epidermis to become an amphibian retina (Halder et al. 1995; Chow et al. 1999).

Even the known differences between major groups seem to be diminishing. Although vertebrates don't express juvenile hormone and insects don't express estrogens, the receptors for these hormones are found in both insects and vertebrates, and in some cases they will bind hormones from the other clade (Harmon et al. 1995; King-Jones and Thummel 2005) as well as binding hormones in their own clade. The fly eye and vertebrate eye are considered to be analogous, not homologous, structures; the vertebrate eye is not compounded of multitudinous ommatidia, nor do vertebrate eyes derive from imaginal discs activated by ecdysterone. Yet the *Pax6* gene *is* homologous in flies and vertebrates. In both groups, this same gene is essential for initiating eye formation, and in both groups it acts through the same pathway.

Duplication and divergence: The Hox genes

So far we have discussed two new principles of evolution as formulated through evo-devo, (1) enhancer modularity and (2) the small toolbox. The third principle is "duplication and divergence" wherein genes are duplicated and can acquire new functions. Susumo Ohno (1970) had noticed that gene families (such as the globin genes) could be traced back to a single original gene. He realized that this provided material for new variation. Many of the genes active in development have duplicated and have taken on new functions.

Studying the Hox gene complex is especially instructive, because Hox gene expression provides the basis for anterior-posterior axis specification throughout the animal kingdom. This is a remarkable example of the small toolkit. The same genes are providing anterior-posterior instructions in the fly, worm, and mouse embryos. The Hox genes encode a family of transcription factors, and the enormous variation of morphological forms among animals is underlain by this single common set of instructions.

The similarity of all the Hox genes is best explained by descent from a common ancestor (Figure 10.9). First, the different Hox genes in the homeotic gene cluster appear to have originated from duplications of an ancestral gene. This would mean that in fruit flies, the *Deformed* (*Dfd*), *Ultrabithorax* (*Ubx*), and *Antennapedia* (*Antp*) genes all emerged as duplications of an original gene in some ancestor. The sequence patterns of these three genes (especially in the homeodomain region) are extremely well conserved. Such tandem gene duplications are thought to be the result of errors in DNA replication, and they are very common. Once duplicated, the gene copies can diverge by random mutations in their coding sequences and regulatory

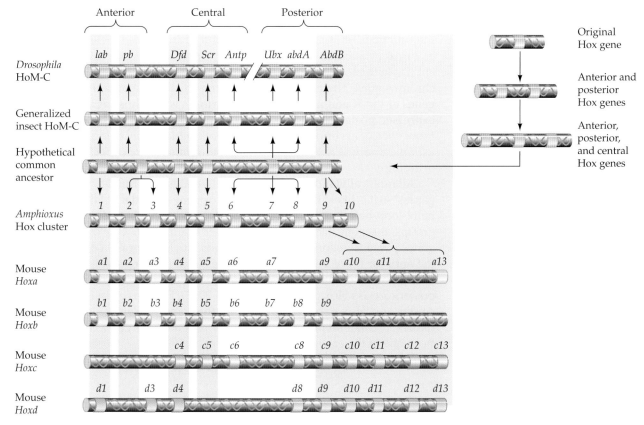

Figure 10.9   Postulated ancestry of the Hox genes from a hypothetical ancestor of both protostomes and deuterostomes. The deuterostome, amphioxus, has only one set of Hox genes, just like the protostomes. Vertebrates, derived from an amphioxus-like organism, have four Hox clusters, none of which is complete. (After Holland and Garcia-Fernández 1996; Gehring et al. 2009.)

regions (such as enhancers), developing different expression patterns and new functions (Lynch and Conery 2000; Damen 2002; Locascio et al. 2002). Susumu Ohno (1970), one of the founders of the gene family concept, likened gene duplication to a method used by a sneaky criminal to circumvent surveillance. While the "police force" of natural selection makes certain that there is a "good" gene performing its function properly, that gene's duplicate—unencumbered by the constraints of selection—can mutate and potentially serve new functions.

This **duplication and divergence** scenario is seen in the Hox genes, the globin genes, the collagen genes, the *Distal-less* genes, and in many paracrine factor families (e.g., the Wnt genes). Each member of such a gene family is homologous to the others (i.e., their sequence similarities are due to descent from a common ancestor and are not the result of convergence for a particular function), and they are called **paralogues**. Thus, the *Antennapedia* Hox gene of *Drosophila* is a paralogue of the *Ultrabithorax* Hox gene of *Drosophila*.

Moreover, each Hox gene in *Drosophila* has a homologue in vertebrates. In some cases, the homologies go very deep and can also be seen in the gene's functions. Not only is the vertebrate *Hoxb4* gene similar in sequence to its *Drosophila* homologue, *Dfd*, but the human *HOXB4* gene can perform the functions of *Deformed* when introduced into *Dfd*-deficient *Drosophila* embryos (Malicki et al. 1992). Not only are the Hox genes of these different phyla homologous, but they are in the same order on their respective chromosomes. Their expression patterns are also remarkably similar: the genes at the 3' end are expressed anteriorly, while those at the 5' end are expressed posteriorly (Figure 10.10). Thus, Hox genes are homologous *between* species (as opposed to members of a gene family being homologous *within* a species). Genes that are homologous *between* species are called **orthologues**.

All multicellular organisms—animals, plants, and fungi—have Hox-like genes,* so it is likely that an ancestral Hox gene existed that encoded a basic helix-loop-helix transcription factor in protozoans. In the earliest animal groups, this gene became duplicated (see Figure 10.9). One of the two Hox genes present in some living cnidarians (such as corals) corresponds to the anterior set of vertebrate Hox genes (and is expressed in the anterior portion of the cnidarian larva), while another sequences corresponds to a posterior-class Hox gene (DuBuc et al. 2012). Perhaps even in the ancient cnidarians, Hox genes distinguished their anterior and posterior larval tissues. In bilateral phyla, the central Hox genes emerged as a duplication from one of the earlier genes (de Rosa et al. 1999).

---

*The Hox genes appear to have originated after the origin of multicellularity. Placozoans (very primitive metazoans) have Hox genes, and sponges have genes similar to Hox genes. However, Hox genes have not been found in the few protists whose genomes have been sequenced (Larroux et al. 2007; King et al. 2008). Interestingly, the choanoflagellate genome contains dozens of genes whose protein products are similar to those used in vertebrate embryos for cell adhesion and intercellular signaling. And, although no Hox genes are present in these protists, homologues of the genes encoding the p53, Myc, and Sox transcription factors have been found.

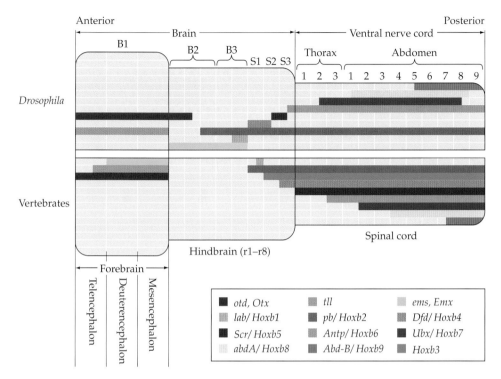

Figure 10.10   Expression of regulatory transcription factors in *Drosophila* and vertebrate embryos along the anterior-posterior axis. The Hox genes specify the anterior-posterior axis and are expressed in similar patterns from the hindbrain posteriorly through the spinal cord. The head regions are defined by homologues of the same three genes in both vertebrates and flies. (After Hirth and Reichert 1999.)

In chordates, however, two large-scale duplications of the entire Hox cluster took place, such that vertebrates have four Hox clusters per haploid genome instead of just one. (Indeed, after the divergence of tetrapods, the fish duplicated their Hox genes yet again, giving themselves eight Hox clusters.) Thus, instead of having a single *Hox4* gene (orthologous to *Deformed* in *Drosophila*), vertebrates have *Hoxa4*, *Hoxb4*, *Hoxc4*, and *Hoxd4*. This constitutes the *Hox4* **paralogue group** in vertebrates. Such large-scale duplications have had several consequences. First, these duplications create much redundancy. It is difficult to obtain a loss-of-function mutant phenotype, since to do so means all copies of these paralogue group genes must be deleted or made nonfunctional (Wellik and Capecchi 2003). However, in some instances, the genes *have* become specialized. *Hoxd11*, for instance, plays an important role in the mammalian limb bud, but not in the reproductive system. Mammalian *Hoxa11*, on the other hand, plays roles in both the limb (where it is critical in specifying the bones) and in the female reproductive tract (where it helps construct the uterus; Wong et al. 2004).

The evolutionary importance of Hox genes in the specification of the body axis can be seen in their evolutionary modification in vertebrates. Gaunt (1994) and Burke and colleagues (1995) have compared the vertebrae of the mouse and the chick. Although the mouse and the chick have similar

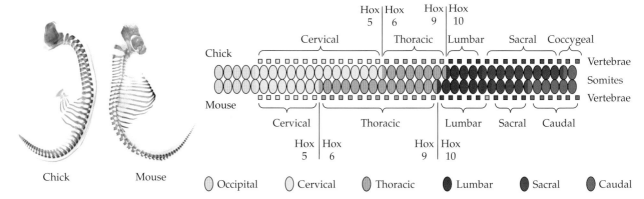

Figure 10.11 Schematic representation of the chick and mouse vertebral patterns along the anterior-posterior axis. The boundaries of expression of certain Hox gene paralogous groups have been mapped onto these domains. The chick has twice as many cervical vertebrae as the mouse. (After Burke et al. 1995, photograph courtesy of M. Kmita and D. Duboule.)

numbers of vertebrae, they apportion them differently. Mice (like almost all mammals, be they giraffes or whales) have only 7 cervical vertebrae. These are followed by 13 thoracic vertebrae, 6 lumbar vertebrae, 4 sacral vertebrae, and a variable (20+) number of caudal vertebrae (Figure 10.11). The chick, on the other hand, has 14 cervical vertebrae, 7 thoracic vertebrae, 12 or 13 (depending on the strain) lumbosacral vertebrae, and 5 coccygeal (fused tail) vertebrae. The researchers asked, "Does the constellation of Hox gene expression correlate with the type of vertebra formed (e.g., cervical or thoracic) or with the relative position of the vertebra (e.g., number 8 or 9)?"

The answer is that the constellation of Hox gene expression predicts the type of vertebra formed. In the mouse, the transition between cervical and thoracic vertebrae is between vertebrae 7 and 8; in the chick, it is between vertebrae 14 and 15. In both cases, the *Hox5* paralogues are expressed in the final cervical vertebra, while the anterior expression boundary of the *Hox6* paralogues extends to the first thoracic vertebra. Similarly, in both animals, the thoracic-lumbar transition is seen at the boundary between expression of the *Hox9* and *Hox10* paralogous groups. It appears there is a code of differing Hox gene expression along the anterior-posterior axis, and that code determines the type of vertebra formed rather than the number of segments.

One of the most radical modifications of the Hox gene patterning is seen in snakes. The forelimb usually forms just anterior to the most anterior expression domain of *Hoxc6*. Posterior to that point, *Hoxc6*, in collaboration with *Hoxc8*, helps specify the vertebrae to become thoracic (ribbed) vertebrae. During early snake development, *Hoxc6* is not expressed in the absence of *Hoxc8*, so the forelimbs do not form.* Rather, the combination

---

*The hindlimbs of snakes are inhibited in a different manner, probably by the loss of *sonic hedgehog* expression (Chiang et al. 1996). Paleontological evidence also suggests that the hindlimbs and forelimbs were lost at different times, the loss of the forelimbs occurring earlier (Caldwell and Lee 1997).

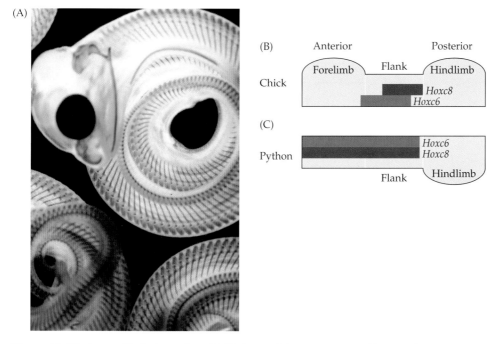

Figure 10.12 Loss of limbs in snakes. (A) Skeleton of the garter snake, *Thamnophis*, stained with alcian blue. Ribbed vertebrae are seen from the head to the tail. (B,C) Hox expression patterns in chick (B) and python (C). (A courtesy of A. C. Burke; B,C after Cohn and Tickle 1999.)

of *Hoxc6* and *Hoxc8* expression is found for most of the length of the snake embryo, specifying nearly all of the vertebrae to be thoracic (Figure 10.12; Cohn and Tickle 1999). In addition, the posterior Hox genes (*Hoxa13* and *Hoxd13*) are not expressed in the snake embryos, causing the new vertebrae to keep forming and to become trunk, not tail vertebrae (Di-Poï et al. 2010; Mallo et al. 2010). Thus, to understand the evolutionary mechanisms by which snakes lost their legs, one has to know the mechanisms of vertebrate axial development.

DUPLICATION AND DIVERGENCE IN HUMAN BIG BRAIN DEVELOPMENT   One of the most interesting of the gene duplications and modifications involves a gene that may promote the remarkable brain growth of human fetuses and infants. There exists in mammals a gene, *SRGAP2*, that encodes a neural growth inhibitor. The protein encoded by this gene is expressed in the mammalian brain cortex and appears to *slow down* cell division and decrease the length and density of dendritic processes. However, humans differ from all other animals (including chimpanzees) by having duplicated this gene twice. Moreover, the second duplication event was not complete, so one of the newly formed genes is only a partial duplicate. This partial gene produces a truncated SRGAP2 protein, SRGAP2C, that is also made in the cerebral cortex and which *inhibits* the activity of normal SRGAP2 made from the complete genes. As a result, cell division in the cerebral cortex continues for longer periods of time, and the dendrites are larger with more

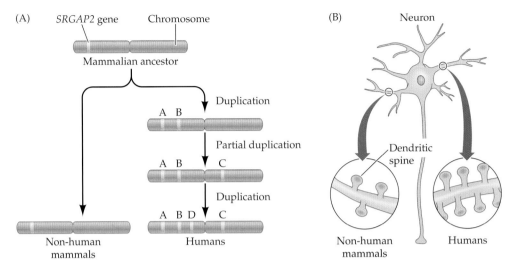

Figure 10.13 Duplication and divergence of human *SRGAP2* gene. (A) The *SRGAP2* gene is found as a single copy in the genomes of all mammals except humans. However, in the lineage giving rise to humans, this gene was duplicated twice, giving rise to four similar versions of the gene, named A–D. (B) The "ancestral" version, *SRGAP2A* (with minor contributions from *SRGAP2B* and *SRGAP2D*), enables the maturation of dendritic spines (protuberances) on the surfaces of neurons. *SGRAP2C* is only partially duplicated, and its product inhibits SRGAP2A, slowing dendritic spine maturation and promoting neuronal migration. This may have allowed for longer maturation time and greater flexibility in the human brain. (After Geschwind and Konopka 2012.)

connections (Charrier et al. 2012; Dennis et al. 2012; Figure 10.13). Based on genomic evidence, this gene duplication event is calculated to have taken place about 2.4 million years ago. This would be about the time of the transition from *Australopithecus* to *Homo*, the increase in primate brain size, and the first known use of tools (Tyler-Smith and Xue 2012).

Homologous pathways of development

One of the most exciting findings of the past two decades has been the discovery not only of homologous regulatory genes, but also of homologous developmental pathways. In different organisms, these pathways are composed of homologous proteins arranged in a homologous manner (Zuckerkandl 1994; Gilbert et al. 1996). In some instances, homologous pathways made of homologous parts are used for the same function in both protostomes and deuterostomes. Conserved similarities in both the pathway and its function over millions of years of phylogenetic divergence are considered to be evidence of "deep homology" between these processes and structures (Brown et al. 2008). One example of deep homology is the chordin/BMP4 pathway for construction of the central nervous system and the patterning of the dorsoventral (back-belly) axis. In the mid 1990s, several investigators found that the same chordin/BMP4 interaction that specifies the formation of the neural tube in vertebrates also specifies the neural tissue in fruit flies (Holley et al. 1995; Schmidt et al. 1995; De Robertis and Sasai 1996; Hemmati-Brivanlou and Melton 1997). In other words, the same set of genes appears to instruct the formation of the vertebrate *dorsal* neural tube and the *Drosophila ventral* neural cord. In both vertebrates and invertebrates, chordin (called short gastrulation, or Sog, protein in insects) inhibits the binding of Bmp4 (called Decapentaplegic, or Dpp, in insects) to its receptors on the ectoderm cells. When Bmp4/Dpp binds to ectodermal cells, it instructs them to become epidermis (skin). However, if chordin/Sog blocks the binding, the ectoderm becomes neural tissue. This blocking of Bmp4/Dpp occurs dorsally in vertebrates and ventrally in insects. These reactions are so similar that *Drosophila* Dpp protein can induce epidermal fates in *Xenopus* (Holley et al. 1995; Figure 10.14). Indeed, the Sog and chordin proteins are themselves regulated by homologous proteins (the Tolloid [fly] and Xolloid [frog] proteins) in arthropods and vertebrates. Thus the arthropod and vertebrate nervous systems, despite their obvious differences, seem to be formed by similar sets of instructions. The plan for specifying the animal nervous system probably evolved only once (Mizutani and Bier 2008).*

Toolkit genes and evolution: A summary

To sum up, three important principles become manifest in the small toolkit for development. First, certain homologous (orthologous) regulatory genes

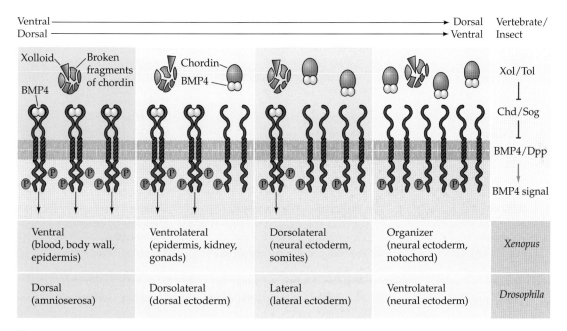

Figure 10.14 Homologous pathways specifying the central nervous system and the dorsal-ventral axes of flies and vertebrates. Both pathways involve a source of chordin (Chd)/Sog protein (the most dorsal region of the vertebrate embryo; the most ventral region of the fly embryo) and a source of BMP4/Dpp (the ventral region of the vertebrate; the dorsal region of the fly). These two regions form antagonistic gradients. Those regions with the most chordin/Sog become central nervous tissue—on the dorsal side of the vertebrate body and on the ventral side of the fly body). The side with the BMP4/Dpp becomes epidermis. In both instances, the gradient is shaped by a constant supply of Xolloid/Tolloid (Xol/Tol) protein, which degrades chordin/Sog. (After Dale and Wardle 1999.)

are highly conserved, critical for the development of all animals (Table 10.1). Thus *Pax6* genes are important for eye development throughout the animal kingdom; *tinman* genes are responsible for heart development in flies, worms, and mammals; and Hox genes are critical for specifying the axes in all animals studied so far.

Second, gene duplication and divergence is an extremely important mechanism for evolution. Duplication allows the formation of homologous

*In 1822, the French anatomist Étienne Geoffroy Saint-Hilaire provoked one of the most heated and critical confrontations in biology when he proposed that the lobster was nothing more or less than a vertebrate upside down. He claimed that the ventral (belly) side of the lobster (with its nerve cord) was homologous to the dorsal (spinal) side of the vertebrate (Appel 1987). It seems that he was correct on the molecular level, if not on the anatomical. Indeed, despite the differences between arthropod and vertebrate neurons, the genes specifying neural cell type are also remarkably similar and play very similar roles in insects and mammals (see Bertrand et al. 2002).

TABLE 10.1 Developmental regulatory genes conserved between protostomes and deuterostomes

Gene	Function	Distribution
achaete-scute group	Cell fate specification	Cnidarians, *Drosophila*, vertebrates
Bcl2/Drob-1/ced9	Programmed cell death	*Drosophila*, nematodes, vertebrates
Caudal	Posterior differentiation	*Drosophila*, vertebrates
delta/Xdelta-1	Primary neurogenesis	*Drosophila*, *Xenopus*
Distal-less/DLX	Appendage formation (proximal-distal axis)	Numerous phyla of protostomes and deuterostomes
Dorsal/NFκB	Immune response	*Drosophila*, vertebrates
forkhead/Fox	Terminal differentiation	*Drosophila*, vertebrates
Fringe/radical fringe	Formation of limb margin (apical ectodermal ridge in vertebrates)	*Drosophila*, chick
Hac-1/Apaf/ced 4	Programmed cell death	*Drosophila*, nematodes, vertebrates
Hox complex	Anterior-posterior patterning	Widespread among metazoans
lin-12/Notch	Cell fate specification	*C. elegans*, *Drosophila*, vertebrates
Otx-1, Otx-2/Otd, Emx-1, Emx-2/ems	Anterior patterning, cephalization	*Drosophila*, vertebrates
Pax6/eyeless; Eyes absent/eya	Anterior CNS/eye regulation	*Drosophila*, vertebrates
Polycomb group	Hox expression/cell differentiation control	*Drosophila*, vertebrates
Netrins, Split proteins, and their receptors	Axon guidance	*Drosophila*, vertebrates
RAS	Signal transduction	*Drosophila*, vertebrates
sine occulus/Six3	Anterior CNS/eye pattern formation	*Drosophila*, vertebrates
sog/chordin, dpp/BMP4	Dorsal-ventral patterning, neurogenesis	*Drosophila*, *Xenopus*
tinman/Nkx 2-5	Heart/blood vascular system	*Drosophila*, mouse
vnd, msh	Neural tube patterning	*Drosophila*, vertebrates

Source: Erwin 1999.

(paralogous) genes, and divergence allows these genes to assume new roles. Numerous transcription factors and paracrine factors are members of such paralogue families. Hox genes are used to pattern the body and limb axes, *Distal-less* genes are used to extend appendages and to pattern the vertebrate skull, and members of the *MyoD* family specify different stages of muscle development.

Third, structural or regulatory modification of these genes can allow them to acquire different functions and to play important roles in evolution. These genes, each family derived from a single ancestral gene, are active in different tissues and provide instructions for the formation of different cell types. So perhaps "the most important difference between the genome of a fruit fly and that of a human is therefore not that the human has new genes but that where the fly only has one gene, our species has multigene families" (Morange 2001, p. 33). Descent with modification is critical for developmental regulatory genes as well as for the organisms they help construct.

Mechanisms of Macroevolutionary Change

As mentioned earlier, there are differences in the agendas of studies of microevolution and macroevolution. Students of microevolution usually stay within a species and ask the question "How did natural selection alter the color of hair in this species?" Students of macroevolution ask questions that involve comparing species, such as "How did mammals get hair, when other vertebrates have scales and feathers?" One can get answers to those questions by seeing how developmental changes link changes in gene expression to anatomical alterations.

In 1977, François Jacob, the Nobel laureate who helped establish the operon model of gene regulation, proposed that evolution rarely creates a new gene. Rather, evolution works with what it has: it combines existing parts in new ways rather than creating new parts. He predicted that such "tinkering" would be most likely to occur in those genes that construct the embryo, not in the genes that function in adults (Jacob 1977). Wallace Arthur (2004) has catalogued four ways in which Jacob's "tinkering" can take place at the level of gene expression:

- Heterotopy (change in location)
- Heterochrony (change in time)
- Heterometry (change in amount)
- Heterotypy (change in kind)

Heterotopy

Changing the *location* of a gene's expression (i.e., altering which cells it is expressed in) can result in enormous amounts of selectable variation. For instance, one of the biggest differences between the duck and the chicken is the webbed feet of the duck. This is an obviously important adaptation to swimming (as are the duck's broad beak and the oil glands of its skin). Vertebrate embryonic limbs develop with their fingers and toes surrounded by

BMP Gremlin Apoptosis Newborn

Chick hindlimb

Duck hindlimb

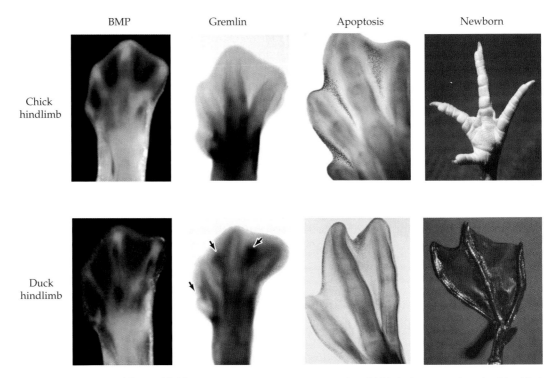

Figure 10.15 Heterotopy exemplified by the role of BMPs in the generation of webbed feet in ducks. BMPs cause apoptosis in the interdigital webbing. Autopods of chick feet (upper row) and duck feet (lower row) are shown at similar stages. The in situ hybridizations show that while BMPs are expressed in both the chick and duck hindlimb webbing, the duck limb shows expression of Gremlin protein (arrows) in the webbing as well. Gremlin is an inhibitor of BMPs. The pattern of cell death (shown by neutral red dye accumulation) becomes distinctly different in the two species. (Photographs courtesy of J. Hurle and E. Laufer.)

a web of connective tissue and skin. In chicks (and humans), the cells that make up the webbing between the digits undergo apoptosis (programmed cell death) initiated by the paracrine factor BMP4. This cell death destroys the webbing and frees the digits. It turns out that ducks also express BMP4 in the webbing of their hindlimbs, but this protein is prevented from signaling cell death by the presence of another protein, the BMP inhibitor Gremlin (Laufer et al. 1997; Merino et al. 1999). Chick and human limbs also express Gremlin around the cartilaginous skeletal elements of the digits, but not in the webbing (Figure 10.15). Thus, the webbing in embryonic duck feet remains intact thanks to the heterotopy—the change in place of gene expression—of the *Gremlin* gene. Indeed, if one adds beads containing the Gremlin protein to the webbing of an embryonic chick limb, the limb will keep its webbing, just as a duck's does (Figure 10.16).

Bat forelimbs present a similar situation. Here the interdigital webbing that forms the wing has been maintained where it has disappeared in other mammals. Again, inhibition of BMP-induced apoptosis is critical. In addition to Gremlin, moreover, a second agent of BMP inhibition in bats appears to be FGF signaling. Unlike other mammals, bats express *Fgf8* in

(A)   (B)

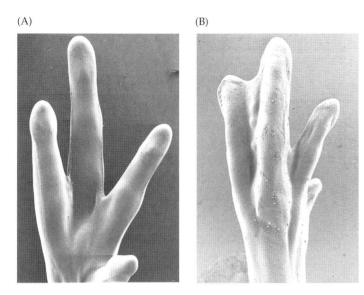

Figure 10.16 Inhibition of cell death by inhibiting BMPs. (A) Control chick limbs show extensive apoptosis in the space between the digits, leading to the absence of webbing. (B) However, when beads soaked with Gremlin protein are placed in the interdigital meso-derm, the webbing persists and generates a duck-like pattern. (After Merino et al. 1999, photographs courtesy of E. Hurle.)

their interdigital webbing, and Fgf8 is critical for maintaining the cells there. If FGF signaling is inhibited in the bat by certain drugs, BMPs will induce the apoptosis of the forelimb webbing, just as in other mammals* (Weatherbee et al. 2006).

Many other evolutionary novelties, including turtle shells, feathers, and teeth, are caused by genes being expressed in new places. Sometimes, though, a new structure is formed when an entire suite of genes, a genetic "subroutine," is transferred into a different cell type. Such transferring is called recruitment or co-option.

RECRUITMENT   Enhancer modularity allows for recruitment. This is seen when a structure usually found in one part of the body is expressed in a new part. This is often because a genetic mutation has formed a new enhancer such that a transcription factor that activates a particular series of genes is activated in a new cell type (Davidson 2011). One of the most beautiful ex-amples of recruitment is seen in the wing structure that defines the beetles. Beetles are the most successful animal order on the planet, accounting for

*Notice that BMP4 is being used in one part of the embryo to specify the ectoderm to become skin and in another part of the embryo to cause cell death. As we will see later, BMP4 is also used to construct the avian beak and to make bone tissue. Classical genetics usually studied proteins that have a single function; hemoglobin, insulin, and collagen, for instance, are each the end product of their cell lineage and each has a single function. Genes active in development, however, often have several different functions. Philosopher John Thorp as called monotelism (the idea that each of nature's structures has but one purpose) "Aristotle's worst idea."

(A)

(B)

Figure 10.17 Elytra are the hardened forewings that are characteristic of Coleoptera, the beetles. Elytra are formed through the recruitment of the genetic module for exoskeleton development into the module for dorsal forewing development. (A) The elytra of a "ladybug" beetle. Its forewings are ornamented with exoskeleton, and its hindwings are extended. (B) These "living jewels" from the Oxford University Museum of Natural History illustrate some of the diversity of beetle elytra.

more than 20% of extant animal species (Hunt et al. 2007). They differ from other insects in forming an elytron—a forewing encased in a hard exoskeleton. This makes them the "living jewels" so beloved of naturalists (Figure 10.17).* In beetles, as in *Drosophila*, the *Apterous* gene is expressed in the dorsal compartment of the wing imaginal discs, and the Apterous transcription factor organizes the tissue to differentiate dorsal wing structures. However, in beetles (and in no other known insect group), Apterous also activates the exoskeleton genes in the forewing while repressing them in the hindwing (Tomoyasu et al. 2009). Thus, a new type of wing emerges from the recruitment of one module (the subroutine of exoskeletal development) into another (the subroutine of dorsal forewing development).

Another example of the recruitment of one organ's gene expression pattern into a new set of cells is the horn of the dung beetle. This prodigious structure was mentioned in Chapter 1. The horns of dung beetles arose from the co-option of the leg-patterning gene expression (especially the genes *homothorax*, *dachshund*, and *Distal-less*) into the head ectoderm (Moczek and Rose 2009). The horn grows like a seventh leg on the head of the male beetle larvae.

Heterochrony

Heterochrony is a shift in the relative *timing* of two developmental processes from one generation to the next. Heterochrony can be seen at any

---

*Although Darwin and Wallace were avid beetle collectors, it was geneticist J. B. S. Haldane (see Gould 1993) whose remark may best reflect the prominence of these insects. When asked by a cleric what the study of nature could tell us about God, Haldane is said to have replied, "He has an inordinate fondness for beetles."

Figure 10.18   Limb development in a flying mammal and a flightless bird. (A) The bat *Rousettus amplexicaudatus* has a large forelimb bud that develops into a wing (orange), while the hindlimb (brown) is underdeveloped relative to other mammals. (B) The flightless kiwi (*Apteryx australis*) has much smaller forelimbs relative to the wings of most birds. (After Richardson 1999.)

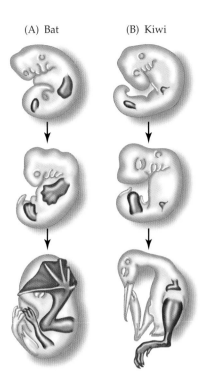

(A) Bat   (B) Kiwi

level of development from gene regulation to adult animal behaviors (West-Eberhard 2003). In heterochrony, one module changes its time of expression or growth rate relative to the other modules of the embryo. Indeed, heterochrony shows some of the best examples of the modularity of development. The bat embryo, for instance, doesn't make a long limb by sustaining limb development for a longer period of time after the bat has been born. Rather, these forelimb buds have started to develop earlier than normal, and by the time the bat is born, the forelimbs are much larger than the hindlimbs (Figure 10.18A). On the other hand (no pun intended), at the same early stage of development, the kiwi has a small forelimb bud and a large hindlimb bud (Figure 10.18B), even though most avian and mammalian embryos at this stage have a forelimb bud that is much larger than the hindlimb bud. In kiwis, the forelimb bud starts to develop *later* than the hindlimb. Thus, although the target of natural selection is an *adult* morphology, arriving at that adaptive phenotype can involve changing the parameters of *early* development. (What allows these early developmental changes is that forelimb development can change independently from that of the rest of the body. The entire body does not have to become small to produce a small limb bud; the limb bud behaves as a developmental module. This mosaic semiautonomy has been called developmental modularity.)

Limb heterochronies are in fact quite common. In marsupials, for instance, the forelimbs develop at a faster rate relative to the embryonic forelimbs of placental mammals, allowing the marsupial embryo to climb into the maternal pouch and suckle (Smith 2003; Sears 2004). The elongated fingers in the dolphin flipper appear to be the result of heterochrony, wherein the region responsible for producing the growth factors for the developing limb (especially Fgf8) is present longer than in other mammals (Figure 10.19; Richardson and Oelschläger 2002). Another digit example of molecular heterochrony occurs in the lizard genus *Hemiergis*, which includes species with three, four, or five digits on each limb. The number of digits is regulated by the length of time the *sonic hedgehog* (*Shh*) gene is active in the limb bud's zone of polarizing activity. The shorter the duration of *Shh* expression, the lower the number of digits (Shapiro et al. 2003).

Heterochrony is an incredibly important mechanism producing selectable developmental changes. The enormous number of ribs formed in embryonic snakes (more than 500 in some species; see Figure 10.12A) is likewise due to heterochrony: the segmentation reactions cycle nearly four times faster relative to tissue growth in snake embryos than they do in related vertebrate embryos (Gomez et al. 2008). One of the most exciting heterochronies that has been discovered is the recent analysis showing

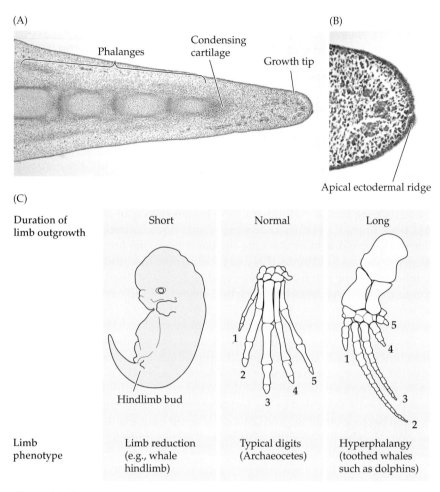

Figure 10.19 Heterochrony in the flipper development of the spotted dolphin (*Stenella attenuata*). (A) Dolphins show "hyperphalangy" of digits 2 and 3. That is to say, digits 2 and 3 continue growing long after the other digits, and they keep adding on new cartilaginous regions. (B) The mechanism for this appears to be the retention of the apical ectodermal ridge that secretes growth factors necessary for the digit growth. (C) A heterochronic hypothesis for cetacean limb development. When limb growth terminates early, the loss of distal structures is seen. This leads to rudimentary appendages in many whales. In Archaeocetes (the progenitors of whales), there appears to a normal amount of limb growth. In toothed whales (such as dolphins), limb development terminates late, giving rise to extra-long digits. (After Richardson and Oelschläger 2002.)

that the bird's skull could have evolved by relatively small heterochronic changes during the development of dinosaur skulls (Bhullar et al. 2012). Just as the human brain (see box) continues certain fetal processes for longer durations, so the bird's brain can be seen as similar to that of an immature dinosaur. It has expanded certain processes that were typical of dinosaur young, especially the large eyes and visual centers of the brain.

Regulatory Alleles May Help Make Us Human

Mary Jane West-Eberhard (2005) noted that "lack of attention to developmental phenomena in relation to speciation promises to change, because genomic studies of speciation can now contemplate gene expression as well as gene frequency data." The synthesis of comparative genomics and developmental expression data is already starting to yield remarkable fruits in looking at that most anthropocentric of questions: "What distinguishes humans from the other great apes?"

If there is one developmental trait that distinguishes humans from the rest of the animal kingdom, it is our *retention of the fetal neuronal growth rate*. Both human and ape brains have a high growth rate before birth. After birth, however, this rate slows greatly in the apes, whereas human brain growth continues at a rapid rate for about 2 years (Martin 1990; see Leigh 2004; Figure A). Portmann (1941), Montagu (1962), and Gould (1977b) have each made the claim that we are essentially "extra-uterine fetuses" for the first year of life.

During early postnatal development, we add approximately 250,000 neurons per minute (Purves and Lichtman 1985). The ratio of brain weight to body weight at birth is similar for great apes and humans, but by adulthood the ratio for humans is literally "off the chart" when compared with that of other primates (Figure B; Bogin 1997). Indeed, if one follows the charts of ape maturity,

actual human gestation is 21 months. Our "premature" birth is an evolutionary compromise based on maternal pelvic width, fetal head circumference, and fetal lung maturity.

In addition to the neurons made after birth, the number of synapses increases by an astronomical number. At the cellular level, no fewer than 30,000 synapses per square centimeter of cortex are formed *every second* during the first few years of human life (Rose 1998; Baringa 2003). It is speculated that these new neurons and rapidly proliferating neural connections enable plasticity and learning, create an enormous storage potential for memories, and enable us to develop skills such as language, humor, and music—that is, those things that help make us human.

So there is a remarkable heterochrony on the level of organ formation. What is responsible for this? King and Wilson (1975) were the first to suggest that regulatory changes in development were responsible for the morphological and behavioral differences between chimps and humans. It seems that on the gene level, gene duplication, heterometry, and heterotypy are very important.

We have already discussed the partial duplication of the *SRGAP2* gene, which inhibits an inhibitor of neuronal migration and dendrite growth (see p. 399). Other human-specific mutations include the deletion of enhancers

(A)

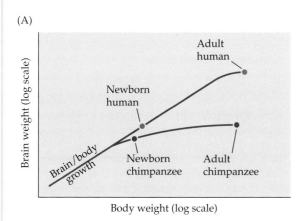

(B)

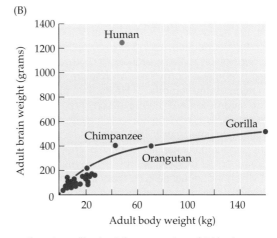

Heterochrony: retention of fetal neuronal growth rate in humans. (A) Whereas other primates (e.g., chimpanzees) complete neuron proliferation around the time of birth, in newborn humans neurons continue to proliferate at the same rate as fetal brain neurons. (B) The brain-to-body weight ratio (encephalization index) of humans is about 3.5 times higher than that of apes. (After Bogin 1997.)

(Continued on next page)

Regulatory Alleles May Help Make Us Human (*continued*)

that would normally activate growth repressors in the brain. The protein encoded by the *GADD45G* gene inhibits neural growth in the ventral forebrains of chimpanzees and mice. However, the forebrain enhancer has been deleted in humans, so this gene is not active there (McLean et al. 2011). Heterometry is implicated as well. Using microarrays, several recent investigations found that while the quantity and types of genes expressed in human and chimp livers and blood were indeed extremely similar, human *brains* produced over five times as much mRNA as the chimp brains (Enard et al. 2002a; Preuss et al. 2004). Some researchers propose heterotypy: Enard and colleagues (2002b) have shown that the *FOXP2* gene is critical for language, and although this gene is extremely well conserved throughout most of mammalian evolution, it has a unique form in humans.

But some of the most interesting ideas are coming from changes in the sequence of regulatory RNAs. Pollard and her collaborators (2006) found this through a very logical and elegant path. First, they asked the DNA databases which areas of DNA (at least 100 base pairs long) are identical between the mouse, the rat, and the chimp. These **evolutionarily conserved regions** would be expected to encode genes whose functions are extremely important, and random mutations in these regions would be expected to have been strongly selected against. Such conserved regions often contain regulatory sequences that control the expression of transcription factors.

Next, the researchers looked at about 35,000 of these conserved sequences and asked whether human homologues of any of these sequences diverge from the common mammalian norm. Their analyses revealed 202 **human accelerated regions**, or **HARs**. Two of these, *HAR1* on chromosome 20 and *HAR2* on chromosome 2, were especially interesting because they are dramatically distinct in humans, with DNA changes more than 10 times greater than changes found in chimp, orangutan, gorilla, spider monkey, or crab-eating macaque, in all of which the sequence is basically the same. The *HAR1* region appears to be adjacent to a region of genes encoding transcription factors, and it encodes a regulatory RNA that is expressed in the cortex of the mammalian brain.

Another region (HARE5) turns out to be an enhancer of a Wnt receptor (Frizzled8) that is found in the neural progenitor cells of the forebrain cortex. The human ortholog of this enhancer contains 16 changes compared to the homologous chimp sequence, and when placed

(C)

Chimpanzee   Human

(C) Molecular evolution among primates. A blue-staining reporter gene was linked to the Human HARE5 enhancer or to its chimp homologue and placed into transgenic mice. The human enhancer is much more active during early brain cortex development than is the chimp enhancer. (After Boyd et al. 2015, photographs courtesy of D. Silver.)

into mice, the human sequence shortens the cell cycle and accelerates the production of neural progenitor cells (Boyd et al. 2015; Figure C).

There are many components to being human, and most of them—the large and cognitive brain, the opposable thumb, the skeleton altered to allow walking, the musculature modified to allow language—arise by changing development. This field of human evo-devo continues to be a fascinating research area, linking development, evolution, and the biological basis of becoming human.

However, our large brain could become a liability in a warm environment. It would overheat if it were not for other changes in human development. While most organs can function within a wide range of temperatures,

Regulatory Alleles May Help Make Us Human *(continued)*

the brain cannot. At 103°F (40°C), delirium begins, and the brain stops functioning at about 106°F (42°C). Most furred animals cannot exert themselves in strenuous activity during hot weather. As sweat builds between their skin and fur, they begin to overheat. Unless they can drink while exerting themselves, the animals will collapse from heat exhaustion (Jablonsky 2013). Evaporation-induced cooling of the skin is critical and is accomplished by two means—hairlessness and eccrine sweat glands. Animals with furry coats and with few eccrine sweat glands produce only about 15% of the sweat that we humans do. Sweating allows us to work in the sunlight (and fossil evidence suggests that the earliest humans ran) without raising our body temperature and overheating our precious brains. As Jablonsky (2013, p. 49) says,

"Behind every large human brain there is a potentially very sweaty human body."

Most mammals have eccrine glands only in their palms and soles. Humans have them throughout their skin (Cabanac and Caputa 1979; Falk 1990; Nelson and Nunnely 1998). Our closest relatives have them over their body, too, but not as densely as humans do. With increasing brain size comes the necessity for a very efficient cooling system: the hairless skin with its eccrine sweat glands. Most other animals don't sweat the amounts of water that humans do, and other primates (and horses) have eccrine sweat glands only in their armpits. It's humans who live by the sweat of their brow.

Heterometry

Heterometry is the change in the *amount* of a gene product. One example within our own species is expression of interleukin-4 (IL-4). IL-4 is a paracrine factor released by cells of the immune system to promote the differentiation of IgE-secreting B cells. These are the cells that initiate the cascade to remove pathogens, including parasitic worms, from our bodies. However, if these B cells recognize nonpathogenic substances, one gets allergies. One regulatory allele of the *IL-4* gene is *–524T* (i.e., in this allele the nucleotide base thymine occupies the position 524 base pairs before the start of transcription). The *–524T* allele is recent and is found only among human beings (Rockman et al. 2003). The wild-type allele, *–524C* (which has cytosine in the –524 position), is found in all primate populations—and indeed in every other mammal tested. The *–524T* allele, therefore, emerged in the lineage separating *Homo* from chimps (and hence from other primates in general). Individuals carrying the *–524T* allele are subject to several disease states, including severe allergies, asthma, contact dermatitis, and subacute sclerosing panencephalitis. However, population genetic statistics show that while the *–524T* allele may cause lowered fitness in many habitats, this new allele appears to be advantageous in human populations exposed to certain parasitic worms. It allows these people to defend themselves better against helminth parasites. What does the *–524T* allele do differently than the *–524C* allele? The C → T mutation creates a new binding site for the transcription factor NFAT (*nuclear factor for activated T cells*) in the IL-4 enhancer (Figure 10.20). The new site at *–524T* leads to more rapid production of IL-4, yielding protein levels of at least threefold the norm. So in areas of filariasis (elephantiasis) and schistosomiasis, this allele appears to provides more fitness than the *–524C* allele and thus may be selected for in those populations.

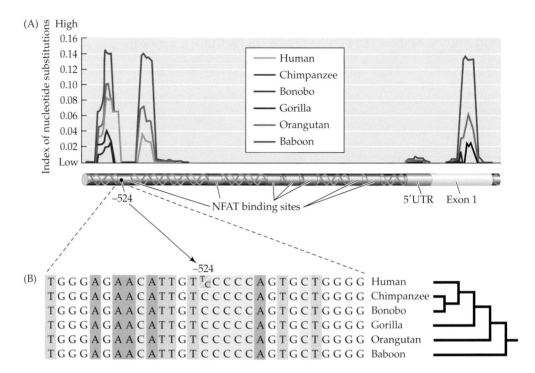

Figure 10.20 Heterometry (a change in the amount of a gene product expressed) can help drive evolution. (A) Regions of variability in the enhancer of the gene for interleukin-4 (IL-4) among primates. Several regions are highly conserved, or invariable, indicating that any change (i.e., mutation) in these nucleotide sequences usually results in decreased fitness. These regions include the enhancers for the *IL-4* genes. (B) At position –524, in the midst of a highly conserved enhancer sequence, a mutation in the human population has created a new binding site for the NFAT transcription factor, enabling *IL-4* to be transcribed in greater amounts. Although this has negative fitness consequences in many populations, it functions positively for the survival of individuals living in places where worm parasites are common. Heterometry is caused by mutation in the enhancer site of the *IL-4* gene. (After Rockman et al. 2003.)

THE BEAKS OF THE FINCHES   One of the best examples of heterometry involves Darwin's celebrated finches (mentioned in Chapter 8), a set of closely related bird species collected by Charles Darwin and his shipmates during his visit to the Galápagos and Cocos Islands in 1835. These birds helped him frame his evolutionary theory of descent with modification, and they still serve as one of the best examples of adaptive radiation and natural selection (see Grant 1999; Weiner 1994; Grant and Grant 2007). One important behavioral and morphological distinction in these birds is that between the cactus finches and the ground finches* (Figure 10.21). The ground finches evolved deep, broad beaks that enable them to crack

*The phylogeny of Darwin's finches in quite complex, with different species adapting different beak morphologies on different islands (Lamichhaney et al. 2015); but for our discussion, cactus and ground finches will be defined by their habitats and not phylogeny.

Figure 10.21   Correlation between beak shape and the expression of *Bmp4* in Darwin's finches. In the genus *Geospiza*, the ground finches (represented here by *G. fuliginosa*, *G. fortis*, and *G. magnirostris*) separated from the cactus finches (represented here by *G. scandens* and *G. conirostris*). The differences in beak morphology correlate to heterochronic and heterometric changes in *Bmp4* expression in the beak. That is, *Bmp4* is expressed earlier and at a higher level in the seed-crushing ground finches. This gene expression difference provides an explanation for the role of natural selection on these birds. (After Abzhanov et al. 2004.)

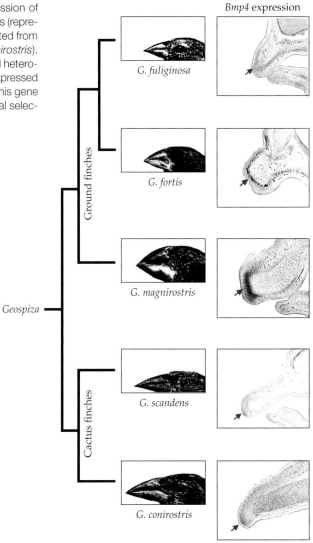

seeds open, whereas the cactus finches evolved narrow, pointed beaks that allow them to probe cactus flowers and fruits for insects.

Research on other birds (Schneider and Helms 2003) has shown that species differences in the beak pattern were caused by changes in the growth of the cranial neural crest–derived cells of the frontonasal process (i.e., those cells that form the facial bones and cartilages). Moreover, it has further been demonstrated that birds use BMP4 as a growth factor promoting cell division in their beaks, and that species-specific differences in beaks (such as those between the chick and the duck) are due to the placement of *Bmp4* expression (Wu et al. 2004, 2006).

Abzhanov and his colleagues (2004) have found a remarkable correlation between the beak shape of the finches and the amount of *Bmp4* expression. No other paracrine factor has shown such differences. The expression of *Bmp4* in the mesenchyme of embryonic *ground* finch beaks starts earlier and is much greater than *Bmp4* expression in *cactus* finch beaks. In all cases, the *Bmp4* expression pattern correlates with the breadth and depth of the beak.

The importance of these expression differences was confirmed experimentally by changing the *Bmp4* expression pattern in chick embryos (Abzhanov et al. 2004; Wu et al. 2004, 2006). When *Bmp4* expression was enhanced in the frontonasal process mesenchyme, the chick developed a broad beak reminiscent of the beaks of the ground finches. Conversely, when BMP signaling was inhibited in this region (by adding a BMP inhibitor to the developing beak primordium), the beak became narrow and pointed, like those of cactus finches. Thus, enhancers controlling the amount of beak-specific BMP4 synthesis may be critically important in the evolution of Darwin's finches.

But this is only the beginning of the story. While the ground finches develop broad, blunt beaks, the cactus finches are known for their slender beaks. Gene chip technology has shown that the cactus finch embryos have more calmodulin in their embryonic beaks than the blunt-beaked ground finches (Figure 10.22). Calmodulin is a protein that combines with many enzymes to make their activity dependent upon calcium ions. In situ hybridization and other techniques have demonstrated that the calmodulin gene is expressed at higher levels in the cactus finch embryonic beaks than in the embryonic beaks of ground finches (see Figure 10.22). When calmodulin was experimentally up-regulated in the embryonic chicken beak, the chicken, too, developed a long beak. Thus, the genes for BMP4

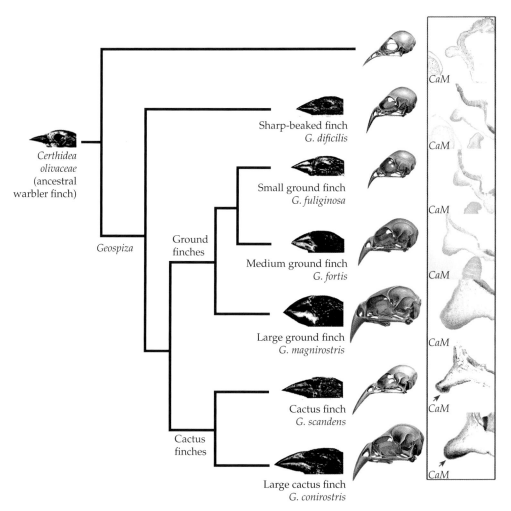

Figure 10.22   Correlation between the length of the beaks of Darwin's finches and the amount of calmodulin (*CaM*) gene expression. The various species of *Geospiza* are a monophyletic group (i.e., all are descended from a common ancestor); the differences in their beak morphologies can be seen skeletally. The *CaM* gene is expressed in a strong distal-ventral domain in the mesenchyme of the upper beak prominence of the large cactus finch *G. conirostris*, at somewhat lower levels in the cactus finch *G. scandens*, and at very low levels in the large ground finch and medium ground finch *G. magnirostris* and *G. fortis*, respectively. Very low levels of *CaM* gene expression were also detected in the mesenchyme of *G. difficilis*, *G. fuliginosa*, and the basal warbler finch *Certhidea olivacea*. (After Abzhanov et al. 2006; skull images from Bowman 1961.)

and calmodulin represent two targets for natural selection (Figure 10.23; Abzhanov et al. 2006). While natural selection will allow certain morphologies to survive, the generation of those morphologies depends upon variations of developmental regulatory genes such as those for BMP4 and calmodulin. It isn't the amount of adult beak components, but the amount of the embryonic construction units, that is being changed.

The molecular analysis of the genomes of these finches has also identified new genes for beak shape (Lamichhaney et al. 2015). One of the conclusions made by comparing the genomes of these finches is that their respective *Bmp4* genes are very similar and that it is unlikely that enhancer differences exist between them. This would mean that some other gene is probably involved in regulating the differing amounts of *Bmp4* transcription. These comparisons also identified the *Alx1* gene as a critical component in at least some of the species-specific differences in beak size. This gene is important in recruiting cranial neural crest cells into the frontonasal

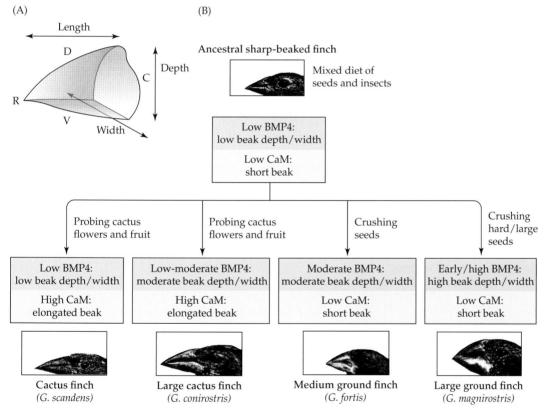

Figure 10.23 Beak evolution among Darwin's finches. (A) Bird beaks are three-dimensional structures that can change along any of the growth axes. C, caudal (tailward); D, dorsal (spineward); R, rostral (forward); V, ventral (belly). (B) A sharp-beaked finch represents the basal morphology for *Geospiza*, as the ancestral species is thought to have had such a morphology. The model for BMP4 and calmodulin involvement explains development of both the elongated and the deep/wide beaks of the derived species of *Geospiza*. (After Abzhanov et al. 2006.)

(A) Surface-dwelling
      populations

(B) Cave-dwelling
      populations

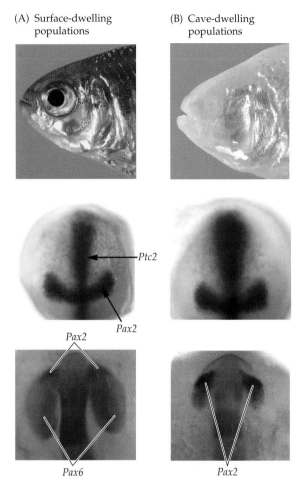

Figure 10.24 Surface-dwelling (A) and cave-dwelling (B) Mexican tetras (*Astyanax mexicanus*). The eye fails to form in the population that has lived in caves for more than 10,000 years (top right). Two genes that respond to Shh proteins, *Ptc2* and *Pax2*, are expressed in broader domains in the cavefish embryos than in those of surface dwellers (center). The embryonic optic vesicles (bottom) of surface-dwelling fish are of normal size and have small domains of *Pax2* expression (specifying the optic stalk). The optic vesicles of the cave-dwelling fishes' embryos (where *Pax6* is usually expressed) are much smaller, and the *Pax2*-expressing region has grown at the expense of the *Pax6* region. (From Yamamoto et al. 2004, photographs courtesy of W. Jeffery.)

process, and humans deficient in this gene have a syndrome of malformations involving the nose and forehead (Uz et al. 2010; Dee et al. 2013).

BLIND CAVEFISH   The Mexican tetra (cavefish), *Astyanax mexicanus*, has forms that live in caves and those that live on the surface. Those populations that live in the almost total darkness of subterranean caves have lost their eyes. This is due to an overexpression of the paracrine factor sonic hedgehog (Shh) from the prechordal plate of the head. Shh from the prechordal plate suppresses *Pax6* expression in the center of the neural tube, dividing the eye field into two districts (Figure 10.24). If the mouse *Shh* gene is mutated, or if the processing of this protein is inhibited, the single median eye field does not split. The result is called "cyclopia"—a single eye in the center of the face, usually below the nose (see Chiang et al. 1996). Conversely, if too much Shh is synthesized by the prechordal plate, *Pax6* is suppressed in too large an area and the eyes fail to form at all. This phenomenon may explain why cave-dwelling fish are blind. Yamamoto and colleagues (2004) demonstrated that the difference between surface populations of *A. mexicanus* and eyeless cave-dwelling populations of the same species is the amount of Shh secreted from the prechordal plate (see Figure 10.24). This eventually leads to the failure of proper lens induction and the degeneration of the eye (Ma et al. 2014). But why would elevated Shh and blindness be selected? The reason for the selection is probably that in caves, sight doesn't matter. What is important is that elevated Shh gives the organism larger nostrils, increased ability to recognize smells, and larger jaws. What was probably selected in cave-dwelling species is the heightened oral sensing for finding prey and larger jaws for consuming it (Yamamoto et al. 2009). In other places, the loss of eyes would be a lethal price to pay for these features, but not in the dark cave environments.

Heterotypy

In heterochrony, heterotopy, and heterometry, the mutations affect the regulatory regions of the gene. The gene's product—the protein—remains the same, although it may be synthesized in a new place, at a different time, or in different amounts. In **heterotypy**, the changes affect the protein that binds to these regulatory regions. The changes of heterotypy affect the actual coding region of the gene, and thus they can change the functional properties of the protein being synthesized. In agriculture, one of the most important evolutionary changes has been in maize (commonly known in the United States as corn).

(A)

(B)

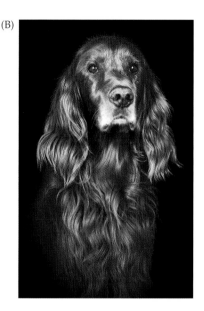

Figure 10.25   Mutations in the structural gene sequence that produces the protein product can yield changes in phenotype. (A) Mutations in the melanocortin receptor (expressed in the pigment cells) give red hair pigmentation rather than black pigmentation. (B) This gene also is mutated in many strains of dogs, including Irish setters.

Here, the transition from ancestral teosinte to maize was promoted by a single mutation of the gene encoding the transcription factor Tga1 (Wang et al. 2005). This mutation exposed the kernels on the cob, allowing them to be readily harvested for human consumption.

In humans, heterotypy helps produce such phenotypes as red hair color (caused by mutations in the gene encoding the receptor for the paracrine factor melanocortin), and increased hair thickness and number of eccrine sweat glands (mutations in the gene for the receptor for the paracrine factor ectodysplasin). Interestingly, both of these examples have to do with biogeography. There does not appear to be positive selection for red hair and the particular alleles of melanocortin receptor genes (Figure 10.25). Rather, there is selection *against* such alleles near the equator, since people having these genes make smaller amounts of protective melanin pigmentation (Harding et al. 2000).

WHY INSECTS HAVE ONLY SIX LEGS   Insects have six legs, whereas most other arthropod groups (spiders, millipedes, centipedes, and crustaceans) have many more. How is it that the insects came to form legs only in their three thoracic segments and have no appendages in their abdominal regions? The answer seems to reside in the relationship between Ultrabithorax (Ubx) protein and the *Distal-less* gene.

The *Ubx* gene is one of the Hox genes. It encodes a transcription factor that is used throughout the arthropods to activate those genes expressed in the posterior thorax and abdomen. In most arthropod groups, Ubx protein does not inhibit the *Distal-less* gene. However, in the insect lineage, a mutation occurred in the *Ubx* gene wherein the original 3′ end of the

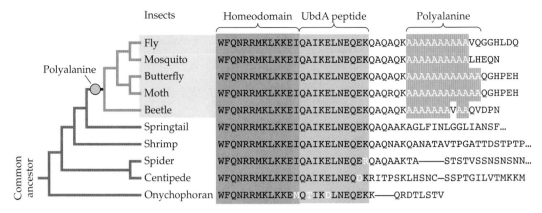

Figure 10.26 Changes in Ubx protein associated with the insect clade in the evolution of arthropods. Of all arthropods, only the insects have Ubx protein that is able to repress *Distal-less* gene expression and thereby inhibit abdominal legs. This ability to repress *Distal-less* is due to a mutation wherein a stretch of polyalanine residues is encoded in the carboxyl terminus of the Ubx protein. This mutation is seen only in the insect *Ubx* gene. (After Galant and Carroll 2002; Ronshaugen et al. 2002.)

protein-coding region was replaced by a group of nucleotides encoding a stretch of about 10 alanine residues (Figure 10.26; Galant and Carroll 2002; Ronshaugen et al. 2002). This polyalanine region represses *Distal-less* transcription. When a shrimp *Ubx* gene is experimentally modified to encode this polyalanine region, its Ubx, too, represses the *Distal-less* gene. The ability of insect Ubx to inhibit *Distal-less* thus appears to be the result of a gain-of-function mutation that characterizes the insect lineage. One cannot explain insect evolution without a developmental explanation of why limb development of this lineage has been so constrained.

GAMETE RECOGNITION AND POSSIBLE SPECIATION EVENTS   One of the most important of developmental mechanisms for speciation is provided by heterotypy of gamete recognition proteins. Here, a reproductive barrier is established within a species by allowing only some of the eggs to pair with only some of the sperm. Proteins that mediate species-specific interactions between the sperm and egg—**gamete recognition proteins**—are the fastest-evolving proteins known in the animal kingdom (Swanson and Vacquier 2002). Even when closely related species have near identity at every other protein, their gamete recognition proteins may have diverged significantly. In such cases, the sperm and egg cannot bind one another, and fertilization is thwarted. Such divergence causes reproductive isolation and speciation (Biermann et al. 2004; Palumbi and Lessios 2005). As Levitan and Ferrell (2006) have concluded, "gamete recognition proteins can evolve at astonishing rates and lie at the heart of reproductive isolation and speciation in diverse taxa." Indeed, the coevolution of the male and female gamete recognition proteins shows the hallmarks of sexual selection and coevolved genes. It is thought that this coevolution between the genes encoding male and female gamete recognition proteins can lead to reproductive barriers

(A)

(B)

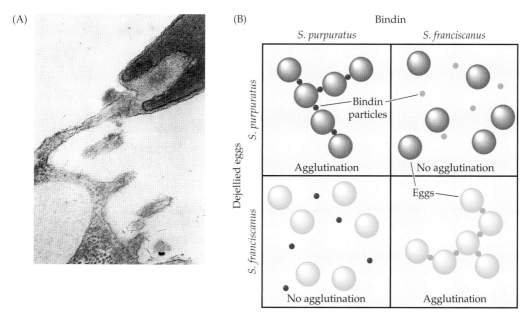

Figure 10.27 Species-specific binding of acrosomal process to egg surface in sea urchins. (A) Actual contact of a sea urchin sperm acrosomal process with an egg microvillus. (B) In vitro model of species-specific binding. The agglutination of dejellied eggs by bindin was measured by adding bindin particles to a plastic well containing a suspension of eggs. After 2–5 minutes of gentle shaking, the wells were photographed. Each bindin bound to and agglutinated only eggs from its own species. (A from Epel 1977, photograph courtesy of F. D. Collins and D. Epel; B based on photographs in Glabe and Vacquier 1977.)

that have the potential to drive speciation by dividing a species into two different mating groups (see Clark et al. 2009; Figure 10.27).

Indeed, the beginnings of speciation might be seen in the separation of a species into two reproductively isolated populations. DNA sequencing has suggested that such gamete-induced genetic isolation has occurred in certain sea stars, abalones, and sea urchins. Here, different alleles can be observed encoding the proteins of the bindin (on the sperm) and bindin receptor (egg) proteins. In the Canadian sea star (starfish) *Patiria miniata*, the genes encoding the gamete recognition proteins in the southern area of its range have different allelic frequencies than in the northern range. What's more, sperm from the southern part of the range has more successful fertilization with the eggs from the southern region, and this has been correlated with the genotypes of the bindin protein and its receptor (Hart et al. 2014). In the sea urchin *Strongylocentrotus purpuratus*, there are numerous alleles of the genes encoding bindin and the bindin receptor. Interestingly, even though the genes for these proteins are not physically linked, there is linkage disequilibrium for these genes (i.e., these genes are found together in a population more often than would be expected from their frequency in the population), indicating that there is assortative mating and that these gamete-recognition genes are likely to be coevolving within a population (Plata Strapper 2013).

In summation, form appears to evolve largely by altering the timing, location, and amount of gene expression. The heterotopy, heterochrony, and heterometry are accomplished by changing the enhancer elements regulating gene expression. Heterotypy can alter the transcription factors that bind to these *cis*-regulatory sequences, giving these proteins different properties. As Carroll (2008) has noted, the importance of the developmental phenomena to evolution "constitutes a developmental genetic theory of morphological evolution which can supplement and extend the Modern Synthesis of evolutionary biology and population genetics."

Developmental Constraints on Evolution

There are only about three-dozen animal phyla, representing all the major body plans of the animal kingdom. One can easily envision other types of body plans and imagine animals that do not exist (science fiction writers do it all the time). So why aren't more body plans found among the animals? To answer this, we have to consider the constraints that development imposes on evolution.

The number and forms of possible phenotypes that can be created are limited by the interactions that are possible among molecules and between modules. These molecular interactions also allow change to occur in certain directions more easily than in others. Collectively, the restraints on phenotype production are called **developmental constraints**. These developmental constraints on evolution fall into three major categories: physical, morphogenetic, and phylogenetic (see Richardson and Chipman 2003).

Physical constraints

The laws of diffusion, hydraulics, and physical support are immutable and will permit only certain physical phenotypes to arise (Forgacs and Newman 2005; von Dassow and Davidson 2008). A vertebrate on wheeled appendages such as Dorothy saw in Oz cannot exist, because in reality blood cannot circulate through a rotating organ; this entire evolutionary avenue is closed off. Similarly, structural parameters and fluid dynamics would prohibit the existence of 6-foot-tall mosquitoes or 25-foot-long leeches.

Morphogenetic constraints

Rules for morphogenetic construction also limit the phenotypes that are possible (Oster et al. 1988). Bateson (1894) and Alberch (1989) noted that when organisms depart from their normal development, they do so in only a limited number of ways. Some of the best examples of these types of constraints come from the analysis of limb formation in vertebrates. Although there have been many modifications of the vertebrate limb over more than 300 million years, some modifications (such as a middle digit shorter than its surrounding digits) are not found (Holder 1983). Moreover, analyses of natural populations suggest that there is a relatively small number of ways in which limb changes can occur (Wake and Larson 1987). If a longer limb is favorable in a given environment, an elongated humerus variant may become favored by selection, but one never sees two smaller humeri joined together in tandem, although one could imagine the selective advantages that such an arrangement might have. This observation suggests a limb construction scheme that follows certain rules.

These rules are now becoming known. The chemical principles of the reaction-diffusion model (see box) appear to govern the architecture of the limb (Newman and Müller 2005). Oster and colleagues (1988) found that this model can explain the known morphologies of the limb, and it could explain why other morphologies are forbidden. The reaction-diffusion equations predicted the observed succession of bone development from stylopod (humerus/femur) to zeugopod (ulna-radius/tibia-fibula) to autopod (hand/foot). If limb morphology is indeed determined by the reaction-diffusion mechanism, then spatial features that cannot be easily generated by reaction-diffusion kinetics will not readily occur.

Phylogenetic constraints

Phylogenetic constraints on the evolution of new structures are historical restrictions based on the history of an organism's development (Gould and Lewontin 1979). Once a structure comes to be generated by inductive interactions, it is difficult to start over again. The notochord, for example, which is functional in adult protochordates such as amphioxus (Berrill 1987), degenerates in adult vertebrates. Yet it is transiently necessary in vertebrate embryos, where it specifies the neural tube. Similarly, Waddington (1938) noted that although the pronephric kidney of the chick embryo is considered vestigial (since it has no ability to concentrate urine), it is the source of the ureteric bud that induces the formation of a functional kidney during chick development.

As genes acquire new functions during the course of evolution, they may become involved in more than one module, making change difficult. This ability of a gene to play multiple roles in different cells is called **pleiotropy**. In a sense, pleiotropy is the *opposite* of modularity, involving interdependent connections between parts rather than their independence. Galis and colleagues (2002) provide evidence that the reason the segment polarity gene network is conserved in all types of insects is that these genes play roles in several different pathways. Such pleiotropy constrains the possibilities for alternative mechanisms, since it makes change difficult.

Pleiotropies may underlie some of the constraints seen in mammalian development. Galis speculates that mammals have only seven cervical vertebrae (while birds may have dozens) because the Hox genes that specify these vertebrae have become linked to cell proliferation in mammals (Galis 1999; Galis and Metz 2001). Thus, changes in Hox gene expression that might facilitate evolutionary changes in the skeleton might also misregulate cell proliferation and lead to cancers. She supports this speculation with epidemiological evidence showing that changes in skeletal morphology correlate with childhood cancer. Selection against embryos having more or fewer than seven cervical ribs appears to be remarkably strong: at least 78% of human embryos with an extra anterior rib (i.e., six cervical vertebrae) die

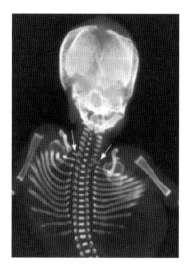

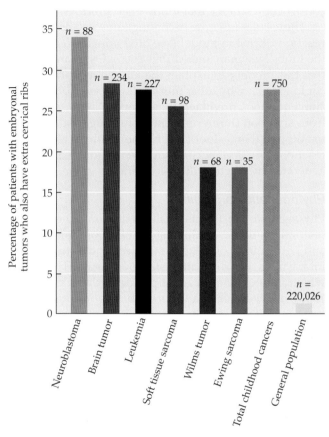

Figure 10.28 Nearly 80% of fetuses with extra cervical ribs (arrows) die before birth. Those surviving have extremely high rates of cancers. The association of this axial abnormality with lower fitness suggests pleiotropy, namely that the specification of the cervical ribs is associated with the regulation of cell division. Thus, it is difficult to change axial specification (in this case, the number of cervical ribs) without causing abnormalities of mitosis that can lead to cancer. (From Galis et al. 2006.)

before birth, and 83% die before they are a year old (Figure 10.28). These deaths appear to be the result of multiple congenital anomalies or cancers (Galis et al. 2006).

Looking Ahead

Traditionally, there have been two major research programs in evolutionary biology. The first is the study of adaptation, looking at the modification of characters through natural selection. The second program has been speciation, looking at how populations split into distinctly separate lineages of descent (Wagner and Lynch 2010). There had also been a third program, but one that had been eclipsed since the Modern Synthesis of population

Reaction-Diffusion Models

Some constraints arise from the mathmatics of chemical diffusion. One of the most important mathematical models in developmental biology was formulated by Alan Turing (1952), one of the founders of computer science (and the mathematician who cracked the German "Enigma" code during World War II). He proposed a model wherein two homogeneously distributed substances would interact to produce stable patterns during morphogenesis. These patterns would represent regional differences in the concentrations of the two substances. Their interactions would produce an ordered structure out of random chaos.

Turing's reaction-diffusion model involves two substances. Substance P promotes the production of more substance P as well as substance S. Substance S, however, inhibits the production of substance P. Turing's mathematics show that if S diffuses more readily than P, sharp waves of concentration differences will be generated for substance P (Figure A,B). These waves have been observed in certain chemical reactions (Prigogine and Nicolis 1967; Winfree 1974).

The reaction-diffusion model predicts alternating areas of high and low concentrations of some substance.

When the concentration of such a substance is above a certain threshold level, a cell (or group of cells) may be instructed to differentiate in a certain way. An important feature of Turing's model is that particular chemical wavelengths will be amplified while all others will be suppressed. As local concentrations of P increase, the values of S form a peak centering on the P peak, but becoming broader and shallower because of S's more rapid diffusion. These S peaks inhibit other P peaks from forming. But which of the many P peaks will survive? That depends on the size and shape of the tissues in which the oscillating reaction is occurring. The mathematics describing which particular wavelengths are selected consists of complex polynomial equations. Such functions have been used to model the spiral patterning of slime molds, the polar organization of the limb, the radiation of mammalian teeth, and the pigment patterns of mammals, fishes, and snails. The pattern of the scutes on turtle shells can be explained by reaction-diffusion mechnisms, and the mathematics of these reactions also explain why nearly all turtle species have the same number and pattern of scutes, independent of whether they inhabit deserts or oceans (Moustakas-Verho et al. 2014).

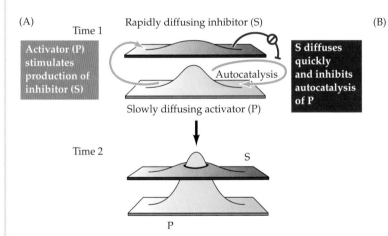

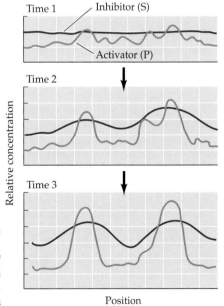

Reaction-diffusion (Turing model) system of pattern generation. Generation of periodic spatial heterogeneity can come about spontaneously when two reactants, S and P, are mixed together under the conditions that S inhibits P, P catalyzes production of both S and P, and S diffuses faster than P. (A) The conditions of the reaction-diffusion system yielding a peak of P and a lower peak of S at the same place. (B) The distribution of the reactants is initially random, and their concentrations fluctuate over a given average. As P increases locally, it produces more S, which diffuses to inhibit more peaks of P from forming in the vicinity of its production. The result is a series of P peaks ("standing waves") at regular intervals.

(Continued on next page)

Reaction-Diffusion Models *(continued)*

(C,D) Computer simulations of chick limb development. A computer model of limb development can be made after providing experimentally derived parameters of limb bud geometry, and the diffusion, synthesis, and degradation of activators and inhibitors. (C) The model predicts the sequential formation of the single stylopod, the double zeugopod, and the multiple cartilagenous rods of the autopod. (D) Interestingly, changing the parameters slightly will also generate the observed limb skeletons found in fossils such as the fishlike aquatic reptile *Brachypterygius* (whose forelimb had a paddle-like shape) and *Sauripterus*, one of the first land-dwelling reptiles. (After Zhu et al. 2010.)

This reaction-diffusion pattern has had special importance in forming the limbs. The proximal-distal axis of the limbs includes a single bone, a double bone, a wrist/ankle, and several digits (fingers or toes). By modulating the kinetics of diffusion and degradation, one can mathematically acquire those forms intermediate between limb and fin, and many of these forms are known from the fossil record (Figure C,D; Zhu et al. 2010). The rules of the reaction-diffusion system mandate that the fingers are never between other bones; nor is there a duplication of the humerus or the paired bones, no matter how advantageous that might be. In other words, the Turing model accounts for why there are certain morphologies we never see. (For example, we never see striped animals with spotted tails, whereas spotted animals with striped tails are rather common.) There are rules of development that allow evolution to go in some directions, but not in others.

Recent experiments (Raspopovic et al. 2014) have identified a possible set of paracrine factors—BMPs and Wnts—that interact through reaction-diffusion to generate precisely the type of patterns found in vertebrate digits, where the expression of the Sox9 gene (which is critical in generating cartilage) becomes amplified in those digit regions and downregulated in the webbing regions. The number of digits permitted is generated by the Hox and Fgf proteins that shape the physical space of the limb bud. Remarkably, the patterns of gene expression are consistent with the fin-to-limb placement of bony elements and predict the correct number of digits in the five-toed mouse, the two-toed camel, and the one-toed horse (Sheth et al. 2012; Cooper et al. 2014; Cooper 2015).

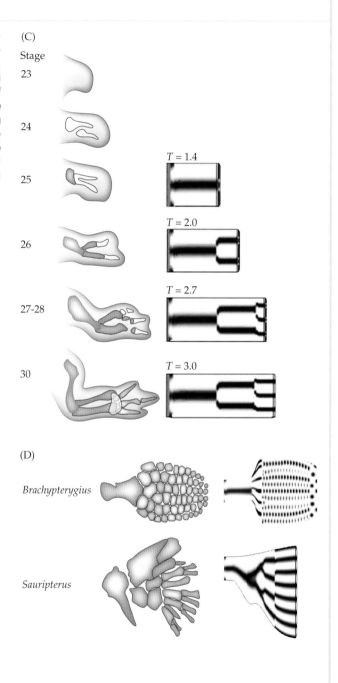

genetics and evolution. This was the study of how novelty arose in evolution. Where the first two programs emphasized the survival of the fittest, this last program combined evolution and development to look at the arrival of the fittest. It was a program championed by Darwin's friend and champion, Thomas Huxley. While Darwin saw evolution as the result of natural selection, Huxley saw evolution in terms of embryonic development. He wrote in 1893 (p. 202), "Evolution is not a speculation but a fact; and it takes place by epigenesis." Darwin was emphasizing a theory of population change; Huxley was emphasizing a theory of body construction. Both theories are correct.

Huxley emphasized (see Huxley 1880) that evolution depends on heritable changes in development. When we say that the contemporary one-toed horse has an ancestor with five toes, we are saying that during equine evolution, there have been changes in the placement and growth of its limb cartilage cells (Huxley 1880; Wolpert 1983). As we've seen, these changes may be caused by alterations in the timing, placement, and amount of gene expression. They can also be caused by changes in the protein being synthesized, and this is especially important when the protein is involved in organ formation or cell differentiation. Modularity and gene duplication have enabled these changes to occur without causing major disastrous changes in the developing organism. However, in some cases, the possible routes of evolutionary change are constrained by developmental and physical parameters. We are on the verge of combining Darwin's and Huxley's insights in several new: (1) the reintegration of population genetics and developmental genetics, (2) the combination of genomics and mathematical models of development to account for the emergence of new morphologies, (3) the reinterpretation of the fossil record in the light of new discoveries in animal development, and (4) the creation of novel cell types through genetic re-integration.

Population genetics of regulatory alleles

Originally the Modern Synthesis looked at genes in a theoretical way, because the original theorists did not know that DNA was the basis of genetic material, nor could they know the DNA sequence differences that formed the alleles. Once protein and DNA sequencing became possible (as mentioned in the previous chapter), the Modern Synthesis focused on the variation produced by differences in the protein-encoding sequences of DNA (such as those for globin and rhodopsin). However, we now know that the variations that natural selection works on are most often produced by differences in the *regulatory regions of genes* active in constructing the embryo. With the ability to sequence entire genomes, we are going to be able to look at something that may provide incredible knowledge of evolution: the population genetics of regulatory alleles. For instance, there are numerous polymorphisms in the enhancer elements around particular genes. This allows members of a population to have different degrees of regulating the gene and to have the gene respond to different stimuli (Hahn et al. 2004; Goering et al. 2009; Kasowski et al. 2010; Garfield et al. 2012). One such new program is underway (Salazar-Ciudad and Jernvall 2010) to sequence the genome of every member of the ringed seal subspecies *Phoca hispida ladogensis* and to correlate the regulatory alleles to the dental morphology of each seal in the population.* The combination of population genetics and developmental genetics may be one of the most important projects in modern evolutionary biology, integrating genotype, development, phenotype, and evolutionary fitness.

Mathematically modeling developmental change

Related to this is the ability to mathematically model the development of organs and see how genetic factors change them. Based on a mathematical model of tooth development predicting certain reaction-diffusion processes, a recent paper was able to "replay" the fossil record of rodent tooth evolution by changing paracrine factor concentrations as the tooth bud developed! Starting with a "generic" tooth that lacked specific cusps and ridges, Harjunmaa and colleagues (2014) were able to restore the evolutionary sequence of events leading to the three-cusped molar by adding increasing concentrations of the paracrine factor ectodysplasin, and they were able to reconstitute the ridge pattern by manipulating expression of sonic hedgehog and fibroblast growth factor 3. Small amounts of ectodysplasin (EDA) caused the formation of the trigonid cusp (the first part of the molar to have evolved and which is used for shearing food), while higher concentrations brought forth the talonid cusp, which evolved more recently and which is also used for shearing food (Figure 10.29).

This brings us to another frontier of evo-devo: *evolvability*, the very capacity of a system to adaptively change (Kirschner and Gerhart 1998; Arthur 2004; Pigliucci 2008). How can mutations transform certain genotypes into particular phenotypes? This involves both the ability to generate novel phenotypes and the restrictions on producing certain phenotypes. Thus, novelties can arise only from those subsets of possibilities that are both permitted by the constraints and performable by the genome. As mentioned earlier, evolutionary explanation requires knowledge of the developmental dynamics between the gene changes and the anatomical changes. Recent studies (Jiménez et al. 2015) demonstrate mathematically that the dynamics of gene networks are critical in getting a phenotype from the genome and that these dynamic parameters both allow and constrain particular types of morphological variation. That is to say, the way a gene regulatory network is constituted is itself a developmental constraint. This may enable us ask another new question—What biases evolution in particular directions?

Rereading the fossil record

Rereading the fossil record in the light of developmental biology has given new insights into evolution. We mentioned earlier that bird skulls may have evolved by heterochronic changes of dinosaur skulls. Another important example has been finding the place of turtles in the history of the Earth. Until 2013, one of the biggest embarrassments of vertebrate paleontology

---

*Unlike other mammals, seals do not have occlusion—the upper teeth do not need to fit into the lower ones. Thus, an entire layer of developmental constraints is lacking in seals. This allows one to look at what morphological variants are possible without having selection for functional occlusion.

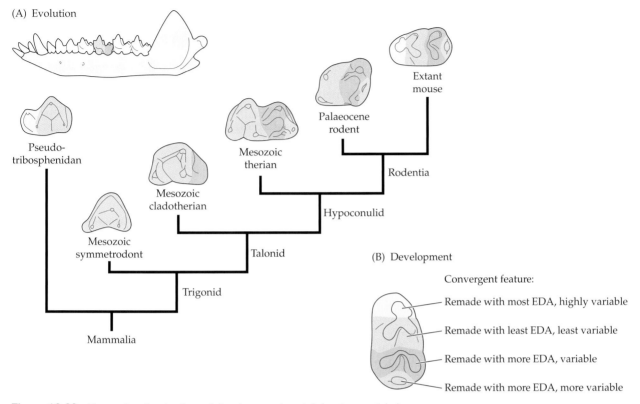

(A) Evolution

Pseudo-
tribosphenidan

Mesozoic
symmetrodont

Mesozoic
cladotherian

Mesozoic
therian

Palaeocene
rodent

Extant
mouse

Rodentia

Hypoconulid

Talonid

Trigonid

Mammalia

(B) Development

Convergent feature:

Remade with most EDA, highly variable

Remade with least EDA, least variable

Remade with more EDA, variable

Remade with more EDA, more variable

Figure 10.29   Reconstructing tooth evolution by experimental developmental change. (A) Scheme of tooth evolution showing that as mammals evolved, their teeth became more complex. Trigonid cusps (tan) developed in early mammals (symmetrodonts in the Mesozoic), while the talonid cusps are seen in a slightly later group of mammals called the cladotherians. Hypoconulids (green) evolved from these, and the anteroconulids are seen only in advanced rodents. (B) A loss of all cusps occurs when teeth are deficient in ecto-dysplasin (EDA), a paracrine factor made in the tooth signaling center. By the incremental addition of EDA to the EDA-deficient tooth, these features are brought back sequentially Moreover, the features that evolved the earliest are the most stable. Further changes (the ridges between cusps) can be made by modifying sonic hedgehog and Fgf3 levels. (After Luo 2014.)

(and one that the creationists constantly took advantage of) was that we didn't know the ancestor of the turtle. Some paleontologists said turtles were related to birds; others said they was related to crocodiles; one group said turtles were related to lizards; and another, large, group of scientists said that turtles arose before the other groups of reptiles and were not related much to any of the others.

But when Lyson and colleagues (2013) looked carefully at the fossil bones of *Eunotosaurus*, they saw evidence that it had the same developmental pattern as turtles. In turtles, the ribs do not form a rib cage; rather they migrate laterally into the dermis. This appears to be due to a heterotopy—Fgf10 signaling in the dermis. As a result, turtle ribs have different shapes than other ribs, and they also lack the attachments for the costal muscles (that make spareribs in beef and pork species). The turtle-specific pattern of rib growth was seen in *Eunotosaurus*, linking this non-turtle reptile with the turtles. As a result of this, turtles are now seen by most biologists as being allied to the birds and crocodiles. This has recently been corroborated of genomic studies (Field et al. 2014; Crawford et al. 2014). The combination of developmental genetic, genomic, and paleontological evidence has finally found a home for the turtles in the fossil record.

As we've seen in this chapter, developmental biology can show how the various new groups of animals could be formed. Thewissen and colleagues (2012) have noted:

> Whereas paleontology strives to determine "What happened in evolution?," developmental genetics uses gene control in embryos to try to answer "How did it happen?" Combined, the two approaches can lead to remarkable insights that benefit both fields.

How does one evolve new cells?

So far, we've been discussing the creation of new organs by tinkering with existing possibilities. Turtle shells, webbed feet, and bat wings are modifications of bones and soft tissue altered by processes common to all vertebrates. They are new structures, but they are made of old material. Let's go to a new level of novelty: new material. How does one make a new cell type? This is a fascinating area, and it concerns recombining genes into new functional pathways (see Wagner and Lynch 2010; Wagner 2014; Schlosser 2015; Simões-Costa and Bronner 2015). The chordates invented several new cell types: the notochord was first seen in the earliest chordates (indeed, defining them as a new group); the neural crest cell and dentine cells were first seen in the vertebrate lineage; and the decidual stromal cells necessary for intrauterine pregnancy were first seen made by placental mammals (Hall and Gillis 2012; Wagner et al. 2014). How do these cells emerge? It appears that the protochordates (such as tunicates) have all the genes necessary to form a neural crest cell, but they are not yet connected together in a way that makes for a stable cell type (Green and Bronner 2014; Simões-Costa et al. 2014). The mechanisms for their successful "rewiring" make for a critically important research program. As Wagner and Lynch (2010) have written, "The bottom line for any study of evolutionary novelties is explaining the origin of the gene regulatory network that executes organ-specific gene expression."

One hint for how such reassociations can be made comes from studies of the decidual stromal cells. These are the uterine cells that regulate mammalian pregnancy, coordinating with the mammalian embryo to make the placenta and the uterine vasculature, as well as suppressing the maternal immune response against the embryo. In this case, transposons (mobile genetic elements originating from viruses) appear to have altered the expression of certain transcription factors, and the evolution of these transcription factors allows them to pair with partners they had not interacted with previously (Lynch et al. 2008; Brayer et al. 2011). Remarkably, thousands of enhancer sequences that mediate the identity and functions of the decidual stromal cells are derived from ancient transposons. These enhancers contain hormone-responsive elements, especially those that bind the receptor of progesterone, the major hormone of pregnancy maintenance. Thus, it appears that decidual stromal cells, a new cell type found in placental mammals and that permits long-term pregnancy, were derived from the uterine fibroblasts of organisms that acquired new enhancers from retroviruses (Emera et al. 2012; Lynch et al. 2015; Kin et al. 2015). Genes that had been expressed in other organs became expressed in the uterine stromal cells, giving them new functions, such as immunosuppression and maternal-fetal signaling. Such transposon-mediated "rewiring" of genetic regulatory networks would provide a rapid mechanism for distributing nearly identical copies of enhancers that would be receptive to the same signal throughout the genome. The origin of novel cell types is an entirely new question that can now be asked, and this idea that evolution be mediated by genetic elements and signals originating from other organisms will be discussed further in the next chapter.

References

Abzhanov, A., W. P. Kuo, C. Hartmann, B. R. Grant, P. R. Grant and C. J. Tabin. 2006. The calmodulin pathway and evolution of elongated beak morphology in Darwin's finches. *Nature* 442: 563–567.

Abzhanov, A., M. Protas, B. R. Grant, P. R. Grant and C. J. Tabin. 2004. *Bmp4* and morphological variation of beaks in Darwin's finches. *Science* 305: 1462–1465.

Alberch, P. 1989. The logic of monsters: Evidence for internal constraints in development and evolution. *Geobios* (Lyon), Mémoires Spécial 12: 21–57.

Amundson, R. 2005. *The Changing Role of the Embryo in Evolutionary Thought: Structure and Synthesis.* Cambridge University Press, Cambridge.

Appel, T. A. 1987. *The Cuvier-Geoffroy Debate: French Biology in the Decades before Darwin.* Oxford University Press, New York.

Arthur, W. 2004. *Biased Embryos and Evolution.* Cambridge University Press, Cambridge.

Barinaga, M. 2003. Newborn neurons search for meaning. *Science* 299: 32–34.

Bates, H. W. 1862. Contributions to an insect fauna of the amazon valley (Lepidoptera: Heliconidae). *Trans. Linn. Soc. Lond.* 23: 495–556.

Bateson, W. 1894. *Materials for the Study of Variation.* Cambridge University Press, Cambridge.

Berrill, N. J. 1987. Early chordate evolution. I. *Amphioxus*, the riddle of the sands. *Int. J. Invert. Reprod. Dev.* 11: 1–27.

Bertrand, N., D. S. Castro and F. Guillemot. 2002. Proneural genes and the specification of neural cell types. *Nature Rev. Neurosci.* 3: 517–530.

Bhullar, B. A., J. Marugán-Lobón, F. Racimo, G. S. Bever, T. B. Rowe, M. A. Norell and A. Abzhanov. 2012. Birds have paedomorphic dinosaur skulls. *Nature* 487: 223–226.

Biermann, C. H., J. A. Marks, A. C. Vilela-Silva, M. O. Castro and P. A. Mourao. 2004. Carbohydrate-based species recognition in sea urchin fertilization: Another avenue for speciation? *Evol. Dev.* 6: 353–361.

Bogin, B. 1997. Evolutionary hypotheses for human childhood. *Yrbk. Phys. Anthropol.* 40: 63–89.

Bolker, J. A. 2000. Modularity in development and why it matters to evo-devo. *Am. Zool.* 40: 770–776.

Bowman, R. I. 1961. Morphological differentiation and adaptation in the Galapagos finches. *Univ. Calif. Publ. Zool.* 58: 1–302.

Boyd, J. L. and 7 others. 2015. Human-chimpanzee differences in a FZD8 enhancer alter cell-cycle dynamics in the developing neocortex. *Curr. Biol.* 25: 772–779.

Brayer K. J., V. J. Lynch and G. P. Wagner. 2011. Evolution of a derived protein-protein interaction between HoxA11 and Foxo1a in mammals caused by changes in intramolecular regulation. *Proc. Natl. Acad. Sci. USA* 108(32): E414–E420.

Brown, F. D., A. Prendergast and B. J. Swalla. 2008. Man is but a worm: Chordate origins. *Genesis* 46: 605–613.

Burke, A. C., A. C. Nelson, B. A. Morgan and C. Tabin. 1995. Hox genes and the evolution of vertebrate axial morphology. *Development* 121: 333–346.

Cabanac, M. and M. Caputa. 1979. Natural selective cooling of the human brain: Evidence for its occurrence and magnitude. *J. Physiol.* 286: 255–264.

Caldwell, M. W. and M. S. Y. Lee. 1997. A snake with legs from the marine Cretaceous of the Middle East. *Nature* 386: 705–709.

Carroll, S. B. 2005. *Endless Forms Most Beautiful: The New Science of Evo Devo.* W. W. Norton, New York.

Carroll, S. B. 2008. Evo-devo and an expanding evolutionary synthesis: A genetic theory of morphological evolution. *Cell* 134: 25–36.

Carroll, S. B. 2009. *Remarkable Creatures: Epic Adventures in the Search for the Origins of Species.* Houghton Mifflin Harcourt, Boston, MA.

Chan, Y. F. and 15 others. 2010. Adaptive evolution of pelvic reduction in sticklebacks by recurrent deletion of a Pitx1 enhancer. *Science* 327: 302–305

Charrier, C. and 10 others. 2012. Inhibition of SRGAP2 function by its human-specific paralogs induces neoteny during spine maturation. *Cell* 149: 923–935.

Chiang, C., Y. Litingtung, E. Lee, K. E. Young, J. L. Cordoen, H. Westphal and P. A. Beachy. 1996. Cyclopia and axial patterning in mice lacking *sonic hedgehog* gene function. *Nature* 383: 407–413.

Chow, R. L., C. R. Altmann, R. A. Lang and A. Hemmati-Brivanlou. 1999. *Pax6* induces ectopic eyes in a vertebrate. *Development* 126: 4213–4222.

Clark, N. L., J. Gasper, M. Sekino, S. A. Springer, C. F. Aquadro and W. J. Swanson. 2009. Coevolution of interacting fertilization proteins. *PLoS Genet.* 5: e1000570.

Cohn, M. J. and C. Tickle. 1999. Developmental basis of limblessness and axial patterning in snakes. *Nature* 399: 474–479.

Colosimo, P. F., C. L. Peichel, K. Nereng, B. K. Blackman, M. D. Shapiro, D. Schluter and D. M. Kingsley. 2004. The genetic architecture of parallel armor plate reduction in threespine sticklebacks. *PLoS Biol.* 2: 635–641.

Colosimo, P. F. and 9 others. 2005. Widespread parallel evolution in sticklebacks by repeated fixation of ectodysplasin alleles. *Science* 307: 1928–1933.

Cooper, K. L. 2015. Self-organization in the limb: A Turing mechanism for digit development. *Curr. Opin. Genet. Dev.* 32: 92–97.

Cooper, K. L. and 8 others. 2014. patterning and post-patterning modes of evolutionary digit loss in mammals. *Nature* 511: 41–45.

Crawford, N. G. and 8 others. 2015. A phylogenomic analysis of turtles. *Mol. Phylogenet. Evol.* 83: 250–257.

Dale, L. and F. C. Wardle. 1999. A gradient of BMP activity specifies dorsal–ventral fates in early *Xenopus* embryos. *Sem. Cell Dev. Biol.* 10: 319–326.

Damen, W. G. 2002. *Fushi tarazu*: A Hox gene changes its role. *BioEssays* 24: 992–995.

Davidson, E. H. 2006. *The Regulatory Genome: Gene Regulatory Networks in Development and Evolution*. Academic Press, New York.

Davidson, E. H. 2011. Evolutionary bioscience as regulatory systems biology. *Dev. Biol.* 357: 35–40.

Dee, C. T., C. R. Szymoniuk, P. E. Mills and T. Takahashi. 2013. Defective neural crest migration revealed by Zebrafish model of Alx1-related frontonasal dysplasia. *Hum. Mol. Genet.* 22: 239–251.

De Robertis, E. M. and Y. Sasai. 1996. A common plan for dorso-ventral patterning in Bilateria. *Nature* 380: 37–40.

de Rosa, R. and 7 others. 1999. Hox genes in brachiopods and priapulids and protostome evolution. *Nature* 379: 772–776.

Dennis, M. Y. and 15 others. 2012. Evolution of human-specific neural *SRGAP2* genes by incomplete segmental duplication. *Cell* 149: 912–922.

Di-Poï, N., J. I. Montoya-Burgos, H. Miller, O. Pourquié, M. C. Milinkovitch and D. Duboule. 2010. Changes in Hox genes' structure and function during the evolution of the squamate body plan. *Nature* 464: 99–103.

Diamond, J. 2002. *The Third Chimpanzee: The Evolution and Future of the Human Animal*. HarperCollins, New York.

Dobzhansky, Th. 1955. *Evolution, Genetics, and Man*. Wiley, New York.

Dobzhansky, Th. 1964. Biology, molecular and organismic. *Amer. Zool.* 4: 443–452 (p. 443).

DuBuc, T. Q., J. F. Ryan, C. Shinzato, N. Satoh and M. Q. Martindale. 2012. Coral comparative genomics reveal expanded Hox cluster in the cnidarian-bilaterian ancestor. *Integr. Comp. Biol.* 52: 835–841.

Emera, D., C. Casola, V. J. Lynch, D. E. Wildman, D. Agnew and G. P. Wagner. 2012. Convergent evolution of endometrial prolactin expression in primates, mice, and elephants through the independent recruitment of transposable elements. *Mol. Biol. Evol.* 29: 239–247.

Enard, W. and 12 others. 2002a. Intra- and interspecific variation in primate gene expression patterns. *Science* 296: 340–343.

Enard, W. and 7 others. 2002b. Molecular evolution of *FOXP2*, a gene involved in speech and language. *Nature* 418: 869–872.

Epel, D. 1977. The program of fertilization. *Sci. Am.* 237(5): 128–138.

Erwin, D. H. 1999. The origin of bodyplans. *Am. Zool.* 39: 617–629.

Falk, D. 1990. Brain evolution in *Homo*: The "Radiator" theory: *Behav. Brain Sci.* 13: 333–381.

Field, D. J., J. A. Gauthier, B. L. King, D. Pisani, T. R. Lyson and K. J. Peterson. 2014. Toward consilience in reptile phylogeny: miRNAs support an archosaur, not lepidosaur, affinity for turtles. *Evol. Dev.* 16: 189–196.

Forgacs, G. and S. A. Newman. 2005. *Biological Physics of the Developing Embryo*. Cambridge University Press, Cambridge.

Galant, R. and S. B. Carroll. 2002. Evolution of a transcriptional repression domain in an insect Hox protein. *Nature* 415: 910–913.

Galis, F. 1999. Why do almost all mammals have seven cervical vertebrae? Developmental constraints, Hox genes, and cancer. *J. Exp. Zool./Mol. Dev. Evol.* 285: 19–26.

Galis, F. and J. A. Metz. 2001. Testing the vulnerability of the phylotypic stage: On modularity and evolutionary conservation. *J. Exp. Zool./Mol. Dev. Evol.* 29: 195–204.

Galis, F., T. J. M. van Dooren and J. A. Metz. 2002. Conservation of the segmented germband stage: Robustness or pleiotropy? *Trends Genet.* 18: 504–509.

Galis, F. and 7 others. 2006. Extreme selection in humans against homeotic transformations of cervical vertebrae. *Evolution* 60: 2643–2654.

Garfield, D., R. Haygood, W. J. Nielsen and G. A. Wray. 2012. Population genetics of *cis*-regulatory sequences that operate during embryonic development in the sea urchin *Strongylocentrotus purpuratus*. *Evol. Dev.* 14: 152–167.

Gaunt, S. J. 1994. Conservation in the Hox code during morphological evolution. *Int. J. Dev. Biol.* 38: 549–552.

Gehring, W. J. 1998. *Master Control Genes in Development and Evolution: The Homeobox Story*. Yale University Press, New Haven, CT.

Gehring, W. 2005. New perspectives on eye development and the evolution of eyes and photoreceptors. *J. Heredity* 96: 171–184.

Gehring, W. J., U. Kloter and H. Suga. 2009. Evolution of the Hox gene complex from an evolutionary ground state. *Curr. Top. Dev. Biol.* 88: 35–61.

Geschwind, D. H. and G. Konopka. 2012. Neuroscience: Genes and human brain evolution. *Nature* 486: 481–482.

Gilbert, S. F. 1996. Cellular dialogues in organogenesis. In M. E. Martini-Neri, G. Neri and J. M. Opitz (eds.), *Gene Regulation and Fetal Development: Proceedings of the Third International Workshop on Fetal Genetic Pathology, June 3–6, 1993*. Wiley-Liss, New York, pp. 1–12.

Gilbert, S. F. 2006. The generation of novelty: The province of developmental biology. *Biol. Theory* 1: 209–212.

Gilbert, S. F. 2013. *Developmental Biology*, 10th Ed. Sinauer Associates, Sunderland, MA.

Gilbert, S. F. and R. Burian. 2003. Development, evolution, and evolutionary developmental biology. In B. K. Hall and W. M. Olson (eds.), *Key Concepts and Approaches in Evolutionary Developmental Biology*. Harvard University Press, Cambridge, MA, pp. 61–68.

Gilbert, S. F., J. M. Opitz and R. A. Raff. 1996. Resynthesizing evolutionary and developmental biology. *Dev. Biol.* 173: 357–372.

Goering, L. M. and 7 others. 2009. Association of orthodenticle with natural variation for early embryonic patterning in *Drosophila melanogaster*. *J. Exp. Zool. B Mol. Dev. Evol.* 312: 841–854.

Gomez, C., E. M. Özbudak, J. Wunderlich, D. Baumann and O. Pourquié. 2008. Control of segment number in vertebrate embryos. *Nature* 454: 335–338.

Gompel, N., B. Prud'homme, P. J. Wittkop, V. A. Kassner and S. B. Carroll. 2005. Chance caught on the wing: *cis*-regulatory evolution and the origin of pigment patterns in *Drosophila*. *Nature* 433: 481–487.

Goodman, M. 1999. The genomic record of humankind's evolutionary roots. *Am. J. Hum. Genet.* 64: 31–39.

Gould, S. J. 1977a. *Ontogeny and Phylogeny*. Harvard University Press, Cambridge, MA.

Gould, S. J. 1977b. *Ever Since Darwin*. W. W. Norton, New York.

Gould, S. J. 1993. A special fondness for beetles. *Natural Hist.* 102(1): 4.

Gould, S. J. and R. C. Lewontin. 1979. The spandrels of San Marcos and the Panglossian paradigm: A critique of the adaptationist programme. *Proc. R. Soc. Lond. B* 205: 581–598.

Grant, P. 1999. *The Ecology and Evolution of Darwin's Finches*, Rev. Ed. Princeton University Press, Princeton, NJ.

Grant, P. R. and B. R. Grant. 2007. *How and Why Species Multiple: The radiation of Darwin's Finches*. Princeton University Press, Princeton, NJ.

Green, S. A. and M. E. Bronner. 2014. The lamprey: A jawless vertebrate model system for examining origin of the neural crest and other vertebrate traits. *Differentiation* 87: 44–51.

Guenther, C. A., B. Tasic, L. Luo, M. A. Bedell and D. M. Kingsley. 2014. A molecular basis for classic blond hair color in Europeans. *Nature Genet.* 46: 748–752.

Hahn, M. W., M. V. Rockman, N. Soranzo, D. B. Goldstein and G. A. Wray. 2004. Population genetic and phylogenetic evidence for positive selection on regulatory mutations at the factor VII locus in humans. *Genetics* 167: 867–877.

Haldane, J. B. S. 1948. The theory of a cline. *J. Genetics* 48: 277–284.

Halder, G., P. Callaerts and W. J. Gehring. 1995. Induction of ectopic eyes by targeted expression of the *eyeless* gene in *Drosophila*. *Science* 267: 1788–1792.

Hall, B. K. 1999. *Evolutionary Developmental Biology*. Kluwer Academic Publishers, Dordrecht.

Hall, B. K. and J. A. Gillis. 2012. Incremental evolution of the neural crest, neural crest cells and neural crest-derived skeletal tissues. *J. Anat.* 222: 19–31.

Harding, R. M. and 10 others. 2000. Evidence for variable selective pressures at MC1R. *Am. J. Hum. Genet.* 66: 1351–1361.

Harjunmaa, E. and 11 others. 2014. Replaying evolutionary transitions from the dental fossil record. *Nature* 512: 44–48.

Harmon, M. A., M. F. Boehm, R. A. Heyman and D. J. Mangelsdorf. 1995. Activation of mammalian retinoid X receptors by the insect growth regulator methoprene. *Proc. Natl. Acad. Sci. USA* 92: 6157–6160.

Harris, A. 1904. Quoted in H. DeVries. 1904. *Species and Varieties: Their Origin by Mutation*. Open Court Publishing, Chicago, p. 401.

Hart, M. W., J. M. Sunday, I. Popovic, K. J. Learning and C. M. Konrad. Incipient speciation of sea star populations by adaptive gamete recognition coevolution. *Evolution* 68: 1294–1305.

Hemmati-Brivanlou, A. and D. A. Melton. 1997. Vertebrate embryonic cells will become nerve cells unless told otherwise. *Cell* 88: 13–17.

Hirth, F. and H. Reichert. 1999. Conserved genetic programs in insect and mammalian brain development. *BioEssays* 21: 677–684.

Holder, N. 1983. Developmental constraints and the evolution of vertebrate limb patterns. *J. Theor. Biol.* 104: 451–471.

Holland, P. W. H. and J. Garcia-Fernández. 1996. Hox genes and chordate evolution. *Dev. Bio.* 173: 382–395.

Holley, S., P. D. Jackson, Y. Sasai, B. Lu, E. M. De Robertis, F. M. Hoffmann and E. L. Ferguson. 1995. A conserved system for dorsal-ventral patterning in insects and vertebrates involving sog and chordin. *Nature* 376: 249–253.

Hughes, C. L. and T. C. Kaufman. 2002. Hox genes and the evolution of the arthropod body plan. *Evol. Dev.* 4: 459–499.

Hunt, T. and 15 others. 2007. A comprehensive phylogeny of beetles reveals the evolutionary origins of a superradiation. *Science* 318: 1913–1916.

Huxley, J. 1942. *Evolution: The Modern Synthesis*. Allen and Unwin, London, p. 8.

Huxley, T. H. 1880. Quoted in "Professor Huxley on Evolution." *Engl. Mechanic World Science* 32: 393 (31 Dec 1880).

Huxley, T. 1893. *Darwiniana: Collected Essays, Vol. II*. Macmillan, London.

Jablonsky, N. G. 2013. *Skin: A Natural History*. University of California Press, Berkeley.

Jacob, F. 1977. Evolution and tinkering. *Science* 196: 1161–1166.

Jernvall, J. 2006. *The curious incident of the seal in the pond*. Plenary session. European Society for Evolutionary Developmental Biology meeting, Prague, August 2006.

Jiménez, A., A. Munteanu and J. Sharpe. 2015. Dynamics of gene circuits shapes evolvability. *Proc. Natl. Acad. Sci. USA* 112: 2103–2108.

Kasowski, M. and 16 others. 2010. Variation in transcription factor binding among humans. *Science* 328: 232–235.

Kin, K., M. C. Nnamani, V. J. Lynch, E. Michaelides and G. P. Wagner. 2015. Cell-type phylogenetics and the origin of endometrial stromal cells. *Cell Rep.* 10: 1398–1409.

King, M.-C. and A. C. Wilson. 1975. Evolution at two levels in humans and chimpanzees. *Science* 188: 107–116.

King, N. and 35 others. 2008. The genome of the choanoflagellate *Monosiga brevicollis* and the origins of metazoans. *Nature* 451: 783–788.

King-Jones, K. and C. S. Thummel. 2005. Nuclear receptors: A perspective from *Drosophila*. *Nature Rev. Genet.* 6: 311–323.

Kirschner, M. and J. Gerhart. 1998. Evolvability. *Proc. Natl. Acad. Sci. USA* 95: 8420–8427.

Kirschner, M. W. and J. C. Gerhart. 2005. *The Plausibility of Life: Resolving Darwin's Dilemma*. Norton, New York.

Kunte, K. and 7 others. 2014. doublesex is a mimicry supergene. *Nature* 507: 229–232.

Lamichhaney, S. and 13 others. 2015. Evolution of Darwin's finches and their beaks revealed by genome sequencing. *Nature* 518: 371–375.

Larroux, C., B. Fahey, S. M. Degnan, M. Adamski, D. S. Rokhsar and B. M. Degnan. 2007. The NK homoeobox gene cluster predates the origin of Hox genes. *Curr. Biol.* 17: 706–710.

Laubichler, M. 2013. Homology as a bridge between evolutionary morphology, developmental evolution and phylogenetic systematics. In A. Hamilton (ed.), *The Evolution of Phylogenetic Systematics*. University of California Press, Berkeley, pp. 63–86.

Laufer, E., S. Pizette, H. Zou, O. E. Orozco and L. Niswander. 1997. BMP expression in duck interdigital webbing: A reanalysis. *Science* 278: 305.

Leigh, S. R. 2004. Brain growth, life history, and cognition in primate and human evolution. *Am. J. Primatol.* 62: 139–164.

Levitan, D. R. and D. L. Ferrell. 2006. Selection on gamete recognition proteins depends on sex, density, and genotype frequency. *Science* 312: 267–269.

Linnen, C. R., Y. P. Poh, B. K. Peterson, R. D. Barrett, J. G. Larson, J. D. Jensen and H. E. Hoekstra. 2013. Adaptive evolution of multiple traits through multiple mutations at a single gene. *Science*. 339: 1312–1316.

Locascio, A., M. Manjanares, M. J. Blanco and M. A. Nieto. 2002. Modularity and reshuffling of *Snail* and *Slug* expression during vertebrate evolution. *Proc. Natl. Acad. Sci. USA* 99: 16841–16846.

Luo, Z.-X. 2014. Tooth structure re-engineered. *Nature* 512: 36–37.

Lynch, M. and J. S. Conery. 2000. The evolutionary fate and consequences of duplicate genes. *Science* 290: 1151–1155.

Lynch, V. J., A. Tanzer, Y. Wang, F. C. Leung, B. Gellersen, D. Emera and G. P. Wagner. 2008. Adaptive changes in the transcription factor HoxA-11 are essential for the evolution of pregnancy in mammals. *Proc. Natl. Acad. Sci. USA* 105: 14928–14933.

Lynch, V. J. and 15 others. 2015. Ancient transposable elements transformed the uterine regulatory landscape and transcriptome during the evolution of mammalian pregnancy. *Cell Rep.* 10: 551–561.

Lyson, T. R., G. S. Bever, T. M. Scheyer, A. Y. Hsiang and J. A. Gauthier. 2013. Evolutionary origin of the turtle shell. *Curr. Biol.* 23: 1113–1119.

Ma, L., A. Parkhurst and W. R. Jeffery. 2014. The role of a lens survival pathway including sox2 and αA-crystallin in the evolution of cavefish eye degeneration. *Evodevo* 5: 28.

Malicki, J., L. C. Cianetti, C. Peschle and W. McGinnis. 1992. Human *HOX4B* regulatory element provides head-specific expression in *Drosophila* embryos. *Nature* 358: 345–347.

Mallo, M., D. M. Wellik and J. Deschamps. 2010. Hox genes and regional patterning of the vertebrate body plan. *Dev. Biol.* 344: 7–15.

Manceau, M., V. S. Domingues, R. Mallarino and H. E. Hoekstra. 2011. The developmental role of Agouti in color pattern evolution. *Science* 331: 1062–1065.

Martin, A. and 10 others. 2012. Diversification of complex butterfly wing patterns by repeated regulatory evolution of a Wnt ligand. *Proc. Natl. Acad. Sci. USA* 109: 12632–12637.

Martin, A., K. J. McCulloch, N. H. Patel, A. D. Briscoe, L. E. Gilbert and R. D. Reed. 2014. Multiple recent co-options of Optix associated with novel traits in adaptive butterfly wing radiations. *Evodevo* 5: 7.

Martin, R. D. 1990. *Primate Origins and Evolution: A Phylogenetic Reconstruction*. Princeton University Press, Princeton, NJ.

Maxam, A. and W. Gilbert. 1977. A new method for sequencing DNA. *Proc. Natl. Acad. Sci. USA* 74: 560–564.

McGregor, A. P., V. Orgogozo, I. Delon, J. Zanet, D. G. Srinivasan, F. Payre and D. L. Stern. 2007. Morphological evolution through multiple *cis*-regulatory mutations at a single gene. *Nature* 448: 587–590.

McLean, C. Y. and 12 others. 2011. Human-specific loss of regulatory DNA and the evolution of human-specific traits. *Nature* 471: 216–219.

Meehan, T. J. and L. D. Martin. 2003. Extinction and re-evolution of similar adaptive types (ecomorphs) in Cenozoic North American ungulates and carnivores reflect van der Hammen's cycles. *Naturwiss* 90: 131–135.

Merino, R., J. Rodríguez-Leon, D. Macias, Y. Ganan, A. N. Economides and J. M. Hurle. 1999. The BMP antagonist Gremlin regulates outgrowth, chondrogenesis and programmed cell death in the developing limb. *Development* 126: 5515–5522.

Mizutani, C. M. and E. Bier. 2008. EvoD/Vo: The origins of BMP signaling in the neuroectoderm. *Nature Rev. Genet.* 9: 663–675.

Moczek, A. P. and D. J. Rose. 2009. Differential recruitment of limb patterning genes during development and diversification of beetle horns. *Proc. Natl. Acad. Sci. USA* 106: 8992–8997.

Moczek, A. and 23 others. 2015. The significance and scope of evolutionary developmental biology: A vision for the 21st century. *Evol. Dev.* 17: 198–219.

Moustakas-Verho, J. E. and 9 others. 2014. The origin and loss of periodic patterning in the turtle shell. *Development* 141: 3033–3039.

Montagu, M. F. A. 1962. Time, morphology, and neoteny in the evolution of man. In M. F. A. Montagu (ed.), *Culture and Evolution of Man*. Oxford University Press, New York.

Morange, M. 2001. *The Misunderstood Gene*. Harvard University Press, Cambridge, MA.

Mullen, L. M. and H. E. Hoekstra. 2008. Natural selection along an environmental gradient: A classic cline in mouse pigmentation. *Evolution* 62: 1555–1570.

Müller, Fritz. 1878. Über die Vortheile der Mimicry bei Schmetterlingen. *Zoologischer Anzeiger* 1: 54–55.

Müller, G. B. 2007. Evo-devo: Extending the evolutionary synthesis. *Nature Rev. Genet.* 8: 943–949.

Mundy, N. I. 2005. A window on the genetics of evolution: MC1R and plumage colouration. *Proc. Biol. Soc.* 272: 1633–1640.

Naisbit, R. E., C. D. Jiggins and J. Mallet. 2003. Mimicry: Developmental genes that contribute to speciation. *Evol. Dev.* 5: 269–280.

Nelson, D. A. and S. A. Nunneley. 1998. Brain temperature and limits of transcranial cooling in humans: Quantitative modeling results. *Eur. J. Appl. Physiol.* 78: 353–359.

Newman, S. A. and G. B. Müller. 2005. Origination and innovation in the vertebrate limb skeleton: An epigenetic perspective. *J. Exp. Zool.* 304B: 593–609.

Nunes, M. D., S. Arif, C. Schlötterer and A. P. McGregor. 2013. A perspective on micro-evo-devo: Progress and potential. *Genetics* 195: 625–634.

Oakley, T. H. and C. W. Cunningham. 2002. Molecular phylogenetic evidence for the independent evolutionary origin of an arthropod compound eye. *Proc. Natl. Acad. Sci. USA* 99: 1426–1430.

Odling-Smee, F. J., K. N. Laland and M. W. Feldman. 2003. *Niche Construction. The Neglected Process in Evolution*. Monographs in Population Biology 37, Princeton University Press, Princeton, NJ.

Ohno, S. 1970. *Evolution by Gene Duplicaiton*. Springer, Berlin.

Oster, G. F., N. Shubin, J. D. Murray and P. Alberch. 1988. Evolution and morphogenetic rules: The shape of the vertebrate limb in ontogeny and phylogeny. *Evolution* 42: 862–884.

Palumbi, S. R. and H. A. Lessios. 2005. Evolutionary animation: How do molecular phylogenies compare to Mayr's reconstruction of speciation patterns in the sea? *Proc. Natl. Acad. Sci. USA* 102(suppl. 1): 6566–6572.

Pigliucci, M. 2008. Is evolvability evolvable? *Nature Rev. Genet.* 9: 75–82.

Plata Stapper, A. 2013. Evolutionary Dynamics of the Interacting Gamete Recognition Proteins Sperm Bindin and Its Egg Receptor Ebr1 in the Purple Sea Urchin *Strongylocentrotus purpuratus*. *Electronic Theses, Treatises and Dissertations*. Paper 7553.

Pollard, K. S. and 15 others. 2006. An RNA gene expressed during cortical development evolved rapidly in humans. *Nature* 443: 167–172.

Portmann, A. 1941. Die Tragzeiten der Primaten und die Dauer der Schwangerschaft beim Menschen: Ein Problem der vergleichen Biologie. *Rev. Suisse Zool.* 48: 511–518.

Preuss, T. M., M. Caceres, M. C. Oldham and D. H.Geschwind. 2004. Human brain evolution: Insights from microarrays. *Nature Rev. Genet.* 5: 850–860.

Prigogine, I. and G. Nicolis. 1967. On symmetry-breaking instabilities in dissipative systems. *J. Chem. Phys.* 46: 3542–3550.

Punnett, R. G. 1915. *Mimicry in Butterflies.* Cambridge University Press, Cambridge.

Purves, D. and J. W. Lichtman. 1985. *Principles of Neural Development.* Sinauer Associates, Sunderland, MA.

Quammen, D. 2007. Quoted in L. Valenti. Wordsmiths of southwestern Montana: David Quammen. *Outside Bozeman* (Winter 2006/7): 21. Available at www.outsidebozeman.com

Raff, R. A. 1996. *The Shape of Life: Genes, Development, and the Evolution of Animal Form.* University of Chicago Press, Chicago.

Raspopovic, J., L. Marcon, L. Russo and J. Sharpe. 2014. Modeling digits. Digit patterning is controlled by a Bmp-Sox9-Wnt Turing network modulated by morphogen gradients. *Science* 345: 566–570.

Reed, R. D. and 12 others. 2011. Optix drives the repeated convergent evolution of butterfly wing pattern mimicry. *Science* 333: 1137–1141.

Richardson, M. 1999. Vertebrate evolution: The developmental origins of adult variation. *BioEssays* 21: 604–613.

Richardson, M. D. and H. H. Oelschläger. 2002. Time, pattern and heterochrony: A study of hyperphalangy in the dolphin embryo flipper. *Evol. Dev.* 4: 435–444.

Richardson, M. K. and A. D. Chipman. 2003. Developmental constraints in a comparative framework: A test case using variations in phalanx number during amniote evolution. *J. Exp. Zool. (MDE) B* 296: 8–22.

Rockman, M. V., M. W. Hahn, N. Soranzo, D. B. Goldstein and G. A. Wray. 2003. Positive selection on a human-specific transcription factor binding site regulating IL4 expression. *Curr. Biol.* 13: 2118–2123.

Ronshaugen, M., N. McGinnis and W. McGinnis. 2002. Hox protein mutation and macroevolution of the insect body plan. *Nature* 415: 914–917.

Rose, S. 1998. *Lifelines: Biology beyond Determinism.* Oxford University Press, Oxford.

Salazar-Ciudad, I. and J. Jernvall. 2010. A computational model of teeth and the developmental origins of morphological variation. *Nature* 464: 583–586.

Sanger, F., S. Nicklen and A. R. Coulson. 1977. DNA sequencing with chain-terminating inhibitors. *Proc. Natl. Acad. Sci. USA* 74: 5463–5467.

Schlosser, G. 2015. Vertebrate cranial placodes as evolutionary innovations: The ancestor's tale. *Curr. Top. Dev. Biol.* 111: 235–300.

Schluter, D. 2000. *The Ecology of Adaptive Radiation.* Oxford University Press, Oxford.

Schmidt, J., V. Francoise, E. Bier and D. Kimelman. 1995. *Drosophila* short gastrulation induces an ectopic axis in *Xenopus*: Evidence for conserved mechanisms of dorsoventral patterning. *Development* 121: 4319–4328.

Schneider, R. A. and J. A. Helms. 2003. The cellular and molecular origins of beak morphology. *Science* 299: 565–568.

Sears, K. E. 2004. Constraints on the evolution of morphological evolution of marsupial shoulder girdles. *Evolution* 58: 2353–2370.

Shapiro, M. D., M. A. Bell and D. M. Kingsley. 2006. Parallel genetic origins of pelvic reduction in vertebrates. *Proc. Natl. Acad. Sci. USA* 103: 13753–13758.

Shapiro, M. D., J. Hanken and N. Rosenthal. 2003. Developmental basis of evolutionary digit loss in the Australian lizard *Hemiergis*. *J. Exp. Zool.* 279B: 48–56.

Shapiro, M. D. and 7 others. 2004. Genetic and developmental basis of evolutionary pelvic reduction in threespine sticklebacks. *Nature* 428: 717–723.

Shbailat, S. J. and E. Abouheif. 2013. The wing-patterning network in the wingless castes of Myrmicine and Formicine ant species is a mix of evolutionarily labile and non-labile genes. *J. Exp. Zool. B Mol. Dev. Evol.* 320: 74–83.

Sheth, R. and 8 others. 2012. Hox genes regulate digit patterning by controlling the wavelength of a Turing-type mechanism. *Science* 338: 1476–1480.

Simões-Costa, M. and M. E. Bronner. 2015. Establishing neural crest identity: A gene regulatory recipe. *Development* 142: 242–257.

Simões-Costa, M., J. Tan-Cabugao, I. Antoshechkin, T. Sauka-Spengler and M. E. Bronner. 2014. Transcriptome analysis reveals novel players in the cranial neural crest gene regulatory network. *Genome Res.* 24: 281–290.

Smith, K. 2003. Time's arrow: Heterochrony and the evolution of development. *Int. J. Dev. Biol.* 47: 613–621.

Sucena, E. and D. Stern. 2000. Divergence of larval morphology between *Drosophila sechellia* and its sibling species caused by *cis*-regulatory evolution of *ovo/shaven-baby*. *Proc. Natl. Acad. Sci. USA* 97: 4530–4534.

Swanson, W. J. and V. D. Vacquier. 2002. The rapid evolution of reproductive proteins. *Nature Rev. Genet.* 3: 137–144.

Thewissen, L. G. M., L. N. Cooper and R. R. Behringer. 2012. Developmental biology enriches paleontology. *J. Vert. Paleo.* 32: 1223.

Tomoyasu, Y., Y. Arakane, K. J. Kramer and R. E. Denell. 2009. Repeated co-options of exoskeleton formation during wing-to-elytron evolution in beetles. *Curr. Biol.* 19: 2057–2065.

Tournamille, C., Y. Colin, J. P. Cartron and C. Le Van Kim. 1995. Disruption of a GATA motif in the *Duffy* gene promoter abolishes erythroid gene expression in Duffy-negative individuals. *Nature Genet.* 10: 224–228.

Turing, A. M. 1952. The chemical basis of morphogenesis. *Philos. Trans. R. Soc. Lond. B* 237: 37–72.

Tyler-Smith, C. and Y. Xue. 2012. Sibling rivalry among paralogs promotes evolution of the human brain. *Cell* 149: 737–739.

Uz, E. and 15 others. 2010. Disruption of ALX1 causes extreme microphthalmia and severe facial clefting: Expanding the spectrum of autosomal-recessive ALX-related frontonasal dysplasia. *Am. J. Hum. Genet.* 86: 789–796.

Vignieri, S. N., J. G. Larson and H. E. Hoekstra. 2010. The selective advantage of crypsis in mice. *Evolution* 64: 2153–2158.

von Dassow, M V. and L. A. Davidson. 2008. Variation and robustness of the mechanics of gastrulation: The role of tissue mechanical properties during morphogenesis. *Birth Def. Res. C: Embryo Today* 81: 253–269.

Waddington, C. H. 1938. The morphogenetic function of a vestigial organ in the chick. *J. Exp. Biol.* 15: 371–376.

Wagner, G. P. 2014. *Homology, Genes, and Evolutionary Innovation.* Princeton University Press, Princeton, NJ.

Wagner, G. P. and V. J. Lynch. 2010. Evolutionary novelties. *Curr. Biol.* 20: R48–R52.

Wagner, G. P., K. Kin, L. Muglia and M. Pavlicev. 2014. Evolution of mammalian pregnancy and the origin of the decidual stromal cell. *Int. J. Dev. Biol.* 58: 117–126.

Wake, D. B. and A. Larson. 1987. A multidimensional analysis of an evolving lineage. *Science* 238: 42–48.

Wang, H. and 8 others. 2005. The origin of the naked grains of maize. *Nature* 436: 714–719.

Weatherbee, S. D., R. R. Behringer, J. J. Rasweiler, IV and L. A. Niswander. 2006. Interdigital webbing retention in bat wings illustrates genetic changes underlying amniote limb diversification. *Proc. Natl. Acad. Sci. USA* 103: 15103–15107.

Weiner, J. 1994. *The Beak of the Finch: A Story of Evolution in Our Time.* Random House, New York.

Wellik, D. M. and M. R. Capecchi. 2003. *Hox10* and *Hox11* genes are required to globally pattern the mammalian skeleton. *Science* 301: 363–367.

West, D. A. 2003. *Fritz Müller: A Naturalist in Brazil.* Pocahontas Press, Blacksburg, VA.

West-Eberhard, M. J. 2003. *Developmental Plasticity and Evolution.* Oxford University Press, New York.

West-Eberhard, M. J. 2005. Developmental plasticity and the origin of species differences. *Proc. Natl. Acad. Sci. USA* 102: 6543–6549.

Winfree, A. T. 1974. Rotating chemical reactions. *Sci. Am.* 230(6): 82–95.

Wolpert, L. 1983. Constancy and change in the development and evolution of pattern. In B. C. Goodwin, N. Holder and C. C. Wylie (eds.), *Development and Evolution.* Cambridge University Press, Cambridge, pp. 47–57.

Wong, K. H., H. D. Wintch and M. R. Capecchi. 2004. Hoxa11 regulates stromal cell death and proliferation during neonatal uterine development. *Mol. Endocrinol.* 18: 184–193.

Wu, P., T. X. Jiang, S. Suksaweang, R. B. Widelitz and C. M. Chuong. 2004. Molecular shaping of the beak. *Science* 305: 1465–1466.

Wu, P., T. X. Jiang, J. Y. Shen, R. B. Widelitz and C. M. Chuong. 2006. Morphoregulation of avian beaks: Comparative mapping of growth zone activities and morphological evolution. *Dev. Dyn.* 235: 1400–1412.

Yamamoto, Y., D. W. Stock and W. R. Jeffery. 2004. Hedgehog signalling controls eye degeneration in blind cavefish. *Nature* 431: 844–847.

Yamamoto, Y., M. S. Byerly, W. R. Jackman and W. R. Jeffery. 2009. Pleiotropic functions of embryonic Sonic hedgehog expression link jaw and taste bud amplification with eye loss during cavefish evolution. *Dev. Biol.* 330: 200–211.

Zhu, J., Y. T. Zhang, M. S. Alber and S. A. Newman. Bare bones pattern formation: A core regulatory network in varying geometries reproduces major features of vertebrate limb development and evolution. *PLoS ONE* 5(5): e10892.

Zuckerkandl, E. 1994. Molecular pathways to parallel evolution. I. Gene nexuses and their morphological correlates. *J. Mol. Evol.* 39: 661–678.

Environment, Development, and Evolution

Toward a New Evolutionary Synthesis

In my opinion, the greatest error which I have committed, has not been allowing sufficient weight to the direct action of the environment, i.e. food, climate, etc., independently of natural selection.

Charles Darwin to Moritz Wagner, 1876

May your symbionts be with you.

Angela Douglas, 2010

To develop is to interact with the environment. To evolve is to alter these interactions in a heritable manner.

Armin Moczek, 2015

The importance of development for evolutionary biology is not limited to the role of developmental regulatory genes discussed in Chapter 10. The studies of developmental plasticity and developmental symbiosis point to something quite unexpected in evolutionary theory: that the environment not only *selects* variation, it helps *construct* and shape variation. This integration of ecological developmental biology into evolutionary biology has sometimes been referred to as **ecological evolutionary developmental biology**, or "eco-evo-devo." Eco-evo-devo has several research projects, but its main goal, as recently stated (Abouheif et al. 2014) is to "uncover the rules that underlie the interactions between an organism's environment, genes, and development and to incorporate these rules into evolutionary theory."

With the application of knowledge gained from the studies now being done in ecological developmental biology, a new and more inclusive evolutionary theory is being forged. So far, eco-devo has contributed at least three components to this nascent evolutionary synthesis. These are the three concepts introduced in the first section of the textbook. The first concept introduced into evolutionary biology is developmental plasticity. This concept forms the basis of **genetic accommodation**, the idea that environmentally induced changes to a phenotype, when adaptive over long periods of time, can become the genetic norm for a species. Developmental plasticity is also the basis of **niche construction**, wherein the developing organism can modify its environment because there are plastic features in its habitat. Second, eco-devo has maintained that epigenetic inheritance systems, such as the epialleles formed by environmental agents, are also important sources of selectable variation. The third of these is the concept of developmental symbiosis. Symbiosis has given rise to processes of symbiont-mediated evolution, including the possibility that the holobiont is a unit of evolutionary selection. Thus, what is added to evolutionary biology are the agencies of the environment (see Moczek 2015). These factors had been hypothesized to be exceptions to the general rules of nature, and thus were only minor areas of evolutionary theory. Now, these same factors are seen as being pervasive throughout nature, even characteristics of natural life on this planet. They therefore need to be incorporated into an extended theory of evolution.

The above-mentioned discoveries of eco-evo-devo have allowed evolutionary biology to expand beyond the constraints of the following four assumptions of the Modern Synthesis:

1. *"Genetic variation in alleles is the only evolutionarily relevant variation."* Ecological developmental biology demonstrates that epigenetic variation can also be transmitted from generation to generation, can have profound phenotypic consequences, and therefore constitutes an important component of selectable variation.

2. *"Individual genotypes are the main target of selection."* Ecological developmental biology shows that organisms may be more like ecosystems, holobionts composed of numerous genotypes that interact with each other. This may allow natural selection to favor "teams" rather than particular individuals and may also privilege "relationships" as a unit of selection.

3. *"The environment is a selective agent of phenotypes but is of no consequence in producing the phenotype."* Ecological developmental biology has shown that developing organisms respond to environmental conditions by altering which phenotypes they produce and that environmental agents help generate particular phenotypes.

4. *"The environment is a given and is unchanged by the organism being selected."* Ecological developmental biology finds that organisms reciprocally shape their immediate environment in ways that match their traits. The environment is altered by the organisms as they develop.

Symbiotic Inheritance

There are several ways that symbiotic inheritance is extremely important for evolution. As described in Chapter 3, the holobiont may be the unit of natural selection (Margulis and Fester 1991; Sapp 1994; Margulis 1999; Douglas 2010; Gilbert et al. 2010, 2012; McFall-Ngai et al. 2013). In addition, developmental symbiosis may be critical for several evolutionary processes, including:

- The production of new selectable variation. This is the result of **symbiopoiesis**, wherein the symbiont is part of the developmental interactions that generate the holobiont.

- The alteration of genomes, constraining and permitting certain ecological ranges

- Reproductive isolation, potentiating new species

- The formation of new types of cells. This can be accomplished by **symbiogenesis**, the acquisition of new genetic material from other organisms into the cell's inheritance systems.

- Evolutionary transitions such as those enabling multicellular life and the origins of complex ecosystems

Holobiont variation produced by symbionts

Symbionts can be passed from generation to generation. Alleles in these symbionts can alter the phenotype of the holobiont and lead to selection of the holobiont based on the alleles of the symbiont. One example of symbionts conferring variation to the entire organism involves pea aphids and their symbiotic bacteria. The pea aphid *Acyrthosiphon pisum* and its bacterial symbiont *Buchnera aphidicola* have a mutually obligate symbiosis. That is, neither the aphids nor the bacteria will flourish without their partner. Pea aphids rely on *B. aphidicola* to provide essential amino acids that are absent from the pea aphids' phloem sap diet (Baumann 2005). In exchange, the pea aphids supply nutrients and intracellular niches that permit the *B. aphidicola* to reproduce (Soler et al. 2001). These bacteria live inside the cells of the aphids (and are therefore called **endosymbionts**). Because of this interdependence, the aphids are highly constrained to the ecological tolerances of *B. aphidicola* (Dunbar 2007).

   Buchnera aphidicola provides more than essential amino acids, though. Certain strains of *B. aphidicola* can give thermotolerance to the holobiont. The heat tolerance of pea aphids and *B. aphidicola* is abrogated by a single nucleotide deletion in the promoter of the heat-shock gene *ibpA*. This microbial gene encodes a small heat-shock protein, and the deletion eliminates the transcriptional response of *ibpA* to heat (Dunbar et al. 2007). In other words, the thermotolerance phenotype of the holobiont depends upon the regulatory allele of the symbiont. Only those aphids containing the wild-type *Buchnera* can produce offspring at high temperatures. Although pea aphids harboring *B. aphidicola* with the short *ibpA* promoter allele suffer from decreased thermotolerance, they experience increased reproductive

Figure 11.1 Fecundity (number of offspring per day) of the pea aphid *Acyrthosiphon pisum* depends upon the temperature at which the juvenile is raised. Those aphids whose *Buchnera aphidicola* symbionts have the wild-type allele (5A-long) for this heat-shock protein have slightly lower fecundity at most temperatures but can have fecundity at high temperatures, whereas those aphids carrying symbionts with the mutant allele (5A-short) do not. (After Dunbar et al. 2007.)

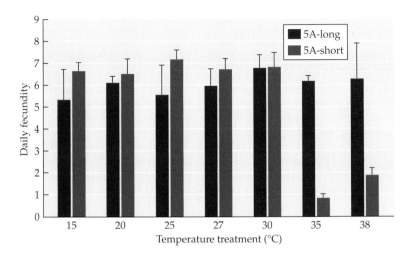

rates under cooler temperatures (15°C–20°C). Aphid lines containing the short-promoter *B. aphidicola* produce more offspring per day during the first 6 days of reproduction compared with aphid lines containing long-allele *B. aphidicola* (Figure 11.1). This trade-off between thermotolerance and fecundity allows the pea aphids and *B. aphidicola* to diversify.

Moran and Yun (2015), taking advantage of the maternal inheritance of *B. aphidocola*, were able to replace a heat-sensitive strain of *Buchnera* with a thermotolerant strain (thus disrupting 100 million years of continuous maternal transmission in its hosts!). The new symbionts were stably transmitted, and the aphids with the new, thermotolerant *Buchnera* displayed a dramatic increase in heat tolerance. This directly demonstrated that the symbiont genotype has a strong effect on holobiont fitness and ecology. Such partnerships, where thermotolerance is provided by the symbiont, are also seen in coral and Christmas cactuses* (see Gilbert et al. 2010).

Another bacterial symbiont of the pea aphid, the bacterium *Rickettsiella*, can alter the *color* phenotype of the holobiont. Those red juvenile aphids that do not inherit the *Rickettsiella* become red adults. However, those aphids that do inherit *Rickettsiella* become green adults, because *Rickettsiella* contains genes that induce the synthesis of quinone compounds that alter the cuticle color (Figure 11.2; Tsuchida et al. 2010). Moreover, a third bacterial symbiont of pea aphid cells, *Hamiltonella*, can provide immunity against parasitoid wasp infection (Oliver et al. 2009). But in this case, the protective variants of *Hamiltonella* result from the incorporation of a specific lysogenic bacteriophage within the bacterial genome. The aphid must be infected with *Hamiltonella*, and the *Hamiltonella* must be infected by phage APSE-3. As Oliver and colleagues (2009, p. 994) write, "In our system, the

---

*As Rosenberg and colleagues (2007) have pointed out, advantageous mutations will spread more quickly in bacterial genomes than in host genomes because of the rapid reproductive rates of bacteria. This may be especially important in species such as aphids that are produced parthenogenetically (without males) and therefore are essentially clonal populations.

(A) Without *Rickettsiella*

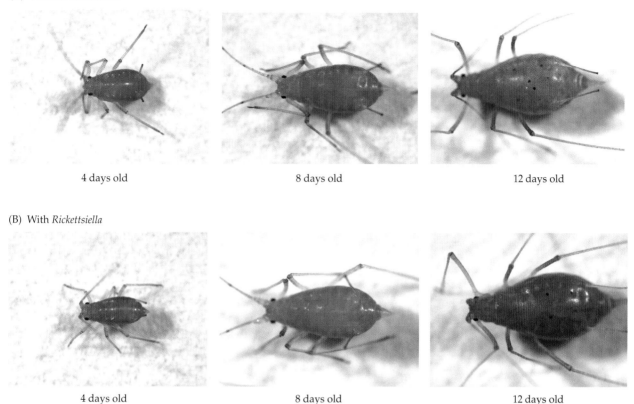

4 days old   8 days old   12 days old

(B) With *Rickettsiella*

4 days old   8 days old   12 days old

Figure 11.2   The color of adult pea aphids depends on whether or not their cells contain *Rickettsiella* bacterial symbionts. (A) Without *Rickettsiella*, red aphid newborns become red adults. (B) With *Rickettsiella*, red aphid newborns become green adults. (After Tsuchida et al. 2010, photographs courtesy of T. Tsuchida.)

evolutionary interests of phages, bacterial symbionts, and aphids are all aligned against the parasitoid that threatens them all. The phage is implicated in conferring protection to the aphid and thus contributes to the spread and maintenance of *H. defensa* in natural *A. pisum* populations." So here, we can see the holobiont, where phage, bacteria, and host are working together for a common phenotype. Again, there is a trade-off for the hosts carrying this protective bacterial "allele." In the absence of parasitoid infection, those aphids carrying the bacteria with lysogenic phage are not as fecund as those lacking this phage.

Thus, the symbionts (and the symbionts' symbionts!) can change the phenotype of the holobiont and provide variations within a population. Are there any examples in mammals? Remarkably, in humans, the gut microbe *Bacteroides plebeius* in Japanese populations differs from the *B. plebeius* of American populations (Hehemann et al. 2010, 2012). The Japanese *B. plebeius* contains two genes that are absent from the American strains. These two genes encode enzymes that enable the bacteria to digest the complex sugars found in seaweed. (Indeed, these genes probably were acquired by lateral gene transmission from related marine bacteria that grow on algae.) Thus, the *B. plebeius* alleles in Japan enable that human population to get more calories from the seaweed in their diet. Hence, the Japanese population appear to be able to digest sushi more completely than European or American populations.

Altering the genome through symbiotic interactions

When the host and the symbiont are necessary for each other's existence, they can afford to get rid of redundant genes that the other partner has. Progressive genome reduction is commonly seen in obligate symbionts (McCutcheon and Moran 2012). The crucial stage of genome reduction is the loss of the symbionts' DNA repair systems, allowing the buildup of mutations and the spread of mobile elements. Those nonfunctional regions of DNA will normally be deleted (Moran 2003). In the symbiotic system of the mealybug *Planococcus citri*, the endosymbiont *Tremblaya princeps* has itself an endosymbiont, *Moranella endobia*. The three organisms together constitute a metabolic team, wherein the synthesis of essential amino acids often utilizes enzymes encoded by the genes of each of the species. The pathway for synthesizing the amino acid phenylalanine, for instance, begins in *T. princeps*, and the products of these enzymes pass into *M. endobia*, and then into the *P. citri* cytoplasm, where phenylalanine is made (McCutcheon and von Dohlen 2011). Having proteins provided by both its endosymbiont and its host, the *T. princeps* genome retains very few enzyme encoding genes, and is basically making ribosomes and little else (Husnik et al. 2013; López-Madrigal et al. 2013). Thus, the holobiont becomes a genetically composite organism in part by having symbiotic genomes reduced so that the organisms comprising it cannot function independently.

This reduction of genomes does not extend only to endosymbionts (i.e., those symbionts living within their host's cells). One of the most intensely specialized symbiotic systems concerns figs and the wasps that pollinate them and live within them. In many cases, the fig has an obligate symbiotic relationship with a particular species of wasp, and neither can exist without the other. The genome of the fig wasp *Ceratosolen solmsi* has undergone a dramatic reduction of those genes involved with environmental sensing and detoxification. Despite the long-range flight of the female to find a fig tree, the wasps have little or no need for environmental protection, since the fig wasps spend almost their entire lives within a benign host (Xiao et al. 2013).

Indeed, as mentioned in Chapter 3, the co-evolution of symbionts can be a double-edged sword (Bennett and Moran 2015). The incorporation of *Buchnera* into the pea aphid holobiont enabled the aphid to use sap as a nutritional resource. But once fully integrated into such mutually obligate symbiosis, the aphid became dependent upon this symbiont for its development.

Reproductive isolation caused by symbionts

Recall that for speciation to occur, one needs both selectable variation and reproductive isolation. Reproductive isolation is the result of mechanisms that prevent members of two species from exchanging genes. In the previous chapter, we discussed one genetic form of reproductive isolation, that which occurs when the gametes fail to recognize each other. Symbionts have been implicated in two other types of reproductive isolation: mate selection and cytoplasmic incompatibility.

Reproductive isolation can arise when mating preferences within a species are altered. If groups within a species fail to mate successfully with one another, they will fission into separate groups, each with its own gene pool. Symbiotic bacteria are able to alter mating behavior by altering the chemicals used in courtship. The mating preference of *Drosophila* appears to be due to the bacteria on the food they eat as larvae (Sharon et al. 2010, 2011). Flies who eat yeast media mate preferentially with other former yeast eaters, while those flies fed molasses as larvae preferentially later mate with other flies who had molasses (Figure 11.3). *Lactobacillus plantarum* is a bacterium that grows abundantly on molasses and less frequently on starch. When antibiotics are put into the fly food to eliminate these bacteria, the matings become random. However, when these flies are then infected with *L. plantarum*, they preferentially mate with each other. It appears that the bacteria are modifying the contact pheromones that flies use while mating, and these differences are abrogated when the bacteria are removed.

In mammals, bacteria may also create conditions for sexual selection by altering pheromones. Theis and colleagues (2013) provide initial evidence that the odors that establish and maintain the social hierarchies of hyena communities are brought about by the bacteria in the scent glands. The ability of different species of bacteria to produce distinctive odors may have an especially important role in mammalian evolution (Archie and Theis 2011).

As mentioned in Chapter 3 (see Figure 3.11), symbionts can also cause reproductive isolation through cytoplasmic incompatibility (Brucker and Bordenstein 2013). When jewel wasps (*Nasonia vitripennis*) become infected by different *Wolbachia* strains, a situation develops where each population has a specific relationship with its own strain of *Wolbachia*. There is reciprocal incompatibility in *both* directions, such that hybrids between the two strains cannot exist. Thus, species can diverge from a common population through having different symbionts.

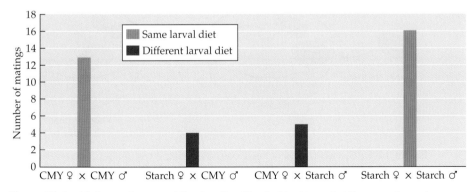

Figure 11.3 Mating preference of flies for other flies that had been fed the same type of food. Antibiotics wiped out these differences, and *Lactobacillus plantarum* restored them. CMY is the molasses-based media. (After Sharon et al. 2010.)

Symbiosis and great evolutionary transitions

Lynn Margulis and Dorian Sagan (2003) speculated that symbionts were behind many, if not all, the major transitions of evolution: "Much more significant [than random mutation] is the acquisition of new genomes by symbiotic merger." Certainly, as Margulis had theorized, the origin of eukaryotic cells came about through a progression, from the symbioses of different bacteria and archaea to their eventual existence as nuclear, mitochondrial, and chloroplast components of eukaryotes (Margulis 1970, 1993; Koonin 2010). The "domestication" of bacteria (probably a proteobacteria similar to *Rickettsia*) into mitochondria enabled eukaryotes to thrive in non-anoxic environments, and it allowed the diversification of all subsequent life. Subsequent domestication of a cyanobacteria into chloroplasts allowed plants to form (Figure 11.4). The process by which new organisms form by acquiring symbionts is called symbiogenesis (Merezhkowsky 1909; Margulis 1993).

As mentioned in Chapter 10, one of the most interesting areas of symbiogenesis currently focuses on whether the origins of certain cell types, including the decidual stromal cell that makes mammalian pregnancy possible, originates from the incorporation of viral genes into the genomes of mammalian ancestors (Dupressoir et al. 2011; Lynch et al. 2015). If viruses

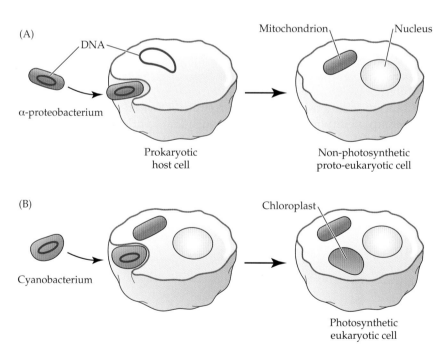

Figure 11.4 Symbiogenesis of the ancestral eukaryotic cell. (A) Sequence homologies indicate that the nuclear DNA of all eukaryotic cells is from an archaean cell, while the mitochondria are derived from bacteria, probably proteobacteria (similar to *Rickettsia*) that it incorporated. (B) When this type of cell absorbed a cyanobacterium, the cyanobacterium eventually became a chloroplast. Both chloroplasts and mitochondria still contain circular DNA reminiscent of bacteria, as well as bacteria-like membranes.

are considered symbionts (see Ryan 2009), then viruses have played a critical role in the evolution of the group called mammals.

Other critically important evolutionary transitions have also been mediated through symbiopoiesis. Mycorrhizal symbiosis was the "key innovation" enabling plants to survive on land (Pirozynski and Malloch 1975; Heckman et al. 2001). But such plants cannot be fully digested by animals until symbiotic bacteria provide the enzymes needed by the vertebrate and invertebrate digestive systems to digest cellulose and lignins. In vertebrates, microbe-mediated herbivory evolved over 50 times during the past 300 million years, enabling them the luxury of plant nutrition (Vermeij 2004). Ants and termites also use such symbiotic bacteria for nutrition, along with other symbiotic partners such as aphids and fungi. The coral reefs that make up some of the most important marine biomes are also products of symbiosis. As mentioned in Chapter 3, the calcareous exoskeletons of the coral animals are made possible by the photosynthesis of the *Symbiodinium* algae inside their cells. The corals only became reef builders when they acquired symbionts in the late Triassic (the same time that saw the origins of mammals and dinosaurs). This is when their isotopic patterns of carbon and oxygen demonstrate photosynthetic activity (see Douglas 2010; Thompson et al. 2015). And, as also mentioned in Chapter 3, the symbioses between legumes and rhizobial bacteria enable the fixation of atmospheric nitrogen, and thereby the synthesis of amino acids, purine, and pyrimidines. Eukaryotic life is made possible by symbiosis. Thus, Margulis and Sagan (2001) have claimed that competition within species plays a relatively minor, and pruning, role compared with symbiotic cooperation between species: "Life did not take over the globe by combat, but by networking."

Symbionts may even have a major share of the responsibility in the development and evolution of multicellularity. First, the atmosphere had to change. The oldest known macroscopic multicellular organisms come from fossils in strata some 2.1 billion years old, soon after the "great oxidation event"—the accumulation of atmospheric oxygen that occurred about 2.4 billion years ago as a result of photosynthesis by bacteria that were probably similar to the modern cyanobacteria ("blue-green algae"). This oxidation event was the most important climate change in evolutionary history, altering the atmosphere from a mixture of ammonia and carbon dioxide to an oxygen-rich mixture that would make aerobic metabolism—and thus metazoan life—possible (El Albani et al. 2010).

Second, the unicellular protists had to relinquish their individuality and become multicellular. The "inertial condition" for a eukaryotic cell is proliferation (see Sonnenschein and Soto 1999). Unicellular protists abound and were the first eukaryotic form of life. What encouraged unicellular organisms to give up their independence and cell division to form multicellular aggregates? The answer appears to include bacteria. Another bacterial boost to metazoan evolution may have been a symbiosis between bacteria and unicellular protists. Recent analyses agree that the metazoans probably arose from a group of protists very much like today's choanoflagellates. Choanoflagellates are single-celled, and their name comes from their resemblance to the choanocytes (collar cells) of sponges. In filtered seawater, one such protist, *Salpingoeca rosetta,* proliferates asexually, forming more

Figure 11.5 Two morphologies of the choanoflagellate protist *Salpingoeca rosetta*. (A) Single-celled form. (B) Colonial form with multiple cells linked by an extracellular matrix. The *Algoriphagus* bacteria, often found with *S. rosetta*, can convert the organism from dividing into individual cells to forming multicellular "rosettes." (From Dayel et al. 2011, photographs courtesy of M. Dayel and N. King.)

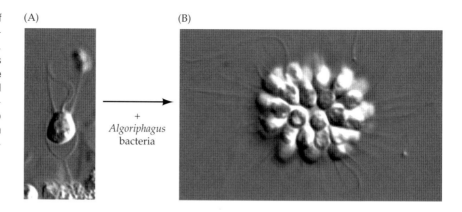

(A)   (B)

+
Algoriphagus
bacteria

single-celled protists. However, when cultured in media containing the bacteria *Algoriphagus machipongonensis* (a *Bacteroides*-like bacteria), the cells do not separate. Rather, they form sheets or rosettes, in which the cells are connected by an extracellular matrix and cytoplasmic bridges (Figure 11.5; Dayel et al. 2011; Alegado et al. 2012). The sphingolipids in the bacterial cell wall are able to effect this transition, and the bacteria are found naturally with colonial forms of this choanoflagellate species. It is thought that the new aggregation increases fluid flow into the colony, thereby increasing the feeding rate of each cell within it (Roper et al. 2013). Thus, multicellularity may have arisen by developmental changes induced by neighboring bacteria.

Symbiosis is a critical, perhaps ubiquitous, component of the way of life in this world, able to facilitate key evolutionary transitions by altering host-symbiont interactions. Therefore, in the study of major processes in evolution, the nature of symbiotic relationships deserves to take center stage.

Epiallelic Inheritance

In addition to the inheritance of nuclear genes, mitochondria, chloroplasts, and symbionts, there is also the inheritance of epialleles (Jablonka and Lamb 2006, 2015; Jablonka and Raz 2009). Recall that traditional evolutionary biology views allelic variation of DNA sequences as the main if not sole source of heritable variation for selection to act upon. However, epiallelic variation has now been described in a vast number of organisms, including humans, forcing a revision of the thinking about the types of variation that matter in evolution, and their origins (see Appendix D; Cabej 2008; Jablonka and Raz 2009; Crews et al. 2014). For example, we have seen how environmental agents can cause alterations in DNA methylation and/or histone modification and that the altered chromatin can be passed through the germline from one generation to the next.

Transgenerational inheritance of environmentally induced phenotypes

Numerous examples of epiallelic heredity have recently been discovered. We saw in Chapter 2 that enzymatic and metabolic phenotypes are established in utero by protein-restricted diets in mice. Protein restriction during a female mouse's pregnancy leads to a specific methylation pattern in her pups and grandpups (Burdge et al. 2007). The endocrine disruptors vinclozolin, methoxychlor, DDT, and bisphenol A also have the ability to alter DNA methylation patterns in the germline, thereby causing illnesses and predispositions to diseases in the grandpups of mice exposed to these chemicals in utero (Anway et al. 2005, 2006a,b; Chang et al. 2006; Newbold et al. 2006; Crews et al. 2007; see Figure 6.23). In the *viable yellow Agouti* phenotype in mice, methylation differences affect coat color and obesity. When a pregnant female is fed a diet that contains substantial levels of methyl donors, the specific methylation pattern at the *Agouti* locus is transmitted not only to the progeny developing in utero (see Figure 2.5), but also to the progeny of those mice and to their progeny (Jirtle and Skinner 2007). These altered chromatin configurations act just as genetic alleles might act and are therefore referred to as epialleles (or, sometimes, epimutations). Such epiallelic transmission has also been seen in invertebrates such as the brine shrimp *Artemia*, where environmentally induced chromatin configurations conferring resistance to heat and bacterial pathogens are transmitted from the stressed generation to at least three successive (and unstressed) generations (Norouzitallab et al. 2014). It does not matter to the developing system whether a gene has been inactivated by a mutation or by an altered chromatin configuration; the effect is the same.

The stress-resistant behavior of rats was also shown to be epiallelic, wherein altered methylation patterns, induced by maternal care, were seen in the glucocorticoid receptor genes. Meaney (2001) found that rats who received extensive maternal care had less stress-induced anxiety and, if female, developed into mothers who gave their offspring similar levels of maternal care. Here, epialleles are combined with maternal behavior to propagate the epiallele (and its behavior) from one generation of rats to another (see Figure 2.16). The maternal behavior causes the formation of a particular epiallele that promotes the hormonal conditions that generate a particular behavior in the pups, and when these pups mature, those who become mothers have a behavior that induces the formation of this epiallele in their pups.

One of the most far-reaching (and therefore controversial) observations of transgenerational epigenetic inheritance induced by the environment concerns memory. Parental olfactory experience was seen to be transmitted from one generation to the next, and this phenomenon was associated with hypomethylated regions of particular olfactory genes in the sperm. Dias and Ressler (2014) conditioned male mice to associate a fruity odor, that of acetophenone, with a mild foot shock. The young mice would come to associate the odor with the shock and to eventually show a startle response to the odor, alone. These mice had changes in the organization of their brain's olfactory bulb neurons, generating more neurons that synthesized the M71 odor receptor that makes the mouse capable of smelling the acetophenone. After this training, the mice were bred to female mice that had never received such training. The pups that were born (the F_1 generation) showed a startle response to acetophenone, even though they had never smelled it previously (and had no reason to fear it.) Other odors did not

(A)

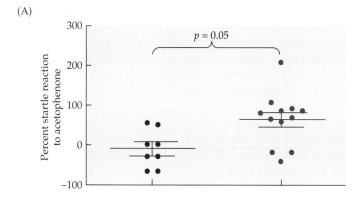

(B)

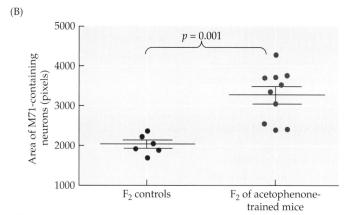

(C)

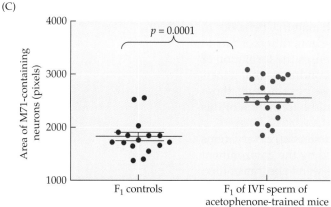

Figure 11.6 Behavioral sensitivity and neuroanatomical changes are inherited in F_2 and in vitro fertilization (IVF) derived generations of odor-trained mice. Mice whose fathers were trained while young to fear acetophenone were able to transfer both the large number of M71-containing neurons and their fear of acetophenone to F_1 and F_2 generations. (A) The large number of M71-containing neurons (that receive the acetophenone stimulus) is inherited in the F_2 generation of mice derived from the trained males. (B) The startle response to the odor was similarly inherited in the grandpups of the trained males. (C) The F_1 mice born from the sperm of a trained male inherited the large number of M71-containing olfactory neurons. Data are presented with black lines showing the average as well as the standard error of the mean. (After Dias and Ressler 2014.)

elicit this response. Moreover, the offspring of the F_1 mice, the F_2 generation, also showed this sensitivity to acetophenone. These F_2 mice, whose grandsires had become sensitive to acetophenone prior to conception, also had brains containing many more M71-containing neurons than animals whose ancestors were so trained (Figure 11.6A,B).

To control for the possibility that fathers who feared the fruity odor treated their offspring differently or that mothers (sensing that their mates were a bit odd, perhaps), might have treated their pups differently, the researchers performed in vitro fertilization experiments in which sperm from the mice that feared acetophenone were sent to another laboratory that artificially inseminated female mice. The offspring from this in vitro fertilization had more M71-containing neurons than control mice, too (Figure 11.6C). (The behavioral studies couldn't be performed due to quarantine regulations.)

The cause of this inheritance of environmentally induced anatomical and behavioral traits is thought to be differential DNA methylation. The sperm DNA of the F_1 mice of fathers that had been trained to acetophenone had a different DNA methylation pattern than that of control mice (exposed and trained to some other chemical). Indeed, the region around the gene encoding the M71 odor receptor appears specifically to be less methylated in the sensitized mice than in control mice. The propagation of this hypomethylation into the brain of the adult mice would help explain the phenotype. However, this is a controversial subject, in that there is no known mechanism to get the signal from the brain into the sperm. These findings are also controversial if applied to humans and children of people suffering from post-traumatic shock disorder (see Hughes 2013).

How epialleles are formed and preserved over several generations are questions just beginning to be studied. Usually, the DNA methylation and histone modifications of the genome are erased during germ cell development and early embryogenesis. However, some genes ("imprinted genes") are able to retain their modified chromatin; and so do certain epialleles induced by environmental agents. Skinner and colleagues (2013) have shown that the major periods for epigenetic programming by vinclozolin are those developmental stages associated with primordial germ cell migration to the gonads (E13 in rats) and with germ cell differentiation (E16). These coincide with the times that most epigenetic marks (such as methylation) are being erased (in the migrating germ cells) and reestablished in a sexually specific manner (gamete differentiation). The vinclozolin-induced methylation appears similar to sex-specific imprinted genes that do not become erased at each generation. The epigenetically marked genes (the "epigenome") are transmitted to all the cells of the developing embryo, such that all the somatic tissues and germline tissues have this altered epigenome. The somatic epigenomic alterations can then produce a cascade of tissue-specific events that become associated with the adult-onset diseases, which in the case of vinclozolin include prostate disease, kidney disease, immune malfunction, and testicular dysgenesis. The altered epigenome is able to get through the erasure of most of the DNA methylation and gets passed on to the next generation.

Epigenetic factors in sexual selection and reproductive isolation

In Chapter 10, we discussed the ability of genes encoding gamete recognition proteins to mediate reproductive isolation, causing one population to split into two or more groups with highly assortative mating. And earlier in this chapter, we discussed the ability of symbiotic microbes to cause such sexual selection and reproductive isolation. Here, we will see that epialleles can also alter mating preference. The fungicide vinclozolin is an antiandrogenic endocrine disruptor, inducing DNA methylation changes that can last for several generations (see Chapter 6). Crews and colleagues (2007) have shown that female rats prefer males who have no history of vinclozolin exposure. The females even recognized males who were three generations removed from the female rat originally exposed to vinclozolin. In other words, males whose parents or grandparents had been exposed to vinclozolin were less attractive to normal females. Indeed, the DNA methylation patterns expressed in the courtship-associated brain regions of these third-generation male rats are altered (Skinner et al. 2014a). Thus, conclude the researchers (Crews et al. 2007), "an environmental factor can promote a transgenerational alteration in the epigenome that influences sexual selection and could impact the viability of a population and evolution of the species."

The concept of epigenetic inheritance—especially symbiopoiesis and epialleles—brings back into evolutionary biology the notion of the inheritance of acquired characters. While this model does not represent the Lamarckian "use and disuse" principle (see Appendix D), the above concepts indicate that if the germline is susceptible to DNA methylation and other epigenetic changes, then mutation is not needed to inactivate these genes, and the inheritance of these environmentally induced chromatin alterations is possible.

Thus epiallelic inheritance (as well as the inheritance of symbionts) must be included in a new evolutionary synthesis. In addition to mutation-driven mechanisms of phenotype production, we find that epialleles and symbionts provide additional hereditary mechanisms of heredity (Table 11.1). Epialleles and symbionts provide sources of selectable variation within species, means of isolating species, and means of producing

TABLE 11.1 Three pathways of evolution: Genetic (Modern Synthesis) and epigenetic (epiallelic and symbiopoietic) mechanisms

Resource	Symbionts	Plasticity	Genes/Function
Origin of variation	Different microbes	Developmental plasticity, epialleles	Mutation, recombination
Fixation of change	Genomic reduction	Genetic assimilation	Drift
Reproductive isolation	Mating preference, cytoplasmic incompatibility	Mating preference	Mating preference, physical barriers, gametes

major evolutionary transitions. Moreover, inheritance from symbionts and epialleles may explain the "missing heritability" wherein many common heritable conditions do not appear to be caused by the inheritance of specific genes. GWAS (genome-wide association studies, where complete genomes of individuals with specific traits are compared with those of the general population) had promised to find the genetic alleles for common conditions—diabetes, schizophrenia, homosexuality, Alzheimer's disease, intelligence, cancers, left-handedness—where there appears to be some familial component. However, the genes associated with these conditions can explain the occurrence of the conditions in only a very small percentage of cases, and the ability of genes to predict who will develop asthma, diabetes, or Alzheimer's disease based on genetic alleles is still wanting (Manolio et al. 2009; Ho 2013).

Plasticity-Driven Adaptation

As discussed in Chapters 1 and 4, plasticity plays major roles in integrating organisms into their environment. Plasticity plays critical roles in defense, predation, sex determination, and even sexual selection (Prudic et al. 2014). In addition, the production of heritable adaptive phenotypes may be made possible by the generation of such phenotypes by developmental plasticity followed by the genetic stabilization of this phenotype into the genetic repertoire of the organism by natural selection. In other words, what had been an environmentally generated phenotype would be a normal genetic phenotype produced irrespective of the environment.

Even before the rediscovery of Mendel's laws in 1900, biologists such as Gulick (1872), Spalding (1873), Baldwin (1896), Lloyd Morgan (1896), and Osborn (1897) had been impressed by developmental and behavioral plasticities. These scientists envisioned ways by which developmental and behavioral responses to environmental stimuli might become genetically fixed, making the environmental inducer unnecessary for the expression of those traits in subsequent generations. This might occur if a developmental response to the environment became adaptive in all situations an organism was likely to meet. These traits would become expressed through genes, and the responses would be initiated within the organism by cell-cell interaction

Parental Effects

Maternal and paternal effects may form another epigenetic inheritance system. These parental effects have been mentioned in Chapter 8 and in Appendix D. We have seen how the environmentally induced phenotype of the mother can be passed down to subsequent generations. In *Daphnia*, for instance, the predator-induced morph is propagated in subsequent generations even when the predator is not present (Agrawal et al. 1999). Similarly, the gregarious and solitary morphs of the migratory locust are transmitted from one generation to another through the foam placed on the eggs by the mother (McCaffery and Simpson 1998). Here, mothers pass on traits to their offspring, some of which become mothers and pass on the traits.

Paternal effects have been less well studied, and we mentioned some earlier in Chapter 7. As mentioned there, sperm can also bring to the egg regulatory RNAs. The first known of these regulatory RNAs explained a strange inheritance of the Kit phenotype. The *Kit* gene encodes a receptor tyrosine kinase necessary for the proliferation and differentiation of the blood, germ cell, and pigment cells. (The ligand for Kit is the paracrine factor discussed earlier in the production of blond hair, on p. 390). When *Kit* heterozygous mice are mated with wild-type mice ($Kit^{+/+}$), about half the offspring are also heterozygous for *Kit* and develop the white paws and tail characteristic of *Kit* heterozygotes. (*Kit* mutant homozygotes die because their blood stem cells fail to mature.) When these heterozygotes are mated to wild-type mice, a fraction of the offspring have wild-type *Kit* genes but develop the heterozygous phenotype (see figure; Rassoulzadegan et al. 2006). These phenotypically mutant mice with wild-type genes are called **paramutants**.

The cause of such paramutants was found to be an RNA within the small amount of sperm cytoplasm. Microinjecting RNA from the sperm of the *Kit* heterozygote into the pronuclei of fertilized mouse eggs produced the heterozygous Kit phenotype in the adult products of those eggs. Moreover the sperm of the paramutant males also had this regulatory RNA. Several small RNAs have been found in sperm, and difference in sperm microRNAs have been seen in males with diet-induced obesity (Kawano et al. 2012; Fullston et al. 2013). Thus, parent-of-origin effects may be another inheritance system. These inherited changes may involve a little-studied phenomenon, where the RNA is methylated (Kiani et al. 2013).

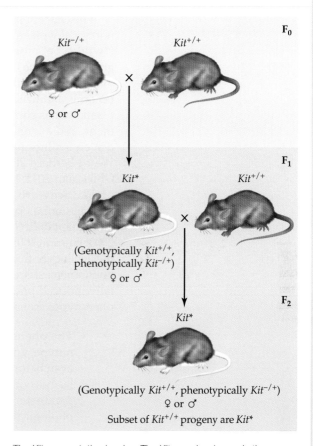

The *Kit* paramutation in mice. The *Kit* gene in mice and other mammals interferes with stem cell migration, and one of its phenotypes is white paws because the pigment cells don't migrate that far. (The homozygotes are completely white and die from anemia, since the blood stem cells don't get into the bone marrow.) A heterozygous mouse for *Kit* genes, when mated to a wild-type mouse, produces heterozygotes (white-pawed) and wild-type mice (dark-pawed). But some of the mice with the wild-type genes have white paws because an RNA in the sperm down-regulated the expression of the *Kit* gene from that allele. A subset of these $Kit^{+/+}$ mice will retain this heterozygous expression pattern. Eventually, the wild-type mice will show wild-type expression. (After Chandler 2007.)

rather than by interaction between the organism's cells and the environment. However, without a theory of gene transmission to undergird such a concept, this line of thinking remained merely interesting speculation.

The concept that one of the environmentally induced morphs of a phenotypically plastic trait could become the genetically transmitted standard ("wild type") for that species has gone under many names. We wish to group under the heading of "plasticity-driven adaptation" all those mechanisms whereby, through selection, environmentally induced phenotypes become stabilized in the genome. During the last decade, developmental plasticity has come to be seen as an important part of normative development, and a large number of studies have indicated that other plasticity-driven evolutionary schemes might also be normative for evolution (see West-Eberhard 2003; Pigliucci et al. 2006; Pfennig et al. 2010; Bateson and Gluckman 2011; Moczek et al. 2011).

Plasticity-driven adaptation has been called "the Baldwin effect," "genetic assimilation," "stabilizing selection," "genetic accommodation," and "the adaptability driver" (see King and Stanfield 1985; Gottlieb 1992; Hall 2003; Bateson 2005, 2014). The differences are generally in the details of how the genetic control can incorporate environment-induced phenotypes. These phenomena share many ideas in common, including that (1) environmentally induced phenotypes are seen first, and (2) there is selection for those phenotypes that are most adaptive. The advantages of such scenarios are:

- **The phenotype is not random.** The environment elicits the phenotype, and the phenotype becomes tested by natural selection even before it is directly produced by genes. As Garson and colleagues (2003) note, although mutation is random, developmental parameters may account for some of the directionality in morphological evolution. Developmental plasticity yields integrated phenotypes biased in certain directions over others.

- **The phenotype already exists in a large portion of the population.** In the Modern Synthesis, one of the problems of explaining new phenotypes is that the bearers of such phenotypes are "monsters" compared with the wild type.* How would such mutations, perhaps present only in one individual or one family, become established and eventually take over a population? The developmental plasticity model solves this problem: this phenotype has been around for a long while, and the capacity to express it is widespread in the population.

Three modes of plasticity-driven adaptation will be considered here and shown to be evolutionarily important. These three overlapping phenomena are phenotypic accommodation, genetic accommodation, and genetic assimilation. **Phenotypic accommodation** concerns the mutual adjustment of parts to one another during development, such that a change in one part creates changes in other parts. This does not typically involve gene mutations. Such phenotypic accommodation can promote **genetic**

---

*Richard Goldschmidt called them "hopeful monsters." What they were hoping for was reproducing.

accommodation wherein environmentally induced phenotypes are selected and subsequently made part of the genetic repertoire of development. Genetic accommodation usually refers to changes in gene frequency that result from environmentally induced phenotypes. What had been induced by the environment becomes either the normative phenotype or an alternative phenotype with a genetically acquired threshold. **Genetic assimilation** is the subset of genetic accommodation wherein the selection is for the environmentally induced phenotype, and cryptic genetic variation (genetic differences that usually do not become phenotypic differences) or new mutations allow this phenotype to be induced through embryonic cell interactions rather than by the environment. Here, plasticity is lost.

Phenotypic accommodation

Phenotypic accommodation has a long history in both embryology and evolutionary biology, but it is only recently that such a notion has been revived in modern evolutionary biology (Riedl 1978; Wagner and Laubischler 2004). Darwin (1868, p. 312) wrote extensively about "correlated variation," noting that "when one part is modified through continued selection, either by man or under nature, other parts of the organization will be unavoidably modified."

In the model for the evolution of new phenotypes proposed by West-Eberhard (2003), phenotypic accommodation, the mutual adjustment of parts during development, plays a central role. In this view, there are four steps by which an environmentally induced trait can come to characterize a species:

1. **Developmental plasticity.** An environmental change produces changes in development leading to the appearance of a new trait.

2. **Phenotypic accommodation.** The regulative ability of the developing organism adapts to this new trait.

3. **Spread of the new variant.** If the initial change is environmentally induced, the variant trait occurs in a large part of the population.

4. **Genetic fixation.** Allelic variation in the population along with natural selection allow for the genetic fixation (assimilation) of the trait so that it is produced irrespective of the environment.

Phenotypic accommodation is due not only to developmental plasticity (i.e., the interactions of the developing organism with its environment), but also to normal embryonic induction (Waddington 1957). Embryologist Hans Spemann (1901, 1907), reviewing the field of "developmental correlations,"* noted that when he put a small piece of *frog* tadpole tissue into a *salamander* embryo, the salamander formed a tadpole jaw, and that the musculature and skeletal elements were all coordinated. Similarly, Twitty (1932) found that when he transplanted the eye-forming region

---

*This concept of correlated development and phenotypic accommodation has gone under many aliases. Frazetta (1975) called this "phenotypic compensation," and Müller (1990) has named it "ontogenetic buffering." Its basis is the normal reciprocal embryonic induction that is responsible for forming complex organs such as the eye or kidney, making certain that the proportion of cells allocated for each structure will result in an appropriately formed organ. This is another excellent example of the concept that it is not possible to explain evolutionary change without knowing the underlying developmental principles.

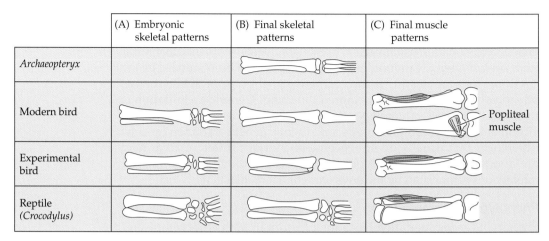

	(A) Embryonic skeletal patterns	(B) Final skeletal patterns	(C) Final muscle patterns
Archaeopteryx			
Modern bird			Popliteal muscle
Experimental bird			
Reptile (*Crocodylus*)			

Figure 11.7 Experimental "atavisms" produced by altering embryonic fields in the limb. Results of Müller's experiments using gold foil to split the chick hindlimb field. (A,B) The embryonic and final bone patterns, indicating that the fibulare structure was retained by the experimental chick limb, as it is in extant reptiles and as it is thought to have been in *Archaeopteryx*, the earliest known bird. (C) Some of the correlated muscle patterns. The popliteal muscle is present in the normal chick limb but is absent from reptile limbs and from the experimental limb. The fibularis brevis muscle, which normally originates from both the tibia and fibula in chicks, takes on the reptilian pattern of originating solely from the fibula in the experimental limb. (After Müller 1989.)

from the embryos of large salamanders into embryos of smaller species, the midbrain region of the side innervated by the larger eye grew bigger to accommodate the larger number of neurons. This coordination of head morphogenesis, where changes in one element cause reciprocal changes on other elements, has also been verified in chimeras made between chick and quail embryos (Köntges and Lumsden 1996).

Phenotypic accommodation has also been shown in the limb, and this accommodation has important evolutionary implications. Repeating the earlier experiments of Hampé (1959), Gerd Müller (1989) inserted a barrier of gold foil into the prechondrogenic hindlimb buds of 3.5-day chick embryos. This barrier separated the regions of tibia formation and fibula formation. The results of these experiments were twofold. First, the tibia was shortened, and the fibula bowed and retained its connection to the fibulare (the distal portion of the tibia). Such relationships between the tibia and fibula are not usually seen in birds, but they are characteristic of reptiles (Figure 11.7). Second, the musculature of the hindlimb underwent changes in parallel with the bones. Three of the muscles that attach to these bones showed characteristic reptilian patterns of insertion. It seems, therefore, that experimental manipulations that alter the development of one part of the mesodermal limb-forming field also alter the development of other mesodermal components. This was crucial in the evolution of the bird hindlimb from the reptile hindlimb. As with the correlated progression seen in facial development, these changes all appear to be due to interactions within a module, in this case, the chick hindlimb field. These changes are not global effects and can occur independently of the other portions of the body.

EVOLUTIONARY CONSEQUENCES OF PHENOTYPIC ACCOMMODATION   The ability of embryos to "improvise" and adjust their development to new conditions allows remarkable changes in anatomy. Each part of an organ has to change independently. The dramatic changes in bone arrangement from agnathans to jawed fishes, from jawed fishes to amphibians, and from reptiles to mammals were coordinated with changes in jaw structure, jaw musculature, tooth deposition and shape, and the structure of the cranial vault and ear (Kemp 1982; Thomson 1988; Fischman 1995). On a less sweeping scale, the results of artificial selection in domestic dogs also demonstrate correlated development (see box). Although the skulls of the many dog breeds range from the pointed snouts of collies to the blunt snouts of bulldogs, in each case the jaw muscles are coordinated with jaw cartilage and bone to allow the dog to properly grasp and chew its food. Goldschmidt (1940) and Schmalhausen (1938, 1949) recognized that the regulative ability of the embryo allowed it to accept changes in its structure, and Schmalhausen (1938) used this notion of integration to criticize evolutionary biologists who thought of organisms as mere mosaics of independent characters (see Levit et al. 2006).

Phenotypic accommodation can be seen in the ability of certain mutants to be functional, despite their limitations. When quadrupeds are born with forelimb deformities, they often learn to become bipedal (and there are several such cases on the Internet).* A handicapped goat originally studied by Slijper (1942a,b; West-Eberhard 2003) was born with paralyzed forelimbs and hopped on its two functional hind legs. Its pelvis had changed shape to accommodate an upright posture, especially in the ischium. This accommodation went along with changes in the skeletal musculature around the pelvis, including an elongated and thickened gluteal tongue whose attachment to the pelvis was mediated by tendons that do not exist in normal goats. The bones of hindlimb, thorax, and sternum were also changed. Thus, a complete set of muscles, skeleton, and tendons coadapted to form an anatomical novelty.

More importantly, such phenotypic accommodation between skeleton, musculature, and behavior may have been essential for the "conquest of land" by aquatic vertebrates. *Polypterus senegalus* is a birchir, a fish that has ventrolaterally positioned pectoral fins and functional lungs, both of which are considered to be traits of the group of fishes that evolved into tetrapods 400 million years ago. Moreover, *P. senegalus* can breathe on land and can perform tetrapod-like walking with its pectoral fins (Figure 11.8). Raising *P. senegalus* in either aquatic (normal) or terrestrial (abnormal) environments showed that the land-acclimated individuals walked faster, held their fins closer to their bodies, and kept their heads raised, all behaviors that were predicted for a transition from water to land (Standen et al. 2014). In addition, the anatomy of the land-reared fish had changed. The bones of the neck and shoulder had changed shape to allow the bracing of the

*One of the most famous is Faith, a dog born with congenitally deformed forelimbs (www.youtube.com/watch?v=emiZpSuz-eA).

(A)

(B)

(C)

(D)

Figure 11.8 Fish trained to walk on land. *Polypterus senegalus* that walk on land show new behaviors and anatomical connections. (A) The fish has its nose raised and plants its left pectoral fin on the ground while its right fin swings forward. (B,C) The head and tail turn toward the left fin, which is grounded. (D) Last, the right pectoral fin is planted on the ground while the left fin is raised. (From Hutchinson 2014, photographs courtesy of H. Larsson and A. Morin; see Baker [2014] for an excellent video on fish learning to walk.)

collarbone and a greater independence of motion. These were some of the same macroevolutionary changes that have been seen in the fossil record of the fishes that evolved into amphibians. Development can provide a mechanism for the ability of organisms to adapt in the direction they do. As Standen and colleagues (2014) conclude:

> Developmental plasticity can be integrated into the study of major evolutionary transitions. The rapid, developmentally plastic response of the skeleton and behaviour of Polypterus to a terrestrial environment, and the similarity of this response to skeletal evolution in stem tetrapods, is consistent with plasticity contributing to large-scale evolutionary change. Similar developmental plasticity in Devonian sarcopterygian fish in response to terrestrial environments may have facilitated the evolution of terrestrial traits during the rise of tetrapods.

Phenotypic accommodation, like enhancer modularity, gene duplication, and the small toolkit, is an essential precondition for macroevolution.

Genetic accommodation

In plasticity-driven models of evolution, new traits often begin as environmentally initiated phenotypic changes and only later become genetically fixed. According to these theories of evolution, some environmentally triggered phenotypes may (by chance) improve an organism's viability under particular conditions. If there is heritable variation among members of a population in their ability to develop this newly favored trait, then selection should favor those alleles or allele combinations that best stabilize, refine, and extend the new trait's expression. This process, where there is evolutionary change due to the selection of the variation in the regulation, form, or effects of an environmentally induced trait, has been called genetic accommodation (West-Eberhard 2003).

In these evolutionary scenarios, a trait that was initially produced in response to an environmental stimulus may eventually either become canalized (so that it is produced independently of the environment, a phenomenon called "genetic assimilation") or become part of a polyphenic system that evolves a "switch point" for the production of alternative phenotypes depending on the environmental circumstances (i.e., "environmental polyphenism"). What had been an environmentally induced phenotype becomes part of the genetic repertoire of the organism. Thus, West-Eberhard (2005) stated her belief that "genes are probably more often followers than leaders in evolutionary change."

There is growing evidence for the importance of genetic accommodation in evolution, and there is a growing recognition of cryptic genetic variation, which makes genetic accommodation possible. Numerous studies of wild populations have shown that an adaptive potential for plasticity can evolve (Nussey et al. 2005; Danielson-François et al. 2006; Gutteling et al. 2007). In one set of studies, Ledon-Rettig and her colleagues (2008) have shown genetic accommodation in the gut morphologies of toads with different diets. Here, data suggest that an ancestral population of spadefoot toads exhibited morphological plasticity in their larval guts in response to eating shrimp. Moreover, this toad species probably had genetic variation in its

Domestication

Phenotypic accommodation may explain one of the most widespread of evolutionary convergences: the phenotypes of domesticated mammals—pigs, cattle, dogs, mice, horses, and so forth. In his observations of plant and animal life, Charles Darwin (1859) discovered that domesticated mammals were different from their wild progenitors. They have a suite of unusual heritable traits that include increased docility and tameness, smaller teeth, coat color changes (reduced pigmentation, spots), ear alterations, and smaller brains. An extensive program of fox domestication, begun in 1959 by Belyaev (1969) and continued by Trut and colleagues (2009), has shown that all these features arise rapidly and can be attained simply by selecting for increased tameness. It appears that the "domestication syndrome" arises together, and not as an assemblage of independent, sequentially acquired traits.

Recently, a developmental hypothesis has been proposed to explain the origin of this domestication syndrome (Wilkins et al. 2014). It begins with selecting for animals that show little fear of humans (Darwin 1875;

Belyaev 1974). This reduction in fear appears to be due to the small size of the adrenal gland, the organ whose hormones regulate stress and fear responses (Künzl and Sachser 1999). The cells of the internal adrenal gland come from the neural crest, and Wilkins and colleagues propose that selecting for tameness selects for animals with a relatively small number of neural crest cells or attenuated migrational capabilities of neural crest cells. What is more interesting, though, is that neural crest cells also give rise to the melanocytes of the coat pigment, the odontoblasts of the teeth, the cells of the middle ear, and much of the cranium (see figure).

So domesticity could arise rapidly through the selection of a behavioral trait (tameness/lack of fear) that reflected a tendency toward a lower number of functional neural crest derivatives. When bred together and selected, these wild animals could give rise to progeny that become rapidly domesticated, and these could show a suite of traits that characterize mild neural crest deficiencies.

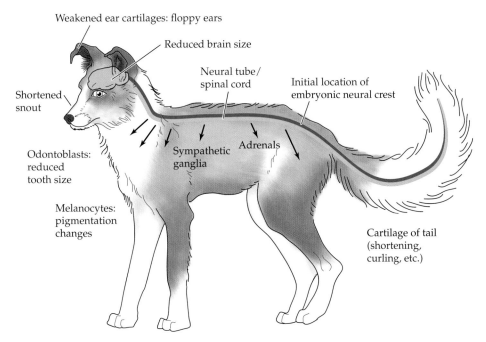

Selection for tameness may enable the domestication system through altered development. The neural crest hypothesis for domestication suggests that in selecting for tameness, one selects for low numbers of neural crest cells in the adrenal gland, reflecting low numbers of functional neural crest cells forming teeth, pigment, and skull. Cells from the neural crest form the stress-response cells of the internal adrenal glands and sympathetic neurons. And they also form pigment cells (melanocytes), tooth cells (odontoblasts), and cranial skeletal cells. (After Wilkins et al. 2014.)

ability to produce such phenotypes. This variation, sometimes called "cryptic genetic variation" because it does not usually show up as a phenotype, can be exposed to natural selection by a variation in the environment, in this case, the diet. The derived spadefoot species had either diet-induced polyphenism (when the environment had both plant detritus and shrimp) or a canalized morphology (when only one resource was available).

Here cryptic genetic variation appears to have enabled the evolutionary transition to carnivory (i.e., shrimp eating) in the spadefoot toad tadpoles (Ledon-Rettig et al. 2010), and such variation might in general facilitate rapid evolutionary transitions to utilize new food sources. Like symbiosis and plasticity, cryptic genetic variation—genetic variation that normally produces no phenotypic difference but which can be uncovered during periods of stress—has been known for a long while, but it's widespread nature and importance to evolution are only recently being appreciated (Ledon-Rettig et al. 2014; Paaby and Rockman 2014).

In a few instances, the basis for the cryptic genetic variation is known. In the nematode *Pristionchus pacificus*, there is polyphenism for the type of mouth. During larval development, it can develop a complex, broad mouth with a tooth, enabling it to eat other nematodes, or it can develop a simple, narrow mouth, which enables it to develop faster but only allows it to eat bacteria. The two phenotypes convey fitness advantages in different environments. When animal prey are available (and induce the broad mouth), the toothed phenotype is more fit. When only bacteria are available, the faster-growing narrow-mouth phenotype takes over the population (Figure 11.9; Ragsdale et al. 2013). Such condition-dependent fitness advantages are predicted by the genetic accommodation model. The gene that executes this irreversible decision, *eud-1*, encodes a sulfatase enzyme (Ragsdale et al. 2013). The expression of this gene is induced by environmental agents (acting through a hormone) and causes the formation of the broad mouth. Loss-of-function mutations of the *eud-1* gene cannot produce broad mouths, whereas those animals overexpressing this gene cannot produce any other type of mouth.

(A)

(B)

(C)

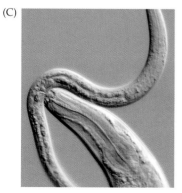

(D)

Figure 11.9 Mouth structures and feeding in the nematode *Pristionchus pacificus*. (A) Broad-mouth phenotype with its clawlike tooth (green). (B) Narrow-mouth phenotype, which lacks the clawlike tooth. (C) Broad-mouth nematodes can eat other nematodes, including larval *Caenorhabditis elegans*, shown here. (D) Narrow-mouth nematodes eat the bacteria that cluster around dead scarab beetles. (From Ragsdale et al. 2013, photographs courtesy of R. Sommer.)

GENETIC ACCOMMODATION IN THE LABORATORY   Suzuki and Nijhout (2006) have shown genetic accommodation in the larvae of the tobacco hornworm moth. By judicious selection protocols, they were able to breed one line in which plasticity decreased (genetic assimilation) and another line in which plasticity increased.

The hornworm moth *Manduca quinquemaculata* has a temperature-dependent color polyphenism that appears to be adaptive. When these caterpillars hatch at 20°C, they have a black phenotype, which enables them to absorb sunlight more efficiently. During the warmer season, with temperatures around 28°C, the hatching caterpillars are green, which affords them camouflage in their environment. A trade-off between cryptic coloration and the need to absorb heat appears to drive this phenotypic plasticity. Wild caterpillars of the related species *M. sexta* (the tobacco hornworm moth), however, are not phenotypically plastic in coloration. The *M. sexta* caterpillar is green no matter what the temperature, although mutant black forms exist (Figure 11.10A). This black mutation is due to reduced levels of juvenile hormone, causing increased levels of melanin production in the caterpillar's skin.

(A)

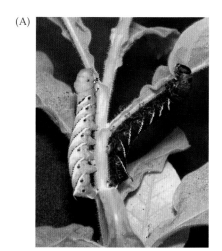

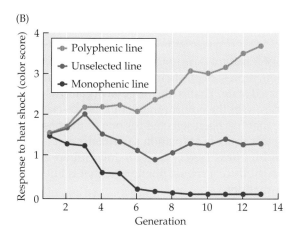

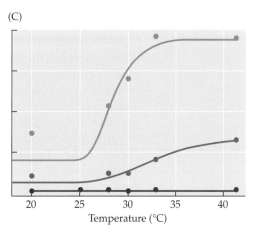

Figure 11.10 Effect of selection on temperature-mediated larval color change in the *black* mutant of the moth *Manduca sexta*. (A) The two color morphs of *M. sexta*. (B) Changes in the coloration of heat-shocked larvae in response to selection for increased (black) and decreased (green) color response to heat-shock treatments, compared with no selection (red). The color score indicates the relative amount of colored regions in the larvae. (C) Reaction norm for generation 13 moth larvae reared at constant temperatures between 20°C and 33°C, and heat shocked at 42°C. Note the steep polyphenism at around 28°C. (After Suzuki and Nijhout 2006, photograph courtesy of Fred Nijhout.)

(B)

(C)

Suzuki and Nijhout tested whether they could induce phenotypic plasticity in *M. sexta* by artificial selection. When green larvae were heat shocked, they remained green. However, when the mutant black larvae were heat shocked at 42°C for 6 hours, they developed phenotypes ranging from black to green. These mutant caterpillars were then subjected to three different regimens. In one case, the adults developing from those black larvae that became green with heat shock were bred to one another. This scheme selected for plasticity. In a second case, a "monophenic" line was selected by mating together those moths resulting from black larvae that did not turned green after heat shock. A third group had no selection applied to it.

Within 13 generations, all individuals selected for the polyphenism (i.e., turning green) always developed a green phenotype after heat shock. Within 7 generations, all the individuals in the monophenic line remained black after heat shock. In the 13th generation, larvae from both lines were subjected not to heat shock but to the mild heat difference (28°C) that would initiate the polyphenism in *M. quinquemaculata*. The monophenic strain stayed black at all temperatures. In other words, this strain had gone from a slightly polyphenic population to a monophenic population by genetic assimilation. Selection for green coloration by heat shock, however, produced a polyphenic population. In this population, the larvae were usually black in temperatures below 28.5°C. Above this temperature, individuals were mostly green. In other words, the selection of this polyphenic line resulted in a temperature-dependent polyphenic switch (Figure 11.10B,C). Although the polyphenic response was not directly selected for (by selection of individuals in different environments), the color response evolved through artificial selection. Moreover, Suzuki and Nijhout showed that the target of this selection was most likely the titer of juvenile hormone, just as in *M. quinquemaculata*.

Thus, at least in the laboratory, genetic accommodation can occur through the interaction of a sensitizing mutation (which brings the phenotype closer to a threshold), environmental change (which can allow the phenotype to cross that threshold), and cryptic genetic changes (which can stabilize this plastic phenotype). Genetic accommodation allows plasticity to persist such that it is expressed only when advantageous, and it demonstrates how the environmental influences become integrated into developmental trajectories.

Genetic assimilation

As mentioned in the studies of *Manduca sexta*, one subset of genetic accommodation, genetic assimilation, represses plasticity, narrowing the range of variation from a plastic to a fixed state. That is, the original organisms have a plastic phenotype that varies with the environment, but the descendant population has a phenotype that has been selected for one of the possible variants. Waddington called this process of fixation **canalization**; it is also known as **developmental robustness** (see Chapter 4 and Gilbert 2002). The processes of canalization produce a robust phenotype that is buffered against environmental perturbations.

In order for an environmentally induced trait to become genetically fixed, the population must be exposed to environmental conditions that repeatedly induce the same phenotype, there must be selective pressure such that the induced phenotype results in higher fitness in that environment, and there must be sufficient genetic variation within the population to stabilize this particular phenotype. In this manner, a phenotype that had been initially part of an environmentally induced polyphenism can be genetically stabilized (Schmalhausen 1949; Waddington 1952, 1953). If a given plastic response to novel environmental challenges happens to be adaptive, and if it continues to be induced by the environment, the ability to mount such a response is predicted to spread through a population. Moreover, if the phenotype could be produced even more effectively without environmental induction but directly through genetically canalized development, then it should become stabilized by genetic means, provided variation for genetic modifiers of the plastic response already exist in the population or become available in the process. If sufficient variation for genetic modifiers does indeed exist, their frequency in the population will increase over generations, ultimately resulting in a situation in which the originally environment-induced phenotype becomes independent of the inducing environment, produced constitutively, and thus genetically assimilated.

GENETIC ASSIMILATION IN THE LABORATORY   Genetic assimilation is readily demonstrated in the laboratory (see Figure 11.10). Waddington (1952, 1953) found that when pupae from a laboratory population of wild-type *Drosophila melanogaster* were exposed to a heat shock of 40°C, the wings of some of the emerging adults had a gap in their posterior crossveins. This gap is not normally present in untreated flies. Two selection regimens were followed, one in which only the aberrant flies were bred to one another, and another in which only nonaberrant flies were mated to one another. After some generations of selection in which only the individuals showing the gap were allowed to breed, the proportion of adults with broken crossveins induced by heat shock at the pupal stage was raised to above 90%. Moreover (and significantly), by generation 14 of such inbreeding, a small proportion of individuals were crossveinless *even among flies of this line that had not been exposed to temperature shock*. When Waddington extended this artificial selection by breeding together only those adults that had developed the abnormality without heat shock, the frequency of crossveinless individuals among untreated flies became very high, reaching 100% in some lines. The phenotypically induced trait had become genetically assimilated into the population.*

Schmalhausen explained such results by saying that the environmental perturbation unmasked genetic heterogeneity for modifier genes which had already existed in the population. In other words, selection on existing genetic variability can stabilize the environmentally induced phenotype.

---

*Note that in these artificial selection experiments, the original phenotype induced by the environment was not intrinsically adaptive. Only the hand of the experimenter choosing which flies mated made it so.

Waddington emphasized that these modifier genes could act by changing the activation threshold for the phenotype, allowing it to become expressed over a wider set of environmental conditions (see Pigliucci et al. 2006). He explained his results by saying that the development of crossveins can be influenced by environmental disturbances above a certain threshold of intensity. But individuals from wild-type populations have a threshold so high that only an unusually strong stimulus, such as a heat shock, can effectively induce a modified expression. Thus, phenotypic variation does not arise if all the fly embryos have a threshold too high to be affected by the disturbances prevailing in the usual environment. However, when an exceptionally severe disturbance occurs, it is that set of individuals in which a phenotypic change is induced (those with the most sensitive genotypes for responding to the environmental stimulus) that are favored under artificial selection. As we will see, Waddington's and Schmalhausen's explanations appear to fit different cases of genetic assimilation.

In addition to finding the crossveinless phenotype upon exposure to heat shock, Waddington showed that his laboratory strains of *Drosophila* had a particular reaction norm in the fly's response to ether. Embryos exposed to ether at a particular stage developed a phenotype similar to the *bithorax* mutation and had four wings instead of two. (The flies' halteres—balancing structures on the third thoracic segment—were transformed into wings.) Generation after generation was exposed to ether, and individuals showing the four-winged state were selectively bred each time. After 20 generations, the selected *Drosophila* strain produced the mutant phenotype even when no ether was applied (Figure 11.11; Waddington 1953, 1956).

Subsequent experiments have borne out Waddington's findings (see Bateman 1959a,b; Matsuda 1982; Ho et al. 1983; Brakefield et al. 1996). In

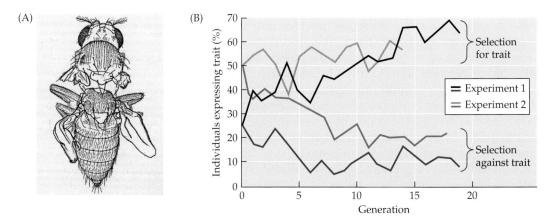

Figure 11.11 Phenocopy of the *bithorax* (four-winged) mutation. (A) Waddington's drawing of a *bithorax* mutant phenotype produced after treatment of the fly embryo with ether. The forewings have been removed to show the aberrant metathorax. This particular individual was in fact taken from "assimilated" stock that produced this phenotype without being exposed to ether. (B) Selection experiments for or against *bithorax*-like response to ether treatment. Two experiments are shown (red and blue lines). In each experiment, one group of flies was selected for the trait while the other was selected against the trait. (From Waddington 1956.)

1996, Gibson and Hogness repeated Waddington's experiments and "saw a steady increase in the frequency of thoracic abnormalities called phenocopies in each generation from 13 percent in the starting population to a plateau of 45 percent" (Gibson 1996). Conversely, when Gibson and Hogness selectively bred nontransformed flies resistant to ether treatment, the frequency of thoracic abnormalities dropped steadily. Gibson and Hogness (1996) found that four alleles of the *Ultrabithorax* (*Ubx*) gene had already existed in the population and were critical for the genetic assimilation of the ether-induced *bithorax* phenotype.* "Waddington's experiment showed some fruit flies were more sensitive to ether-induced phenocopies than others, but he had no idea why," Gibson said. "In our experiment, we show that differences in the *Ubx* gene are the cause of these morphological changes." In other words, the phenocopies were stabilized by alleles of regulatory genes such as those described in Chapter 10. In this case, then, Schmalhausen's model of preexisting regulatory alleles within the population appears to be the correct explanation.

GENETIC ASSIMILATION IN NATURE: MECHANISMS, MODELS, AND INFERENCES   Although it is difficult to document genetic assimilation in nature, there are several instances where it appears that phenotypic variation due to developmental plasticity was later fixed by genes. The first involves pigment variations in butterflies (see Hiyama et al. 2012). As early as the 1890s, scientists used heat shock to disrupt the pattern of butterfly wing pigmentation. In some instances, the color patterns that develop after temperature shock mimic the normal genetically controlled patterns of races (or related species) living at different temperatures. A race (subspecies) whose phenotype is characteristic of the species in a particular geographic area is called an **ecotype** (Turesson 1922). Standfuss (1896) demonstrated that a heat-shocked phenocopy of the Swiss subspecies of *Iphiclides podalirius* resembled the normal form of the Sicilian subspecies of that butterfly, and Richard Goldschmidt (1938) observed that heat-shocked specimens of the central European subspecies of *Aglais urticae* produced wing patterns that resembled those of the Sardinian subspecies (Figure 11.12). Conversely, cold-shocked individuals of the central European ecotype of *A. urticae* developed the wing patterns of the subspecies from northern Scandinavia.

Further observations on the mourning cloak butterfly (*Nymphalis antiopa*; Shapiro 1976), the buckeye butterfly (*Precis coenia*; Nijhout 1984), and the lycaenid butterfly (*Zizeeria maha*; Otaki et al. 2010) have confirmed the view that temperature variation can induce phenotypes that mimic genetically controlled patterns of related races or species existing in colder or warmer conditions. Chilling the pupa of *Pieris occidentalis* will cause it to have the short-day phenotype (Shapiro 1982), which is similar to that of the northern subspecies of pierids. Even "instinctive" behavioral phenotypes associated with these color changes (such as mating and flying) are phenocopied (see Burnet et al. 1973; Chow and Chan 1999).

(A)

(B)

(C)

Figure 11.12 Temperature shocking *Aglais urticae* produces phenocopies of geographic variants shown in Goldschmidt's original illustrations. (A) Usual central European variant. (B) Heat-shocked phenocopy resembling the Sardinian form. (C) The Sardinian form of the species. (From Goldschmidt 1938.)

*This finding confirmed genetic mapping studies, which had already shown that mutant alleles of the *Ubx* gene were responsible for producing the bithorax phenotype.

Thus, an *environmentally* induced phenotype might become the standard *genetically* induced phenotype in one part of the range of that organism.

Yet another case of natural genetic assimilation concerns the tiger snake (*Notechis scutatus*), which, like many fish species, has a head structure that can be altered by diet. The tiger snake can develop a bigger head to ingest bigger prey. This plasticity is seen when its diet includes both large and small mice. However, on some islands the diet contains only large mice, and here the snakes are born with large heads, and there is no plasticity (Figure 11.13). Thus, Aubret and Shine (2009) conclude that their data may provide empirical evidence of genetic assimilation, "with the elaboration of an adaptive trait shifting from phenotypically plastic expression through to canalization within a few thousand years."

GENETIC ASSIMILATION AND NATURAL SELECTION   Both Waddington and Schmalhausen independently proposed genetic assimilation to explain how some species have evolved rapidly in particular directions (see Gilbert 1994). For instance, both scientists were impressed by the calluses of the ostrich. Most mammalian and avian skin has the ability to form calluses on areas that are abraded by the ground or some other surface. The skin cells respond to friction by proliferating. While such examples of environmentally induced callus formation are widespread, the ostrich is

Figure 11.13 Genetic assimilation proposed in tiger snakes. The tiger snakes on the right-hand side of the cage are from mainland populations. They are born with small heads, and they achieve large heads through their plasticity, eating larger prey items (such as rodents and birds). The snakes on the left are from populations that have emigrated to islands where there are no small prey species. These island snakes are born with larger heads. (From Aubret and Shine 2009, courtesy of F. Aubret.)

The Heat-Shock Protein Hypothesis

While genetic assimilation is readily shown to occur in the laboratory, the idea that it could occur in nature remained controversial until Rutherford and Lindquist (1998) demonstrated a possible molecular mechanism unmasking cryptic genetic variation through stress. They were looking at the effects of mutations of the *Drosophila* heat-shock protein gene *Hsp83* and the inactivation of its protein product, Hsp90. When this gene or its protein product was inactivated, a whole range of phenotypes appeared. Several mutations that were preexisting in the population were allowed to become expressed. Moreover, the resulting mutants could be bred such that, within several generations, almost all the flies expressed the mutant phenotype, and some of the flies had that phenotype even if they contained a functional *Hsp83* gene. This phenomenon looked a great deal like genetic assimilation.

Hsp90 is a molecular "chaperone" that is required to keep many proteins in their proper three-dimensional conformation. It is especially important for the structure of several signaling molecules. There are numerous genetic mutations that are usually not expressed as mutant phenotypes because Hsp90 can correct the small changes in mutant protein structure. However, when heat shock (or ether) is applied, there are numerous other proteins that need the attention of Hsp90, and there is not enough Hsp90 to go around. Thus, the variation that has existed but has not been expressed then becomes expressed. In this way, Hsp90 appears to be a "capacitor" for evolutionary change, allowing genetic changes to accumulate until environmental stress releases them to reveal their effects on phenotype. This allows a wide range of genetic mixtures to become possible and allows these combinations to accumulate. When the environment changes and these combinations are expressed, most of them are predicted to be neutral or detrimental. But some might be beneficial and selected in the new environment. Continued selection would enable the genetic fixation of the adaptive physiological response (Rutherford and Lindquist 1998; Queitsch et al. 2002).

Sangster and colleagues (2008a,b) demonstrated that Hsp90-buffered variation is so widespread through the plant species *Arabidopsis thaliana* that every quantitative trait can be predicted to have at least one major component buffered by Hsp90. Moreover, they found that relatively slight environmental changes result in stress conditions that lower the levels of Hsp90 available and thereby cause the expression of hitherto hidden variations. Similarly, Rohner and colleagues (2014) showed that Hsp90 masked cryptic genetic variation affecting the eye size in surface relatives of blind cavefish. The potential for small eyes was found to exist in the natural

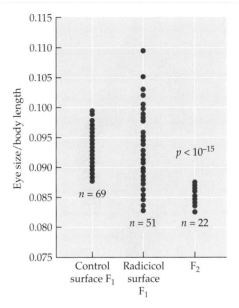

Genetic assimilation of the cryptic variation uncovered by HSP90 inhibition. Treatment of surface fish with the drug radicicol results in a population of offspring that have far more phenotypic variation than the unexposed population. Both larger and small eyes were seen. When these small-eyed fish (from one end of the population distribution) were bred together, the resulting generation had eyes smaller than those of the smallest fish of the original population. No new genes were involved in this change. (After Rohner et al. 2014.)

surface populations, and it can be observed when Hsp90 is inhibited. Inhibiting Hsp90 caused a greater standard deviation in the sizes of the eye orbits. Moreover, raising the surface fish in cave-like situations taxed the Hsp90 and allowed the cryptic variation to become expressed. It was even possible to observe genetic assimilation by selecting for small eye size in the cavefish. Here, by selecting the smallest eyes, one could generate a population that had eyes so small that they were not in the range of the parent population (see figure). It appears, then, that Hsp90 is a major cause of the canalization of phenotypes, enabling the same "wild-type" phenotype to be displayed across a range of genetic and environmental conditions (see Chapter 4). Environmental stress can cause the appearance of variants that can then be challenged by natural selection and selected such that they occur even when the environmental stressor is not present. Sangster and colleagues conclude that "Hsp90 appears to fulfill Waddington's concept of a developmental buffer or molecular canalization mechanism; one which can lead to the assimilation of novel traits on a large scale."

Figure 11.14 Ventral side of an ostrich; arrows mark calluses that are present on the animal from the time it is hatched. (From Waddington 1942.)

born with calluses already present where it will touch the ground (Figure 11.14; Duerden 1920). Waddington and Schmalhausen hypothesized that since the skin cells are already competent to be induced environmentally by friction, they could be induced by other things as well. As ostriches evolved, a mutation or a particular combination of alleles appeared that enabled the skin cells to respond to some substance *within* the embryo. Waddington (1942) wrote:

> Presumably its skin, like that of other animals, would react directly to external pressure and rubbing by becoming thicker.... This capacity to react must itself be dependent upon genes.... It may then not be too difficult for a gene mutation to occur which will modify some other area in the embryo in such a way that it takes over the function of external pressure, interacting with the skin so as to "pull the trigger" and set off the development of callosities.

As Waddington (1961) pointed out, a combination of orthodox Darwinism and orthodox embryology can enable the inheritance of what had been an environmentally induced phenotype.

West-Eberhard (2003) has noted that environmentally induced novelties may be extremely important in evolution. Mutationally induced novelties would occur in only a family of individuals, whereas environmentally induced novelties would occur throughout a population. Moreover, the inducing environment would most often also be a selecting environment. Thus, there would be selection immediately for this trait, and the trait would be continuously induced. Computer modeling has shown that such an environmentally induced genetic assimilation can produce rapid evolutionary change as well as speciation (Kaneko 2002; Behara and Nanjundiah 2004). Indeed, West-Eberhard (2005) has proposed that studying gene expression and plasticity will help us understand the genetic divergence that gives rise to speciation. "Contrary to common belief, environmentally initiated novelties may have greater evolutionary potential than mutationally induced ones."

Thus, genetic assimilation is a mechanism whereby environmentally induced traits can become internally induced during embryonic development through genomic influences. In this process, previously hidden (cryptic) genetic variation becomes important for the stabilization of that phenotype or for the regulation of that expression after an environmental stimulus overcomes the threshold for the expression of these phenotypes. Selection in the presence of this environmental factor enriches the gene pool for the cryptic alleles that would determine this trait, and eventually these alleles become so frequent that the trait appears even in the absence of the environmental stimulus. In this way, a phenotypically plastic trait can be converted into a genetically fixed trait that is constantly produced under a wide range of environmental conditions.

Niche construction

Phenotypic accommodation is predicated on the embryonic interactions called **reciprocal embryonic induction**. Reciprocal embryonic induction between two adjacent tissues is the fundamental principle for organ formation. For instance, the presumptive retina of the mammalian eye is a bulge of cells from the forebrain, while the presumptive lens is a subset of epithelial cells in the surface ectoderm of the head. When the two tissues meet, a complex dialogue is established whereby the presumptive lens cells tell the brain bulge to become the retina, and the presumptive retinal cells tell the placodal epithelium of the head to become the lens (Figure 11.15). Similarly, when the ureteric bud grows into and contacts the nephrogenic mesenchyme, the nephrogenic mesenchyme is told to become the nephron, and at the same time it tells the ureteric bud to become the renal pelvis and collecting ducts of the kidney.

But, as we have seen, the external environment can also be a source of signals that affect development. As environmental signals have been found to be critical in induction, it is not surprising that such environmental signals can also be accommodated. But what if these developmental interactions seen in reciprocal embryonic induction extended to interactions

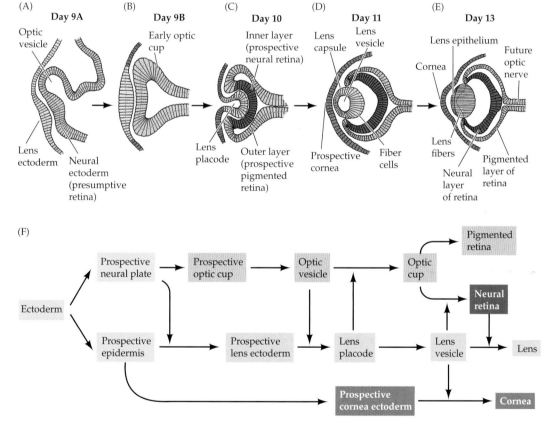

Figure 11.15 Reciprocal embryonic induction forms the eye of a mouse. (A,B) At embryonic day 9, a region of surface ectoderm (destined to become the lens and cornea of the eye) comes into contact with a bulge of neuroectoderm from the forebrain, which is induced to form an optic cup. (C) By day 10, the lens-forming cells have been induced to invaginate, and two layers of retinal cells have been distinguished. (D) By day 11 of gestation, contact with the optic vesicle has induced differentiation of the lens from the presumptive corneal cells. (E) By day 13, two types of lens cells—cuboidal cells and elongated fiber cells—have been established and the cornea has developed in front of them. (F) Summary of some of the inductive interactions during mammalian eye development. (After Cvekl and Piatigorsky 1996.)

between organisms? This question is being addressed by a relatively new branch of evolutionary biology called **niche construction** (see Laland et al. 2008).

Niche construction is an evolutionary idea that emphasizes the capacity of organisms to modify their environments and thereby act as codirectors of their own evolution and that of other species. The importance of this perspective to evolutionary biology was recognized by Richard Lewontin (1982, 1983), who noted that "organisms fit the world so well because they have constructed it." He argued that the organism was not just a passive entity being acted upon by selective forces in the environment, but also an active agent capable of constructing an environment suited to its own ends. This idea has recently been extended by Lewontin (2000) and by John Odling-Smee (1988) and has become strengthened by theoretical population genetic and experimental findings that show niche construction to be an important factor in an organism's fitness (Laland et al. 1996; Odling-Smee et al. 1996; Laland 1999; Odling-Smee 2003; Donohue 2005). Niche construction can even counteract natural selection in instances where adult organisms can modify the environment to constrain or overcome selective forces. For instance, earthworms (*Lumbricus terrestris*) make burrows that provide an aqueous niche within the terrestrial environment. As a result, this species has evolved very little since migrating onto land more than 50 million years ago (Turner 2000).

Niche construction takes on an even more important role when it is linked to development. Developmental environments become coupled to developing organisms by the niche-constructing activities of their organisms. Here, the niche construction seen during co-development (see Chapter 3) becomes the macroscopic analogue to reciprocal embryonic induction (Waddington 1953, 1957; Gilbert 2001; Laland et al. 2008). Niche construction similarly extends the dialogues of organ formation from within the embryo to outside it. The "circuitry diagrams" of organ formation within the embryo become united with the "circuitry diagrams" of organisms within the ecosystem.

We have already mentioned some interesting cases of niche construction. Some of the symbiotic microbes in the mouse intestine, for instance, induce gene expression in the gut epithelia not only to help the host, but to help themselves. The normal gut microbes, such as *Bacteroides*, induce gene expression in the Paneth cells of the intestine, instructing these cells to produce two compounds—angiogenin-4 and RegIII—that prevent the colonization of the intestine by *other* species of microbes. *Bacteroides, Escherichia coli*, and other symbiotic species are impervious to this compound, while several pathogenic Gram-positive bacteria (*Enterococcus faecalis* and *Listeria monocytogenes*) are wiped out by it (Hooper et al. 2003; Cash et al. 2006). Thus, the microbial species is modifying its niche, causing its environment to change in such a way that it can better survive.

Such niche-constructing behavior is also seen in relationships between symbiotic bacteria and their arthropod hosts. Indeed, sometimes the interaction is more one-sided than in the mammalian example. We saw in Chapter 3 how the bacterium *Wolbachia* can skew the sex ratio of the progeny of its host insects by feminizing genetically male embryos, resulting in all or mostly female progeny (see Figures 3.9 and 3.10). This feminization of insect populations has a clear advantage for the infecting *Wolbachia* because it directly increases the number of females, the only sex capable of passing the bacteria on to the insect progeny.

Another example of niche-constructing behavior is found in the goldenrod gall fly (*Eurosta solidaginis*). The female fly lays her eggs inside the goldenrod's stem, within which the eggs hatch into larvae (caterpillars). As a caterpillar eats the stem, proteins in the larval saliva induce cell proliferation in the goldenrod, resulting in the formation of a gall (Figure 11.16). The caterpillar enters the gall and continues eating from within it. As winter approaches, the larva, in danger of freezing to death, begins to produce sorbitol and trehalose sugars that act to prevent ice formation inside the cells. The trigger for synthesis of this "antifreeze" is not temperature but aromatic substances produced by the desiccating plant tissues of the gall (Williams and Lee 2005).

Here we see reciprocal induction on the ecological level. The fly larva creates a niche (the gall) by causing the plant to change its development. The niche provides not only nutrition but also the signal that allows the larva to change its development as winter approaches. Reviewing the literature on gall development, West-Eberhard (2003) concludes, "In gall

(A)

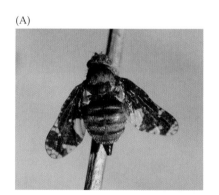

(C)

(B)

Figure 11.16 Niche-constructing behavior of *Eurosta solidaginis*, the goldenrod gall fly. (A) A female fly deposits her eggs inside the goldenrod's stem, within which the eggs hatch. (B) As the larva feeds, proteins in its saliva induce the formation of a gall on the plant's stem. (C) An opened gall reveals the fly larva at the center. (A,C photographs by K. and J. M. Storey, www.carleton.ca/~kbstorey.)

Creating Each Other's Environment: Reciprocal Developmental Plasticity

Niche construction results when the traits of an organism influence the habitat that it experiences. Sometimes this can result in strange consequences. In Chapter 1, we discussed predator-induced polyphenism, where an organism's plasticity enables it to respond to a predator (developmentally, physiologically, or behaviorally) by becoming less readily eaten. However, if this strategy is successful, one would expect that the predator should adapt to it. If the predator can't evolve a strategy to circumvent this new adaptation in the prey, it must find another food source or it will die. This is sometimes called "the evolutionary arms race." Here, predator-induced plasticity can be coupled with prey-induced plasticity in the predator.

Defensive structures are often difficult to make, which is why their induction only in the continued presence of predators is such a good strategy. In northern Japan, the melting snow forms ephemeral ponds in which both salamanders (*Hynobius retardatus*) and frogs (*Rana pirica*) lay their eggs. Thus, the salamander larvae and the tadpoles (frog larvae) inhabit the ponds in early spring (Figure A–D). The salamander larvae are carnivores and will consume the tadpoles. When such salamander predators are present, they unknowingly signal the tadpoles to develop a "bulgy" phenotype with enlarged bodies and tails. This prevents their immediate ingestion by the salamanders. Conversely, when the large tadpoles are abundant, they inadvertently signal the salamander larvae to develop a broad, gaping mouth, which allows them to swallow tadpoles more easily (Figure E; Michimnae and Wakahara 2002; Kishida et al. 2006). Thus, both predator and prey use developmental plasticity to increase their fitness. Moreover, the timing of salamander larva metamorphosis is also plastic and is influenced by the number and morphology of the prey. This causes strange population fluctuations that depend on how rapid the speed of adaptation is between predator and prey (Mougi 2012; Kishida et al. 2013). In this way, there is a mutual inductive response, wherein each organism changes the environment of the other.

(A)   (B)   (C)   (D)

(E)

Reciprocal plasticity. (A–D) Offensive and defensive morphs of the salamander larvae (*Hynobius retardatus*) and frog larvae (*Rana pirica*) that inhabit the same ephemeral ponds. (A) Nondefensive frog tadpole. (B) Defensive "bulgy" frog tadpole. (C) Nonpredatory salamander larva. (D) Predaceous salamander larva. (E) Salamander larva consuming frog tadpole in natural pond. (From Dr. Kishida Osamu, forestcsv.ees.hokudai.ac.jp/en/research.)

production, the plant is a crucial and specific influential element of the insect's development, and the insect is a specific form-inducing element of the plant's developmental environment." There is a co-developmental relationship much like that of the organs in the embryo.

In addition, pollinators and flowers must coevolve developmental patterns in which the eclosion of pollinators is in synchrony with the emergence of floral organs, and the timing of seed production with the foraging habits of seed dispersers coordinates the growth of forests. Thus reciprocal developmental interactions can coordinate ecological rhythms.

In short, when the organism alters its development in response to environmental cues such that the organism is more fit in the particular environment, it is called "adaptive developmental plasticity"; when a developing organism induces changes in its physical environment in ways that make the environment more fit for the organism, it is called "niche construction."* Moreover, there can be reciprocity between these two sets of inductive phenomena.

Looking Ahead

The fields of developmental symbiosis, developmental plasticity, and epialleles are fields of the twenty-first century. In 1990, there were no articles published that had "gut microbiome" in their titles or abstracts. In 2000, there were eight; and in 2014, there were 1687 such articles. Articles mentioning developmental plasticity in their titles or abstracts increased from 57 to 659 in the same timeframe. And the numbers keep rising.

Until very recently, developmental symbiosis, developmental plasticity, and epialleles were thought to be exceptions to the rules. This book has documented that they are not exceptional cases. They are the rules. Developmental plasticity appears to be ubiquitous. Developmental symbiosis appears to be ubiquitous. Cryptic genetic variation appears to be ubiquitous. Whether epialleles are similarly so widespread is just being studied, but they, too, seem to be more prevalent than had been thought even a year ago. Data from ecological developmental biology add entirely new features to evolutionary biology. The evolutionary biology of the Modern Synthesis did not see an organism as a multilineage individual, a holobiont consortium of many species. Nor did it see transposable elements and plasticity as a force in generating novelty. The idea that plasticity and symbiosis could be critical for evolutionary transitions and speciation events was not mainstream. The ability to form a "team" involves both competition and cooperation, and the organism as an ecosystem has not been part of the thinking of mainstream evolutionary biology.

Evolutionary theory tends to develop like a pearl, adding concentric layers as it grows. At its core—the grain that initiates the pearl—is the Darwinian synthesis of anatomy, biogeography, taxonomy, and paleontology. Around that core is the synthesis of Fisher and Haldane, which brought evolutionary biology into line with theoretical population genetics. On

---

*As Stella says in *Silverado*, "The world is what you make of it, friend. If it doesn't fit, you make alterations."

top of that layer, Dobzhansky and Mayr showed that population genetics could also model natural populations and speciation. The outer core of traditional evolutionary theory is presently the syntheses of Maynard Smith, Price, Hamilton and others who used population genetic models to explain social behaviors. Now a new layer is being added to this pearl. One can see the incorporation into evolutionary biology of ideas such as enhancer modularity, regulatory genes, gene duplication, and the need to have a more complete theory of evolution through a genetic theory of body construction. The first wave of evolutionary developmental biology is becoming part of evolutionary theory. This keeps with the thoroughly gene-centered Modern Synthesis view of evolution, but it adds a critically important developmental dimension. Evolution can be seen as heritable changes in development. Developmental genetics is being added on to population genetics, quantitative genetics merging with molecular genetics. Indeed, as was predicted in the early days of evolutionary developmental biology, the study of evolution is beginning to emphasize the population genetics of regulatory alleles (see, for instance, Gilbert et al. 1996; Rockman and Wray 2002; Boyd et al. 2015; Manceau et al. 2011).

But starting around 2000, developmental biologists came to realize that the environment plays an important role in development. Symbionts, diet, predators, conspecifics, and temperature were critical for forming the phenotype of animals. If this were the case, then a developmental approach to animal evolution must also contain the newly appreciated elements of developmental plasticity, symbiogenesis, and environmentally derived epialles (West-Eberhart 2003; Jablonka and Lamb 2005; Gilbert and Epel 2009; Pigliucci and Müller 2010; Gissis and Jablonka 2011). This second wave of evo-devo, this eco-devo-evo, is where the debate is currently centered.

This debate was starkly represented in a well publicized article, entitled "Does evolutionary theory need a rethink?" (Laland et al. 2014). Those on the affirmative side said that an extended evolutionary theory was needed "in which the processes by which organisms grow and develop are recognized as causes of evolution." The negative side in the debate countered that there was no compelling evidence for the importance of non-genetic approaches to evolution. Interestingly, several of the scientists debating against the need to expand evolutionary biology are at the vanguard of incorporating the *first* wave of evo-devo (regulatory alleles, paracrine factors) into evolutionary theory. But they are convinced that there was no evidence (and not convinced that there was evidence) that symbiosis, plasticity, genetic accommodation, epialleles, or niche construction mandated a dramatic rethink of evolutionary theory.

This book has attempted to provide such evidence. And if such evidence is accepted into evolutionary theory, the way we perceive nature changes. And this is the subject of the Coda.

References

Abouheif, E., M. J. Favé, A. S. Ibarrarán-Viniegra, M. P. Lesoway, A. M. Rafiqi and R. Rajakumar. 2014. Eco-evo-devo: The time has come. *Adv. Exp. Med. Biol.* 781: 107–125.

Agrawal, A. A., C. Laforsch and R. Tollrian. 1999. Transgenerational induction of defences in animals and plants. *Nature* 401: 60–63.

Alegado, R. A. and 7 others. 2012. Bacterial sulfonolipid triggers multicellular development in the closest living relatives of animals. *Elife* 1: e00013. doi: 10.7554/eLife.00013.

Alegado, R. A., J. D. Grabenstatter, R. Zuzow, A. Morris, S. Y. Huang, R. E. Summons and N. King. 2013. *Algoriphagus machipongonensis* sp. nov. co-isolated with a colonial choanoflagellate. *Int. J. Syst. Evol. Microbiol.* 63(Pt.1): 163–168.

Anway, M. D., A. S. Cupp, M. Uzumcu and M. K. Skinner. 2005. Epigenetic transgenerational actions of endocrine disruptors and mate fertility. *Science* 308: 1466–1469.

Anway, M. D., C. Leathers and M. K. Skinner. 2006a. Endocrine disruptor vinclozolin induced epigenetic transgenerational adult-onset disease. *Endocrinology* 147: 515–5523.

Anway, M. D., M. A. Memon, M. Uzumcu and M. K. Skinner. 2006b. Transgenerational effect of the endocrine disruptor vinclozolin on male spermatogenesis. *J. Andrology* 27: 868–879.

Archie, E. A. and K. R. Theis. 2011. Animal behavior meets microbial ecology. *Anim. Behav.* 82: 425–436.

Aubret, F. and R. Shine. 2009. Genetic assimilation and the post-colonization erosion of phenotypic plasticity in island tiger snakes. *Curr. Biol.* 19: 1932–1936.

Baker, N. 2014. How fish can learn to walk. www.nature.com/news/how-fish-can-learn-to-walk-1.15778

Baldwin, J. M. 1896. A new factor in evolution. *Am. Nat.* 30: 441–451; 536–553.

Bateman, K. G. 1959a. Genetic assimilation of the *dumpy* phenocopy. *J. Genet.* 56: 341–352.

Bateman, K. G. 1959b. Genetic assimilation of four venation phenocopies. *J. Genet.* 56: 443–474.

Bateson, P. 2005. The return of the whole organism. *J. Biosci.* 30: 31–39.

Bateson, P. 2014. Evolution, epigenetics, and cooperation. *J. Biosci.* 38: 1–10.

Bateson, P. and P. Gluckman. 2011. *Plasticity, Robustness, Development, and Evolution*. Cambridge University Press, Cambridge.

Baumann, P. 2015. Biology bacteriocyte-associated endosymbionts of plant sap-sucking insects. *Annu. Rev. Microbiol.* 59: 155–189.

Behera, N. and V. Nanjundiah. Phenotypic plasticity can potentiate rapid evolutionary change. *J. Theor. Biol.* 226: 177–184.

Belyaev, D. K. 1969. Domestication of animals. *Science J.* 5: 47–52.

Belyaev, D. K. 1974. Domestication, plant and animal, pp. 936–942 in *Encyclopaedia Britannica*, Ed. 15, edited by H. H. Benton. Encyclopedia Britannica–Helen Hemingway Benton Publishing, Chicago. (Cited in Wilkins et al., op. cit.)

Bennett, G. M. and N. A. Moran. 2015. Heritable symbiosis: The advantages and perils of an evolutionary rabbit hole. *Proc. Natl. Acad. Sci. USA* pii: 201421388.

Boyd, J. L. and 7 others. 2015. Human-chimpanzee differences in a FZD8 enhancer alter cell-cycle dynamics in the developing neocortex. *Curr. Biol.* 25: 772–779.

Brakefield, P. M., F. Kesbeke and P. B. Koch 1996. The regulation of phenotypic plasticity of eyespots in the butterfly *Bicyclus*. *Am. Nat.* 152: 853–860.

Brucker, R. M. and S. R. Bordenstein. 2013. The hologenomic basis of speciation: gut bacteria cause hybrid lethality in the genus *Nasonia*. *Science*. 341: 667–669.

Burdge, G. C., J. Slater-Jefferies, C. Torrens, E. S. Phillips, M. A. Hanson and K. A. Lillycrop. 2007. Dietary protein restriction of pregnant rats in the F_0 generation induces altered methylation of hepatic gene promoters in the adult male offspring in the F_1 and F_2 generations. *Br. J. Nutr.* 97: 435–439.

Burnet, B., K. Connolly and B. Harrison. 1973. Phenocopies of pigmentary and behavioural effects of the *Yellow* mutant in *Drosophila* induced by α-dimethyltyrosine. *Science* 181: 1059–1060.

Cabej, N. 2008. *Epigenetic Principles of Evolution*. Albenet, Dumont, NJ.

Chandler, V. 2007. Paramutation: From maize to mice. *Cell* 128: 641–645.

Cash, H. L., C. V. Whitman, C. L. Benedict and L. V. Hooper. 2006. Symbiotic bacteria direct expression of an intestinal bactericidal lectin. *Science* 313: 1126–1130.

Chang, H. S., M. D. Anway, S. S. Rekow and M. K. Skinner. 2006. Transgenerational epigenetic imprinting of the male germline by endocrine disruptor exposure during gonadal sex determination. *Endocrinology* 147: 5524–5541.

Chow, K. L. and K. W. Chan. 1999. Stress-induced phenocopy of *C. elegans* defines functional steps of sensory organ differentiation. *Dev. Growth Diff.* 41: 629–637.

Crews, D., R. Gillette, I. Miller-Crews, A. C. gore and M. K. Skinner. 2014. Nature, nurture, and epigenetics. *Mol. Cell Endocrinol.* 398: 42–52.

Crews, D. and 7 others. 2007. Transgenerational epigenetic imprints on mate preference. *Proc. Nat. Acad. Sci. USA* 104: 5942–5946.

Cvekl, A. and J. Piatigorsky. 1996. Lens development and crystallin gene expression: Many roles for Pax6. *BioEssays* 18: 621–630.

Danielson-François, A. M., J. K. Kelly and M. D. Greenfield. 2006. Genotype × environment interaction for mate attractiveness in an acoustic moth: Evidence for plasticity and canalization. *J. Evol. Biol.* 9: 532–542.

Darwin, C. 1859. *On the Origin of Species*. John Murray, London

Darwin, C. 1868. *The Variation of Animals and Plants under Domestication*. Appleton, New York.

Darwin, C. 1875. *The Variation of Animals and Plants under Domestication*. John Murray, London.

Dayel, M. J., R. A. Alegado, S. R. Fairclough, T. C. Levin, S. A. Nichols, K. McDonald and N. King. 2011. Cell differentiation and morphogenesis in the colony-forming choanoflagellate *Salpingoeca rosetta*. *Dev. Biol.* 357: 73–82.

Dias, B. G. and K. Ressler. 2014. Prental olfactory experiences influences behavior and neural structure in subsequent generations. *Nature Neurosci.* 17: 89–96.

Donohue, K. 2005. Niche construction through phonological plasticity: Life history dynamics and ecological consequences. *New Phytologist* 166: 83–92.

Douglas, A. E. 2010. *The Symbiotic Habit*. Princeton University Press, Princeton, NJ.

Duerden, J. E. 1920. Inheritance of callosities in the ostrich. *Am. Nat.* 54: 289–312.

Dunbar, H. E., A. C. C. Wilson, N. R. Ferguson and N. A. Moran. 2007. Aphid thermal tolerance is governed by a point mutation in bacterial symbionts. *PLoS Biol.* 5(5): e96.

Dupressoir, A., C. Vernochet, F. Harper, J. Guégan, P. Dessen, G. Pierron and T. Heidmann. 2011. A pair of co-opted retroviral envelope syncytin genes is required for formation of the two-layered murine placental syncytiotrophoblast. *Proc. Natl. Acad. Sci. USA* 108: E1164–E1173.

El Albani, A. and 19 others. 2010. Large colonial organisms with coordinated growth in oxygenated environments 2.1 Gyr ago. *Nature* 466: 100–104.

Fischman, J. 1995. Why mammalian ears went on the move. *Science* 270: 1436.

Frazetta, T. H. 1975. *Complex Adaptations in Evolving Populations*. Sinauer Associates, Sunderland, MA.

Fullston, T. and 8 others. 2013. Paternal obesity initiates metabolic disturbances in two generations of mice with incomplete penetrance to the F_2 generation and alters the transcriptional profile of testis and sperm microRNA content. *FASEB J.* 27: 4226–4243.

Garson, J., L. Wang and S. Sarkar. 2003. How development may direct evolution. *Biol. Philos.* 18: 353–370.

Gibson, G. 1996. Quoted in S. Pobojewsky, *The University Record*, January 16.

Gibson, G. and D. S. Hogness. 1996. Effect of polymorphism in the *Drosophila* regulatory gene *Ultrabithorax* on homeotic stability. *Science* 271: 200–203.

Gilbert, S. F. 1994. Dobzhansky, Waddington, and Schmalhausen: Embryology and the modern synthesis. In M. B. Adams (ed.), *The Evolution of Theodosius Dobzhansky*. Princeton University Press, Princeton, NJ, pp. 143–154.

Gilbert, S. F. 2001. Ecological developmental biology: Developmental biology meets the real world. *Dev. Biol.* 233: 1–12.

Gilbert, S. F. 2002. Canalization. In M. Pagel (ed.), *Encyclopedia of Evolution*, Vol. 1. Oxford University Press, New York, pp. 133–135.

Gilbert, S. F. and D. Epel. 2009. *Ecological Developmental Biology*, 1st Ed. Sinauer Associates, Sunderland, MA.

Gilbert, S. F., J. Opitz and R. A. Raff. 1996. Resynthesizing evolutionary and developmental biology. *Dev. Biol.* 173: 357–372.

Gilbert S. F., J. Sapp and A. I. Tauber. 2012. A symbiotic view of life: We have never been individuals. *Q. Rev Biol.* 87: 325–341.

Gilbert, S. F. and 7 others. 2010. Symbiosis as a source of selectable epigenetic variation: Taking the heat for the big guy. *Philos. Trans. R. Soc. Lond. B. Biol. Sci.* 365: 371–378.

Gissis, S. and E. Jablonka. 2011. *Transformations of Lamarckism: From Subtle Fluids to Molecular Biology*. MIT Press, Cambridge, MA.

Goldschmidt, R. B. 1938. *Physiological Genetics*. McGraw-Hill, New York.

Goldschmidt, R. B. 1940. *The Material Basis of Evolution*. Yale University Press, New Haven, CT.

Gottlieb, G. 1992. *Individual Development and Evolution*. Oxford University Press, New York.

Gulick, J. T. 1872. *Nature* 6: 222–224. Quoted in B. K. Hall, 2006, Evolutionist and missionary, the Reverend John Thomas Gulick (1832–1923). II. Coincident or Ontogenetic Selection: The Baldwin Effect. *J. Exp. Zool. MDB* 306B: 489–495.

Gutteling, E. W., J. A. G. Riksen, J. Bakker and J. E. Kammenga. 2007. Mapping phenotypic plasticity and genotype-environment interactions affecting life history traits in *Caenorhabidis elegans*. *Heredity* 98: 28–37.

Hall, B. K. 2003. Baldwin and beyond: Organic selection and genetic assimilation. In B. H. Weber and D. J. Depew (eds.), *Evolution and Learning: The Baldwin Effect Reconsidered*. MIT Press, Cambridge, MA, pp. 141–168.

Hampé, A. 1959. Contribution à l'étude du développement et la régulation des déficiences et excédents dans la patte de l'embryon de poulet. *Arch. Anat. Microsc. Morphol. Exp.* 48: 347–479.

Heckman, D. S., D. M. Geiser, B. R. Eidell, R. L. Stauffer, N. L. Kardos and S. B. Hedges. 2001. Molecular evidence for the early colonization of land by fungi and plants. *Science* 293: 1129–1133.

Hehemann, J. H., G. Correc, T. Barbeyron, W. Helbert, M. Czjzek and G. Michel. 2010. Transfer of carbohydrate-active enzymes from marine bacteria to Japanese gut microbiota. *Nature* 464: 908–912.

Hehemann, J. H., A. G. Kelly, N. A. Pudlo, E. C. Martens and A. B. Boraston. 2012. Bacteria of the human gut microbiome catabolize red seaweed glycans with carbohydrate-active enzyme updates from extrinsic microbes. *Proc. Natl. Acad. Sci. USA* 109: 19786–19791.

Hiyama, A., W. Taira and J. M. Otaki. 2012. Color-pattern evolution in response to environmental stress in butterflies. *Front. Genet.* 3: 15.

Ho, M.-W. 2013. No genes for intelligence in the fluid genome. *Adv. Child Dev. Behav.* 45: 67–92.

Ho, M.-W., E. Bolton and P. T. Saundres. 1983. The bithorax phenocopy and pattern formation. 1. Spatiotemporal characteristics of the phenocopy response. *Exp. Cell Biol.* 51: 282–290.

Hooper, L. V., T. S. Stappenbeck, C. V. Hong and J. I. Gordon. 2003. Angiogenins: A new class of microbicidal proteins involved in innate immunity. *Nature Immunol.* 4: 269–273.

Hughes, V. 2013. Mice inherit specific memories, because epigenetics? phenomena.nationalgeographic.com/2013/12/01/mice-inherit-specific-memories-because-epigenetics/

Husnik, F. and 12 others. 2013. Horizontal gene transfer from diverse bacteria to an insect genome enables a tripartite nested mealybug symbiosis. *Cell* 153: 1567–1578.

Hutchinson, J. 2014. Dynasty of the plastic fish. *Nature* 513: 37–38.

Jablonka, E. and M. J. Lamb. 2005. *Evolution in Four Dimensions: Genetic, Epigenetic, Behavioral, and Symbolic Variation in the History of Life*. MIT Press, Cambridge, MA.

Jablonka, E. and M. J. Lamb. 2006. *Evolution in Four Dimensions: Genetic. Epigenetic, Behavioral, and Symbolic Variation in the History of Life*. MIT Press, Cambridge, MA.

Jablonka, E. and M. J. Lamb. 2015. The inheritance of acquired epigenetic variations. *Int. J. Epidemiol.* pii: dyv020.

Jablonka, E. and G. Raz. 2009. Transgenerational epigenetic inheritance: Prevalence, mechanisms, and implications for the study of heredity. *Q. Rev. Biol.* 84: 131–176.

Jirtle, R. L. and M. K. Skinner. 2007. Environmental epigenomics and disease susceptibility. *Nature Rev. Genet.* 8: 253–262.

Kaneko, K. 2002. Symbiotic sympatric speciation: Consequence of interaction–driven phenotype differentiation through developmental plasticity. *Pop. Ecol.* 44: 71–85.

Kawano, M., H. Kawaji, V. Grandjean, J. Kiani and M. Rassoulzadegan. 2012. Novel small noncoding RNAs in mouse spermatozoa, zygotes and early embryos. *PLoS ONE* 7(9): e44542.

Kemp, T. S. 1982. *Mammal-Like Reptiles and the Origin of Mammals*. Academic Press, New York.

Kiani, J. and 7 others. 2013. RNA-mediated epigenetic heredity requires the cytosine methyltransferase Dnmt2. *PLoS Genet.* 9: e1003498.

King, R. C. and W. D. Stanfield. 1985. *A Dictionary of Genetics*, 3rd Ed. Oxford University Press, Oxford.

Kishida, O., Y. Mizuta and K. Nishimura. 2006. Reciprocal phenotypic plasticity in a predator-prey interaction between larval amphibians. *Ecology* 87: 1599–1604.

Kishida, O., Z. Costa, A. Tezuka and H. Michimae. 2013. Inducible offences affect predator-prey interactions and life-history plasticity in both predators and prey. *J. Anim. Ecol.* 83: 899–906.

Köntges, G. and A. Lumsden. 1996. Rhombencephalic neural crest segmentation is preserved throughout craniofacial ontogeny. *Development* 122: 3229–3242.

Künzl, C. and N. Sachser. 1999. The behavioral endocrinology of domestication: A comparison between the domestic guinea pig (*Cavia aperea* f. porcellus) and its wild ancestor, the cavy (*Cavia aperea*). *Horm. Behav.* 35: 28–37.

Koonin, E. V. 2010. The origin and early evolution of eukaryotes in the light of phylogenomics. *Genome Biol.* 11: 209.

Laland, K. N. 1999. Evolutionary consequences of niche construction and their implications for ecology. *Proc. Natl. Acad. Sci. USA* 96: 10242–10247.

Laland, K. N., F. J. Odling-Smee and M. W. Feldman. 1996. On the evolutionary consequences of niche construction. *J. Evol. Biol.* 9: 293–316.

Laland, K. N., J. Odling-Smee and S. F. Gilbert. 2008. Evo-devo and niche construction: Building bridges. *J. Exp. Zool. B.* 310: 549-566.

Laland, K. and 14 others. 2014. Does evolutionary theory need a rethink? *Nature* 514: 161–164.

Ledon-Rettig, C. C., D. W. Pfennig, A. J. Chunco and I. Dworkin. 2014. Cryptic genetic variation in natural populations: A predictive framework. *Integr. Comp. Biol.* 54: 783–793.

Ledon-Rettig, C. C., D. W. Pfennig and E. J. Crespi. 2010. Diet and hormonal manipulation reveal cryptic genetic variation: Implications for the evolution of novel feeding strategies. *Proc. Biol. Sci.* 277: 3569–3578.

Ledon-Rettig, C. C., D. W. Pfennig and N. Nascone-Yoder. 2008. Ancestral variation and the potential for genetic accommodation in larval amphibians: Implications for the evolution of novel feeding strategies. *Evol. Devel.* 10: 316–325.

Levit, G. S., U. Hossfeld and L. Olsson. 2006. From the "modern synthesis" to cybernetics: Ivan Ivanovich Schmalhausen (1884–1963) and his research program for a synthesis of evolutionary and developmental biology. *J. Exp. Zool.* 306B: 89–106.

Lewontin, R. C. 1982. Organism and environment. In H. C. Plotkin (ed.), *Learning, Development and Culture*. Wiley, New York.

Lewontin, R. 1983. Gene, organism, and environment. In D. S. Bendall (ed.), *Evolution from Molecules to Men*. Cambridge University Press, Cambridge.

Lewontin, R. 2000. *The Triple Helix: Gene, Organism, and Environment*. Harvard University Press, Cambridge, MA.

Lloyd Morgan, C. 1896. *Habit and Instinct*. Arnold, London.

López-Madrigal, S., S. Balmand, A. Latorre, A. Heddi, A. Moya and R. Gil. 2013. How does *Tremblaya princeps* get essential proteins from its nested partner *Moranella endobia* in the mealybug *Planoccocus citri*? *PLoS ONE* 8(10): e77307.

Lynch, V. J., R. D. Leclerc, G. May and G. P. Wagner. 2011. Transposon-mediated rewiring of gene regulatory networks contributed to the evolution of pregnancy in mammals. *Nature Genet.* 43: 1154–1159.

Manolio, T. A. and 26 others. 2009. Finding the missing heritability of complex diseases. *Nature* 461: 747–753.

Manceau, M., V. S. Domingues, R. Mallarino and H. E. Hoekstra. 2011. The developmental role of Agouti in color pattern evolution. *Science* 331: 1062–1065.

Margulis, L. 1970. *Origin of Eukaryotic Cells*. Yale University Press, New Haven, CT.

Margulis, L. 1993. *Symbiosis in Cell Evolution*. W. H. Freeman, New York.

Margulis, L. 1999. *Symbiotic Plane*t. Basic Books, New York.

Margulis, L. and R. Fester, eds. 1991. *Symbiosis as a Source of Evolutionary Innovation*. MIT Press, Cambridge, MA.

Margulis, L. and D. Sagan. 2001. Marvellous microbes. *Resurgence* 206: 10–12.

Margulis, L. and D. Sagan. 2003. *Acquiring New Genomes: A Theory of the Origins of Species*. Basic Books, New York.

Matsuda, R. 1982. The evolutionary process in talitrid amphipods and salamanders in changing environments, with a discussion of "genetic assimilation" and some other evolutionary concepts. *Can. J. Zool.* 60: 733–749.

McCaffery, A. and S. J. Simpson. 1998. A gregarizing factor present in the egg pod foam of the desert locust *Schistocerca gregaria*. *J. Exp. Biol.* 201: 347–363.

McCutcheon, J. P. and N. A. Moran. 2011. Extreme genome reduction in symbiotic bacteria. *Nature Rev. Microbiol.* 10: 13–26.

McCutcheon, J. P. and C. D. von Dohlen. 2011. An interdependent metabolic patchwork in the nested symbiosis of mealybugs. *Curr. Biol.* 21: 1366–1372.

McFall-Ngai, M. and 25 others. 2013. Animals in a bacterial world, a new imperative for the life sciences. *Proc. Natl. Acad. Sci. USA* 110: 3229–3326.

Meaney, M. J. 2001. Maternal care, gene expression, and the transmission of individual differences in stress reactivity across generations. *Annu. Rev. Neurosci.* 24: 1161–1192.

Merezhkowsky, C. 1909 *The Theory of Two Plasms as Foundation of Symbiogenesis, New Doctrine on the Origin of Organisms*,

Michimae, H. and M. Wakahara. 2002. A tadpole-induced polyphenism in the salamander *Hynobius retardatus*. *Evolution* 56: 2029–2038.

Moczek, A. 2015. Re-evaluating the environment in developmental evolution. *Front. Ecol. Evol.* 3: 1.

Moczek, A. P. and 7 others. 2011. The role of developmental plasticity in evolutionary innovation. *Proc. R. Soc. Lond. B.* 278: 2705–2713.

Moran, N. A. 2003. Tracing the evolution of gene loss in obligate bacterial symbionts. *Curr. Opin. Microbiol.* 6: 512–518.

Moran, N. A. and Y. Yun. 2015. Experimental replacement of an obligate insect symbiont. *Proc. Natl. Acad. Sci. USA* 112: 2093–2096.

Mougi, A. 2012. Unusual predator-prey dynamics under reciprocal phenotypic plasticity. *J. Theor. Biol.* 305: 96–102.

Müller, G. B. 1989. Ancestral patterns in bird limb development: A new look at Hampé's experiment. *J. Evol. Biol.* 1: 31–47.

Müller, G. B. 1990. Developmental mechanisms at the origin of morphological novelty: A side-effect hypothesis. In M. H. Nitecki (ed.), *Evolutionary Innovations*. University of Chicago Press, Chicago.

Newbold, R. R., E. Padilla-Banks and W. N. Jefferson. 2006. Adverse effects of the model environmental estrogen diethylstilbestrol are transmitted to subsequent generations. *Endocrinology* 147: S11–S17.

Nijhout, H. F. 1984. Color pattern modification by cold shock in Lepidoptera. *J. Embryol. Exp. Morphol.* 81: 287–305.

Norouzitallab, P. and 8 others. 2014. Environmental heat stress induces epigenetic transgenerational inheritance of robustness in the parthenogenetic Artemia model. *FASEB J.* 28: 3552–3563.

Nussey, D. H., E. Postma, P. Gienapp and M. E. Visser. 2005. Selection on hereditable phenotypic plasticity in a wild bird population. *Science* 310: 304–306.

Odling-Smee, F. J. 1988. Niche constructing phenotypes. In H. C. Plotkin (ed.), *The Role of Behavior in Evolution*. MIT Press, Cambridge, MA, pp. 73–132.

Odling-Smee, F. J. 2003. *Niche Construction. The Neglected Process in Evolution*. Monographs in Population Biology 37, Princeton University Press, Princeton, NJ.

Odling-Smee, F. J., K. N. Laland and M. W. Feldman. 1996. Niche construction. *Am. Nat.* 147: 641–648.

Oliver, K. M., P. H. Degnan, M. S. Hunter and N. A. Moran. 2009. Bacteriophages encode factors required for protection in a symbiotic mutualism. *Science* 325: 992–994

Osborn, H. F. 1897. Organic selection. *Science* 15: 583–587.

Otaki, J. M., A. Hiyama, M. Iwata and T. Kudo. 2010. Phenotypic plasticity in the range-margin population of the lycaenid butterfly *Zizeeria maha*. *BMC Evol. Biol.* 10: 252.

Paaby, A. B. and M. V. Rockman. 2014. Cryptic genetic variation: Evolution's hidden substrate. *Nature Rev. Genet.* 15: 247–258.

Pfennig, D., M. A. Wund, E. C. Snell-Rood, T. Cruickshank, C. D. Schlichting and A. P. Moczek. 2010. Phenotypic plasticity's impacts on diversification and speciation. *Trends Ecol. Evol.* 25: 459–467.

Pigliucci, M. and G. B. Müller. 2010. *Evolution: The Extended Synthesis*. MIT Press, Cambridge, MA.

Pigliucci, M., C. J. Murren and C. D. Schlichting. 2006. Phenotypic plasticity and evolution by genetic assimilation. *J. Exp. Biol.* 209: 2362–2367.

Pirozynski, K. A. and D. W. Malloch. 1975. The origin of land plants: A matter of mycotrophism. *Biosystems* 6: 153–164.

Prudic, K., C. Jeon and A. Monteiro. 2014. Developmental plasticity in sexual roles of butterfly drives mutual sexual ornamentation. *Science* 331: 73–75.

Queitsch, C., T. A. Sangster and S. Lindquist. 2002. Hsp90 as a capacitor of phenotypic variation. *Nature* 417: 618–624.

Ragsdale, E. J., M. R. Müller, C. Rödelsperger and R. J. Sommer. 2013. A developmental switch coupled to the evolution of plasticity acts through a sulfatase. *Cell* 45: 922–933.

Rassoulzadegan, M., V. Grandjean, P. Gounon, S. Vincent, I. Gillot and F. Cuzin. 2006. RNA-mediated non-mendelian inheritance of an epigenetic change in the mouse. *Nature* 441: 469–474.

Riedl, R. 1978. *Order in Living Systems: A Systems Analysis of Evolution*. John Wiley and Sons, New York.

Rockman, M. V. and G. A. Wray. 2002. Abundant raw material for *cis*-regulatory evolution in humans. *Mol. Biol. Evol.* 19: 1991–2004.

Rohner, N. and 7 others. 2014. Cryptic variation in morphological evolution: HSP90 as a capacitor for loss of eyes in cavefish. *Science* 342: 1372–1375.

Roper, M., M. J. Dayel, R. E. Pepper and M. A. Koehl. 2013. Cooperatively generated stresslet flows supply fresh fluid to multicellular choanoflagellate colonies. *Phys. Rev. Lett.* 110: 228104.

Rosenberg, E., O. Koren, L. Reshef, R. Efrony and I. Zilber-Rosenberg. 2007. The role of microorganisms in coral health, disease and evolution. *Nature Rev. Microbiol.* 5: 355–362.

Rutherford, S. L. and S. Lindquist. 1998. Hsp90 as a capacitor for morphological evolution. *Nature* 396: 336–342.

Ryan, F. 2009. *Virolution*. Collins, NY.

Sangster, T. A., N. Salathia, S. Uneurraga, K. Schellenberg, S. Lindquist and C. Queitsch. 2008a. Hsp90 affects the expression of genetic variation and developmental stability in quantitative traits. *Proc. Natl. Acad. Sci. USA* 105: 2963–2968.

Sangster, T. A. and 9 others. 2008b. Hsp90-buffered genetic variation is common in *Arabidopsis thaliana*. *Proc. Natl. Acad. Sci. USA* 105: 2969–2974.

Sapp, J. 1994. *Evolution By Association: A History of Symbiosis*. Oxford University Press, New York.

Schmalhausen, I. I. 1938. Organizm kak tseloje v individual'nom I istoricheskom razvitii. ("The Organism as a Whole in its Individual and Historical Development"). AN SSR, Leningrad.

Schmalhausen, I. I. 1949. *Factors of Evolution: The Theory of Stabilizing Selection*. Blakiston, Philadelphia.

Shapiro, A. M. 1976. Seasonal polyphenism. *Evol. Biol.* 9: 259–333.

Shapiro, A. M. 1982. Redundancy in pierid polyphenisms: Pupal chilling induces vernal phenotype in *Pieris occidentalis* (Pieridae). *J. Lepidopt. Soc.* 36: 174–177.

Sharon, G., D. Segal, J. M. Ringo, A. Hefetz, I. Zilber-Rosenberg and E. Rosenberg. 2010. Commensal bacteria play a role in mating preference of *Drosophila melanogaster*. *Proc. Natl. Acad. Sci. USA* 107: 20051–20056.

Sharon, G., D. Seal, I. Zilber-Rosenberg and E. Rosenberg. 2011. Symbiotic bacteria are responsible for diet-induced mating preference in *Drosophila melanogaster*, providing support for the hologenome concept of evolution. *Gut Microbes* 2: 190–192.

Skinner, M. K., C. G. Haque, E. Nilsson, R. Bhandari and J. R. McCarrey. 2013. Environmentally induced transgenerational epigenetic reprogramming of primordial germ cells and the subsequent germ line. *PLoS ONE* 8(7): e66318.

Skinner, M. K., M. I. Savenkova, B. Zhang, A. C. Gore and D. Crews. 2014a. Gene bionetworks involved in the epigenetic transgenerational inheritance of altered mate preference: environmental epigenetics and evolutionary biology. *BMC Genomics* 15(1): 377.

Skinner, M. K. and 6 others. 2014b. Epigenetics and the evolution of Darwin's finches. *Genome Biol. Evol.* 6: 1972–1989.

Slijper, E. J. 1942a. Biologic-anatomical investigations on the bipedal gait and upright posture in mammals, with special reference to a little goat, born without forelegs. *I. Proc. Konink Ned. Akad. Wet.* 45: 288–295.

Slijper, E. J. 1942b. Biologic-anatomical investigations on the bipedal gait and upright posture in mammals, with special reference to a little goat, born without forelegs. *II Proc. Konink Ned. Akad. Wet.* 45: 407–415.

Soler, T., A. Latorre, B. Sabater and F. J. Silva. 2000. Molecular characterization of the leucine plasmid from *Buchnera aphidicola*, primary endosymbiont of the aphid *Acyrthosiphon pisum*. *Curr. Microbiol.* 40: 264–268.

Sonnenschein, C. and A. Soto. 1999. *A Society of Cells: Cancer and Control of Cell Proliferation*. Oxford University Press, Oxford.

Spalding, D. 1873. Instinct with original observations on young animals. *MacMillan's Magazine* 27: 282–293.

Spemann, H. 1901. Über Correlationen in der Entwicklung des Auges. *Verhand. Anat. Ges.* 15: 61–79.

Spemann, H. 1907. Zum Problem der Correlation in der tierischen Entwicklung. *Verhadl. Deutsche Zool. Gesell.* 17: 22–49.

Standen, E. M., T. Y. Du and H. C. Larsson. 2014. Developmental plasticity and the origin of tetrapods. *Nature* 513: 54-58.

Standfuss, M. 1896. *Handbuch der palearctischen Gross-Schmetterlinge für Forscher und Sammler*. Gustav Fischer, Jena.

Suzuki, Y. and H. F. Nijhout. 2006. Evolution of a polyphenism by genetic assimilation. *Science* 311: 650–652.

Theis, K. R. and 7 others. 2013. Symbiotic bacteria appear to mediate hyena social odors. *Proc. Natl. Acad. Sci. USA* 110: 19832–19837.

Thompson, J., H. Rivers, C. J. Closek and M. Medina. 2015. Microbes in the coral holobiont: Partners throughout evolution, development, and ecological interactions. *Front. Cell. Infect. Microbiol.* 4: 176.

Thomson, K. S. 1988. *Morphogenesis and Evolution*. Oxford University Press, New York.

Trut, L., I. Oskina and A. Kharlamova. 2009. Animal evolution during domestication: The domesticated fox as a model. *Bioessays* 31: 349–360.

Tsuchida, T. and 7 others. 2010. Symbiotic bacterium modifies aphid body color. *Science* 330: 1102–1104.

Turesson, G. 1922. The geotypical response of a plant species to the habitat. *Hereditas* 3: 211–350.

Turner, J. S. 2000. *The Extended Organism: The Physiology of Animal-Built Structures*. Harvard University Press, Cambridge, MA.

Twitty, V. C. 1932. Influence of the eye on the growth of its associated structures, studied by means of heteroplastic transplantation. *J. Exp. Zool.* 61: 333–374.

Vermeij, G. J. 2004. *Nature: An Economic History*. Princeton University Press, Princeton, NJ.

Waddington, C. H. 1942. The canalization of development and the inheritance of acquired characters. *Nature* 150: 563.

Waddington, C. H. 1952. Selection of the genetic basis for an acquired character. *Nature* 169: 278.

Waddington, C. H. 1953. Epigenetics and evolution. In *Symposia for the Society for Experimental Biology VII: Evolution*. Cambridge University Press, pp. 186–189.

Waddington, C. H. 1956. Genetic assimilation of the *bithorax* phenotype. *Evolution* 10: 1–13.

Waddington, C. H. 1957. *The Strategy of the Genes*. Allen and Unwin, London.

Waddington, C. H. 1961. Genetic assimilation. *Adv. Genet.* 10: 257–290.

Wagner, G. P. and M. D. Laubichler. 2004. Rupert Riedl and the re-synthesis of evolutionary and developmental biology: Body plans and evolvability. *J. Exp. Zool. B Mol. Dev. Evol.* 302: 92–102.

West-Eberhard, M. J. 2003. *Developmental Plasticity and Evolution*. Oxford University Press, Oxford.

West-Eberhard, M. J. 2005. Phenotypic accommodation: Adaptive innovation due to developmental plasticity. *J. Exp. Zool. (MDE)* 304B: 610–618.

Wilkins, A. S., R. W. Wrangham and W. T. Fitch. 2014. The "domestication syndrome" in mammals: A unified explanation based on neural crest cell behavior and genetics. *Genetics* 197: 795–808.

Williams, J. B. and R. E. Lee Jr. 2005. Plant senescence cues entry into diapause in the gall fly *Eurosta solidaginis*: Resulting metabolic depression is critical for water conservation. *J. Exp. Biol.* 208: 4437–4444.

Xiao, J. H. and 51 others. 2013. Obligate mutualism within a host drives, the extreme specialization of a fig wasp genome. *Genome Biol.* 14(12): R141.

Philosophical Concerns Raised by Ecological Developmental Biology

In considering the study of physical phenomena ... we find its noblest and most important result to be a knowledge of the chain of connection, by which all natural forces are linked together, and made mutually dependent upon each other; and it is the perception of these relations that exults our views and ennobles our enjoyments.

Alexander von Humboldt, Kosmos, 1844

This book has attempted to present an integrated perspective on the origins of animal diversity, the origins of certain diseases, and the integration of living communities through co-development. Emphasizing the evidence that developmental symbiosis and plasticity are normative for eukaryotic life, it proposes a revised view of the organismal world, its evolution, its maintenance, and its health.

The common denominator of all the components of this perspective on life is development: the emergence of form, the processes of becoming. In this view of living beings (note that both these words are gerunds), the processes of development percolate through evolution, genetics, and ecology. In effect, evolution is the *outcome* of development, while genetics is the set of processes through which similar developmental trajectories are transmitted between generations (Wilson 1896; Griesemer 2002; Burian 2005). Ecology becomes a science focusing on developmental and adult interactions in time and space, which in turn produce and sustain the patterns of interspecies interactions at particular locations.

So how (if at all) does this relational perspective change our views concerning the living world and how it might be appropriately studied? Philosophy is often subdivided into *ontology* (a theory of what is real), *epistemology* (a theory of how we perceive and organize data), and *ethics* (a theory of how to live a righteous life). The applications of these branches of philosophy include, respectively, pedagogy (what and how we teach), methodology (how we acquire data), and public policy (how we make social decisions consistent with our values). The revised view of nature presented in this volume impacts each of these areas. Indeed, we expect that this developmental view of nature will elicit new questions from philosophers, or at least cause them to ask questions that haven't been asked for a while.

Ontology

What is an "individual" in terms of its developmental and ecological history?

Ecological developmental biology has some extremely significant propositions concerning ontology. At the very least, the prevalence of symbioses, polyphenisms, and reaction norms instructed by environmental agents abolishes any notion of a genetic determinism. We can give a definite answer to the question posed by Lewis Wolpert (1994):

> Will the egg be computable? That is given a total description of the fertilized egg—the total DNA sequence and location of all proteins and RNA—could we predict how the embryo will develop?

The answer is a resounding "No—and thank goodness." The trajectory of development depends significantly on environmental agents. As Michael Skinner (2014) has recently summarized:

> One of the most predominant paradigms in the biological sciences today is "genetic determinism." The concept is that the DNA sequence alone is the building block for biology and that mutations in this sequence are the primary causal factors for most biological phenomena from disease development to evolutionary biology. This paradigm is the basis for most of our current education programs and theories in biology. The problem is that many phenomena cannot be easily explained with classic genetics or DNA sequence mutation mechanisms alone. An example is the numerous genome wide association studies (GWAS) that have generally shown less than 1% of a specific disease population have a correlated DNA sequence mutation (Visscher et al. 2012; Zhao and Chen 2013).

If we are seriously considering the "nature of nature," then we must realize that the major biological paradigm of "The Century of the Gene" (Keller 2002) has to be seen as a subset of a much larger theory of biological phenotype and individuality.

Moreover, ecological developmental biology challenges our notions of individuality. We are not individuals in the biological ways we thought we were. We are not individuals by anatomical, physiological, or genetic criteria, and we even develop as multispecies entities. The two major sensory networks through which we apprehend the outside world, the immune system and the nervous system, are brought into existence, in part, by symbiotic bacteria, and even behaviors appear to be conditioned by commensal gut microbes (Gilbert et al. 2012; McFall-Ngai et al. 2013).

Developmental symbioses are no longer relegated to marginal exceptional cases. Rather, they are the norm, and this changes the way we think about what an organism is. In their groundbreaking theory of symbiosis, Sagan and Margulis (1991) saw symbiosis as occurring between autopoietic (self-developing) entities. They wrote, "What is remarkable is the tendency of autopoietic entities to interact with other autopoietic entities." However, if developmental symbioses are the rule, then even the notion of an "autopoietic individual" has to be questioned. We are not adults entering into symbioses with other consenting adults or microbes. Rather, the processes that construct a particular adult organism involve symbioses between *developing* organisms (Gilbert 2002). We are talking "co-construction" on an enormous scale. Developmental symbiosis and phenotypic plasticity demonstrate that development is *co-development*. Epigenesis is between species as well as between the embryonic cells produced by the zygote.

These ideas concerning the importance of the environment in phenotype production have been fundamental platforms of *developmental systems theory* (DST), a school of philosophy that takes developmental biology very seriously and that incorporated the epigenetic ideas of Conrad Waddington, Alfred North Whitehead, and J. M. Baldwin at about the same time that evolutionary developmental biology was being formed (see Gottlieb 1992; Robert 2004). Susan Oyama (1985, p. 123), one of the founders of DST, argued that the unit of development is not the organism but, rather, a developmental system encompassing "not just genomes with cellular structures and processes, but intra- and inter-organismic relations, including relations with members of other species and interactions with the inanimate surround as well."

Perhaps even more, developmental symbiosis forms the ontology of philosophers such as Donna Haraway (2003, 2008). Haraway proposes that reality consists of "becoming with," where the "self" does not exist prior to interactive relations with others. Indeed, Haraway specifically uses developmental symbiosis both as an actual, nonmetaphorical, account of who we are, and as a model for how other "bodies" (bodies of knowledge, societies, etc.) are co-constructed. She contrasts "symbiopoiesis" (developmental symbiosis) against "autopoiesis" (self-realization), that is, the collectively reproducing systems as opposed to the self-reproducing system, at every stage of life. In her view, developmental symbiosis means that information and control are distributed throughout the collectively developing system, rather than being concentrated in a single or small number of entities within a single self-reproducing system.

Philosophers John Dupré and Maureen O'Malley (2009) similarly see living entities as interactive collections. "Life, we claim, is typically found at the collaborative intersections of many lineages, and we even suggest that collaboration should be seen as a central characteristic of living matter…" They make the case that evolution without microbes is a sterile discipline. Symbiosis is what needs to be examined and explained.

What is an "individual" in terms of its evolutionary history?

The ability of species to directly induce the development of other species into more fit trajectories has implications for evolutionary biology. Evolutionary biologists, at least in Britain and the United States, have tended to find in nature the ontology of Locke, Mill, and Hobbes—philosophers of individual rights and of a nature characterized by Hobbes's "war of all against all." Moreover, Darwin's and A. R. Wallace's pioneering insights into natural selection were historically and philosophically grounded in Thomas Malthus's political economy of individual competition, which they transferred from European economics into the natural world. As historian Silvan Schweber (1985, p. 38) noted, in *The Origin of Species* "biology joined hands with Scottish political economy, sociology, and historiography, and with English philosophy of science. The political economy was that of Adam Smith and his disciples." This philosophy assumes that each individual is a unit of self-interest that interacts with other units of self-interest to yield cooperation and wealth. Thus Darwin concluded the *Origin* with praise to a nature that can form a "tangled bank" of harmonious interactions from competition, death, and destruction.

Modern evolutionary theorists have taken evolutionary biology from metaphors of mercantile capitalism to metaphors of investment capitalism, positing genes as the unit of currency (see Haraway 1979; Young 1985). Indeed, most evolutionary biology has accepted the individualistic premises by which each unit is competing for limited resources against all others. Thus, for classical Darwinism, the harmonious interactions one sees in the biosphere are the result of underlying competitive interactions. Just as the wooden pencil was created from self-interested individuals, so was the tree from which the wood came.*

Embryologists, however, have had different philosophical inputs. Centered on the European continent, classical embryology had as its underpinnings the philosophies of emergent form found in Immanuel Kant and Johann Wolfgang Goethe. These philosophies allowed a scientist to hold both mechanistic and teleological views (Lenoir 1982; Gilbert and Faber 1996). Lenoir (1982) and Huneman (2006) argue that the founders of modern embryology—Dollinger, Pander, von Baer, and Rathke—subscribed to the organicism set forth in Kant's *Critique of Judgment*:

> The first principle required for the notion of an object conceived as a natural purpose is that the parts, with respect to both form and being, are only possible through their relationship to the whole.... Secondly, it is required that the parts bind themselves mutually into the unity of a whole in such a way that they are mutually cause and effect of one another.

According to this view, the parts not only determine the properties of the whole, but the whole reciprocally determines the properties of the parts. This is similar to the relationship between words and sentences. The

---

*Several of Darwin's contemporaries, especially those in Russia, felt that Darwin had made an error in borrowing so strongly from British economic theorists and philosophers, making evolution a particularly British rather than universal theory. Chernyshevskii (1888), for instance, wrote that Darwin's view of nature was that to be expected "if Adam Smith had taken it upon himself to write a course in zoology." Darwin was also partial to the philosophy of Gottfried Leibniz, who also saw cooperation coming from competition (it is from Leibniz that Darwin got the idea that nature doesn't make jumps). Leibniz went on to say that not all things can be present simultaneously. While Darwin thought of this in terms of adult competition, this idea, which Leibniz called "compossibility," can be even more important when thinking in terms of ecological developmental biology.

meaning of the sentence (whole) is obviously determined by the meanings of the words (parts), but also (and less obviously) the meaning of each word is determined by its interaction with other words in the sentence. Thus, the meanings of the words of the sentences "The party leaders were split on the platform" or (in American baseball) "The pitcher was driven home on a sacrifice fly" depend on the entire sentences. The words "become with" each other, and the meaning of each word is suspended until the sentence is completed.*

These philosophical views of causation, where the parts help form one another and are interdependent upon each other, emphasize cooperation and integration rather than competition. Embryologists took them to be the natural way of life. Writing about experimental embryology in Weimar Germany, Hans Holtfreter (1968, p. xi) claimed:

> We managed more or less successfully to keep our work undisturbed
> by humanity's strife and struggle around us and proceeded to study
> the plants and animals, and particularly, the secrets of amphibian
> development. Here, at least, in the realm of undespoiled Nature,
> everything seemed peaceful and in perfect order. It was from our
> growing intimacy with the inner harmony, the meaningfulness,
> the integration, and the interdependence of the structures and
> functions as we observed them in dumb creatures that we derived
> our own philosophy of life. It has served us well in this continuously
> troublesome world.

Holtfreter's American contemporary Ernest E. Just, who quoted Goethe and who came from a tradition of observing rather than experimenting, also claimed, "Whether we study atoms or stars or that form of matter, known as living, always must we reckon with inter-relations" (Just 1939, p. 368). Indeed, Just routinely saw developing organisms as being integrated into their ecological contexts (see Byrnes and Eckberg 2006). Interdependence, harmony, and integration promote a different perspective on nature than the autonomous, competitive nature of classical evolutionary biology. In addition to Locke and Hobbes, we have Kant and Goethe.[†]

These developmental views do not invalidate Charles Darwin's or Thomas Huxley's competitive perceptions of nature. (It was Huxley who compared nature to gladiatorial combat and who claimed Hobbes to be one of the three greatest philosophers.) Nature abounds in cruel and "heartless" behaviors. Anyone who has studied in the Galápagos Islands can discourse on birds that throw their younger sibs out of the nest to starve (with their mothers' complicity) and on female lizards that fight viciously for limited nesting sites. Darwin (see Gould 1982; Gillespie 1979) wrote that a "devil's chaplain" could write a great book concerning the "horribly cruel works of nature," and he felt that the life cycles of parasitoid wasps (see Chapter 3)

---

*And the meanings of sentences, themselves, often depend on other sentences. The co-emergence of meaning, biological and linguistic, is not the exception but the norm, at every level of a system.

[†]Perhaps it is through ecological evolutionary developmental biology that we can best appreciate Haeckel's (1882) intuition that Darwin had solved the great conundrum posed by Kant, namely, "how a purposefully directed form of organization can arise without the aid of a purposefully effective cause."

were incompatible with a deity who was both benevolent and omnipotent. The eco-evo-devo model of evolution does not claim that competition is unimportant in the animal kingdom any more than it could claim the absence of predation. It does, however, claim that alongside this competitive model of evolution there belongs a cooperative model of evolution.

Such an account of evolution through cooperative interactions has been popularized by Ashley Montague (1955) and more recently by Lynn Margulis and Dorian Sagan (2002), who emphasize symbiosis as demonstrating the cooperative interconnectedness of life. "Life," they claim, "did not take over the globe by combat, but by networking" (Sagan and Margulis 1986). They argue that symbiosis is responsible for the origin of life as we know it as well as for speciation events. The grand symbioses of nitrogen-fixing bacteria and legumes, of myccorhizal fungi and roots, of reefs and tidewater ecosystems structured by invertebrate-microbe interactions show that our terrestrial and marine ecosystems are predicated on symbiotic relationships. Likewise, our cells and genes show vestiges of ancient symbioses that have become the norm. We have argued here that the symbiosis between developing organisms shows that each organism develops as an interconnected community and that the environment (including other organisms) provides normative cues for embryogenesis. Thus, developmental symbioses and environmentally induced polyphenisms support Margulis' (1998) notions of a "symbiotic planet" and that symbiosis is the signature of life on Earth.

INTERSPECIES EPIGENESIS   Evolutionary biologist and embryologist Conrad Waddington noted (1953) that evolution takes place in two areas—in the embryo and in the environment—and that the parts of an embryo have to function cooperatively to form the organism that can then compete.* This notion of cooperation now extends beyond the embryo to the realms of "interspecies epigenesis," where development is coordinated between two or more species. Therefore, instead of demanding that the external (competitive) mode of evolution serve as a model for the embryo (see Buss 1987; cf. Gilbert 1992), ecological developmental biology suggests that the internal, cooperative embryonic mode of evolution be seen between species. With signals coming from outside the embryo, the choreography of embryonic induction is also danced between organisms, not just within a single organism derived from a zygote. Symbiotic bacteria help form our guts and immune systems; predators, nutrition, and competitors alter developmental trajectories by the evolved mechanisms of developmental plasticity; signals from the warmth of the sun or from the number of hours the sun is above the horizon determine the sex of some vertebrates and the pigmentation of others. In some cases, the sympoietic induction is

---

*In this view, Waddington reflects T. H. Huxley's ideas about inner unity and external struggle. Postulating an early version of group selection, Huxley (1894) noted that a society must cooperate internally to be strong enough to successfully compete against other societies. Huxley's Russian contemporary Petr Kropotkin (1902) noted that other animals form societies and that those that cooperate best would survive. We will return to these concepts later.

reciprocal, allowing each organism to find its place within the ecosystem as it develops.

This has important implications for evolutionary theory. If the environment is giving instructive information as well as selective pressures, and if relationships, rather than individuals, are being selected, then selection may be on the holobiotic "individual," that is, the nature might be selecting a multispecies consortium or team, rather than a monogenetic "individual" whose genes only come from the zygote.

The data of ecological developmental biology also have profound philosophical implications for the question of how evolution works. As mentioned earlier, one of the major insights of twentieth-century philosophy was that we are "defined by the 'other.'" However, this relationship has been depicted as being a confrontation between the "self" and the "other." In this way, it is similar to the traditional evolutionary view of competition between separate autonomous entities. Both traditional evolutionary biology and existentialism grow from the soil of Hobbesian competition and individuality. While ecological developmental biology similarly postulates that we are defined, in part, by the "other," it depicts our identities as *becoming with* the "other." The relationship between "self" and "other" is not that of autonomous and antagonistic individuals, but that of non-autonomous agents-in-the-making that inter- and intra-act to form both themselves and other novel patterns (Haraway, 2008; see Barad 2007 for similar conclusions emanating from quantum physics). The older idea of evolution as intraspecies competition may be a small, fine-tuning, part of a larger view of evolution, wherein natural selection favors certain relationships between embryonic cells (to generate novelty), between cells and symbionts (to form organisms), and between species (to stabilize ecosystems). The biological data of the twenty-first century have been enriched by the discovery of worlds of molecular relationships. These relationships are revolutionizing developmental biology, ecology, evolutionary biology (Shapiro 2011), physiology (Noble 2008, 2013), molecular biology (Mattick 2012), and psychology (Moore 2015) and are creating a new perspective on the relationship of organisms to their environments. The biological models of the twenty-first century should be richer than those of the twentieth.

Integrative philosophical traditions

Several philosophical systems of ontology have also come to this conclusion of mutually dependent co-development. Historically, two of these important philosophies have been dialectical materialism and process philosophy. Dialectical materialism was prominent in Russian biology even before the advent of Communism, and it is not an accident that so many of the early researchers mentioned in this book were Russian. The Russian school of evolutionary biology thought that competition played only a relatively minor role in the origin of species, and one of the leading books on the history of Russian evolutionary thought (Todes 1989) is provocatively titled *Darwin without Malthus*. Several British embryologists, notably Joseph and Dorothy Needham, also thought that the interactivism of dialectical materialism provided a framework for animal development (Haraway 1976; Werskey 1978; Winchester 2008).

In the 1930s, another philosophy of emergence, the "process philosophy" of Alfred North Whitehead, became popular among embryologists (see Haraway 1976; Gilbert 1991). Although the totality of Whitehead's philosophy and theology was not taken up by the scientists, several embryologists were impressed by Whitehead's (1929) concepts of developing systems, processes, and the creative advance into novelty. Speaking for a group of such embryologists, Waddington (1975, pp. 3, 5) reflected, "As far as scientific practice is concerned, the lessons to be learned from Whitehead were his replacement of 'things' by processes which have an individual character which depends upon the 'concrescence' into a unity of very many relations with other processes." Indeed, Waddington reflected, "I tried to put the Whiteheadian outlook to actual use in particular experimental situations."

Affinities between Whitehead's ontology and those of classical Asian philosophers have long been noted (see, for instance, Streng 1975; Odin 1982). Chief among these ancient Asian thinkers was the Buddhist sage Nagarjuna. Nagarjuna is regarded as one of the most important interpreters of what the Buddha said, and he emphasized that "things derive their being and nature by mutual dependence and are nothing in themselves" (quoted in Murti 1955, p. 138). This doctrine, called *Pratityasamutpada*, is variously translated as "dependent origination," "conditioned genesis," "interdependent co-origination," and "interdependent arising." Thus, any phenomenon comes into being ("be-comes") only because of the coming into existence of other phenomena in a network of mutual cause and effect. Indeed, according to Nagarjuna, what the Buddha awakened *to* (*bodhi* means "to awaken") was the truth of interdependent co-origination.

In the twentieth century, another group of philosophers, who became known as phenomenologists, believed that the physical sciences had become too abstract—that they had sacrificed the physical nature of the perceived universe for particles too small to be observed. Atoms and protons had become more real than chairs and trees, the parts more real than the whole. The phenomenologists proposed that there was a reciprocal interaction between the knower and the known. The sensing body was seen as not being preprogrammed by its genes or by its brain but being constantly in the act of improvising its relationship to other factors in the world. According to Merleau-Ponty (1962, p. 52), the body and its environment are in constant and reciprocal exchange, each creating and changing itself and the other. Indeed, in his *Phenomenology of Perception*, Merleau-Ponty (1962, p. 317) uses symbiosis as a metaphor for the reciprocity he is trying to express. Ecological philosophers such as David Abram (1996) and Bruce Foltz (1995) have applied phenomenological analysis to biology, claiming that twentieth-century biology had made genes more real than organisms. As developmental biology begins to forms its relationships with ecology, phenomenology might prove to be a good starting point for a philosophy that can organize a theory of reality around co-development. Here the organism can be seen as the concrete, fleshy nexus integrating (in time and space) the internal networks of developmental genetic interactions with the external networks of ecological interactions.

But the phenomenological approach to development should neither supersede nor eclipse the molecular. Interdependent origination and the

context-dependent co-construction of identity are to be found at each biological level, from the molecular through the ecological. Interactions within
and between each of these levels should reinforce rather than exclude one
another. Whitehead (1920, p. 152) chided philosophers for dividing reality
into phenomenal and causal realms, saying that they were all concretions
of processes. And indeed, developmental biologists (including the most
molecularly oriented) have found that even the properties of proteins depend upon their contexts: A repressor of gene transcription in one cell
may activate the same gene in another cell; the same gene that produces
an enzyme in the liver may encode a structural protein in the lens. BMP4,
a protein that changes cartilage into bone in one part of the limb, tells cells
to die in another part of the limb, tells the cells in the face to divide, and
tells the heart to form in a particular region of the embryo. What BMP4
"is" depends on its context (see Gilbert 2006).

Haraway (2003, 2008) fuses phenomenological analyses with dialectic materialism, process philosophy, and developmental systems theory.
Trained in both developmental biology and ecology, Haraway has kept up
with epigenetic biology and has used it extensively in her philosophical
approaches to both nature and society. More than 20 years ago, she used the
term *Myxotricha paradoxica* as "an entity that interrogates individuality
and collectivity at the same time" (Haraway 1991), and she has extended
this interactive epigenetic account to all creatures. "Organisms are ecosystems of genomes, consortia, partly digested dinners, mortal boundary
formations," Haraway writes (2008, p. 31), and she also quotes Margulis
and Sagan's notion that the organism "is a co-option of strangers."

Haraway uses embryogenesis and developmental symbioses in her on-
tologies. "Reciprocal induction," she argues (Haraway 2008, p. 228), "is the
name of the game," and it's "reciprocating complexity all the way down."
She cites this reciprocal induction, along with developmental symbioses,
as examples of "becoming with." Haraway (2003, pp. 20, 24) has written,
"The relation is the smallest unit of analysis," but in 2008, she preferred
not to use the word "unit" because she felt that this word misled one into
thinking that there is an "atom" that instigates identity. She later writes
(Haraway 2008, p. 26) that "relationships are the smallest possible pattern
for analysis." This is a concept that seems fundamental to ecological developmental biology at all levels: the combinatorial relationship of transcription factors to DNA; the relationship of different cells forming an organ; the
relationship of bacteria (both beneficial and harmful) to the host organism;
the relationship of organisms to other developing organisms, etc. Haraway
doesn't so much ground her beliefs in nature as use natural examples to
visualize principles and ground them in biological data, especially the notion of "becoming with." These principles are not mere metaphors; they
have substance, and these substances interact.

Emergence

Emergence is normative in developmental biology and has been an important concept since the late 1800s (see Gilbert and Sarkar 2000). The
fertilized egg has no hemoglobin, tooth enamel, liver enzymes, gut capillaries, or even the rudiments of a heart. Rather, these entities emerge. And

they emerge through reciprocal embryonic induction and symbiopoiesis. Classical embryology has insisted that an entity is defined both by its parts and by its context. The genes → organ approach must be complemented by the organ → gene approach. As we've seen, the structure and function of a liver or mammary gland epithelial cell depend not only upon the properties of its genetic products, but also on the properties of the organ in which it resides and on the history of that organ. A teratocarcinoma stem cell can become a tumor in one tissue and a set of organs in another. Experimental embryologist Oskar Hertwig (1894), who catalogued the ways in which the environment controlled development (see Chapter 1), claimed explicitly that "the parts of the organism develop in relation to each other, that is, the development of the part is dependent on the development of the whole." And embryologist Hans Spemann (1943, p. 219) remarked, "We are standing and walking with parts of our body which could have been used for thinking had they developed in another part of the embryo."

When reciprocal induction is applied between and among organisms, this concept expands into evolution. Entities need to be thought of in terms of several geometries at the same time. They are defined by the braiding of down-top and top-down (as well as lateral and temporal) networks built from patterns of reciprocal causation. Evolutionary biologist Ernst Mayr (1988) recognized that the characteristics of living wholes "cannot be deduced (even in theory) from the most complete knowledge of the components, new characteristics of the whole emerge that could not have been predicted from a knowledge of the constituents." How they emerge can now be seen in terms of reciprocal inductions.

Some of the most interesting work involving reciprocal causation and the emergence of organization comes from studies of self-organizing regulatory processes (Kauffman et al. 2007; Deacon 2012). Here, not only is there the ability of interacting matter to self-organize, but the organization makes its own constraints, which allow for order to persist and for new types of order (such as living matter) to emerge. It is tempting to speculate that this relates to the embryological situation mentioned above, where the mammary gland cells make the mammary gland; but the architecture of the mammary gland they make causes the differentiation of the cells specifically in the mammary direction (see Chapter 8; see also Bissell 2007; Xu et al. 2009). Indeed, Longo and colleagues (2012) see evolutionary biology as the analysis of the ongoing constraints to the two main principle characteristics of life, proliferation with variation and motility. This would provide the mechanisms for the different patterns of morphogenesis.

Pedagogy

A discipline can be defined both by the set of questions it hopes to answer and by the content of its textbooks and journals. Pedagogical issues thus reflect, as well as shape, views of scientific reality. It should be obvious that this book makes a sustained argument that the present disciplinary boundaries need to be altered in order to apprehend the realities that science is discovering. Developmental biology needs to be expanded to overlap and to include both evolutionary and ecological dimensions. An important part of ecological developmental biology is to go outdoors to see the world in its natural and cultural complexities. As Louis Agassiz said (and this quotation is above the library entrance at the Marine Biological Laboratory in Woods Hole!), "Study Nature, Not Books." Student field trips should not be only the province of ecologists; developmental biology students need to see frog eggs in ponds as well as in the laboratory. Indeed, one of the first papers in ecological developmental biology (Mead and Epel 1995) reminded developmental biologists that to study sea urchin fertilization properly, one has to take into account that most sea urchin eggs are fertilized in oceans, not in laboratories. Moreover, developmental biology courses should include discussions of philosophical issues and public policy issues attendant upon new discoveries. Social and scientific dimensions are interactive in a host of ways, and awareness of these interactions is critical for better science.*

Evolutionary biology needs to be similarly expanded to intersect developmental biology. Indeed, the last three chapters of this book attempt to show the importance and relevance of such an expansion. The *explananda* (things which need to be explained) of evolutionary biology must include both a theory of change *and* a theory of body construction. And ecology must see itself not as an isolated discipline but as providing wider bridges connecting itself to both developmental and evolutionary biology. Harkening back to Van Valen's (1973) notion that evolution is development controlled by ecology, ecology must embrace life cycles and cell cycles as well as nutrient cycles. The arrows of causation must extend reciprocally from the environment through the cell nucleus. None of these disciplines should subsume the other, for their questions are, for the most part, different ones. Yet, together, they constitute Waddington's "diachronic biology" as well as Severtsov's (1935; see Chapter 11) recipe for a complete evolutionary science.

Ecological developmental biology, when it focuses on symbiosis and plasticity, highlights different questions in evolution, *different explananda*. What does evolution look like when the proper unit of analysis is not the individual but the relationship (at each different level)? What does evolution look like when selection may be on "teams" of organisms and on the relationships between these teams? What does natural selection mean when the environment not only is an agent that selects adaptive phenotypes but also contains agents that help instruct the formation of adaptive phenotypes (and may undergo changes itself because of it)? Moreover, how do we revise our views about the environment and evolution when germline DNA methylation can effect the transmission of environmentally induced characters from one generation to the next? And given the data mentioned in this book, how are developmental plasticity and developmental symbiosis related to some of the great transitions in the evolution of biodiversity.

---

*This is not a new idea. The first meeting of the Growth Society (later the Society for Developmental Biology), organized by N. J. Berrill in 1939, had representatives from numerous areas of biology, as well as the philosopher J. H. Woodger, who spoke on what it all meant. This notion has been revised and revitalized (see Gilbert and Fausto-Sterling 2003) for the inclusion of social issues concerning stem cell research, abortion, cloning, and environmental justice in science classes.

These should be highly provocative and productive questions during the next decades.

Even the metaphors we use to explain and describe evolution should change. Evolution is real, to be sure; but the metaphors we use to describe it need revision. In the type of reciprocity seen throughout this volume, metaphors and models both shape the way we do science and also are shaped by what science shows the world to be like. Darwin used the metaphor of metallic wedges. Only so many wedges fit into a piece of wood, and when one fits better, it forces another out. "Fit" is the root metaphor here, and it is similar to the lock-and-key model of enzyme activity that was being proposed at a similar time in biochemistry (see Gilbert and Greenberg 1984). However, the lock-and-key model in biochemistry doesn't work. It doesn't have a good "fit" to nature. Rather, the current model is that of "induced fit," wherein there is an interaction between the substrate and the catalytic site. Developmental plasticity indicates that there are interactions between the animal and its environment such that the phenotype can become fit for the environment. It is a kind of induced fitness.

Another metaphor concerns the environment as a selector. It is sometimes thought of as a policeman, a filter, or a sieve, but we now see that the environment is both selective *and* instructive. Although "instructive" is in fact a technical term that describes certain developmental processes, one can use it in its pedagogic sense as well. The environment can be likened to an instructor. The instructor has two roles: one is to impart information; the other is to test the student. So it is with the environment. It both imparts information to the developing organisms and simultaneously challenges them to see if they have integrated that information in a useful way. Our ways of visualizing and narrating evolution will have to change to fit these new perceptions.

Our metaphors of gene expression must also change. The gene is sometimes the initiator, but as Mary Jane West-Eberhard has shown (see Chapter 11), the gene is often the follower that stabilizes an existing environmentally determined phenotype. We have been saddled (a metaphor indicating someone's riding us) with the notion of "selfish genes" controlling development and evolution through an informational script that has to be decoded (i.e., transcribed and translated). We take for granted the programming language of computer engineers with its genetic instructions (Nijhout 1990; Boudry and Pigliucci 2013). But there are other possibilities. One alternative is to see each holobiotic organism as a performance. We inherit a *score* (the genome), a means of *interpreting* the score, and a means of *improvisation* should the score be incomplete (about 50% of induced gene mutations produce no perceptible phenotypic difference). Each performance will be different, even if the DNA is the same (Gilbert and Bard 2014). Identical twins are different performances of the same score.

In addition to ecology, evolutionary biology, and developmental biology, numerous medical disciplines should be affected by ecological developmental biology. Certainly oncology must include development as a major component. Not only are epigenetic mechanisms capable of causing cancers, but numerous tumors have now been shown to arise from the misregulation of relationships between cells (see Chapter 8). Not only the somatic mutation theory of cancer, but also "evolutionary medicine" and "evolutionary psychology" are each predicated on the Modern Synthesis view of evolution by competition between individuals having different genetic alleles. Changing these sciences to include the expanded view of evolution coming from ecological developmental biology (and highlighting plasticity and epigenetics) would change each of these disciplines (Stotz 2014).

Ecological developmental biology has provided some metaphors and models for restructuring disciplinary boundaries. Indeed, its existence is a model of interdisciplinarity, taking material from nearly every biological discipline. It also provides metaphors for philosophers who want to depict knowledge as not disciplinary but "rhizomatic," characterized by interacting pathways, webs, knots, and routes of connectivity rather than by distinct borders. The rhizome metaphor comes from Deleuze and Guattari's use of the notion of the plant root system, connected through interactions of symbiotic fungi. This rhizomatic network of knowledge is ceaselessly establishing connections between scientific conclusions, networks of meaning, the arts, organizations of power, and processes relating to social struggles (Deleuze and Guattari 1987, p. 7). Different configurations of topics should therefore be emerging in our curriculum, and old boundaries should be changed or made porous (Root-Bernstein and Root-Bernstein 1999; Root-Bernstein et al. 2008). One can imagine, for instance, the fusion of medical embryology and human genetics courses to include not only the developmental genetic mechanisms of human disease, but also the mechanisms of teratogenesis, endocrine disruption, and public policy that regulates toxic waste and assisted reproductive technology.*

And in addition to forming some new categories, the developmental perspective can be used to redefine and enlarge some old disciplines. These include parasitology (see Collins et al. 2013), paleontology (see Davis 2013; Lyson et al. 2013, 2014; Schneider and Shubin 2013; Standen et al. 2014; Werneburg and Sanchez-Villagara 2014), and natural history. Natural history is the study of how organisms come to live where they live, eat what they eat, attune their life cycles to certain environmental cues, and become situated within specific communities and ecosystems. That is to say, natural history combines paleontology, developmental biology, physiology, and ecology. It isn't so much a discipline as a way of looking at nature. Like any other science, it contains both facts and an aesthetic that organizes them (Gilbert and Faber 1996). Again, returning to Thomas Huxley (1854):

> To a person uninstructed in natural history, his country or sea-side stroll is a walk through a gallery filled with wonderful works of art, nine-tenths of which have their faces turned to the wall. Teach him

---

*Developmental symbiosis, for instance, abolishes the border between developmental biology and ecology, allowing new perspectives. Human childbirth, for example, has been traditionally depicted as the culmination of a strictly developmental process. However, eco-devo has also allowed it to be described as an ecological process involving the maintenance of a holobiont symbiotic community (Gilbert 2014; Chiu and Gilbert 2015).

something of natural history, and you place in his hands a catalogue of those which are worth turning around. Surely our innocent pleasures are not so abundant in this life, that we can afford to despise this or any other source of them.

Natural history has been disparaged as being "old-fashioned" because it is merely an interactive inventory of nature. It is not experimental. In the twentieth century, biology had to be experimental to become a "real" (and fundable) science. Historians (see Allen 1979; Benson 1988; Nyhart 1994) have shown that evolutionary and developmental biology were changed from being outdoor disciplines to being indoor disciplines (first through physiology and then through genetics) during this time. In this process, the material form lost its importance, and theoretical concepts ascended. This view is a holdover from the late nineteenth-century perspective that only physics is a real science and that everything else, as Nobel Prize–winning physicist Ernest Rutherford said, "is stamp collecting." But this can't be the whole answer, because genomics, transcriptomics, and proteomics are also merely interactive inventories of nature—just at the gene level. What had happened during the twentieth century is the ascendancy of the gene over the organism as the explanatory level for biology and medicine (Keller 2002). As in so much of Western art, music, and literature, the abstract has triumphed over the figurative. This decline is reflected in the number of courses taught in natural history, as well as in funding to zoological and botanical collections in museums. In textbooks, the pages devoted to natural history have declined as the pages devoted to molecular biology have expanded (Tewksbury et al. 2014).

But as the biomes of the world grow closer together and organisms and products from one environment are entering all others, natural history, surprisingly, turns out to be one of the most "relevent" disciplines. For instance, cholera appears to fluctuate with climate-induced algal blooms, where photosynthetic organisms fill the water. Why should this be so? Scientists discovered a symbiosis between *Vibrio cholerae* (the bacterium that causes cholera) and certain copepods (small crab-like crustaceans) that permits the bacteria to survive. Copepods are dependent on algae, and when there are algal blooms, the number of copepods expands, along with the number of their symbiotic cholera microbes. This allows algal blooms to be used as an early detection marker for cholera (Huq et al. 1996; Colwell and Huq 1994; Constantine de Magny and Colwell 2009), and it also explains how Indian women prevent cholera by filtering polluted water through their saris (which can filter out copepods to which the bacteria remain attached).

Ignorance of natural history has created numerous ecological disasters that could have been prevented* (LaBastille 1984; Spencer et al. 1991; Bailey 2011; Tewksbury et al. 2014). For instance, in the Pacific Northwest of the United States and Canada, tree stumps were removed from rivers to help

---

*The attempt to do what is helpful without knowledge of natural history can lead to devastating consequences, as in our earlier discussion (see p. 88) of the British blue butterfly. Cartoonist Gary Larson's 1999 book *There's a Hair in My Dirt* is a playful and sardonic lesson of such attempts to do good without the requisite scientific knowledge.

adult salmon migration. However, tree stumps and other large woody debris form pools that are critical for the survival of *juvenile* salmon over the winter (Fausch and Northcote 1992; Watanabe et al. 2005). Ecological developmental biology has a great deal to say about how we teach biology in the twenty-first century.

Epistemology and Methodology

How we study development

The demands of eco-devo require that developmental biologists get beyond the easily accessible model systems in order to study the effects of the environment on development (and more broadly, the interactions of development and environment). The six model organisms of animal developmental biology (the fly *Drosophila*, the frog *Xenopus*, the nematode *Caenorhabditis*, the mammal *Mus*, the bird *Gallus*, and the zebrafish *Danio*) were all selected in part because of the absence of major environmental factors in their early development; the absence of significant plasticity makes these model organisms easier to study because environmental variables are not important. Model organisms play a large role in defining the reality of a discipline, and these model organisms make it appear that the fertilized egg might contain everything needed for phenotype production. This is because the model organisms derived in the 1970s were constructed for looking at the roles of genes during development* (Bolker 1995, 2014). Newer model systems are needed for "characterizing and understanding the interconnectedness between an organism's environment, its development responses, and its ecological interactions in natural populations, [where] such research promises to clarify further the role of the environment in not only selecting among diverse phenotypes, but also creating such phenotypes in the first place" (Ledón-Retting and Pfennig 2011). Getting away from "model systems" will take a great deal of effort and a change in funding priorities (see NSF 2005; Collins et al. 2007; Jenner and Wills 2007).

Although the development of these model organisms provides an excellent approximation of development and a needed first step in figuring out the proteins and genes involved in forming complex structures, the model system approach does not give complete answers. As McFall-Ngai (2002) has noted, our epistemology is flawed. "The implicit assumption that has accompanied the study of animal development is that only 'self' cells (i.e., those containing the host genome) communicate to induce developmental pathways." The presence of other organisms and abiotic signals can cause the phenotypes that develop in nature to differ significantly from those obtained in the laboratory. Relyea and Mills (2001) found that when tadpoles were raised in isolation, they were relatively insensitive to the herbicide

---

*The model organisms of the earlier part of the twentieth century (the chick, the flatworm, sea urchins, and salamanders) had been constructed for the ease of tissue transplantation and the explication of inducing centers and morphogenetic fields. The model organisms of evolutionary developmental biology and ecological developmental biology are not single organisms but collections of organisms that allow comparisons along phylogenetic or ecological axes (Collins et al. 2007; Gilbert 2009).

carbaryl. However, when they were simultaneously exposed to that herbicide and made to undergo predator-induced polyphenism by the placement of a dragonfly nymph in the water, the same concentration of the herbicide could wipe out the population. What had seemed a harmless chemical in laboratory tests could wipe out the frog population under more natural conditions. We now have to include the environment in our concepts of development. Moreover, environments themselves vary—their dynamic histories affect and are affected by developing organisms in multispecies patterns (Odling-Smee et al. 2003). This has not been the standard way of studying development, nor have we usually looked at developmental differences within populations or between habitats. Developmental biologists have tended to look at species as a whole rather than looking at the variations within a species or variations within and among interacting species that are crucial to the development of each partner. This is a more evolutionary approach, and it is one that developmental biologists are just beginning to adopt (see Kopp et al. 2000).

Another epistemological change may concern the autonomy of developmental biology. The past decades has seen the emergence and rise in popularity of a "new" field called "systems biology"—but this approach already had a long-standing history in developmental biology. Systems biology views an organism as an integrated and interacting network of genes, gene products, cellular components, physiological coordinating systems, and ecological agents (biotic and abiotic). The interactions themselves are seen as temporally patterned entities. Starting in the 1920s, several developmental biologists, notably Paul Weiss (1971) and Ludwig von Bertalanffy (1968), put forth the notion of systems biology. That Weiss and von Bertalanffy both trained in Vienna was probably not incidental to their coming up with such ideas (see Drack et al. 2007), since the research program of Vienna's Prater Vivarium has been seen as a precursor to eco-evo-devo (Wagner and Laubichler 2004).

The systems biology approach gained momentum only after the various genome projects failed to deliver the expected goal of reading the "Book of Life" from the accumulated DNA sequences. It was then realized that epigenetic approaches were also required. Ecological developmental biology provides one kind of systems biology approach, through which the neuroendocrine and paracrine signaling cascades integrate the gene wiring diagrams of developmental biology with the biotic and abiotic flow diagrams of ecology. Historically dynamic processes such as cell cycles, life cycles, nitrogen cycles, and solar cycles interact to create a new systems biology wherein developmental physiology is linked to ecology. And when one has a theory of development that involves signal transduction both within and between embryos, one can ask new questions. Marc Kirschner (2005) highlights some new, big questions raised by the systems approach:

> The big question to understand in biology is not regulatory linkage but the nature of biological systems that allows them to be linked together in many nonlethal and even useful combinations.... These circuits may have certain robustness, but more important they have adaptability and versatility. The ease of putting conserved processes under regulatory control is an inherent design feature of the processes themselves.

> Among other things, it loads the deck in evolutionary variation and makes it more feasible to generate useful phenotypes upon which selection can act.

However, the nascent field of systems biology is still basing its modeling on concepts of autopoiesis and homeostasis (Noble 2006; Boudry and Pigliucci 2013). But autopoiesis in multicellular organisms is probably very rare, if it exists at all, and homeostasis is underlain by metabolism, a temporally dynamic set of processes. Because metabolism enables the stabilization of an "individual" by permitting the organism to retain constancy while constantly changing its component parts, "individuals" are always relational processes in time (Jonas 1966; Gilbert 1982). Individuals are always in the making, and are always in relation with other becoming entities over many simultaneous wavelengths of time—metabolic, developmental, ecological, and evolutionary. The questions of developmental biology will be placed into new contexts, and the *explananda* of the field are expected to change accordingly. Eco-evo-devo is progressively proposing models based on symbiopoiesis and dynamic relational adaptation within and among holobionts—such as dung beetles, squids, toads—that can model an evolutionary nature that includes developmental symbiosis and plasticity as normative.

How we study evolution and ecology

Ecological developmental biology emphasizes phenotype. Genes and environments are both major contributors to the production of the phenotype. In evolution, phenotypic plasticity may produce a morphology or behavior that later becomes fixed into the genome. Certainly in any new evolutionary synthesis, genetic assimilation, phenotypic accommodation, and genetic accommodation have to be taken seriously, as do epialleles that can be transmitted between generations. Phenotypic plasticity can also hide genetic variations, which can express themselves under different environmental conditions (Sultan 2011), and predator-induced polyphenism in prey species can even trigger prey-induced polyphenism in the predator (Kishida et al. 2006; Mougi 2012; Kishida et al. 2013). Modeling species interactions and possible extinction events necessitates understanding the roles and limitations of plasticity (Botero et al. 2014).

Plasticity and its limits must be a major issue, and one that unites developmental biology, ecology, and evolution. In an era of record-breaking droughts and fires, rapidly melting glaciers, and extremes of temperature, conservation biology demands knowledge of how holobionts, groups, and species meet—and fail to meet—the threats of extinction. Indeed, if the data of this volume are accurate, then plasticity across the entire life cycle may be key to such survival. The interaction of climate and life cycle produces conditions that can retard or enhance the ability of a species to adapt to new conditions. And while the road to extinction is paved with remarkable adaptations to existing environments, flexibility may provide the means of surviving climate changes (Janzen 1967; Ghalambor et al. 2006; Deutsch et al. 2008; McCain 2009).

And, as mentioned in our discussion of ontology, the units of selection should be seen as symbiotic organisms. As documented in Chapter 11, the protagonists of "individual selection" may be holobionts, multi-lineage "individuals." So the interactions that create fitness and health within the holobiont have to become paramount for discussions of population and ecosystem changes. Also, since symbionts may play important roles during evolutionary transitions (origins of eukaryotic cells, origins of multicellularity, origin of mammals, etc.), one cannot confine evolution to the changes in allele frequency over time. The roles of symbionts in sympatric reproductive isolation may be critical for the origins of new species. Thus, ecological developmental biology and especially the roles of plasticity, epialleles, and symbiosis are critical in future studies of evolution and ecology.

Ethics for the Anthropocene

In February 2008, a group of 21 scientists of the Geological Society of London concluded that we and the other organisms with whom we share this planet, and indeed, the planet itself, are no longer in the Holocene epoch. The interglacial age is over, and we are now living in the Anthropocene, an epoch characterized by a human-dominated environment (Zalasiewicz et al. 2008).

Multispecies flourishing

While the "Anthropocene epoch" is a useful term for calling attention to the effects of human activity on the planet, there are numerous critiques of the term. One is that not all humans are part of the "Anthropos" that supposedly runs and ruins the affairs of the planet. Secondly, the idea that this constitutes a geological epoch of thousands of years produces a hopelessness about solving these problems. More likely, the Anthropocene is not a geological "epoch," but a geological "event," analogous to the K/Pg extinction event (that exterminated the dinosaurs 65 million years ago) and the Permian "Great Dying" (which destroyed 95% of marine animal species 250 million years ago). But instead of the iridium that marks the K/Pg event, the Anthropocene may be recognized by a layer of ferric oxide rust.* According to Lynn Margulis (2011):

> You know what the index fossil of *Homo sapiens* in the recent fossil record is going to be? The squashed remains of the automobile. There will be a layer in the fossil record where you're going to know people were here because of the automobiles. It will be a very thin layer.

The ethical goals are to make the Anthropocene event as short as possible and to have a more biodiverse and healthier equilibrium (for wildlife as well as humans) on the other side of it. The first thing that needs to be done is to understand that we are part of nature and that our existence depends on the other-than-human parts of it. That means letting go of our

*One must realize that if the extent of life is represented by a human forelimb from shoulder to thumb, the region from the shoulder to wrist comprises an entirely microbial realm. The Cambrian Explosion of eukaryotic multicellularity occurs in the wrist, dinosaurs come and go within the thumb, and mammals appear at the base of the thumbnail. "All human existence would be confined to the extreme tip of the thumbnail, and could be erased with a single stroke of a nail-file" (Lee 2014).

desire to be manager of nature and becoming responsive and responsible partners in natural alliances. Throughout this book, we have continually found that the rules of the evolutionary process include the crucial notion of "becoming with." Co-development is normative, and evolution is not merely the battle of each against all. As Chris Cuomo (1998, p. 62) concludes, sustainability is not enough, and stewardship is at best a beginning (and does not call into question human hegemony over nature). Cuomo proposes that the ethical starting point of ecological theory is "a commitment to the *flourishing* or well-being, of individuals, species, and communities." This ethic of multispecies flourishing and well-becoming are perhaps visualized best in that very union we mentioned in Chapter 3—the orchid seed. The beauty of the orchid is made possible by the invasion of its seed coat by a symbiotic fungus. Alliances are strange and wonderful things in the living kingdom. We "become with" the world, and that is our enormous responsibility.

Philosophers of ecology have recognized this reciprocity of humans and the more-than-human world. David Abram (1996) writes:

> Our bodies have formed themselves in delicate reciprocity with the manifold textures, sounds, and shapes of an animate earth—our eyes have evolved in subtle interaction with other eyes, as our ears are attuned by their very structure to the howling of wolves and the honking of geese. To shut ourselves off from these other voices, to continue by our lifestyles to condemn these other sensibilities to the oblivion of extinction, is to rob our own senses of their integrity, and to rob our minds of their coherence. We are human only in contact, and conviviality, with what is not human.

Attention to life cycles

Ecological developmental biology adds to this recognition of reciprocity two other principles: (1) a focus on juvenile populations and (2) the importance of integrating social webs with natural webs. Nature does not exist anymore apart from human endeavors (and vice versa), and most of the animal population in the world is embryonic or larval. Both these new ways of thinking have been highlighted in conservation efforts to sustain one of the most iconic and awe-inspiring butterflies, the monarch (*Danaus plexippus*).

The monarch butterfly undergoes a multigenerational migratory life cycle that, for the eastern part of the population, includes spending the winter in the mountains of central Mexico. Millions of these butterflies spend summers in the northern United States and fly to the mountains guided by light-sensitive magnetosensors on their antennae that integrate light signals with signals from the Earth's magnetic field (Brower et al. 2006; Guerra et al. 2014). Over the past 20 years, the monarch population has declined 80%, and the question is "What is killing the butterflies?" There have been three major hypotheses: (1) loss of the overwintering habitat due to land-use changes in the Mexican mountains, (2) extreme weather events, and (3) destruction of habitat for the adult butterflies in the United States.

It has turned out that none of these is probably the major factor. Although each of these has contributed to the population decline, the major cause of the monarch butterfly's destruction is the loss of the milkweed plant (*Asclepias*), which is the food plant of the butterfly *larvae*. The adult monarch butterfly can get nectar from many types of plants, but the monarch *caterpillar* is a specialist, eating only milkweed. Without milkweed, the adult has no place to lay her eggs. If she finds milkweed, there are now often other eggs laid on it, and the caterpillars must compete for limited food when they hatch. So the problem is not the destruction or starvation of adults. It's the limitation of food for the larvae (Flockhart et al. 2014).

And what has stopped milkweed from growing? Here is where human social and technological networks must be added to biological networks. The culprit seems to be the glyphosate herbicides mentioned in Chapter 5. Here, they are acting as predicted, as herbicides, killing plants. But the herbicides are now being sprayed in high concentrations and indiscriminately, because the large-scale farms are growing genetically modified glyphosate-resistant soy and corn. The herbicide is sprayed over an entire area of plants, leaving only the genetically modified crops surviving. This has led to a 58% decrease in milkweed plants and a catastrophic loss of monarch butterflies* (Pleasants and Oberhauser 2013; Smith 2014).

Similarly, insect development is impeded by a new generation of pesticides (neonicotinoids) that are killing insects quite well, but indiscriminately. These neonicotinoids are thought to be responsible, at least in part, for the collapse of bee colonies. Those bee colonies whose forage plants had been treated with neonicotinoids developed hardly any queen bees compared with those whose plants had not been treated (Whitehorn et al. 2012; Sandrock et al. 2014). Without the insects, the symbioses perfected over millions of years perish, and so do our crop plants. In addition, the bird and fish populations that depend on insect prey plummet (Hallmann et al. 2014). If we are to conserve our planet's species, we must know about their development and how human activities can impact their entire life cycles.

The Anthropocene and its critics

To some researchers who study cultural norms, the term "Anthropocene" is appropriate in highlighting our biopolitical problems, but it should be complemented by other terms, especially terms that suggest more adequate modes of thinking and acting in our current planetary dilemmas. Certainly, the ruling "Anthropos" in "Anthropocene" does not include all of humanity, nor is it particularly descriptive. The term "Capitalocene" has been used to emphasize our world-encircling network of hierarchical

---

*Federal protection has been sought for the monarch butterfly (Xerces 2014a,b). And this is something that can be remedied in part by planting the right milkweed plants in one's gardens. The nonnative, tropical milkweeds flower most of the year and can actually be deleterious, preventing the monarchs from migrating, and promoting the spread of a parasite that kills the butterflies (Satterfield et al. 2015). Other milkweed plants might not flower at the appropriate time (see nativeplantwildlifegarden.com/help-monarchs-with-the-right-milkweeds/). Another helpful website is monarchjointventure.org/get-involved/create-habitat-for-monarchs, which includes a link to the Xerces Society website for information on where to buy local, native milkweed seeds. Further, the Buddleia plant that is frequently called the "butterfly bush" does nothing helpful for the monarch migration. While it attracts butterflies, it is not a milkweed, and it provides no food for the larvae of any American butterfly.

economic relationships as the hallmark of our times, developing over the past few hundred years. Here, global capitalism is understood "as a world-ecology, joining the accumulation of capital, the pursuit of power, and the co-production of nature in dialectic unity" (Moore 2013). "Capitalocene" can do important work to focus attention on dynamic historical economic processes that must be reworked. But it, too, is a limited term. Another term, "Plantationocene" has been recently been suggested by a group of biologists, sociologists, and anthropologists (AURA 2014a; Haraway 2015) to focus on past and present mass migrations of human and nonhuman genes and organisms into Earth-changing patterns of organized and mechanized means of mass production. This includes, for example, the massive replacement of mixed-species forests with monoculture palm oil plantations in recent and current practices, and the historical movements of peoples and organisms in the slave-based sugar and cotton plantations of earlier periods of the Plantationocene and Capitalocene. "Plantationocene" focuses on the fact that multispecies rearrangements in agriculture preceded and supported the mechanized factory with its fossil fuel–burning apparatus; plantations were the first modern factories and the models for regimenting both human and nonhuman beings.

Both the Capitalocene and Plantationocene cause, and are caused by, the human population explosion. The enormous increase in population causes the attendant crises of food production, multispecies habitat destruction, and threatened wide-scale ecological collapse. The increasing number of humans to feed reciprocally propels and is argued to justify the accelerating mechanization of agriculture, which further displaces species, uses huge amounts of herbicides (against weeds), pesticides (against insects), and synthetic fertilizers (which require large amounts of energy to produce). Only when these economic relationships are seen as containing and being contained in natural relationships do the data make sense and can intervention be accomplished (Figure 1).

For instance, genetically modified crops may indeed be doing what they were intended to do—reduce chemical pesticide use, increase crop yields, and increase profits (see Klümper and Qain 2014). But this is at the cost of higher herbicide use, plant monoculture, corporate control of farms, the displacement of local crops for those that can be exported for higher prices, and the death of many nonpest insects such as the monarch butterflies and honeybees.

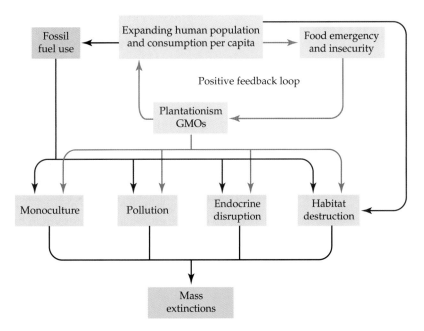

Figure 1 Positive feedback loop of elevated human population growth, food emergency, and plantationism leads to the mass extinctions of the Anthropocene/Capitalocene event. (After discussions of AURA 2014a.)

Human beings are certainly not the only players in this story. Our stories must include not only other species, but also the global and local economics of a nature and culture that cannot be separated (Tsing 2004; Haraway 2008; Kirksey 2014).

Environmental justice

Ecological developmental biology changes the stories that scientific data allow us to tell. And these stories are critically important. Anthropologist Marilyn Strathern (1992, p. 10) has commented, "It matters what ideas we use to think other ideas." Ecological developmental biology gives us new ideas to think with.

As can be seen in Chapters 5 and 6, ecological developmental biology highlights areas of environmental justice, where there is exposure to chemicals that damage the developing embryo. Chemicals such as bisphenol A and PCBs are associated with increased risk of human miscarriages (Chen et al. 2013; Lathi et al. 2014) and may cause more infertility and abortion than elective procedures. Indeed, this is a place where sciences and diverse religions can become allies to forge policy (Gilbert 2013). The antiabortion bishop and the physician who prescribes contraceptives both claim to want healthy babies in healthy families. To this end, both can promote and support legislation to ban endocrine-disrupting chemicals. We have a science-based ethic that has caught up with Dr. Martin Luther King's 1963 letter from the Birmingham, Alabama, jail: "We are caught in an inescapable network of mutuality, tied in a single garment of destiny. Whatever affects one directly, affects all indirectly."

Environmental justice involves who and what benefits from new technologies (and the new natural and social arrangements emerging from their use) and who or what suffers. Nature must come into the stories in a major way. Who benefits from fracking, and who suffers infertility, illness, and abortion? Who benefits from incinerating toxic waste, and who suffers having the sick children and impaired larval development? Who gains from the disposal of plastic in the ocean, and which organisms (both human and nonhuman) suffer from the Great Pacific Garbage Patch, an area of plastic refuse larger than the state of Texas? The existence of this enormous plastic island is made possible by the largely unconscious choice we have made "to base our way of life on things that are intended to be used once and thrown away, and to manufacture those things out of material that will last for thousands of years" (Seidenberg 2015). Surveys show that certain cosmetics containing endocrine-disrupting chemicals are used by one ethnic group more than by others (Tiwari and Ward 2003; James-Todd et al. 2012), and this conclusion becomes even more important if the diseases or conditions caused by such endocrine-disrupting chemicals are transmitted from one generation to the next.

Epigenetic factors may also play an important role in decisions about early opportunities and experience. If environmental enrichment either in utero or soon after birth are critical for developing learning skills, then a polity would benefit by increasing the possibilities of mothers to raise their children in such environments. Similarly, if the nutrition a fetus receives is critical for its adult health, governments may get a healthier and less-expensive-to-care-for citizenry by providing food supplements to expectant mothers. Ecological developmental biology may be extremely important in public health and environmental health policy decisions.

A third area of ecological justice concerns the genocide that involves the elimination of entire species. Within the next century, we may witness the extinction of nearly half the species of life on this planet (Barnosky et al. 2011; Pearson 2011; Hansen et al. 2013a,b). This is due in large part to the economic relationships between people and also between people and that part of nature that is mistakenly seen to be separate from humans and exchangeable for money (see Moore 2014). Do humans have the right to destroy an entire species? Or even more, to destroy entire complexes of interacting species? Indeed, this is another place where science and religion can/must become allies. In science, it's called preservation of biodiversity. In various religious traditions, it's called stewardship or guardianship for Creation, or the doctrine of doing no harm.*

A fourth area of environmental justice is called "intergenerational justice" (Weiss 1989; Weston 2008, 2012; Hansen 2013a). Intergenerational justice demands that one generation not use up the resources and the beauty of the world such that other generations cannot use or perceive it. Each generation is seen to be entitled to the diversity of natural and cultural resources (and the access to these resources) that had been enjoyed by previous generations.

Advocates for intergenerational environmental justice, such as the Eco-Justice Collaborative (EJC 2014) and UNICEF (2014), claim that the generation of adults alive today, through their decisions and lifestyles, are determining to an unprecedented degree the future that later generations will inherit. As mentioned here and in several chapters, the stability of ecosystems depends on the complex developmental, symbiotic, and plastic interactions between species. Should we be allowed to bequeath to the next generation a world devoid of butterflies, songbirds, or coral? It takes knowledge, such as that from eco-devo, to understand what policies could best protect environments and what decisions individuals can make to preserve resources and sources of wonder.

Population equilibrium

The human population exploded 2.5 times from 1950 to 2008, reaching a total of more than 7 billion persons, with many living in desperately unsustainable extremes of both excess and deprivation. This growth created emergency conditions that have reciprocally driven the green revolutions, first with artificial fertilizers and secondly with genetically modified high-yield and herbicide-resistant plants. Such emergency conditions have also changed farming practices such that billions of domesticated animals spend most of their lives indoors, in industrial "animal containment facilities" modeled after prisons (see Richmond et al. 2003 for such rationales). The alternatives to enormous human population expansion and continuing

*Note that in this ecological developmental biology story, the religion-science dispute over the *origin* of species becomes a sideshow. The real action, the real story, is in the *preservation* and *flourishing* of species (see Gilbert 2013).

vastly unjust and unequal exploitation and consumption of resources are (1) the stark Malthusian correctives of disease, war, and malnutrition; (2) a strictly hierarchical management of nature and humanity, such as those envisioned in the planetary plantation-like structures of dystopian science fiction; or (3) more collaborative and likely more effective policies of encouraging major changes in models of wealth, consumption, progress, and good living. This latter solution most likely means not only new economic models, but also promoting and supporting non-biological families of many kinds, including those consisting of numerous parents per child, innovative adoption practices for old and young people, and other arrangements of durable, supportive non-biogenetic kin and community. This will take imagination if it is to be done in a voluntary, non-coercive and non-eugenic manner. Adequate models and practices do not yet exist, and creative thinking within and about eco-devo may help to envision these new systems (AURA 2014b). To go further, we need to tell more and better stories and to send our imaginations into areas explored mostly by speculative fiction writers.

Better stories

Storytelling matters. If there is anything that is appreciated by all cultures, it is a good story, and storytelling is a critical part of the scientific enterprise. Scientists have a moral and scientific imperative to tell accurate stories. Scientific stories must not contradict existing data.* Ecological developmental biology has just expanded the realm in which scientific stories can be told. Moreover, these stories are important to us. The account of our evolution has ethical consequences. Abraham J. Heschel (1965, pp. 7–8) noted that whereas "a theory about the stars never becomes part of the being of the stars," a theory about human beings enters our consciousness, determines our self-image, and modifies our very existence. "The image of man affects the nature of man.... We become what we think of ourselves." If we think of ourselves as killer apes, certain behavioral phenotypes are acceptable that would not be socially allowed if we viewed humans as the product of sympoietic and plastic evolution that has enabled us to voluntarily cooperate to an extent unimaginable by other vertebrates.

Ecological developmental biology, in particular, focuses on the notion that genes do not control destiny. Thus, the notion of improving the world by selecting people with "good genes" is not so much good science as bad fantasy. Geneticist Theodosius Dobzhansky pointed out that the world's problems are not caused by people with inferior genomes, but by people with perfectly good genomes using their talents for improper ends. Cor van der Weele (1999) has noted that polyphenisms and epigenetics mean that

*Some of Thomas Huxley's attempts to change biological stories should be mentioned here. Huxley was among the first to point out the need for a developmental account of evolution (Chapter 9), and he was among the first to criticize religious writers for freely ignoring science when it conflicted with their views. He also wrote (Huxley 1894) that the story being told about subsistence-level struggle for existence in humans (and that humans improve because of it) was wrong. "It is a contest, not for the means of existence, but for the means of enjoyment." Earlier in his career, Huxley (1870) codified the notion that one must "learn what is true in order to do what is right," and he personally felt this meant combining research, teaching, writing, and public service (see Desmond 1999).

genes do not determine our lives, and she calls for an "ethics of attention." Science, she points out, helps determine what we see and what we hope to understand. For example, since 88% of breast cancers are thought to be caused by "the environment" (American Cancer Society 2006) whereas the "cancer genes" *BRCA1* and *BRCA2* account for only 5%–10% of breast tumors, our emphasis on the genetic component of this disease appears to be dangerously misplaced.

Evolutionary narratives are the most critical stories in biology, in science, and perhaps in Western civilization, so we had better get them in line with the biological data.* The competitive story of evolution has given rise to stories of hierarchy and lethal competition. Our narratives tell us what science thinks is normal, and we have been told that selfishness is normal and adaptive. Michael Ghiselin (1974) compares social and biological evolution:

> The evolution of society fits the Darwinian paradigm in its most
> individualistic form. Nothing in it cries out to be otherwise explained.
> The economy of nature is competitive from beginning to end....
> No hint of genuine charity ameliorates our vision of society, once
> sentimentalism has been laid aside.

The stories usually told about evolution are stories of a nature characterized by self-interest and competition to the death. Science is told as a story wherein nature has no fundamental cooperative tendencies. That narrative leaves out almost everything that this book shows about cooperativeness and evolution. It gives an incomplete story and partial account of nature that makes humans and all other organisms seem like completely self-interested, atomic individuals. So one underappreciated moral imperative is highlighted here: we must tell better stories. Ecological developmental biology has expanded the range of our storytelling possibilities. We must tell much richer stories about interactive co-shaping relationships at every level. Eco-evo-devo narratives, concerned with providing for large, interactive, and flourishing human and non-human populations, might lead to more distributed, flexible, collaborative models and research programs for restoring and repairing ecosystems and economies.

Unfortunately, the language to tell these stories has not yet evolved. None of the available categories—Anthropocene, Capitalocene, or Plantationocene—offers adequate models and metaphors for our needed rethinking and acting. Moreover, our stories are usually framed by the opposition of competition and cooperation, a binary system that we know does not reflect nature. For instance, in the *Euprymna-Vibrio* symbiosis, the squid first kills all the attached bacteria that are not functioning *Vibrio fischeri*. Then, the vibrios initiate the symbiotic construction of the light organ via cell death. The vibrios and squid finally generate a light organ that allows the squid to more successfully compete against squid-eating predators. And this competition, in turn, stabilizes the ecosystem. We find such interpenetrations of competition and cooperation throughout nature, and

*The stories promulgated by intelligent design proponents, for example, are not bounded by scientific data, are frequently based on nonscientific assumptions, and freely ignore known science when it conflicts with previously held views (see Gilbert et al. 2007).

it is inherent in our metaphor of "teams." One tries out for a team in a competitive manner. Then, the best particular mix of people get to become the team, and that team cooperates to compete against other teams. And in that way, the league flourishes. No simple binary pair of terms (such as "competition" and "cooperation") can do justice to the emergent interactive complexity.

Metaphors and models for telling such stores traditionally come from nonscientific sources. In evolutionary biology, economics (mercantile capitalism, investment capitalism, Gini coefficient) has provided some of the most important models. So has literature, from whence we get the "Red Queen Strategy" (from Lewis Carroll's *Through the Looking Glass*). Each of these metaphors privileges the competitive adaptationist view of evolution. We are still looking for sources of new metaphors with which to tell our new stories. Thomas Huxley tried some speculative fiction in his *Prolegomena to Evolution and Ethics* (1894), where he formulated two competing human ecosystems (including pets and domesticated animals and crops) that must cooperate internally to survive external competition. However, he found that all his just systems crumbled before the onslaught of human population expansion. That was, however, before modern birth control, so there might be more hope now. More recent literary experimentation to make such societies of interspecies responsibility, cooperation, and competition include Ursula Le Guin's Hain and Kesh stories, as well as the Earthseed stories of Octavia Butler. The recent sports book *The Boys in the Boat* is an excellent example of the mixture of competition and cooperation at multiple levels that makes a team possible.

Ariew and Lewontin (2004) remind us that important evolutionary stories have been re-written before. Darwin's notion of fitness was an ecological metaphor based on the image of items physically "fitting" into a pre-existing pattern. Those individuals matching the pattern would fit better than those that didn't. Population geneticists changed this ecological metaphor into a story of quantitative allelic variation wherein those individuals in a population that provided more progeny to the next generation (either in absolute or relative numbers) would be considered more fit. This resulted in "specifically *genetical* theories of evolution... that have essentially no natural historical elements or even internal causal narratives (with the exception of genetic mechanisms) and in which Darwinian 'fit' plays no role." One of the exciting projects of ecological evo-devo would be to construct a new measure of fitness based on sustained and active ecological integration. In such new measures, differential procreation would not be the sole measure of success. This has obvious parallels to economists looking for measures other than the gross domestic product for assessing economic welfare.

But there are very few traditions in the West where multispecies interactions get beyond the binaries of cooperation/competition, self/other, internal/external. To do so in the civilized countries is to flirt with animism and naive sentimentality. One place to look for multispecies interactions and reciprocity is in crafts traditions. First, it is a place where intimate knowledge of the natural entity is critical for creating the artifact. Thomas Huxley saw science in the craft tradition, too, and many scientists have had the "feeling for the organism" aid their work. Second, craft traditions include a variety of models that differ from the hierarchical ones that predominate capitalistic economics. Certain craft traditions dealing directly with living organisms include artisanal food-based industries. In trying to stay free of agribusiness, many artisans have adopted a model of "rivalry" that includes both cooperation and competition (Kirksey 2014; Verhal and Hoskins 2014). These groups include microbrewers, artisanal cheese producers, non-farmed mushroom hunters, and organic food restaurateurs.

Another food-based "artisanal" movement with relevance to ecological developmental biology is "grass farming," an ecologically efficient approach to livestock production. This management-intensive program is an ecological developmental approach, based on synchronizing the development of grass, insects, poultry, and mammals in a manner that takes advantage of their mutual interactions (Ekarius 1999; Pollan 2006; Nation 2007; Wilkins 2008). Thus, when the grass reaches a certain height, cows are allowed to graze. The cows then defecate in the field, and insect larvae grow in the cow feces. At a certain time, the larvae will be at the appropriate stage for chickens to feed on them. By rotation of the placement of the cows and the chickens, livestock grows with minimal investment in chemical fertilizers, antibiotics, or pesticides. It integrates knowledge of life cycles into farming practices. Like most artisanal ventures, however, it is labor-intensive. It is also like many of the other artisanal food movements, not only cognizant of life cycles, but also crucially associated with symbiotic microbes.

New vocabularies are beginning to bridge the gap between craft and science. For example, "terroir" is the French term for a "sense of place" that refers to the effects that the local climate, geography, and geology have on the production of a product. It expresses the idea that a given organism will react differently to different environmental conditions, and it was first used by the French winegrowers to distinguish the wines of one region from another (see Paxson 2014). Some of the terroir may indeed come from transient microbes whose fermentation products distinguish particular realms and times. Celebrity chef David Chang recently collaborated with Harvard University microbiologists on a study showing that fungi indigenous to particular regions may impart to a cuisine its own "microbial terroir." Indeed, the goal of that project (Felder et al. 2012) was "to create truly indigenous products, through stewardship of our native microorganisms."

Ecological developmental biology gives some great new stories. It forms a view of nature that is immediate and intimate: Why preserve ecosystems? The first answer is that we are part of them. As much as we have separated ourselves from nature, we are intimately connected to the rest of the biological world through symbioses. We are built by symbiotic relationships, and our air and food are produced by such relationships. We are intimately connected to nature, and even the algae in the ocean are essential to our living.

The second answer relates to who we are as people. Heschel (1955) noted that we are human because we are able to experience wonder. Wonder has a short half-life, and it decays rapidly into awe and curiosity. Awe, he points out, is the font of religion, while curiosity gives rise, as Plato and Aristotle said it did, to philosophy and science (see Gilbert 2013). So humans have a vested interest in preserving sources of wonder. Preserving sources of wonder is where religion, philosophy, and science—the "grand-children of wonder"—have a natural alliance. In this sense, retaining the mountain forests, the meadows, and rivers is indeed valuable to humans, for they maintain our humanity.

We must also recognize that nature, as we know it, is rarely untouched. While we should try to preserve the few expanses of "pristine" nature left, it is critical to remember that just as we are constantly being touched by "external" nature, nature is constantly touched by everything we place into the water and air (see McKibben 1989). Keeping sources of wonder alive does not necessitate their being pristine; indeed, by the very nature of the interactions between organisms, we could hardly expect them to be. Eco-logical developmental biology can integrate both selfishness and otherness into an ethic of "becoming with." It is an ethic of wonder and connectivity where nature and culture create each other, where each organism is a blend of different species, and where each organism is linked to a nature-culture matrix that both generates it and is made by it. We biologists have a great new story to offer the world.

References

Abram, D. 1996. *The Spell of the Sensuous: Perception and Language in a More-than-Human World*. Vintage Books, New York.

Allen G. 1979. Naturalists and experimentalists: The genotype and the phenotype. *Stud. His. Bio*. 3: 179–209.

American Cancer Society. 2006. Frequently requested statistics in cancer. www.cancer.org

Ariew, A. and R. C. Lewontin. 2004. The confusions of fitness. *Br. J. Philos. Sci*. 55: 347–363; quote on p. 351, emphasis in original.

AURA (Aarhus University Research on the Anthropocene). 2014a. 30 October 2014.

AURA (Aarhus University Research on the Anthropocene). 2014b. Santa Cruz. anthropocene.au.dk/arts-of-living-on-a-damaged-planet; vimeo.com/97663518

Bailey, K. M. 2011. An empty donut hole: The great collapse of a North American fishery. *Ecol. Soc*. 16: 28.

Barad, K. 2007. *Meeting the Universe Half-Way: Quantum Physics and the Entanglement of Matter and Meaning*. Duke University Press, Durham, NC.

Barnosky, A. D. and 11 others. 2011. Has the earth's sixth mass extinction already arrived? *Nature* 471: 51–57.

Benson, K. R. 1988. From museum research to laboratory research: The transformation of natural history into academic biology. In R. Rainger, K. R. Benson and J. Maienschein. *The American Development of Biology*. U. Pennsylvania Press, PA., pp. 49–83.

Bissell, M. J. 2007. Architecture is the message: The role of extracellular matrix and 3-D structure in tissue-specific gene expression and breast cancer. *Pezcoller Found J*. 16(29): 2–17.

Bolker, J. A. 1995. Model systems in developmental biology. *BioEssays* 17: 451–455.

Bolker, J. A. 2014. Model species in evo-devo: A philosophical perspective. *Evol. Dev*. 16: 49–56.

Botero, C. A., F. J. Weissing, J. Wright and D. R. Rubenstein. 2014. Evolutionary tipping points in the capacity to adapt to environmental change. *Proc. Natl. Acad. Sci. USA* 112: 184–189.

Boudry, M. and M. Pigliucci. 2013. The mismeasure of machine: Synthetic biology and the trouble with engineering metaphors. *Stud. Hist. Philos. Biol. Biomed. Sci*. 44(4 Pt B): 660–668.

Brower, L. P., L.S. Fink and P. Walford. 2006. Fueling the fall migration of the monarch butterfly. *Integr. Comp. Biol*. 46: 1123–1142.

Burian, R. 2005. *The Epistemology of Development, Evolution, and Genetics*. Cambridge University Press, Cambridge, New York.

Buss, L. W. 1987. *The Evolution of Individuality*. Princeton University Press, Princeton, NJ.

Byrnes W. M. and W. R. Eckberg. 2006. Ernest Everett Just (1883–1941): An early ecological developmental biologist. *Dev. Biol*. 296: 1–11.

Chen, X. and 12 others. 2013. Parental phenols exposure and spontaneous abortion in Chinese population residing in the middle and lower reaches of the Yangtze River. *Chemosphere* 93: 217–222.

Chernyshevskii, N. G. 1888. Quoted in Todes, op. cit. 1989.

Chiu, L. and S. F. Gilbert. 2015. The birth of the holobiont: Multi-species birthing through mutual scaffolding and niche construction. *Biosemiotics* March: 1–20.

Collins, J. J., 3rd, B. Wang, B. G. Lambrus, M. E. Tharp, H. Iyer and P. A. Newmark. 2013. Adult somatic stem cells in the human parasite *Schistosoma mansoni*. *Nature* 494: 476–479.

Collins, J. P., S. F. Gilbert, M. D. Laubichler and G. B. Müller. 2007. How to integrate development, evolution, and ecology: Modeling in evo-devo. In M. D. Laubichler and G. B. Müller (eds.), *Modeling Biology: Structures, Behaviors, Evolution*. MIT Press, Cambridge, MA, pp. 355–378.

Colwell, R. R. and A. Huq. 1994. Vibrios in the environment: Viable but nonculturable *Vibrio cholerae*. In I. K. Wachsmuth, P. A. Blake and O. Olsvik (eds.), Vibrio cholerae *and Cholera: Molecular to Global Perspectives*. American Society for Microbiology, pp. 117–133.

Constantin de Magny, G. and R. R. Colwell. 2009. Cholera and climate: A demonstrated relationship. *Trans. Am. Clin. Climatol. Assoc.* 120: 119–128.

Cuomo, C. 1998. *Feminism and Ecological Communities: An Ethic of Flourishing*. Routledge, New York.

Davis, M. C. 2013. The deep homology of the autopod: Insights from hox gene regulation. *Integr. Comp. Biol.* 53: 224–232.

Deacon, T. 2012. *Incomplete Nature: How the Mind Emerged from Matter*. W. W. Norton, New York.

Deleuze, G. and F. Guattari. 1987. *A Thousand Plateaus: Capitalism and Schizophrenia*. B. Massumi (trans.). Continuum Press, NY.

Desmond, A. 1999. *Huxley: From Devil's Disciple to Evolution's High Priest*. Addison-Wesley, Reading, MA.

Deutsch, C. A., J. J. Tewksbury, R. B. Huey, K. S. Sheldon, C. K. Ghalambor, D. C. Haak and P. R. Martin. 2008. Impacts of climate warming on terrestrial ectotherms across latitude. *Proc. Natl. Acad. Sci. USA* 105: 6668–6672.

Drack M., W. Apfalter and D. Pouvreau. 2007. On the making of a system theory of life: Paul A. Weiss and Ludwig von Bertalanffy's conceptual connection. *Q. Rev. Biol.* 82: 349–373.

Dupré, J. and M. A. O'Malley. 2009. Varieties of living things: Life at the intersection of lineages and metabolism. *Philos. Theor. Biol.* 1: e003.

EJC (Eco-Justice Collaborative). 2014. ecojusticecollaborative.org/issues/climate-change/intergenerational-justice/

Ekarius, C. 1999. *Small-Scale Livestock Farming: A Grass-Based Approach for Health, Sustainability, and Profit*. Storey Books, North Adams, MA.

Fausch, K. D. and T. G. Northcote. 1992. Large woody debris and salmonid habitat in a small coastal British Columbia stream. *Can. J. Fish. Aqua. Sci.* 49: 682–693.

Felder, D., D. Burns and D. Chang. 2012. Defining microbial terroir: The use of native fungi for the study of traditional fermentation processes. *Int. J. Gastron. Food Sci.* 1: 64–69.

Flockhart, D. T., J. B. Pichancourt, D. R. Norris and T. G. Martin. 2014. Unravelling the annual cycle in a migratory animal: Breeding-season habitat loss drives population declines of monarch butterflies. *J. Anim. Ecol.* 84: 155–165.

Foltz, B. V. 1995. *Inhabiting the Earth: Heidegger, Environmental Ethics and the Metaphysics of Nature*. Prometheus Books, Amherst, NY.

Foyston, J. 2014. Cooperation, not competition: Brit brewer teams with craft brewers to revive ailing English pubs. *The Oregonian* 23 October, 2014.

Ghalambor, C. K., R. B. Huey, P. R. Martin, J. J. Tewksbury and G. Wang. 2006. Are mountain passes higher in the tropics? Janzen's hypothesis revisited. *Integr. Comp. Biol.* 46: 5–17.

Ghiselin, M. T. 1974. *The Economy of Nature and the Evolution of Sex*. University of California Press, Berkeley.

Gilbert, S. F. 1982. Intellectual traditions in the life sciences: Molecular biology and biochemistry. *Persp. Biol. Med.* 26: 151–162.

Gilbert, S. F. 1991. Induction and the origins of developmental genetics. In S. F. Gilbert (ed.), *A Conceptual History of Modern Embryology*. Plenum Press, NY, pp. 181–206.

Gilbert, S. F. 1992. Cells in search of community: Critiques of Weismannism and selectable units of ontogeny. *Biol. Phil.* 7: 473–487.

Gilbert, S. F. 2002. The genome in its ecological context: Philosophical perspectives on interspecies epigenesis. *Ann. NY Acad. Sci.* 981: 202–218.

Gilbert, S. F. 2006. *Developmental Biology*, 8th Ed. Sinauer Associates, Sunderland, MA.

Gilbert, S. F. 2008. When "personhood" begins in the embryo: Avoiding a syllabus of errors. *Birth Defects Res. C: Embryo Today*. 84: 164–173.

Gilbert, S. F. 2009. The adequacy of model systems for evo-devo. In A. Barberousse, T. Pradeu and M. Morange, (eds.) *Mapping the Future of Biology: Evolving Concepts and Theories*. Springer Press, New York, pp. 57–68.

Gilbert, S. F. 2013. Wonder and the necessary alliances between science and religion. *Euresis* 4: 7–30.

Gilbert, S. F. 2014. A holobiont birth narrative: The epigenetic transmission of the human microbiome. *Front. Genet.* 5: 282.

Gilbert, S. F. and J. Bard. 2014. Formalizing theories of development: A fugue on the orderliness of change. In S. Minelli and T. Pradeu (eds.), *Towards a Theory of Development*. Oxford University Press, Oxford, pp. 129–143.

Gilbert, S. F. and M. Faber. 1996. Looking at embryos: The visual and conceptual aesthetics of emerging form. In A. I. Tauber (ed.), *The Elusive Synthesis: Aesthetics and Science*. Kluwer, Dordecht, pp. 125–151.

Gilbert, S. F. and A. Fausto-Sterling. 2003. Educating for social responsibility: Changing the syllabus of developmental biology. *Int. J. Dev. Biol.* 47: 327–244.

Gilbert, S. F. and J. Greenberg. 1984. Intellectual traditions in the life sciences. II. Stereospecificity. *Persp. Biol. Med.* 28: 18–34.

Gilbert, S. F. and S. Sarkar. 2000. Embracing complexity: Organicism for the twenty-first century. *Dev. Dynamics* 219: 1–9.

Gilbert, S. F. and the Swarthmore College Evolution and Development Seminar. 2007. The aerodynamics of flying carpets: Why biologists are loath to "teach the controversy." In N. C. Comfort (ed.), *The Panda's Black Box: Opening Up the Intelligent Design Controversy*. Johns Hopkins University Press, Baltimore, pp. 40–62.

Gilbert, S. F., J. Sapp and A. Tauber. 2012. A symbiotic view of life: We have never been individuals. *Q. Rev. Biol.* 87: 325–341.

Gillespie, N. C. 1979. *Charles Darwin and the Problem of Creation*. University of Chicago Press, Chicago.

Gottlieb, G. 1992. *Individual Development and Evolution: The Genesis of Novel Behavior*. Oxford University Press, New York.

Gould, S. J. 1982. Nonmoral Nature. *Natural History* 91(Feb.): 19–26.

Griesemer, J. 2002. What is "Epi" about epigenetics? *Ann. NY Acad. Sci.* 981: 97–110.

Guerra, P. A., R. J. Gegear and S. M. Reppert. 2014. A magnetic compass aids monarch butterfly migration. *Nature Commun.* 5: 4164.

Haeckel, E. 1882. "Ueber die Naturanschauung von Darwin, Göthe und Lamarck," Tageblatt der 55. Versammlung Deutscher Naturforscher und Aerzte in Eisenach, von 18. bis 22. September 1882 (Eisenach: Hofbuchdruckerei von H. Kahle,), pp. 82.

Hallmann, C.A., R. P. Foppen, C. A. van Turnhout, H. de Kroon and E. Jongejans. 2014. Declines in insectivorous birds are associated with high neonicotinoid concentrations. *Nature* 511: 341–343.

Hansen, J. and 17 others. 2013a. Scientific case for avoiding dangerous climate change to protect young people and nature. Available at arxiv.org/abs/1110.1365v3.

Hansen, J. and 17 others. 2013b. Assessing "dangerous climate change": Required reduction of carbon emissions to protect young people, future generations and nature. *PLoS ONE* 8(12): e81648.

Haraway, D. J. 1976. *Crystals, Fabrics, and Fields: Metaphors of Organicism in Twentieth-Century Biology*. Yale University Press, New Haven, CT.

Haraway, D. J. 1979. The biological enterprise: Sex, mind and profit from human engineering to sociobiology. *Rad. Hist. Rev.* 20: 206–37.

Haraway, D. J. 1991. Otherworldly conversations, Terran topics, local terms. *Sci. Cult. (Lond.)* 3: 64–98.

Haraway, D. J. 2003. *The Companion Species Manifesto: Dogs, People, and Significant Otherness*. Prickly Paradigm Press, Chicago.

Haraway, D. J. 2008. *When Species Meet*. University of Minnesota Press, Minneapolis.

Haraway, D. J. 2015. Anthropocene, Capitalocene, Plantationocene, Chthulocene: Making kin. *Environ. Humanit.* 6. environmentalhumanities.org

Hertwig, O. 1894. Zeit- und Streitfragen der Biologie I. Präformation oder Epigenese? Grundzüge einer Entwicklungstheorie der Organismen. Gustav Fischer Jena. Translated by P. C. Mitchell as *The Biological Problem of Today: Preformation or Epigenesis?* Macmillan, New York.

Heschel, A. J. 1955. *God in Search of Man*. Harper, New York.

Heschel, A. J. 1965. *Who Is Man?* University of California Press, Berkeley.

Holtfreter, J. 1968. Address in honor of Viktor Hamburger. In M. Locke (ed.), *The Emergence of Order in Developing Systems*. Academic Press, New York.

Huneman, P. 2006. Naturalising purpose: From comparative anatomy to the "adventure of reason." *Stud. Hist. Phil. Biol. Biomed. Sci.* 37: 649–674.

Huq, A., B. Xu, M. A. Chowdhury, M. S. Islam, R. Montilla and R. R. Colwell. 1996. A simple filtration method to remove plankton-associated Vibrio cholerae in raw water supplies in developing countries. *Appl. Environ. Microbiol.* 62: 2508–2512.

Huxley, T. H. 1854. On the value of the natural history sciences. Van Voorst, Lonon (Reprinted in *Lay Sermons, Addresses, and Reviews*. Macmillan, 1870.)

Huxley, T. H. 1870. On Descartes' "Discourse touching the method of using one's reason rightly and of seeking scientific truth." *Macmillan's Magazine* 24 March 1870; reprinted in T. H. Huxley 1877. *Lay Sermons, Addresses, and Reviews*. Appleton, New York.

Huxley, T. H. 1893. Evolution in biology. In *Darwiniana: Collected Essays* II. Macmillan, London.

Huxley, T. H. 1894. Prolegomena. In T. H. Huxley, 1896. *Evolution and Ethics and Other Essays*. Appleton, New York, pp. 1–45.

James-Todd, T., R. Senie and M. B. Terry. 2012. Racial/ethnic differences in hormonally-active hair product use: a plausible risk factor for health disparities. *J. Immigr. Minor Health.* 14: 506–511.

Janzen, D. H. 1967. Why mountain passes are higher in the tropics. *Am. Nat.* 101: 233–249.

Jenner, R. and M. Wills. 2007. The choice of model organisms in evo-devo. *Nature Rev. Genet.* 8: 311–319.

Jonas, H. 1966. *The Phenomenon of Life: Toward a Philosophical Biology*. University of Chicago Press, Chicago.

Just, E. E. 1939. *The Biology of the Cell Surface*. Blakiston Press, Philadelphia.

Kauffman, S., R. K. Logan, R. Este, R. Goebel, D. Hobill and I. Shmulevich. 2007. Propagating organization: An enquiry. *Biol. Philos.* 23: 27–45.

Keller, E. F. 2002. *Century of the Gene*. Harvard University Press, Cambridge, MA.

King, M. L., Jr. 1963. Letter from Birmingham Jail. May 19, 1963 New York Post Sunday Magazine. abacus.bates.edu/admin/offices/dos/mlk/letter.html

Kirksey, E. (ed.). 2014. *The Multispecies Salon*. Duke University Press, Durham, NC.

Kirschner, M. W. 2005. The meaning of systems biology. *Cell* 121: 503–504.

Kishida, O., Y. Mizuta and K. Nishimura. 2006. Reciprocal phenotypic plasticity in a predator-prey interaction between larval amphibians. *Ecology* 87: 1599–1604.

Kishida, O., Z. Costa, A. Tezuka and H. Michimae. 2013. Inducible offences affect predator-prey interactions and life-history plasticity in both predators and prey. *J. Anim. Ecol.* doi: 10.1111/1365-2656.12186.

Klümper, W. and M. Qain. 2014. A meta-analysis of the impacts of genetically modified crops. *PLoS ONE* 9(11): e111629.

Kopp, A., I. Duncan, D. Godt and S. B. Carroll. 2000. Genetic control and evolution of sexually dimorphic characters in *Drosophila*. *Nature* 408: 553–559.

Kropotkin, P. 1902/1955. *Mutual Aid: A Factor of Evolution*. Porter-Sargent, Boston.

LaBastille, A. 1984. Drastic decline in Guatemala's Giant Piedbilled Grebe population. *Envir. Conserv.* 11: 346–348.

Larson, G. 1999. *There's a Hair in my Dirt! A Worm's Story*. Harper-Collins, NY.

Lathi, R. B., C. A. Liebert, K. F. Brookfield, J. A. Taylor, F. S. vom Saal, V. Y. Fujimoto and V. L. Baker. 2014. Conjugated bisphenol A in maternal serum in relation to miscarriage risk. *Fertil. Steril.* 102: 123–128.

Ledón-Rettig, C and D. W. Pfenning. 2011. Emerging model systems in eco-evo-devo: The environmentally responsive spadefoot toad. *Evol. Dev.* 13: 391–400.

Lee, M. S. Y. 2014. In M. R. Sánchez-Villagara and N. MacLeod (eds.), *Issues in Paleopbiology: A Global View*. Scidinge Hall, Zurich, pp. 143–149.

Lenoir, T. 1982. *The Strategy of Life: Teleology and Mechanics in Nineteenth Century German Biology*. D. Reidel, Boston.

Longo, G., P.-A. Miquel, C. Sonnenschein and A. M. Soto. 2012. Is information a proper observable for biological organization? *Prog. Biophys. Mol. Biol.* 109: 108–114.

Lyson, T. R. and 7 others. 2014. Origin of the unique ventilatory apparatus of turtles. *Nature Commun.* 5: 5211.

Lyson, T. R., G. S. Bever, T. M. Scheyer, A. Y. Hsiang and J. A. Gauthier. 2013. Evolutionary origin of the turtle shell. *Curr. Biol.* 23: 1113–1119.

Margulis, L. 1998. *Symbiotic Planet: A New Look at Evolution*. Basic Books, New York.

Margulis, L. 2011. In D. Teresi 2011. Discover Interview: Lynn Margulis says she's not controversial, she's right. *Discover*, April 2011. discovermagazine.com/2011/apr/16-interview-lynn-margulis-not-controversial-right

Margulis, L. and D. Sagan. 2002. *Acquiring Genomes: A Theory of the Origins of Species*. Basic Books, New York.

Mattick, J. S. 2012. Rocking the foundations of molecular genetics. *Proc. Nat. Acad. Sci. USA* 109: 16400–16401.

Mayr, E. 1988. *Toward a New Philosophy of Biology: Observations of an Evolutionist*. Harvard University Press, Cambridge, MA.

McCain, C. M. 2009. Vertebrate range sizes indicate that mountains may be "higher" in the tropics. *Ecol. Lett.* 12: 550–560.

McFall-Ngai, M. J. 2002. Unseen forces: The influence of bacteria on animal development. *Dev. Biol.* 242: 1–14.

McFall-Ngai, M. and 25 others. 2013. Animals in a bacterial world: A new imperative for the life sciences. *Proc. Nat. Acad. Sci. USA* 110: 3229–3236.

McKibben, B. 1989. *The End of Nature.* Anchor Books, New York.

Mead, K. S. and D. Epel. 1995. Beakers versus breakers: How fertilisation in the laboratory differs from fertilisation in nature. *Zygote* 3: 95–99.

Merleau-Ponty, M. 1962. *Phenomenology of Perception.* Translated by C. Smith. Humanities Press, New York.

Montague, A. 1955. Foreword. In P. Kropotkin's *Mutual Aid: A Factor of Evolution.* Porter-Sargent, Boston.

Moore, D. 2015. *The Developing Genome.* Oxford University Press, Oxford.

Moore, J. 2013. Anthropocene or capitalocene? jasonwmoore. wordpress.com/2013/05/13/anthropocene-or-capitalocene/

Moore, J. W. 2014. The capitalocene/ I. on the nature and origins of our ecological crisis. Available at jasonwmoore.wordpress. com/2013/05/13/anthropocene-or-capitalocene/

Mougi, A. 2012. Unusual predator-prey dynamics under reciprocal phenotypic plasticity. *J. Theor. Biol.* 305: 96–102.

Murti, T. R. V. 1955. The *Central Philosophy of Buddhism.* Allen and Unwin, London.

Nation, A. 2007. Getting started in grass farming. The Stockman Grassfarmer. stockmangrassfarmer.net

National Science Foundation (NSF). 2005. *An Integrated Developmental Biology*: Workshop Report. Booklet 08053.

Nijhout, H. F. 1990 Metaphors and the role of genes in development. *Bioessays* 12: 441–446.

Noble, D. 2006. *The Music of Life.* Oxford University Press, Oxford.

Noble, D. 2008. *The Music of Life: Biology Beyond Genes.* Oxford University Press, Oxford.

Noble, D. 2013. Physiology is rocking the foundations of evolutionary biology. *Exp. Physiol.* 98: 1235–1243.

Nyhart, L. K. 1994. *Biology Takes Form: Animal Morphology and the German Universities, 1800–1900.* University of Chicago Press, Chicago.

Odin, S. 1982. *Process Metaphysics and Hua-Yen Buddhism.* SUNY Press, Albany, NY.

Odling-Smee, F. J., K. N. Laland and M. W. Feldman. 2003. *Niche Construction: The Neglected Process in Evolution.* Princeton University Press, Princeton, NJ.

Oyama, S. 1985. *The Ontogeny of Information.* Cambridge University Press, Cambridge.

Paxson, H. 2014. Microbiopolitics. In Kiksey, E. (op. cit.) pp. 115–121.

Pearson, R. 2011. *Driven to Extinction: The Impact of Climate Change on Biodiversity.* Sterling, NY.

Pleasants, J. M. and K. S. Oberhauser. 2013. Milkweed loss in agricultural fields due to herbicide use: Effect on the Monarch Butterfly population. *Insect Conserv. Divers.* 6: 135–144.

Pollan, M. 2006. *The Omnivore's Dilemma: A Natural History of Four Meals.* Penguin Press, New York.

Relyea, R. A. and N. Mills 2001. Predator-induced stress makes the pesticide carbaryl more deadly to gray treefrog tadpoles (*Hyla versicolor*). *Proc. Natl. Acad. Sci. USA* 98: 2491–2496.

Richmond, J. Y., R. H. Hill, R. S.Weyant, S. L. Nesby-O'Dell and P. E.Vinson. 2003. What's hot in animal biosafety. *ILAR J.* 44: 20–27.

Robert, J. S. 2004. *Embryology, Epigenesis, and Evolution: Taking Development Seriously.* Cambridge University Press, Cambridge.

Root-Bernstein, R. and M. Root-Bernstein. 1999. *Sparks of Genius: The Thirteen Thinking Tools of the World's Most Creative People,* Houghton & Mifflin, New York.

Root-Bernstein, R. and 14 others. 2008. Arts foster scientific success: Comparison of Nobel prizewinners, Royal Society, National Academy, and Sigma Xi members. *J. Psychol. Sci. Technol.* 1: 51–63.

Sagan, D. and L. Margulis. 1986. *Origins of Sex: Three Billion Years of Genetic Recombination.* Yale University Press, New Haven, CT.

Sagan, D. and L. Margulis. 1991. Epilogue: The uncut self. In A. I. Tauber (ed.), *Organism and the Origins of Life.* Kluwer, Dordrecht, pp. 361–364.

Sandrock, C., M. Tanadini, L. G. Tanadini, A. Fauser-Misslin, S. G. Potts and P. Neumann. 2014. Impact of chronic neonicotinoid exposure on honeybee colony performance and queen supersedure. *PLoS ONE* 9(8): e103592.

Satterfield, D. A., J. C. Maerz and S. Altizer. 2015. Loss of migratory behaviour increases infection risk for a butterfly host. *Proc. Biol. Sci.* 282: 20141734.

Schneider, I. and N. H. Shubin. 2013. The origin of the tetrapod limb: From expeditions to enhancers. *Trends Genet.* 29: 419–426.

Schweber, S. S. 1985. The wider British context in Darwin's theorizing. In D. Kohn (ed.), *The Darwinian Heritage.* Princeton University Press, Princeton, NJ.

Seidenberg, D. M. 2015. *Kabbalah and Ecology: God's Image in the More-Than Human World.* Cambridge University Press, Cambridge.

Severtsov, A. N. 1935. *Modes of Phyloembryogenesis.* Quoted in M. B. Adams, 1980. Severtsov and Schmalhausen: Russian morphology and the evolutionary synthesis, In E. Mayr and W. B. Provine (eds.). *The Evolutionary Synthesis: Perspectives on the Unification of Biology.* Harvard University Press, Cambridge, MA, pp. 193–225 (p. 217).

Shapiro, J. A. 2011. *Evolution: A View from the 21st Century.* FT Press, Saddle River, NJ.

Skinner, M. K. 2014. Endocrine disruptor induction of epigenetic transgenerational inheritance of disease. *Mol. Cell. Endocrinol.* 398: 4–12.

Smith, L. 2014. Monarch butterfly's reign threatened by milkweed decline. Use of herbicides in U.S. causing a decline in plants essential for the golden icon. National Geographic. news. nationalgeographic.com/news/2014/08/140819-monarch-butterfly-milkweed-environment-ecology-science/

Spemann, H. 1943. *Forschung und Leben.* Quoted in Spemann and the organizer, 1986. In T. J. Horder, J. A. Witkowski and C. C. Wylie (eds.). *A History of Embryology.* Cambridge University Press, New York.

Spencer, C. N., B. R. McClelland and J. A. Stanford. 1991. Shrimp stocking, salmon collapse and eagle displacement: Cascading interactions in the food web of a large aquatic ecosystem. *Bioscience* 41: 14–21.

Standen, E. M., T. Y. Du and H. C. Larsson. 2014. Developmental plasticity and the origin of tetrapods. *Nature* 513: 54–58.

Stotz, K. 2014. Extended evolutionary psychology: The importance of transgenerational developmental plasticity. *Front. Psychol.* 5: 908.

Strathern, M. 1992. *Reproducing the Future.* Manchester University Press.

Streng, F. J. 1975. Metaphysics, negative dialectic, and the expression of the inexpressible. *Philos. East West* 25: 429–447.

Sultan , S. E. 2011. Evolutionary implications of individual plasticity. In S. B. Gissis and E. Jablonka (eds.), *Transformations of Lamarckism: From Subtle Fluids to Molecular Biology*. MIT Press, Cambridge, MA, pp. 193–204.

Tewksbury, J. J. and 16 others. 2014. Natural history's place in science and society. *BioScience* 64: 300–310.

Tiwary, C. M. and J. A. Ward. 2003. Use of hair products containing hormone or placenta by US military personnel. *J. Pediatr. Endocrinol. Metab.* 16: 1025–1032.

Todes, D. P. 1989. *Darwin without Malthus: The Struggle for Existence in Russian Evolutionary Thought*. Oxford University Press, New York.

Tsing, A. L. 2004. *Friction: An Ethnography of Global Connection*. Princeton University Press, Princeton, NJ.

UNICEF. 2014. *The Challenges of Climate Change: Children on the Front Line*. UNICEF Office of Research, Florence, Italy.

van der Weele, C. 1999. *Images of Development: Environmental Causes in Ontogeny*. SUNY Press, Albany, NY.

Van Valen, L. 1973. Festschrift. *Science* 180: 488.

Verhaal, C. and J. Hoskins. 2014. Rivalry in cooperative environments: Collective organization and the emergence of competitive dynamics in craft-based industries. Strategies in a World of Networks Meeting, Madrid, 2014.

Visscher, P.M., M. A. Brown, M. I. McCarthy and J. Yang. 2012. Five years of GWAS discovery. *Am. J. Hum. Genet.* 90: 7–24.

von Bertalanffy, L. 1968. *General System Theory: Foundations, Development, Applications*. George Braziller, New York.

Waddington, C. H. 1953. Epigenetics and evolution. In R. Brown and J. F. Danielli (eds.). *Evolution* (SEB Symposium VII). Cambridge University Press, Cambridge, pp. 186–199.

Waddington, C. H. 1975. The practical consequences of metaphysical beliefs on a biologist's work: An autobiographical note. In *The Evolution of an Evolutionist*, Cornell University Press, Ithaca, NY.

Wagner, G. P. and M. D. Laubichler. 2004. Rupert Reiedl and the re-synthesis of evolutionary and developmental biology: Body plans and evolvability. *J. Exp. Zool. (MDB)*, 302B: 92–102.

Watanabe, M. and 8 others. 2005. Toward efficient riparian restoration: integrating economic, physical, and biological models. *J. Environ. Management* 75: 93–104.

Weiss, E. B. 1989. In *Fairness to Future Generations: International Law, Common Patrimony, and Intergenerational Equity*. Transnational Publications. Dobbs Ferry, NY.

Weiss, P. A. (ed.). 1971. *Hierarchically Organized Systems in Theory and Practice*. Hafner, New York.

Werneburg, I. and M. R. Sánchez-Villagra. 2014. Skeletal heterochrony is associated with the anatomical specializations of snakes among squamate reptiles. *Evolution* 69: 254–263.

Werskey, G. 1978. *The Visible College: A Collective Biography of British Scientists and Socialists of the 1930s*. Viking Press, New York.

Weston, B. H. 2008. Climate Change and Intergenerational Justice: Foundational Reflections. *Vermont Journal of Environmental Law* 9: 375.

Weston, B. H. 2012. Foundations of intergenerational Ecological Justice: An Overview. *Human Rights Quart.* 34: 251–256.

Whitehead, A. N. 1920. *The Concept of Nature*. Cambridge University Press, Cambridge.

Whitehead, A. N. 1929. *Process and Reality*. Macmillan, New York.

Whitehorn, P. R., S. O'Connor, F. L. Wackers and D. Goulson. 2012. Neonicotinoid pesticide reduces bumble bee colony growth and queen production. *Science* 336: 351–352.

Wilkins, R. J. 2008. Eco-efficient approaches to land management: a case for increased integration of crop and animal production systems. *Philos. Trans. Roy. Soc. Lond. B Biol. Sci.* 363: 517–525.

Wilson, E. B. 1896. *The Cell in Development and Inheritance*. Macmillan, New York.

Winchester, S. 2008. *The Man Who Loved China*. Harper Collins, New York.

Wolpert, L. 1994. Do we understand development? *Science* 266: 571–572.

Xerces Society. 2014a. Partnership with Monarchs. www.xerces.org/monarchs/

Xerces Society. 2014b. Federal Petition. www.xerces.org/after-90-percent-decline-federal-protection-sought-for-monarch-butterfly-2/

Xu, R., A. Boudreau and M. J. Bissell. 2009. Tissue architecture and function: dynamic reciprocity via extra- and intra-cellular matrices. *Cancer Metastasis Rev.* 28: 167–176.

Young, R. M. 1985. *Darwin's Metaphor*. Cambridge University Press, New York.

Zalasiewicz, J. and 20 others. 2008. Are we now living in the Anthropocene? *GSA Today* 18(2): 4–8.

Zhao, H. and Z. J. Chen. 2013. Genetic association studies in female reproduction: from candidate-gene approaches to genome-wide mapping. *Mol. Hum. Reprod.* 19: 644–654.

Lysenko, Kammerer, and the Truncated Tradition of Ecological Developmental Biology

The notions that (1) a particular embryo could develop different phenotypes depending upon the particular environmental circumstances it was in, and (2) that such specific environmentally induced phenotypes could sometimes be inherited from one generation to the next were important parts of late nineteenth-century and early twentieth-century biology. It formed a major part of the biology of central Europe (see Herbst 1894), although it was of lesser importance in Great Britain and the United States.

This Continental synthesis of ecology, development, and evolution was largely demolished in the 1940s. The Nazis dismantled the Prater Vivarium of Vienna, which had been one of its leading research centers, and many of its scientists either died in concentration camps or committed suicide. The German biological enterprise was almost completely wiped out, and many of its major embryologists fled to Canada and the United States. Meanwhile, in the Soviet Union, the concept of phenotypic plasticity became so malignantly hegemonic that its application took it far beyond the realm of scientific practice. The Soviet biological establishment, headed by Trofim Lysenko, considered the phenotype to be totally at the mercy of the environment and asserted that the genes did nothing. The firing, exile, and imprisonment of Soviet geneticists caused a huge reaction in other countries, ultimately leading to the dismissal of environmental causes of phenotype production in most of Western biology.

In the eyes of Western geneticists, two related scandals severely weakened the hypothesis that phenotypic plasticity was an important source of heritable variation: Lysenko's failed attempts to vernalize wheat, and the disastrous outcome of Kammerer's experiments on developmental plasticity in the midwife toad.

Lysenko's Vernalized Wheat

A major reason for the marginalization of phenotypic plasticity in the study of evolution and development is almost certainly the specter of Trofim Denisovich Lysenko. Lysenko was an opportunistic plant breeder whose anti-genetics rhetoric resonated with Josef Stalin's views on the malleability of human nature. Lysenko rose through the ranks of Soviet biology, largely because of the patronage of particular Soviet officials and his ability to obtain the support of influential scientists who thought Lysenko had the potential for greatness. (He would turn on those scientists later.) Lysenko capitalized on his peasant background, castigating the academic scientists as neither knowing nor caring about the practical value of science. How can one study eye color in fruit flies when people are suffering from food shortages? He cast his agricultural argument against genetics as a struggle of true, useful, Soviet science against false, racist, capitalist metaphysics (see Medvedev 1969; Joravsky 1970; Soyfer 1994). From the 1930s to the early 1960s, Lysenko's manifestos declared that genetics was a capitalist product and that the gene was a metaphysical entity with little, if any, role in the production of an organism's traits. Rather, the organism was plastic and the environment determined its specific traits.

In the early 1930s, the Soviet Union had one of the most advanced genetics programs in the world, with luminaries such as Chetverikov, Severtsov, Philipchenko, Karpechenko, Timofeev-Ressovsky, and Vavilov. Lysenko's campaign against these geneticists was waged both scientifically and politically, and his techniques were vicious in both contexts. In 1940, the leading plant geneticist in the Soviet Union, Nikolai Vavilov (one of the scientists who at first promoted Lysenko) was seized on a field trip and charged with sabotaging Soviet agriculture, spying for England, and belonging to a right-wing conspiracy. Sentenced to life imprisonment, Vavilov, a man who was trying to end world hunger, died of starvation in the Gulag in 1943 (see Pringle 2008). Other scientists were dismissed from their posts. Initially, some biologists felt secure enough to combat Lysenko, but in 1948, Lysenko announced at a scientific congress that he had Stalin's full support. He and his followers removed the remaining geneticists from their positions, imprisoning some and causing others to flee (see Medvedev 1969; Krementsov 1998; Pringle 2008).

Lewontin and Levins (1976) have shown that Lysenkoism was nothing less than an aborted cultural revolution and that no analysis of Lysenkoism can be made without understanding Russia's ecology. Nearly all of Russia lies above 40°N (the latitude of Minneapolis), so its climate is more similar to that of Canada than that of the United States. However, Russia has even more extreme weather conditions than Canada, and the temperature gap between summer and winter varies enormously from year to year. The growing season in its chief agricultural belt is short and prone to unpredictable fluctuations in rainfall and frosts.

The big problem for Soviet wheat production was that the crop had to be planted late enough in the spring to avoid chilling frosts, but early enough to get the full benefit of the comparatively short growing season. Rather than following Vavilov's lead in breeding hybrid wheat strains that might be fit for particular environments, Lysenko claimed that he could adapt Russian wheat to the harsh climate through a process called vernalization.

Vernalization is the acquisition of the ability of a plant to flower in the spring by the exposure of the plant to cold conditions. Many plants have natural vernalization, and in some cases, treating a plant to cold condtions can induce early flowering. Vernalization is a phenomenon that is well known to plant biologists, and the cold-sensitive transcription of the *VRN1* gene (which encodes a transcription factor necessary for the flowering of cereal crops) has been well studied (see Chapter 2; Baulcombe and Dean 2014; Deng et al. 2015). Lysenko was far from the first to describe the phenomenon of vernalization, although he did coin the name. Plant physiologists of the 1920s and 1930s were especially interested in how light and temperature could direct plant development in different directions. In particular, there was widespread interest in the notion of "aftereffects," cases in which specific temperature or light regimes focused on seeds or seedlings would cause changes in development weeks or months later (see Roll-Hansen 2005, 2011). For instance, temperature had long been used to cause plants to flower "out of season." Winter cereals need a period of cold in order to flower. They usually are sown in late autumn and flower the next summer. If sown in the spring, they will stay in vegetative growth and not flower until after the winter. But Gassner (1918) had shown that if given a pulse of low temperature immediately after germination in the spring, winter cereals can flower that season. Gassner initiated a research program to find the physiological mechanism by which the norms of reaction (the ability to flower early or late) became directed into one path or the other by temperature.

Lysenko claimed to have done this on a grand scale and for the sake of Soviet agriculture. He said that his results showed that if seeds are chilled and wetted, these "vernalized" seeds can be planted in the spring, rapidly develop, and complete their growth cycle before the winter frosts. In other words, Lysenko claimed that he was able to turn a winter grain into a spring grain through chilling the seed. In a letter of recommendation for Lysenko, Vavilov (1933) praised Lysenko:

> For the first time, with such penetrating profoundness and scope, Lysenko managed to find ways to cope with vegetation control, shift vegetation phases and transform winter crops into spring ones, or late ripening into early. His work is a discovery of primary importance, as it is opening a new sphere for research, and quite an attainable sphere. Undoubtedly Lysenko's work will entail development of the whole branch of plant physiology; this discovery would provide for an opportunity of wide-scale utilization of the world's plant diversity in hybridization for shifting their areas to more remote northern territories. Even the current phase of Lysenko's dicoveries is of paramount interest.

So even though Lysenko later made exaggerated claims for his work, it started off within the realm of science. Similarly, Lysenko's antagonism to genetics was not outside of scientific debate, especially in the Soviet Union. In 1920s and 1930s, many biologists did not think that classical genetics gave natural selection enough variation to work on. Among taxonomists, agronomists, systematists, ecologists, and paleontologists, there was a widespread dissatisfaction with the genetic approach to evolution (Roll-Hansen 2005).

Moreover, in the 1920s, genetics had become increasingly tied to eugenics, the "science" of breeding better people. As a social movement that had a great deal of popular support in the West, eugenics had become increasingly tied to conservative and even fascist politics (Ludmerer 1972; Kevles 1998; Selden 1999; Carlson 2001). Consequently, the value of genetics in the Soviet Union diminished, especially in the botanical community, which did not see the gene as a particularly valid explanation of variation. Indeed, when mass selection experiments were performed on populations of plants (where all plants surviving selection were able to breed rather than only the most outstanding individuals being allowed to mate), they usually failed to select anything. Leading plant ecologist Vladimir Komarov (1944, quoted in Roll-Hansen 2005) noted that botanists saw that a large amount of variability was due to the plants' responses to environmental factors. Such adaptive plasticity, he noted, could be heritable if it was maintained through the plants' germ cells. The germ cells are not as rapidly segregated in plants as they were thought to be in animals, suggesting that traits acquired by the young plant could be transmitted to its progeny.

However, Lysenko took this idea of vernalization and phenotypic plasticity far beyond the bounds of science. His well-publicized experiments were criticized as not having been well controlled. For instance, Konstantinov and colleagues (1936) complained that Lysenko threw out data that did not confirm his ideas, and that he gave vernalized seeds better growing conditions than nonvernalized seeds. Lysenko responded that such criticism betrayed the capitalist notions of its authors. In 1935, Lysenko and I. I. Prezent published a textbook that contradicted genetics in favor of an environmental view wherein the genotype was described as a set of possibilities for development in different directions, dependent upon the conditions of the environment. In defiance of the plant-breeding experiments of Mendel and the work being done with *Drosophila* in Thomas Hunt Morgan's laboratory, this book claimed that the genome did not contain instructive information. When well-known plant breeders pointed out the deficiency of Lysenko's views and the critical nature of genes as shown by careful experiments, Lysenko responded with theoretical comments that often had no basis in fact (such as the notion that early ripening was always dominant over late ripening).

Outside the botanical community, there were many evolutionary geneticists in the Soviet Union who felt that the environment could provide guidance for the phenotype through its interaction with genes. These geneticists tried to find a common dialogue between Lysenko and the rest of the genetics community. However, Lysenko would have no compromise. In the late 1930s, Lysenko and Prezent initiated a campaign in Soviet newspapers, linking geneticists to traitors, and calling their scientific opponents "Trotskyites-Bukharinites." Here, historian Roll-Hansen notes (2005), "The links to eugenics and race theories became fatal for genetics. It got caught up in the campaign against Fascism." In 1939, Lysenko and Prezent succeeded in blocking the election of two eminent biologists (N. K. Koltsov and M. M. Zavadovskii) to the Academy of Science while assuring Lysenko's election. In this campaign, they mentioned the mild support for eugenics that these geneticists had shown decades earlier. In 1940, with the arrest of Vavilov and other geneticists (and with Koltsov dying of a heart attack), Lysenko had consolidated his position as the head of biology in the Soviet Union.

Eventually, when Lysenko was unable to show significant gains in Soviet wheat production and his followers (who were not the brightest of scientific lights) fumbled at meetings, Lysenko's position grew more unstable. Some scientists, including the brilliant geneticist Ivan I. Schmalhausen (see Appendix D), worked to push Lysenko and his followers out and fill their positions with competent scientists; but it wasn't until 1964, when Nikita Khrushchev was ousted from power, that the Lysenkoists began to be removed from their posts.

Kammerer's Midwife Toads

The idea that evolution involves the inheritance of acquired characteristics was formulated in the early 1800s by Jean Baptiste Lamarck (see Appendix D), and the early part of the twentieth century was "the Golden Age of Lamarckianism" (Gliboff 2011). Before that time, genetics had not risen to prominence, and the motor for evolution was seen to be development. Darwin used "Lamarckian" notions of inheritance, and the energetic particles and transformative fluids of Lamarck were recast as Darwin's "gemmules." Darwin and Lamarck were seen as allies. Eventually, genetics would separate them, with the growing acceptance of Mendel's particulate theory of inheritance (Berrill and Liu 1948; Gilbert 1998). Morgan (1932) was adamant that the best explanation of evolution was genetics. The last bastion of phenotypic plasticity research in Western biology was probably the Prater Vivarium in Vienna. In the 1920s, one of its biologists, Paul Kammerer, attempted to synthesize Darwinism and chromosome theory while highlighting (1) phenotypic plasticity as a major source of variation and (2) the inheritance of acquired characteristics as an experimentally demonstrable effect. He claimed that animal form could be modified by environmental influences, and that these variations were passed on from generation to generation. According the Kammerer, this is how striped salamanders became spotted and blind cave salamanders developed functional eyes.

Kammerer's most famous (or infamous) experiment involved changing the morphology and behaviors of midwife toads (genus *Alytes*). These toads normally mate on land, and the eggs become attached to the hindlimbs of the male. Kammerer kept the toads in water and induced them to mate there (where the eggs floated away). From the few eggs that survived, Kammerer claimed to have bred a line of midwife toads that habitually bred in the water. Moreover, these toads developed "nuptial pads" on the male forelimbs. These were rough dark patches of skin that enabled the now aquatic male to grab his slippery female partner. A scandal broke in 1926 when one of the last remaining water-habituated midwife toads was demonstrated to have had India ink injected where the nuptial pads were supposed to be. Kammerer denied faking the nuptial pad, but a few weeks later he committed suicide. Numerous scientists had seen the toads and attested to the presence of nuptial pads, but G. K. Noble and William Bateson doubted their existence.

There have been numerous attempts to explain this work and the scandal. Arthur Koestler (1971) suggests that an enemy of Kammerer (and he had several) framed him or that a well-meaning laboratory assistant touched up a deteriorating specimen. He claims that Kammerer's suicide was not due to his despondency over a supposed fraud, but to his failure to win the beautiful Alma Mahler as his mistress. More recently, Sander Gliboff (2006) provided evidence that Kammerer had a difficult time photographing his slimy subjects, and that he may well have inked the nuptial pad to make it more visible on film:

> I agree with Iltis that the pad was not fabricated. Kammerer worked with highly plastic organisms that probably did exhibit the traits he reported, and there was more evidence for the existence of the nuptial pad than just the doctored specimen. However, I suspect Kammerer of injecting the specimen anyway, only to enhance, not create, its appearance.

Whatever the reason, the discovery that Kammerer's celebrated toads had inked forelimbs, along with Kammerer's suicide coming so soon afterward, created a sense of scientific fraud that severely damaged the notion of developmental plasticity.* The toad's case was one of the few publicly known examples of non-Mendelian inheritance, and with its demise another objection to the rule of genetics was demolished.

The Lysenkoist revolution and Kammerer's toad had a major effect on Western biology. First, they were seen as object lessons: this is what happens when you let science be subservient to ideology and politics. (Kammerer was an outspoken Socialist who was thinking about moving his laboratory to Moscow; see Goldschmidt 1949 for a view of these episodes as warnings to the West.) This was a somewhat ironic twist, since the generation of geneticists in America and Great Britain in the 1940s and 1950s were trying to free Western genetics from its own political involvement with the eugenics and anti-immigration movements. Second, the reaction against Lysenkoism led to a hardening of a genetic position that did not include phenotypic plasticity. The American Society for Genetics hired a public relations firm and began a Golden Jubilee Program for 1950, to celebrate the 50 years of genetics since the rediscovery of Mendel's paper (and enshrine Mendel as the founder of the new science; see Sapp 1987; Wolfe 2002; Gormley 2007). But in placing itself in opposition to Lysenkoism, Western genetics had also placed itself in opposition to phenotypic plasticity. As historian Jan Sapp (1987) has documented, the West also became ideologically committed. The biologists' anti-Communism became manifest in anti-plasticity.

Biology in the Context of Ideological Struggle

Biology, as constructed after World War II, has seen the dominance of American traditions. Much of what we now consider to be the basic structure

---

*Most of Kammerer's experiments, including the nuptial pad investigations, have never been re-attempted. However, embryologist J. R. Whittaker (1985) was unable to reproduce Kammerer's results when he redid the experiments on lengthening in tunicate siphons.

of the biological sciences (and indeed the basic structure of the natural world) was formulated in the context of the Cold War. Nowhere is this more clearly seen than in the formulation of contemporary genetics in the context of Lysenkoism. Genetics, in the 1930s, was America's and Britain's contribution to biology. Moreover, the geneticists provided a mechanism for inheritance that was seen as being as independent of development and also independent of the environment. The genotype directed the phenotype, and each character was seen as individually inherited (see Appendix C). As Lewontin and Levins (1976) have pointed out, the reduction of character to gene and the atomicity (separateness) of each trait was consistent with the Anglo-American notions of individuality and analytic philosophy.

Another reason for the marginalization of developmental plasticity was that it did not fit into the paradigms of evolutionary biology or developmental biology. Evolutionary biology was defined by variation that was inherited through genes. Environmentally induced variation was not considered heritable, and those schemes that proposed models for the inheritance of environmentally induced variation (such as Waddington's genetic assimilation or Schmalhausen's stabilizing selection) were looked upon as either bad science, politically motivated science, or a combination of both (see Appendix D).

One of the few Western evolutionary biologists who did refer to norms of reaction was Theodosius Dobzhansky. He came from the Russian school of evolution, but he stayed in the United States while Lysenkoism eclipsed Soviet evolutionary genetics. In 1926 (while still in Russia), he equated the reaction norm with the genotype (Sarkar 1999), and his 1937 book, *Genetics and the Origin of Species*, reintroduced the idea of the reaction norm to Western science. However, rather than seeing it as a developmental phenomenon, Dobzhansky (1955; Dobzhansky and Spassky 1963) redefined the norm of reaction to refer to the adaptive plasticity of the entire population's collection of genomes. As Sarkar notes, "Dobzhansky, unlike Schmalhausen, and like a true geneticist from that period, generally ignored embryology." Thus the notion of developmental plasticity in individuals became lost from genetics and evolutionary biology.

It also became lost from developmental biology. As developmental biology became molecularized, this branch of science became more dependent on model systems—those organisms from which a large community of scientists could obtain detailed information. In the field of animal developmental biology spanning the 1960s to the present, research on *Xenopus laevis*, *Drosophila melanogaster*, *Caenorhabditis elegans*, and *Mus musculus* has been ascendant. As Bolker (1995) pointed out, these model organisms converged on a particular suite of phenotypes that allowed them to be used for the genetic analysis of development: small size, fast growth to sexual maturity, large litter size, and a "rapid, highly canalized development" that reduces the effects of environment on phenotype production. "These biases," she notes, "influence both data collection and interpretation, and our views of how development works and which aspects of it are important." Indeed, they led to an erroneous perception that the organism, in the words of Jacques Monod (1971, p. 87), "came into being spontaneously and autonomously, without outside help and without the injection of additional information."

The Return of Plasticity

The return of developmental plasticity into discussions of evolution, development, and health has several sources. First, of course, it was never totally absent from discussions. Scientists such as C. H. Waddington and Richard Lewontin (one of Dobzhansky's students) kept alive discussions of the possible roles of plasticity in evolution. The controversy over IQ testing made plasticity a fundamental concern for Lewontin (1972, 1976). When some social and physical scientists (Arthur Jensen and William Shockley being the most prominent) wrote popular books claiming that intelligence was a genetically fixed trait that was distributed unequally among races, Lewontin used the concept of developmental plasticity to show that genetics and environment interact to make phenotypes. (Indeed, this argument had been used previously by Lancelot Hogben to counter eugenic proposals claiming that intelligence was hereditary.) Lewontin continued to bring this line of reasoning into evolutionary theory and to popularize it as well (Lewontin 1993, 2001, 2002).

Lewontin was also important in keeping alive a reaction against the view that genetics (and the Human Genome Project in particular) could explain all the important parts of one's phenotype. In the latter part of the twentieth-century, genetic reductionism—the view that the complete phenotype of an organism can be traced back solely to the genome—became part of the rhetoric of evolutionary biology and later, the Human Genome Project (see Burian 1982; Keller 1992; Nelkin and Lindee 1995). Dawkins (1979) extended this rhetoric to the conclusion that the genome was the book of life and that our bodies are merely the transient vehicles for the survival and propagation of our immortal DNA. This "genocentric" view of life came under fire from several biologists and philosophers (see Malacinski 1990). Some critics were scientists who were actively pursuing research into phenotypic plasticity or epigenesis. David Nanney (1957, 1985) viewed the cytoplasm as critical in heredity and saw in the rhetoric of DNA the image of a totalitarian state. H. Frederik Nijhout (1990a,b) and Richard Strathmann (1985) criticized the genocentric view of inheritance from their studies of the hormonal control of larval development. Brian Goodwin (1982) and Stuart Newman (1979, 1990), developmental biologists concerned with organizational patterning, also formulated critiques of genetic reductionism. In 1983, Newman co-founded the Council for Responsible Genetics, a major source of information against genetic reductionism.

Other critiques of genetic reductionism and determinism came from feminist scientists who saw that the genes-are-destiny rhetoric was being used against women. Donna Haraway (1979), Evelyn Fox Keller (1985), Anne Fausto-Sterling (1985), and Bonnie Spanier (1995), used their scientific training to criticize the genetic determinism in the field. The Biology and Gender Study Group (1988) was likewise constituted of scientists and science students. Philosophers of biology (e.g., Tauber and Sarkar 1992) and developmental systems theorists starting with Gilbert Gottlieb (1971, 1992) and Susan Oyama (1985) made the critique of genetic determinism a central part of their analyses. This provided another cause for the rebirth of interest in phenotypic plasticity. By 1999, developmentally plastic phenomena (seasonal polyphenisms, temperature-sensitive and context-dependent sex determination, nutritional polyphenisms, predator-induced polyphenisms, and neural plasticity) were linked to criticisms of genetic determinism to provide a positive alternative to the genocentric approach (see Gilbert 2001, 2002).

Another positive alternative came from newly discovered epigenetic phenomena. One major contribution to the re-emergence of plasticity was a set of studies in mammalian molecular biology that demonstrated the contextual nature of genetics. In particular, the discovery of chromosome imprinting demonstrated that developmental as well as genetic parameters determined whether a gene functioned (Barton et al. 1984; McGrath and Solter 1984; DeChiara et al. 1991). Indeed, as early as 1991, a "Human Epigenome Project" was mooted to balance the genetic deterministic rhetoric of the Human Genome Project (Gilbert 1991a; Bradbury 2003).

Conservation biology provided another goad for studying plasticity. Conservation programs require information not just about adult survival in limited or compromised environments, but also about the development and vulnerabilities of embryos and larvae. Such data can have profound practical implications. For example, Morreale and coworkers (1982) revealed that sea turtle conservation programs had failed to take into account the temperature-dependent mechanism of turtle sex determination and as a result were releasing thousands of hatchling turtles—all of the same sex. In addition, Mead and Epel (1995) showed that fertilization phenomena studied in the laboratory may not be adequate descriptions of what occurs in nature. Indeed, few criticisms of developmental biology are as strong as "It doesn't tell us about how animals develop in nature." To correct this, we must forge an ecological developmental biology. A related reason for the increase in interest in the ecological aspects of development has been the increasing recognition of (and alarm about) environmental contamination by anthropogenic teratogens. Environmental chemicals that we once assumed were harmless (at least to adults) may be dangerous to developing organisms and may also threaten adult fertility. Some recent studies have been popularized by the media (Colburn et al. 1996; Hayes et al. 2002), and several others (Relyea and Mills 2001) demonstrate the need to study these compounds in the "real world" as opposed to just the laboratory.

For these reasons, and perhaps other reasons as well, phenotypic plasticity is coming back to play important roles in developmental biology, ecology, genetics, and evolutionary developmental biology (evo-devo). Since 2000, numerous books concerning plasticity have been published, and Massimo Pigliucci's *Phenotypic Plasticity* (2001) and Mary Jane West-Eberhard's *Developmental Plasticity and Evolution* (2003) brought developmental plasticity into the heart of evolutionary theory. The first symposium in ecological developmental biology (Dusheck 2002; Gilbert and Bolker 2003) was conducted at the Society for Integrative and Comparative Biology in January 2002.

References

Barton, S. C., M. A. Surani and M. L. Norris. 1984. Role of paternal and maternal genomes in mouse development. *Nature* 311: 374–376.

Baulcombe, D. C. and C. Dean. 2014. Epigenetic regulation in plant responses to the environment. *Cold Spring Harb. Perspect. Biol.* 6(9): a019471.

Berrill, N. J. and C. K. Liu. 1948. Germplasm, Weismann, and hydrozoa. *Quart. Rev. Biol.* 23: 124–132.

Biology and Gender Study Group. 1988. The importance of feminist critique for contemporary cell biology. *Hypatia* 3: 172–186.

Bolker, J. 1995. Model systems in developmental biology. *Bioessays* 17: 451–455.

Bradbury, J. 2003. Human Epigenome Project—Up and running. *PLoS Biol.* 1: e82.

Bradshaw, A. D. 1965. Evolutionary significance of phenotypic plasticity in plants. *Adv. Genet.* 13: 115–155.

Burian, R. 1982. Human sociobiology and genetic determinism. *Philos. Forum* 13: 40–66.

Carlson, E. A. 2001. *The Unfit: A History of a Bad Idea.* Cold Spring Harbor Press, Cold Spring Harbor, NY.

Collins, J. P., S. F. Gilbert, M. D. Laubichler and G. B. Müller. 2007. How to integrate development, evolution, and ecology: modeling in evo-devo. In M. D. Laubichler (ed.), *Roots of Theoretical Biology: The Prater Vivarium Centenary.* MIT Press, Cambridge, MA, pp. 355–378.

Colburn, T., D. Dumanoski and J. P. Myers. 1996. *Our Stolen Future.* Dutton, New York.

Dawkins, R. 1976. *The Selfish Gene.* Oxford University Press, New York.

DeChiara, T. M., E. J. Robertson and A. Efstratiadis. 1991. Parental imprinting of the mouse insulin-like growth factor II gene. *Cell* 64: 849–859.

Deng, W., M. C. Casao, P. Wang, K. Sato, P. M. Hayes, E. J. Finnegan and B. Trevaskis. 2015. Direct links between the vernalization response and other key traits of cereal crops. *Nature Commun.* 6: 5882.

Dobzhansky, Th. 1955. *Evolution, Genetics, and Man.* John Wiley, New York.

Dobzhansky, Th. and B. Spassky. 1963. Genetics of natural populations. XXXIV. Adaptive norm, genetic load, and genetic elite in *Drosophila pseuodoobscura. Genetics* 48: 1467–1485.

Dusheck, J. 2002. It's the ecology, stupid. *Nature* 418: 578–579.

Fausto-Sterling, A. 1985. *Myths of Gender: Biological Theories about Men and Women.* Basic Books, New York.

Gassner, G. 1918. Beiträge zur physiologischen Charakterisik sommer-und winterannueller Gewächse, insbesondere der Getreidepflanzen. *Zeit. Botan.* 10: 417–480.

Gilbert, S. F. 1991a. Cytoplasmic action in development. *Q. Rev. Biol.* 66: 309–316.

Gilbert, S. F. 1991b. Induction and the origins of developmental genetics. In S. F. Gilbert (ed.), *A Conceptual History of Modern Embryology.* Plenum, New York, pp. 181–206.

Gilbert, S. F. 1998. Bearing crosses: The historiography of genetics and embryology. *Amer. J. Med. Genet.* 76: 168–182.

Gilbert, S. F. 2000. Diachronic biology meets evo-devo: C. H. Waddington's approach to evolutionary developmental biology. *Am. Zool.* 40: 729–737.

Gilbert, S. F. 2001. Ecological developmental biology: Developmental biology meets the real world. *Dev. Biol.* 233: 1–12.

Gilbert, S. F. 2002. Genetic determinism: The battle between scientific data and social image in contemporary developmental biology. In A. Grunwald, M. Gutmann and E. M. Neumann-Held (eds.), *On Human Nature: Anthropological, Biological, and Philosophical Foundations.* Springer-Verlag, New York, pp. 121–140.

Gilbert, S. F. and J. Bolker. 2003. Ecological developmental biology: Preface to the symposium. *Evol. Dev.* 5: 3–8.

Gliboff, S. 2006. The case of Paul Kammerer: Evolution and experimentation in the early twentieth century. *J. Hist. Biol.* 39: 525–563.

Gliboff, S. 2011. The golden age of Lamarckism: 1866–1926. In S. Gissis and E. Jablonka (eds.), *Transformations of Lamrckianism.* MIT Press, Cambridge, pp. 45–55.

Goldschmidt, R. B. 1949. Research and politics. *Science* 109: 219–227.

Goodwin, B. C. 1982. Genetic epistemology and constructionist biology. *Rev. Int. Philo.* 36: 527–548.

Gormley, M. 2007. L. C. Dunn and the reception of Lysenkoism in the United States. Ph.D. thesis, Oregon State University.

Gottlieb, G. 1971. *Development of Species Identification in Birds: An Inquiry into the Prenatal Determinants of Perception.* University of Chicago Press, Chicago.

Hall, B. K. 1992a. Waddington's legacy in development and evolution. *Amer. Zool.* 32: 113–122.

Hall, B. K. 1992b. *Evolutionary Developmental Biology.* Chapman and Hall, New York.

Haraway, D. J. 1979. The biological enterprise: Sex, mind, and profit from human engineering to sociobiology. *Rad. Hist. Rev.* 20: 206–237.

Hayes, T. B., A. Collins, M. Lee, M. Mendoza, N. Noriega, A. A. Stuart and A. Vonk. 2002. Hermaphroditic, demasculinized frogs after exposure to the herbicide atrazine at low ecologically relevant doses. *Proc. Natl. Acad. Sci. USA* 99: 5476–5480.

Hertwig, O. 1894. *The Biological Problem of Today: Preformationism or Epigenesis ?* (P.C. Mitchell, transl.) Macmillan, New York.

Joravsky, D. 1970. *The Lysenko Affair.* Harvard University Press, Cambridge, MA.

Keller, E. F. 1985. *Reflections on Gender and Science.* Yale University Press, New Haven.

Kevles, D. 1998. *In the Name of Eugenics.* Harvard University Press, Cambridge, MA.

Koestler, A 1971. *The Case of the Midwife Toad.* Random House, New York.

Konstantinov, P. N., P. I. Lisitsyn and D. Kostov. 1936. Neskol'koslov o rabotakh odesskogo instituta seleksii I genetikii. *SRSKh* 11: 121–130.

Krementsov, N. 1998. *Stalinist Science.* Princeton University Press, Princeton, NJ.

Lewontin, R. C. 1972. The apportionment of human diversity, *Evolutionary Biology* 6: 381–398.

Lewontin, R. C. 1976. The fallacy of biological determinism. *The Sciences* 16: 6–10.

Lewontin, R. C. 1993. *Biology as Ideology: The Doctrine of DNA.* Harper, New York.

Lewontin, R. C. 2001. *It Ain't Necessarily So.* Granta Press, Cambridge University Press, Cambridge.

Lewontin, R. C. 2002. *The Triple Helix: Gene, Organism, and Environment.* Harvard University Press, Cambridge, MA.

Lewontin, R. C. and R. Levins 1976. The problem of Lysenkoism. In H. Rose and S. Rose (eds.), *The Radicalisation of Science. Ideology of/in the Natural Sciences*. Macmillan, London, pp. 32–64.

Ludmerer, K. M. 1972. *Genetics and American Society*. Johns Hopkins University Press, Baltimore.

Lysenko, T. D. and I. I. Prezent. 1935. *Seleksiia i Teoriia Stadiinogo Razvitiia Rasteniia*. Sel'khozgiz, Moscow.

Malacinski, G. 1990. *Cytoplasmic Organization Systems*. McGraw Hill, New York.

McGrath, J. and D. Solter. 1984. Completion of mouse embryogenesis requires both the maternal and paternal genomes. *Cell* 37: 179–183.

Mead, K. S. and D. Epel. 1995. Beakers versus breakers: How fertilization in the laboratory differs from fertilization in nature. *Zygote* 3: 95–99.

Medvedev, Z. 1969. *The Rise and Fall of T. D. Lysenko*. Columbia University Press, New York.

Monod, J. 1971. *Chance and Necessity*. Knopf, New York.

Morreale, S. J., G. J. Ruiz, J. R. Spotila and E. A. Standora. 1982. Temperature-dependent sex determination: Current practices threaten conservation of sea turtles. *Science* 216: 1245–1247.

Morgan, T. H. 1932. *The Scientific Basis of Evolution*. W. W. Norton, NY.

Nanney, D. L. 1957. The role of the cytoplasm in heredity. In W. D. McElroy and B. Glass (eds.), *The Chemical Basis of Heredity: A Symposium*. Johns Hopkins University Press, Baltimore, pp. 134–166.

Nanney, D. L. 1985. Heredity without genes: Ciliate explorations of clonal heredity. *Trends Genet.* 1: 295–298.

Nelkin, D. and M. S. Lindee. 1995. *The DNA Mystique: The Gene as a Cultural Icon*. W. H. Freeman, New York.

Newman, S. A. and W. D. Comper. 1990. "Generic" physical mechanisms of morphogenesis and pattern formation. *Development* 110: 1–18.

Newman, S. A. and H. L. Frisch. 1979. Dynamics of skeletal pattern formation in developing chick limb. *Science* 205: 662–668.

Nijhout, H. F. 1990a. Metaphors and the role of genes in development. *BioEssays* 12: 441–446.

Nijhout, H. F. 1990b. A comprehensive model for color pattern formation in butterflies. *Proc. Roy. Soc. B* 239: 81–113.

Nyhart, L. 1995. *Biology Takes Form*. University of Chicago Press, Chicago.

Oyama S. 1985. *The Ontogeny of Information: Developmental Systems and Evolution*. Duke University Press, Durham, NC.

Pigliucci, M. 2001. *Phenotypic Plasticity: Beyond Nature and Nurture*. Johns Hopkins University Press, Baltimore.

Pringle, P. 2008. *The Murder of Nikolai Vavilov*. Simon and Schuster, New York.

Relyea, R. A. and N. Mills. 2001. Predator-induced stress makes the pesticide carbaryl more deadly to grey treefrog tadpoles (*Hyla versicolor*). *Proc. Natl. Acad. Sci. USA* 98: 2491–2496.

Roll-Hansen, N. 2005. *The Lysenko Effect: The Politics of Science*. Humanity Books, Amherst, NY.

Roll-Hansen, N. 2011. Lamarck and Lysenkoism revisited. In S. Gissis and E. Jablonka (eds.), *Transformations of Lamrckianism*. MIT Press, Cambridge, pp. 77–88.

Sapp, J. 1987. *Beyond the Gene: Cytoplasmic Inheritance and the Struggle for Authority in Genetics*. Oxford University Press, Oxford.

Sarkar, S. 1999. From the *Reaktionsnorm* to the Adaptive Norm: The norm of reaction, 1906–1960. *Biol. Philos.* 14: 235–252.

Selden, S. 1999. *Inheriting Shame: The Story of Eugenics and Racism in America*. Teachers College Press, New York.

Soyfer, V. 1994. *Lysenko and the Tragedy of Soviet Science*. Transl. L. Gruliow and R. Gruliow. Rutgers University Press, New Brunswick, NJ.

Spanier, B. 1995. *Im/Partial Science: Gender Ideology in Molecular Biology*. Indiana University Press, Bloomington, IN.

Strathmann, R. R. 1985. Feeding and non-feeding larval development and life history evolution in marine invertebrates. *Annu. Rev. Ecol. Syst.* 16: 339–361.

Tauber, A. I. and S. Sarkar. 1992. The human genome project: Has blind reductionism gone too far? *Perspect. Biol. Med.* 35: 220–235.

Vavilov, N. K. 1933. Quoted in Ni. I. Vavilov Institute of Plant Industry. Lysenko's role in the development of agricultural science in the USSR. www.vir.nw.ru/history/lysenko.htm

Waddington, C. H. 1968. The basic ideas of biology. In C. H. Waddington (ed.), *Towards a Theoretical Biology*. Edinburgh University Press, Edinburgh, pp. 1–32.

Weismann A. 1875. Über den Saison-Dimorphismus der Schmetterlinge. In *Studien zur Descendenz-Theorie*. Engelmann, Leipzig.

West-Eberhard, M. J. 2003. *Developmental Plasticity and Evolution*. Oxford University Press, Oxford.

Whittaker, J. R. 1985. Paul Kammerer and the suspect siphons. *MBL Science*; www.mbl.edu/publications/pub_archive/Ciona/Kammerer/index.html

Wolfe, A. J. 2002. Speaking for Nature and Nation: Biologists as public intellectuals in cold war culture. Ph.D. thesis, University of Pennsylvania.

Woltereck, R. 1909. Weitere experimentelle Untersuchungen über Artveränderung, speziel über das quantitativer Artunterschiede bei Daphnien. *Verhandlungen der Deutschen zoologischen Gessellschaft* 19: 110–173.

The Molecular Mechanisms of Epigenetic Change

Epigenetics has been redefined many times over the past half century. Originally, C. H. Waddington coined the term to refer to what today we would call developmental genetics—how the genotype gives rise to the phenotype through developmental interactions. Waddington modified the term "epigenesis" (development) by melding it with the word "genetics." However, the "epi" in "epigenesis" could also mean "that which is above" (as in *epi*dermis or *epi*glottis). So Robin Holliday (1990) defined epigenetics as "the study of the mechanisms of temporal and spatial control of gene activity during the development of complex organisms." Thus "epigenetic" can be used to describe anything other than DNA sequence that influences the development of an organism. In recent years, epigenetics has sometimes been used to mean that which is "above" the genetic sequences—in other words, the "formatting" of the genome.

As Eccleston and colleagues (2007) wrote, "Epigenetics is typically defined as the study of heritable changes in gene expression that are not due to changes in DNA sequence." However, this definition depends on what one means by "heritable." If one is thinking in terms of cells, it means that the gene expression pattern of one cell is transmitted accurately to that cell's descendants. This is important in the stability of differentiated states. We don't want our skin cells turning into gut cells while we sleep. But, as this book demonstrates (see Chapters 2, 6, and 11), we are coming to realize how agents causing epigenetic changes in germ cells can create stable patterns of gene expression that become transmitted from an organism to its progeny.

Understanding epigenetics requires that we understand how the chromatin becomes altered to activate or suppress transcription (gene expression). We also have to understand how these alterations in chromatin structure become stabilized. That is to say, how is each gene region given a "transcriptional memory" that it can rely on after replication so that alterations in the chromatin are repeated and gene expression patterns are stable? Although we lack anything near a complete understanding of these events, we do know that both of these involve modifications of the nucleosomes.

Three integrated systems may be involved in stabilizing chromatin conformations between cell generations. It is possible that each progresses to the same end from different starting points. One system starts with methylation of cytosine nucleotides in the organism's DNA. The other two start with nucleosome alterations (histone modification and substitution).

DNA Methylation

In many animals (including mammals, but not usually in flies), the nucleotide sequence CG has the potential to become methylated (i.e., to have methyl groups attached) at its C (cytosine) residue. (In plants, methylation is often on a CNG sequence cytosine.) But how is the methyl group restored each time a new DNA strand is replicated during mitosis? The answer lies in the fact that the other C of the opposing CG pair would be methylated, and a particular enzyme (DNA methyltransferase-1, or Dnmt1) recognizes hemimethylated (half-methylated) sites and "completes" them so that each methyl-CG has a corresponding GC-methyl group. Dnmt1 does not recognize unmethylated Cs. In that way, the pattern of methylated cytosines is repeated from cell generation to generation.

This copying of methylated sites is far from exact. While DNA replication is so accurate (due to the editing subunit of DNA polymerase) that only 1 base in 10^8 is copied incorrectly, the error rate for DNA methylation is about 1 in every 25 cytosines. This means that methylated sites are often bunched close together, so the effect of errors is reduced. Even so, random errors in methylation can create changes in gene expression within clonal populations of cells (Laird et al. 2004; Bird 2007).

The importance of methylation in stably inhibiting transcription was originally shown in two mammalian conditions in which genes are silenced during development. First, in mammalian dosage compensation, one X chromosome is "turned off" in every female (XX) cell (which contains two X chromosomes). Once this is accomplished, that same X chromosome—either the one from the father or the one from the mother—remains inactive in all the progeny of that cell (see Brockdorff and Turner 2007; Migeon 2007). This is why calico cats, in which a major pigmentation gene is located on the X chromosome, have separate areas of black and orange pigmentation; which color is expressed depends on which X chromosome was inactivated in the progenitor pigment cell. It also explains why calico cats are always females (or sterile XXY males). The maintenance (but not the origin) of this pattern depends on DNA methylation of clustered CG pairs (Venolia et al. 1982; Wolf et al. 1984).

The second phenomenon explained by DNA methylation is genomic imprinting, where the chromosomes from the male and the female are not equivalent. At about 100 known loci, the genes contributed by the sperm differ in activity from the genes contributed by the egg. In some of these cases, the nonfunctioning gene has been rendered inactive by DNA methylation.* As mentioned before, methylated DNA is associated with stable

*This means that mammals must have both a male and a female parent. Unlike in sea urchins, flies, or frogs, parthenogenesis—virgin birth—cannot happen naturally in mammals.

DNA silencing, either (1) by interfering with the binding of gene-activating transcription factors or (2) by recruiting repressor proteins that stabilize nucleosomes in a restrictive manner along the gene. The presence of a methyl group can prevent transcription factors from binding to the DNA, thereby preventing the gene from being activated. For instance, during early embryonic development in mice, the *Igf2* (insulin-like growth factor-2) gene is active only from the father's chromosome 7; the egg-derived *Igf2* gene does not function. This is because in the female-derived chromosome 7, the CTCF protein—an inhibitor that can block a gene's promoter region from receiving activation signals from enhancers—binds to a region near the maternal *Igf2* gene. Once bound, it prevents the maternally derived *Igf2* gene from functioning. But it is only in the female-derived homologue that this binding region is unmethylated and available. In the sperm-derived chromosome 7, the region where CTCF would bind is methylated. CTCF cannot bind and thus the gene is not inhibited and expresses a functional protein (Figure 1; Bartolomei et al. 1993; Ferguson-Smith et al. 1993; Bell and Felsenfeld 2000).

The second mechanism is the converse of the first. Instead of impeding proteins from binding, the methyl-CG actually is bound by specific repressor proteins. Proteins such as MeCP2, MBD1, and Kaiso have a

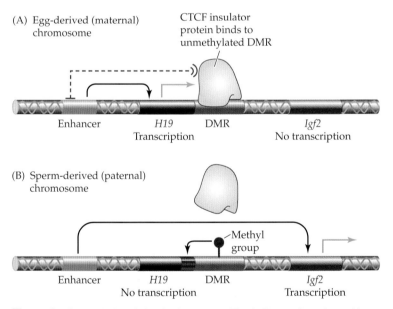

Figure 1   Genomic imprinting in the mouse. The *Igf2* gene is activated by an enhancer element it shares with the *H19* gene. A differentially methylated region (DMR) lies between the enhancer and the *Igf2* gene, adjacent to the promoter region of the *H19* gene. (A) In the egg-derived chromosome, the DMR is unmethylated and binds the CTCF insulator protein, thus blocking the enhancer signal and preventing *Igf2* expression. (B) In the sperm-derived chromosome, the DMR is methylated, with two consequences. First, the CTCF insulator protein cannot bind to the DMR and the enhancer can activate *Igf2*. Second, the methylated DMR alters the chromatin of the *H19* promoter, preventing its activation by the enhancer. Therefore, the maternal chromosome expresses *H19* but not *Igf2*, whereas the paternally derived chromosome expresses *Igf2* but not *H19*.

methyl-CpG-binding domain (MBD) that recognizes methyl-CpG sequences on the DNA (Bird and Wolffe 1999). When these proteins bind to DNA, they can associate with histone deacetylases and histone methyltransferases that stabilize nucleosomes and prevent transcription (Jones et al. 1998; Nan et al. 1998; Sarraf and Stancheva 2004). Methylated DNA may also bind (probably indirectly) histone H1, the "linker" histone subunit found between the nucleosomes, thus providing another mechanism through which the nucleosomes become tightly packed on methylated DNA (Rupp and Becker 2005).

Histone Modification and Histone Substitution

Histone modification is a way of getting nucleosomes into a conformation that will either permit or block RNA polymerase from either initiating or elongating messenger RNA at a particular gene. Some modifications appear to be critical for initiating transcription—that is, for getting RNA polymerase poised on the promoter (see Figure 2.1). These modifications often include a trimethylated lysine on the fourth position of histone H3 (H3K4me3) and the acetylation of histone H3 at lysines on positions 9 and 14. Other histone modifications, such as H3K36me3 (translated as "trimethylated lysine at position 36 of histone H3") appear to be necessary within the gene to allow the RNA polymerase to pass through. Both modifications might be required to achieve transcription of the specific gene (Guenther et al. 2007). Figure 2 shows a nucleosome and its H3 tail, which contains certain amino acids whose modification can regulate transcription.

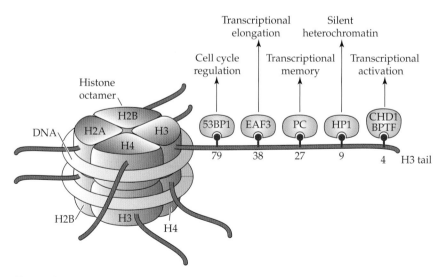

Figure 2   Methylation of histone H3. The tail of histone H3 sticks out from the nucleosome and is capable of being methylated. In this diagram, methylated lysine residues are recognized by particular proteins. The methylated residues at positions 4, 38, and 79 are associated with gene activation, whereas methylated lysines at positions 9 and 27 are associated with repression. The specific proteins binding these sites, represented above the methyl groups, are not shown to scale. (After Kouzarides and Berger 2007.)

Many transcription factors work by binding one site to the DNA and using another site on the protein to recruit histone-modifying enzymes. For instance, the Pax7 transcription factor that activates muscle-specific genes binds to the enhancer region of these genes within the muscle precursor cells and recruits a histone methyltransferase to that region. The histone methyltransferase methylates the lysine in the fourth position of histone H3 (H3K4), resulting in trimethylation of this lysine and the activation of gene transcription (McKinnell et al. 2008). The repair and displacement of nucleosomes along the DNA make it possible for the transcription factors to find their binding sites (Adkins et al. 2004; Li et al. 2007).

Protein recruitment

Histone modifications can also signal the recruitment of the proteins that can retain the memory of transcriptional state from generation to generation through mitosis. These are the proteins of the Trithorax and Polycomb families. Trithorax proteins, when bound to nucleosomes of active genes, keep these genes active; Polycomb proteins, which bind to condensed nucleosomes, keep the genes in an inactive state.

The Polycomb proteins fall into two categories that act sequentially in repression. The first set has histone methyltransferase activities that methylate lysines H3K27 and H3K9 to repress gene activity. In many organisms, this repressive state is stabilized by the activity of a second set of Polycomb factors that bind to the methylated tails of histone H3 and keep the methylation active and that also methylate adjacent nucleosomes, thereby forming tightly packed repressive complexes* (Grossniklaus and Paro 2007).

The Trithorax proteins act to counter the effect of the Polycomb proteins and are necessary to retain the memory of activation. They can either modify nucleosomes or alter the positions of nucleosomes on the chromatin. Thus, some of the Trithorax proteins become parts of complexes that use the energy from ATP hydrolysis to move nucleosomes along the chromatin or to remodel them, allowing transcription factors to bind. Other Trithorax group proteins keep the H3K4 lysine trimethylated (Kingston and Tamkun 2007).

Reinforced repression

Reinforcement between repressive chromatin and repressive DNA has also been observed. Just as methylated DNA is able to attract proteins that deacetylate histones and attract H1 linker histones (both of which stabilize nucleosomes), so repressive states of chromatin are able to recruit enzymes that methylate DNA. DNA methylation patterns during gametogenesis depend in part on the DNA methyltransferase Dnmt3L. Although this protein has lost its enzymatic activity, it can still bind to the amino end of

---

*Indeed, the maintenance of active and repressed states of the *FLC* gene involved in vernalization (see the box on p. 46) involves chromatin modification by Polycomb proteins. Cold temperature (or artificial vernalization) results in the inactivation of the *FLC* gene by the cold-induced activation of the genes producing VIN3 and Polycomb group proteins. These proteins deacetylate histone H3 and methylate it on specific regions. This methylation recruits the HP1 proteins that aggregate nucleosomes together through histone H3. This nucleosome aggregation is maintained in the nuclei of the apical meristem cells through mitotic divisions. In this way, when warm weather returns, winter wheat is ready to flower (see Sung and Amasino 2005).

histone H3. Once bound, it will recruit and/or activate the DNA methyl-transferase Dnmt3a2 to methylate the cytosines on nearby CG pairs (Fan et al. 2007; Ooi et al. 2007). However, if the lysine at H3K4 is methylated, Dnmt3L will not bind.

Histone substitution

The epigenetic memory of an active gene state can also be maintained by histone substitution. When nuclei from differentiated tadpole cells are placed into enucleated oocytes, they can form complete, cloned tadpoles; however, some of the cloned tissues can retain transcription characteristic of the original cell from which the nucleus was taken. For example, Ng and Gurdon (2008) found that when the donor nuclei came from muscle cells, the muscle-specific gene *MyoD* was expressed in nonmuscle cell lineages—even after the cell nuclei had gone through 28 replication cycles in which *MyoD* transcription had been suppressed. The marker of this transcriptional memory was neither DNA methylation nor a covalent modification of the histones. Rather, it was the substitution of a variant form (H3.3) of histone H3. In several species studied, the H3.3 variant appears to be enriched in active gene loci (Ahmad and Henikoff 2002; McKittrick et al. 2004).

Epigenetic Repression by Noncoding RNAs

In addition to proteins, small, noncoding RNAs can also inhibit transcription. Whatever a protein can do, a ribonucleic acid can also do, and this includes catalysis, scaffolding, translational repression, and transcriptional regulation (Cech and Steitz 2014). In yeasts, heterochromatin (large regions of repressed chromatin) is induced by the modification of histones through RNA-induced transcriptional silencing complex. The small interfering RNA (siRNA) forms a duplex with complementary DNA where it recruits proteins that stabilize nucleosomes and that degrade any new transcripts. In female mammals, a similar mechanism may initiate X chromosome inactivation (see above), where the noncoding *Xist* RNA appears to bind proteins that maintain the methylated state of chromatin throughout most of the inactivated chromosome (Chang et al. 2006; Migeon 2007).

Recent studies have demonstrated that siRNAs can regulate transcription from mammalian germline cells by initiating methylation in these regions (Kuramochi-Migawa et al. 2008; Watanabe et al. 2008). Moreover, in mammals and nematodes, noncoding RNAs can be inherited through the germline, and these RNAs maintain an inhibited pattern of gene expression from one generation to the next (see Rassoulzadegan et al. 2006; Rechavi et al. 2011, 2014; Liebers et al. 2014).

Cell-Cell Interactions and Chromatin Remodeling

One of the significant recent breakthroughs of developmental biology has been the linking of cell-cell interactions with chromatin remodeling. These two phenomena had always been considered separate areas. The first was concerned with how changes at the cell surface cause cells to differentiate along particular pathways. The second concerned the mechanisms by which transcription factors in the nucleus cause new patterns of gene expression. We now have dozens of examples of the following paradigm that links the two (see Gilbert 2006).

Paracrine factors from one cell bind to receptors on another cell. Once these receptors have bound the factor, they gain enzymatic activity (often becoming kinases), which then activates certain enzymes within the cytoplasm. These cytoplasmic enzymes trigger a signal transduction cascade that ends in the activation of a dormant transcription factor. This transcription factor, now active, can bind to DNA and recruit chromatin-modifying enzymes that will regulate gene transcription. Thus, signal transduction cascades bring together events occurring at the cell membrane with events occurring in the cell nucleus.

Metabolism and Chromatin Remodeling

In Chapters 1 and 7 we see that changes in nutriton can cause changes in gene expression in offspring. In addition, paternal diet in rats is important in regulating insulin production in the pancreatic β-cells of female pups. High-fat diets in the fathers will alter the expression of those genes involving glucose tolerance and insulin sensitivity (Ng et al. 2010). The father's high-fat diet induces gene expression changes in the white fat of his daughters (Ng et al. 2014). How is the chromatin altered by metabolism? Histone acetylation may play an important role (Gut and Verdin 2013; Figure 3). Acetylation of hundreds of cellular proteins depends on the cytosol levels of acetyl-CoA, sometimes called "active acetate." This depends on a host of metabolic factors, including diet. Recently, it was found that this is also reflected in the nucleus, where the acetylation of histones is dependent on nuclear concentrations of acetyl-CoA. Wellen (2009) demonstrated in mammalian cells that when acetyl-CoA is abundant in the nucleus, histone acetylation increases. This, in turn, allows transcription from genes that promote fat formation and the production of adipocytes. When glucose concentration drops critically (as during starvation or fasting), ketone bodies, containing such short-chain fatty acids as β-hydroxybutyrate, form. This compound has numerous protective functions, and it also acts as an inhibitor of histone deacetylases (Shimazu et al. 2013).

(A)

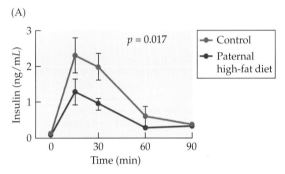

(B)

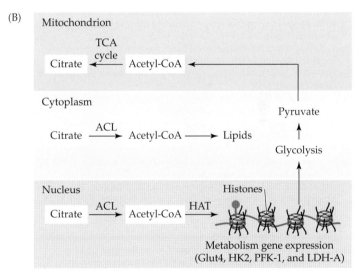

Figure 3   Links between metabolism and gene regulation. (A) Insulin secretion during a glucose tolerance test given to control female mice (open circles) and to female mice that were the daughters of fathers fed high-fat diets. (B) A proposed scheme (Wellen et al. 2009) wherein high glucose can create high levels of citrate in the mitochondria through the TCA cycle. The citrate would enter the cytoplasm where it could make lipids, and it could enter into the nucleus, where the enzyme ACL would make intranuclear acetyl-CoA, which serves as a source for histone acetyltransferases (HAT) to acetylate histones. This would enable the transcription of genes associated with adipocyte (fat cell) development and fat storage. (A after Ng et al. 2010; B after Rathmell and Newgard 2009.)

The mechanisms of epigenetic control restrict which genes are able to become active in each cell and also determine which genes remain active through rounds of mitosis. In this way, cell types are specified and stabilized. These mechanisms largely involve DNA methylation and the modification of nucleosomes by enzymes that can methylate or acetylate particular histone residues. The epigenetic changes in DNA are crucial to our understanding of the mechanisms by which the environment can alter organismal phenotypes, the origin of cancerous cellular phenotypes, the origin of the aging syndrome, and the production of selectable traits in evolution. Indeed, since the Human Genome Project has not clarified how the various genetic components are regulated, a Human Epigenome Project has been conceived (Beck et al. 1999). This project has recenty identified and catalogued the sites of variable human DNA methylation and histone modifications in normal and cancerous cells (Kundaje et al. 2015).

References

Adkins, C. C., S. R. Howar and J. K. Tyler. 2004. Chromatin disassembly mediated by histone chaperone Asf1 is essential for transcriptional activation of the yeast *PHO5* and *PHO8* genes. *Mol. Cell* 14: 657–666.

Ahmad, K. and S. Henikoff. 2002. The histone variant H3.3 marks active chromatin by replication-independent nucleosome assembly. *Mol. Cell* 9: 1191–1200.

Bartolomei, M. S., A. L. Webber, M. E. Brinow and S. M. Tilghman. 1993. Epigenetic mechanisms underlying the imprinting of the mouse *H19* gene. *Genes Dev.* 7: 1663–1673.

Beck, S., A. Olek and J. Walter. 1999. From genomics to epigenomics: A loftier view of life. *Nature Biotechnol.* 17: 1144.

Bird, A. 2002. DNA methylation patterns and epigenetic memory. *Genes Dev.* 16: 6–21.

Bird, A. and A. P. Wolffe. 1999. Methylation-induced repression: Belts, braces, and chromatin. *Cell* 99: 451–454.

Brockdorff, N. and B. M. Turner. 2007. Dosage compensation in mammals. In C. D. Allis, T. Jenuwein and D. Reinberg (eds.), *Epigenetics*. Cold Spring Harbor Press, New York, pp. 321–340.

Cech, T. T. and J. A. Steitz. 2014. The noncoding RNA revolution: Trashing old rules to forge new ones. *Cell* 157: 77–94.

Chang, S. C., T. Tucker, N. P. Thoprogood and C. J. Brown. 2006. Mechanisms of X-chromosome inactivation. *Frontiers Biosci.* 11: 852–866.

Eccleston, A., N. DeWitt, C. Gunter, B. Marte and D. Nath. 2007. Epigenetics. *Nature* 447: 395.

Fan, L. and 7 others. 2007. Recognition of unmethylated histone H3 lysine 4 links BHC80 to LSD1-mediated gene repression. *Nature* 448: 718–722.

Ferguson-Smith, A. C., H. Sasaki, B. M. Cattanach and M. A. Surani. 1993. Paternal-origin-specific epigenetic modification of the mouse *H19* gene. *Nature* 362: 751–755.

Grossniklaus, U. and R. Paro. 2007. Transcriptional silencing by Polycomb group proteins. In C. D. Allis, T. Jenuwein and D. Reinberg (eds.), *Epigenetics*. Cold Spring Harbor Press, NY, pp. 211–230.

Guenther, M. G., S. S. Levine, L. A. Boyer, R. Jaenisch and R. A. Young. 2007. A chromatin landmark and transcription initiation at most promoters in human cells. *Cell* 130: 77–88.

Gut, P. and E. Verdin. 2013. The nexus of chromatin regulation and intermediary metabolism. *Nature* 502: 489–498.

Holliday, R. 1990. DNA methylation and epigenetic inheritance. *Philos. Trans. R. Soc. Lond. B Biol. Sci.* 326: 329–338.

Jones, P. L. and 7 others. 1998. Methylated DNA and MeCP2 recruit histone deacetylase to repress transcription. *Nature Genet.* 19: 187–191.

Kingston, R. E. and J. W. Tamkun. 2007. Transcriptional regulation by Trithorax group proteins. In C. D. Allis, T. Jenuwein and D. Reinberg (eds.), *Epigenetics*. Cold Spring Harbor Press, New York, pp. 231–248.

Kouzarides, T. and S. L. Berger. 2007. Chromatin modifications and their mechanisms of action. In C. D. Allis, T. Jenuwein and D. Reinberg (eds.), *Epigenetics*. Cold Spring Harbor Press, NY, pp. 191–209.

Kundaje, A. and 94 others. 2015. Integrative analysis of 111 reference human epigenomes. *Nature* 18: 317–330.

Kuramochi-Miyagawa, S. and 17 others. 2008. DNA methylation of retrotransposon genes is regulated by Piwi family members MILI and MIWI2 in murine fetal testes. *Genes Dev.* 22: 908–917.

Laird, C. D. and 9 others. 2004. Hairpin-bisulfite PCR: Assessing epigenetic methylation patterns on complementary strands of individual DNA molecules. *Proc. Natl. Acad. Sci. USA* 101: 204–209.

Li, B., M. Carey and J. L. Workman. 2007. The role of chromatin during transcription. *Cell* 128: 707–719.

Liebers, R., M. Rassoulzadegan and F. Lyko. 2014. Epigenetic regulation by heritable RNA. *PLoS Genet.* 10(4): e1004296

McKinnell, I. W. and 7 others. 2008. Pax7 activates myogenic genes by recruitment of a histone methyltransferase complex. *Nature Cell Biol.* 10: 77–84.

McKittrick, E., P. R. Gafken, K. Ahmad and S. Henikoff. 2004. Histone H3.3 is enriched in covalent modifications associated with active chromatin. *Proc. Natl. Acad. Sci. USA* 101: 1525–1530.

Migeon, B. R. 2007. *Females Are Mosaics: X-inactivation and Sex Differences in Disease*. Oxford University Press, NY.

Nan, X., H.-H. Ng, C. A. Johnson, C. D. Laherty, B. M. Turner, R. N. Eisenman and A. Bird. 1998. Transcriptional repression by the methyl-CpG-binding protein MeCP2 involves a histone deacetylase complex. *Nature* 393: 386–389.

Ng, S.-F., R. C. Y. Lin, D. R. Laybutt, R. Barres, J. A. Owens and M. J. Morris. 2010. Chronic high-fat diet in fathers programs β-cell dysfunction in female rat offspring. *Nature* 467; 963–966.

Ooi, S. K. T. and 11 others. 2007. DNMT3L connects unmethylated lysine 4 histone H3 to de novo methylation of DNA. *Nature* 448: 714–717.

Rassoulzadegan, M., V. Grandjean, P. Gounon, S. Vincent, I. Gillot and F. Cuzin. 2006. RNA-mediated non-Mendelian inheritance of an epigenetic change in the mouse. *Nature* 441: 469–474.

Rathwell, J. C. and C. B. Newgard. 2009. Biochemistry: A glucose-to-gene link. *Science* 324: 1021–1022.

Rechavi, O., G. Minevich and O. Hobert. 2011. Transgenerational inheritance of an acquired small RNA-based antiviral response in *C. elegans*. *Cell* 147: 1248–1256.

Rupp, R. A. W. and P. B. Becker. 2005. Gene regulation by histone H1: New links to DNA methylation. *Cell* 123: 1178–1179.

Sarraf, S. A. and I. Stancheva. 2004. Methyl-CpG-binding protein MBD1 couples histone H3 methylation at lysine 9 by SETDB1 to DNA replication and chromatin assembly. *Mol. Cell* 15: 595–605.

Shimazu, T. and 15 others. 2013. Suppression of oxidative stress by β-hydroxybutyrate, an endogenous histone deacetylase inhibitor. *Science* 339: 211–214.

Sung, S. and R. M. Amasino. 2005. Remembering winter: Toward a molecular understanding of vernalization. *Annu. Rev. Plant Biol.* 56: 491–508.

Venolia, L., S. M. Gartler, E. R. Wasserman, P. Yen, T. Mohandas and L. J. Shapiro. 1982. Transformation with DNA from 5-aza-cytidine reactivated X chromosomes. *Proc. Natl. Acad. Sci. USA* 79: 2352–2354.

Watanabe, T. and 12 others. 2008. Endogenous siRNAs from naturally formed dsRNAs regulate transcripts in mouse oocytes. *Nature* 453: 539–543.

Wellen, K. E., G. Hatzivassilliou, U. M. Sachdeva, T. V. Bui, J. R. Cross and C. B. Thompson. 2009. ATP-citrate lyase links cellular metabolism to histone acetylation. *Science* 324: 1076–1080.

Wolf, S. F., D. J. Jolly, K. D. Lunnen, T. Friedman and B. R. Migeon. 1984. Methylation of the hypoxanthine guanosine phosphoribosyltransferase locus on the human X chromosome: Implications for X-chromosome inactivation. *Proc. Natl. Acad. Sci. USA* 81: 2806–2810.

Writing Development Back into Evolutionary Theory

As described in Chapter 9, two disciplines left out of the Modern Synthesis were ecology and embryology (Harrison 1937; Hamburger 1980; Amundson 2000, 2001). Embryology is the study of how the zygote gives rise to the interconnected organs of the body, and for many embryologists, no theory of evolution made sense without a theory of body construction. For instance, Lillie (1927) could say that the methods and theories used by geneticists "have no place among their categories for the ontological process and *a fortiori* for the phylogenetic." Thus, for Lillie (and for many other embryologists, especially N. J. Berrill and Gavin de Beer), if genetics could not explain development, then it could not explain evolution. De Beer (1954) would write that "the processes whereby the structures are formed are as important as the structures themselves from the point of view of evolutionary morphology and homology."

But until recently, embryologists themselves did not have an adequate theory of body construction. There was no coherent theory of differentiation or morphogenesis until the 1980s, and to geneticists, this was reason enough for excluding embryology from the Modern Synthesis. In this sense, it was important for geneticists that evolution be a theory of change, independent of a theory of body construction. Most geneticists (with a few exceptions such as Sewall Wright) did not think that development was needed to explain evolution. Development happened, and that was enough. The genotype was expressed as a phenotype, and development was merely the way it got there. In contradistinction to de Beer, Ernst Mayr (1980) would write, "The clarification of the biochemical mechanism by which the genetic program is translated into the phenotype tells us absolutely nothing about the steps by which natural selection had built up the particular genetic program."*

Related to this issue was that genetics and developmental biology explained different things. The explanandum (that which requires explanation) of the Modern Synthesis was natural selection; the population geneticists did not need knowledge of development to get their answers. The explanandum

*Mayr's "genetic program" metaphor had a lot to do with taking development out of the modern synthesis (see, for instance, Wallace 1986).

for developmental biologists interested in evolution was anatomical change; natural selection was secondary in that it allowed some of these changes to persist. Rudy Raff, one of the earliest advocates of a developmental genetic approach to evolution, said, "They're interested in species; we're interested in bodies" (Amundson 2005, p. 253). Indeed, most early embryologists were not interested in variations within species, and this disinterest allowed the evolutionary biologists to ignore developmental considerations in their work.

Goldschmidt: Macromutations and "Hopeful Monsters"

Some biologists thought development and evolution should be reconnected and so began constructing their own theories. In the 1940s, Richard Goldschmidt criticized the Modern Synthesis, writing that their paradigm of the accumulation of small genetic changes was not sufficient to explain the generation of evolutionarily novel anatomical structures such as teeth, feathers, cnidocysts (the stinging cells of cnidarians), or the shells of snails. He claimed that evolution could only occur through heritable changes in those genes that regulated development (Goldschmidt 1940).

In *The Material Basis of Evolution*, Goldschmidt (1940) presented two models relating gene activity, development, and evolutionary dynamics. The first model proposed that new species might originate as "hopeful monsters" that were the result of mutations in developmentally important loci ("developmental macromutations"). In the second model, Goldschmidt argued that chromosomal rearrangements ("systemic mutations") would have the effect of many cumulative developmental macromutations and cause even larger phenotypic changes. Goldschmidt's view of systemic mutations did not win much favor at the time, nor did his idea that "a single mutational step affecting the right process at the right moment can accomplish everything, providing that it is able to set in motion the ever present potentialities of embryonic regulation" (1940, p. 297). However, evolutionary developmental biology has since shown that single mutational steps can may in fact be critical in the generation of novel structures, and several studies (see Bateman and DiMichele 2002) have shown that stress to the nucleus can cause gene duplications, the movement of transposable DNA sequences, and an increase in recombination rates (i.e., the rate of the "systematic mutations") that can result in rapid and extensive phenotypic changes.

However, Goldschmidt's presentation of these ideas went against the grain of the genetic science of the 1940s. To Goldschmidt, the gene wasn't a locus or an allele, but a unit of development (Goldschmidt 1940, p. 197). For Goldschmidt, the regulatory processes of development relieved the need for thousands of modifier genes, and for this reason, he attempted "to convince evolutionists that evolution is not only a statistical genetical problem but also one of the developmental potentialities of the organism" (Goldschmidt et al. 1951). However, in the absence of molecular data, Goldschmidt's ideas were generally ignored (see Gould 1980).

Waddington: A Theory of Bodies, a Theory of Change

Goldschmidt's ideas were very different form the genetic ideas of Morgan's school of genetics. However, other scientists were trying to synthesize the orthodox genetics of Morgan's group with the orthodox embryology. Two of the most far-reaching syntheses of evolution, genetics, and development were attempted by Ivan Ivanovich Schmalhausen and Conrad Hal Waddington (whose work is discussed in Chapter 11 and in Appendix D). Waddington was trained in genetics, experimental embryology, paleontology, and evolutionary biology and was able to appreciate the links between them. In his 1953 essay "Epigenetics and Evolution," Waddington analyzed the shortcomings of the population genetic account of evolution. He noted (1953, p. 187) that the genetic approach to evolution had culminated in the Modern Synthesis, but he also noticed that "it has been primarily those biologists with an embryological background who have continued to pose questions."

Waddington's essay claimed that the Modern Synthesis failed to work in at least three areas. First (as has been stressed throughout this book), much variation appears to be regulated by the environment, not the inherited genotype. Second, as Goldschmidt had noted, the large taxonomic divisions of animals encompass an extent of physical differences that is not compatible with local races branching off. That is, accumulations of small mutations in a local group would not be sufficient to separate, say, amphibians from fish or reptiles from amphibians. Waddington noted that Goldschmidt's own hypotheses about how such divisions might take place were so unconvincing to geneticists that they obscured the very real cogency of Goldschmidt's arguments for the existence of large phenotypic gaps that could not be bridged by the accumulation of small allelic changes. Third, Waddington noted the different rates of evolution seen in the paleontological record.

Waddington (1953, p.190) claimed that in conventional evolutionary studies, the animal is considered either as a genotype (and is studied by geneticists) or as a phenotype (and is studied by ecologists and taxonomists). What is needed, he said, is an evolutionary study of those processes that transform genotype into phenotype—the epigenetics of development. Following Goldschmidt, Waddington (1953, pp. 190–191) declared:

> Changes in genotypes only have ostensible effects in evolution if they bring with them alterations in the epigenetic processes by which phenotypes come into being; the kinds of change possible in the adult form of an animal are limited to the possible alterations in the epigenetic system by which it is produced.

Waddington proposed two levels of evolution. First, there would be an internal evolution whereby changes in gene expression *in the embryo* would create the possibility for new variation. Such changes would have to be met by other embryonic changes in order for the organism to survive to birth. Second, there would be the natural selection on the organism itself.

Thus, Waddington hoped to combine evolution as a theory of bodies with evolution as a theory of change. He also hoped to combine the nascent field of developmental genetics (which he was helping to create) with the established science of population genetics. This respect for both traditions was characteristic of both Waddington and Schmalhausen and provided a paradigm for today's evolutionary developmental biology.

Evolution without Development

While the theoretical issues raised by Goldschmidt, Waddington, and Schmalhausen were being debated, the founders of the Modern Synthesis were reacting strongly against developmental views. Most of them argued that a theory of evolutionary change did not need to be underwritten by a theory of body construction (and, unfortunately, at that time the science of embryology was in no condition to provide such a theory). Moreover, to the architects of the Modern Synthesis, developmental biology not only was asking the wrong questions, but was structurally incapable of asking the right ones. Morgan (1932) had rewritten the *history* of evolutionary biology in such a way that embryology was excluded, and then Ernst Mayr rewrote the *philosophy* of evolutionary biology so as to exclude embryology from having anything at all to say about evolutionary studies (see Amundson 2005; Levit and Meister 2006).

In her analyses of the origins of the Modern Synthesis, Polly Winsor (2003, 2006) points out that while Mayr proposed some of the most insightful and far-reaching ideas in evolutionary biology, his history wasn't always the best. In order to keep evolutionary biology focused on intraspecific variation, she writes, Mayr invented a typological dichotomy that in fact did not exist. On one hand, Mayr postulated that a semimystical, scientifically discredited mode of thinking called "essentialism" was characteristic of embryologists and paleontologists. On the other hand, he felt there was a scientifically rigorous and philosophically valid way of approaching life called "population thinking," which is what geneticists do. Darwin, Mayr said, had shown that essentialism was wrong and founded the latter perspective.

But as Winsor points out, the type of essentialism Mayr evoked in the 1950s had no adherents in contemporary biology, and in fact biology won the battle against essentialism decades before Darwin. She further points out that this dichotomy was historically false and, despite its popularity, "little more than a myth." Thus, even though Mayr did not believe that evolution was solely a matter of allele frequency changes, his philosophical analysis demanded that the only scientific focus of evolutionary biology be evolution within species.

Another fundamental assumption of the Modern Synthesis was that macroevolution (the origin of new species and higher taxa) could be explained by natural selection. Natural selection was thought to be so strong a force that the reason unrelated organisms looked alike was that they had similar selection pressures, and not that they inherited genes from a common ancestor.* In fact, Mayr (1966) and Dobzhansky (1955) explicitly

---

*As Amundson (2005) points out, the need of genetic studies to have interbreeding species prevented the discovery of homologous genes between species. He writes that Mayr's and Dobzhansky's belief in the strength of natural selection far overreached the data and the methodology. "Both Mayr and Dobzhansky were perfectly willing to predict that common characters among species were the result of adaptive convergence, not homologous genes, even though they both knew that the Mendelian blind spots prevented direct evidence for (or against) that prediction. They had faith in the power of natural selection to be able to produce any phenomenon that homologous genes could produce. … It illustrates an aspect of scientific commitment that extends well beyond the data immediately at hand" (2005, p. 217).

rejected the possibility that widely shared characteristics were caused by homologous genes. As Mayr (1966) wrote:

> Much that has been learned about gene physiology makes it evident that the search for homologous genes is quite futile except in very close relatives (Dobzhansky 1955). If there is only one efficient solution for a certain functional demand, very different gene complexes will come up with the same solution, no matter how different the pathway by which it is achieved.

Thus, Mayr believed that "eyes evolved independently at least 40 times in different groups of animals" (Mayr 1988, p. 72) and had converged on a common phenotype from many sources (Salvini-Plawen and Mayr 1977). As discussed in Chapter 10, however, evolutionary developmental biology has a much more parsimonious explanation for eye development.

Not only is the non-developmental perspective an incomplete view of evolution, it becomes an inaccurate one when the variances are not linear. Brandon and Nijhout (2007; Nijhout 2003, 2007) have shown that $G \times E$ (genotype with environment) interactions are not linear and are dependent on developmental history (i.e., the temporal and spatial contexts in which they occur). Quantitative genetics (because it is statistical) is a linear theory that can work with nonlinear relationships between the phenotypes of parent and offspring only if there is an equation that linearizes the relationship. In certain developmental systems there does not seem to be any linear relationship; rather, nonlinearity is an important component of the phenotype's development. Moreover, Brandon and Nijhout see quantitative genetics as largely descriptive (and not explanatory) because it essentially says that the mean trait value in the next generation will be the same as that of the current generation, modified only by its selection differential (roughly: difference of parental mean from population mean) and its heritability.

As mentioned in Chapter 9, evolutionary theory appears to expand in concentric layers. The exclusion of embryology from the Modern Synthesis may have been a necessary step in the formation of an evolutionary theory. Until the 1970s, embryology did not have much to say about the role of genes in development. Meanwhile, the fusion of genetics and population biology allowed the formation of a theory of evolutionary change that could explain and predict the maintenance of diversity within populations. There were reasons for the focus on strictly genetic mechanisms for evolution. First, genetics was much better understood than developmental biology. Second, genetics provided a mechanism for the transmission of phenotypic traits from one generation to the other, as well as mathematical rules for understanding the transmission of phenotypes from one generation to the next. Third, genetics provided a fundable way to study evolution, especially in the United States (Beatty 2006). Fourth, neo-Darwinism linked evolution with population genetics, and to get a good genetic story, one does not want to deal with organisms whose phenotypes are significantly controlled by the environment. As Sonia Sultan (2003) points out, "neo-Darwinian botanists were often quite frustrated in their attempts to discern genetically based local adaptations through this 'environmental noise,'" and this led them to overlook the adaptive nature of these plastic responses. And fifth, the fear of Lysenkoism caused a backlash against any type of hereditary variation that was not linked directly to allelic variation (Lindegren 1966; Lewontin and Levins 1976; Sapp 1987).

There was also the view (see Wilkins 2011) that to accept "soft" inheritance would mean undermining the concept that complex traits could be explained by the cumulative effects of numerous Mendelian genes. The data of genetics was, at the time, a far better way of explaining variation than the inheritance of acquired characteristics. And it should be recalled that many famous evolutionary geneticists (such as Mayr and Maynard Smith) started their careers as Lamarckians and came to Darwinism only after the data of genetics convinced them to abandon their former ways (Wilkins 2011). In embracing one, they denied the other. Later, there was a general feeling (voiced by Mayr, 2002; see Wilkins 2011) that Crick's "central dogma" that information could flow from nucleic acid to protein, but not vice versa, made genes the only possible route of inheritance and variation.

The molecular biology techniques that have been invented since the 1950s now allow us to study evolution outside of interbreeding populations, thus removing a major constraint of the Modern Synthesis. We now know that this earlier evolutionary theory made assumptions that were unwarranted and have been shown to be false. Similarities in extremely divergent organisms can be explained by homologous genes, and embryology need not be essentialist. Moreover, the data from ecological developmental biology show that phenotype is not the mere unrolling of genotype.

Development is now being considered much more seriously in evolutionary theory than it has since the early days of the Modern Synthesis. Alleles of regulatory genes are being found and studied to determine if they can account for evolutionary change both within and between species. Evo-devo is in the process of becoming part of evolutionary theory. *Eco-evo-devo* is still in the process of proving its merits. The past decade, though, have been remarkable in showing the critical importance of symbionts, plasticity, and epiallic variation to the phenomena of variation, inheritance, reproductive isolation, and major transitional changes.

References

Amundson, R. 2000. Embryology and evolution 1920–1960: Worlds apart? *Hist. Philos. Life Sci.* 22: 35–352.

Amundson, R. 2001. Adaptation and development: On the lack of common ground. In S. Orzack and E. Sober (eds.), *Adaptation and Optimality*. Cambridge University Press, Cambridge, pp. 303–334.

Amundson, R. 2005. *The Changing Role of the Embryo in Evolutionary Thought: Structure and Synthesis*. Cambridge University Press, Cambridge.

Bateman, R. M. and W. A. DiMichele. 2002. Generating and filtering major phenotypic novelties: neoGoldschmidtian saltation revisited. In Q.C.B. Cronk, R. M. Bateman and J. A. Hawkins (eds.), *Developmental Genetics and Plant Evolution*. Taylor and Francis, London, pp. 109–159.

Beatty, J. 2006. Masking disagreement among experts. *Episteme* 3: 52–67.

Brandon, R. N. and H. F. Nijhout. 2007. The empirical non-equivalence of genic and genotypic models of selection: A decisive refutation of genic selectionism and pluralistic genic selectionism. *Philos. Sci.* 73: 277–297.

De Beer, G. R. 1954. *Embryos and Ancestors*, Rev. Ed. Oxford University Press, Oxford.

Dobzhansky, Th. 1955. *Evolution, Genetics, and Man*. Wiley, New York.

Goldschmidt, R. B. 1940. *The Material Basis of Evolution*. Yale University Press, New Haven.

Goldschmidt, R., A. Hannah and I. Piternick. 1951. The podoptera effect in *Drosophila melanogaster*. *Univ. Calif. Publ. Zool.* 55: 67–294.

Gould, S. J. 1980. The return of hopeful monsters. *Natural History* 86 (June/July): 22–30. Reprinted in *The Panda's Thumb*. 1992. W. W. Norton, New York.

Hamburger, V. 1980. Embryology and the Modern Synthesis in evolutionary theory. In E. Mayr and W. Provine (eds.), *The Evolutionary Synthesis: Perspectives on the Unification of Biology.* Cambridge University Press, New York, pp. 97–112.

Harrison, R. G. 1937. Embryology and its relations. *Science* 85: 369–374.

Levit, G. S. and K. Meister. 2006. The history of essentialism vs. Ernst Mayr's "Essentialism story": A case study of German idealistic morphology. *Theory Biosci.* 124: 281–307.

Lewontin, R. C. and R. Levins 1976. The problem of Lysenkoism. In H. Rose and S. Rose (eds.), *The Radicalisation of Science. Ideology of/in the Natural Sciences.* Macmillan, London, pp.32–64.

Lillie, F. R. 1927. The gene and the ontogenetic process. *Science* 64: 361–368.

Lindegren, C. C. 1966. *The Cold War in Biology.* Planarian Press, Ann Arbor, MI.

Mayr, E. 1966. *Animal Species and Evolution.* Harvard University Press, Cambridge, MA.

Mayr, E. 1980. Prologue: Some thoughts on the history of the evolutionary synthesis. In E. Mayr and W. B. Provine (eds.), *The Evolutionary Synthesis: Perspectives on the Unification of Biology.* Harvard University Press, Cambridge, MA, pp. 9–10.

Mayr, E. 1988. *Toward a New Philosophy of Biology.* Harvard University Press, Cambridge, MA.

Mayr, E. in A. Wilkins 2002. Interview with Ernst Mayr. *BioEssays* 24: 960–973.

Morgan, T. H. 1932. *The Scientific Basis of Evolution.* W. W. Norton, New York.

Nijhout, H. F. 2003. On the association between genes and complex traits. *J. Invest. Derm. Symp.* 8: 162–163.

Nijhout, H. F. 2007. Complex traits: Genetics, development, and evolution. In R. Sansom and R. Brandon (eds.), *Integrating Evolution and Development: From Theory to Practice.* MIT Press, Cambridge, pp. 93–112.

Salvini-Plawen, L. V. and E. Mayr. 1977. On the evolution of photoreceptors and eyes. *Evol. Biol.* 10: 207–263.

Sapp, J. 1987. *Beyond the Gene: Cytoplasmic Inheritance and the Struggle for Authority in Genetics.* Oxford University Press, Oxford.

Sultan, S. E. 2003. Phenotypic plasticity in plants: A case study in ecological development. *Evol. Dev.* 5: 25–33.

Waddington, C. H. 1953. Epigenetics and evolution. In R. Brown and J. F. Danielli (eds.), *Evolution* (SEB Symposium VII). Cambridge University Press. Cambridge, pp. 186–199.

Wallace, B. 1986. Can embryologists contribute to an understanding of evolutionary mechanisms? In W. Bechtel (ed.), *Integrating Scientific Disciplines.* Martinus Nijhoff, Dordrecht, pp. 149–163.

Wilkins, A. 2011. Why did the Modern Synthesis give short shrift to "Soft Inheritance"? In S. B. Gissis and E. Jablonka (eds.), *Transformations of Lamarckism: From Subtle Fluids to Molecular Biology.* MIT Press, Cambridge, MA, pp. 127–132.

Winsor, M. P. 2003. Non-essentialist methods of pre–Darwinian taxonomy. *Biol. Philos.* 18: 387–400.

Winsor, M. P. 2006. The creation of the essentialism story: An exercise in metahistory. *Hist. Philos. Life Sci.* 28: 149–174.

Epigenetic Inheritance Systems
The Inheritance of Environmentally Induced Traits

Ecological evolutionary developmental biology has the data to bring three controversial alternative inheritance systems back into the discussion of evolutionary biology. The first idea concerns the inheritance of environmentally acquired traits, an ancient idea usually associated with Jean-Baptiste Lamarck (1744–1829) but which was also used by Charles Darwin and many other Victorian naturalists. The second controversial idea usually goes by the name "genetic assimilation," and it concerns the genetic fixation of an adaptive, plastic response into the genome. In this hypothesis, a response that was once part of a phenotypically plastic repertoire is now part of the normative genetic "program." The third controversial idea is symbiotic inheritance.

But if eco-evo-devo is going to shed this far-too-cumbersome moniker and become mainstream evolutionary biology, it has to deal with issues of the inheritance of acquired traits and show why they were considered wrong and why new data suggest that these theories need to be reconsidered as containing some important insights into evolution.

The Ghost of Lamarck

Epigenetic inheritance systems recall the specter of a banished ghost—Lamarckian inheritance. The year 2009 was not only the bicentenary of Darwin's birth and the centenary of the Woltereck and the Johannsen papers described in Chapter 1, it is also the bicentenary of Lamarck's *Philosophie Zoologique*. Lamarckian inheritance was based on physiology, behavior, and phenotypic plasticity: if you used your muscles, they grew bigger. Moreover, such muscular changes would be passed on to subsequent generations, so the offspring of runners would have stronger leg muscles. In the absence of any notion of genes, this physiological model seemed appropriate. Lamarck's two "laws" of evolution were:

- *First law*: In every animal that has not passed the limit of its development, more frequent and continuous use of any organ gradually strengthens, develops, and enlarges that organ and gives it a power proportional to the length of time it has been so used; whereas the permanent disuse of any organ weakens and deteriorates it by imperceptible increments, progressively diminishing its functional capacity until it finally disappears.

- *Second law*: All the acquisitions or losses wrought by nature on individuals through the influence of their environment and the predominant use or permanent disuse of any organ, are preserved by reproduction, provided that the acquired modifications are common to both sexes, or at least to the individual that produces the young.

Lamarck concludes:

Nature has produced all the species of animals in succession, beginning with the most imperfect or simplest, and ending her work with the most perfect, so as to create a gradually increasing complexity in their organisation; these animals have spread at large throughout all the habitable regions of the globe, and every species has derived from its environment the habits that we find in it and the structural modifications which observation shows us.

Here, Lamarck proposes plasticity as a model for inheritance, in which the organism, influenced by its environment, alters its body (and those of its descendants) through use and disuse.

Darwin also believed that it might be possible for acquired characteristics to be inherited. He speculated that the body's tissues produced "gemmules" that circulated through the body and entered into the germ cells (Darwin 1883). Thus, if a tissue were bigger due to more usage in a more demanding environment, it could produce more gemmules, causing the phenotype to be transmitted into germ cells. The amount and the type of gemmule produced would determine its inheritance.

This notion of diffusible gemmules was disproved by Darwin's cousin, Francis Galton. Galton assumed that Darwin's postulated gemmules had to be transferred through the blood.* So he transfused blood between gray rabbits and lop-eared rabbits until he felt the lop-eared rabbits were totally transfused with gray rabbit blood and vice versa. However, each type of rabbit continued to breed true to its type—no influence of blood-borne gemmules was ever observed. A little later, August Weismann proposed that only the germline counted in heredity, and that the germline was separate from the somatic lineages of cells that formed the body. Therefore, anything that affected the individual could not influence heredity if the germline were not affected. Weismann cut off the tails of mice for nineteen generations and showed that a tailless race did not develop. (Commenting on these experiments, Thomas Huxley [1884] wryly noted that Weismann should have known that circumcision among Jews and Muslims had not produced men without foreskins.)

---

*Recall that before scientists knew about genes, blood was considered the backbone of heredity. We still speak of bloodlines, blood-ties, and blood being thicker than water. There are numerous cultures where semen is thought to be purified blood. Indeed Aristotle, Galen, and Avicenna, three of the leading sources of Western medicine, held that view.

However, numerous examples of inheritance of acquired characteristics *had* been reported. By rearing butterfly pupae at different temperatures, Standfuss and Fischer (quoted in Thomson 1908, p. 215) produced butterflies that resembled related species, and the temperature-induced phenotype was transmitted to their progeny. Even into the twentieth century, reports of inheritance of acquired characteristics occasionally were published (Lewontin and Levins 1976). For instance, geneticist Viktor Jollos (1934) reported that heat resistance induced by heat shock could be transmitted to offspring. As mentioned in Appendix A, numerous examples of the inheritance of acquired characteristics were proposed by the Viennese biologists of the Prater Vivarium. In 1920, Austrian biologists Paul Kammerer and Eugen Steinach reported morphological changes in offspring and grandoffspring of rats that were exposed to high temperatures (Gliboff 2005; Logan 2007).

As genetic studies found more and more examples of Mendelian inheritance, examples of inherited acquired traits became suspect. Indeed, Lindegren (1949), surveying inheritance in the fungus *Neurospora*, wrote that about two out of every three newly discovered variants did not show Mendelian segregation. Therefore, he said, such variants were usually discarded. A similar assessment was made by Benkemoun and Saupe (2006), who suggested that these "anomalies" to Mendelian segregation might be caused by epigenetic inheritance, and that "we should have a closer look before putting them in the autoclave." So it is possible that human bias about what can be best studied has prevented us from seeing further examples of epigenetic inheritance (see Jablonka and Raz 2009).

Plasticity-Driven Evolution

With the rediscovery of Mendel's laws, Johannsen (1909) made the distinction between genotype and phenotype, claiming that phenotype was formed through the combination of genotype and environmental conditions. Also in 1909, Woltereck, working with parthenogenetically pure lines of *Daphnia*, introduced the concept of the reaction norm to illustrate the phenotypic differences that a given genotype could display in response to environmental cues. Five years later, the Swedish biologist H. Nilsson-Ehle (1914) placed this into the larger concept of phenotypic plasticity to define the observed variations exhibited by a single genotype due to either environmental or stochastic factors. This view that the genotype represented a reaction norm became the dominant view of the Russian school of evolutionary biology that coalesced around Chetverikov in the 1920s (Adams 1980a). Indeed, in 1926, the young Dobzhansky equated the reaction norm with the gene, claiming that the reaction norm was the actual Mendelian unit of inheritance (see Sarker and Fuller 2003).

But such plasticity was not a major part of the British and American schools of evolutionary biology. When T. H. Morgan (1932) claimed that genetics provided the only scientific way to study evolution, he meant the *Drosophila* genetics that his laboratory had founded, not the more plastic notions of heredity operating in Europe. When Morgan's genetics merged with the mathematical models of Fisher, Haldane, and Wright to form the

Modern Synthesis, the phenotype was thought to be a direct readout of the genotype. Any phenotypic variation due to environmental factors was given a "genetic" explanation by the use of such terms as "penetrance" and "expressivity" (Vogt 1926; Sarkar 1999). Although Dobzhansky originally stressed the importance of the reaction norm concept in evolution, by 1937 he had abandoned this approach, defining evolution as a "change in the genetic composition of populations" and describing the reaction norm as a property of the population rather than of the individual (see Sarkar and Fuller 2002).

One of the first people to realize the importance of plasticity for evolution was Lancelot Hogben. Hogben (1933) looked at the reaction norms in flies having two genotypes. He found that when he graphed the number of eye facets versus temperature for these two strains, he did not obtain identical or parallel lines. From this, he concluded that there was an interdependence of genes and environment and that phenotype production was a complex relationship between these factors. Although this analysis was largely overlooked in favor of Haldane's linear models, this appreciation of genome-environment (G × E) interactions was further developed in the 1970s and became important in newer models of genome-environment interactions (see, for example, Lewontin 1974).

Another person who championed a plasticity-driven approach to evolution was Jean Piaget. He began his career as a biologist and only later became an educator who applied much of his biological knowledge to educational theory. Piaget noted that the pond snail *Limnaea stagnalis* had two phenotypes. There were snails with elongated shells and snails with flattened shells. A flatter shell developed in those snails that were raised in turbulent water, as this shape was better adapted for the rough environment. Moreover, after several generations, this environmentally induced phenotype became genetically fixed and would continue to be formed even if the snails were transferred to still water (Piaget 1929a,b).

Meanwhile, Russian biologists continued to integrate the notion of phenotypic plasticity into their accounts of evolution. In one of his last publications, Alexsei Nikolaevich Severtsov (1935), the founder of the Russian school of evolutionary morphology, wrote "At the present time, we morphologists do not have the full theory of evolution. It seems to us that in the near future, ecologists, geneticists, and developmental biologists must move forward to create such a theory, using their own investigations, based on ours." To Severtsov, a complete theory of evolution must causally explain the morphological changes seen in paleontology through the mechanisms of genetics, ecology, and embryology. He felt that genetics alone could not provide the mechanism because it did not involve the "how" of evolution (Adams 1980b). Only ecology and embryology could do that. To this end, Severtsov and others worked on evolutionary syntheses that included developmental and ecological considerations. Georgii Gause, best known for his competitive exclusion principle in ecology, wrote one of the first statements of the genetic fixation of a plastic phenotype. He claimed that adaptive phenotypes can emerge as physiological responses to the environment and that these modifications can become genotypically fixed by mutations if there is a fitness advantage for selection (Gause 1947).

This integration of embryology, development, and ecology became the project of the Institute of Evolutionary Morphology, headed by Severtsov's student Ivan Ivanovich Schmalhausen. For Schmalhausen (1949), such plasticity was critical for evolution: "Every genotype is characterized by its own specific norm of reaction, which includes adaptive modifications of the organism to different environments." His book, *Factors of Evolution: The Theory of Stabilizing Selection*, provides numerous examples of polyphenisms that appear to have been genetically fixed. Independently of Schmalhausen's work, Conrad Waddington (1942, 1952) linked the fixation of developmental plasticity to genetic variation and tried to incorporate this idea into evolutionary biology. Schmalhausen called his idea "stabilizing selection"; Waddington named his notion "genetic assimilation." Waddington had to present his ideas against a Western background that was suspicious or ignorant of developmental plasticity, while Schmalhausen had to present his ideas against a Soviet background that was suspicious of genes* (see Gilbert 1994).

In the West, a reaction against Lysenkoism made geneticists very wary of the notion of phenotypes being produced by the environment (Lindegren 1966; Sapp 1987). As mentioned in Appendix A, attempts to look at nongenomic contributions to development were a casualty of the Cold War. Indeed, the founders of the Modern Synthesis were unimpressed by genetic assimilation/genetic accommodation and other such theories based on plasticity. George Gaylord Simpson (1953) dismissed genetic assimilation as a "relatively minor" contribution to evolution, and Ernst Mayr (1963) and Theodosius Dobzhansky (1970) interpreted genetic assimilation as a failed attempt to support a model of Lamarckian/Lysenkoist inheritance. Certainly, the title of Waddington's 1953 paper, "Genetic Assimilation of an Acquired Character," would raise a number of red flags.

Epialleles and Germline Transmission of Alternative Chromatin Structures

This book has tried to show the evidence that plasticity, as well as symbionts and epialleles, might be playing an enormously important role in evolution. First, developmental plasticity can provide the material bases for genetic accommodation, through which developmentally induced phenotypes may be selected for and assimilated into the germline. Second,

---

*Lysenkoism (see Appendix A) was critical to the negative reception of both hypotheses. Just as Western geneticists thought that Waddington, a British socialist, was making concessions to the Marxists, the Soviet scientists thought Schmalhausen was making concessions to the capitalists. Schmalhausen was one of the very few Soviet biologists (along with D. A. Sabinin and A. N. Formozov) who openly disagreed with Lysenko in 1947. However, his Order of the Red Banner did not save Schmalhausen from Lysenko's vitriolic public attack; indeed Lysenko crafted much of his infamous 1948 speech at the All-Union Academy of Agricultural Sciences against Schmalhausen (Adams 1980b). Schmalhausen was stripped of all his appointments and assistants, after which he retreated to his dacha and independently studied vertebrate embryology. (Sabinin committed suicide.) As Lysenko lost power, Schmalhausen was gradually brought back into Russian biology. His last years were spent (like those of Waddington) reformulating his evolutionary theories into a cybernetic framework (see Wake 1986; Levit et al. 2006).

important evolutionary transitions have been shown to occur through the transmission of symbionts. And third, as mentioned in Chapter 2, epialleles can be induced in the chromatin, and these chromatin changes (DNA methylation, histone conformations) can be transmitted through the germline.

The basis of the barrier against the concept of inheritance of acquired characteristics has been the idea that changes to the body did not affect the genes in the germ cells. We now know this to be false for some traits, even if true for most. Environmentally induced DNA methylation and other chromatin alterations can be passed through the germline. Therefore, environmentally induced phenotypes can be passed from generation to generation by chromatin modifications. Table 1 lists some of the documented epigenetic inheritance systems in animals.

The diet of worms

In addition to examples mentioned in the body of the text, there is more evidence pointing to the importance of small noncoding RNAs in transmitting stable phenotypes. As mentioned earlier (see Chapter 2 and Appendix B), small noncoding RNAs can be propagated through the germline and restrict gene expression. Recently published reports (Rassoulzadegan et al. 2006; Cuzin and Rassoulzadegan 2010; Cossetti et al. 2014) strongly suggest that RNA-mediated epigenetic inheritance in mice may be much more prevalent than originally thought. In roundworms, it appears that acquired characteristics can be transmitted for over 100 generations through the amplification of small germline RNAs (Burton et al. 2011; Rechavi et al. 2011).

A new series of studies published by Rechavi and colleagues (2014) demonstrates an epigenetic inheritance of a starvation-induced pattern of gene expression wherein endogenous small RNAs silence particular genes and result in increased longevity for many generations. In addition to the ability of caloric restriction to form dauer larvae during the second larval instar period of *Caenorhabditis elegans*, starvation can also reversibly arrest *C. elegans* development at the first (L1) larval stage. While this arrest occurs without the morphological changes seen in the dauer larva, the expression of fully 27% of the protein-coding genes is altered, and the L1-arrested larva has increased stress resistance and increased longevity (Baugh 2013). This arrest is mediated by the production of small interfering RNAs (siRNAs) that are amplified by an RNA-dependent RNA polymerase, once they are bound to their target mRNA (Gu et al. 2012; Grishok 2013). These siRNAs bind to exonic regions of mRNA and prevent the mRNA from being translated. Rechavi and coworkers (2014) found that starvation-induced developmental arrest in L1 generated siRNAs that were transgenerationally inherited for at least three generations (Figure 1). Moreover, these siRNAs targeted mRNAs with roles in nutrition. This transgenerational effect is thought to occur because amplification of siRNAs occurs in the germline of *C. elegans* as well as in the somatic lineages, and the siRNAs keep being amplified each generation. The extended longevity given to the original nematode by starvation is seen in its third-generation offspring.

Here we have a condition where (1) an environmental factor (starvation) directly causes heritable alterations, and (2) the induced changes are seen only in a subset of traits that are responding to the environmental

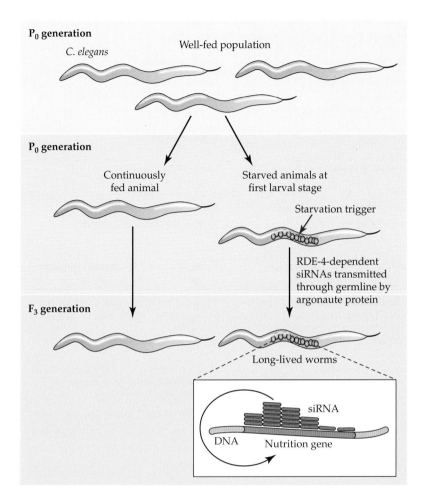

P₀ generation

C. elegans   Well-fed population

P₀ generation

Continuously
fed animal

Starved animals at
first larval stage

Starvation trigger

RDE-4-dependent
siRNAs transmitted
through germline by
argonaute protein

F₃ generation

Long-lived worms

siRNA

DNA   Nutrition gene

Figure 1 Caloric restriction during the first larval stage of *Caenorhabditis elegans* induces the expression of siRNAs that silence expression of particular genes involved in nutrition. The germline of the nematode has an RNA-dependent RNA polymerase (RDE-4) that makes more of the siRNAs that bind to exon regions of mRNA, even in the absence of the environmental trigger (starvation) that induced them originally. (After Rechavi et al. 2014.)

condition. There is also a mechanism (although not Lamarck's physiological "use and disuse") that can mediate this inheritance. Thus, Koonin (2014) claims that "the siRNA-based starvation response clearly comes across as inheritance of an acquired trait, or Lamarckian evolution." So, in addition to the component of our evolution provided by Mendelian alleles, there are epigenetic components that allow for the inheritance of certain acquired traits (see Gissis and Jablonka 2009). This may be very important in those lineages where there is not a strict and early segregation of the germline (Buss 1987). Epigenetic inheritance is not uncommon in plants, and this is because their germline is normally derived from somatic lineages (Heard and Martienssen 2014). Indeed, much (if not most) human variation cannot be explained by allelic differences (Visscher et al. 2012), and Mattick (2012) has suggested that the inherited component of complex traits (which has not been explained by genome-wide association studies) may be due to such epigenetic inheritance. Exploration of the relative importance of epialleles and alleles in producing variation is just beginning. In a provocative paper, Skinner and colleagues (2014b) show that the epiallelic variation between five species of Darwin's finches included genes of the

TABLE 1 Some examples of transgenerational epigenetic inheritance in animals

Organism	Phenotype	Gene involved
Caenorhabditis elegans (nematode)	Small and dumpy appearance	RNAi of *ceh13*
	Silencing of green fluorescent protein (GFP)	RNAi of GFP transgene
	Various effects, not reported	RNAi of 13 genes
Daphnia pulex (water flea)	Expression of *G6PD S* and *N* variants	*G6PD* or its regulator
Drosophila melanogaster (fruit fly)	Modifying ability of Y chromosome	Imprintor gene interaction
	Ectopic outgrowth in eyes	*Kr* (*Kr*$^{\text{If-1}}$ allele), (*TrxG* mutation)
	Eye color	Transgenic *Fab7* flanking *lacZ* and mini-*white* reporter transgenes
	(a) Eye color (due to derepression of mini-*white* reporter gene cloned downstream to transgenic *Fab7*)	(a) Activation state of endogenous and transgenic *Fab7* elements containing CMM
	(b) Suppression of wing deformations	(b) Derepression of *sd*
	Susceptibility to tumors	Probably several loci, including heritable epigenetic variation in the *ftz* promoter
Ephestia kuehniella (moth)	Reversion of shortened antennae and associated mating disadvantage	Suppressor of *sa* (*sa*^{WT})
Schistocerca gregaria (locust)	Gregarious behavior; body color	Not known
Myzus persicae (peach potato aphid)	Loss of insecticide resistance	Probably amplified resistance genes
Vulpes vulpes (fox)	Piebald spotting	Activation state of *Star* gene
Mus musculus (mouse)	Probability of developing yellow coat color, obesity, susceptibility to diabetes and cancer	*A*^{vy} (see Chapter 2)
	Reduced body weight, reduced level of proteins involved in sexual recognition and possibly higher mortality between birth and weaning	Not specified, but connected to major urinary protein (*Mup*) and olfactory marker protein (*Omp*) genes
	Probability of kinked tail shape	Axin-fused
	White-spotted tail and feet	*Kit*
	Repression of recombination of the *LoxP* element and concomitant methylation	Transgenic *LoxP* and surrounding chromosomal sequences

Transgenerational persistance	Cause	Mechanism
Over 40 generations	Feeding with bacteria expressing dsRNA targeting ceh13	Chromatin remodeling; RNAi-mediated
At least 40 generations	Feeding with bacteria expressing dsRNA targeting GFP	Chromatin remodeling; RNAi-mediated
At least 10 generations	Feeding with bacteria expressing dsRNA targeting the 13 genes	Chromatin remodeling; RNAi-mediated
Spontaneous reversion rate between the two forms 1 in 10 and 1 in 2	Spontaneous and glucose induced. Presence of S form related to stressful conditions	Not known
11 generations	Transient effect of imprintor gene	Chromatin marking
At least 13 generations	Geldanamycin treatment given to Kr^{If-1} strain, or transient presence of TrxG mutation vtd^3	Chromatin marking involved
At least 4 generations	Transient presence of GAL4 protein	Probably chromatin inheritance
(a) At least 4 generations (more than 4 years) (b) Not specified (stability lower and hard to detect) (b) Not specified (stability lower and hard to detect)	(a) High temperature (b) High temperature	Probably chromatin inheritance
Increased tumorigenicity, 2 generations; modified ftz methylation, at least 1 generation	Crossing with hop^{Tum1} and Kr^1 mutants	DNA methylation; chromatin inheritance
Up to 5 generations; incompletely inherited from mother but almost fully inherited from father	Exposure of larva and pupa to lithium ions; alternate electrical field; or 25°C at late 5th instar larval and pupal phases	Probably chromatin inheritance
Several generations	Material in egg foam secreted by females	Not known
Stable inheritance of lost resistance in clones which have amplified DNA	Induced by DNA amplification	DNA methylation involved
Star (semidominant allele) activated in ~ 1% of domesticated animals; inherited for more than 2 generations	Spontaneous in foxes raised on fur farms; hormonal stress suggested	Possibly heritable chromatin modification
At least 2 generations of agouti epigenotype	Spontaneous, but affected by diet	Chromatin marking, including DNA methylation
Preliminary results suggest transmission of the traits to the F_2 generation	Induced by transfer of mouse pronuclei at the one-cell stage to eggs of a different genotype; traits are transmitted to most offspring through the male germline	DNA methylation assumed to be involved
Spontaneous rate of inactivation 6%; rate of reactivation 1%	Spontaneous; influenced by diet. Injection of hydrocortisone during spermiogenesis reduces penetrance	Chromatin marking including DNA methylation
2 generations of outbred crossing; 6 generations of inbred crossing between paramutants	Transient presence of Kit^{tm1Alf} mutation	RNA inheritance; RNAi involved
Methylated state maintained for at least 3 generations	Transient presence of Sycp1-Cre; exposure of wild type to recombinase activity	DNA methylation

(Continued on next page)

TABLE 1 **Some examples of transgenerational epigenetic inheritance in animals** (*continued*)

Organism	Phenotype	Gene involved
Mus musculus (mouse) (*continued*)	Genome stability	Many
	Glucose intolerance; fat storage	*PPARα* (see Chapter 7); *HGR*
	Tendency to develop tumors	Elevated expression of gene coding for LF (an estrogen-responsive protein) and *c-fos*
	Odor sensitivity and discrimination	*Olfr151*
Rattus norvegicus (rat)	Modified serotonin content in immune cells	Not specified
	Increased expression of genes coding for metabolic factors	Promoters of *PPARα* and *GR* in liver; increased expression of other RNAs
	Decreased spermatogenic capacity; elevated incidence of tumor, prostate and kidney diseases, serum cholesterol levels, immune system abnormalities; premature aging and male mating disadvantage	Methylation state of 15 different DNA sequences. Reduced expression of *ankyrin28*, *Ncstn*, *Rab12*, *Lrrn6a*, and *NCAM1* found in vinclozolin group as well as increased expression of *Fadd*, *Pbm1b*, *snRP1c*, and *Waspip*
	Altered glucose homeostasis	Not specified
Artemia (crustacean)	Thermotolerance	*Hsp70*
Homo sapiens (human)	Cardiovascular mortality and diabetes susceptibility	Imprinted tandem repeat upstream of *INS-IGF2-H19* region
	Predisposition for tumor formation and hereditary nonpolyposis colorectal cancer	*MSH2*
	Angelman and Prader-Willi syndromes	15q11-q13

Source: Jablonka and Raz 2009 (quoted with the permission of E. Jablonka), supplemented by M. Skinner 2014.

BMP signaling pathway that are known to alter beak morphology, as well as genes in the pigment pathway.

There is a large ecological literature on "indirect genetic effects," where the expression of a gene in one individual affects the phenotype of another individual of the same species (see Wolf et al. 1998). This is especially the case for parental effects. It would be interesting if symbionts, siRNAs, paramutations, and epialleles contribute to this phenomenon. Future research will need to ascertain how these genetic and epigenetic systems interact in animals, how abundant such epigenetic systems are, and how natural selection mediates which one predominates in a particular circumstance.

Transgenerational persistance	Cause	Mechanism
At least 3 generations	Irradiation	Chromatin methylation
2 generations; some effect in the third generation	Diet; endocrine disruptors	DNA methylation
Apparent in the F_1 and F_2 generations	Induced by diethylstilbestrol during pregnancy	DNA methylation probably involved
At least 2 generations		
2 generations	Intramuscular administration of β-endorphin during 19th day of pregnancy	Not specified
At least 2 generations	Protein-restricted diet during pregnancy	DNA methylation
At least 4 generations; transmission through the male germline	Vinclozolin or methoxychlor treatment during gestation	DNA methylation
Parental and 3 generations	Low-protein diet in mother, from day 1 of pregnancy through lactation	Not specified; DNA methylation probably involved in F_1 animals
3 generations		
At least 2 generations	Food availability during childhood growth period	Possibly methylation; transmitted through male germline
3 generations	Not known	DNA methylation
Inherited from paternal grandmother (no imprint erasure in the father)	Spontaneous	DNA methylation

It is likely that all of these mechanisms function during evolution; indeed, philosopher of biology Richard Burian (1988) has proposed that this plurality of mechanisms is itself a product of evolution:

> I conclude that the lack of a single dominant disciplinary matrix in evolutionary biology is a consequence of the nature of evolutionary phenomena, and particularly of the role of historical accidents in affecting the evolutionary success, failure, and transformation of lineages. For this reason, I submit, we would be foolish to expect the unification of evolutionary theory within a single paradigm—and, what's more, we should count the failure to achieve such unification as a Good Thing.

This does not mean that there are many theories of evolution that all have the same scientific support. Evolution by natural selection, the mainstay of the Darwinian synthesis, is not in question. But we have realized that animals have been able to evolve in many different ways. Most of this diversification involves the mechanisms used for the generation of variation, and that is where development plays its enormous roles. Natural selection works on the array of variants that development offers it. Evolution by natural selection is not going away. It's just getting a whole lot richer.

References

Adams, M. B. 1980a. Sergei Chetverikov, the Koltsov Institute, and the evolutionary synthesis. In E. Mayr and W. B. Provine (eds.), *The Evolutionary Synthesis: Perspectives on the Unification of Biology*. Harvard University Press, Cambridge, MA, pp. 242–278.

Adams, M. B. 1980b. Severtsov and Schmalhausen: Russian morphology and the evolutionary synthesis, In E. Mayr and W. B. Provine (eds.), *The Evolutionary Synthesis: Perspectives on the Unification of Biology*. Harvard University Press, Cambridge, MA, pp. 193–225.

Baugh, L. R. 2013. To grow or not to grow: Nutritional control of development during *Caenorhabditis elegans* L1 arrest. *Genetics* 194: 539–555.

Benkemoun, L. and S. J. Saupe. 2006. Prion proteins as genetic material in fungi. *Fungal Genet. Biol.* 43: 789–803.

Burian, R. M. 1988. Challenges to the Evolutionary Synthesis. *Evol. Biol.* 23: 247–269.

Burton, N. O., K. B. Burkhart and S. Kennedy. 2011. Nuclear RNAi maintains heritable gene silencing in *Caenorhabditis elegans*. *Proc. Nat. Acad. Sci. USA* 108: 19683–19688.

Buss, L. W. 1987. *The Evolution of Individuality*. Princeton University Press, Princeton, NJ.

Cossetti, C., L. Lugini, L. Astrologo, I. Saggio, S. Fais and C. Spadafora. 2014. Soma-to-germline transmission of RNA in mice xenografted with human tumour cells: Possible transport by exosomes. *PLoS ONE* 9(7): e101629.

Cuzin, F. and M. Rassoulzadegan. 2010. Non-Mendelian epigenetic heredity: Gametic RNAs as epigenetic regulators and transgenerational signals. *Essays Biochem.* 48: 101–106.

Darwin, C. 1883. *The Variation of Animals and Plants under Domestication*, 2nd ed., Vol. 2. Appleton, New York.

Dobzhansky, Th. 1970. *Genetics of the Evolutionary Process*. Columbia University Press, New York, pp. 210–211.

Gause, G. F. 1947. Problems of evolution. *Transact. Ct. Acad. Sci.* 37: 17–68.

Gilbert, S. F. 1994. Dobzhansky, Waddington and Schmalhausen: Embryology and the Modern Synthesis. In M. B. Adams (ed.), *The Evolution of Theodosius Dobzhansky: Essays on His Life and Thought in Russia and America*. Princeton University Press, Princeton, NJ, pp. 143–154.

Gissis, S. B. and E. Jablonka (eds.). 2011. *Transformations of Lamarckianism: From Subtle Fluids to Molecular Biology*. MIT Press, Cambridge, MA.

Gliboff, S. 2005. "Protoplasm ... is soft wax in our hands": Paul Kammerer and the art of biological transformation. *Endeavor* 29: 162–167.

Grishok, A. 2013. Biology and mechanisms of short RNAs in *Caenorhabditis elegans*. *Adv. Genet.* 83: 1–69.

Gu, S. G., J. Pak, S. Guang, J. M. Maniar, S. Kennedy and A. Fire. 2012. Amplification of siRNA in *Caenorhabditis elegans* generates a transgenerational sequence-targeted histone H3 lysine 9 methylation footprint. *Nature Genet.* 44: 157–164.

Heard, E. and R. A. Martienssen. 2014. Transgenerational epigenetic inheritance: Myths and mechanisms. *Cell* 157: 95–109.

Hogben, L. 1933. *Nature and Nurture*. W. W. Norton, New York.

Huxley, T. H. 1884. Quoted in *The Medical Record*, May 31, 1884, p. 628.

Jablonka, E. and M. J. Lamb. 1995. *Epigenetic Inheritance and Evolution: The Lamarckian Dimension*. Oxford University Press, Oxford.

Johannsen, W. 1909. *Elemente der exacten Erblichkeitslehre*. Gustav Fischer, Jena.

Jollos, V. 1934. Inherited changes produced by heat-treatment in *Drosophila melanogaster*. *Genetica* 16: 476–494.

Koonin, E. V. 2104. Calorie restriction à Lamarck. *Cell* 158: 237–238.

Lamarck, J. B. 1809. *Philosophie zoologique, ou exposition des considérations relatives à l'histoire naturelle des animaux*. Transl. H. Eliot. Macmillan 1914 (p. 113; 126). Online at www.ucl.ac.uk/taxome/jim/Mim/lamarck6.html.

Levit, G. S., U. Hossfeld and L. Olsson. 2006. From the "Modern Synthesis" to cybernetics: Ivan Ivanovich Schmalhausen and his research program for a synthesis of evolutionary and developmental biology. *J. Exp. Zool.* 306B: 89–106.

Lewontin, R. C. 1974. The analysis of variance and the analysis of causes. *Am. J. Human Genet.* 26: 400–411.

Lewontin, R. C. and R. Levins. 1976. The problem of Lysenkoism. In H. Rose and S. Rose (eds.), *The Radicalisation of Science*. Macmillan, London, pp. 32–65.

Lindegren, C. C. 1949. *The Yeast Cell, its Genetics and Cytology*. Educational Publishers, St. Louis.

Lindegren, C. C. 1966. *The Cold War in Biology*. Planarian Press, Ann Arbor, MI.

Lysenko, T. 1948. Quoted in Lindegren 1966, op. cit., p. 57.

Logan, C. A. 2007. Overheated rats, race, and the double gland: Paul Kammerer, endocrinology, and the problem of somatic induction. *J. Hist. Biol.* 40: 683–725.

Mattick, J. S. 2012. Rocking the foundations of molecular genetics. *Proc. Nat. Acad. Sci. USA* 109: 16400–16401.

Mayr, E. 1963. *Animal Species and Evolution*. Harvard University Press, Cambridge, MA.

Morgan, T. H. 1932. *The Scientific Basis of Evolution*. W. W. Norton, New York.

Nilsson-Ehle, H. 1914. Vilka erfarenheter hava hittills vunnits rörande möjligheten av växters acklimatisering? *Kunglig Landtbruksakademiens. Handlingar och Tidskrift* 53: 537–572.

Piaget, J. 1929a. Les races lacustres de la *Limnaea stagnalis* L.: Recherches sur les rapports de l'adaptation hereditaires avec le milieu. *Bull. Biol. Fr. Belg.* 63: 424–455.

Piaget, J. 1929b. L'adaptation de la *Limnaea stagnalis* au milieu lacustres de la Suisse Romande: Etude biométrique et génétique. *Rev. Suisse Zool.* 36: 263–531.

Rassoulzadegan, M., V. Grandjean, P. Gounon, S. Vincent, I. Gillot and F. Cuzin. 2006. RNA-mediated non-Mendelian inheritance of an epigenetic change in the mouse. *Nature* 441: 469–474.

Rechavi, O., G. Minevich and H. Hobert. 2011. Trans-generational inheritance of an acquired smallRNA-based anti-viral response in *C. elegans*. *Cell* 147: 1248–1256.

Rechavi, O., L. Houri-Ze'evi, S. Anava, W. S. Goh, S. Y. Kerk, G. J. Hannon and O. Hobert. 2014. Starvation-induced transgenerational inheritance of small RNAs in *C. elegans*. *Cell* 158: 277–287.

Sapp, J. 1987. *Beyond the Gene*. Oxford University Press, New York.

Sarkar, S. 1999. From the *Reaktionsnorm* to the adaptive norm: The norm of reaction 1909–1960. *Biol. Philos.* 14: 235–252.

Sarkar, S. and T. Fuller. 2002. Generalized norms of reaction for ecological developmental biology. *Evol. Dev.* 5: 106–115.

Schmalhausen, I. I. 1949. *Factors of Evolution: The Theory of Stabilizing Selection*. Blakiston, Philadelphia.

Severtsov, A. N. 1935. *Modes of Phyloembryogenesis*. Quoted in Adams 1980b, p. 217.

Simpson, G. G. 1953. The Baldwin Effect. *Evolution* 7: 110–117.

Skinner M. K., M. I. Savenkova, B. Zhang, A. C. Gore and D. Crews. 2014a. Gene bionetworks involved in the epigenetic transgenerational inheritance of altered mate preference: Environmental epigenetics and evolutionary biology. *BMC Genomics* 15: 377.

Skinner, M. K., C. Gurerrero-Bosagna, M. M. Haque, E. E. Nilsson, J. A. Koop, S. A. Knutie and D. H. Clayton. 2014b. Epigenetics and the evolution of Darwin's finches. *Genome Biol. Evol.* 6: 1972–1989.

Thomson, J. A. 1908. *Heredity*. Putnam, New York.

Visscher, P. M., M. A. Brown, M. I. McCarthy and J. Yang. 2012. Five years of GWAS discovery. *Am. J. Hum. Genet.* 90: 7–24.

Vogt, O. 1926. Psychiatrisch wichtige Tatsachen der zoologisch-botanischen Systematik. *J. Psychol. Neurologie* 101: 805–832.

Waddington, C. H. 1942. The canalization of development and the inheritance of acquired characters. *Nature* 150: 563.

Waddington, C. H. 1952 Selection of the genetic basis for an acquired character. *Nature* 169: 278.

Wake, D. B. 1986. Foreword to *Factors of Evolution. The Theory of Stabilizing Selection* by I. I. Schmalhausen. University of Chicago Press, Chicago, pp. v–xii.

Wolf, J. B., E. D. Brodie, III, J. M. Cheverud, A. J. Moore and M. J. Wade. 1998. Evolutionary consequences of indirect genetic effects. *Trends Ecol. Evol.* 13: 64–69.

Woltereck, R. 1909. Weitere experimentelle Untersuchungen über Artveränderung, speziell über das Wesen quantitativer Artunderscheide bei Daphniden. *Versuch. Deutsch. Zool. Gest.* 1909: 110–172.

Illustration Credits

Chapter 10

Opener: © Jochen Tack/Alamy. **10.17A:** © F1online digitale Bildagentur GmbH/Alamy. **10.17B:** © Jochen Tack/Alamy. **10.25A:** © Paffy69/istock. **10.25B:** © DragoNika/Shutterstock.

Chapter 11

Opener: Courtesy of Dr. Kishida Osamu, forestcsv.ees.hokudai.ac.jp/en/research. **11.16B:** David McIntyre.

Index

Italic type indicates the information will be found in an illustration.
The designation "n" indicates the information will be found in a footnote.